Lignocellulosic Polymer Composites

Scrivener Publishing
100 Cummings Center, Suite 541J
Beverly, MA 01915-6106

Polymer Science and Plastics Engineering

The series combines the two interdisciplinary fields of polymer science and plastics engineering to publish state of the art advances in the science and engineering of polymers and plastics. The book series publishes both short and standard length monographs, textbooks, edited volumes, handbooks, practical guides, and reference works related to all aspects of polymer science and plastics engineering including, but not limited to, renewable and synthetic polymer chemistry and physics, compositions (e.g. blends, composites, additives), processing, characterization, testing, design (materials and equipment), and applications. The books will serve a variety of industries such as automotive, food packaging, medical, and plastics as well as academia.
Proposals or enquiries should be sent to the series editor Dr. Srikanth Pilla at: spilla@clemson.edu

Publishers at Scrivener
Martin Scrivener(martin@scrivenerpublishing.com)
Phillip Carmical (pcarmical@scrivenerpublishing.com)

Lignocellulosic Polymer Composites

Processing, Characterization, and Properties

Edited by

Vijay Kumar Thakur

WILEY

Co-published by John Wiley & Sons, Inc. Hoboken, New Jersey, and Scrivener Publishing LLC, Salem, Massachusetts.
Published simultaneously in Canada.

For general information on our other products and services or for technical support, please contact our Customer Care Department within the United States at (800) 762-2974, outside the United States at (317) 572-3993 or fax (317) 572-4002.

Wiley also publishes its books in a variety of electronic formats. Some content that appears in print may not be available in electronic formats. For more information about Wiley products, visit our web site at www.wiley.com.

For more information about Scrivener products please visit www.scrivenerpublishing.com.

Cover design by Russell Richardson

Library of Congress Cataloging-in-Publication Data:

ISBN 978-1-118-77357-4

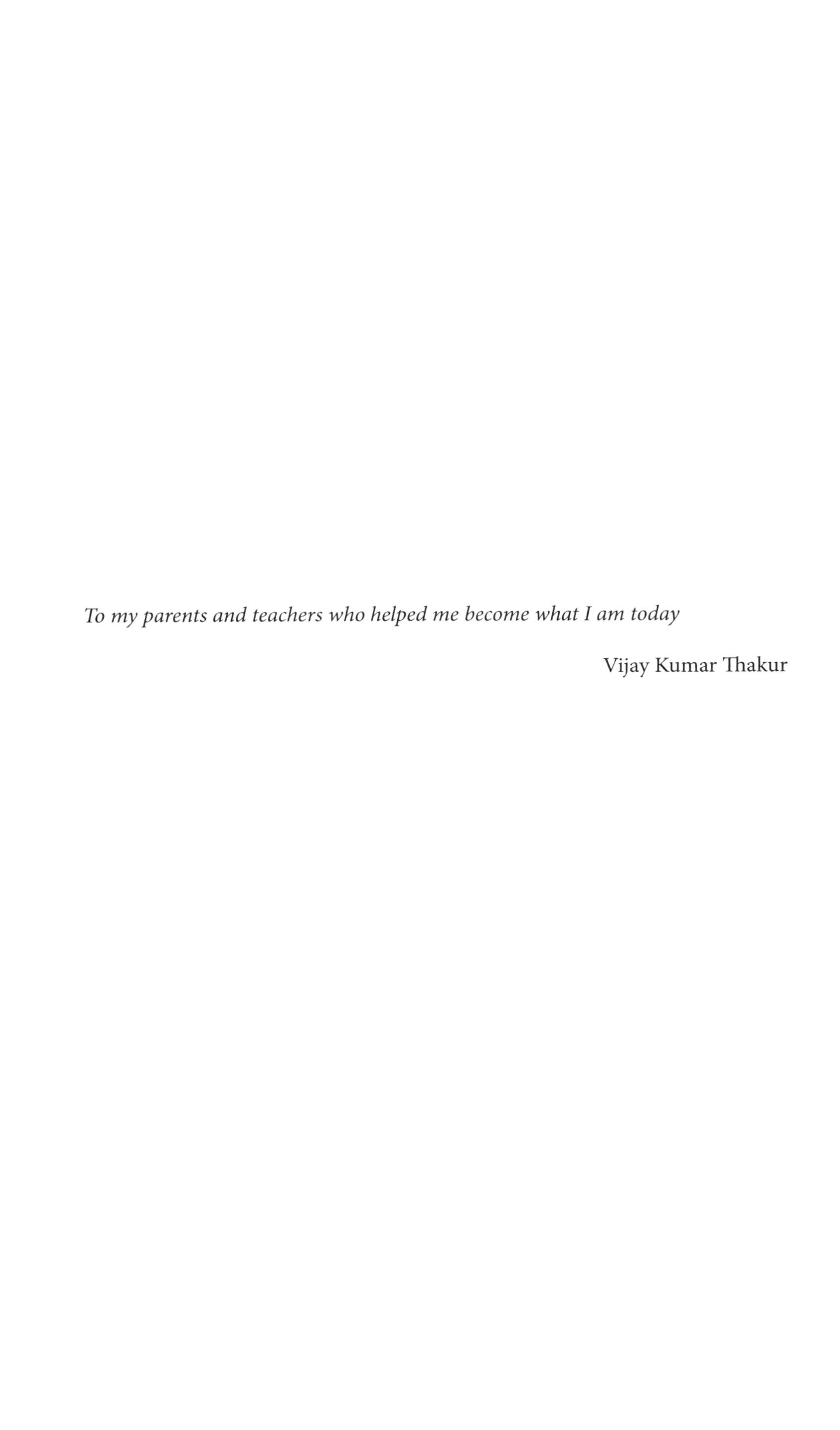

To my parents and teachers who helped me become what I am today

Vijay Kumar Thakur

Contents

Preface **xvii**

Part I: LIGNOCELLULOSIC NATURAL POLYMERS BASED COMPOSITES

1 Lignocellulosic Polymer Composites: A Brief Overview **3**
Manju Kumari Thakur, Aswinder Kumar Rana and Vijay Kumar Thakur
1.1 Introduction 3
1.2 Lignocellulosic Polymers: Source, Classification and Processing 4
1.3 Lignocellulosic Natural Fibers: Structure, Chemical Composition and Properties 8
1.4 Lignocellulosic Polymer Composites: Classification and Applications 10
1.5 Conclusions 13
References 13

2 Interfacial Adhesion in Natural Fiber-Reinforced Polymer Composites **17**
E. Petinakis, L. Yu, G. Simon, X. Dai, Z. Chen and K. Dean
2.1 Introduction 17
2.2 PLA-Based Wood-Flour Composites 18
2.3 Optimizing Interfacial Adhesion in Wood-Polymer Composites 20
2.3.1 Chemical Modification 21
2.3.2 Physical Modification 27
2.4 Evaluation of Interfacial Properties 30
2.4.1 Microscopic Characterisation 31
2.4.1.1 Scanning Electron Microscopy 31
2.4.1.2 Atomic Force Microscopy 32
2.4.2 Spectroscopic Techniques 33
2.4.2.1 Acoustic Emission Spectroscopy (AES) 33
2.4.3 Other Techniques 34
2.5 Conclusions 34
References 35

3 Research on Cellulose-Based Polymer Composites in Southeast Asia **41**
Riza Wirawan and S.M. Sapuan
3.1 Introduction 42
3.2 Sugar Palm (*Arenga pinnata*) 44
3.3 Oil Palm (*Elaeis Guineensis*) 46

3.4 Durian (*Durio Zibethinus*) 49
3.5 Water Hyacinth (*Eichhornia Crassipes*) 51
3.6 Summary 57
References 58

4 Hybrid Vegetable/Glass Fiber Composites 63
Sandro C. Amico, Jose R. M. d'Almeida, Laura H. de Carvalho and Mª Odila H. Cioffi
4.1 Introduction 63
4.1.1 The Hybrid Concept 65
4.2 Vegetable Fiber/Glass Fiber Thermoplastic Composites 67
4.3 Intra-Laminate Vegetable Fiber/glass Fiber Thermoset Composites 69
4.4 Inter-Laminate Vegetable Fiber/glass Fiber Thermoset Composites 71
4.5 Concluding Remarks 75
Acknowledgement 76
References 76

5 Flax-Based Reinforcement Requirements for Obtaining Structural and Complex Shape Lignocellulosic Polymer Composite Parts 83
Pierre Ouagne and Damien Soulat
5.1 Introduction 84
5.2 Experimental Procedures 86
5.2.1 Flax Tow Testing 86
5.2.2 Flax Fabric Testing 86
5.2.2.1 Biaxial Tensile Test 87
5.2.3 Sheet Forming Device for Dry Textile Reinforcement 87
5.3 Results and Discussion 90
5.3.1 Tensile Behavior of Reinforcement Components: Flax Tow Scale 90
5.3.1.1 Flax Tow Tensile Behavior 90
5.3.1.2 Effect of Gauge Length on Tensile Properties 91
5.3.1.3 Evolution of Failure Behavior 91
5.3.2 Tensile Behavior of Reinforcement Components: Scale of Fabric 94
5.3.3 Global Preform Analysis 94
5.3.4 Analysis of Tensile Behavior of Tows During Forming 95
5.4 Discussions 97
5.5 Conclusions 98
References 98

6 Typical Brazilian Lignocellulosic Natural Fibers as Reinforcement of Thermosetting and Thermoplastics Matrices 103
Patrícia C. Miléo, Rosineide M. Leão, Sandra M. Luz, George J. M. Rocha and Adilson R. Gonçalves
6.1 Introduction 104
6.2 Experimental 105
6.2.1 Preparation of cellulose and lignin from sugarcane bagasse 106
6.2.2 Surface Treatment for Coconut Fibers 106
6.2.3 Chemical Characterization of Fibers and Lignin 106
6.2.3.1 Carbohydrates and Lignin Determination 106
6.2.3.2 Determination of Ashes Content in Lignin 107

6.2.3.3 Elemental Analysis of Lignin 107
6.2.3.4 Total Acid Determination in Lignin 107
6.2.3.5 Total Hydroxyls in Lignin 107
6.2.3.6 Phenolic Hydroxyls in Lignin 107
6.2.3.7 Determination of Carbonyl Groups in Lignin 108
6.2.3.8 Analysis of the Molecular Weight Distribution of Lignin 108
6.2.4 Infrared Spectroscopy (FTIR) Applied to Fibers and Lignin 108
6.2.5 Preparation of Thermosetting and Thermoplastic Composites Reinforced with Natural Fibers 108
6.2.6 Scanning Electron Microscopy (SEM) 109
6.2.7 Thermogravimetric Analysis (TGA) 109
6.2.8 Differential Scanning Calorimetry (DSC) Characterization 109
6.3 Results and Discussion 110
6.3.1 Chemical Composition and Characterization of Sugarcane Bagasse and Coconut Fibers 110
6.3.2 Chemical Characterization of Lignin Extracted from Sugarcane Bagasse 111
6.3.3 Modification of Coconut Fibers by Chemical Treatment 112
6.3.4 Fourier Transform Infrared Spectrometry Applied to Coconut Fibers 113
6.3.5 Composites with Thermoplastic and Thermosetting as Matrices 113
6.3.5.1 Coconut Fibers 113
6.3.6 Morphological Characterization for Composites Reinforced with Cellulose and Lignin from Sugarcane Bagasse and Coconut Fibers 114
6.3.7 Thermogravimetric Analysis for Composites and Fibers 117
6.3.8 Differential Scanning Calorimetry Studies for Composites and Fibers 120
6.4 Conclusions 122
Acknowledgements 123
References 123

7 Cellulose-Based Starch Composites: Structure and Properties 125
Carmen-Alice Teacă, Ruxanda Bodîrlău and Iuliana Spiridon
7.1 Introduction 125
7.2 Starch and Cellulose Biobased Polymers for Composite Formulations 126
7.3 Chemical Modification of Starch 127
7.4 Cellulose-Based Starch Composites 129
7.4.1 Obtainment 129
7.4.1.1 Preparation of Starch Microparticles (StM) and Chemically Modified Starch Microparticles (CStM) 129
7.4.1.2 Determination of the Molar Degree of Substitution of CMSt 130
7.4.1.3 Preparation of CMSt/St/cellulose Filler Composite Films 131
7.4.2 Characterization of Starch Polymer Matrix 133
7.4.2.1 FTIR Spectroscopy Investigation 133
7.4.2.2 X-ray Diffraction Analysis 134

7.4.3 Properties Investigation 136
7.4.3.1 Opacity Measurements 136
7.4.3.2 Water Sorption Properties 137
7.4.3.3 Mechanical Properties 138
7.4.3.3 Thermal Properties 139
7.5 Conclusions/Perspectives 139
References 140

8 Spectroscopy Analysis and Applications of Rice Husk and Gluten Husk Using Computational Chemistry 147
Norma-Aurea Rangel-Vazquez, Virginia Hernandez-Montoya and Adrian Bonilla-Petriciolet
8.1 Introduction 148
8.1.1 Computational Chemistry 148
8.1.1.1 Molecular Mechanics Methods 149
8.1.1.2 Semi-Empirical Methods 150
8.1.2 Lignocellulosic Materials 153
8.1.2.1 Rice Husk 154
8.1.2.2 Wheat Gluten Husk 155
8.1.3 Benzophenone 158
8.1.4 Glibenclamide 159
8.1.4.1 Mechanism of Action 159
8.1.4.2 Medical Uses 160
8.2 Methodology 160
8.2.1 Geometry Optimization 160
8.2.2 FTIR 160
8.2.3 Electrostatic Potential 160
8.3 Results and Discussions 161
8.3.1 Geometry Optimization 161
8.3.2 FTIR Analysis 161
8.3.3 Electrostatic Potential 163
8.3.4 Absorption of Benzophenone 164
8.3.4.1 Geometry Optimization 164
8.3.4.2 FTIR 164
8.3.4.3 Electrostatic Potential 168
8.3.5 Absorption of Glibenclamide 169
8.3.5.1 Geometry Optimization 169
8.3.5.2 FTIR 169
8.3.5.3 Electrostatic Potential 172
8.4 Conclusions 172
References 172

9 Oil Palm Fiber Polymer Composites: Processing, Characterization and Properties 175
S. Shinoj and R. Visvanathan
9.1 Introduction 176
9.2 Oil Palm Fiber 177
9.2.1 Extraction 177

9.2.2 Morphology and Properties 178
9.2.3 Surface Treatments 181
9.3 Oil Palm Fiber Composites 184
9.3.1 Oil Palm Fiber-Natural Rubber Composites 185
9.3.1.1 Mechanical Properties 185
9.3.1.2 Water Absorption Characteristics 187
9.3.1.3 Thermal Properties 187
9.3.1.4 Electrical Properties 187
9.3.2 Oil Palm Fiber-Polypropylene Composites 189
9.3.2.1 Mechanical Properties 189
9.3.2.2 Water Absorption Characteristics 191
9.3.2.3 Degradation/weathering 192
9.3.3 Oil Palm Fiber-Polyurethane Composites 192
9.3.3.1 Mechanical Properties 192
9.3.3.2 Water Absorption Characteristics 193
9.3.3.3 Degradation/weathering 194
9.3.4 Oil Palm Fiber-Polyvinyl Chloride Composites 194
9.3.4.1 Mechanical Properties 194
9.3.4.2 Thermal Properties 195
9.3.5 Oil Palm Fiber-Polyester Composites 196
9.3.5.1 Physical Properties 196
9.3.5.2 Mechanical Properties 196
9.3.5.3 Water Absorption Characteristics 197
9.3.5.4 Degradation/weathering 198
9.3.6 Oil Palm Fiber-Phenol Formaldehyde Composites 198
9.3.6.1 Physical Properties 199
9.3.6.2 Mechanical Properties 199
9.3.6.3 Water Absorption Characteristics 200
9.3.6.4 Thermal Properties 201
9.3.6.5 Degradation/weathering 201
9.3.7 Oil Palm Fiber-Polystyrene Composites 202
9.3.7.1 Mechanical Properties 202
9.3.8 Oil Palm Fiber-Epoxy Composites 202
9.3.8.1 Mechanical Properties 203
9.3.9 Oil Palm Fiber-LLDPE Composites 203
9.3.9.1 Physical Properties 204
9.3.9.2 Electrical Properties 205
9.3.9.3 Mechanical Properties 205
9.3.9.4 Thermal Properties 207
9.4 Conclusions 208
References 208

10 Lignocellulosic Polymer Composites: Processing, Characterization and Properties 213
Bryan L. S. Sipião, Lais Souza Reis, Rayane de Lima Moura Paiva, Maria Rosa Capri and Daniella R. Mulinari
10.1 Introduction 213

10.2 Palm Fibers 214
10.2.1 Effect of Modification on Mechanical Properties of Palm Fiber Composites 215
10.2.2 Alkali Treatment and Coupling Agent 216
10.3 Pineapple Fibers 220
10.3.1 Alkali Treatment 221
10.3.2 Acid Hydrolysis 223
Acknowledgements 227
References 227

Part II: CHEMICAL MODIFICATION OF CELLULOSIC MATERIALS FOR ADVANCED COMPOSITES

11 Agro-Residual Fibers as Potential Reinforcement Elements for Biocomposites 233
Nazire Deniz Yılmaz
11.1 Introduction 233
11.2 Fiber Sources 235
11.2.1 Wheat Straw 235
11.2.2 Corn Stalk, Cob and Husks 235
11.2.3 Okra Stem 236
11.2.4 Banana Stem, Leaf, Bunch 236
11.2.5 Reed Stalk 237
11.2.6 Nettle 237
11.2.7 Pineapple Leaf 238
11.2.8 Sugarcane 238
11.2.9 Oil Palm Bunch 238
11.2.10 Coconut Husk 239
11.3 Fiber Extraction methods 239
11.3.1 Biological Fiber Extraction Methods 240
11.3.2 Chemical Fiber Separation Methods 241
11.3.3 Mechanical Fiber Separation Methods 241
11.4 Classification of Plant Fibers 246
11.5 Properties of Plant Fibers 247
11.5.1 Chemical Properties of Plant Fibers 247
11.5.1.1 Cellulose 247
11.5.1.2 Hemicellulose 248
11.5.1.3 Lignin 248
11.5.1.4 Pectin 249
11.5.1.5 Waxes 249
11.6. Properties of Agro-Based Fibers 249
11.6.1 Physical Properties 249
11.6.2 Mechanical Properties 251
11.6.3 Some Important Features of Plant Fibers 252
11.6.3.1 Insulation 252
11.6.3.2 Moisture Absorption 252
11.6.3.3 Dimensional stability 254

11.6.3.4 Thermal Stability 255
11.6.3.4 Photo Degradation 257
11.6.3.5 Microbial Resistance 257
11.6.3.6 Variability 257
11.6.3.7 Reactivity 258
11.7 Modification of Agro-Based Fibers 258
11.7.1 Physical Treatments 258
11.7.2 Chemical Treatments 260
11.7.2.1 Alkalization 260
11.7.2.2 Acetylation 263
11.7.2.3 Silane Treatment 263
11.7.2.4 Bleaching 263
11.7.2.5 Enzyme Treatment 264
11.7.2.6 Sulfonation 265
11.7.2.7 Graft Copolymerization 265
11.8 Conclusion 266
References 266

12 Surface Modification Strategies for Cellulosic Fibers 271
Inderdeep Singh, Pramendra Kumar Bajpai
12.1 Introduction 271
12.2 Special Treatments during Primary Processing 273
12.2.1 Microwave Curing of Biocomposites 274
12.2.2 Chemical Treatments of Fibers During Primary Processing of Biocomposites 274
12.2.2.1 Alkaline Treatment 275
12.2.2.2 Silane Treatment 276
12.3 Other Chemical Treatments 277
12.4 Conclusions 278
References 279

13 Effect of Chemical Functionalization on Functional Properties of Cellulosic Fiber-Reinforced Polymer Composites 281
Ashvinder Kumar Rana, Amar Singh Singha, Manju Kumari Thakur and Vijay Kumar Thakur
13.1 Introduction 282
13.2 Chemical Functionalization of Cellulosic Fibers 283
13.2.1 Alkali Treatment 283
13.2.2 Benzoylation 283
13.2.3 Composites Fabrication 283
13.3 Results and Discussion 284
13.3.1 Mechanical Properties 284
13.3.1.1 Tensile Strength 284
13.3.1.2 Compressive Strength 286
13.3.1.3 Flexural Strength 288
13.3.2 FTIR Analysis 288
13.3.3 SEM Analysis 289

13.3.4 Thermogravimetric Analysis 290
13.3.5 Evaluation of Physico-Chemical Properties 290
13.3.5.1 Water Absorption 290
13.3.5.2 Chemical Resistance 292
13.3.5.3 Moisture Absorption 293
13.3.6 Limiting Oxygen Index (LOI) Test 295
13.4 Conclusion 297
References 297

14 Chemical Modification and Properties of Cellulose-Based Polymer Composites 301
Md. Saiful Islam, Mahbub Hasan and Mansor Hj. Ahmad @ Ayob
14.1 Introduction 302
14.2 Alkali Treatment 303
14.3 Benzene Diazonium Salt Treatment 306
14.4 *o*-hydroxybenzene Diazonium Salt Treatment 310
14.5 Succinic Anhydride Treatment 313
14.6 Acrylonitrile Treatment 317
14.7 Maleic Anhydride Treatment 318
14.8 Nanoclay Treatment 318
14.9 Some other Chemical Treatment with Natural Fibers 320
14.9.1 Epoxides Treatment 320
14.9.2 Alkyl Halide Treatment 320
14.9.3 β- Propiolactone Treatments 320
14.9.4 Cyclic Anhydride Treatments 321
14.9.5 Oxidation of Natural Fiber 321
14.10 Conclusions 321
References 322

Part III: PHYSICO-CHEMICAL AND MECHANICAL BEHAVIOUR OF CELLULOSE/ POLYMER COMPOSITES

15 Weathering of Lignocellulosic Polymer Composites 327
Asim Shahzad and D. H. Isaac
15.1 Introduction 328
15.2 Wood and Plant Fibers 329
15.3 UV Radiation 330
15.3.1 Lignocellulosic Fibers 332
15.3.2 Polymer Matrices 333
15.3.3 Methods for Improving UV Resistance of LPCs 334
15.4 Moisture 335
15.4.1 Lignocellulosic Fibers 336
15.4.2 Polymer Matrices 339
15.4.3 Methods for Improving Moisture Resistance of LPCs 340
15.5 Testing of Weathering Properties 342
15.6 Studies on Weathering of LPCs 345
15.6.1 Lignocellulosic Fibers 345

15.6.2 Lignocellulosic Thermoplastic Composites 346
15.6.2.1 Effects of Photostabilizers and Surface Treatments 352
15.6.3 Lignocellulosic Thermoset Composites 359
15.6.4 Lignocellulosic Biodegradable Polymer Composites 360
15.7 Conclusions 362
References 363

16 Effect of Layering Pattern on the Physical, Mechanical and Acoustic Properties of Luffa/Coir Fiber-Reinforced Epoxy Novolac Hybrid Composites 369
Sudhir Kumar Saw, Gautam Sarkhel and Arup Choudhury
16.1 Introduction 369
16.2 Experimental 373
16.2.1 Materials 373
16.2.2 Synthesis of Epoxy Novolac Resin (ENR) 373
16.2.3 Fabrication of Composite Materials via Hot-pressing 373
16.3 Characterization of ENR-Based Luffa/Coir Hybrid Composites 374
16.3.1 Dimensional Stability Test 374
16.3.2 Mechanical Strength Analysis 375
16.3.3 Sound Absorption Test 375
16.3.4 Scanning Electron Microscopy (SEM) 375
16.4 Results and Discussion 376
16.4.1 Water Absorption Test 376
16.4.2 Thickness Swelling Test 377
16.4.3 Effect of Different Configurations on Mechanical Properties 378
16.4.4 Sound Absorption Performances 380
16.4.5 Study of Hybrid Composite Microstructure 381
16.5 Conclusions 383
Acknowledgements 383
References 383

17 Fracture Mechanism of Wood-Plastic Composites (WPCS): Observation and Analysis 385
Fatemeh Alavi, Amir Hossein Behravesh and Majid Mirzaei
17.1 Introduction 385
17.1.1 Fracture Behavior of Particulate Composites 386
17.1.1.1 Particle Size, Volume Fraction, and Fillers Orientation 386
17.1.1.2 Fillers & Polymers Characteristics 389
17.1.1.3 Loading 391
17.1.1.4 Temperature 391
17.1.1.5 Interface 393
17.2 Fracture Mechanism 396
17.3 Toughness Characterization 398
17.4 Fracture Observation 400
17.5 Fracture Analysis 402
17.5.1 Macroscale Modeling 402
17.5.2 Multi-scale Modeling 403
17.5.3 Cohesive Zone Model (CZM) 404
17.5.4 Other Numerical Methods 407
17.5.5 Inverse Method 408

17.6 Conclusions 409
References 410

18 Mechanical Behavior of Biocomposites under Different Operating Environments 417
Inderdeep Singh, Kishore Debnath and Akshay Dvivedi
18.1 Introduction 417
18.2 Classification and Structure of Natural Fibers 419
18.3 Moisture Absorption Behavior of Biocomposites 421
18.4 Mechanical Characterization of Biocomposites in a Humid Environment 423
18.5 Oil Absorption Behavior and Its Effects on Mechanical Properties of Biocomposites 424
18.6 UV-Irradiation and Its Effects on Mechanical Properties of Biocomposites 425
18.7. Mechanical Behavior of Biocomposites Subjected to Thermal Loading 426
18.8 Biodegradation Behavior and Mechanical Characterization of Soil Buried Biocomposites 428
18.9 Conclusions 429

Part IV: APPLICATIONS OF CELLULOSE/ POLYMER COMPOSITES

19 Cellulose Composites for Construction Applications 435
Catalina Gómez Hoyos and Analía Vazquez
19.1 Polymers Reinforced with Natural Fibers for Construction Applications 435
19.1.1 Durability of Polymer-Reinforced with Natural Fibers 438
19.1.2 Classification of Polymer Composites Reinforced with Natural Fibers 439
19.2 Portland Cement Matrix Reinforced with Natural Fibers for Construction Applications 440
19.2.1 Modifications on Cement Matrix to Increase Durability 441
19.2.1.1 Pozzolanic Aditions 441
19.2.1.2 Carbonation of Cement Matrix 442
19.2.2 Modifications on Natural Fibers to Increase Durability of Cement Composites 443
19.2.3 Application of Cement Composites Reinforced with Cellulosic Fibers 445
19.2.4 Celllulose Micro and Nanofibers Used to Reinforce Cement Matrices 446
References 448

20 Jute: An Interesting Lignocellulosic Fiber for New Generation Applications 453
Murshid Iman and Tarun K. Maji
20.1 Introduction 453
20.2 Reinforcing Biofibers 455

20.2.1 Chemical Constituents and Structural Aspects of Lignocellulosic Fiber 457
20.2.2 Properties of Jute 458
20.2.3 Cost Aspects, Availability and Sustainable Development 460
20.2.4 Surface Treatments 461
20.2.5 Processing 461
20.2.5.1 Compression Molding 462
20.2.5.2 Resin Transfer Molding 462
20.2.5.3 Vacuum-Assisted Resin Transfer Molding (VARTM) 463
20.2.5.4 Injection Molding 464
20.2.5.5 Direct Long-Fiber Thermoplastic Molding (D-LFT) 464
20.3 Biodegradable Polymers 465
20.4 Jute-Reinforced Biocomposites 466
20.5 Applications 468
20.6 Concluding Remarks 468
Acknowledgement 469
References 469

21 Cellulose-Based Polymers for Packaging Applications 477
Behjat Tajeddin
21.1 Introduction 477
21.1.1 Packaging Materials 479
21.1.2 Plastics 479
21.1.3 Problems of Plastics 480
21.2 Cellulose as a Polymeric Biomaterial 481
21.2.1 Cellulose Extraction 482
21.2.2 Cellulosic Composites (Green Composites) 483
21.2.3 Cellulose Derivatives Composites 486
21.2.3.1 Esterification 486
21.2.3.2 Etherification 487
21.2.3.3 Regenerated Cellulose Fibers 488
21.2.3.4 Bacterial Cellulose (BC) 489
21.3 Cellulose as Coatings and Films Material 490
21.3.1 Coatings 491
21.3.2 Films 492
21.4 Nanocellulose or Cellulose Nanocomposites 492
21.5 Quality Control Tests 493
21.6 Conclusions 495
References 496

22 Applications of Kenaf-Lignocellulosic Fiber in Polymer Blends 499
Norshahida Sarifuddin and Hanafi Ismail
22.1 Introduction 499
22.2 Natural Fibers 500
22.3 Kenaf: Malaysian Cultivation 505
22.4 Kenaf Fibers and Composites 508

22.5 Kenaf Fiber Reinforced Low Density Polyethylene/Thermoplastic Sago Starch Blends 509
22.6 The Effects of Kenaf Fiber Treatment on the Properties of LDPE/TPSS Blends 512
22.7 Outlook and Future Trends 517
Acknowledgement 517
References 517

23 Application of Natural Fiber as Reinforcement in Recycled Polypropylene Biocomposites 523
Sanjay K Nayak and Gajendra Dixit
23.1 Introduction 523
23.1.1 Natural Fibers – An Introduction 525
23.1.2 Chemical Composition of Natural Fiber 526
23.1.3 Classification of Natural Fibers 529
23.1.4 Surface Modification of Natural Fibers 530
23.1.4.1 Alkali Treatment 530
23.1.4.2 Silane Treatment (SiH4) 530
23.1.4.3 Acetylation of Natural Fibers 531
23.1.5 Properties of Natural Fibers 532
23.2 Recycled Polypropylene (RPP) - A matrix for Natural Fiber Composites 533
23.3 Natural Fiber-Based Composites – An Overview 534
23.3.1 Sisal Fiber–Based Recycled Polypropylene (RPP) Composites 535
23.3.1.1 Mechanical and Dynamic Mechanical Properties of Sisal RPP Composites 536
23.3.1.2 Thermal Properties Sisal RPP Composites 539
23.3.1.3 Weathering and Its Effect on Mechanical Properties of Sisal RPP Composites 541
23.3.1.4 Fracture Analysis of RPP and its Composites 543
23.4 Conclusion 545
References 545

Index 551

Preface

The development of science and technology is aimed to create a better standard of life for the benefit of human beings all over the world. Among the various materials used in present day life, polymers have substituted many of the conventional materials, especially metals, in various applications due to their advantages. However, for some specific uses, some mechanical properties, e.g. strength and toughness, of polymer materials are found to be inadequate. Various approaches have been developed to improve such properties. In most of these applications, the properties of polymers are modified using fillers and fibers to suit the high strength/ high modulus requirements. Generally, synthetic fibers such as carbon, glass, kevlar etc., are used to prepare the polymer composites for high-end sophisticated applications due to the fact that these materials have high strength and stiffness, low density, and high corrosion resistance. Despite having several good properties, these materials (both the reinforcement and polymer matrices) are now facing problems due to their shortcomings especially related to health and biodegradability. Moreover, these fibers are not easy to degrade and results in environmental pollution. On the economic side, making a product from synthetic fiber reinforced polymer composites is a high cost activity associated with both manufacturing process and the material itself. The products engineered with petroleum-based fibers and polymers suffer severely when their service life meets the end. The non-biodegradable nature of these materials has imposed a serious threat to the environment when ecological balance is concerned. These are some of the issues which have led to the reduced utilization of petroleum-based non-biodegradable composites and the development of bio-based composite materials in which at least one component is from biorenewable resources.

Indeed, the concerns about the environment and the increasing awareness around sustainability issues are driving the push for developing new materials that incorporate renewable sustainable resources. Researchers all around the globe have been prompted to develop more environmentally-friendly and sustainable materials as a result of the rising environmental awareness and changes in the regulatory environment. These environmentally-friendly products include biodegradable and bio-based materials based on annually renewable agricultural and biomass feedstock, which in turn do not contribute to the shortage of petroleum sources. Biocomposites, which represents a group of biobased products, are produced by embedding lignocellulosic natural fibers into polymer matrices and in these composites at least one component (most frequently lignocellulosic natural fibers as the reinforcement) is from green biorenewable resources. For the last two decades, lignocellulosic natural fibers have started to be considered as alternatives to conventional man-made fibers in the academic as well as commercial arena, for a number of areas including transportation, construction, and packaging applications. The use of lignocellulosic fibers and their components as raw material in the production of polymer composites has been considered as technological progress in the context of sustainable development. The interest in lignocellulosic polymer composites is mainly driven by the low cost of lignocellulosic natural fibers, as well for their other unique advantages, such as the lower environmental pollution due to their bio-degradability, renewability, high specific properties, low density, lower specific gravity, reduced tool wear, better end-of-life characteristics, acceptable specific strength and the control of carbon dioxide emissions.

Keeping in mind the advantages of lignocellulosic polymers, this book primarily focuses on the processing, characterization and properties of lignocellulosic polymer composites. Several critical issues and suggestions for future work are comprehensively discussed in this book with the hope that the book will provide a deep insight into the state-of-the-art of lignocellulosic polymer composites. The principal credit of this goes to the authors of the chapters for summarizing the science and technology in the exciting area of lignocellulosic materials. I would also like to thank Martin Scrivener of Scrivener Publishing along with Dr. Srikanth Pilla (Series Editor) for their invaluable help in the organisation of the editing process.

Finally, I would like to thank my parents and wife Manju for their continuous encouragement and support.

Vijay Kumar Thakur, Ph.D.
Washington State University, U.S.A.
August 5, 2014

Part I

LIGNOCELLULOSIC NATURAL POLYMERS BASED COMPOSITES

1

Lignocellulosic Polymer Composites: A Brief Overview

Manju Kumari Thakur[*,1], **Aswinder Kumar Rana**[2] **and Vijay Kumar Thakur**[*,3]

[1]*Division of Chemistry, Government Degree College, Sarkaghat, Himachal Pradesh University, Summer Hill, Shimla, India*
[2]*Department of Chemistry, Sri Sai University, Palampur, H.P., India*
[3]*School of Mechanical and Materials Engineering, Washington State University, Washington, U.S.A.*

Abstract
Due to their environmental friendliness and several inherent characteristics, lignocellulosic natural fibers offer a number of advantages over synthetic fibers such as glass, carbon, aramid and nylon fibers. Some of the advantages of lignocellulosic natural fibers over synthetic fibers include biodegradability; low cost; neutrality to CO_2 emission; easy processing; less leisure; easy availability; no health risks; acceptable specific properties and excellent insulating/noise absorption properties. Due to these advantageous properties, different kinds of lignocellulosic natural fibers are being explored as indispensable components for reinforcement in the preparation of green polymer composites. With these different advantageous properties in mind, this chapter provides a brief overview of different lignocellulosic natural fibers and their structure and processing, along with their applications in different fields.

Keywords: Lignocellulosic natural fibers, structure, processing and applications

1.1 Introduction

Different kinds of materials play an imperative role in the advancement of human life. Among various materials used in present day life, polymers have been substituted for many conventional materials, especially metals, in various applications due to their advantages over conventional materials [1, 2]. Polymer-based materials are frequently used in many applications because they are easy to process, exhibit high productivity, low cost and flexibility [3]. To meet the end user requisitions, the properties of polymers are modified using fillers and fibers to suit the high strength/high modulus requirements [4]. Generally synthetic fibers such as carbon, glass, kevlar, etc., are used to prepare polymer composites for high-end, sophisticated applications due to the fact that these materials have high strength and stiffness, low density and high corrosion

**Corresponding author*: shandilyamn@gmail.com; vktthakur@hotmail.com

Vijay Kumar Thakur (Ed.), Lignocellulosic Polymer Composites, (3–16) 2015 © Scrivener Publishing LLC

resistance [5]. Fiber-reinforced polymer composites have already replaced many components of automobiles, aircrafts and spacecrafts which were earlier used to be made by metals and alloys [6]. Despite having several good properties, these materials (both the reinforcement and polymer matrices) are now facing problems due to their shortcomings especially related to health and biodegradability [7]. As an example, synthetic fibers such as glass and carbon fiber can cause acute irritation of the skin, eyes, and upper respiratory tract [8]. It is suspected that long-term exposure to these fibers causes lung scarring (i.e., pulmonary fibrosis) and cancer. Moreover, these fiber are not easy to degrade and results in environmental pollution [9]. On the economic side, making a product from synthetic fiber-reinforced polymer composites is a high cost activity associated with both the manufacturing process and the material itself [10]. The products engineered with petroleum-based fibers and polymers suffer severely when their service life meets their end [11]. The non-biodegradable nature of these materials has imposed a serious threat for the environment where ecological balance is concerned [12]. Depletion of fossil resources, release of toxic gases, and the volume of waste increases with the use of petroleum-based materials [13]. These are some issues which have led to the reduced utilization of petroleum-based non-biodegradable composites and development of biobased composite materials in which at least one component is from biorenewable resources [14].

Biobased composites are generally produced by embedding lignocellulosic natural fibers into polymer matrices, and in these composites at least one component (most frequently natural fibers as the reinforcement) is from green biorenewable resources [15]. This book is primarily focused on the effective utilization of lignocellulosic natural fibers as an indispensable component in polymer composites. The book consists of twenty-three chapters and each chapter gives an overview of a particular lignocellulosic polymer composite material. Chapter 2 focuses on natural fiber-based composites, which are the oldest types of composite materials and are the most frequently used. The book has been divided into three parts, namely: (1) Lignocellulosic natural polymer-based composites, (2) Chemical modification of cellulosic materials for advanced composites, and (3) Physico-chemical and mechanical behavior of cellulose/polymer composites. In the following section a brief overview of lignocellulosic fibers/polymer composites will be presented.

1.2 Lignocellulosic Polymers: Source, Classification and Processing

Different kinds of biobased polymeric materials are available all around the globe. These biobased materials are procured from different biorenewable resources. Chapters 2–10 primarily focus on the use of different types of lignocellulosic fiber-reinforced composites, starting from wood fibers to hybrid fiber-reinforced polymer composites. Chapter 3 summarizes some of the recent research on different lignocellulosic fiber-reinforced polymer composites in the Southeast region of the world, while Chapter 6 summarizes the research on some typical Brazilian lignocellulosic fiber composites. The polymers obtained from biopolymers are frequently referred to as biobased

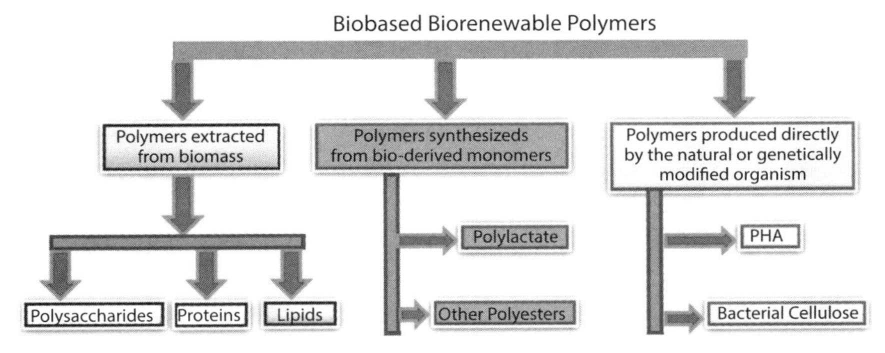

Figure 1.1 (a) Classification of biobased polymers [11, 13, 16].

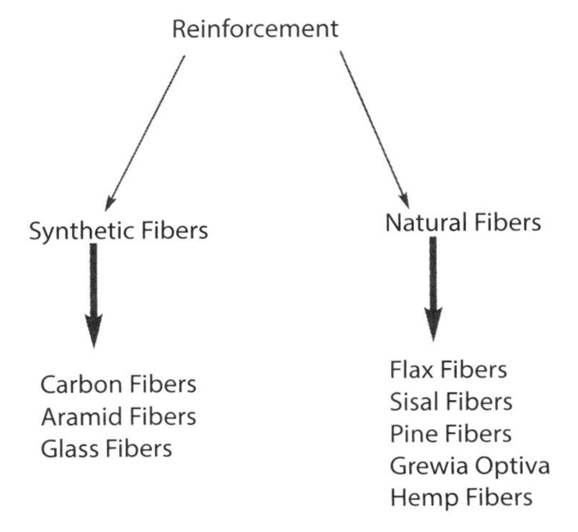

Figure 1.1 (b) Types of fiber reinforcement used in the preparation of polymer composites[11, 13, 16].

biorenewable polymers and can be classified into different categories depending upon their prime sources of origin/production. Figure 1.1(a) shows the general classification of biobased biorenewable polymers [11, 13, 16].

For the preparation of polymer composites, generally two types of fibers, namely synthetic and natural fibers, are used as reinforcement. Figure 1.1(b) shows different types of natural/synthetic fibers frequently used as reinforcement in the polymer matrix composites.

Natural fibers can further be divided into two types: plant fibers and animal fibers. Figure 1.2 shows the detailed classification of the different plant fibers. These plant fibers are frequently referred to as lignocellulosic fibers.

Among biorenewable natural fibers, lignocellulosic natural fibers are of much importance due to their inherent advantages such as: biodegradability, low cost, environmental friendliness, ease of separation, recyclability, non-irritation to the skin, acceptable specific strength, low density, high toughness, good thermal properties, reduced tool wear, enhanced energy recovery, etc. [11,13,16,17]. Different kinds of lignocellulosic materials are available all around the world. These lignocellulosic materials are procured from different biorenewable resources. The properties of the lignocellulosic materials depend upon different factors and growing conditions. Lignocellulosic

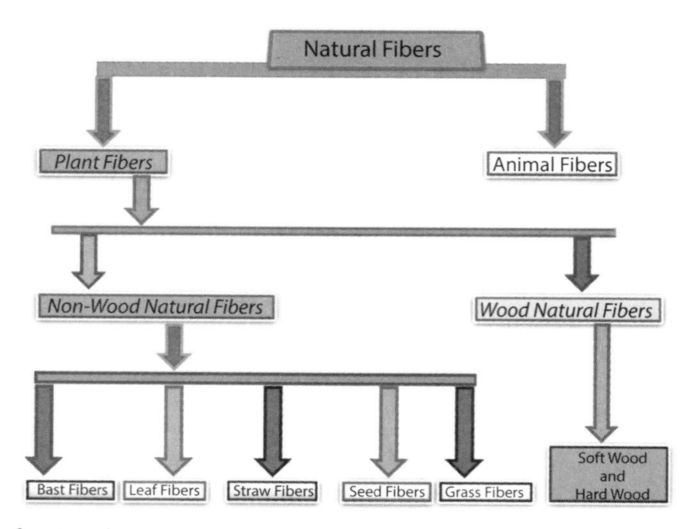

Figure 1.2 Classification of natural fibers [11, 13, 16].

Table 1.1 Factors effecting fiber quality at various stages of natural fiber production. Reprinted with permission from [18]. Copyright 2012 Elsevier.

Stage	**Factors effecting fiber quality**
Plant growth	Species of plant
	Crop cultivation
	Crop location
	Fiber location in plant
	Local climate
Harvesting stage	Fiber ripeness, which effects: – Cell wall thickness – Coarseness of fibers – Adherence between fibers and surrounding structure
Fiber extraction stage	Decortication process
	Type of retting method
Supply stage	Transportation conditions
	Storage conditions
	Age of fiber

natural fibers are generally harvested from different parts of the plant such as stem, leaves, or seeds. [18]. A number of factors influence the overall properties of the lignocellulosic fibers. Table 1.1 summarizes some of the factors affecting the overall properties of lignocellulosic fibers. The plant species, the crop production, the location, and the climate in which the plant is grown significantly affect the overall properties of the lignocellulosic fibers [18].

The properties and cost of lignocellulosic natural fibers vary significantly with fiber type. Figure 1.3(a-c) shows the comparison of potential specific modulus values of

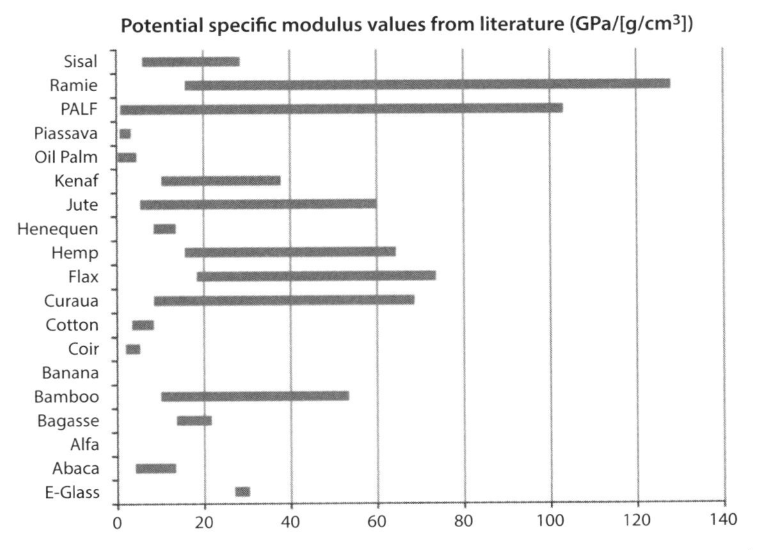

Figure 1.3 (a) Comparison of potential specific modulus values and ranges between natural fibers and glass fibers. Reprinted with permission from [18]. Copyright 2012 Elsevier.

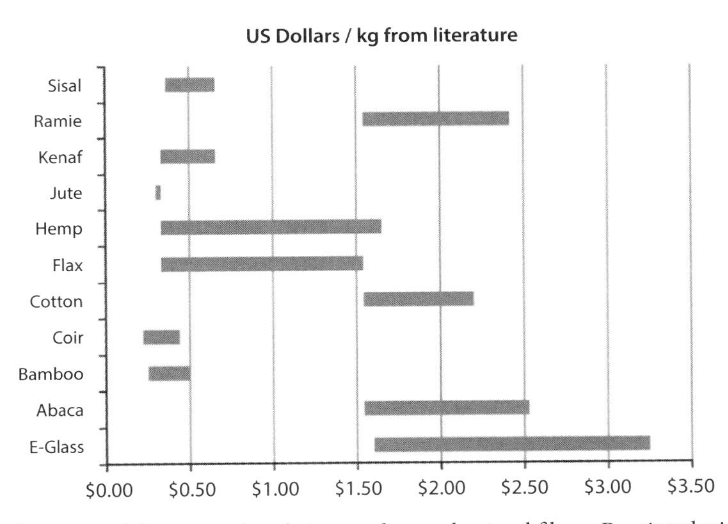

Figure 1.3 (b) Cost per weight comparison between glass and natural fibers. Reprinted with permission from [18]. Copyright 2012 Elsevier.

natural fibers/glass fibers; cost per weight comparison between natural fibers and glass and cost per unit length respectively.

Chapters 2–10 discuss in detail the different properties of natural lignocellulosic fibers, their processing and fabrication of polymer composites. Chapter 11 summarizes the structure, chemistry and properties of different agro-residual fibers such as wheat straw; corn stalk, cob and husks; okra stem; banana stem, leaf, bunch; reed stalk; nettle; pineapple leaf; sugarcane; oil palm bunch and coconut husk; along with their processing.

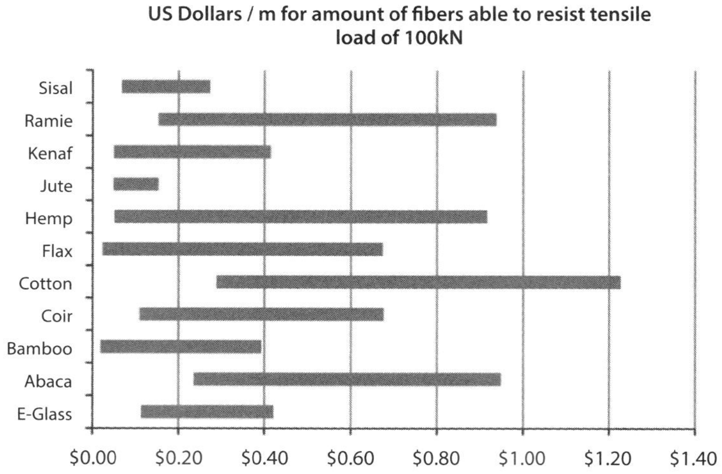

Figure 1.3 (c) Cost per unit length (capable of resisting 100 KN load) comparison between glass and natural fibers. Reprinted with permission from [18]. Copyright 2012 Elsevier.

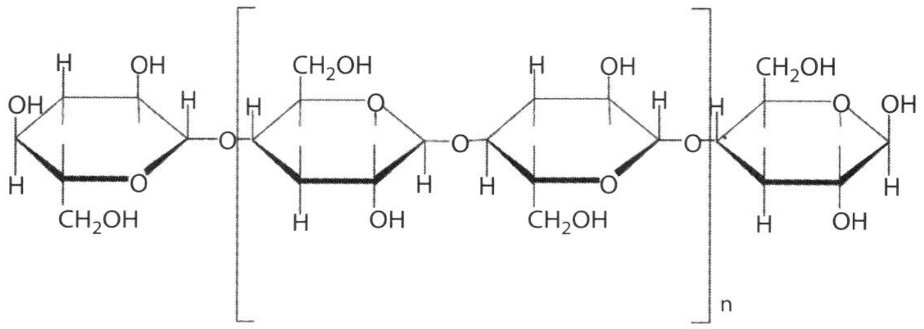

Figure 1.4 (a). Chemical structure of cellulose. Reprinted with permission from [19]. Copyright 2011 Elsevier.

1.3 Lignocellulosic Natural Fibers: Structure, Chemical Composition and Properties

Lignocellulosic natural fibers are primarily composed of three components, namely cellulose, hemicellulose and lignin. Figure 1.4 (a, b) shows the structure of cellulose and lignin. Cellulose contains chains of variable length of 1-4 linked β-d-anhydroglucopyranose units and is a non-branched macromolecule[19]. As opposed to the structure of cellulose, lignin exhibits a highly branched polymeric structure [17-19]. Lignin serves as the matrix material to embed cellulose fibers along with hemicellulose, and protects the cellulose/hemicellulose from harsh environmental conditions [1, 13, 16]. Chapter 11 discusses in detail the chemical composition of lignocellulosic natural fibers.

The plant cell wall is the most important part of lignocellulosic natural fibers. Figure 1.4(c) shows the schematic representation of the natural plant cell wall [19]. The cell wall of lignocellulosic natural fibers primarily consists of a hollow tube with four different layers [19]. The first layer is called the primary cell wall, the other three, the secondary cell walls, while an open channel in the center of the microfibrils is called the lumen

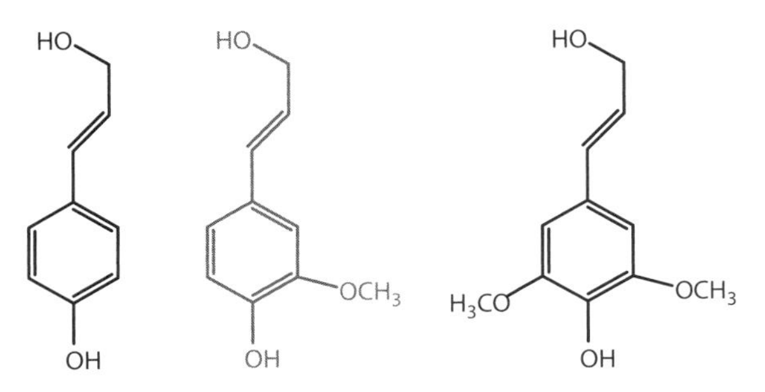

Figure 1.4 (b) Structure of Lignin [1, 13, 16].

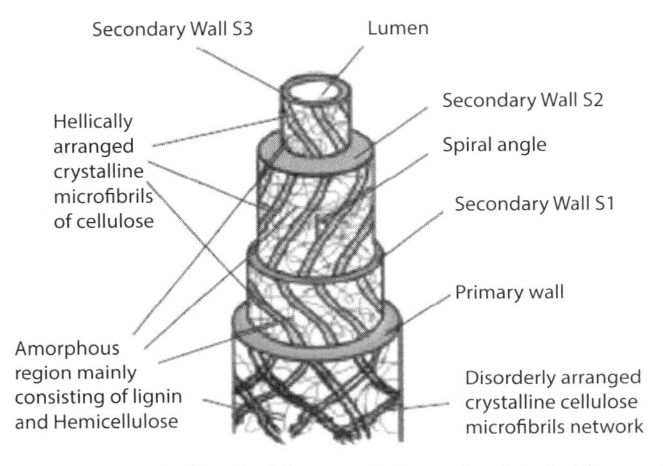

Figure 1.4 (c) Schematic picture of cell wall of the natural plants. Reprinted with permission from [19]. Copyright 2011 Elsevier.

[1, 13, 16]. These layers are composed of cellulose embedded in a matrix of hemicellulose and lignin. In lignocellulosic natural fibers, cellulose components provide the strength and stiffness to the fibers via hydrogen bonds and other linkages. On the other hand, hemicellulose has been found to be responsible for moisture absorption, biodegradation, and thermal degradation of the fibers [1, 13, 16].

Table 1.2 summarizes some of the advantages/disadvantages of lignocellulosic natural fibers[20].

Chapter 5 summarizes the investigation of lignocellulosic flax fiber-based reinforcement requirements to obtain structural and complex shape polymer composites. This chapter discusses in detail the possibility of forming complex shape structural composites which are highly desirable for advanced applications. Chapter 7 focuses on the structure and properties of cellulose-based starch polymer composites, while Chapter 8 focuses on the spectroscopic analysis of rice husk and wheat gluten husk-based polymer composites using computational chemistry. Chapter 9 summarizes the processing, characterization and properties of oil palm fiber-reinforced polymer composites. In this chapter, the use of oil palm as reinforcement in different polymer matrices such as natural rubber, polypropylene, polyurethane, polyvinyl chloride, polyester, phenol formaldehyde, polystyrene, epoxy and LLDPE is discussed. Chapter 10 also focuses on

Table 1.2 Advantage and disadvantages of natural fibers cellulosic/synthetic fiber-reinforced polymer hybrid composites [20]. Copyright 2011 Elsevier.

Advantages	Disadvantages
Low specific weight results in a higher specific strength and stiffness than glass	Lower strength especially impact strength
Renewable resources, production require littleenergy and low CO_2 emission	Variable quality, influence by weather
Production with low investment at low cost	Poor moisture resistant which causes swelling of the fibers
Friendly processing, no wear of tools andno skin irritation	Restricted maximum processing temperature
High electrical resistant	Lower durability
Good thermal and acoustic insulating properties	Poor fire resistant
Biodegradable	Poor fiber/matrix adhesion
Thermal recycling is possible	Price fluctuation by harvest results or

the processing and characterization of oil palm- and pine apple-reinforced polymer composites.

1.4 Lignocellulosic Polymer Composites: Classification and Applications

Lignocellulosic polymer composites refer to the engineering materials in which polymers (procured from natural/petroleum resources) serve as the matrix while the lignocellulosic fibers act as the reinforcement to provide the desired characteristics in the resulting composite material. Polymer composites are primarily classified into two types: (a) fiber-reinforced polymer composites and (b) particle-reinforced polymer composites. Figure 1.5 (a) shows the classification of polymer composites depending upon the type of reinforcement.

Depending upon the final application perspectives of the polymer composite materials, both the fibers as well as particle can be used as reinforcement in the polymer matrix.

Polymer composites are also classified into renewable/nonrenewable polymer composites depending upon the nature of the polymer/matrix [1, 13, 16]. Figure 1.5 (b) show the classification of polymer composites depending upon the renewable/nonrenewable nature. Polymer composites in which both components are obtained from biorenewable resources are referred to as 100% renewable composites, while composites in which at least one component is from a biorenewable resource are referred to as partly renewable polymer composites[1, 13, 16]. Chapter 4 of the book presents a review on the state-of-the-art of partly renewable polymer composites with a particular focus on the hybrid vegetable/glass fiber composites. This chapter summarizes the hybridization effect on the properties of the final thermoplastic and thermoset polymer matrices

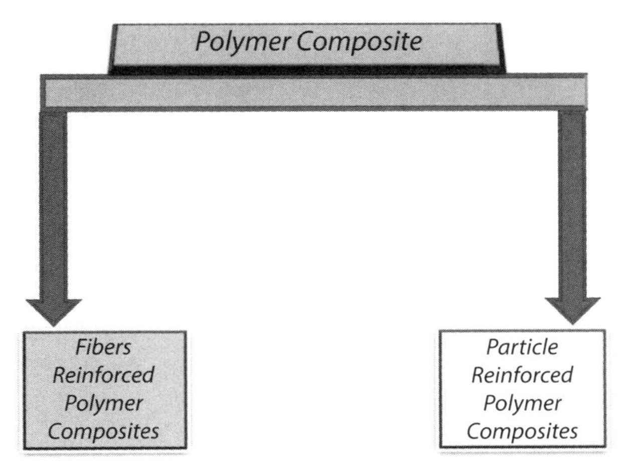

Figure 1.5 (a) Classification of polymer composites, depending upon the reinforcement type [1, 13, 16].

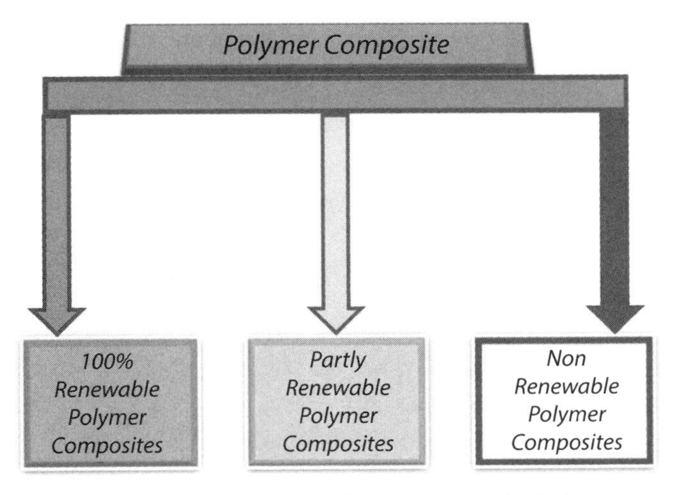

Figure 1.5 (b) Classification of polymer composites depending upon both the polymer matrix and reinforcement type [1, 13, 16].

composites. On the other hand, the polymer composites in which none of the parts are from biorenewable resources are referred to as nonrenewable composites.

Although lignocellulosic natural fibers and their respective polymer composites offer a number of advantages over their synthetic counterparts, these lignocellulosic fibers/polymer composites also suffer from a few drawbacks [21][22][23]. One of the biggest drawbacks of these fibers and their composites is their sensitivity towards the moisture and water, which ultimately deteriorates the overall properties of these materials [24][25][26]. These lignocellulosic fiber/polymer composites also show a poor chemical resistance [27–29]. In addition to these drawbacks, another main disadvantage encountered during the addition of lignocellulosic natural fibers into a polymer matrix, is the lack of good interfacial adhesion between the two components [2][14][19][30]. Chapter 2 primarily focuses on the adhesion aspects of natural fiber-based polymer composites. Different characterization techniques for the evaluation of the interfacial properties of the polymer composites are described in this chapter. The hydroxyl groups present on the lignocellulosic fibers are also incompatible with most of the matrices, especially the thermoplastic polymer matrices [2][14] [19][30]. A number of methods are presently being explored to improve the surface characteristics of these lignocellulosic fibers. Some of the most common techniques used to increase the physico-chemical characteristics of the lignocellulosic natural fibers include mercerization, silane treatment and graft copolymerization [9][12][31–35].

Chapters 11–14 of this book focus solely on the different chemical modification techniques used to improve the physico-chemical properties of the lignocellulosic fibers. In addition to these chapters, other chapters also briefly focus on some selected chemical modification techniques. For example, Chapter 10 briefly discusses the effect of alkali treatment on the properties of oil palm- and pine apple-reinforced polymer composites. Chapter 11 discusses the effect of both the physical and chemical modification techniques on the properties of lignocellulosic polymer composites. The chemical modification techniques summarized in this chapter include alkalization treatment, acetylation, silane treatment, bleaching, enzyme treatment, sulfonation and graft copolymerization. Chapter 12 also focuses on the different chemical treatments for cellulosic fibers carried out during the primary processing of polymer composites. Chapter 13 summarizes the effect of mercerization and benzoylation on different physico-chemical properties of the lignocellulosic *Grewia optiva* fiber-reinforced polymer composites. Chapter 14 focuses on the effect of chemical treatments, namely alkali treatment, benzene diazonium salt treatment, o-hydroxybenzene diazonium salt treatment, succinic anhydride treatment, acrylonitrile treatment, maleic anhydride treatment, and nanoclay treatment; along with several other chemical treatments on different cellulosic fibers.

Chapters 15–18 focus on the weathering/mechanical study of lignocellulosic fiber-reinforced polymer composites. The effect of different environmental conditions on the physico-chemical and mechanical properties of the polymer composites is discussed in detail in these chapters. Chapter 15 mainly focuses on the effect of weathering conditions on the properties of lignocellulosic polymer composites. Most of the focus of this chapter is the effect of UV radiation on different properties of composites. Chapter 16 describes the effect of layering pattern on the physical, mechanical and acoustic properties of luffa/coir fiber-reinforced epoxy novolac hybrid composites, and Chapter 17 summarizes the fracture mechanism of wood plastic composites. Chapter 18 focuses on the mechanical behavior of biocomposites under different environmental conditions.

Lignocellulosic polymer composites are mainly fabricated using the following processes: (a) Compression Molding, (b) Injection Molding, (c) EXPRESS Process (extrusion-compression molding), and (d) Structural Reaction Injection Molding (S-RIM). Among these processes, compression molding is most frequently used and sometime combined with the hand lay-up method.

The use of natural fiber-reinforced composites is increasing very rapidly for a number of applications ranging from automotive to aerospace. Chapters 19–23 describe the different applications of lignocellulosic polymer composites. Chapter 19 focuses on the applications of lignocellulosic fibers in construction, while Chapter 20 summarizes the use of lignocellulosic jute fibers for next generation applications. Chapter 21 discusses in detail the use of cellulosic composites for packaging applications. Lignocellulosic fiber-reinforced polymer composites are seriously being considered as alternatives to

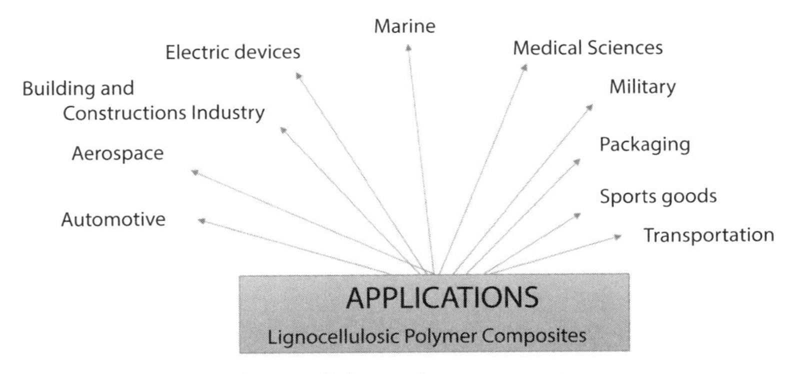

Figure 1.6 Potential Applications of lignocellulosic polymer composites.

synthetic fiber-reinforced composites as a result of growing environmental awareness. Figure 1.6 summarizes some of the recent applications of lignocellulosic natural fiber-reinforced polymer composites in different fields.

1.5 Conclusions

Among different composite materials, lignocellulosic polymer composites have a bright future for several applications due to their inherent eco-friendliness and other advantages. The effective utilization of different lignocellulosic fibers as one of the components in the polymer composites have immense scope for future development in this field. For the successful development of low-cost, advanced composites from different lignocellulosic materials, comprehensive research on how to overcome the drawbacks of lignocellulosic polymer composites, along with seeking new routes to effectively utilize these composites, is of utmost importance.

References

1. R. Prasanth, R. Shankar, A. Dilfi, V. Thakur, and J.-H. Ahn, Eco-friendly fiber-reinforced natural rubber green composites: A perspective on the future, in *Green Composites from Natural Resources,* CRC Press, Boca Raton, FL (2013).
2. V.K. Thakur, A.S. Singha, and M.K. Thakur, Green Composites from Natural Fibers: Mechanical and Chemical Aging Properties. *Int. J. Polym. Anal. Charact.* 17, 401–407 (2012).
3. V.K. Thakur, M.K. Thakur, and R. Gupta, Eulaliopsis binata: Utilization of Waste Biomass in Green Composites, in *Green Composites from Natural Resources,* pp. 125-130, CRC Press, Boca Raton, FL (2013).
4. V.K. Thakur, M.K. Thakur, and R. Gupta, Green Composites from Functionalized Renewable Cellulosic Fibers, in *Green Composites from Natural Resources, pp. 307-318,* CRC Press, Boca Raton, FL (2013).
5. M.A. Pinto, V.B. Chalivendra, Y.K. Kim, and A.F. Lewis, Effect of surface treatment and Z-axis reinforcement on the interlaminar fracture of jute/epoxy laminated composites. *Eng. Fract. Mech.* 114, 104–114 (2013).

6. M.A. Pinto, V.B. Chalivendra, Y.K. Kim, and A.F. Lewis, Valuation of surface treatment and fabrication methods for jute fiber/epoxy laminar composites. *Polym. Compos.* 35, 310–317 (2014).
7. Y. Srithep, T. Effingham, J. Peng, R. Sabo, C. Clemons, L.-S. Turng, and S. Pilla, Melt compounding of poly (3-hydroxybutyrate-co-3-hydroxyvalerate)/nanofibrillated cellulose nanocomposites. *Polym. Degrad. Stab.* 98, 1439–1449 (2013).
8. A.S. Singha and V.K. Thakur, Fabrication of Hibiscus sabdariffa fibre reinforced polymer composites. *Iran. Polym. J.* 17, 541–553 (2008).
9. V.K. Thakur, M. K. Thakur, and R. K. Gupta, Graft copolymers from cellulose: Synthesis, characterization and evaluation. *Carbohydr. Polym.* 97, 18–25 (2013).
10. V.K. Thakur, M.K. Thakur, and R.K. Gupta, Synthesis of lignocellulosic polymer with improved chemical resistance through free radical polymerization. *Int. J. Biol. Macromol.* 61, 121–126 (2013).
11. N. Dissanayake and J. Summerscales, Life cycle assessment for natural fiber composites, in *Green Composites from Natural Resources*, pp. 157-181, CRC Press, Boca Raton, FL (2013).
12. V.K. Thakur, M.K. Thakur, and R.K. Gupta, Graft copolymers from natural polymers using free radical polymerization. *Int. J. Polym. Anal. Charact.* **18**, 495–503 (2013).
13. V.K. Thakur and A.S. Singha, *Biomass-based biocomposites, iSmithers Rapra,* (2013).
14. V.K. Thakur, A.S. Singha, and M.K. Thakur, Fabrication and physico-chemical properties of high-performance pine needles/green polymer composites. *Int. J. Polym. Mater.* 62, 226–230 (2013).
15. V.K. Thakur, A.S. Singha, and M.K. Thakur, Biopolymers Based Green Composites: Mechanical, Thermal and Physico-chemical Characterization. *J. Polym. Environ.* **20**, 412–421 (2011).
16. V.K. Thakur, M.K. Thakur, R. Gupta, R. Prasanth, and M. Kessler, Green composites: an introduction, *in Green Composites from Natural Resources, pp. 1-10*CRC Press, Boca Raton, FL (2013).
17. V.K. Thakur, M.K. Thakur, and R.K. Gupta, Review: Raw natural fibers based polymer composites, *Int. J. Polym. Anal. Charact.*.
18. D.B. Dittenber and H.V.S. GangaRao, Critical review of recent publications on use of natural composites in infrastructure. *Compos. Part Appl. Sci. Manuf.* 43, 1419–1429 (2012).
19. H.M. Akil, M.F. Omar, A.A.M. Mazuki, S. Safiee, Z.A.M. Ishak, and A. Abu Bakar, Kenaf fiber reinforced composites: A review. *Mater. Des.* 32, 4107–4121 (2011).
20. M. Jawaid and H.P.S.A. Khalil, Cellulosic/synthetic fibre reinforced polymer hybrid composites: A review. *Carbohydr. Polym.* 86, 1–18 (2011).
21. A.S. Singha and V.K. Thakur, Synthesis and characterization of grewia optiva fiber-reinforced PF-based composites. *Int. J. Polym. Mater.* 57, 1059–1074 (2008).
22. A.S. Singha and V.K. Thakur, Synthesis, characterisation and analysis of hibiscus sabdariffa fibre reinforced polymer matrix based composites. *Polym. Polym. Compos.* 17, 189–194 (2009).
23. A.S. Singha, V.K. Thakur, and B.N. Mishra, Study of Grewia optiva fiber reinforced urea-formaldehyde composites. *J. Polym. Mater.* 26, 81–90 (2009).
24. A.S. Singha and V.K. Thakur, Synthesis and characterization of pine needles reinforced RF matrix based biocomposites. *J. Chem.* 5, 1055–1062 (2008).
25. A.S. Singha and V.K. Thakur, Chemical resistance, mechanical and physical properties of biofibers-based polymer composite. *Polym. Plast. Technol. Eng.* 48, 736–744 (2009).
26. A.S. Singha and V.K. Thakur, Fabrication and characterization of H. sabdariffa fiber-reinforced green polymer composites. *Polym. Plast. Technol. Eng.* 48, 482–487 (2009).

27. A.S. Singha and V.K. Thakur, Physical, chemical and mechanical properties of Hibiscus sabdariffa fiber/polymer composite. *Int. J. Polym. Mater.* 58, 217–228 (2009).
28. A.S. Singha and V.K. Thakur, Morphological, thermal, and physicochemical characterization of surface modified pinus fibers. *Int. J. Polym. Anal. Charact.* 14, 271–289 (2009).
29. V.K. Thakur and A.S. Singha, Physicochemical and mechanical behavior of cellulosic pine needle-based biocomposites. *Int. J. Polym. Anal. Charact.* 16, 390–398 (2011).
30. V.K. Thakur, A.S. Singha, and M.K. Thakur, Natural cellulosic polymers as potential reinforcement in composites: Physicochemical and mechanical studies. *Adv. Polym. Technol.* 32, E427–E435 (2013).
31. V.K. Thakur, A.S. Singha, and M.K. Thakur, Synthesis of natural cellulose-based graft copolymers using methyl methacrylate as an efficient monomer. *Adv. Polym. Technol.* 32, E741–E748 (2013).
32. V.K. Thakur, A.S. Singha, and M.K. Thakur, Pressure induced synthesis of EA grafted Saccaharum cilliare fibers. *Int. J. Polym. Mater. Polym. Biomater.* 63, 17–22 (2014).
33. V.K. Thakur, A.S. Singha, and M.K. Thakur, Graft copolymerization of methyl methacrylate onto cellulosic biofibers. *J. Appl. Polym. Sci.* 122, 532–544 (2011).
34. V.K. Thakur, M.K. Thakur, and R.K. Gupta, Rapid synthesis of graft copolymers from natural cellulose fibers. *Carbohydr. Polym.* 98, 820–828 (2013).
35. V.K. Thakur, M.K. Thakur, and R.K. Gupta, Graft copolymers of natural fibers for green composites. Carbohydr. *Polym.* 104, 87–93 (2014).
36. R.M. dos Santos, W.P. Flauzino Neto, H.A. SilvÉrio, D.F. Martins, N.O. Dantas, and D. Pasquini, Cellulose nanocrystals from pineapple leaf, a new approach for the reuse of this agro-waste. *Ind. Crops Prod.* 50, 707–714 (2013).

2

Interfacial Adhesion in Natural Fiber-Reinforced Polymer Composites

E. Petinakis [*,1,2], L. Yu [2], G. P. Simon [2], X. J. Dai[3], Z. Chen[3] and K. Dean [2]

[1]CSIRO, Manufacturing Flagship, Melbourne, Australia
[2]Department of Materials Engineering, Monash University, Melbourne, Australia
[3]Institute for Frontier Materials, Deakin University, Melbourne, Australia

Abstract
Concerns about the environment and increasing awareness about sustainability issues are driving the push for developing new materials that incorporate renewable sustainable resources. This has resulted in the use of natural fibers for developing natural fiber-reinforced polymer composites (NFRPCs). A fundamental understanding of the fiber-fiber and fiber-matrix interface is critical to the design and manufacture of polymer composite materials because stress transfer between load-bearing fibers can occur at the both of these interfaces. Efficient stress transfer from the matrix to the fiber will result in polymer composites exhibiting suitable mechanical and thermal performance. The development of new techniques has facilitated a better understanding of the governing forces that occur at the interface between matrix and natural fiber. The use of surface modification is seen as a critical processing parameter for developing new materials, and plasma-based modification techniques are gaining more prominence from an environmental point of view, as well as a practical approach.

Keywords: Natural fibers, biocomposites, interfacial adhesion, impact strength, morphology, atomic force microscopy

2.1 Introduction

In order to develop natural fiber-reinforced polymer composites, the primary focus should be given to the nature of the interface between the natural fiber and the polymer matrix. A fundamental understanding of fiber-fiber and fiber-matrix interface is critical to the design and manufacture of polymer composite materials because stress transfer between load-bearing fibers can occur at the fiber-fiber interface and fiber-matrix interface. In wood-polymer composite systems there are two interfaces that exist, one between the wood surface and the interphase and one between the polymer and interphase [1]. Therefore, failure in a composite or bonded laminate can occur as follows; (i) adhesive failure in the wood–interphase interface, (ii) the interphase-polymer

**Corresponding author*: steven.petinakis@csiro.au

Vijay Kumar Thakur (Ed.), Lignocellulosic Polymer Composites, (17–40) 2015 © Scrivener Publishing LLC

interface, or (iii) cohesive failure of the interphase. The major role of fibers is to facilitate the efficient transfer of stress from broken fibers to unbroken fibers by the shear deformation of the resin at the interface. Therefore, the degree of mechanical performance in natural fiber composites is dictated entirely by the efficiency of stress transfer through the interface. In general, increased fiber-matrix interactions will result in a composite with greater tensile strength and stiffness, while the impact properties of the composite will be reduced. The final properties of the composite will entirely be controlled by the nature of the interface. The interfacial shear strength (IFSS) in fiber-reinforced composites is controlled primarily by mechanical and chemical factors, as well as surface energetics. The mechanical factors include thermal expansion mismatch between the fibers and the resin, surface roughness and resulting interlocking of the fiber in the resin, post-debonding fiber resin friction, specific surface area, and resin microvoid concentration adjacent to the fibers. Most fiber-reinforced composites are processed above room temperature, so as the composites cool down following processing, differences in the coefficient of thermal expansion of the fiber and resin will result in resin shrinkage, causing circumferential compressive forces acting on the fiber, resulting in a strong grip by the resin on the fiber. Fiber surface roughness also improves the IFSS since the surface roughness can increase mechanical interlocking between well-aligned fibers and polymer matrix.

The nature of the interface/interphase in polymer composites incorporating natural fibers is still not well understood. Most natural fibers consist of cellulose, hemicelluloses, lignin and other low molecular weight compounds [2]. The properties of natural fibers are dictated by their growing circumstances and processing. The heterogeneous nature of natural fibers can lead to composites with variations in interfacial interactions, which can impact negatively or positively on the mechanical and thermal performance of the resulting final composites. Therefore, the aim of the research in this chapter will be to attempt to address this issue and provide insight into the surface and adhesion properties of polymer composite systems incorporating natural fibers and biopolymers as well as conventional polymers. Special attention will be given to Poly(lactic acid)-based composites, as this is the subject of ongoing research by the authors of this chapter [3, 4].

2.2 PLA-Based Wood-Flour Composites

Polymer composites incorporating natural fibers can result in materials with improved mechanical and impact performance, which is well-documented [5-7], and is of particular interest for enhancing the properties of biodegradable polymeric materials such as PLA [8-10]. The benefits of using natural fiber instead of conventional reinforcing agents, such as glass fibers, talc, or carbon fibers, for improving the performance of biodegradable polymers include the retention of the biodegradability of the composite, as well as lower density and improved performance. Furthermore, these natural fibrous fillers are also usually lower cost because they are typically industrial bio-waste, for example wood-flour byproduct from the timber industry.

PLA composites incorporating a range of cellulosic fibers have been reported, and include the use of flax [11], sisal [12], bamboo [13], short abaca [14], jute [15], lyocell

[16, 17], paper and pulp [18, 19], and microcrystalline cellulose [20]. The extent of particle-matrix interactions in polymer composites comprising natural polymeric fillers strongly influences the mechanical properties of the final composite, with poor interfacial adhesion between the particle and the matrix generally leading to composites having inferior mechanical properties [21]. PLA/cellulose-fiber composites containing less than 30%w/w fiber have been shown to have increased tensile modulus and reduced tensile strength compared with PLA, and this has been attributed to factors that include the weak interfacial adhesion between the less polar PLA matrix and the highly polar surface of the cellulose fibers [22], and lack of fiber dispersion due to a high degree of fiber agglomeration [21]. According to the literature, the surface tension, (γ_s^d_) of PLA is typically between 20-30mJ/m^2[23] and wood-flour is typically around 40mJ/m2 [24].

In order to improve interfacial properties of PLA based composites incorporating natural fibers and/or fillers, surface modification techniques can be deployed by (a) modification of natural fibers using conventional wet-based chemistry techniques or addition of functional compatibilizers and/or coupling agents or (b) modification of the matrix through the addition of impact modifiers, with reactive end groups. It is well known that the use of compatibilizers can improve interfacial adhesion between polymer and natural fiber in polymer composites, and this approach has been reported to improve the compatibility between PLA and carbohydrate materials such as starch [25, 26]. This improved compatibility is attributed to the introduction of reactive groups at the interface between the PLA matrix and the surface of the more polar starch particles, where the formation of this interphase strengthens the chemical and physical interaction between the polar surface of the starch particles and the less polar PLA host matrix. The accompanying increase in the overall polarity of the host polymer matrix resulting from the polar nature of the introduced reactive sites is also likely to promote more uniform dispersion of the polar starch particles in the host matrix.

Compatibilizers utilized in PLA-based composites include maleic anhydride-polypropylene (MAPP) [27, 28] and methylenediphenyl-diisocyanate (MDI). The use of MDI as a coupling agent in renewable polymer composite materials has been studied extensively for blends of starch and PLA [29], and has been shown to yield materials having enhanced mechanical properties, as a consequence of the reaction between isocyanate moieties of MDI and free hydroxyl groups in starch, resulting in the *in situ* formation of isocyanate groups on the surface of the starch particles. Free isocyanate groups can then act to compatibilize the starch and the PLA by reacting with hydroxyl end groups of PLA. We have demonstrated in a previous paper that MDI has been shown to compatibilize composites comprising PLA and wood-flour in a similar fashion (see Scheme 2.1).

The use of poly (ethylene–acrylic acid) (PEAA), as toughening agents (also known as impact modifiers) have shown to improve the flexibility and impact performance of composites of polyolefins and wood fibers [30-32]. The improved toughness is attributed to the incorporation of the rubbery polyethylenic chains into the polymer matrix, which can assist in dissipating or absorbing the energy during crack propagation. However, since the acrylic acid functionality of PEAA can, in principle, react with the cellulosic hydroxyl groups located at the surface of wood-flour particles to form ester linkages, as illustrated in Scheme 2.2, it is possible that enhanced compatibilisation between the wood-flour particles and PLA, might also be achieved through the

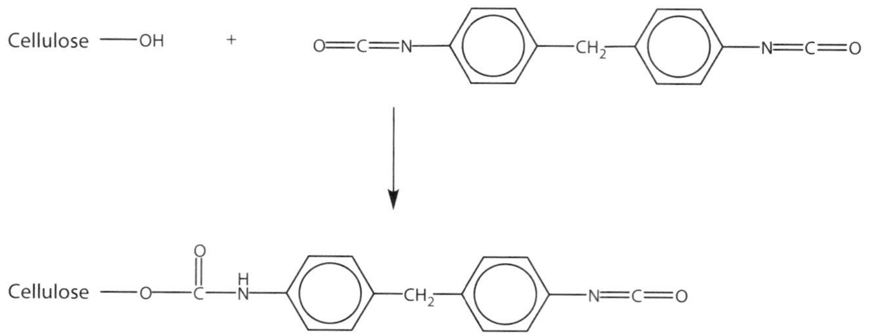

Cellulose —OH + O=C=N–C$_6$H$_4$–CH$_2$–C$_6$H$_4$–N=C=O

↓

Cellulose –O–C(=O)–NH–C$_6$H$_4$–CH$_2$–C$_6$H$_4$–N=C=O

Scheme 2.1 (Petinakis *et al.* 2009).

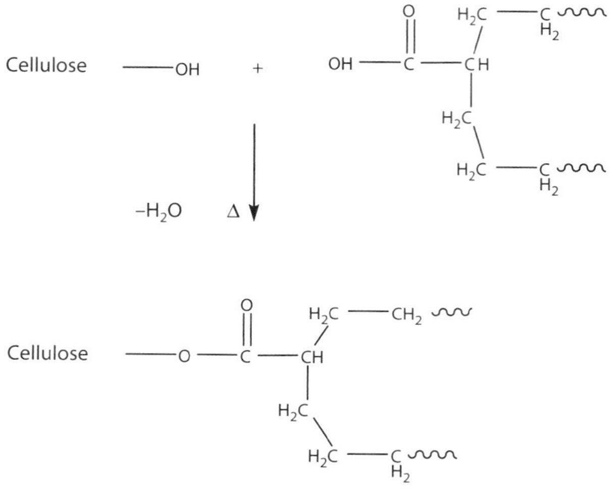

Cellulose —OH + OH–C(=O)–CH(H$_2$C–C(H$_2$)~)–H$_2$C–H$_2$C–C(H$_2$)~

$-H_2O$ Δ ↓

Cellulose –O–C(=O)–CH(H$_2$C–CH$_2$~)–H$_2$C–H$_2$C–C(H$_2$)~

Scheme 2.2 (Petinakis *et al.* 2009).

addition of PEAA. However, in our work on modification of PLA [3], it was found that the addition of PEAA actually modified the PLA matrix, through the dispersion of fine rubbery particles within the PLA matrix. Hence, the impact properties of the resulting composites were improved at the expense of the interfacial adhesion.

2.3 Optimizing Interfacial Adhesion in Wood-Polymer Composites

Optimization of interfacial adhesion in the development of natural fiber-reinforced polymer composites has been the subject of extensive research of the past two decades. Many techniques have been developed and tested and the principal aim of various modification strategies has been to reduce the fiber-fiber interaction through aiding improved wetting and dispersion, as well as improving interfacial adhesion and the resulting stress transfer efficiency from the matrix to the fiber.

2.3.1 Chemical Modification

Alkaline treatment is one of the most widely used chemical treatments for natural fibers for use in natural fiber composites. The effect of alkaline treatment on natural fibers is it disrupts the incidence of hydrogen bonding in the network structure, giving rise to additional sites for mechanical interlocking, hence promoting surface roughness and increasing matrix/fiber interpenetration at the interface. During alkaline treatment of lignocellulosic materials, the alkaline treatment removes a degree of the lignin, wax and oils which are present, from the external surface of the fiber cell wall, as well as causing chain scissioning of the polymer backbone, resulting in small crystallites. The treatment exposes the hydroxyl groups in the cellulose component to the alkoxide. Beg *et al.* studied the effect of the pre-treatment of radiate pine fiber with NaOH and coupling with MAPP in wood fiber-reinforced polypropylene composites. It was found that fiber pre-treatment with NaOH resulted in an improvement in the stiffness of the composites (at 60% fiber loading) as a function NaOH concentration, however at the same time, a decrease was observed in the strength of the composite [33]. The reason for a reduction in the tensile strength was attributed to a weakening of the cohesive strength of the fiber as a result of alkali treatment. The use of alkali treatment in conjunction with MAPP was found to improve the fiber/matrix adhesion. However, it seems that only small concentrations of NaOH can be used to treat fibers, otherwise the cohesive strength can be compromised. Ichazo et. al also studied the addition of alkaline treated wood flour in polypropylene/wood flour composites. It was shown that alkaline treatment only improved fiber dispersion within the polypropylene matrix, but not the fiber-matrix adhesion. This was attributed to a greater concentration of hydroxyl groups present, which increased the hydrophilic nature of the composites. As a result, no significant improvement was observed in the mechanical properties of the composites and a reduction in the impact properties [34]. From previous studies it is shown that the optimal treatment conditions for alkalization must be investigated further in order to improve mechanical properties. Care must be taken in selecting the appropriate concentration, treatment time and temperature, since at certain conditions the tensile properties are severely compromised. Islam *et al.* studied the effect of alkali treatment on hemp fibers, which were utilized to produce PLA biocomposites incorporating hemp fibers. This study showed that crystallinity in PLA was increased due to the nucleation of hemp fibers following alkaline treatment. The degree of crystallinity had a positive impact on the mechanical and impact performance of the resulting composites with alkaline treated hemp fibers, as opposed to the composites without treated hemp fibers.

Qian *et al.* conducted a study on bamboo particles (BP) that were treated with low-concentrations of alkali solution for various times and used as reinforcements in PLA based composites [35]. Characteristics of BP by composition analysis, scanning electron microscopy, Brunauer-Emmett Teller test, and Fourier transform infrared spectroscopy, showed that low-concentration alkali treatment had a significant influence on the microstructure, specific surface area, and chemical groups of BP. PLA/treated-BP and PLA/untreated-BP composites were both produced with 30 wt% BP content. Mechanical measurements showed that tensile strength, tensile modulus, and elongation at break of PLA/BP composites increased when the alkali treatment time reached

3.0 h with maximal values of 44.21, 406.41MPa, and 6.22%, respectively. The maximum flexural strength and flexural modulus of 83.85MPa and 4.50 GPa were also found after 3.0-h alkali treatment. Differential scanning calorimetric analysis illustrated that PLA/BP composites had a better compatibility and larger PLA crystallinity after 3.0-h treatment.

Silane coupling agents have been used traditionally in the past in the development of conventional polymer composites reinforced with glass fibers. Silane is a class of silicon hydride with a chemical formula SiH_4. Silane coupling agents have the potential to reduce the incidence of hydroxyl groups in the fiber-matrix interface. In the presence of moisture, hydrolysable alkoxy groups result in the formation of silanols. Silanols react with hydroxyl groups of the fiber, forming a stable, covalently-bonded structure with the cell wall. As a result, the hydrocarbon chains provided by the reaction of the silane produce a crosslinked network due to covalent bonding between fiber and polymer matrix. This results in a hydrophobic surface in the fiber, which in turn increases the compatibility with the polymer matrix. As mentioned earlier, silane coupling agents have been effective for the treatment of glass fibers for the reinforcement of polypropylene. Silane coupling agents have also been found to be useful for the pre-treatment of natural fibers in the development of polymer composites. Wu *et al.* demonstrated that wood fiber/polypropylene composites containing fibers pre-treated with a vinyl-tri methoxy silane significantly improved the tensile properties. It was discovered that the significant improvement in tensile properties was directly related to a strong interfacial bond caused by the acid/water condition used in the fiber pre-treatment [36].

In a study by Bengtsson *et al.*, the use of silane technology in crosslinking polyethylene-wood flour composites was investigated [37]. Composites of polyethylene with wood-flour were reacted *in-situ* with silanes using a twin screw extruder. The composites showed improvements in toughness and creep properties and the likely explanation for this improvement was that part of the silane was grafted onto polyethylene and wood, which resulted in a crosslinked network structure in the polymer with chemical bonds occurring at the surface of wood. X-ray microanalysis showed that most of the silane was found within close proximity to the wood-flour. It is known that silanes can interact with cellulose through either free radical or condensation reaction but also through covalent bonding by the reaction of silanol groups and free hydroxyl groups at the surface of wood, however the exact mechanism could not be ascertained. In a study by González *et al.* focused on the development of PLA based composites incorporating untreated and silane treated sisal and kraft cellulose fibers [38]. The tensile properties of the resulting composites did not present any major statistical difference between composites with untreated cellulose fibers and silane treated cellulose fibers, which suggested that silane treatment of the cellulose fibers did not contribute to further optimization in the reinforcing affect of the cellulose fibers. The analysis of the high resolution C1s spectra (XPS) indicates that for C_1 (C-C, C-H), the percentage of lignin in the intreated sisal fibers was higher, in comparison with kraft fibers. But after modification with silanes, the C_1 signal decreases for sisal fibers, which shows that attempted grafting with the silane has resulted in removal of lignin and exposed further cellulose. The higher C_1 signal reported for kraft fibers suggested some grafting with silane as a result of the contribution from the alkyl chain of the attached silanol, but no further characterisation was provided to support grafting of silanes to kraft fibers.

A study by Petinakis *et al.* focused on the suitability of a peel adhesion test as a macro-scale, exploratory technique for assessing the effectiveness of chemical modification techniques for improving adhesion between PLA film and a model wood substrate [4]. The study was conducted in order to gain insight into the nature of wood surface following chemical modification and how the resulting adhesion with PLA film could provide insight into failure mechanisms in PLA based biocomposites wood-flour as a natural filler. In this study, three different silanes were used to modify the surface of pine with- and without alkaline pre-treatment, namely 3-aminopropyltriethoxysilane (APTES), vinyltrimethoxysilane (VTMS) and 3-(Triethoxysilyl)propyl methacrylates (TESPM), and were compared with alkaline treatment of pine alone. The improvement observed in the peel fracture energy of the pine-PLA laminate following modification with APTES was possibly due to the possible reaction of amino-silane groups with the pine surface, since the results from the XPS study demonstrated that silicon and nitrogen atoms were attached to the surface. The bonding mechanism resulting in improved adhesion can be attributed to the ethoxy group (-CH_3CH_2) being hydrolysed to produce silanol groups, and these silanol groups can then react the free hydroxyl groups on the pine surface [38]. PLA can also bond with the amine group of the amino-silane treated pine through an acid-base interaction. The XPS of the PLA fracture surface indicated the presence of trace quantities of nitrogen, which support this mechanism. The XPS of the peel fracture surfaces followed silane modification with VTMES, indicating that grafting did not occur. XPS showed very little silicon on the surface, which indicated that silanol groups were not produced for reaction with hydroxyl groups on the pine surface. This is not uncommon, as previous studies have shown that the amount of water bound to cellulose fibers is a pre-requisite for silane reactions [38]. The XPS showed that attempted grafting with VTMES actually led to further exposure of the cellulose and increased the surface roughness. Figure 2.1 (b) clearly shows increased surface roughness following silane modifaction, which supports the argument that improvement in peel fracture energy achieved with VTMES is not due to chemical bonding but instead due to exposure of active sites and to negatively charged OH groups in cellulose that can physically interact with the PLA film.

The improvement in the peel fracture energy observed with TESPM can, however, be attributed to the formation of silanol groups, which have formed by the hydrolisation of the ethoxy group (-CH_3CH_2) in TESPM. The silanol groups can readily react with free hydroxyl groups associated with the cellulose component attached to the pine surface. The modified pine surface can then interact favourably with PLA, by molecular entanglement (interdiffusion) between the methacryl end group of the silane and the PLA matrix.

Sis *et al.* prepared composites based on poly (lactic acid) (PLA)/poly (butylene adipate-co-terephthalate) (PBAT)/kenaf fiber using a melt blending method [39]. A PLA/PBAT blend with the ratio of 90:10 wt%, and the same blend ratio reinforced with various amounts of kenaf fiber were prepared and characterized. The addition of kenaf fiber reduced the mechanical properties sharply due to the poor interaction between the fiber and polymer matrix. Modification of the composite by (3-aminopropyl)trimethoxysilane (APTMS) showed improvements in mechanical properties, increasing up to 42.5, 62.7 and 22.0% for tensile strength, flexural strength and impact strength, respectively. The composite treated with 2% APTMS successfully exhibited optimum

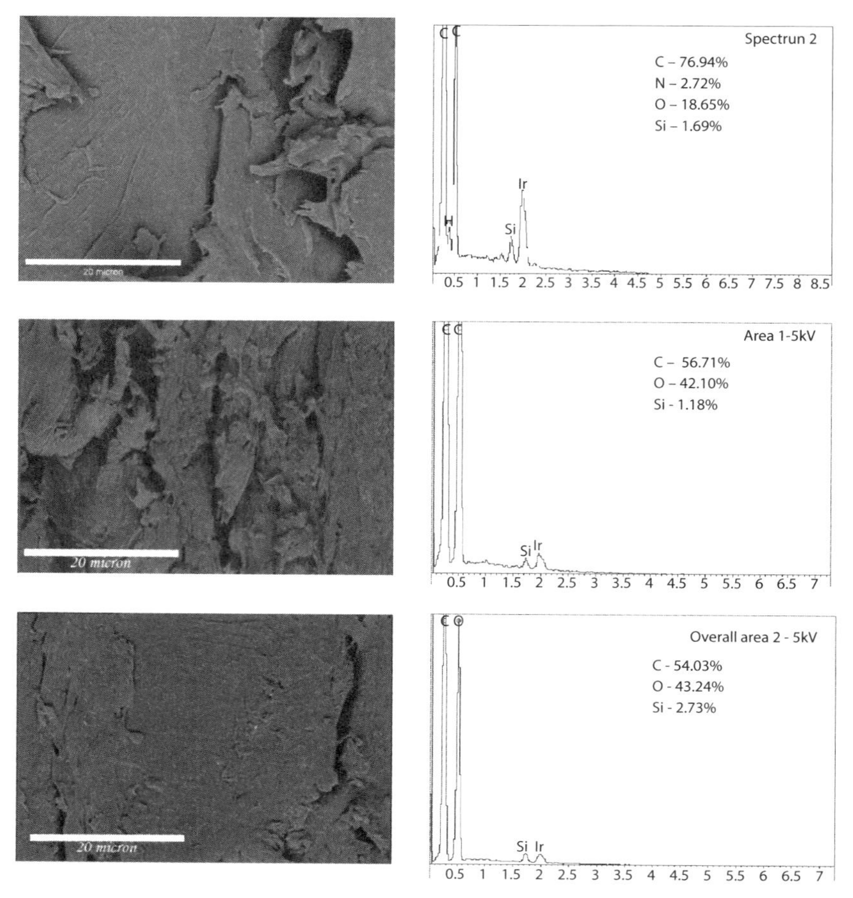

Figure 2.1 5000x images of pine surfaces following silane modification (without pre-treatment) with EDAX spectrums with elemental compositions (a) Pine-1 wt% APTES (b) Pine-1 wt% VTMES and (c) 1 wt% TESPM [4].

tensile strength (52.27 MPa), flexural strength (64.27 MPa) and impact strength (234.21 J/m). Morphological interpretation through scanning electron microscopy (SEM) revealed improved interaction and interfacial adhesion between PLA/PBAT blend and kenaf fiber and the fiber was well distributed. DMA results indicated lower storage modulus (E') for PLA/PBAT/kenaf fiber blend and an increase after modification by 2 wt% APTMS. Conversely, the relative damping properties decreased. Based on overall results, APTMS can be used as coupling agent for the composite since APTMS can improve the interaction between hydrophilic natural fibers and non-polar polymers.

Chemical modification through esterification of natural fibers through reaction with organic acid anhydrides has also been the subject of ongoing research in the field. Anhydrides can be classified into two major groups: non-cyclic anhydrides (i.e., Acetic) and cyclic anhydrides (i.e., Maleic). Of the non-cyclic anhydrides, acetylation with acetic anhydride is the most widely reported [28, 40, 41]. The reaction involves the conversion of a hydroxyl group to an ester group by virtue of the carboxylic group of the anhydride, with the free hydroxyl groups in cellulose. Reactions involving non-cyclic anhydrides are quite problematic as there are several steps involved during the treatment. These reactions also require the use of strong bases or catalysts to facilitate the reaction. Although the use of non-cyclic anhydrides can generally lead to good yields, a large proportion of the treated cellulose can contain free anhydride, which cannot be easily removed from the treated cellulose. Generally, the modified cellulose may be comprised of a distinct odor, which suggests the presence of free anhydride. The other issue with the use of non-cyclic anhydrides is the formation of acid by-products, which are generally present in the modified cellulose. Pyridine, a catalyst used in the reaction, acts by swelling the wood and extracting lignin to expose the cellular structure of the cellulose. This facilitates the exposure of the free hydroxyl groups in cellulose to the anhydride. However, due to the aggressive nature of pyridine, it can also degrade and weaken the structure of the cell wall, which may not allow efficient modification. The effect of esterification on natural fibers is it imparts hydrophobicity, which makes them more compatible with the polymer matrix.

Tserki *et al.* investigated the reinforcing effect of lignocellulosic fibers, incorporating flax, hemp and wood, on the mechanical properties of Bionolle, an aliphatic polyester [42]. The use of acetic anhydride treatment of the fibers was proven not to be effective for improving the matrix tensile strength, compared with other techniques such as compatibilisation; however it did reduce the water absorption of the fibers. Lower tensile strengths were reported for composites reinforced with wood fiber, compared with flax and hemp. This may be attributed to the nature of the fibers, since flax and hemp are fibrous, whereas wood fiber is more flake-like in nature, with an irregular size and shape. The type and nature of lignocellulose fibers (chemical composition and structure) is of paramount importance in the development of polymer composites. It is shown that different fibers behave differently depending on treatment. On the other hand, reactions of cellulose with cyclic anhydrides have also been performed [43]. Reactions involving cyclic anhydrides generally do not result in the formation of by-products and reactions can be performed with milder solvents, which don't interfere with the cell wall structure of cellulose. In order to facilitate reactions of wood flour with cyclic anhydrides it is important that the wood flour be pre-treated. Pre-treatment requires immersion of the wood flour in a suitable solvent, such as NaOH. This process is otherwise known as Mercerization, which is thought to optimize fiber-surface characteristics, by removing natural impurities such as pectin, waxy substances and natural oils. It is widely reported that the wood alone does not readily react with esterifying agents, since the hydroxyl groups required for reaction are usually masked by the presence of these natural impurities.

Gregorova *et al.* prepared PLA film composites with 30 wt% wood-flour and/or mica through melt blending and compression molding [44]. Semi-crystalline PLA was plasticized with poly(ethylene glycol) and filled with 30 wt% of wood-flour and mica. The degree of crystallinity was purposely increased by annealing. The filler/polymer matrix interface was modified through the addition of 4, 4 - Methylenediphenyl diisocyanate (MDI). The results showed that the increase in crystallinity had a strong impact on the mechanical performance of the composites. Another interesting observation was that MDI had a preference to react first with the plasticising agent and then with the wood-flour and this has been observed in previous studies where the MDI has not compatibilised matrix with filler 29].

Baltazar-y-Jimenez *et al.* prepared Poly (D-, L-lactic acid) (PDLA) and PDLA-wood pulp fiber injection molded composites, which were modified with very small amounts (< 1 wt%) of N'-(o-phenylene)dimalemide and 2,2'-dithiobis (benzothiazole) by reactive extrusion [45]. The modification produced an increase in the percent crystallinity (Xc), heat deflection temperature (HDT), impact energy, tensile strength, and modulus in PDLA. A significant reduction in the melting temperature (Tm) and an increase in the thermal resistance (Tmax) were also found. Fourier-Transform infrared spectroscopy (FTIR) suggests the creation of hydrogen bonds, a thiol ester and/or ester bond during the modification. Reactive extrusion of commercially available poly (lactic acid) (PLA) by means of N'-(o-phenylene)dimalemide and 2,2'-dithiobis (benzothiazole) provides a low cost and simple processing method for the enhancement of the properties of this biopolymer.

Altun *et al.* studied the effect of surface treatments and wood flour (WF) ratio on the mechanical, morphological and water absorption properties of poly(lactic acid) (PLA)-based green composites [46]. WF/PLA interfacial adhesion was promoted by means of alkaline treatment and pre-impregnation with dilute solution of matrix material. The mechanical data showed that incorporation of WF without any surface treatment caused high reduction in tensile strength in spite of incremental increase in tensile modulus. As the amount of alkaline treated WF increased, both modulus and tensile strength also increased. Both alkaline treatment and pre-impregnation further increased the mechanical properties including tensile strength, tensile modulus and impact strength. According to dynamic mechanical analysis (DMA) test results, the glass transition temperature of PLA increased with the addition of WF and the highest increment was obtained when pre-impregnated WF was used. The optimization in the glass transition temperature due to the addition of pre-impregnated WF was attributed to the lack of polymer chain motions as a result of interaction with adhered polymer segments.

Csizmadia *et al.* reported on the application of a resol type phenolic resin that was used for the impregnation of wood particles, for the reinforcement of PLA [47]. A preliminary study showed that the resin penetrates wood with rates depending on the concentration of the solution and on temperature. Treatment with a solution of 1 wt% resin resulted in a considerable increase of composite strength and decrease of water absorption. Composite strength improved as a result of increased inherent strength of the wood, but interfacial adhesion might be modified as well. When wood was treated with resin solutions of greater concentration, the strength of the composites decreased, first slightly, then drastically to a very small value. A larger amount of resin results in a thick coating on wood with inferior mechanical properties. At large resin contents the mechanism of deformation changes; the thick coating fails very easily leading to the catastrophic failure of the composites at very small loads. It seems the limiting factor in producing PLA-based natural fiber composites with superior mechanical properties is the inherent strength of the natural fibers. Pre-impregnation of natural fibers with a resin may present as a useful approach to improving the inherent strength of natural fibers, for use as reinforcing elements in biopolymer composites and could also improve interfacial adhesion with polymer matrix.

2.3.2 Physical Modification

Physical methods reported in the literature involve the use of corona or plasma treaters for modifying cellulose fibers for conventional polymers [48-51]. In recent years the use of plasma for treatment of natural fibers has gained more prominence as this provides a "greener" alternative for the treatment of natural fibers for the development of polymer composites. It is of particular interest to polymer composites incorporating biopolymer matrices, since this technique provides a further impetus to the whole notion of "green materials. Sustainability and end of life after use are important considerations when developing polymer composites from renewable resources, as is the toxicity and environmental impact of using various chemical or physical methods for improving the properties of these materials. Some chemical techniques may be toxic, e.g., isocyanates are carcinogenic, and therefore, the use of such agents may not be appropriate for the development of polymer composites from renewable resources. Physical methods involving plasma treatments have the ability to change the surface properties of natural fibers by the formation of free radical species (ions, electrons) on the surfaces of natural fibers [52]. During plasma treatment, surfaces of materials are bombarded with a stream of high energy particles within the stream of plasma. Properties such as wettability, surface chemistry and surface roughness of material surfaces can be altered without the need for employing solvents or other hazardous substances. Alternative surface chemistries can be produced with plasmas, by altering the carrier gas and depositing different reactive species on the surfaces of natural fibers [53]. This can then be further exploited by grafting monomeric and/or polymeric molecules on to the reactive natural fiber surface, which can then facilitate compatibilisation with the polymer matrix.

In a study by Morales *et al.*, low energy glow discharge plasma were used to functionalize cellulose fibers for improving interfacial adhesion between the fibers and polystyrene film [54]. The micro-bond technique was used to study the effect of the plasma treatment on the fiber matrix interface. The results showed that the adhesion in the fiber-matrix interface increased as a function of treatment time up to 4 minutes, however longer treatment times resulted in degradation of the fibers leading to poor interfacial adhesion. Baltazar-y-Jimenez *et al.* conducted a study evaluating the effect of atmospheric air pressure plasma treatment (AAPP) of lignocellulosic fibers on the resulting properties and adhesion to cellulose acetate butyrate[51]. The impact of AAPP treatment on abaca, flax, hemp and sisal fibers was studied by SEM and single fiber pull-out tests. The results of the study showed that interfacial shear strength (τ_{IFSS}) increased marginally for flax, hemp and sisal fibers after 1 minute AAPP, but decreases with prolonged exposure for abaca and sisal, which may be attributed to the formation of weak boundary layers (WBL). The reduction in the interfacial shear strength of the various fibers tested was most likely the result of fiber degradation from exposure to the AAPP, which resulted from the formation of a mechanical weak boundary layer.

In a study conducted by Yuan *et al.*, argon and air-plasma treatments were used to modify the surface of wood fibers in order to improve the compatibility between the wood fibers and a polypropylene matrix [55]. The improvement in the mechanical properties of the resulting composites, as depicted by SEM, was attributed to an increase in the surface roughness of the wood fibers following plasma treatment. The increase in surface roughness can facilitate better mechanical interlocking, but the increase in surface roughness can also expose more reactive cellulosic groups, which can interact favourably with the polymer matrix. The increase in surface roughness and improved O/C ratio can improve interfacial adhesion, which results in improved mechanical performance of the resulting composites.

Byung-Sun *et al.* reported on the use of an atmospheric glow discharge (AGD) plasma for depositing hexamethyl-disiloxane (HMDSO) on wood-flour using helium as the carrier gas [56]. Contact angles of various monomers were measured by a goniometer to calculate surface energies in order to select the monomer with highest surface energy. The highest surface energy was achieved with hexamethyl-disiloxane and was used for plasma coating of wood flour to improve its bonding and dispersion with the polypropylene (PP). The mechanical test results and SEM observations indicated that good dispersion and positive compatibility between the wood flour and the PP could be achieved. A similar study was conducted by Kim *et al* [57].

Plasma induced grafting has been demonstrated as a useful approach to enhance paper hydrophobicity. Song *et al.* conducted a study that involved graft polymerisation of butyl acrylate (BA) and 2-ethylhexyl acrylate (2-EHA) on paper via plasma induced grafting [58]. Contact angle, FTIR, XPS and SEM were conducted in order to gain insight into the level of grafting and the morphological changed following the plasma induced grafting. The results demonstrate that hydrophobicity of the modified paper sheet could be attained following modification. A similar approach could be used to modify the surfaces of natural fibers, with a particular emphasis to green chemistry and modifying the surfaces of natural fibers using green chemicals, which has been reported previously by Gaiolas *et al.* [59]. Deposition of green monomers through the use of plasma-induced surface grafting could present a useful approach for producing natural fiber-reinforced polymer composites that are completely benign and non-toxic

The authors of this chapter conducted a recent study, which evaluated the efficacy of plasma treatment on a model wood substrate. In order to understand the effect of plasma treatment on wood-based fibers, clear pine wood veneers with an identical chemical composition was used as a model substrate. Plasma treatment was conducted using a custom built 13.56MHz inductively coupled plasma chamber in continuous wave mode (CW) and continuous wave mode plus pulse plasma mode [60]. During continuous wave mode the plasma is always "on", so the formation of functional groups is accompanied by their destruction due to the continuous ion bombardment or surface ablation. During the pulse mode, the plasma source is on for a specific period and the plasma generated is similar to the continuous wave mode, where the plasma reactive species interact with the surface of the wood substrate. During the period when the

Table 2.1 Initial plasma treatment conditions conducted on a clear pine wood veneer substrate.

Argon Plasma	Oxygen Plasma
CW	CW
Power: 75W	Power: 100W
Pressure: $5X10^{-2}$	Pressure: $1x10^{-3}$
Time: 60sec	Time: 120sec

plasma source is off, the charged particles and UV radiation will disappear, which prevents further ion bombardment and surface ablation[61].

The initial plasma treatment conditions that were trialled are outlined in Table 2.1. Clear pine wood veneer samples were placed directly under the plasma source and exposed to an argon plasma followed treatment with oxygen (O_2) plasma. The purpose of the initial plasma treatment with argon was performed in order to ablate the surface of the wood veneer to remove surface impurities and contaminants. The SEM images of the control and plasma treated clear pine wood veneers surfaces are shown in Figure 2.2. The control sample (Figure 2.2A) clearly shows evidence of surface contamination or low molecular weight material that appears crystalline in nature and the appearance of less pronounced corrugations, which is indicative of the underlying virgin wood surface. Following plasma treatment (Figure 2.2B), the wood surface appears cleaner since the surface contaminant layer has been removed by ablation due to the argon plasma. The surface also appears rougher than the control surface, which suggests that the oxygen plasma has etched the surface. Further evidence of the positive effect of plasma treatment can be observed from the results of the XPS analysis in Figure 2.3. The main effect of plasma treatment is an increase in the surface oxygen content, specifically in a decrease of the CH_x signal (C 1s at 285 eV) and an increase in the C-O signal (C 1s at 286.5 eV) and the C = O, O-C-O signal (C 1s at 288 – 288.5 eV). This is consistent with an increase in the surface concentration of the main wood components such as cellulose and lignin, particularly cellulose. A likely mechanism is the removal of hydrocarbon-based compounds from the surface by plasma oxidation, exposing the underlying cellulose. Contact angle analysis was also conducted on the pine veneer surface prior to and after plasma modification and shown in Figure 2.4. The contact angle of the water droplet on the control surface was around 40-50°, but following plasma modification instantaneously became zero, which indicated that the effect of plasma has removed the surface contaminant layer and exposing more reactive surface hydroxyl groups, which suggests that the surface has been rendered more hydrophilic due to the removal of the surface contaminant and increase in the surface hydroxyl groups, as supported by the XPS analysis.

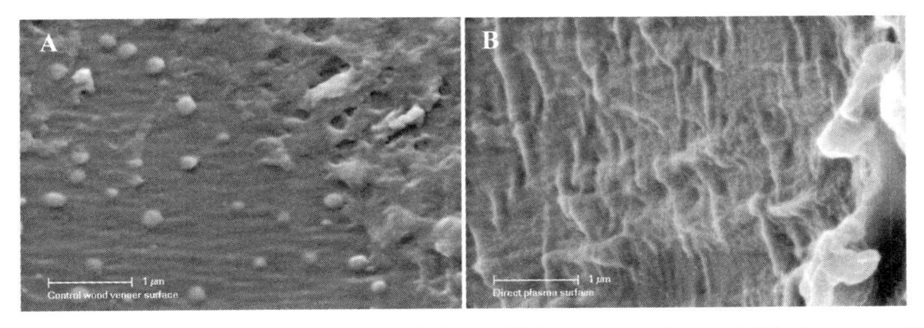

Figure 2.2 SEM images depicting surface morphology of (A) control surface and (B) plasma modified surface. (Petinakis *et al.* 2014).

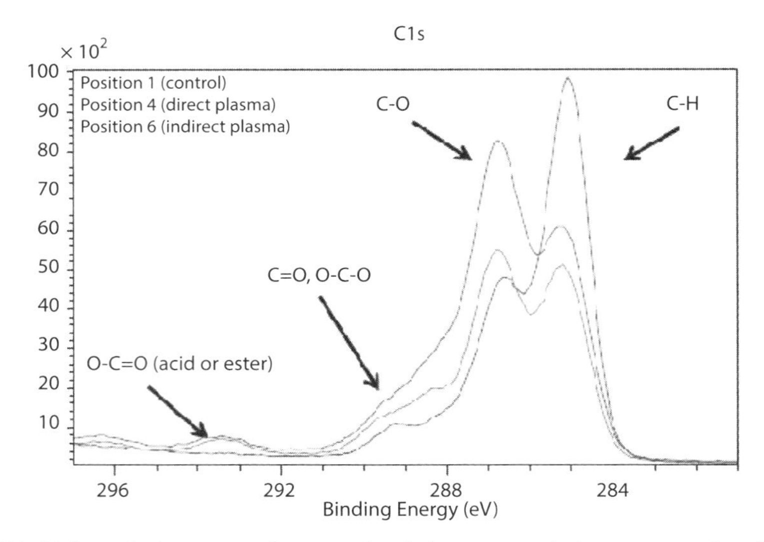

Figure 2.3 C 1s high resolution spectra for control and plasma treated pine veneer surface (Petinakis *et al.* 2014).

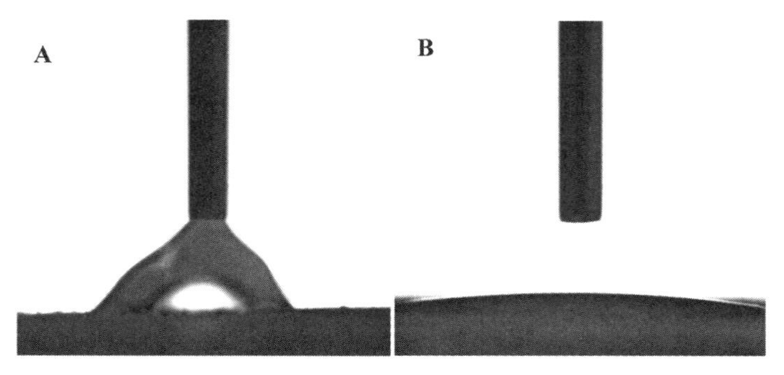

Figure 2.4 Water contact angle on (A) control pine veneer and (B) pine veneer following plasma modification (Petinakis *et al.* 2014).

2.4 Evaluation of Interfacial Properties

One of the most important focus areas of research in the development of natural fiber-reinforced polymer composites is characterisation of the fiber-matrix interface, since the interface alone can have a significant impact on the mechanical performance of the resulting composite materials, in terms of the strength and toughness. The properties of all heterogeneous materials are determined by component properties, composition, structure and interfacial interactions [62]. There have been a variety of methods used to characterize interfacial properties in natural fiber-reinforced polymer composites, however, the exact mechanism of the interaction between the natural fiber and the polymeric matrix has not been clearly studied on a fundamental level and is presently the major drawback for widespread utilization of such materials. The extent of interfacial adhesion in natural fiber-reinforced polymer composites utilizing PLA as the polymer matrix has been the subject of several recent investigations, hence the focus in this section will be on PLA-based natural fiber composites.

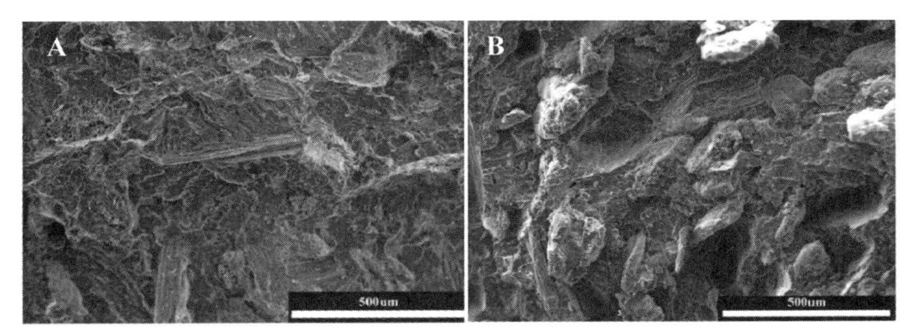

Figure 2.5 Effect of modification techniques on interfacial properties (a) wood-flour modification with MDI and (b) impact modification of PLA matrix with addition of PEAA. (Petinakis *et al.* 2009).

2.4.1 Microscopic Characterisation

2.4.1.1 Scanning Electron Microscopy

SEM is one of the most widely used techniques to evaluate interfacial properties of natural fiber-reinforced polymer composites. SEM can provide insight into the fracture surface morphology in terms of the quality of the fiber-matrix interfacial adhesion. A good example of how SEM is a useful technique for evaluating the quality of the interface in PLA-based composites is depicted in Figure 2.5 and Figure 2.6. The fracture surface morphology of a PLA/wood-flour composite produced with the addition of MDI (Figure 2.5(a)) appears homogeneous with the wood-flour particles not being clearly visible because they are firmly embedded and coated in the PLA matrix due to the improved interfacial interaction. The higher magnification image shown in Figure 2.6(a) shows more clearly the substantially increased adhesion at the matrix-particle interface, with no evidence of the particle deformation or voids that would normally be associated with poor matrix-particle adhesion. In comparison, the fracture surface morphology of the PLA-wood-flour composite modified through impact modification (Figure 2.5b) shows a higher incidence of particle pull-out and void formation compared to the composite with MDI-mediated wood flour. The higher magnification image of the same composite (Figure 2.6b) depicts the fracture surface morphology of the same composite but on this occasion it shows a gap in the interface between the matrix and wood-flour particle that is indicative of poor interfacial adhesion.

In a recent study, Le Moigne *et al.* studied interfacial properties in PLA biocomposites reinforced with flax fibers modified with organosilanes. The origins of the reinforcement at the fiber/matrix interface were investigated at the macromolecular and the microstructural levels by physico-chemical and mechanical cross-analyses. It was shown that interfacial interaction results from both modified chemical coupling and mechanical interlocking at the fiber/matrix interface [63]. Dynamic mechanical thermal analysis revealed a decrease in damping for treated biocomposites due to the formation of a layer of immobilized polymer chains resulting from strong interactions at the interface. *In situ* observations of crack propagation by scanning electron microscopy illustrated clearly that the treated biocomposites show a cohesive interfacial failure at much higher loads, highlighting the enhanced load transfer from the matrix to the fibers.

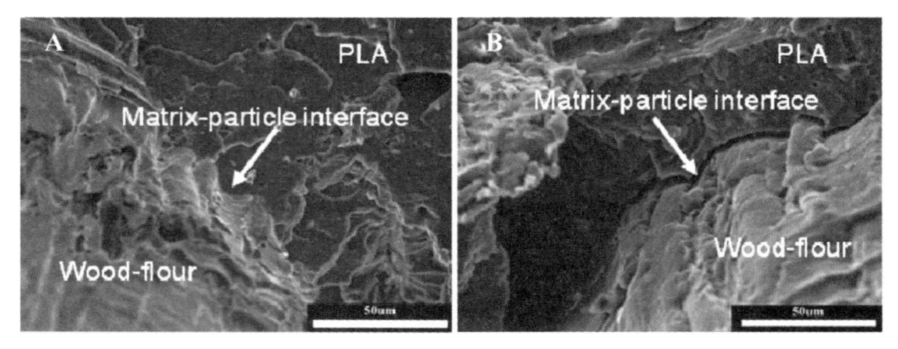

Figure 2.6 Higher magnification images demonstrating (a) good adhesion achieved through chemical modification of wood-flour and (b) poor adhesion through impact modification (Petinakis *et al.* 2009).

2.4.1.2 Atomic Force Microscopy

Atomic force microscopy (AFM) is a non-destructive imaging technique that can be used in the development of natural fiber-reinforced polymer composites to evaluate the efficacy of surface treatment techniques on natural fibers. Avramidis *et al.* conducted a study involving plasma treatment at atmospheric pressure using a dielectric barrier discharge for modifying the hydrophilicity of wood and wood-based materials with a monomer, such as HMDSO [64]. AFM imaging was carried out on the untreated and plasma coated wood surfaces in order to investigate the surface topography. The morphology of untreated wood surface appears to be completely covered by the deposited layer, whereas the macrostructure is still visible. A characteristic feature that can be observed by the AFM image is the nodule-like fine structure of the layer (Figure 2.7).

The authors of this chapter have recently conducted some work involving application of AFM for evaluating surface morphological changes following plasma modification of a wood substrate. Due to the heterogeneous nature of wood fibers it was found more convenient to carry out plasma modification on a model wood veneer substrate that replicated the chemical composition of wood-flour. Plasma modification was conducted using a combination of a continuous wave and pulse wave mode. The effect of plasma modification on the wood surface can be observed in the amplitude images depicted in Figure 2.8. The plasma treated surface (right) exhibits a nodule-like fine structure on a nanometre scale that is indicative of the resulting surface roughness and that is not observed in the untreated wood surface (left)

In a recent study Raj *et al.* presented the first direct study of adhesion forces, by colloidal force microscopy, between smooth PLA films representing the polymer matrix, and a microbead of cellulose that mimics the cellulose material in flax fibers [65]. Normalized adhesion force measurements demonstrated the importance of capillary forces when experiments were carried out under ambient conditions. Experiments, conducted under dry air allowed for the deduction of the contribution of pure van der Waals forces, and the results, through the calculation of the Hamaker constant, show that these forces, for the PLA/cellulose/air system, were lower than those obtained for the cellulose/cellulose/air system and hence underlined the importance of optimizing the interface among these materials. The study demonstrated the capacity of AFM to probe direct interactions in complex systems by adjusting the nature of the surface and

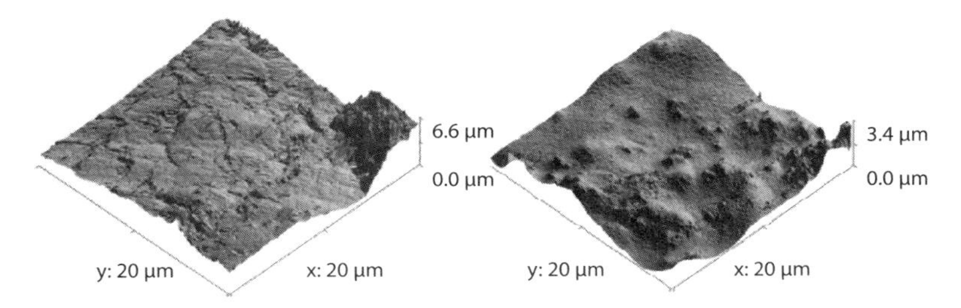

Figure 2.7 Atomic force microscopy images of uncoated (left) and plasma coated Beech veneers (Avramidis *et al.* 2009).

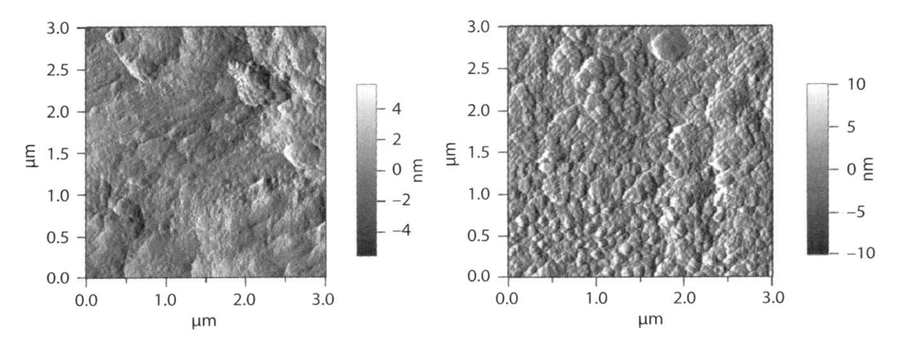

Figure 2.8 AFM amplitude images of (a) untreated wood (b) plasma treated wood (Petinakis *et al.* 2014).

the probe to mimic the contacting materials. Furthermore, the technique can be operated in a variety of environments, consistent with composite making and provides particular opportunity for testing and improving the properties of biopolymer composites.

2.4.2 Spectroscopic Techniques

2.4.2.1 Acoustic Emission Spectroscopy (AES)

The integrity of the interfacial region is of particular interest in the development of natural fiber-reinforced polymer composites. The lack of widespread application of natural fibers in polymer composites has been primarily due to lack of adhesion to most polymeric matrices. The cornerstone to the successful application of natural fiber-reinforced polymer composites in the market is a strong interfacial bond, which can ensure long term durability of the material [66]. In order to provide insight into the interfacial charatersitics of natural fiber-reinforced polymer composites, AES can be used for the detection and localization of fiber failure in polymer composites. In a study conducted by Finkenstadt *et al.*, a combination of AES and confocal microscopy were used to evaluate the mechanical properties of plasticized sugar beet pulp (SBP) and PLA green composites. AES studies showed correlated debonding and fracture mechanisms up to 20 wt% plasticizer content and uncorrelated debonding and fracture for 30-40 wt% sorbitol and 30 wt% glycerol in the SBP-PLA composites [67]. Baltazar-y-Jimenez *et al.* conducted AES on PLA composites incorporating wood pulp fibers. AES studies provided insight into the nature of the interface in the composite, which showed that strong interaction between PLA and modified wood pulp lead to an increase in the speed of sound through the composite. In comparison, unmodified wood pulp exhibited poor sound attenuation through the composite [68].

Faludi *et al.* conducted a study that involved the production PLA biocomposites were using three corn cob fractions and a wood fiber as a reference [69]. The composites were characterized by tensile testing, scanning electron microscopy (SEM) and polarization optical microscopy (POM). Micromechanical deformation processes were followed by acoustic emission measurements. The different strength of the components were proved by direct measurements. Two consecutive micromechanical deformation processes were detected in composites containing the heavy fraction of corn cob, which were attributed to the fracture of soft and hard particles, respectively. The study showed that the fracture of soft particles did not necesarrily result in the failure of the composites that is initiated either by the fracture of hard particles or by matrix cracking. Very large particles debond easily from the matrix resulting in catastrophic failure at very low stresses. At sufficiently large shear stresses large particles break easily during compounding, thus reinforcement depending on interfacial adhesion was practically the same in all composites, irrespective of the initial fiber characteristics.

2.4.3 Other Techniques

In a study by Bax *et al.* the mechanical properties (tensile and impact characteristics) of injection-molded flax and Cordenka-reinforced polylactide (PLA) composites with fiber mass fractions between 10 and 30% were evaluated [70]. The results in indicated that by increasing the fiber content from 10 to 30% resulted in an increase in tensile characteristics resulted, but it was found that a reinforcement with 10% flax fibers led to poorer tensile strength as compared to the neat PLA matrix. The poor result achieved with flax fiber was not anticipated and a recent study by Graupner *et al.* set out to evaluate this observation. Therefore, test specimens with a fiber content of 10 and 30% were examined for their fiber orientations and void content. For the investigations, a new technique involving microcomputer tomography images were created by monochromatic synchrotron radiation [71]. Fiber orientation angles of these micrographs were determined with an adapted measuring mask of the Fiber Shape software. It could be was demonstrated that the fiber orientation in the composite is dependent on the fiber mass fraction and the type of fiber. No voids were found in all the composites investigated. The average fiber orientation angle of 10% flax/PLA showed a larger deviation from the longitudinal axis of the test specimen than the other samples, and is made primarily responsible for the lower tensile strength of this composite.

2.5 Conclusions

Concerns about the environment and increasing awareness around sustainability issues are driving the push for developing new materials that incorporate renewable sustainable resources. This has resulted in the use of natural fibers for developing all natural fiber-reinforced polymer composites. A fundamental understanding of fiber-fiber and fiber-matrix interface is critical to the design and manufacture of polymer composite materials because stress transfer between load-bearing fibers can occur at the fiber-fiber interface and fiber-matrix interface and this is the subject of ongoing investigations. The development of new techniques has facilitated a better understanding of the governing forces that occur at the interface between matrix and natural fiber. The use of surface modification is seen as a critical processing parameter for developing new materials, and plasma-based modification techniques are gaining more prominence, from an environmentally point of view, as well as from a practical approach.

References

1. S. Konaar, Interactions between wood and synthetic polymers, *in Wood-Polymer Composites,* S. KONaM (Ed.), pp. 41–71, Woodhead Publishing, Cambridge. (2008).
2. W.T.Y. Tze, D.J. Gardner, C.P. Tripp, and S.C. O'Neill, Cellulose fiber/polymer adhesion: effects of fiber/matrix interfacial chemistry on the micromechanics of the interphase. *J. Adhes. Sci. Technol.* 20(15), 1649–1668 (2006).
3. E. Petinakis, L. Yu, G. Edward, K. Dean, H. Liu, and A. Scully, Effect of matrix–particle interfacial adhesion on the mechanical properties of poly (lactic acid)/wood-flour micro-composites. *J. Polym. Environ.* 7(2), 83–94 (2009)
4. E. Petinakis, L. Yu, G.P. Simon, G.H. Edward, and K. Dean, Evaluation on effect of surface modification on the adhesion between wood and poly (lactic acid). *J. Biobased Mater. Bioenergy* 6(4), 388–398 (2012).
5. D. Puglia, J. Biagiotti, and J.M. Kenny, A review on natrual fibre-based composites - part II: Application of natural reinforcements in composite materials for automotive industry. *J. Nat. Fibres* 1(3), 23–65. (2005)
6. A.K. Mohanty, M. Misra, and G. Hinrichsen, Biofibres, biodegradable polymers and biocomposites: An overview. *Macromol. Mater. Eng.* 276(277), 1–24 (2000).
7. I.S. Arvanitoyannis, Totally and partially biodegradable polymer blends based on natural and synthetic macromolecules: Preparation, physical properties, and potential as food packaging materials. *J. Macromol. Sci. Rev. Macromol. Chem. Phys.* C39(2), 205–71 (1999).
8. L. Yu, K. Dean, and L. Li, Polymer blends and composites from renewable resources. *Prog. Polym. Sci.* 31(6), 576–602 (2006).
9. B.L. Shah, S.E. Selke, M. B. Walters, and P.A. Heiden, Effects of wood flour and chitosan on mechanical, chemical, and thermal properties of polylactide. *Polym. Compos* 9(6), 655–663 (2008).
10. S. Pilla, S. Gong, E. O'Neill, R. M. Rowell, and A.M. Krzysik, Polylactide-pine wood flour composites. *Polym. Eng. Sci.* 48(3), 578–87 (2008).
11. K. Oksman, M. Skrivars, and J.F. Selin, Natural fibres as reinforcement in polylactic acid (PLA) composites. *Compos. Sci. Technol.* 63(9), 1317–1324 (2003).
12. V.A. Alvarez, R.A. Ruscekaite, and A. Vazquez, Mechanical properties and water absorption behaviour of composites made from a biodegradable matrix and alkaline treated fibres. *J. Compos. Mater.* 37(17),1575–1588 (2003).
13. S.H. Lee, and S. Wang, Biodegradable polymers/bamboo fiber biocomposite with bio-based coupling agent. *Compos. A.* 37(1), 80–91 (2006).
14. M. Shibat, K. Ozawa, N. Teramoto, R. Yosomiya, and H. Takeishi, Biocomposites made from short abaca fiber and biodegradable polyesters. *Macromol. Mater. Eng.* 288 (1), 35–43 (2003).

15. D. Plackett, T.L. Anderson, W.B. Pederson, and L. Nielsen, Biodegradable composites based on L-polylactide and jute fibres. *Compos. Sci. Technol.* 63(9), 1287–1296 (2003).
16. M. Shibata, S. Oyamada, S.I. Kobayashi, D. Yaginuma, Mechanical properties and biodegradability of green composites based on biodegradable polyesters and lyocell fabric. *J. Appl. Poly. Sci.* 92(6), 3857–3863 (2004).
17. N. Graupner, and J. Mussig, A comparison of the mechanical characteristics of kenaf and lyocell fibre reinforced poly (lactic acid) (PLA) and poly (3-hydroxybutyrate) (PHB) composites. *Compos. A Appl. Sci. Manuf.* 42(12), 2010–2019 (2011).
18. M.S. Huda, L.T. Drzal, M. Misra, A.K. Mohant, K. Williams, D.F. Mielewsk, A study on biocomposites from recycled newspaper fiber and poly (lactic acid). *Ind. Eng. Chem. Res.* 44(15), 5593–5601 (2005).
19. M.S. Huda, A.K. Mohanty, L.T. Drzal, E. Schut , and M. Misra, "Green" composites from recycled cellulose and poly (lactic acid): Physico-mechanical and morphological properties evaluation. *J. Mater. Sci.* 40, 4221–4229 (2005).
20. A.P. Mathew, K. Oksman, and M. Sain, Mechanical properties of biodegradable composites from poly lactic acid (PLA) and microcrystalline cellulose (MCC). *J. Appl. Polym. Sci.* 97(5), 2014–2025 (2005).
21. J. George, M.S. Sreekal, and S. Thoma, A review on interface modification and characterization of natural fiber reinforced plastic composites. *Polym. Eng. Sci.* 41(9), 1471–1485 (2001).
22. V. Tserki, P. Matzinos, and C. Panayiotou, Novel biodegradable composites based on treated lignocellulosic waste flour as filler Part II. Development of biodegradable composites using treated and compatibilized waste flour. *Composites. A. Appl. Sci. Manuf.* 37, 1231–1238 (2006).
23. D. Cava, R. Gavara, J.M. Lagarón, and A. Voelkel, Surface characterization of poly (lactic acid) and polycaprolactone by inverse gas chromatography. *J. Chromatogr. A* 1148(1), 86–91 (2007).
24. Z. Dominkovics, L. Dányád, and B. Pukánszk, Surface modification of wood flour and its effect on the properties of PP/wood composites. *Compos. A Appl. Sci. Manuf.* 38(8), 1893–1901 (2007).
25. H. Wang, X. Sun, and P. Seib, Mechanical properties of poly (lactic acid) and wheat starch blends with methylenediphenyl diisocyanate. *J. Appl. Polym. Sci.* 84(6), 1257–1262 (2002).
26. H. Wang, X. Sun, and P. Seib, Properties of poly (lactic acid) blends with various starches as affected by physical aging. *J. Appl. Polym. Sci.* 90(13), 3683–3689 (2003).
27. M. Kazayawoko, J.J. Balatinecz, and L.M. Matuana, Surface modification and adhesion mechanisms in woof fibre-polypropylene composites. *J. Mater. Sci.* 34, 11 (1999).
28. G. Bogoeva-Gaceva, M. Avella, M. Malinconico, A. Buzarovska, A. Grozdanov, G. Gentile, and M.E. Errico, Natural fiber eco-composites. *Polym. Compos.* 28(1), 98–107 (2007).
29. L. Yu, K. Dean, Q. Yuan, L. Chen, and X. Zhang, Effect of compatibilizer Distribution on the Blends of Starch/Biodegradable Polyesters. *J. Appl. Polym. Sci.* 103, 812–818 (2006).
30. K. Oksman, and H. Lindberg, Influence of thermoplastic elastomers on adhesion in polyethylene-wood flour composites. *J. Appl. Polym. Sci.* 68(11), 1845–1855 (1998).
31. Y. Wang, F-C. Yeh, S-M. Lai, H-C. Chan, and H-F. Shen, Effectiveness of functionalized polyolefins as compatibilizers for polyethylene/wood flour composites. *Polym. Eng. Sci.* 43(4), 933–945 (2003).
32. N. Sombatsompop, C. Yotinwattanakumtorn, and C. Thongpin, Influence of type and concentration of maleic anhydride grafted polypropylene and impact modifiers on mechanical properties of PP/wood sawdust composites. *J. Appl. Polym. Sci.* 97(2), 475–484 (2005).
33. M.D.H. Beg, and K.L. Pickerin, Fiber pretreatment and Its effects on wood fiber reinforced polypropylene composites. *Mater. Manuf. Processes.* 21, 303–307 (2006).
34. M.N. Ichazo, C. Albano, J. Gonzalez, R. Perera, and M.V. Candal, Polypropylene/wood flour composites: treatments and properties. *Compos. Struct.* 54, 207–214 (2001).

35. S.P. Qian, H.L. Mao, K.C. Sheng, J. Lu, Y.F. Luo, and C.Y. Hou, Effect of low-concentration alkali solution pretreatment on the properties of bamboo particles reinforced poly (lactic acid) composites. *J. Appl. Polym. Sci.* 130(3), 1667–1674 (2013).
36. J. Wu, D. Yu, C-M. Chan, J. Kim, and Y-W. Mai, Effect of fiber pretreatment condition on the interfacial strength and mechanical properties of wood fiber/PP composites. *J. Appl. Polym. Sci.* 76(7), 1000–1010 (2000).
37. M. Bengtsson, and K. Oksma, The use of silane technology in crosslinking polyethylene/wood flour composites. *Compos. A Appl. Sci. Manuf.* 37, 752–765 (2006).
38. D. González, V. Santos, and J.C. Parajó, Silane-treated lignocellulosic fibers as reinforcement material in polylactic acid biocomposites. *J. Thermoplast. Compos. Mater.* 25, 1005–1022 (2012).
39. A.L.M. Sis, N.A. Ibrahim, and W.M.Z.W. Yunus, Effect of (3-aminopropyl)trimethoxysilane on mechanical properties of PLA/PBAT blend reinforced kenaf fiber. *Iranian Polym. J.* 22(2), 101–108 (2013).
40. C.A.S. Hill, Chemical modification of wood (I): Acetic anhydride modification, *in Wood Modification: Chemical, Thermal and Other Processes,* C.A.S. Hill (Ed.), pp. 45–76, John Wiley and Sons, New Jersey. (2006).
41. X. Li, L. Tabil, and S. Panigrahi, Chemical treatments of natural fiber for use in natural fiber-reinforced composites: A review. *J. Polym. Environ.* 15(1), 25–33 (2007).
42. V. Tserki, P. Matzinos, N.E. Zafeiropoulos, and C. Panayiotou, Development of biodegradable composites with treated and compatibilized lignocellulosic fibres. *J. Appl. Polym. Sci.* 100, 4703–4710 (2006).
43. B. Doczekalska, M. Bartkowiak, and R. Zakrzewski, Modification of sawdust from pine and beech wood with the succinic anhydride. *Holz als Roh- und Werkstoff* 65(3), 187–191 (2007).
44. A. Gregorova, V. Sedlarik, M. Pastorek, H. Jachandra, and F. Stelzer, Effect of compatibilizing agent on the properties of highly crystalline composites based on poly (lactic acid) and wood flour and/or mica. *J. Polym. Environ.* 9(2), 372–81 (2011).
45. A. Baltazar-y-Jimenez, and M. Sain, Effect of bismaleimide reactive extrusion on the crystallinity and mechanical performance of poly (lactic acid) green composites. *J. Appl. Polym. Sci.* 24(4), 3013–3023 (2012).
46. Y. Altun, M. Dogan, and E. Bayramli, Effect of alkaline treatment and pre-impregnation on mechanical and water absorbtion properties of pine wood flour containing poly (lactic acid) based green-composites. *J. Polym. Environ.* 21(3), 850–856 (2013).
47. R. Csizmadia, G. Faludi, K. Renner, J. Moczo, and B. Pukanszky, PLA/wood biocomposites: Improving composite strength by chemical treatment of the fibers. *Compos. A Appl. Sci2. Manuf.* 53, 46–53 (2013).
48. S. Dong, S. Saphieha, and H.P. Schreiber, Mechanical properties of corona-modified cellulose/polyethylene composites. *Polym. Eng. Sci.* 33(6), 343–346 (1993).
49. S. Dong, S. Saphieha, and H.P. Schreiber, Rheological properties of corona modified cellulose/polyethylene composites. *Polym. Eng. Sci.* 32(22), 1734–1739 (1992).
50. N. Olaru, L. Olaru, and G. Cobillac, Plasma modified wood fibres as fillers in polymeric materials. *Rom. J. Phys.* 50(9), 1095–1101 (2005).
51. A. Baltazar-y-Jimenez, M. Bistritz, E. Schulz, and A. Bismarck, Atmospheric air pressure plasma treatment of lignocellulosic fibres: Impact on mechanical properties and adhesion to cellulose acetate butyrate. *Compos. Sci. Technol.* 68(1), 215–227 (2008).
52. K-Y. Lee, A. Delille, and A. Bismarck, Greener surface treatments of natural fibres for the production of renewable composite material, in *Cellulose Fibers: Bio- and Nano-Polymer Composites,* S. Kalia, B.S. Kaith, and I. Kaur, (Ed.), pp. 155–78, Springer, Berlin Heidelberg. (2011)

53. M.H. Nguyen, B.S.Kim, J.R. Ha, and J.I. Song. Effect of plasma and NaOH treatment for rice husk/PP composites. *Adv. Compos. Mater.* 20(5), 435–42 (2011).
54. J. Morales, M.G. Olayo, G.J. Cruz, P. Herrera-Franco, and R. Olayo. Plasma modification of cellulose fibers for composite materials. *J. Appl. Polym. Sci.* 101(6), 3821–3828 (2006).
55. X. Yuan, K. Jayaraman, and D. Bhattacharyya. Effects of plasma treatment in enhancing the performance of woodfibre-polypropylene composites. *Compos. A Appl. Sci. Manuf.* 35(12), 1363–1374 (2004).
56. B-S. Kim, B-H. Chun, I.L. Woo, and B-S.Hwang. Effect of plasma treatment on the wood flour for wood flour/PP composites. J. Thermoplast. Compos. Mater. 22(1), 21–28 (2009).
57. M. Kim, H.S. Kim, J.Y. Lim. A Study on the Effect of Plasma Treatment for Waste Wood Biocomposites. Journal of Nanomaterials. 2013.
58. Z. Song, J. Tang, J. Li, and H. Xiao. Plasma-induced polymerization for enhancing paper hydrophobicity. *Carbohydr. Polym.* 92(1), 928–933 (2013).
59. C. Gaiolas, A.P. Costa, M. Nunes, M.J.S. Silva, and M.N. Belgacem. Grafting of paper by silane coupling agents using cold-plasma discharges. *Plasma Proces. Polym.* 5(5), 444–452 (2008).
60. Li, L. Dai, X. J. Xu, H. S. Zhao, J. H. Yang, P. Maurdev, G. Plessis, J. d. Lamb, P. R. Fox, B. L. Michalski, W. P., Combined continuous wave and pulsed plasma modes: For more stable interfaces with higher functionality on metal and semiconductor surfaces. *Plasma Process. Polym* 2009, 6 (10), 615–619.
61. Z. Chen, X.J. Dai, P.R. Lamb, D.R. de Celis Leal, B.L. Fox, Y. Chen, J.D. Plessis, M. Field, and X. Wang. Practical amine functionalization of multi-walled carbon nanotubes for effective interfacial bonding. *Plasma Proces. Polym.* 9(7), 733–741 (2012).
62. N.E. Zafeiropoulos, D.R. Williams, C.A. Baillie, and F.L. Matthews, Engineering and characterisation of the interface in flax fibre/polypropylene composite materials. Part I. Development and investigation of surface treatments. *Composites A Appl. Sci. Manuf.* 33(8), 1083–1093 (2002).
63. N. ILe Moigne, M. Longerey, J.M. Taulemesse, J.C. Bénézet, and A. Bergere, Study of the interface in natural fibres reinforced poly (lactic acid) biocomposites modified by optimized organosilane treatments. *Ind. Crop Prod.* 52, 481–494 (2014).
64. G. Avramidis, E. Hauswald, A. Lyapin, H. Militz, W. Viöl, and A. Wolkenhauer, Plasma treatment of wood and wood-based materials to generate hydrophilic or hydrophobic surface characteristics. *Wood Mater. Sci. Eng.* 4(1–2), 52–60 (2009).
65. G. Raj, E. Balnois, M.A. Helias, C. Baley, and Y. Grohens, Measuring adhesion forces between model polysaccharide films and PLA bead to mimic molecular interactions in flax/PLA biocomposite. *J. Mater. Sci.* 47(5), 2175–2181 (2012).
66. A. Dufresne, and M.N. Belgacem, Cellulose-reinforced Composites: From Micro-to Nanoscale. *Polimeros-Ciencia E Tecnologia* 23(3), 277–286 (2013).
67. V.L. Finkenstadt, C.K. Liu, P.H. Cooke, L.S. Liu, and J.L. Willett, Mechanical property characterization of plasticized sugar beet pulp and poly (lactic acid) green composites using acoustic emission and confocal microscopy. *J. Polym. Environm.* 16(1), 19–26 (2008).
68. A. Baltazar-y-Jimenez, I. Seviaryna, M. Sain, and E.Y. Maeva, Acoustic, tomographic, and morphological properties of bismaleimide-modified PLA green composites. *J. Reinf. Plast. Compos.* 30(16), 1329–1340 (2011).
69. G. Faludi, G. Dora, K. Renner, J. Moczo, and B. Pukanszky, Biocomposite from polylactic acid and lignocellulosic fibers: Structure-property correlations. *Carbohydr. Polym.* 92(2), 1767–1775 (2013).

70. B. Bax, and J. Müssig, Impact and tensile properties of PLA/Cordenka and PLA/flax composites. *Compos. Sci. Technol.* 68(7–8), 1601–1607 (2008).
71. N. Graupner , F. Beckmann, F. Wilde, and J. Mussig, Using synchroton radiation-based micro-computer tomography (SR mu-CT) for the measurement of fibre orientations in cellulose fibre-reinforced polylactide (PLA) composites. *J. Mater. Sci.* 49(1), 450–460 (2014).

3

Research on Cellulose-Based Polymer Composites in Southeast Asia

Riza Wirawan[*,1] **and S.M. Sapuan**[*,2]

[1]*Department of Mechanical Engineering, Universitas Negeri Jakarta, Jakarta, Indonesia*
[2]*Department of Mechanical and Manufacturing Engineering, Universiti Putra Malaysia, Selangor, Malaysia*

Abstract
Environmental awareness and changes in the regulatory environment have prompted many researchers to develop more environmentally-friendly and sustainable materials. In this context, cellulose-based materials are of particular interest. Besides their environmental friendliness, the advantages of cellulose-based materials include their high availability, renewability and low density in comparison with synthetic or mineral materials. Additionally, the use of agricultural products could help reduce social problems in tropical countries, which contribute to agricultural exodus. Nowadays, people tend to go to big cities to look for jobs that are not suitable for their qualifications or experience. Using agricultural products for cellulose-based polymer composites may help to increase incomes from agriculture, hence leading to people being interested in working in the agricultural field, thereby reducing agricultural exodus.

In this chapter, the discussion is focused on the research on cellulose-based polymer composites conducted in some tropical countries in Southeast Asia. Southeast Asia is a subregion of Asia located in the tropical zone, where a variety of plants grow. Hence, there is a high availability of cellulose-based materials found in this area. This situation enhances the attractiveness of research on cellulose-based composites. The reported findings of researchers in Southeast Asia about the processing, properties and application of composites made of commonlyused cellulose-based fibers such as sisal, flax, hemp, ramie, jute, kenaf, etc., are discussed along with many kinds of cellulose-based fibers of interest which have not been reported elsewhere. This chapter is a combination of reviews on some unique cellulose-based polymer composites in Southeast Asia together with a report on the research conducted by the authors.

Keywords: Cellulose-based, polymer, composite, Southeast Asia

**Corresponding authors*: rwirawan@unj.ac.id; sapuan@eng.upm.edu.my

Vijay Kumar Thakur (Ed.), Lignocellulosic Polymer Composites, (41–62) 2015 © Scrivener Publishing LLC

3.1 Introduction

The development of technology is aimed at creating a better standard of living for the benefit of mankind. Mankind is seeking new products, materials and tools to serve this purpose. Materials play an important role in the advancement of life. In the past, the influence of materials was felt, so much so that eras were named after the materials themselves such as Stone Age, Bronze Age, etc. It is difficult to associate the currrent era with a particular type of material, and if it can be given such a name, it is more appropriate to call it "the age of various materials or advanced materials." Polymers have been substituted for many conventional materials, especially metals, in various applications, due to the advantages of polymers over conventional materials. They are used in many applications because they are easy to process, are highly productive, low cost and flexible. However, for some specific uses, some mechanical properties, e.g., strength and toughness, of polymer materials, are found to be inadequate. Various approaches have been developed to improve such properties. In most of these applications, the properties of polymers are modified using fillers and fibers to suit the high strength/high modulus requirements. Fiber-reinforced polymers have better specific properties compared to conventional materials and find applications in diverse fields, ranging from appliances to spacecraft [1]. Composite materials, or composites, are one of the major developments in material technology in recent decades. Many authors define composite materials in different ways. Daniel and Ishai [2] define a composite as "a material system consisting of two or more phases on a macroscopic scale, whose mechanical performance and properties are designed to be superior to those of the constituent materials acting independently." Meanwhile, Gay and Hoa [3] stated, "a composite material currently refers to material having strong fibers surrounded by a weaker matrix material which serves to distribute the fibers and also to transmit the load to fibers." Composite material is also defined as "a multiphase material that is artificially made, as opposed to one that occurs or forms naturally, chemically dissimilar and separated by a distinct interface" [4]

Roesler et al. (2006) characterize composite materials by the following properties:

- A strengthening second phase is embedded in a continuous matrix.
- The strengthening second phase and the matrix are initially separate materials and are joined during processing—the second phase is thus not produced by internal processes like precipitation.
- The particles of the second phase have a size of at least several micrometers.
- The strengthening effect of the second phase is at least partially caused by load transfer.
- The volume fraction of the strengthening second phase is at least approximately 10%.

These properties, however, cannot be used as a definition for composites in further developments. Some of the new composites (e.g., in nanotechnology) do not posses some of these listed properties.

From all of the definitions, it can be stated that, in general, composite material is considered to be a material that is composed of at least two physically distinct phases of materials. The first phase is a continuous material, called matrix, which surrounds the dispersed phase. The dispersed materials can be in the form of particle as well as fiber. The fibers and particles usually act as reinforcements to carry load or to control strain, whereas the matrix acts as a bonding medium to transfer load and to provide continuity and structural integrity. The matrix also protects the fibers and retains them in position to form the desired shape for a finished article (Owen, 2000a).

Historically, straw, one of the cellulose-based fibers, was used as reinforcement for mud to form bricks for building construction. The process of brickmaking was depicted on ancient Egyptian tomb paintings. However, engineered composites were reported to be introduced in the 1940s, when glass fiber was used as reinforcement for polymers in the aerospace, marine and automotive industries. During that period, a majority of the resins used included epoxy and polyester. Later, in the 1960s, carbon fibers were introduced as reinforcement for composites, particularly in aerospace and sporting goods. In these applications, the prime concern was performance rather than cost, and this was the beginning of the development of advanced composites.

In recent years, it has been found that there are many drawbacks to using glass and carbon fibers. Glass and carbon fibers can cause acute irritation of the skin, eyes, and upper respiratory tract. It is suspected that long-term exposure to these fibers causes lung scarring (i.e., pulmonary fibrosis) and cancer. Moreover, glass fiber is not easily degradable and results in environmental pollution. On the economic side, making a product from carbon fiber-reinforced composites is a high cost activity associated with both the manufacturing process and the material itself. That is why many products made of carbon fiber are very expensive.

Due to the drawbacks of glass and carbon fiber-reinforced composites, cellulose-based fiber-reinforced polymer composites have attracted great interest among material scientists and engineers. The interest is also affected by the need to develop an environmentally-friendly material, and partly replacing currently used glass or carbon fibers in fiber-reinforced composites. The optimum utilization of a renewable resource will provide a positive image for a sustainable "green" environment. New environmental regulations and uncertainty about petroleum and timber resources have also triggered much interest in developing composite materials from cellulose-based fibers. For some tropical countries like Southeast Asia, the development of composite materials from cellulose-based fiber may create new employment in the villages and help to attract people working in the agricultural field, thereby reducing the desire to go to big cities to look for jobs. Hence, it may reduce one of the social problems, which is agricultural exodus.

Like other tropical countries, there are areas of biodiversity in Southeast Asian countries. There is a large variety of species living in the area. Hence, there are many unique species that cannot be found in other places like four-season countries. This situation leads to a greater interest of researchers in Southeast Asia to explore many kinds of natural material resources, including cellulose-based materials. In this chapter, a review of some of the research on unique cellulose-based polymer composites in Southeast Asian countries is presented. The discussed resources include sugar palm

(*Arenga pinnata*), oil palm (*Elaeis guineensis*), durian (*Durio zibethinus*), and water hyacinth (*Eichhornia crassipes*).

The major challenge in utilizing cellulose-based polymer as reinforcing material in synthetic polymer composite is the compatibility issue. Due to the presence of hydroxyl and other polar groups on the surface and throughout the cellulosed-based polymer, moisture absorption can be high. This leads to poor wettability by the polymers and weak interfacial bonding between fibers and hydrophobic polymers, as the matrix [5]

3.2 Sugar Palm (*Arenga pinnata*)

Sugar palm (*Arenga pinnata*) is called by different names such as *kabung*, or *enau* in Malaysia. Sugar palm fiber is a kind of natural fiber that comes from the *Arenga pinnata* plant, a forest plant that can be widely found in Southeast Asia countries like Indonesia and Malaysia. This fiber seems to have properties like other natural fibers, but the detailed properties are not yet generally known. Generally, sugar palm has desirable properties like strength and stiffness, and its traditional applications include paint brushes, septic tank base filters, clear water filters, doormats, carpets, ropes, chair/sofa cushions, and for nests to hatch fish eggs. In certain regions, traditional applications of sugar palm fiber includes handcraft for *kupiah* (Acehnese typical headgear used in prayer) and roofing for traditional houses in Mandailing, North Sumatra, Indonesia. The sugar derived from the sugar palm tree is called palm sugar and it is one of the local delicacies widely consumed by Asians for making cakes, desserts, food coatings or to mix with drinks. It is produced by heating the sap derived from the sugar palm tree [6].

Sugar palm is known as a multipurpose species of palm. From this palm, we can produce many kinds of products, including food and beverages, timber, fibers, biopolymers, biocomposites, and other cellulose-based products [7, 8]. In some places of Indonesia, there are many old traditions using products based on sugar palm cultivation [8]. For the purpose of biobased composites, its fiber is of interest. It is a brown to black fiber with a diameter of 50–800 μm [9]. The fiber is called gomuti. A picture of gomuti is shown in Figure 3.1.

Sastra *et al.* [10], Leman *et al.* [11] and Suriani *et al.* [12] studied the tensile properties of gomuti-reinforced epoxy composites. They hand layed-up the gomuti with epoxy at 10, 15, and 20 weight percentages with different orientations, including random (long and chopped) and woven roving. After curing in room temperature (25–30°C), it was reported that the maximum tensile strength and modulus were achieved by the 10 wt% woven roving gomuti, which are 52 MPa and 1256 MPa, respectively. The scanning electron micrograph showed that woven roving gomuti has better fiber-matrix adhesion compared to the others [12].

Bachtiar *et al.* [13] then continued the study by adding an alkali treatment during the processing of the composite. They studied the effect of alkaline treatment on the tensile properties of gomuti/epoxy composites. Although there is an inconsistent result obtained for tensile strength, it was found that the tensile modulus of treated specimen was higher than that of untreated specimen. It is indicated that the treatment is effective to improve the interfacial bonding between fiber and matrix. The effect of

Figure 3.1 Sugar palm fiber (gomuti).

alkaline treatment on the impact properties of gomuti/epoxy composites was also studied. It was reported that the impact properties of gomuti/epoxy composites improved more than 12% after alkaline treatment at 0.5M NaOH solution with 8 hours soaking time [14].

Beside an alkaline treatment, an environmental treatment was also studied as reported by Leman *et al.,* in which [15] the surface properties of the gomuti improved after seawater and freshwater treatments. Treatment with seawater for 30 days increases the tensile strength of the gomuti/epoxy composites more than 67% as compared to untreated composites.

In term of aging, the gomuti/epoxy performs with higher tensile strength after aging. The aging process increases the tensile strength up to 50%. The impact strength, however, slightly decreases (6.33%) after aging, as reported by Leman *et al.* [16]. The results show that the strength of gomuti/epoxy composites increases during aging, while the ductility decreases. Moreover, as the fiber is hydrophilic, the moisture absorption of the composites increases with the increase of fiber content [17].

Beside gomuti/epoxy composite, studies on gomuti composites with other matrices were also reported [18]. Razak and Ferdiansyah studied the toughness of gomuti/cement composite. It is reported that the gomuti/concrete composites performed with higher toughness in comparison with the plain concrete.

The properties of gomuti itself were studied by Ishak *et al.* [19] and Bachtiar *et al.* [20]. From the studies, it is found that the average tensile strength of the gomuti is 190 Mpa, the Young's Modulus is 4 GPa, and the strain at failure is 20%. Moreover, the source of gomuti affected both tensile and thermal properties. Gomutis obtained from different heights of sugar palm trees were characterized. It has been confirmed that gomuti obtained from sugar palm trees of lower height has higher tensile properties. The tensile properties are lower at the higher part of the tree. This result is associated with the optimum chemical composition, especially cellulose, hemicellulose, and lignin. In terms of thermal stability, fibers from the lower part of the tree perform with higher thermal stability.

In order to obtain better propeties from the gomuti, Sapuan *et al.* [21] impregnated the fiber with phenol formaldehyde (PF) and unsaturated polyester (UP). The effect of impregnation time on physical and tensile properties of the gomuti was studied. As a result, the physical properties were significantly improved, especially after 5 min impregnation with PF. The moisture content and the water absorption of the gomuti decreased without any significant change in specific gravity. The tensile strength and toughness improved after 5 min impregnation with UP. A different result is obtained when the fiber is impregnated with PF. Both tensile strength and toughness of the composites were brought lower after the impregnation with PF. Moreover, there was no significant difference obtained after the extension of impregnation time.

Beside the mechanical properties, other properties were also studied to explore the benefit of gomuti. Ismail *et al.* [22] studied the potential of gomuti as soundproof material. They reported that the sound absorption coefficients of Arenga Pinnata were from 2000 Hz to 5000 Hz within the range of 0.75–0.90 and the optimum sound absroption coefficient was 40 mm. Hence, the gomuti has the potential to be used as a sound absorbing raw material, together with having other benefits such as being cheap, lightweight, and biodegradable.

3.3 Oil Palm (*Elaeis guineensis*)

Oil palm (*Elaeis guineensis*), also known as African Tree, originates from the tropical rain forest region of Africa [23]. It is a high-yielding source of edible and technical oils. Hence, oil palm has been grown as a plantation crop in most tropical countries within 10° of the equator. The international trade of palm oil started in the nineteenth century. At that time, Africa led the world in production and export of oil palm. However, Indonesia and Malaysia surpassed Africa's total palm oil production by 1966. Moreover, among the five top oil palm producing nations, three of them are Southeast Asian countries. They are Indonesia, Malaysia and Thailand.

As in many other agricultural products, oil palm generates abundant amounts of biomass. Properly used, the biomass is not only able to solve environmental problems, but also generates a new income since the wastes may turn into more useful products. On the contrary, if the waste is not used, its disposal may lead to environment problems [24]. This situation attracted the attention of researchers in oil palm producing countries to explore the use of the biomass.

The wood flour of oil palm has been studied for use as filler in polypropylene (PP) [25]. It was found that larger-sized filler obtained higher tensile modulus, tensile strength and impact strength in comparison with the smaller-sized filler. However, without any treatment, the mechanical properties of the composites decreases with the increase of oil palm flour content.

Ismail *et al.* [26] reported the effects of concentration and modification of surface in oil palm fiber-reinforced rubber composite. After modification of the fiber surface, the physical properties of the oil palm fiber/rubber composites increased. Although the tensile strength decreased with the increase of concentration of oil palm fiber, the modulus and hardness increased. Moreover, modification of surface improved the adhesion between fiber and rubber, as observed under scanning electron microscope.

Similar result were obtained when the oil palm wood flour (OPWF) was mixed with epoxidized natural rubber (ENR). The increase of OPWF content resulted in the decrease of tensile strength and elongation at break of the OPWF/ENR composites. However, it increased tensile modulus, tear strength and hardness. Moreover, the cure (t_{90}) and scorch time decreased when the OPWF content increased. Larger particle size of OPWF resulted in shorter t_{90} and scorch time, while the highest fiber content with the smallest particle size resulted in the highest torque [27].

Continuing research then studied the fatigue life of OPWF/natural rubber [28]. Beside the reduction of tensile strength, tear strength and elongation, the increase of OPWF loading in natural rubber compounds resulted in the reduction of fatigue life. When the OPWF loading increases, the interfacial adhesion between the OPWF filler and matrix gets poorer. This is due to poor wetting of the OPWF.

Beside the wood flour, an attempt to use composite from empty fruit brunch (EFB) has also been reported. Rozman *et al.* [29] compounded three sizes of EFB at different filler loading with high-density polyethylene (HDPE) in a single-screw compounder. As a result, the modulus of elasticity of EFB/HDPE increased when the filler loading increased. On the contrary, the modulus of rupture decreased when the filler loading increased. In terms of particle size, sample with smaller-sized particles performed higher in both modulus of elasticity and modulus of rupture in comparison to that of larger-sized particles. Moreover, flexural toughness, tensile strength and impact strength decreased when the filler content increased.

Yusoff *et al.* [30] reported the mechanical properties of oil palm fiber/epoxy. The fiber used was in the form of short random fibers. A hand lay-up technique was conducted to produce 5, 10, 15, and 20% fiber by volume. A decreasing trend of tensile and flexural properties was reported as fiber loading increased.

Hill *et al.* [31] manufactured EFB/polyester composites after various chemical treatments. The composites were then tested by exposure to decay fungi in unsterile soil for up to 1 year. It was found that such exposure resulted in deterioration of mechanical properties. The chemical treatments, however, significantly protected the composites from severe deterioration.

Rozman *et al.* [32] modified the EFB filler by maleic anhydride (MAH). The reaction of EFB and MAH (dissolved in dimethylformamide) was conducted at 90°C. Higher flexural and impact strength was obtained after the treatment. It is an indication that the MAH treatment enhanced the adhesion between EFB filler and matrix. Fourier transform infrared (FTIR) analysis showed that before the reaction occurred, there was an absorption peak at 1630 cm^{-1}, indicating C = C groups inside MAH. After the reaction, the absorption was reduced. The reduction of the absorption indicated the reduction number of C = C groups inside MAH due to reaction with PP.

Beside adding the coupling agent, irradiation of the EFB composites was also conducted [33]. The EFB-filled poly(vinyl chloride)/epoxidized natural rubber (PVC/ENR) blends were irradiated using a 3.0 MeV electron beam at doses ranging from 0 to 100 kGy in air and at room temperature. Mechanical tests were performed to measure the tensile strength, Young's modulus, elongation at break and gel fraction. Moreover, the composites were also compared to poly(methyl acrylate)-grafted EFB. As a result, electron beam irradiation increased the tensile strength, Young's modulus and gel fraction. The elongation at break, however, decreased. Moreover, although there was an improvement in the adhesion between the EFB and the matrix, there was no significant effect of the grafting EFB with methyl acrylate on the tensile properties and gel fraction of the composites upon irradiation.

Since the mechanical properties of the EFB/polymer composite are not very high, Rozman *et al.* [34] tried to mix the EFB with the conventional glass fiber (GF) and PP matrix to form a hybrid composite. The incorporation of the EFB and GF into PP matrix, however, still decreased the flexural and tensile strength, while both flexural and tensile modulus increased. The flexural and tensile properties improved after the addition of coupling agents such as maleic-anhydride-modified PP (known as Epolene, E-43) and 3-(trimethoxysilyl)-propylmethacrylate (TPM). However, insignificant improvement was obtained after the addition of polymethylenepolyphenyl isocyanate (PMPPIC). Hence, E-43 and TPM are suitable coupling agents for EFB-GF/PP hybrid composites.

Another study was conducted on the tensile and impact behavior of oil palm fiber/glass fiber hybrid bilayer laminate composites [35]. The epoxy resin was impregnated in the fiber mats and cured at 100°C for 1 h before post-curing at 105°C. An increment of impact strength was obtained after the addition of glass fibers. When the glass fiber parts were impacted, the impact strength of the composite was higher than when the oil palm layers were impacted. In other words, the glass/epoxy parts showed higher impact strength compared to the oil palm/epoxy parts.

Khalil *et al.* [36] studied the combination of oil palm fiber and glass fiber as hybrid composite in polyester matrix. Hybrid laminate composites with different weight ratio were prepared to study the hybrid effect of glass and EFB fibers on the physical and mechanical properties of the composites. Generally, the hybrid composites showed better properties as compared to EFB/polyester composites without glass fiber.

Beside the glass fiber hybrid composite, hybrid of EFB and jute fibers was also studied. Two kinds of trilayer hybrid composites were prepared and compared. The composite with EFB as skin material and jute as the core material was compared to the composite with jute as skin material and EFB as the core material. It was reported that all the composites were resistant to various chemicals. Moreover, the tensile properties of the composite having jute as skin and EFB as core material showed higher tensile properties as compared to the other layering pattern.

Following the trend of using PVC as matrix, research on the use of EFB as composite in PVC matrix was also reported. Bakar *et al.* [37] used the EFB as filler in unplasticized poly(vinyl chloride) (PVC-U). They studied the effects of extracted EFB on the processability, impact and flexural properties of EFB/PVC-U composites. PVC-U resin, EFB and other additives were first dry-blended using a heavy-duty laboratory mixer and then milled into sheets on a two-roll mill before being hot-pressed and cut into impact and flexural test specimens. There were two kinds of EFB used in this experiment, which were extracted and unextracted. The FTIR showed that the unextracted EFB contained oil residue, while the extracted one contained less oil residue. The results showed that both extracted and unextracted EFB decreased the fusion time and melt viscosity. However, the fusion time increased with the increase of extracted EFB content. Meanwhile, there was no significant difference in both the impact and flexural properties of extracted and unextracted EFB.

In order to explore more properties beside mechanical properties, Yousif *et al.* [38] determined the tribological performance of oil palm fiber/polyester composites. It was reported that the existence of oil palm fiber in the matrix improved the wear properties by up to four times compared to unfilled polyester. The friction coefficient of the composite was less by about 23% than that of unfilled polyester. In terms of wear mechanisms, there was debonding, bending and tear of fibers, and high deformation in the resinous region was observed.

We can see that there have been many efforts to study oil palm fibers with various matrices and various treatments. However, the properties are still limited and cannot compete with the synthetic fibres. Further development is still needed before the composites can face the competition in the market. Some issues that have not been well studied and need further work regarding fiber and its composites are [24]:

- Thermal degradation at low temperatures;
- Water absorption, which is related to mechanical properties;
- Various mechanical properties due to various factors;
- Further application with current development.

This presents both a challenge and opportunity for researchers who are working toward a conclusive outcome regarding the properties of oil palm fiber and its composites.

3.4 Durian (*Durio zibethinus*)

Durian actually refers to the fruit of several species of trees belonging to the genus *Durio*. There are about 30 recognized *Durio* species. However, *Durio zibethinus* is the only species available on the international market. Hence, in this chapter, durian refers to *Durio zibethinus*. It is a fruit that originates from Southeast Asia. The name "durian" itself comes from an Indonesian and Malaysian word. *Duri* means thorn. It describes the fruit, which is surrounded by many thorns on its skin. Durians are pictured in Figure 3.2 ,where their sharp thorn-covered skin can be seen.

Nowadays, more than 50% of the fruit of one durian, including skin, seed, etc., is considered as waste [39]. Hence, as in the case of oil palm, using durian skin as composite is an attempt to convert a waste into a value-added product. Although not as extensively used as sugar palm and oil palm, there are several reports on the usage of durian skin for composite materials, including paperboard.

Khedari *et al.* [40] reported an initial investigation on the use of lightweight construction material composed of durian fiber, coconut fiber, cement and sand. The effects of the addition of fiber on thermal conductivity, compressive strength and bulk density were studied. As a result, it was found that the addition of the fibers reduces the thermal conductivity of the composite. Moreover, the composite yieldis lightweight and satisfies the basic requirements of construction materials. It can be used for walls and roofs. Hence, the potential use of durian as filler in cement composite seems promising. The development of the composite may lead to an energy-efficient building, while converting a waste into an added-value product.

Figure 3.2 Durian, "the king of fruits."

The durian peel and coconut coir mixture was then used to develop low thermal conductivity particleboards. It was reported that the mixture ratio of durian peel and coconut coir was optimum at 10:90 by weight. In comparison with durian only particleboards and coconut only particleboards, the mixture particleboards showed better properties, except for the modulus of elasticity. The mixture particleboard is of lower thermal conductivity, which is suitable for ceiling and wall insulating materials. With more development, it will not be impossible to use this material for furniture applications.

A performance simulation test of durian fiber-based lightweight construction materials was conducted by Charoenvai *et al.* [41]. Using commercial research software (WUFI 2D), the heat and moisture transfer through a durian fiber-based lightweight construction material was calculated. In the simulation, the materials were exposed to a climate condition of Bangkok when the hygrothermal characteristics of the material were investigated. From the investigation, it was observed that the weekly mean water content of the surface of the composite was moderately low. The apparent thermal performance of the composite was highly affected by the moisture. When the water was absorbed in the pore structure, the thermal conductivity increased, as the thermal conductivity of water is higher than that of air. Although the mean value of thermal conductivity was still low, coating of the surface is needed to decrease the moisture content. After all, the results confirmed that the composite satisfied the requirement of the construction materials.

Charoenvai *et al.* then tried to replace the cement with rice husk ash (RHA) [42, 43]. The investigation included oxide analysis, X-ray diffraction, surface area, fineness, and size distribution measurements, together with scanning electron microscopy. The study concluded that the RHA can be used as a partial replacement for ordinary Portland cement type I in mortar mixes. The replacement of 30% ordinary Portland cement with ground RHA performed with maximum strength. Moreover, the fineness of RHA affected the compressive strength of the composites. Finer RHA resulted in higher compressive strength. In addition, increasing the size of the sand decreased the compressive strength, bulk density and thermal conductivity, due to more empty spaces or voids.

Durian seed is also available in the form of flour. Durian seed flour has been compounded with PP and HDPE to form durian seed flour (DSF)/thermoplastic composites[44]. The presence of DSF was found to increase the stabilization torque during mixing and to reduce the tensile strength and elongation at break. The only increased mechanical property was the modulus of elasticity. For the matrix comparison, the tensile strength and modulus of elasticity of DSF-filled PP were higher than those of DSF-filled HDPE at the same fiber content. Meanwhile, the elongation at break of DSF-filled PP was lower than that of DSF-filled HDPE at the same fiber content. It showed a lower effect in the mechanical properties; there was evidence that filler improved the thermal properties of the composites. Thermogravimetric analysis results showed that the initial degradation temperature of both DSF/PP and DSF/HDPE composites increased with the incorporation of DSF. Positive results were also reported on biodegradation. There was a significant increase in the carbonyl and hydroxyl index of the composite after a simple biodegradability test. It is evidence that the composites are subjected to biodegradation.

3.5 Water Hyacinth (*Eichhornia crassipes*)

Water hyacinth (*Eichhornia crassipes)* is an aquatic plant native to the Amazon basin. However, it can easily reproduce and hence has spread all around the world. As a result, we can now find water hyacinth in most tropical and subtropical countries. The water hyacinth is one of the fastest growing plants known. As a result, the availability of this plant, especially in tropical countries, is high. In many places this plant may act as a pernicious invasive plant. As it is easy to grow, uncontrolled water hyacinth will cover entire lakes or ponds. As a result, it blocks sunlight and oxygen, often killing fish and other aquatic animals [45]. Figure 3.3 shows water hyacinth covering a river in Makassar, Sulawesi, Indonesia. In it can be seen a large area of water covered by water hyacinth due to the rapid growth of the plants. It depicts the abundance of this resource. Hence, utilizing water hyacinth (WH) as filler or a reinforcing agent in a polymeric composite is an interesting concept. It may contribute not only to economic but also environmental development. Currently the water hyacinth stalk is a raw material used for many kinds of handmade crafts. However, the usage of this material is low compared to its availability.

Supri and Ismail [46] prepared modified and unmodified low-density polyethylene (LDPE) and mixed it with water hyacinth fiber (WHF) composites by melt blending. Tensile test, differential scanning calorimetry (DSC), thermogravimetric analysis (TGA), and water absorption behavior test of the composites were conducted. The NCO-polyol-modified LDPE/WHF showed higher tensile strength, modulus of elasticity, and water absorption resistance as compared to the unmodified LDPE/WHF composites. However, the elongation at break was better when the LDPE was unmodified. Moreover, the modified LDPE/WHF offered better thermal properties in comparison to the unmodified LDPE/WHF. The NCO-polyol was reported to create better dispersion of WHF in the LDPE matrix.

Figure 3.3 Water hyacinth covering a river.

As a continuation of the above research, Supri and Ismail used diisocyanate-polyhydroxyl groups as a coupling agent on low-density polyethylene/acrylonitrile butadiene styrene/water hyacinth fiber (LDPE/ABS/WHF) composites [47]. The coupling agent was reported to enhance the tensile strength and modulus. Moreover, it also improved the thermal stability of the composites.

Further development was conducted by modifying the water hyacinth fiber with poly(methyl methacrylate) (PMMA) [48]. The PMMA-modified WHF was blended with LDPE and natural rubber (NR) to produce a composite in a LDPE/NR blended matrix. The results showed that the PMMA-modified WHF composites showed higher tensile strength and modulus, glass transition temperature, melting temperature and degree of crystallinity compared to the unmodified WHF composites. The PMMA was reacted with the fiber and it was shown in FTIR analysis as the presence of ester carbonyl group and C-O ester group in PMMA-modified WHF. Better interfacial adhesion between fiber and matrix was also found when the WHF was modified with PMMA. This indicates a lower value of interparticle spacing in modified WHF composites as compared to that in unmodified WHF composites.

Another chemical modification studied was poly(vinyl alcohol) (PVA). The modification was conducted in water hyacinth fiber before mixing with low-density polyethylene and natural rubber to form a LDPE/NR/WHF composite. The tensile strength and modulus, melting temperature and water absorption resistance of the LDPE/NR/WHF composites are reported to increase after PVA modification. However, the elongation at break decreased after the modification. Lower value of interparticle spacing was also found in PVA-modified WHF composites, enhancing the interparticle interaction between WHF and the matrix.

Furthermore, Abral *et al.* [49] studied the characteristics of water hyacinth fiber and its composite with unsaturated polyester. An alkali treatment was used to modify the surface of the WHF. The WHF was treated in various alkali concentrations for 1 h

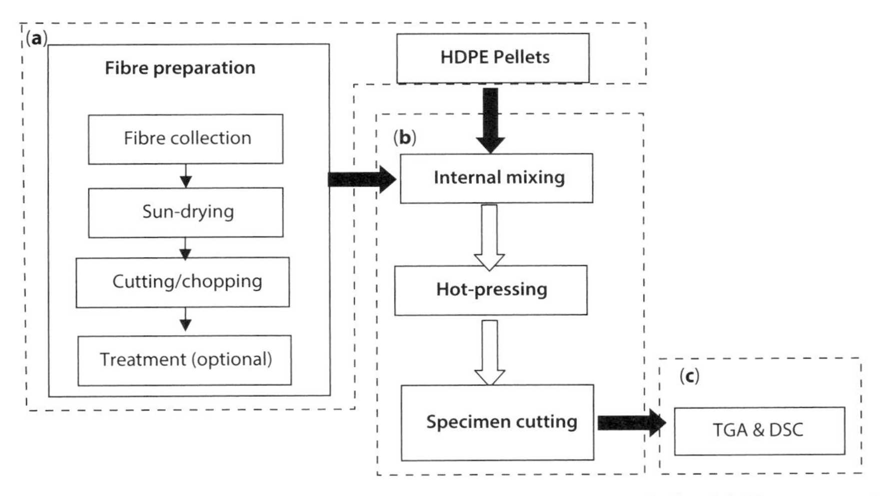

Figure 3.4 Flow chart of the water hyacinth-filled HDPE experiments includes: (a) fiber preparation, (b) material mixing, and (c) material testing.

immersion time before being mixed with the unsaturated polyester. From the experiments, it was found that better mechanical properties were obtained after the treatment with 7% NaOH for 1 h. However, higher alkali concentrations damaged the cellulosic structure and decreased the mechanical properties of the composites.

Wirawan *et al.* [50] and Sutrisno *et al.* [51] reported the thermal properties of water hyacinth-filled high-density polyethylene (HDPE). The flow chart of the experiment is shown in Figure 3.4.

The water hyacinth was sun-dried to remove the water for 24 h. The sun-dried water hyacinth stalks were then chopped to obtain no more than 3 cm of short fibers. After the chopping process, the fibers are ready to be mixed with the HDPE matrix. The fibers and the HDPE pellets were then introduced to a Haake Reomix OS at a temperature of 140°C for a thermal mixing process at a rotor speed of 50 rpm. The HDPE pellets were fed into the chamber of the thermal mixer to form a melted homogeneous matrix. About 5 min later, the fibers were also fed into the chamber. The total mixing process took 20 minutes. Various compositions were made as described in Figure 3.5. The outputs of the internal mixer were then hot-pressed to form a platen composite. The composites were then cut to prepare several tests. In this chapter, reports on the results of thermogravimetric analysis and differential scanning calorimetry are reported. The thermogravimetric analysis result is shown in Figure 3.6. It can be seen that the weight of WHF, HDPE and the composite commonly decreased significantly in a temperature range of 200 to 400°C. The temperature at the initial weight decrease corresponded to the thermal stabilization of the material. It indicates the temperature at which the material started to decompose.

There is a small initial weight loss observed at a temperature of around 100°C for WHF, which corresponds to a moisture uptake. It is known that natural fibers, including WHF, are hydrophilic in nature, and hence there is always moisture content inside. Sometimes, the sun-drying process is not enough to remove the moisture. At a higher temperature, two main stages of degradation were shown. The degradation is related to the decomposition of the main substance of lignocelluloses, which are cellulose,

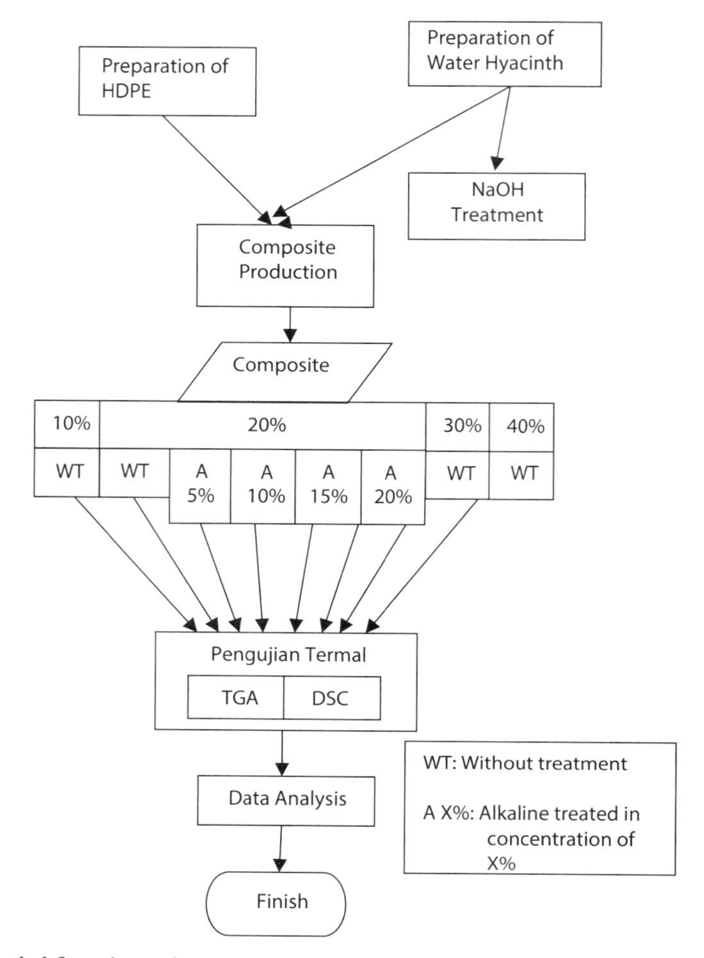

Figure 3.5 Detailed flow chart of experiment.

hemicelluloses and lignin [52]. It was reported that the hemicelluloses and cellulose degraded at a lower temperature than lignin [53]. Based on the literature, the degradation of cellulose and hemicelluloses is represented by the first stage of degradation. The higher temperature of degradation corresponds to the degradation of lignin. Meanwhile, the HDPE performed single stage of degradation, indicating the homogeneity of the material in terms of thermal degradation.

Moreover, in Figure 3.6 we can see that the curve was shifted to the lower temperature when the fiber content of WHF/HDPE composites increased. This corresponds to the higher degradation temperature of the matrix as compared to the first stage degradation of the fiber. In other words, the thermal stability of the fiber was lower than that of the matrix. As a result, the first stage of degradation temperature of the composite decreases when the fiber increases. This phenomenon indicates that less (or no) chemical reaction occurred between WHF and the matrix. Thus, the mixing of WHF and HDPE did not change the physical properties of each material.

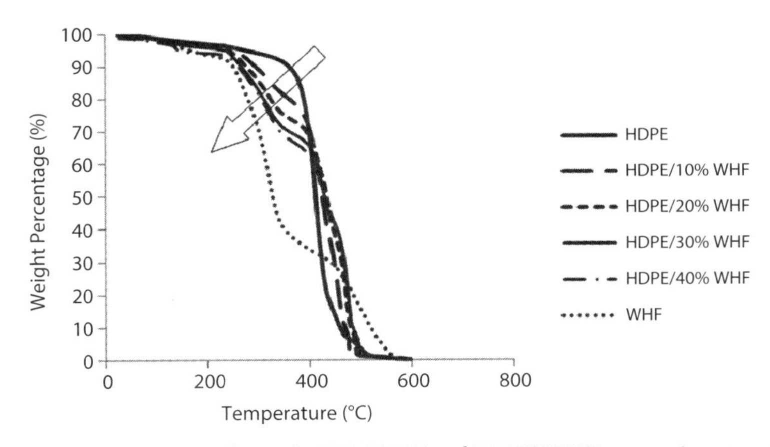

Figure 3.6 Thermogravimetric analysis of WHF, HDPE and WHF/HDPE composites at various WHF contents.

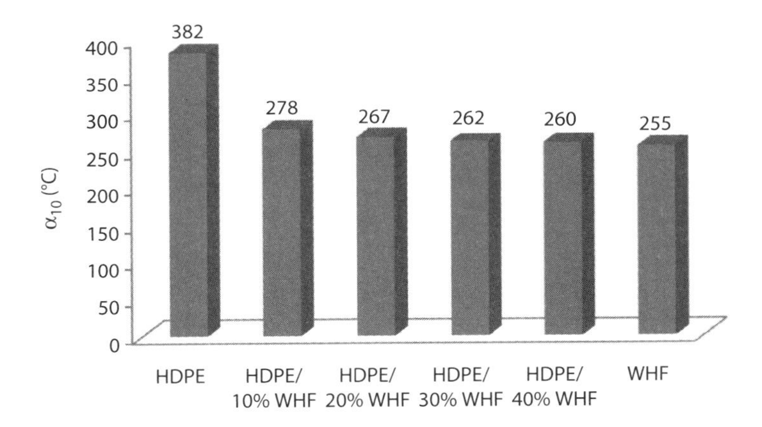

Figure 3.7 Thermal stability value of HDPE, WHF and composites.

The value of thermal stability can be identified based on the temperature of the sample when it is 10% decomposed (α_{10}). Figure 3.7 clearly shows that the thermal stability values are slightly lower with the fiber content.

The thermogravimetric analysis of WHF/HDPE composites after various treatments is depicted in Figure 3.8. It can be seen that 5% alkali treatment caused the decrease of thermal stability of the WHF/HDPE composite. However, when the concentration of alkali increased up to 20%, the thermal stability increased. This can be explained by the fact that alkali treatment removes some particles at the outer part of the fiber skin. The particles are believed to have higher thermal stability. Hence, when the particles are removed, the overall thermal stability of the WHF decreased. However, at the higher concentration, the alkali removes hemicelluloses, which are of lower thermal stability. As a result, the overall thermal stability of the fiber and the composite increased. Figure 3.9 shows the comparison of the thermal stability values.

The effect of fiber content on the DSC curve is depicted in Figure 3.10. It can be seen that there are uniform curves beside the WHF. This happened because at the

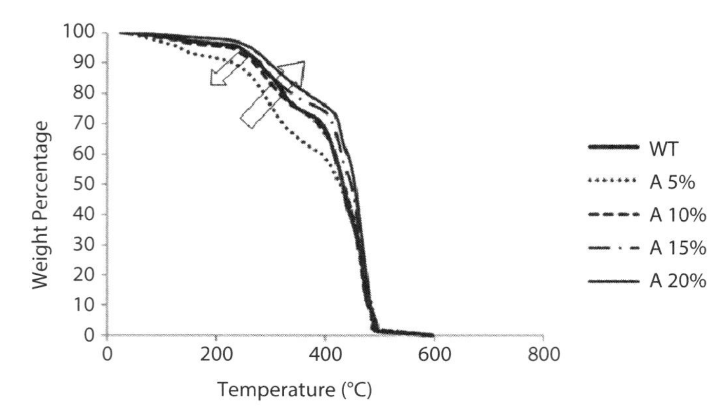

Figure 3.8 Thermogravimetric analysis of WHF/HDPE composites after various treatments.

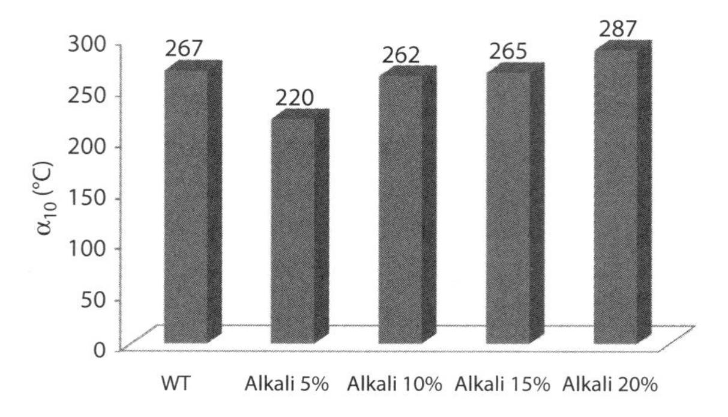

Figure 3.9 Thermal stability values of the WHF/HDPE composites after various treatments.

temperature of 130–145°C HDPE melts. The melting of HDPE occurred not only in the unfilled HDPE but also the HDPE in the composite. The peaks shown in the DSC curves indicate the melting point of the HDPE.

The interesting result here is that the incorporation of the WHF slightly moved the curve to the left. This indicated that there is a decreasing of the melting point of HDPE after the incorporation of WHF. Higher WHF content resuted in a lower melting point of HDPE. The comparison of melting points is depicted in Figure 3.11.

Figure 3.12 shows the effect of concentration of NaOH on the DSC curve. The melting point of the HDPE in the composites slightly increased after 5% NaOH treatment. However, the significant increase of the melting point was only achieved after 10% of NaOH treatment. For the NaOH treatment at the concentration of more than 10%, the increase of melting point was not significant. The comparison of the melting point is depicted in Figure 3.13.

The significant increase of melting point after 10% alkali treatments shows that, at that concentration, the NaOH has a role to improve fiber-matrix interface. When the fiber is untreated, the fiber and matrix is not mixed well due to the hydrophobic nature of the matrix and hydrophilic nature of the fiber. Hence, the adhesion between fiber and matrix is poor and there is no heat transfer between fiber and matrix [54, 55]. As a result, the melting point of the composite is not higher than that of the matrix. After 10% alkali treatment, the adhesion gets better and then the heat transfer occurs. Hence,

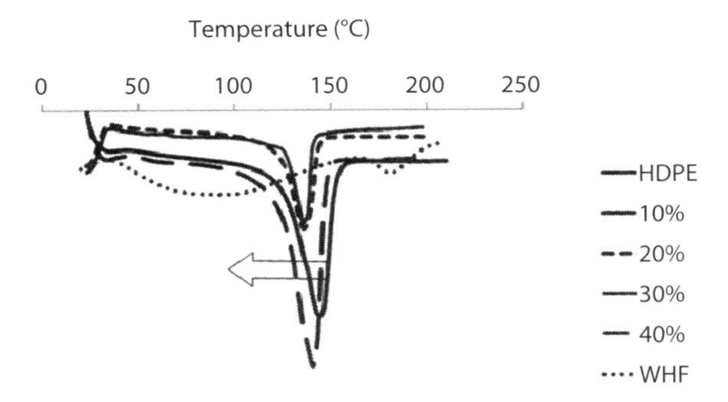

Figure 3.10 Differential scanning calorimetry of HDPE, WHF and their composites.

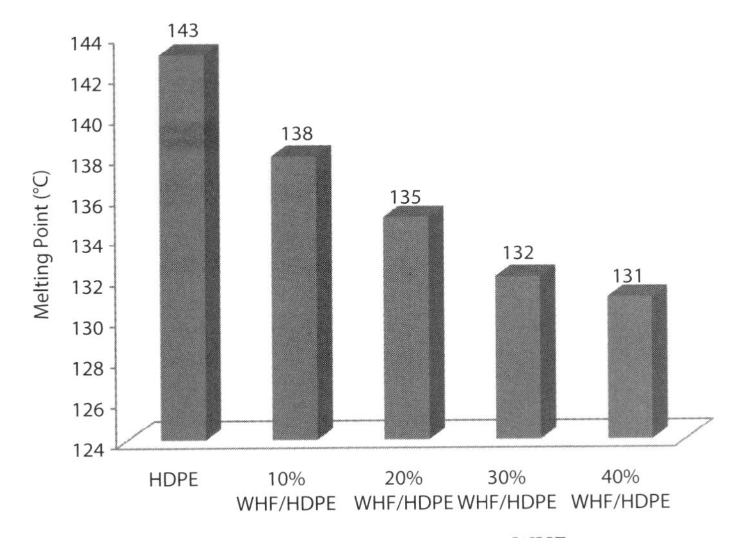

Figure 3.11 Melting points of WHF/HDPE composites at various WHF content.

the fiber, which is of higher thermal stability, also carries the thermal load. As a result, the melting point of the HDPE in the the composite is enhanced.

Alkali treatment in the concentration of 5% NaOH, is not enough to make a better fiber-matrix adhesion. As a result, there is no significant increment in thermal stability. On the other hand, the 15% and 20% alkali treatment cannot make a fibre-matrix adhesion better than that of 10% alkali treatment, as the optimum fiber-matrix adhesion is already reached.

3.6 Summary

This chapter has shown the great efforts of researchers in Southeast Asia to develop cellulose-based polymer composite, especially from the abundant resources in Southeast Asia. In terms of mechanical properties, several treatments, from the cheapest alkali treatment to the expensive coupling agents, have been reported. However, developing a cellulose-based polymer composite that can compete with conventional fiber

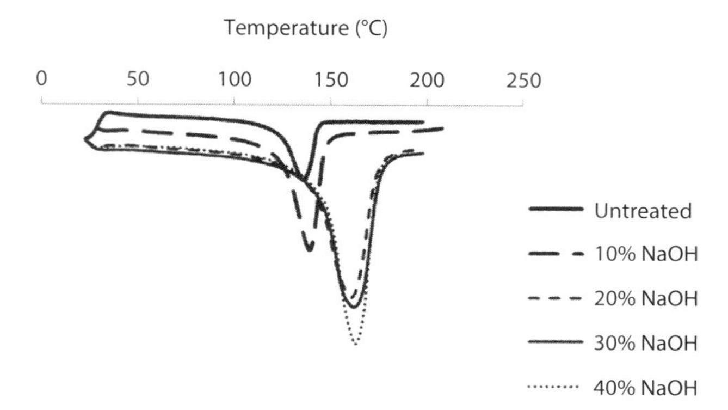

Figure 3.12 Differential scanning calorimetry of WHF/HDPE composites after various treatments.

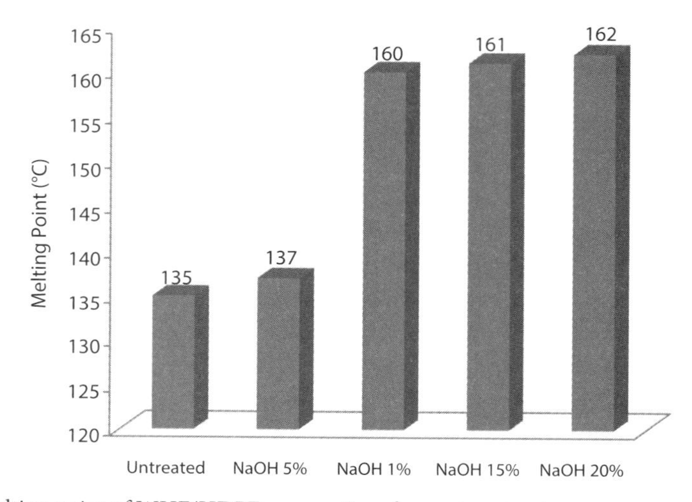

Figure 3.13 Melting point of WHF/HDPE composites after various treatments.

composites is still challenging. Many factors are related to the mechanical properties, from the varied nature of the fiber to the fiber-matrix adhesion. It seems that research on cellulose-based polymer composites will continue to explore some new methods and materials for developing new cellulose-based polymer composites. The Southeast Asian countries have an opportunity to explore more new resources for cellulose-based materials, as they have unlimited cellulose-based polymer resources.

References

1. D.N. Saheb, and J.P. Jog, Natural fiber polymer composites: A review. *Adv. Polym. Technol.* 18(4), 351–363 (1999).
2. I.M. Daniel, and O. Ishai, *Engineering Mechanics of Composite Materials,* Oxford University Press, New York. (1994).
3. D. Gay, and S.V. Hoa, *Composite Materials: Design and Applications. 2* ed., CRC Press, New York. (2007)

4. W.D. Callister, *Materials Science and Engineering: an Introduction*. 7 ed, John Wiley & Sons, New York. (2007)
5. L.Drzal, A.K. Mohanty, R. Burgueno, and M. Misra, Biobased Structural Somposite Materials for Housing an Infrastructure Applications: Opportunities and Challenges, Proceedings of the NSF Housing Research Agenda Workshop, Volume 2, pp. 129–140 (2004).
6. C. W. Ho, W.M. Aida, M.Y. Maskat, and H. Osman, Changes in volatile compounds of palm sap (Arenga pinnata) during the heating process for production of palm sugar. *Food Chem.* 102(4), 1156–1162 (*2007*).
7. M.R. Ishak, S.M. Sapuan, Z. Leman, M.Z.A. Rahman, U.M.K. Anwar, and J.P. Siregar, Sugar palm (Arenga pinnata): Its fibres, polymers and composites. *Carbohydr. Polym.* 91(2), 699–710 (*2013*).
8. J. Mogea, B. Seibert, and W. Smits, Multipurpose palms: the sugar palm (Arenga pinnata (Wurmb) Merr.). *Agrofor. Syst.* 13(2), 111–129 (1991).
9. A. Ticoalu, T. Aravinthan, and F. Cardona. A Review of Current Development in Natural Fiber Composites for Structural and Infrastructure Applications, in *Proceedings of the 2010 Southern Region Engineering Conference (SREC 2010)*. Australia (2010).
10. H. Sastra, J.P. Siregar, S.M. Sapuan, Z. Leman, and M.M. Hamdan Flexural properties of Arenga Pinnata fibre reinforced epoxy composites. *Am. J. Appl. Sci.* 21–24 (2005).
11. Z. Leman, H.Y. Sastra, S.M. Sapuan, M.M.H.M. Hamdan, and M.A. Maleque, Study on impact properties of Arenga pinnata fibre reinforced epoxy composites. *J. Appl. Technol.* 3(1), 14–19 (2005).
12. M. Suriani, M.M. Hamdan, H.Y. Sastra, and S.M. Sapuan, Study of interfacial adhesion of tensile specimens of Arenga pinnata fiber reinforced composites. *Multidiscip. Modeling Mater. Struct.* 3(2), 213–224 (2007).
13. D. Bachtiar, S. Sapuan, and M. Hamdan, The effect of alkaline treatment on tensile properties of sugar palm fibre reinforced epoxy composites. *Mater. Des.* 29(7), 1285–1290 (2008).
14. D. Bachtiar, S. Sapuan, and M. Hamdan, The influence of alkaline surface fibre treatment on the impact properties of sugar palm fibre-reinforced epoxy composites. *Polym. Plast. Techn ol. Eng.* 48(4), 379–383 (2009).
15. Z. Leman, S.M. Sapuan, M. Azwan, M.M.H.M. Ahmad, Maleque, The effect of environmental treatments on fiber surface properties and tensile strength of sugar palm fiber-reinforced epoxy composites. *Polym. Plast. Technol. Eng.* 47(6), 606–612 (2008).
16. A. Ali, A. Sanuddin, and S. Ezzeddin, The effect of aging on Arenga pinnata fiber-reinforced epoxy composite. *Mater. Des.* 31(7), 3550–3554 (2010).
17. Z. Leman, S.M. Sapuan, A.M. Saifol, M.A. Maleque, and M.M. H.M. Ahmad, Moisture absorption behavior of sugar palm fiber reinforced epoxy composites. *Mater. Des.* 29(8),1666–1670 (2008).
18. H.A. Razak,and T. Ferdiansyah, Toughness characteristics of Arenga pinnata fibre concrete. *J. Nat. Fibers* 2(2), 89–103 (2005).
19. M. Ishak, S.M. Sapuan, Z. Leman, M.Z.A. Rahman, and U.M. K. Anwar, Characterization of sugar palm (Arenga pinnata) fibres. *J. Therm. Anal. Calorim.* 109(2), 981–989 (2012).
20. D. Bachtiar, S.M. Sapuan, E.S. Zainudin, A. Khalina, and K.Z.M. Dahlan, The tensile properties of single sugar palm (Arenga pinnata) fibre, in *IOP Conference Series: Materials Science and Engineering*. Volume 11, Issue 1, pp. 012012, IOP Publishing. (2010).
21. S. Sapuan, Z. Leman, S.M. Sapuan, M.Z.A. Rahman, Anwar, Effects of impregnation time on physical and tensile properties of impregnated sugar palm (Arenga pinnata) fibres. *Key Eng. Mater.* 471, 1147–1152 (2011).
22. L. Ismail, M.I. Ghazali, S. Mahzan, and A.M.A.Zaidi, Sound absorption of Arenga pinnata natural fiber. *World Acad. Sci. Eng. Technol.* 67, 804–806 (2010).

23. K. Poku, Small-scale palm oil processing in Africa. *FAO Agric. Serv. Bull.* 148 (2002).
24. A. Hassan, A.A. Salema, F.N. Ani, and A.A. Bakar, A review on oil palm empty fruit bunch fiber-reinforced polymer composite materials. *Polym. Compos.* 31(12), 2079–2101 (2010).
25. M. Zaini, M.A. Fuad, Z. Ismail, M.S. Mansor, J. Mustafah, The effect of filler content and size on the mechanical properties of polypropylene/oil palm wood flour composites. *Polym. Int.* 40(1), 51–55 (1996).
26. H. Ismail, N. Rosnah, and U. Ishiaku, Oil palm fibre-reinforced rubber composite: effects of concentration and modification of fibre surface. *Polym. Int.* 43(3), 223–230 (1997).
27. H. Ismail, H.D. Rozman, R.M. Jaffri, and Z.A. Ishak, Oil palm wood flour reinforced epoxidized natural rubber composites: The effect of filler content and size. *Eur. Polym. J.* 33(10), 1627–1632 (1997).
28. H. Ismail, and R. Jaffri, Physico-mechanical properties of oil palm wood flour filled natural rubber composites. *Polym. Test.* 18(5), 381–388 (1999).
29. H. Rozman, H. Ismail, R.M. Jaffri, A. Aminullah, and Z.A. Mohd Ishak, Mechanical properties of polyethylene-oil palm empty fruit bunch composites. *Polym. Plast. Technol. Eng.*37(4), 495–507 (1998).
30. M.Z.M. Yusoff, M. Zuhri, M.S. Salit, N. Ismail, and R. Wirawan, Mechanical properties of short random oil palm fibre reinforced epoxy composites. *Sains Malays.* 39(1), 87–92 (2010).
31. C. Hill, and H. Abdul, The effect of environmental exposure upon the mechanical properties of coir or oil palm fiber reinforced composites. *J. Appl. Polym. Sci.* 77(6), 1322–1330 (2000).
32. H. Rozman, M. Saad, and Z. Mohd Ishak, Flexural and impact properties of oil palm empty fruit bunch (EFB)–polypropylene composites—the effect of maleic anhydride chemical modification of EFB. *Polym. Test.* 22(3), 335–341 (2003).
33. C.T. Ratnam, G. Raju, and W.M.Z.W. Yunus, Oil palm empty fruit bunch (OPEFB) fiber reinforced PVC/ENR blend-electron beam irradiation. *Nucl. Instr. Meth. Phys. Res. B* 265(2), 510–514 (2007).
34. H. Rozman, G.S. Tay, R.N. Kumar, A. Abusamah, H. Ismail, and Z.A. Mohd Ishak, Polypropylene–oil palm empty fruit bunch–glass fibre hybrid composites: a preliminary study on the flexural and tensile properties. *Eur. Polym. J.* 37(6), 1283–1291 (2001).
35. A.B.A. Hariharan, and H.A. Khalil, Lignocellulose-based hybrid bilayer laminate composite: part I-studies on tensile and impact behavior of oil palm fiber-glass fiber-reinforced epoxy resin. *J. Compos. Mater.* 39(8), 663–684 (2005).
36. H.A. Khalil, S. Hanida, C.W. Kang, and N.N. Fuaad, Agro-hybrid composite: The effects on mechanical and physical properties of oil palm fiber (EFB)/glass hybrid reinforced polyester composites. *J. Reinf. Plast. Compos.* 26(2), 203–218 (2007).
37. A.A. Bakar, A. Hassan, and A.F.M. Yusof, The effect of oil extraction of the oil palm empty fruit bunch on the processability, impact, and flexural properties of PVC-U composites. *Int . J. Polym. Mater.* 55(9), 627 – 641 (2006).
38. B. Yousif, and N. El-Tayeb, The effect of oil palm fibers as reinforcement on tribological performance of polyester composite. *Surf. Rev. Lett.* 14(6), 1095–1102 (2007).
39. H. Anuar,M.M. Romainor, M.N. Nuraimi, and W.N. WANPotential of durian skin fiber as reinforcing agent in PLA composite, *in Composite Science and Technology: 2020-Scientific and Technical Challenges: Proceedings of the 9th International Conference on Composite Science and Technology (ICCST-9),* M. Meo, (Ed.), pp. 430–433, DEStech Publications, In c, U.S.A., (2013).
40. J. Khedari, B. Suttisonk, N. Pratinthong, and J. Hirunlabh, New lightweight composite construction materials with low thermal conductivity. *Cem. Concr. Compos.* 23(1), 65–70 (2001).

41. S. Charoenvai, J. Khedari, J. Hirunlabh, C. Asasutjarit, B. Zeghmati, D. Quenard, and N. Pratintong, Heat and moisture transport in durian fiber based lightweight construction materials. *Sol. Energ.* 78(4), 543–553 (2005).
42. S. Charoenvai, J. Khedari, J. Hirunlabh, M. Daguenet, and D. Quenard, Impact of Rice Husk Ash on the Performance of Durian Fiberbased Construction Materials, in *Tenth DBMC International Conference on Durability of Building Materials and Components*, Lyon, France, April 17–20 (2005).
43. S. Charoenvai, J. Khedari, J. Hirunlabh, and C. Asasutjarit, Development of Durian Fiber-based Composite Material, *The Second TSME International Conference on Mechanical Engineering*, (2011).
44. H. Osman, and M.H. Zakaria, Effects of durian seed flour on processing torque, tensile, thermal and biodegradation properties of polypropylene and high density polyethylene composites. *Polym. Plast. Technol. Eng.* 51(3), 243–250 (2012).
45. F. Yirefu, A. Tafesse, T. Gebeyehu, and T. Tessema, Distribution, impact and management of water hyacinth at Wonji-Showa sugar factory. Ethiop. *J. Weed Manage.* 1,41–52 (2007).
46. A.G. Supri, and H. Ismail, The effect of NCO-polyol on the properties of low-density polyethylene/water hyacinth fiber (Eichhornia crassiper) composites. *Polym. Plast. Technol. Eng.* 49(8), 766–771 (2010).
47. A.G. Supri, and H. Ismail, The effect of isophorone diisocyanate-polyhydroxyl groups modified water hyacinth fibers (Eichhornia crassiper) on properties of low density polyethylene/acrylonitrile butadiene styrene (LDPE/ABS) composites. *Polym. Plast. Technol. Eng.* 50(2), 113–120 (2011).
48. A.G. Supri, S.J. Tan, H. Ismail, Teh, Effect of poly (methyl methacrylate) modified water hyacinth fiber on properties of low density polyethylene/natural rubber/water hyacinth fiber composites. *Polym. Plast. Technol. Eng.* 50(9), 898–906 (2011).
49. H. Abral, H. Putra, S.M. Sapuan, and M.R. Ishak, Effect of alkalization on mechanical properties of water hyacinth fibers-unsaturated polyester composites. *Polym. Plast. Technol. En g.* 52(5), 446–451 (2013).
50. R. Wirawan, , R. Pasaribu, and A. Kholil, Thermogravimetric Analysis of Untreated Water Hyacinth (Eichhornia Crassipes)/HDPE Composites. Composite Science and Technology: 2020-Scientific and Technical Challenges, *Proceedings of the 9th International Conference on Composite Science and Technology (ICCST-9), p.* 418 (2013).
51. H. Sutrisno, R. Wirawan, B. Wibawa, and D. Permatasari, Differential Scanning Calorimetry of Alkali-Treated Water Hyacinth (Eichhornia Crassipes)/HDPE Composites. Composite Science and Technology: 2020-Scientific and Technical Challenges, *Proceedings of the 9th International Conference on Composite Science and Technology (ICCST-9), p.* 407 (2013).
52. E.S. Zainudin, S.M. Sapuan, K. Abdan, and M.T.M. Mohamad, Thermal degradation of banana pseudo-stem filled unplasticized polyvinyl chloride (UPVC) composites. *Mater. Des.* 30(3), 557–562 (2009).
53. R.M.Rowell, Chemical modification of agro-resources for propertyenhancement, in *Paper and Composites from Agro-based Resources*, R.M. Rowell, R.A. Young, and J.K. Rowell, (Ed.), Lewish Publisher, London, Volume 54, pp. 351–375 (1997).
54. N,Z.X. Wang, X. Ma, and J. Fang, Influence of carbon black on the properties of plasticized poly (lactic acid) composites. *Polym. Degrad. Stabil.* 93(6), 1044–1055 (2008).
55. E.T.N. Bisanda, The effect of alkali treatment on the adhesion characteristics of sisal fibres. *Appl. Compos. Materi.* 7(5), 331–339 (2000).

4

Hybrid Vegetable/Glass Fiber Composites

Sandro C. Amico*[,1], José R. M. d'Almeida[2], Laura H. de Carvalho[3] and Mª. Odila H. Cioffi[4]

[1]PPGEM, Federal University of Rio Grande do Sul, Porto Alegre, Brazil
[2]Materials Engineering Department, Pontifical Catholic University of Rio de Janeiro, Rio de Janeiro, Brazil
[3]Materials Engineering Department, Federal University of Campina Grande, Campina Grande, Brazil
[4]Fatigue and Aeronautic Materials Research Group - UNESP – Univ Estadual Paulista, Guaratinguetá/SP, Brazil

Abstract
There is an ever-growing worldwide interest in the use of lignocellulosic fibers as reinforcement in either thermoset (mainly unsaturated polyester) or thermoplastic (mainly polyolefins) composites. However, the wider use of these fibers for replacing synthetic ones is limited by disadvantages like their comparatively poorer mechanical properties, higher moisture absorption, lower compatibility to polymers, fiber heterogeneity, inferior durability and also flammability. Among the ways of minimizing these drawbacks, the concomitant use of vegetable and synthetic fibers, i.e., producing hybrid composites, is among the most promising strategies. For instance, hybridization with glass fiber, the most used synthetic fiber, is an unquestionable way of incrementing overall mechanical and thermal properties. In this context, this chapter presents a review of the state-of-the-art of hybrid vegetable/glass fiber composites, focusing on the hybridization effect on the properties of thermoplastic and thermoset polymer matrices composites.

Keywords: Hybrid composites, glass fiber, vegetable fibers, thermoplastic, thermoset

4.1 Introduction

The market for polymer composites is growing continuously mainly based on the fact that these materials display a combination of properties not attainable by their constituents alone, and that can be tuned to target specific applications by adequately choosing matrix, reinforcement, modifiers, processing methods and conditions [1, 2]. There is also a worldwide interest in the use of lignocellulosic, also known as vegetable fibers, as reinforcement for these composites. Some of the advantages of using these fibers include generally lower cost, lower environmental impact, lower specific gravity, lower abrasion during processing and better end-of-life characteristics. Their use in less strict

**Corresponding author*: amico@ufrgs.br

Vijay Kumar Thakur (Ed.), Lignocellulosic Polymer Composites, (63–82) 2015 © Scrivener Publishing LLC

applications is attracting increasing attention in many sectors, especially civil/building, automotive, packaging, and for consumer goods and acoustic insulation components (e.g., ceiling paneling, partition boards).

Properties of composites are mostly dependent on their constituent's identity and amount, shape, size, orientation, aspect ratio, and matrix/filler adhesion, which is particularly crucial as stresses are transmitted to the reinforcement through the interface [3, 4]. In the case of vegetable fibers, which are hydrophilic, their interface with the polymer, mostly hydrophobic, is poor, which in turn leads to reduced mechanical properties of the composites [5]. In order to overcome this issue, chemical modifications (alkali, acetilation, bleaching, benzylation, silanization, monomer grafting, among others) of the fibers or even of the matrix may be employed. According to Belgacem and Gandini [6], the best overall results are obtained when the chemical modification generates covalent bonds between fiber surface and matrix. Fiber surface modifications also include physical and physico-chemical treatments such as solvent extraction, corona and plasma discharges, laser, X-ray and UV bombardments [7-9].

Other quicker and simpler solutions include pre-drying and addition of a compatibilizer [10, 11]. Pre-drying yields surface moisture removal, and improved fiber wetting and fiber/polymer adhesion, whereas the addition of a compatibilizer, i.e., a component containing both hydrophobic and hydrophilic moieties in its structure, allows the formation of chemical bonds (covalent or van der Waals) between fibers and matrix. However, mixed success has been reported in the literature for the use of adhesion promoters/compatibilizers with results greatly dependent on the matrix employed, the type and amount of the filler, the amount and characteristics (e.g., chemical nature, molecular weight, etc.) of the promoter, the incorporation method and the composite processing technique, among others. In general, although the use of adhesion promoters leads to reduced water absorption, better tensile properties and processing, the elongation at break, toughness and impact properties are often either not enhanced or reduced. In fact, if toughness and impact, rather than tensile properties are desired, the use of elastomeric impact modifiers may be an alternative.

It should be noted, however, that the thermal and mechanical properties of vegetable fiber reinforced polymer composites are notoriously lower than those of similar composites reinforced with synthetic fibers (e.g., carbon, glass, aramid) [1, 2, 12]. The above-mentioned techniques, i.e., fiber drying and surface treatment or the addition of a compatibilizer, are mostly not enough to adjust the properties of vegetable fiber reinforced polymers to the desired level. Moreover, even though these treatments enhance adhesion, there is some controversy in the literature about their effect on the mechanical properties of the fiber itself and even when a more pronounced gain is noticed after treatment, the improvement for the composite is often within the scatter of the results. In addition, the cost and environmental impact of some of these treatments, especially of those more elaborated, often prevent their industrial scale applications.

Indeed, a wider use of vegetable fibers in composites replacing synthetic fibers, or more realistically, glass fiber, in either thermoset (mainly unsaturated polyester) or thermoplastic (mainly polyolefins) composites is mainly limited by their comparatively poorer mechanical properties, higher moisture absorption and lower compatibility to polymers. Inferior durability and even flammability may become an issue. Fiber heterogeneity, translated by the large bounds of properties presented by them due to their very own nature, is also a concern.

Even with those limitations, vegetable fibers have been successfully employed in a range of applications [13], in some of them replacing glass fibers. For instance, Cicala *et al.* [14] observed similar mechanical behavior and notable weight and cost reduction in pipes as a consequence of replacing glass mats by hemp mats. La Rosa *et al.* [15] carried out life cycle impact assessment on these hybrid composites and reported on the benefits of using hemp focusing on estimated damage to human health, to ecosystem quality and to resources. They concluded that damage on resources are reduced because hemp is renewable. Also, its improves several indexes, for instance, related to ozone layer depletion, human toxicity, particulate matter formation, ionizing radiation, climate change, terrestrial acidification, freshwater and marine ecotoxicity, urban land occupation, natural land transformation, and metal and fossil depletion. Only the agricultural land occupation index was considered detrimental. In addition, lignocellulosic fibers require much less energy in their production than any synthetic fiber, they are biodegradable and neutral in respect to CO_2 emission [16, 17].

Even so, the use of glass fibers on composites has several advantages worth mentioning. The intrinsic properties of glass fibers are higher and the variability in their properties is far smaller than those of lignocellulosic fibers; and glass fibers originate stronger fiber/matrix interface and their properties are not affected by the temperatures employed for processing thermoplastic matrices. The water uptake of glass fibers is minimal (1% for glass and 5-15% for lignocellulosic fibers), which is of importance because moisture causes dimensional variation in composites and ultimately affects their mechanical properties. Also, during manufacturing, moisture yields voids and affects fiber/matrix bonding with a detrimental effect on mechanical properties.

4.1.1 The Hybrid Concept

Hybrid composites may be defined as systems in which one kind of reinforcing material is incorporated into a mixture of different matrices (blends) [18], or when two or more reinforcing/filling materials are present in a single matrix [19, 20], or also if both approaches are combined [12].

The concept of hybridization provides even greater flexibility to the design engineer to tailor the material properties according to particular requirements. In a hybrid composite containing two or more fibers, which is the most common type of hybrid, the properties of one fiber could complement the other, resulting in cost and performance balance through proper material design [21], considering that the characteristics of these composites are, in general, a weighed sum of the individual components.

The final performance of hybrid composites will be dictated by the same factors that affect ordinary single-fiber composites, such as matrix and fiber type, length and shape of the fibers, fiber/matrix interfaces, overall fiber content and orientation. But in the case of hybrid composites, their properties also depend on fiber volumetric ratio, failure strain of individual fibers and how they are arranged in the composite, i.e., the hybrid design. Regarding the hybrid design, reinforcements may be combined in various ways, such as fiber-by-fiber mixtures ("intimate" or "intralaminate" hybrids), tow-by-tow mixtures ("discrete" or "zebra" hybrids), layer-by-layer mixtures ("interlaminate"), skin-core-skin structures (i.e., sandwich structures), internal ribs and external ribs [22]. Interlaminate, or simply laminate, consists in depositing layers produced with different fibers, whereas in intralaminates, both fibers are entangled (i.e., intermingled) within a single layer. Especially for the latter, optimum hybrid results are obtained when the fibers are strain compatible [23].

As previously discussed, considerable research work has been devoted to ways of minimizing drawbacks associated with of vegetable fibers (e.g., hydrophilic character, high water absorption and temperature limitations), and the concomitant use of vegetable and synthetic fibers, i.e., one type of hybrid composite, is among the most promising strategies. For instance, hybridization with glass fiber, the most used synthetic fiber, is an unquestionable way of achieving desired mechanical properties and physical characteristics. The individual characteristics of the fibers may be transferred to the final composite, originating a material with generally intermediate properties with potential broadening of the original range of applications.

In addition, since lignocellulosic fibers are commonly less expensive than glass ones, replacing a certain amount of glass by vegetable fibers can increase, on a cost basis, the competitiveness of these hybrid composites. For example, cost savings ranging from 10 to 20% are reported for hybrid composites where glass fibers were replaced by hemp (*Cannabis sativa*), flax (*Linum usitatissimum*) or kenaf (*Hibiscus cannabinus*) fibers, even using a very low volume fraction of these natural fibers—within 8-11% [16]. Also very important is the fact that vegetable fibers are less abrasive than glass fibers to molds and processing equipment, which will translate into long-term cost savings since maintenance or replacement of parts will be reduced.

Recent literature may be found on the hybridization of sisal/glass, bamboo/glass, hemp/glass, palm fiber/glass, jute/glass, curaua/glass, kapok/glass, pineapple/glass, ridge gourd/glass, mostly using hybrid mats, hybrid fabrics, or a stack of mats or fabrics. On this context, this chapter presents a brief review on the state-of-the-art of vegetable/glass fiber composites focusing on the effect of hybridization on their final properties.

Both thermoplastic and thermoset polymer matrices are addressed below and some of the advantages associated with them are discussed. Nevertheless, it is worth mentioning that hybrid fibrous fillers have also been added to elastomeric matrices. For instance, Anuar *et al.* [24] worked with glass/oil palm empty fruit bunch fibers as reinforcement for natural rubber. Interfacial adhesion, and properties were determined as a function of fiber content and modification. Their results showed that the properties of natural rubber were enhanced with fiber incorporation, adhesion was not poor and that both untreated and treated (using silane and maleic anhydride grafted polypropylene) composites displayed a good set of properties. The optimum tensile and impact properties were achieved with composites having 10% each empty fruit bunch and glass fibers. Also, Wan Busu *et al.* [25] reported the properties of kenaf/glass fiber reinforced natural rubber composites compounded by the melt blending method. Composites of various fiber contents (0-20 vol%) were prepared and the authors found that optimized properties were obtained with a hybrid composite having a 3:1 kenaf/glass ratio.

4.2 Vegetable Fiber/Glass Fiber Thermoplastic Composites

Although thermoplastics, thermosets and elastomers can be used as matrix of polymer composites, thermosets are more widely used mainly due to the processing techniques employed in their manufacture, that allow for the easy use of long fibers, mats and fabrics as reinforcement. Thermosets are also generally more thermally stable than thermoplastics, and these two reasons lead to products with significantly higher mechanical properties which are less affected by thermal variations than thermoplastic composites [1, 2, 12].

However, there are also drawbacks associated with thermoset composite manufacturing, which is broadly considered slow and non-versatile and, also, impossible to easily reprocess. Indeed, recycling of polymer composites, particularly of thermosets, is rather difficult since their constituents are very difficult to isolate in an economically viable way [1].

On the other hand, thermoplastic composites, especially those with short fibers, are more complacent as they allow for primary recycling and the use of common processing equipment such as extruders and injection machines. However, polymers, particularly thermoplastics, tend to degrade during processing and exposure, which may significantly affect their mechanical properties with reprocessing and reuse. Reusing these materials may become such a burden that, despite economic and environmental considerations, it is sometimes preferred to discard or to incinerate rather than reprocess them [12, 26].

Nevertheless the use of thermoplastics for polymer composites, although less thermally stable and not as strong as thermosets, has steadily grown. Fibers, particularly glass, but also carbon, are widely used, generating products with much improved mechanical properties, which are attainable even with short fibers [1, 2, 12]. It is important to add that there are also thermoplastic composites with long fibers, but they are much less common, usually being found in specific high-technology applications.

The reasons for manufacturing vegetable fiber reinforced thermoplastic composites include easier processing, reprocessing possibilities and partial substitution of non-renewable (i.e., the polymer) for renewable resources, minimizing environmental impact. These composites have been used in a variety of less demanding applications such as panels, packaging, gardening, utilities and furniture. Despite the fact that their mechanical properties are not as good as their ceramic-filled counterparts, vegetable fiber reinforced thermoplastics may display other interesting characteristics such as thermal and acoustic insulation.

The lignocellulosic materials mostly used as fillers in thermoplastic composites include wood flour, starch, rice husk and a wide variety of vegetable fibers available such as jute, sisal, flax, hemp, coir, banana, pineapple, among others. And whenever vegetable fiber reinforced thermoplastic composites with higher properties are needed, possible solutions include improved adhesion, better fiber orientation, and filler hybridization with synthetic fibers or mineral fillers. The latter solution is an intermediate alternative regarding environmental friendliness, cost, weight and performance compared to an all synthetic composite [12, 26].

Most of the literature related to thermoplastic hybrid composites with at least one lignocellulosic fiber addresses their mechanical properties as a function of fiber length and content, extent of fiber intermingling, hybrid design, fiber chemical treatment, and use of coupling agents to improve fiber/matrix bonding. The matrices employed in these thermoplastic composites are mainly based on polypropylene (PP), polyethylene (low- and high-density), polystyrene, poly (vinyl chloride) and polycarbonate. Although there are cellulosic/cellulosic and cellulosic/synthetic fibers reinforced thermoplastic hybrid composites, the focus here is on hybrid cellulosic/glass fiber thermoplastic composites only, which is further discussed below.

a) Polypropylene (PP) based-hybrid composites:

Several researchers have studied the influence of vegetable fiber/glass fiber hybridization on the properties of PP composites. The vegetable fibers employed included bamboo, empty palm fruit bunch, sisal, jute, flax and hemp, among others. Rozman *et al.* [27] investigated the effect of fiber loading levels and compatibilizer addition on the properties of PP-empty palm fruit bunch-glass fiber composites. They obtained reduced flexural and tensile strengths with the addition of both fibers to the PP matrix but these properties increased with overall fiber loading and the presence of coupling agents. In a follow-up study [28], they investigated the effect of oil extraction of the empty palm fruit bunch on the same properties and obtained significantly higher tensile and flexural properties with this procedure. Scanning electron microscopy (SEM) study revealed that oil extraction had resulted in the formation of continuous interfacial regions between EFB and PP matrix, and in an increase in matrix ductility.

Thwe and Liao [29] investigated the effect of environmental aging on the properties of bamboo/glass fiber reinforced PP composites and found that aging in water for 1600 h was reduced nearly two fold with glass fiber incorporation, which shows that durability of bamboo fiber reinforced PP can be enhanced by hybridization with a small amount of glass fibers. They stated that water immersion leads to a decrease in mechanical properties which is more prominent at longer times and high temperatures and that compatibilization strongly reduces these deleterious effects. The authors concluded that the hybrid approach of blending more durable glass fiber with bamboo fiber is an effective way to improve durability of natural fiber composites under environmental aging. These bamboo/glass fiber reinforced PP systems were also studied in [18, 30, 31], who reported an increase in mechanical properties of the systems and a decrease in water uptake with glass fiber hybridization and glass content. Thermal properties also tended to improve with both hybridization and compatibilization.

Panthapulakkal and Sain [32] investigated the mechanical, water absorption and thermal properties of injection molded hemp/glass fiber-reinforced PP composites and concluded that hybridization with glass fibers improves the mechanical properties of short hemp fiber composites. Reis *et al.* [33] studied the flexural behavior of hand-manufactured hybrid laminate composites with a hemp fiber/PP core and two glass fibers/PP surface layers at each side of the sample. When compared with glass fibers reinforced PP laminates, the hybrid composites showed economical, ecological and recycling advantages and also specific fatigue strength benefits.

Jarukumjorn and Suppakarn [34] incorporated glass fiber into sisal/PP composites, compatibilized with PP-g-MA, and found enhanced tensile, flexural, and impact strength but no significant effect on tensile and flexural moduli. In addition, glass fiber incorporation improved thermal properties and water resistance of composites.

Samal *et al.* [35] described the properties of PP/banana-glass fiber composites and concluded that glass fiber addition led to improved mechanical performance of the composite and decreased water uptake. They also reported better interfacial adhesion and thermal properties with compatibilizer addition and hybridization.

Shakeri and Ghasemian [36] investigated moisture absorption of recycled newspaper-glass fiber reinforced PP composites targeting outdoor applications. They observed that the incorporation of glass fibers and particularly the addition of maleated PP significantly reduced water absorption and thickness swelling, making these composites suitable for use in damp places, such as the interior of bathrooms, wood decks, food packaging, among others.

b) Polyethylene (PE) based hybrid composites:

A number of papers [37-45] on vegetable/glass fiber hybrid PE composites have also been published. The basic findings are similar to those reported for PP/vegetable/glass composites, i.e. mechanical and thermal properties of the composite improve with fiber content and compatibilizer incorporation, the properties are more sensitive to the relative amount of glass fibers, and water uptake is reduced and durability increases with glass fiber addition. In other words, hybridization of vegetable fibers such as sisal, jute, flax with glass fibers leads to composites with a good combination of properties, which, in many cases, are a weighted sum of the contributions of the individual reinforcements.

c) Other matrices:

A number of other thermoplastics such as polystyrene (PS) [46] and poly (vinyl chloride) (PVC) [47-49] have also been used as matrices for the manufacture of hybrid vegetable/glass fiber thermoplastic composites. Properties vary with the nature of the components, their relative amounts, particle size, affinity and properties, but here again, the main conclusions point out that overall properties increase with fiber content, particularly that of glass fibers, appropriate selection of processing conditions and fiber sizing, and the use of a compatibilizer. Polar matrices usually do not require chemical modifications of the fibers or of the matrix, and may even preclude the use of a compatibilizer, although these substances may also act as processing aids and help fiber dispersion, which in turn enhances mechanical performance of these composites.

4.3 Intra-Laminate Vegetable Fiber/glass Fiber Thermoset Composites

As discussed above, hybridizing vegetable fibers with synthetic fiber such as glass fiber, significantly improves strength, stiffness, moisture and fire-resistance behavior of the original lignocellulosic fiber composite [40]. The increase in fiber volume fraction also has a positive influence on composite mechanical properties such as stiffness and strength [50, 51]. The fiber length and the consequent fiber aspect ratio should also be considered when dealing with non-continuous fibers [48].

In the case of intralaminate vegetable fiber/glass fiber thermoset composites, the manufacturing possibilities are numerous. In many cases, compression molding is chosen, as in Romanzini *et al.* [52], Mishra *et al.* [40], Ahmed *et al.* [53], Almeida Jr. *et al.* [54], or Hand Lay-up, as in Biswas and Satapathy [55], Ramesh *et al.* [56], Ramnath *et al.* [57]. For instance, Cicala *et al.* [14] produced curved pipes in glass fabric/natural mat fiber hybrid composites by hand lay-up. The type of fabric used to produce the composites was considered an important issue and various were evaluated, namely, E-glass woven, E-glass random mat and C-glass liner from HP-Textiles (for E-glass woven and mat), and mats of hemp, flax and kenaf.

Curaua, a vegetable fiber from the Northern Region of Brazil, has been used to produce hybrid thermoset composites with glass fiber. As the intended application in their case was automotive components, commonly submitted to shear loading, composites were Iosipescu shear tested according to ASTM D5379 and also evaluated using the impulse excitation technique [54]. The obtained hybrid composites were considered successful, especially for the 30% fiber volume case, and a curaua/glass fiber ratio of 30/70, which showed a good balance between low weight and high mechanical properties. They produced hybrid vegetable/glass fiber epoxy intralaminar composite using curaua fiber rope and chopped glass fibers, been both randomly mixed and compressed. The overall fiber volume fraction ranged within 20-40% and the mat was dried prior to molding. The authors identified some factors that significantly influenced the results, including fiber type and content. In a complementary work, Almeida Jr. *et al.* [58] carried out thermal, mechanical and dynamic mechanical analyses of the same hybrid intralaminate composites and identified higher thermal stability and higher impact strength and hardness for the glass reinforcement. An increase in storage modulus was noticed whereas the glass transition temperature showed no significant trend with glass incorporation.

In both studies shown above, the authors dried the hybrid fiber mat prior to molding. The importance of the drying stage lies in the fact that the high content of moisture of natural fibers is detrimental for the properties of the composite [2], since, in combination with the hydrophobic character of the polymer matrix, is responsible for void formation and weak adhesion [59].

Polyester resin composites with pineapple leaf mat-glass fibers and sisal mat-glass fibers were studied by Mishra *et al.* [40], who reported the influence of fiber content on tensile, flexural and impact properties. The use of natural fiber was justified based on the moderately high specific strength and stiffness of these fibers which could prove suitable for thermoset polymer composites. Due to the expectedly poor fiber/matrix adhesion, the authors washed both natural fibers with detergent solution and deionized water and later dried them. The sisal fibers were also submitted to alkali treatment to remove waxes, followed by cyanoethylation and also acetylation. Optimum glass fiber loadings for pineapple leaf/glass and sisal/glass hybrid polyester composites were found for 8.6 and 5.7 wt.% of glass fibers, respectively, for an overall fiber content of 25 wt.%

and 30 wt.%, respectively. They observed decrease in water absorption due to hybridization and after surface treatment of the natural fiber.

Atiqah *et al.* [60] carried out mercerization treatment of kenaf fiber and a mat of this material was used to produce hybrid polyester composites with woven glass fiber. Since the intention was to obtain high mechanical strength with good moisture resistance, mercerization was considered important to achieve that. Higher tensile strength and modulus were obtained for composites when treated kenaf was used, which was justified by the more suitable fiber matrix interface that improved load carrying capacity and fiber/matrix load transfer. The authors also studied the Izod impact properties, and an increase of 11% in this property for the treated kenaf was reported in hybrid kenaf/glass fiber composite. The kenaf/glass arrangement was also found to contribute for the observed mechanical behavior which assisted the composite in withstanding the impact force. A limit in kenaf fiber volume of 15% was established since Izod impact strength decreased for higher kenaf content, and the treated kenaf-glass hybrid composite (15/15 v/v) showed the highest flexural, tensile and impact strength of the studied formulations.

Davoodi *et al.* [61] also studied a kenaf/glass reinforced epoxy composite, in their case focusing on passenger car bumper beam application and characterized the material based on tensile and Izod impact testing according to ASTM D3039 and D256 standard, respectively, among other evaluations. The hybrid composite was considered able to be utilized in automotive structural components such as the bumper beam but conditioned to an improvement in impact behaviour, that could be achieved by optimizing structural design parameters (thickness, beam curvature and strengthening ribs) or through material improvement such as epoxy toughening to modify the ductility behavior and improve energy absorption.

Velmurugan and Manikandan [62] prepared a hybrid composite using palmyra/glass fiber reinforcement and rooflite resin, which is known to produce a strong bond with glass-fiber mat. The fiber fraction weight was kept in 55%, but the amount of palmyra (28-55%) and glass (27-0%) fiber were varied. The composites were characterized under tensile, flexural, impact and shear testing. The best glass/palmyra fibers composition regarding mechanical properties was 20/35 wt%. The inclusion of a small amount of synthetic fiber improved the overall strength with a consequent increase in barrier properties, allowing them to conclude that natural/synthetic fiber hybridization was an optimal approach to improve mechanical characteristics and durability of natural composites.

4.4 Inter-Laminate Vegetable Fiber/glass Fiber Thermoset Composites

Hybrid glass fiber/vegetable fiber composites are a fine way of combining the advantages of both kinds of fibers, and to minimize their disadvantages. A balance of properties of both fibers is usually obtained when hybrid composites' properties are compared to those properties of single-fiber systems [63]. Regarding the development of vegetable fiber/glass fiber thermoset composites a major decision is on how they can be suitably manufactured. The usual manufacturing methods are hand lay-up, compression molding, resin transfer molding and associated processes such as RIM, and filament winding [63-66]. Apart from the FW technology, where fiber roving is the raw material, the other processes mostly use fabrics, chopped fibers and mats. In fact, a useful way to manufacture hybrid composites is by using fabrics and mats instead of unidirectional or chopped fibers. Fabrics are easier to handle and drape during the manufacture of parts, and *prepregs* can be more easily manufactured with fabrics than with roving, for example.

Because of that, hybrid woven fabrics with glass, carbon or aramid fibers are largely used in advanced polymer composites [67], and the same approach can be used when glass/vegetable hybrid composites are required [68]. For example, hybrid polyester matrix composites produced with glass and jute (*Corchorus capsularis*) fabrics with the plain weave configuration presented flexural strength comparable to that of glass fiber composites [68]. Flexural loading is, in fact, a convenient load configuration to evaluate hybridization, since the stronger and more expensive glass fiber can be placed at the outer layers of the composite to withstand the higher stress loading at that position [69]. The use of glass with jute fibers also minimizes the effect of fiber surface treatments, intended to improve the vegetal fiber/matrix interfacial strength, upon the overall mechanical behavior of a hybrid composite. Therefore, the use of untreated fibers could become feasible and this is a relevant aspect since the use of chemicals will both increase the cost of the final composite as also reduce the "*green*" approach of the vegetable fibers. The use of an outer layer composed of glass fiber and a core of jute fibers could also enhance the aging behavior of the composites, since the outer layers can protect the jute core from weathering [70].

As a matter of fact, many authors consider that the behavior of hybrid composites is mainly influenced by the properties of their outer fiber layers, and that optimum mechanical properties can be obtained by placing high strength fibers at the outer layers [71]. However, this is not a general rule, and depending on the external load being applied to the composite, the use of at least one vegetal fiber outer layer can be of interest.

Another important aspect in the use of hybrid glass fiber-vegetable fiber composites is related to their toughness. Clark and Ansell [72] showed that the use of random jute fibers mat and woven glass fabrics can increase the fracture toughness and the impact resistance of composites. They used several layer stacking sequences and obtained maximum fracture toughness when jute layers were at the core of the composite, sandwiched between glass layers. The increase in fracture toughness was attributed to the additional jute mat to glass fabric interfacial ply separation, and the consequent blunting of the crack propagation. Recent analysis using design of experiments showed, indeed, that a careful design of the layer stacking sequence can be a powerful tool to optimize the mechanical performance of hybrid jute/glass composites [73].

The use of a symmetric laminate, which has only zero elements in its $[B_{ij}]$ coupling matrix, favors tensile and flexural behavior [74]. The impact strength of jute/glass hybrid laminates was recently analyzed by Yu *et al.* [75], who compared the asymmetric jute/jute/glass laminate and the symmetric jute/glass/jute laminate. The results showed higher impact strength of the hybrids in respect to an all-jute composite and also that the asymmetric configuration presented as much as 66.7% higher impact strength than the symmetric one. This behavior is similar to the one described when asymmetric glass/aramid hybrid composites were tested under impact [76]. The increase in energy absorption capacity of these hybrid was attributed to the presence of energy consuming mechanisms due to the non-zero coupling elements A_{16} and A_{26} at the laminate extensional stiffness matrix, A_{ij}, and D_{16} and D_{26} at the laminate bending stiffness matrix, D_{ij}.

Besides jute, several other hybrid composites combining glass and vegetal fibers were investigated, such as sisal, flax, curaua (*Ananas erectifolius*), abaca (*Musa textilis*) and loofa or sponge gourd (*Luffa cyllindrica*), to cite a few of the fibers being evaluated [16, 77]. Although properties vary when a particular vegetal fiber is used or when different thermoset matrices are used, the overall behavior of the majority vegetal fiber/glass fiber thermoset composites follows the common trend summarized above for jute/glass hybrid composites.

For example, Amico *et al.* [77] analyzed the influence of the stacking sequence of sisal/glass hybrid composites. They found that the mechanical properties of the hybrid composite can be very similar to those of the pure glass composite depending on the loading being applied and of the stacking sequence. The optimal flexural behavior was obtained when the sisal fiber layers were sandwiched between the strong glass layers. In addition, Mishra *et al.* [40] observed that the use of glass fiber greatly reduced water uptake in sisal/glass and pineapple (*Ananas comosus)*/glass hybrid composites.

The use of a sandwich configuration with outer layers of glass fibers is indeed the approach chosen by several authors. Salleh *et al.* [78] tested several layer-arrangement configurations of glass and kenaf fibers. They varied the kenaf fiber aspect ratio and used chopped strand or woven glass fiber, reporting an increase in tensile strength with the aspect ratio of the kenaf fibers. It was also found an increase in tensile properties for the hybrid composites with respect to the pure-kenaf fiber composites.

Toughness increase due to hybridization was also discussed and presented by several authors. Santulli *et al.* [79] observed that the substitution of glass fibers by flax fibers at the composite's core enhanced impact damage tolerance and hindered damage propagation. These authors also observed that substitution of glass by flax fiber brought a significant weight reduction when the hybrid composite was compared to the pure-glass fiber one. Similar results for flax/glass fiber composites were reported by Zhang *et al.* [80]. Adekunle *et al.* [81] also studied glass/flax composites and stressed a weight reduction when flax was used instead of glass fibers. Since they used a different stacking sequence, with sandwiched glass fiber mats between woven and non-woven flax fibers, the effect of the more resistant glass fibers was only observed when tensile properties were measured. Under flexural load, the laminates of distinct construction showed similar performance, what was attributed to the particular stress distribution and to the proximity of the glass fiber to the neutral axis of the test specimen.

Morye and Wool [82] used symmetric and non-symmetric stacking sequences with glass/flax hybrid composites and also varied the glass/flax fiber ratio, namely, 100/0, 80/20, 60/40, 40/60, and 0/10. For non-symmetric composites, flexural and impact tests were performed with the top face of the composite being either glass fibers or flax fibers. The mechanical properties of the composites were found to depend on the fiber layer arrangement in the composite, and the non-symmetrical laminates with flax at the loaded top face presented superior performance under flexure or impact.

When the impact behavior is under consideration, the performance of the hybrid composites, regardless of the glass/flax ratio, was even better than that of both single-fiber composites.

Sreekala *et al.* [23] observed an increase in impact strength when a small amount of glass fiber was used along with oil palm (*Elaeis guineensis*) fiber in composites. In fact, the impact behavior of some of the hybrids was greater than that of the pure-glass fiber composite. This behavior was partially attributed to oil palm fiber debonding and pull-out mechanisms. Also, density of the composite decreased for higher volume fraction of oil palm fiber.

The increase in tensile and impact strength due to glass-fiber incorporation was also obtained when oil palm/glass fibers hybrids were investigated [83]. Non-symmetric laminates were used, and the impact strength was also dependent on the face under loading. However, differently from the work of Morye and Wool [82], the best performance was obtained when the glass fiber layers were at the top loaded face. In this case, the variation in impact strength with the glass fiber content was fairly well modeled using a rule of mixtures approach. The increase in impact strength was attributed to extensive delamination between the glass fiber layers and also at the glass fiber/oil palm fiber layer interfaces.

The optimization of the mechanical behavior of palmyra (*Borassus flabellifer*) waste fibers composites was reported when these fibers were used in hybrid composites with glass fibers [84]. The hybrid composites were prepared with a total fiber weight of 60%, and with glass fibers as the outer layers. The mechanical properties increased with the amount of glass fibers, and, for a particular amount of glass fibers (7 wt%), impact strength increased with the amount of palmyra fibers.

The increase in fracture toughness brought by delamination of weak interfaces was also observed when sponge gourd fibers were used in hybrid glass fiber/sponge polyester composites [85]. Sponge gourd fibers are in fact a natural fiber mat that causes crack path deviation, leading to a controlled fracture mode of the composite and increasing composite's toughness. The fracture mode changed from abrupt to a controlled one when sponge gourd fibers were sandwiched between the glass fiber layers.

Hybrid composites using other less common fibers have being investigated by several researchers, for instance, Mahesh *et al.* [86] studied elephant grass *(Miscanthus sinensis)*-glass fiber hybrid composites. Abiy [87] studied bamboo fiber (*Bambusa sp.*)-glass fiber composites, varying bamboo:glass fiber ratio and the layer stacking sequences. The tensile properties of the $[0/90/0/90]_s$ laminates improved for higher glass fiber content, and an optimum result was obtained in respect to tensile modulus when the bamboo:glass ratio was 50:50. The three point bending testing of the $[90_2/0_2/-45/45]_s$ laminates yielded maximum flexural strength for the hybrids with bamboo:glass fiber ratio of 15:85, whereas the 30:70 and 50:50 ratios produced comparable strength in relation to the pure glass fiber composite.

Kapok (*Ceiba pentandra*)/glass fiber composites were manufactured using untreated and alkali treated kapok fibers [88]. The glass and kapok fibers were used as fabrics and the relative kapok fiber fraction was varied within 0-100%. The tensile strength and modulus of the composites increased with the glass fiber content as usual, and the hybrids showed greater properties compared to the pure matrix, even when only 25% of kapok fibers were used. The chemical treatment of the fibers did not produce a significant effect regarding tensile strength, although an increase was obtained for the composites with higher kapok fiber relative content. Tensile modulus, on the other side, showed a marked increase when alkali treated fibers were used, indicating an enhancement in fiber/matrix adhesion. In the work of Priya and Rai [89], a similar trend was found for silk (*Bombyx mori*)/glass fabric hybrid composites, with an increase in mechanical properties even when just a small amount of glass was incorporated into silk fiber composites. The incorporation of glass also reduced water uptake of the hybrid composite compared to that of the pure-silk fiber composite.

Ramnath *et al.* [57] went one step further and combined three simultaneous fiber reinforcements, namely abaca, jute and glass. Abaca fiber, which belongs to the banana family, are usually extracted from the plant trunk, but in this study, they were extracted from the base of the banana leaf. Jute is a natural fiber produced by retting process and can be spun into coarse or strong threads. The composites were manufactured in a way that jute fibers were flanked by abaca fibers on both sides, and glass fibers sandwiched this assembly on both top and bottom layers. Mechanical characterization was conducted based on impact (ASTM D256), tensile (ASTM D638), flexural (ASTM D790) and double shear (ASTM D5379) testing. The abaca/jute/glass hybrid composite showed better properties than the abaca/glass or the jute/glass hybrids, and an intra-fiber delamination process was identified in the abaca mat, which was responsible for a reduced strength.

4.5 Concluding Remarks

An overview of the work being performed on hybrid vegetable/glass fiber composites was presented. The use of natural fibers in industrial applications provides several challenges for researchers and product developers as differences in elastic and polar characteristics between the different fibers and matrices need to be taken into consideration.

The studies in the literature indicate that several different vegetable fibers have been mixed with glass fibers and incorporated into a number of polar and non-polar polymers, including thermoplastics, thermosets and even natural rubber. The influence of fiber content and relative amounts, fiber length, modification of both fibers and matrices and the use of compatibilizers on the mechanical and thermal behaviour of these systems was described.

A wide variety of research work is being conducted worldwide with sometimes conflicting ideas. Focus usually lies on physical, especially water absorption and degradation, and mechanical properties of hybrid composites. The research on electrical, thermal, acoustic and dynamic mechanical properties of hybrid vegetable/glass thermoplastic composites, however, is more restrict.

As expected, the nominal values of the various properties vary widely from system to system and are also dependent on processing and testing conditions. Even though a very different group of fibers and polymer matrices were presented here, some general trends can be identified. Perhaps, the main one is that fiber hybridization even with a small amount of glass fibers can significantly improve the physical and mechanical properties of natural fiber polymer composites. Moreover, these properties can be further enhanced by the use of a compatibilizer and proper processing conditions as they improve fiber/matrix adhesion and promote filler dispersion. It was found that the thermal stability of the composites also increased with glass fiber addition and that the properties of the systems can be reasonably estimated by a weighted sum of their individual properties.

In short, hybrid composites are cost effective and can be used in a number of indoor and outdoor applications. Future research on hybrid composites, currently driven mostly by their possible automotive applications, needs to explore other sectors, such as building, rural, biomedical and perhaps even in non-structural aircraft components. Challenges still exist in suitable analytical modelling on most of the published results and they could not only help in interpreting the experimental results but also in optimization, targeting specific, more demanding, applications.

Despite the fact that these composites are not as ecofriendly as their pure-vegetable reinforced counterparts, they do help reducing the amount of synthetic non-readily recyclable material on these composites and result in products with better mechanical and thermal properties, with a positive hybrid effect reported for several glass/natural fiber composites. Hybrid glass fiber/natural fiber composites usually display better toughness than pure-glass fiber composites and they can also advantageously substitute surface chemical treatments of natural fibers. In all, hybridization with glass fiber, the most used synthetic fiber, is an unquestionable way of making natural fiber composites achieve noteworthy mechanical properties and physical characteristics.

Acknowledgement

The authors would like to acknowledge the continuous support of CNPq and CAPES (Brazil).

References

1. F.P. La Mantia and M. Morreale,Green composites: A brief review. *Compos. A* 42, 579–588 (2011).
2. O. Faruk, A.K. Bledzki, H.P. Fink, and M. Sain, Biocomposites reinforced with natural fibers: 2000–2010. *Prog. Polym. Sci.* 37, 1552–1596 (2012).
3. R.F. Gibson, *Principles of Composite Materials Mechanics*, CRC Press, Boca Raton, FL (2012).
4. L.E. Nielsen, and R.F. Landel, *Mechanical properties of polymers and composites*, Marcel Dekker, New York (1994).
5. M.J. John, and R.D. Anandjiwala, Recent developments in chemical modification and characterization of natural fiber-reinforced composites. *Polym. Compos.* 29, 187–207 (2008).
6. M.N. Belgacem, and A. Gandini, The surface modification of cellulose fibres for use as reinforcing elements in composite materials. *Compos. Interfaces* 12, 41–75 (2005).
7. V. Cech, R. Prikryl, R. Balkova, J. Vanek, and A. Grycova, The influence of surface modifications of glass on glass fiber/polyester interphase properties. *J. Adhes. Sci. Technol.* 17, 1299–1320 (2003).
8. R. Li, L. Ye, and Y.W. Mai, Application of plasma technologies in fibre-reinforced polymer composites: a review of recent developments. *Compos. A Appl. Sci. Manuf.* 28, 73–86 (1997).

9. J. Morales, M.G. Olayo, G.J. Cruz, P. Herrera-Franco, and R. Olayo, Plasma modification of cellulose fibers for composite materials. *J. Appl. Polym. Sci.* 101, 3821–3828 (2006).
10. S. Kalia, B.S. Kaith, and I. Kaur, Pretreatments of natural fibers and their application as reinforcing material inpolymer composites-A review. *Polym. Eng. Sci.* 49, 1253–1272 (2009).
11. Y. Xie, C.A.S. Hill, Z. Xiao, H. Militz, and C. Mai, Silane coupling agents used for natural fiber/polymer composites: A review. *Compos. A Appl. Sci. Manuf.* 41, 806–819 (2010).
12. M. Jawaid, and H.P.S. Abdul Khalil, Cellulosic/synthetic fibre reinforced polymer hybrid composites: A review. *Carbohydr. Polym.* 86, 1–8 (2011).
13. A.K. Mohanty, M. Misra and L.T. Drzal, Sustainable bio-composites from renewable resources: opportunities and challenges in the green materials world. *J. Polym. Environ.* 10, 19–26 (2002).
14. G. Cicala, G. Cristaldi, G. Recca, G. Ziegmann, A. El-Sabbagh, and M. Dickert, Properties and performances of various hybrid glass/natural fibre composites for curved pipes. *Mater. Des.* 30, 2538–2542 (2009).
15. A.D. La Rosa, G. Cozzo, A. Latteri, A. Recca, A. Björklund, E. Parrinello, and G. Cicala, Life cycle assessment of a novel hybrid glass-hemp/thermoset composite. *J. Clean. Prod.* 44, 69–76 (2013).
16. G. Cristaldi, A. Latteri, G. Recca, G. Cicala, and P.D. Dubrovski, *InTech Europe*, p. 317, Rijeka, Croatia, (2010).
17. A. Shahzad, Hemp fiber and its composites - a review. *J. Compos. Mater.* 46, 973–986 (2011).
18. M. M. Thwe, and K. Liao, Durability of bamboo-glass fiber reinforced polymer matrix hybrid composites. *Composites Science and Technology*, 63, 375–387 (2003).
19. S.Y. Fu, G. Xu, and Y.W. Mai, On the elastic modulus of hybrid particle/short-fiber/polymer composites. *Compos. B Eng.* 33, 291–299 (2002).
20. J. Karger-Kocsis, Reinforced polymer blends, in *Polymer Blends Performance*, D.R. Paul, and C.B. Bucknall, Volume 2, Academic press inc., Wiley, New York (2000).
21. M.J. John, and S. Thomas, Biofibres and biocomposites. *Carbohydr. Polym.* 71, 343–364 (2008).
22. D. Short, and J. Summerscales, Hybrids-a review. *Composites*, 10, 215–221 (1979).
23. M.S. Sreekala, J. George, M.G. Kumaran, and S. Thomas, The mechanical performance of hybrid phenol-formaldehyde-based compositesreinforced with glass and oil palm fibres. *Compos. Sci. Technol.* 62, 339–353 (2002).
24. H. Anuar, S.H. Ahmad, R. Rasid, and N.S. Nik Daud, Tensile and impact properties of thermoplastic natural rubber reinforced short glass fiber and empty fruit bunch hybrid composites. *Polym. Plast. Technol. Eng.* 45, 1059–1063 (2006).
25. W.N. Wan Busu, H. Anuar, S.H. Ahmad, R. Rasid, and N.A. Jamal, The mechanical and physical properties of thermoplastic natural rubber hybrid composites reinforced with hibiscus cannabinus , L and short glass fiber. *Polym. Plast. Technol. Eng.* 49, 1315–1322 (2010).
26. Y. Li, Y.M. Mai, and L. Ye, Sisal fibre and its composites: a review of recent developments. *Compos. Sci. Technol.* 60, 2037–2055, (2000).
27. H.D. Rozman, G.S. Tay, R.N. Kumar, A. Abusamah, H. Ismail, and Z.A. Mohd, Polypropylene–oil palm empty fruit bunch–glass fibre hybrid composites: a preliminary study on the flexural and tensile properties. *Eur. Polym. J.* 37, 1283–1291 (2001).
28. H.D. Rozman, G.S. Tay, R.N. Kumar, A. Abusamah, H. Ismail, and Z.A. Mohd, Smart materials and constructions. *Polym. Plast. Technol. Eng.* 40, 703–714 (2001).
29. M.M. Thwe, and K. Liao, Effects of environmental aging on the mechanical properties of bamboo–glass fiber reinforced polymer matrix hybrid composites. *Compos. A Appl. Sci. Manuf.* 33, 43–52 (2002).

30. S.K. Samal, S. Mohanty, and S.K. Nayak, Polypropylene--bamboo/glass fiber hybrid composites: Fabrication and analysis of mechanical, morphological, thermal, and dynamic mechanical behavior. *J. Reinf. Plast. Compos. 28,* 2729–2747 (2009).
31. S.K. Nayak, S. Mohanty, and S. K. Samal, Influence of short bamboo/glass fiber on the thermal, dynamic mechanical and rheological properties of polypropylene hybrid composites. *Mater. Sci. Eng.* 523, 32–38 (2009).
32. S. Panthapulakkal, and M. Sain, Injection-molded short hemp fiber/glass fiber-reinforced polypropylene hybrid composites—Mechanical, water absorption and thermal properties. *J. Appl. Polym. Sci.* 103, 2432–2441, (2007).
33. P.N.B. Reis, J.A.M. Ferreira, F.V. Antunes, and J.D.M. Costa, Flexural behaviour of hybrid laminated composites. *Compos. A Appl. Sci. Manuf.* 38, 1612–1620 (2007).
34. K. Jarukumjorn, and N. Suppakarn, Effect of glass fiber hybridization on properties of sisal fiber–polypropylenecomposites. *Compos. B Eng.* 40, 623–627 (2009).
35. S.K. Samal, S. Mohanty, and S.K. Nayak, Banana/glass fiber-reinforced polypropylene hybrid composites: Fabrication and performance evaluation. *Polym. Plast. Technol. Eng.* 48, 397–414 (2009).
36. A. Shakeri, and A. Ghasemian, Water absorption and thickness swelling behavior of polypropylene reinforced with hybrid recycled newspaper and glass fiber. *Appl. Compos. Mater.* 17, 183–193 (2010).
37. G. Kalaprasad, S. Thomas, C. Pavithran, N.R. Neelakantan, S. Balakrishnan, Hybrid effect in the mechanical properties of short sisal/glass hybrid fiber reinforced low density polyethylene composites. *J. Reinf. Plast. Compos.* 15, 48–73 (1996).
38. G. Kalaprasad, K. Joseph, and S. Thomas, Influence of short glass fiber addition on the mechanical properties of sisal reinforced low density polyethylene composites. *J. Compos. Mater.* 31, 509–527 (1997).
39. G. Kalaprasad, P. Pradeep, G. Mathew, C. Pavithran, and S. Thomas, Thermal conductivity and thermal diffusivity analyses of low-density polyethylene composites reinforced with sisal, glass and intimately mixed sisal/glass fibres. *Compos. Sci. Technol.* 60, 2967–2977 (2000).
40. S. Mishra, A.K. Mohanty, L.T. Drzal, M. Misra, S. Parija, S.K Nayak, S.S. Tripathy, Studies on mechanical performance of biofibre/glass reinforced polyester hybrid composites. *Compos. Sci. Technol.* 63, 1377–1385 (2003).
41. G. Kalaprasad, B. Francis, S. Thomas, C.R. Kumar, C. Pavithran, G. Groeninckx, and S. Thomas, Effect of fibre length and chemical modifications on the tensile properties of intimately mixed short sisal/glass hybrid fibre reinforced low density polyethylene composites. *Polym. Int.* 53, 1624–1638 (2004).
42. K. John, and S.V. Naidu, Sisal fiber/glass fiber hybrid composites: the impact and compressive properties. *J. Reinf. Plast. Compos.* 23, 1253–1258 (2004).
43. J. Rizvi, and G.M. Semeralul, Glass-fiber-reinforced wood/plastic composites. *J. Vinyl Addit. Technol.* 14, 39–42 (2008).
44. M. Valente, F. Sarasini, F. Marra, J, Tirilo, and G. Pulci, Hybrid recycled glass fiber/wood flour thermoplastic composites: Manufacturing and mechanical characterization. *Compos. A Appl. Sci. Manuf.* 42, 649–657 (2011).
45. B. Kord, Studies on mechanical characterization and water resistance of glass fiber/thermoplastic polymer bionanocomposites. *J. Appl. Polym. Sci.* 123, 2391–2396 (2012).
46. A. Haneefa, P. Bindu, I. Aravind, and S. Thomas, Studies on tensile and flexural properties of short banana/glass hybrid fiber reinforced polystyrene composites. *J. Compos. Mater.* 42, 1471–1489 (2008).

47. H. Jiang, D. Pascal Kamdem, and B. Bezubic, Mechanical properties of poly(vinyl chloride)/wood flour/glass fiber hybrid composites. *J. Vinyl Technol.* 9, 138–145 (2003).
48. O. Faruk, and A.K. Bledzki, *Wood Plastic Composite: Present and Future*, Wiley Encyclopedia of Composites, (2012).
49. S. Jeamtrakull, A. Kositchaiyong, T. Markpin, V. Rosarpitak, and N. Sombatsompop, Effects of wood constituents and content, and glass fiber reinforcement on wear behavior of wood/PVC composites. *Compos. B* 43, 2721–2729 (2012).
50. G. George, K. Joseph, B. Abderrahim, and S. Thomas, Recent advances in green composites. *Key Eng. Mater.* 425, 107–166 (2010).
51. A. Arbelaiz, B. Fernandez, J. A. Ramos, R. Retegi, R. Llano-Ponte, and I. Mondragon, Mechanical properties of short flax fibre bundle/polypropylene composites: Influence of matrix/fibre modification, fibre content, water uptake and recycling. *Compos. Sci. Technol.* 65, 1582–1592 (2005).
52. D. Romanzini, A. LavorattiI, H.L. Ornaghi Jr, S.C. Amico, and A.J. Zattera, Influence of fiber content on the mechanical and dynamic mechanical properties of glass/ramie polymer composite. *Mater. Des.* 47, 9–15 (2013).
53. K.S. Ahmed, S. Vijayarangan, and A.C.B. Naidu, Elastic properties, notched strength and fracture criterion in untreated woven jute–glass fabric reinforced polyester hybrid composites. *Mater. Des.* 28, 2287–2294 (2007).
54. J.H.S. Almeida Jr, S.C. Amico, E.C. Botelho, F.D.R. Amado, Hybridization effect on the mechanical properties of curaua/glass fiber composites. *Compos. B*, 55, 492–497 (2013).
55. S. Biswas, and A. Satapathy, A comparative study on erosion characteristics of red mud filled bamboo–epoxy and glass–epoxy composites. *Mater. Des.* 31, 1752–1767 (2010).
56. M. Ramesh, K. Palanikumar, and K.H. Reddy, Comparative evaluation on properties of hybrid glass fiber-sisal/jute reinforced epoxy composites. *Procedia Eng.* 51, 745–750 (2013).
57. B.V. Ramnath, S.J. Kokan, R.N. Raja, R. Sathyanarayanan, C. Elanchezhian, A.R. Prasad, and V.M Maickavasagam, Evaluation of mechanical properties of abaca–jute–glass fibre reinforced epoxy composite. *Mater. Des.* 51, 357–366 (2013).
58. J.H.S. Almeida Jr, H.L. Ornaghi Jr, S.C. Amico, F.D.R. Amado, Study of hybrid intralaminate curaua/glass composies. *Mater. Des.* 42, 111–117 (2012).
59. C. Milanese, M.O.H. Cioffi, and H.C.J. Voorwald, Thermal and mechanical behaviour of sisal/phenolic composites. *Compos. B Eng.* 43, 2843–2850 (2012).
60. A. Atiqah, M.A. Maleque, M. Jawaid, and M. Iqbal, Development of kenaf-glass reinforced unsaturated polyester hybrid composite for structural applications. *Compos. B* 56, 68–73 (2014).
61. M.M. Davoodi, S.M. Sapuan, D. Ahmad, A. Ali, A. Khalina, and M. JonoobiI, Mechanical properties of hybrid kenaf/glass reinforced epoxy composite for passenger car bumper beam. *Mater. Des.* 31, 4927–4932 (2010).
62. R. Velmurugan, and V. Manikandan, Mechanical properties of palmyra/glass fiber hybrid composites. *Compos. A* 38, 2216–2226 (2007).
63. G. Mehta, L.T. Drzal, A.K. Mohanty and M. Misra, Effect of fiber surface treatment on the properties of biocomposites from nonwoven industrial hemp fiber mats and unsaturated polyester resin. *J. Appl. Polym. Sci.* 99, 1055–1068 (2006).
64. D. Romanzini, H.L Ornaghi Junior, S.C. Amico and A.J. Zattera, Influence of fiber hybridization on the dynamic mechanical properties of glass/ramie fiber-reinforced polyester composites. *J. Reinf. Plast. Compos.* 31, p. 1652–1661 (2012).
65. D. Romanzini, H.L. Ornaghi Jr, S.C. Amico and A.J. Zattera, Preparation and characterization of ramie-glass fiber reinforced polymer matrix hybrid composites. *Mater. Res.* 15, 415–420 (2012).

66. P. Lehtiniemi, K. Dufva, T. Berg, M. Skrifvars and Pentti Järvelä, Natural fiber-based reinforcements in epoxy composites processed by filament winding. *J. Reinf. Plast. Compos.* 30, 1947–1955 (2011).
67. K.K. Chawla, *Composite Materials - Science and Engineering*, Springer-Verlag, New York (1987).
68. A.L.F.S. d'Almeida and J.R.M. d'Almeida, Evaluation of the Tensile and Flexural Mechanical Behavior of Hybrid Glass Fiber – Jute Fiber Fabric Reinforced Composites, *Proceedings of the XII International Macromolecular Colloquium and 7th International Symposium on Natural Polymers and Composites*, Gramado – Brazil, p. 129 (2010).
69. M. de Rosa, C. Santulli, F. Sarasini and M. Valente, Effect of loading-unloading cycles on impact-damaged jute/glass hybrid laminates. *Polym. Compos.* 30, 1879–1887 (2009).
70. R. Mohan, and Kishore, Jute-Glass Sandwich Composites. *J. Reinf. Plast. Compos.* 4, 186–194 (1985).
71. S. Nunna, P.R. Chandra, S. Shrivastava and A.K. Jalan, A review on mechanical behavior of natural fiber based hybrid composites. *J. Reinf. Plast. Compos.* 31, 759–769 (2012).
72. R.A. Clark and M.P. Ansell, Jute and glass fibre hybrid laminates. *J. Mater. Sci.* 21, 269–276 (1986).
73. B. Sutharson, M. Rajendran and A. Karapagaraj, Optimization of natural fiber/glass reinforced polyester hybrid composites laminate using Taguchi methodology. *Int. J. Mater. Bio mater. Appl.* 2, 1–4 (2012).
74. R.M. Jones, *Mechanics of Composite Materials*, 2nd Ed., Taylor & Francis, Philadelphia, (1999).
75. Y. Yu, Y. Yang and H. Hamada, Impact Properties of Jute and Jute Hybrid Reinforced Composites, *Proceedings of the 18thInternational Conference On Composite Materials - ICCM18*, Jeju Island, South Korea, paper T44–T43 (2012).
76. M.V.de Souza, S.N. Monteiro and J.R.M. d'Almeida, Effect of the shear and coupling elements of the Aij and Dij matrices on the impact behavior of glass, aramid and hybrid fabric composites. *Compos. Struct.* 76, 345–351 (2006).
77. S.C. Amico, C.C. Angrizani and M.L. Drummond, Influence of the stacking sequence on the mechanical properties of glass/sisal hybrid composites. *J. Reinf. Plast. Compos.* 29, 179–189 (2010).
78. Z. Salleh, M.N. Berhan, K.M. Hyie and D.H. Isaac, In-core sensors readings diagnostics based on neuro-fuzzy techniques. *World Acad. Sci. Eng. Technol.* 71, 969–974 (2012).
79. C. Santulli, M. Janssen and G. Jeronimidis, Partial replacement of E-glass fibers with flax fibers in composites and effect on falling weight impact performance. *J. Mater. Sci.* 40, 3581–3585 (2005).
80. Y. Zhang, Y. Li, H. Ma and T. Yu, Tensile and interfacial properties of unidirectional flax/glass fiber reinforced hybrid composites. *Compos. Sci. Technol.* 88, 172–177 (2013).
81. K. Adekunle, S.-W.Cho, R. Ketzscher and M. Skrifvars, Mechanical properties of natural fiber hybrid composites based on renewable thermoset resins derived from soybean oil, for use in technical applications. *J. Appl. Polym. Sci.* 124, 4530–4541 (2012).
82. S.S. Morye and R.P. Wool, Mechanical properties of glass/flax hybrid composites based on a novel modified soybean oil matrix material. *Polym. Compos.* 26, 407–416 (2005).
83. A. B. A. Hariharan and H. P. S. Abdul Khalil, Lignocellulose-based hybrid bilayer laminate composite: Part I - Studies on tensile and impact behavior of oil palm fiber-glass fiber-reinforced epoxy resin. *J. Compos. Mater.* 39, 663–684 (2005).
84. R. Velmurugan and V. Manikandan, Mechanical properties of glass/palmyra fiber waste sandwich composites. *Indian J. Eng. Mater. Sci.* 12, 563 (2005).

85. A. Boynard and J.R.M. d'Almeida, Morphological characterization and mechanical behavior of sponge gourd (Luffa cylindrica)–polyester composite materials. *Polym. Plast. Technol. Eng.* 39, 489–499 (2000).
86. M. Mahesh, K. Poornima and M.V. Rao, Optimization of effective parameters of jatropha-biosesel using taguchi method and performance analysis using CI engine. *Int. J. Mech. Ind. Eng.* 3, 46–52 (2013).
87. A. Abiy, Design and analysis of bamboo and e-glass fiber reinforced epoxy hybrid composite for wind turbine blade shell, Addis Ababa University, M. Sc. Dissertation, (2013).
88. G.V. Reddy, S.V. Naidu and T.S. Rani, Kapok/glass polyester hybrid composites: tensile and hardness properties. *J. Reinf. Plast. Compos.* 27, 1775–1787 (2008).
89. S.P. Priya, and S.K. Rai, Mechanical performance of biofiber/glass-reinforced epoxy hybrid composites. *J. Ind. Text.* 35, 217–226 (2006).

5

Flax-Based Reinforcement Requirements for Obtaining Structural and Complex Shape Lignocellulosic Polymer Composite Parts

Pierre Ouagne[*,1] **and Damien Soulat**[2]

[1]*Laboratoire PRISME, University of Orleans, Orleans, France*
[2]*GEMTEX-ENSAIT, Roubaix, France*

Abstract

This study concentrates on the possibility of forming complex shapes using the sheet forming process; a technique that could be used by the automotive industry due to interesting cost/cadencies ratio. Particularly investigated is the possibility of realizing a technical complex preform without defect from two different architecture commercial reinforcements not specially designed. Several defects such as tow buckles and too high strains in tows need to be prevented. For this reason a specially designed blank holder set was used to form complex tetrahedron shapes. The defects previously encountered using a non-optimized blank holder set have been suppressed to a great extent. Tow buckles were suppressed by reducing the tension in the vertical tows of the shape and by increasing the tension of the tows exhibiting the buckles. For the plain weave fabric, the buckles were totally suppressed at the end of the forming process, whereas an additional compression stage would probably be necessary in the case of the twill weave to get rid of the tow buckles. By reducing the tension in the vertical tows of the shape in the zone where tow buckles may take place, the too high strain defect was also reduced to such an extent that it was not a problem anymore.

This study therefore shows it is possible to form complex shapes using untwisted flax commercial reinforcements not particularly optimized for complex shape forming by well designed geometry of blank holders. This result is particularly interesting because it is possible to use fabrics that do not necessitate as much energy for their production as twisted yarn reinforcement, and because the composite parts manufactured using such reinforcements show higher performance because higher fiber volume fraction can be obtained.

Keywords: Flax, composite, manufacturing processes, fabric/textile, fabric mechanical behavior, preforming, tow, forming defect, tow buckling

**Corresponding author*: pierre.ouagne@univ-orleans.fr

Vijay Kumar Thakur (Ed.), Lignocellulosic Polymer Composites, (83–102) 2015 © Scrivener Publishing LLC

5.1 Introduction

Natural fibers have long been considered as potential reinforcing materials or fillers in thermoplastic or thermoset composites. Numerous studies deal with the subject [1-6]. Natural fibers are particularly interesting because they are renewable, have low density and exhibit high specific mechanical properties. They also show non-abrasiveness during processing, and more importantly, biodegradability. A large amount of work has been devoted to identify the tensile behavior of individual fibers or group of few fiber of different nature and origin [7-10]. However, few studies deal with the subject of the mechanical behavior of fiber assemblies and particularly analyze the deformability of these structures.

To maximize the reinforcement performance in structural application, fiber orientation needs to be considered and this implies the use of continuous reinforcements. As a consequence, aligned fiber architectures such as unidirectional sheets, non-crimped fabrics and woven fabrics (bidirectional) are usually used as reinforcement. However, natural fibers are not continuous and may show large discontinuity in properties. In order to avoid long considerations about the variability of the fiber properties, it may be interesting to consider for some manufacturing processes such as filament winding [11] or pultrusion [12] the scale of the tow or the scale of the yarn. The scale of the composite (natural fiber reinforcement combined to a polymeric resin) is also interesting if one wants to avoid considering the variability of the fiber properties [13-15]. The homogenized behavior at the composite scale depends on the reinforcement type (mat, woven fabric, non-crimped fabric), the resin used and the process chosen to manufacture the composite. The study of composite samples is also used to analyze the impact of the composite part all along its life cycle [16,17]. The energetic record to produce flax fibers for composite materials has been analyzed by Dissanayake *et al.* [18,19]. They showed, in the case of traditional production of flax mats, with the use of synthetic fertilizers and pesticides associated to traditional fiber extraction such as dew retting and hackling, that the energy consumption linked to the production of a flax mat is comparable to the energy consumed during the production of a glass mat. They also showed that spinning to produce yarns is an energy intensive operation, and in that case, the glass woven fabric may show a lower impact on the environment than an equivalent flax woven fabric if one considers an environmental energy viewpoint. As a consequence, it is recommended to avoid the use of spun yarns to produce natural fiber-based woven fabrics. An alternative to spinning fibers into yarns is the use of a binder to provide cohesion between fibers in the form of tows. Despite studies providing the mechanical characteristics and potential of natural fiber yarns [20-22], few studies focus on that of the flax tows.

When dealing with composite materials, weight reduction is often an issue, and it appears that the best gain can be obtained on structural thick or complex shape parts. However, the possibility to realize these shapes in composite materials is still a problem to be solved. For example, only 25% of the Airbus A380 is constituted of composite materials. Several low-scale manual manufacturing processes exist to realize these complex shape composite parts, particularly for the military or the luxury car industries. Filament winding or pultrusion could be used to manufacture thick structural parts.

The sheet forming of dry or commingled (reinforcement and matrix fibers mixed in a same tow) composite materials can be considered as a solution for the industrial scale manufacture of complex shape composite parts as this process shows a good production rate/cost ratio. Numerical approaches have been used to determine the process parameters to be used [23-24]. However, few of these studies dealt with complex shape parts for which specific defects such as tow buckles may appear [25]. The appearance of such defects may prevent the qualification of the part and indicate the limit of the reinforcement material behavior under a single or a combination of deformation modes. It is therefore important to quantify and understand the mechanisms controlling the appearance of defects so that the numerical tools developed in the literature for complex shape forming [26]can simulate them.

In the liquid composite moulding (LCM) family, the resin transfer moulding (RTM) process has received great attention in the literature [27] , particularly the second stage of the process dealing with the injection of resin in preformed dry shapes and the permeability of the reinforcements [28,29]. The first stage of this process consists of forming dry reinforcements. In the case of specific double curved shapes, woven fabrics are generally used to allow the in-plane strain necessary for forming without dissociation of the tows.

The modification of the tow orientation and local variations of fiber volume fraction have a significant impact on the resin impregnation step as the local permeabilities (in-plane and transverse) of the reinforcement may be affected [30,31]. In the most severe cases, the ply of fabric can wrinkle or lose contact with the mould, hence severely reducing the quality of the finished product [32]. Another defect called tow buckling has also been reported for flax woven fabrics [33,34]. As the quality of the preform is of vital importance for the final properties of the composite parts, it is important when forming of complex shape is considered to prevent the appearance of such defects.

Several experimental devices have been set up to investigate the deformation modes and the possible occurrence of defects during forming of textile reinforcements. Hemispherical punch and die systems were particularly studied because the shape is rather simple, it is doubled curved, and because it leads to large shear angles between the tows [35-37]. In this paper, an experimental device is presented to form severe shapes. As an example, tetrahedron geometry is considered as it is much more difficult to form than hemispherical shapes, especially if the radiuses of curvature are small.

The objective of this study consists mainly in investigating the potential of the flax tow architecture for composite processing. Tensile tests on specimens at different gauge lengths and strain rates will be presented. The identification of failure modes using digital image correlation is also investigated. Finally, in relation to composite processes such as wet filament winding, the evolution of the tensile properties of the tow in the wet state after impregnation will also be studied. This study also proposes to analyze the feasibility of forming the mentioned complex shape with natural fibers-based woven fabric reinforcements. Special attention is given to the defects that may appear during forming. In particular, the high strain and tow buckling defect are discussed and a discussion about ways to prevent their appearance is presented.

5.2 Experimental Procedures

5.2.1 Flax Tow Testing

The flax tow used was manufactured by Groupe Depestele (France) [38] in the form of spools with a linear density of 500 tex. The tow consists of an agglomeration of flax fiber bundles along the tow axis, which overall cohesion is given by a natural binder. The architecture of the tow can be qualified as flat and unidirectional (Figure 5.1); the tow is 0.16 ± 0.03 mm thick and 2.20 ± 0.39 mm wide.

Samples were randomly cut from the spools for specimen preparation. The cut samples were glued to aluminium sheets with an epoxy Araldite glue for the tensile tests (Figure 5.2). To study the effect of the gauge length, six batches of specimens were prepared with gauge length values of 4, 25, 50 125, 250 and 500 mm and were tested under quasi static loading. For the study of the strain rate effect, four batches of specimens were prepared with a gauge length of 250 mm and tested at four loading rates, going from 1.33x10-5 to 1.33x10-2 s-1 by a step factor of 10. To investigate the potential of this tow architecture for composite preforming processes such as filament winding, a batch of specimens of 250 mm gauge length were impregnated in an epoxy resin bath prior to tensile tests (wet state).

An INSTRON 4507 tensile machine with a 10kN load cell (± 2.5N accuracy) was used for all the tests. The quasi-static strain rate was taken as 6.66x10-5 s-1. For short gauge lengths (≤125 mm), the specimens were marked along their axis (Figure 5.2) and the displacement during the tensile tests were monitored using the mark tracking technique [39]. Furthermore, for local analysis of the behavior, some specimens were fitted with a speckle pattern for digital image analysis.

5.2.2 Flax Fabric Testing

Two different flax fabrics (Figure 5.3) are used in this study. The first one is a plain weave fabric with an areal weight of 280 ± 19 g/m^2, manufactured by Groupe Depestele (France) [38]. The fabric is not balanced, as the space between the weft tows (1.59 ± 0.09 mm) is different to the one between the warp tows (0.26 ± 0.03 mm). The widths

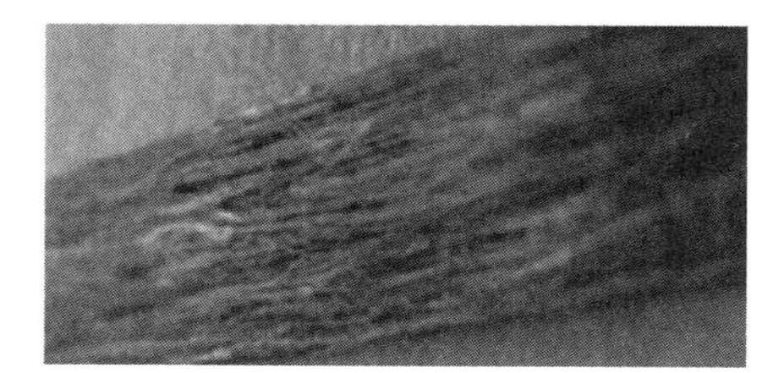

Figure 5.1 Tow architecture.

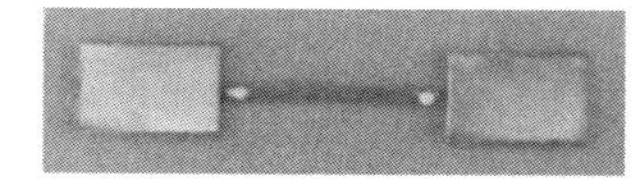

Figure 5.2 Tow tensile test specimen.

of the warp and the weft tows are respectively 2.53 ± 0.12 mm and 3.25 ± 0.04 mm. As a consequence, there are 360 warp tows and 206 weft tows per metre of fabric. The linear mass of the warp and the weft tows is the same and is equal to 494 ± 17 g/km. The second fabric is a 2x2 twill weave with an areal weight of 273 ± 5 g/m^2, also manufactured by Groupe Depestele [38]. This fabric is more balanced than the plain weave fabric but not completely. Between the weft tows, no space can be observed whereas a space between the warp tows of 0.9 ± 0.2 mm takes place. Moreover, the widths of the warp and weft tows are respectively 2.4 ± 0.4 mm and 2 ± 0. 4 mm. As a consequence, there are 312 warp tows and 500 weft tows per metre of fabric. The linear mass of the warp and the weft tows are respectively 275 ± 12 g/km and 293 ± 32 g/km.

5.2.2.1 *Biaxial Tensile Test*

A biaxial tension device (Figure 5.4) has also been used to characterize the tensile behavior of the reinforcements.

Biaxial and uniaxial tension tests as well as tensile test conducted on individual tows can be performed using this device. For synthetic carbon or glass fabrics, the limit to failure is not reached during forming, and the tensile test are designed to analyze the possible non-linearity of the stress-strain curves due to the 2D assembly of the woven textiles generally used. For natural fiber fabrics, the tensile limit of the fabric becomes particularly interesting as the tows used to elaborate the woven fabrics are manufactured from finite length fibers slightly entangled and held together by a natural binder and are not expected to show comparable tensile resistance. The tensile strains for each considered tows are measured using a 2D version of the mark tracking device described previously. The detailed description of the device as well as the procedure of the test may be found in reference [40].

5.2.3 Sheet Forming Device for Dry Textile Reinforcement

A device presented in Figure 5.5 was especially designed to analyze the possibility to form reinforcement fabrics. Particularly, the device was developed to examine the local deformations during the forming process [41]. The device is the assembly of a mechanical part and an optical part. The mechanical part consists of a punch/open die system coupled with a classical blank-holder system. The punch used in this

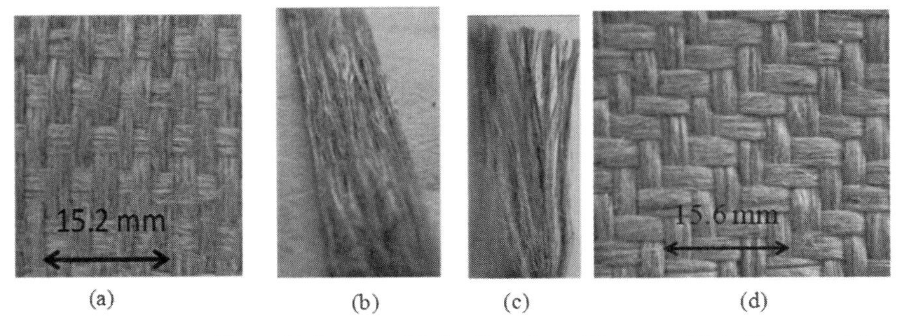

Figure 5.3 (a) Plain weave fabric; (b) flax tow; (c) detailed view of flax tow; (d) twill weaves fabrics.

study (Figure 5.5b) is a tetrahedron form with 265 mm sides. Its total height is 128 mm and the base height is 20mm. The edges and vertices possess 10 mm radius for the punch and 20 mm for the die. As the punch possesses low edges radiusses, it is expected that large shear strains take place during forming. A triangular open die (314x314x314mm^3) is used to allow the measurement of the local strains during the process with video cameras associated to a marks tracking technique [39]. A piloted electric jack is used to confer the motion of the punch. Generally, the punch velocity is 30 mm/min and its stroke 160 mm. The maximum depth of the punch is 160 mm. A classical multi-part blank-holder system is used to prevent the appearance of wrinkling defects during the preforming tests by introducing tension on the fabric. It is composed of independent blank-holders actuated by pneumatic jacks that are

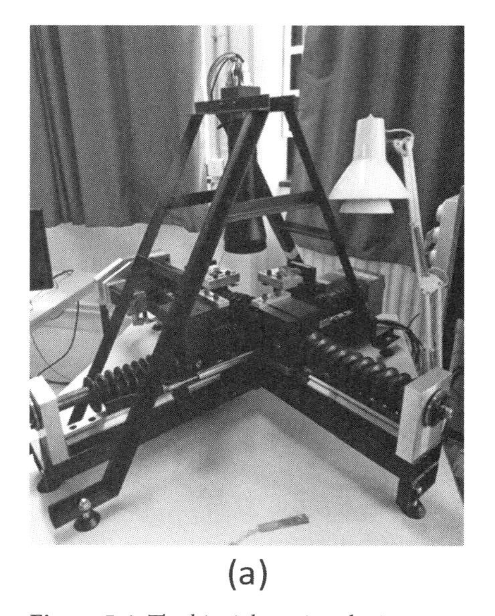

(a) (b)

Figure 5.4 The biaxial tension device.

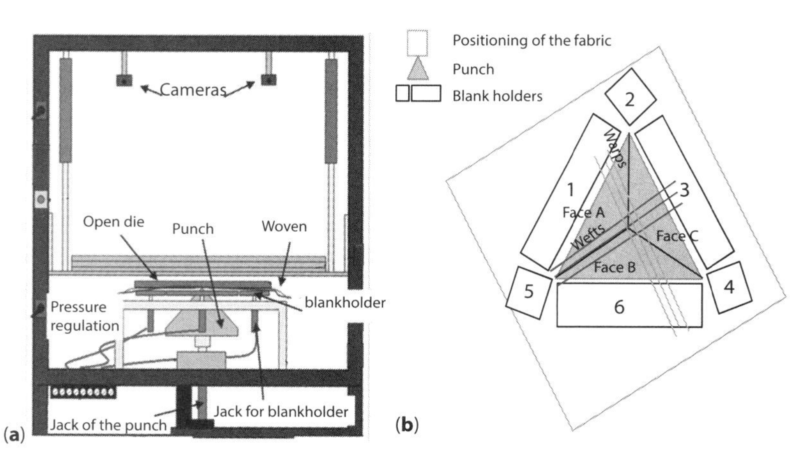

Figure 5.5 (a) The sheet forming device (b) Initial positioning of the fabric and position of the blank holders.

able to impose and sense independently a variable pressure. The quality of the final preform may depend on several process parameters such as the dimensions, positions, and the pressure applied by each of the blank-holders. They can be easily changed to investigate their influence on the quality of the final preform. Before starting the test, a square piece of fabric is positioned between the die and the blank-holders. The initial positioning of the fabric is of particular importance as it partly conditions the final tow orientations within the part. However, it is not possible to establish before the test their final position at the end of the forming and as a consequence their mechanical stiffness. So, for the tests presented below, it was chosen to align the warp or the weft tows with an edge of the tetrahedron (the opposite edge of Face C (Figure 5.5b)). To avoid bending of the fabric under its own mass, a drawbead system is used to apply low tensions at the tow extremities. At the end of the performing test, the dry preform can be fixed by applying a spray of resin on its surface so that the preform can be removed from the tools and kept in its final state.

As mentioned in the Introduction part, several defects may appear during the sheet forming of complex shapes such as a tetrahedron. Both the tow buckles and the excessive tensile strain in tows are localized in zones close to the tows passing by the top of the tetrahedron as indicated in Figure 5.5b. by the lines passing by the top of the pyramid [42]. In this zone, the vertical tows passing by the triple point are too tight in the three faces and on the edge opposed to Face C. The perpendicular tows may show the presence of tow buckles partly resulting of the bending in their plane of tows and also probably because of a too low tension of these tows. In the case of Figure 5.5b the buckles are localized on Face C of the tetrahedron and on the opposite edge. No buckles are observed on Faces A and B. This is linked to the un-balanced architecture of the fabric used [43]. The basis of the new blank holder generation therefore consists in reducing pressure in the vertical tows and to increase the tension of the horizontal ones. A schematic diagram of the blank holder is presented in Figure 5.6. Instead of the 6 initial blank holders (Figure 5.6a), the new blank holder generation consist of 4 blank holders with specific geometries. The blank holders impose tensions to the membrane and particularly to the bent tows exhibiting tow buckles. Between the 4 blank holders, empty zones have been left to release the tension of the tows showing too high tensile strains. It can be noted that small blank holders could be used to fill up the spaces and impose a local pressure to this zone if necessary.

At the end of the preforming test, several analyses at different scales can be performed. A first global analysis at the macroscopic scale concerning the final state of the preform before removing it from the tool can be performed. It consists in analysing if the shape is obtained and if the shape shows defects. Another analysis, at the mesoscopic scale, consists in analysing the evolution of the local strains (shear, tension) during forming.

Using this device, an experimental study to analyze with the tetrahedron shape, the generation of defects (wrinkles, tow buckles, tow sliding, vacancies, etc.) can be performed. The influence of the process parameter and particularly the blank-holder pressures on the generation and the magnitude of defects is also commented and analyzed.

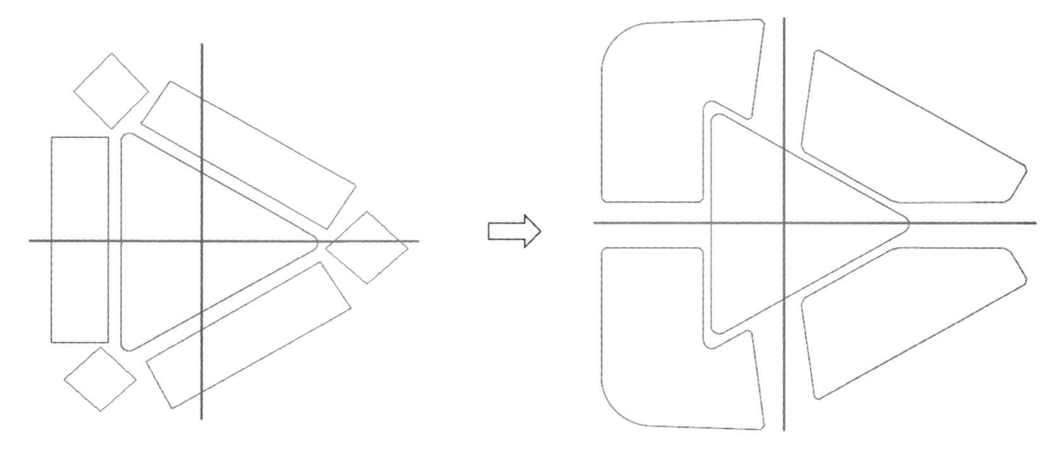

Figure 5.6 Blank holders' designs.

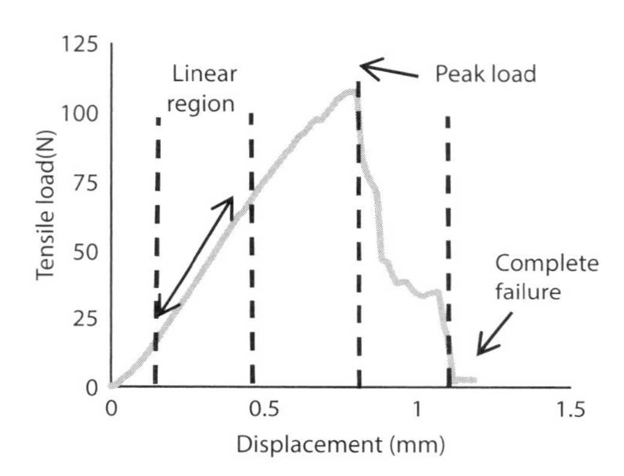

Figure 5.7 Typical tensile curve.

5.3 Results and Discussion

5.3.1 Tensile Behavior of Reinforcement Components: Flax Tow Scale

5.3.1.1 Flax Tow Tensile Behavior

The typical tensile curve of the flax tow can be seen on Figure 5.7. The curve starts with a non-linear region caused by the rearrangement of the fiber bundles within the tow with the loading axis. Then, a linear region is observed. The slope of this linear region is used to calculate the modulus of the tow. This linear region is followed by another non-linear region probably due to random damage of interfaces between the fiber bundles occurring within the tow. A peak load is then reached. The strain at break corresponds to the peak load strain. After the peak load, a non-progressive failure of the tow is observe till complete failure.

To calculate the apparent tensile strength and modulus, the cross-sectional area of the tow was estimated by dividing its linear density (500 tex) by that of its constituent fiber, flax, taken as 1.53 g/cm^3[4].

5.3.1.2 *Effect of Gauge Length on Tensile Properties*

The tensile properties measured at six different gauge lengths are given on Figure 5.8. The markers indicate the mean values and the vertical bars indicate the maximum and minimum measured for each batch. The trend of the mechanical properties shows a decrease with the increasing gauge length. This tendency is similar to the results found on fiber bundles [44-47].

The tow consists of agglomerated flax fiber bundles themselves being an assembly of elementary flax fibers. Therefore, for small gauge lengths, more elementary flax fibers within the bundles are likely to be clamped to both ends of the specimen. As a consequence, the mechanical properties tend to that of the elementary flax fiber (800-1135 MPa for the tensile strength and 50-60 GPa for the modulus [48]) as the gauge length decreases. Inversely, as the gauge length increases, the proportion of elementary flax fibers likely to be clamped to both side of the specimen decreases and the fiber bundles get loaded in exchange. At higher gauge length, the failure takes place at the interfaces between and within the bundles (pectin) [44] which have a relatively lower tensile strength than that of the elementary fibers. The tensile strength of the tow decreases till an almost constant plateau as the gauge length increases. The slight decrease in the tensile strength curve between 250 and 500 mm gauge length is probably due to increasing distribution of flaws in the tow. Concerning the modulus, the curve reaches a constant value as from 250 mm since the constitution of the tow has stabilized. Indeed, since the elementary flax fiber's length ranges between 4-77 mm [49], at 250 mm gauge length, no more elementary fibers are likely to be clamped to both sides of the specimen. Hence, the measured modulus corresponds to that of the combination of the fiber bundles' and binder's modulus. The evolution of the curves of Figure 5.8 suggests that a characteristic length as from which the mechanical behavior is comparatively stable can be fixed at 250 mm.

5.3.1.3 *Evolution of Failure Behavior*

The failure behavior of the tow varies as a function of the gauge length (Figure 5.9). The evolution of the failure behavior of the tow can also be related to its constitution at

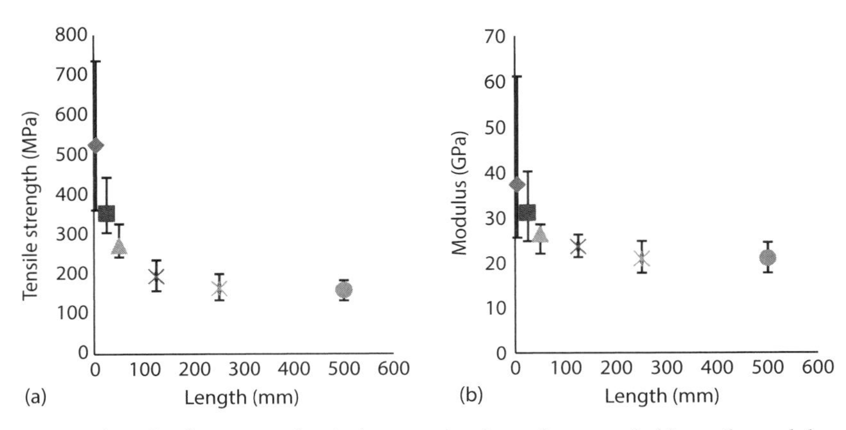

Figure 5.8 Gauge length effect on mechanical properties a) tensile strength, b) tensile modulus.

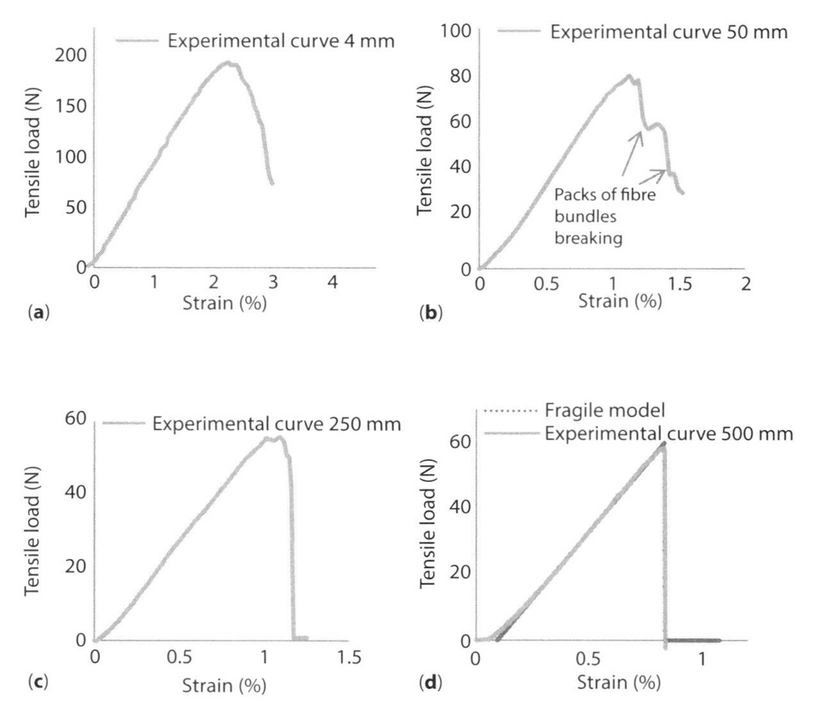

Figure 5.9 Evolution of the tensile failure behavior with the gauge length.

different gauge lengths. At 4 mm gauge length (Figure 5.9 a), a rather progressive failure can be observed after the peak load. For this small gauge length, the proportion of continuous elementary fibers heading for one end of the specimen to the other is high. The failure is governed predominantly by that of the elementary fibers breaking successively within the specimen. However, when the gauge length increases, the proportion of the specimen consisting of continuous elementary fiber decreases, and therefore the failure is more likely to occur at the interfaces [44]. For the 50 mm, it can be seen that after the peak load, the tensile load falls down step by step, Figure 5.9 b. This failure mode in steps is attributed to packs of fiber bundles breaking. The term pack of fiber bundle refers to an agglomeration of fiber bundles held together by the binder. It can represent in terms of size more than 400 μm as compared to typical size of flax fiber bundle, viz. 50-100 μm [44]. The failure in steps is attributed to a delayed activation (slack) of the packs of fiber bundles during the tensile tests, causing them to break successively. This slack occurrence tends to decrease when the gauge length increases [50], as seen on Figure 5.9c for a 250 mm gauge length test and moreover on Figure 5.9 d for a 500 mm gauge length test for which the failure is brittle.

The failure mode by packs of fiber bundles can be observed by digital image correlation. Figure 5.10 a shows the tensile curve of a speckled specimen (Figure 5.10 b) and the states where the image correlation analysis are presented. The analysis area on Figure 5.10 b corresponds to the middle region of the tow. The mm/pixel ratio is 1.39×10^{-2}. The 2D displacement field obtained by image correlation is given on Figure 5.10c & d, the labels on the left-hand side are given in pixel. Before the peak load, state A and B, a 2D displacement field gradient can be seen as expected in a tensile test. After the peak load, at state C, an axial debonding on the left, propagating along the tow's axis can be seen. This debonding leads to a drop in the tensile load. A slight stabilization of the load is observed at this state before the failure of a second pack of fiber bundles at state D. The stabilization of the load is attributed to a delayed activation of the second pack of fiber bundles. The breakage of the second pack of fiber bundles seen at state D corresponds to a major drop of the tensile load. The size of this pack is approximately 700 μm. Further axial debonding can also be seen on the right-hand side of the tow. Nonetheless, the failure of the tow is not total at this state as a displacement gradient can still be seen in the middle region of the tow. The complete failure occurs at state E

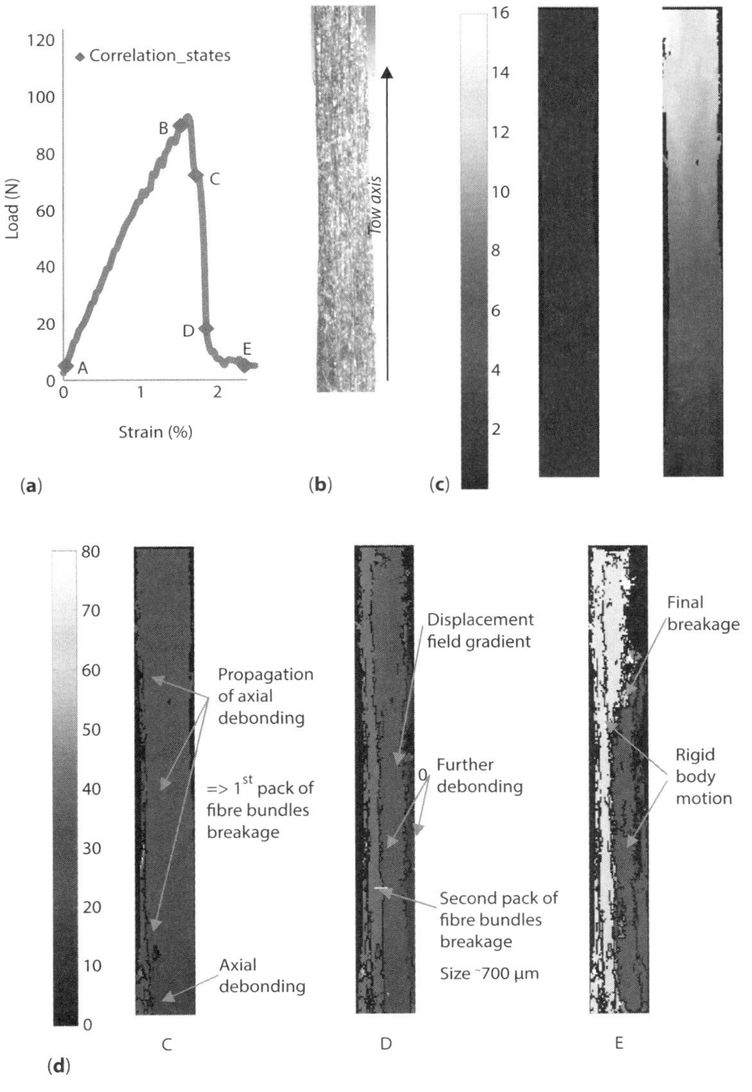

Figure 5.10 Local analysis by digital image correlation, a) tensile curve, b) speckled specimen, c) displacement field before peak load, d) displacement field after peak load.

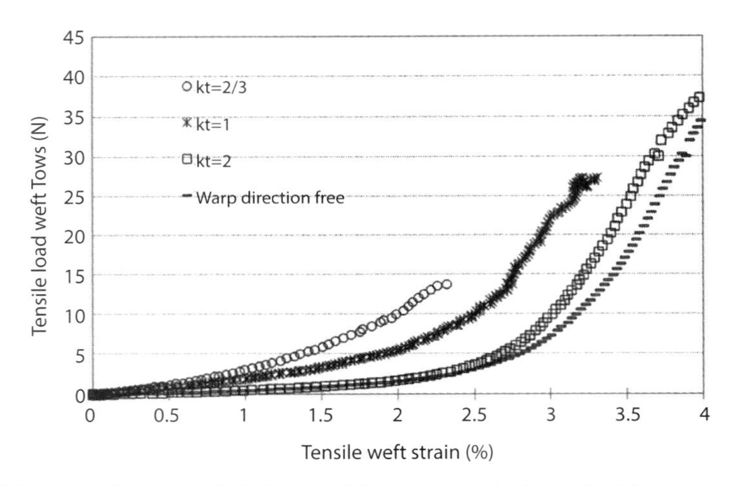

Figure 5.11 Plain weave fabric: tensile behavior of the weft tows during a biaxial test.

with the breakage of the last pack of fiber bundle. The rigid body displacement of the two remaining parts of the tow can thus be observed.

5.3.2 Tensile Behavior of Reinforcement Components: Scale of Fabric

Biaxial tests were carried out to study the tensile behavior of the flax fabric. The results showing the biaxial behaviour of the fabric in the weft direction are presented for the plain weave fabric in Figure 5.11 for different values of the parameter *kt*. The parameter $kt = e_{we}/e_{wa}$ is defined as the ratio between the strain in the weft direction (e_{we}) over the one in the warp direction (e_{wa}).

The results show that, an increasing value of *kt* leads to higher "failure/limiting" strain in the weft tows. This means that the crimp effect decreases in this case. The crimp effect is the lowest when the warp direction is not loaded and therefore left free. As the warp tows are left free, these ones are not tight and therefore the crimp effect and the associated non linearity zone is more pronounced.

The strain values observed at the maximum load point during a biaxial test are lower than in the case of the uniaxial test performed with same dimension samples and this for the two directions. As a consequence, the strains at which the load is maximum are lower than 4.5% in the weft direction during the uniaxial test. These strain values represent the limit from which a loss of fiber density takes place in the tows and this phenomenon should probably be avoided by keeping the tensile strain in the tetrahedron preform lower than these values.

Figure 5.12 indicates that the tensile strain limit in the case of the warp tows left free is about 2% for the twill weave fabric.

5.3.3 Global Preform Analysis

An initial square specimen of the flax fabric is positioned with six blank holders placed on specific places around the tetrahedron punch. On each of them a pressure of one bar is applied. The maximum depth of the punch is 150 mm. At the end of the forming

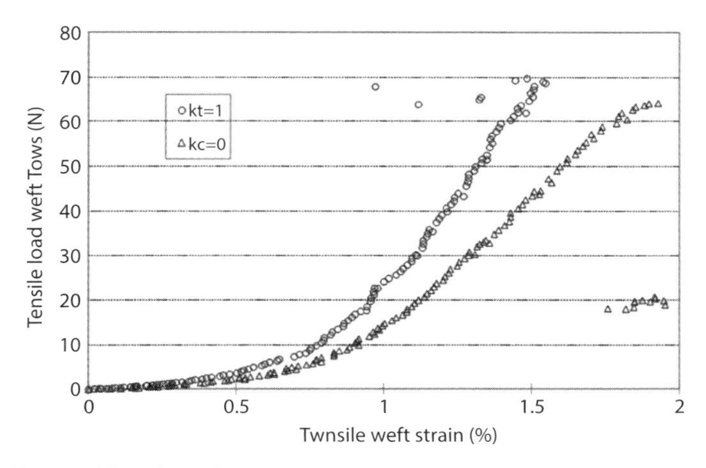

Figure 5.12 Twill weave fabric: biaxial tension curves.

process, an epoxy resin spray is applied to the preform so that the shape is fixed in its deformed state. The preform in its final state is presented in Figure 5.13.a. At the scale of the preform the obtained shape is in good agreement with the expected tetrahedron punch. The fabric is not un-weaved on faces or edges. Some wrinkles appear (Figure5.13.a) at the surrounding of the useful part of the preform.

The position and the size of these wrinkles depend on the blank holder position and on the pressure they apply on the fabric. The process parameters (number and position of blank holders, choice of the punch, etc.) and the initial positioning of the fabric have a significant influence on the final shape. These aspects will be presented in future works. At the local scale, it is possible to analyze during the process the evolution of the shear angle between tows and the longitudinal strain along the tows. During the forming stage, the woven textile is submitted to biaxial tensile deformation, in plane shear deformation, transverse compaction and out-of-plane bending deformations. If all these components can be significant, the feasibility to obtain the expected shape is largely dependent on the in-plane shear behavior. On the formed tetrahedron faces, values of the measured shear angle are relatively homogeneous [33]. These values do not reach the locking angle above which defects such as wrinkles appear.

5.3.4 Analysis of Tensile Behavior of Tows During Forming

If no apparent defects are observed on the faces and on the edges of the tetrahedron formed using the new blank holder geometry, it is still necessary to investigate if the tensile strains of the tows passing by the triple point (top of the pyramid) are higher than the strain at which local failure may happen. Figure 5.14 shows a comparison of the tensile strains measured using both set of blank holder for the plain weave fabric and at the same location on the preform (on the weft tow passing by the triple point of Face C using 2 bar of blank holder pressure).

Figure 5.14 shows that a large decrease in tensile strain is observed, particularly at the end of the forming process. The tensile strain values are lowered from 8.8% to 4.7% indicating that the new set of blank holder has an important beneficial effect on the

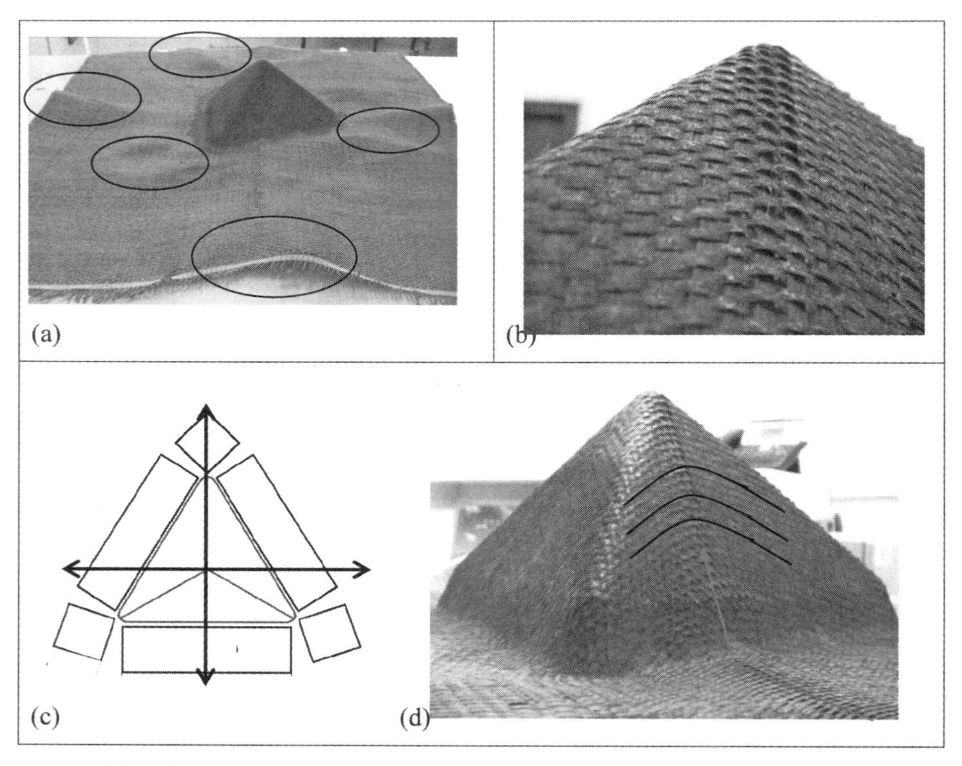

Figure 5.13 (a) Preform and Wrinkles. (b) Zoom on buckles. (c) Position of buckles. (d) tow orientation.

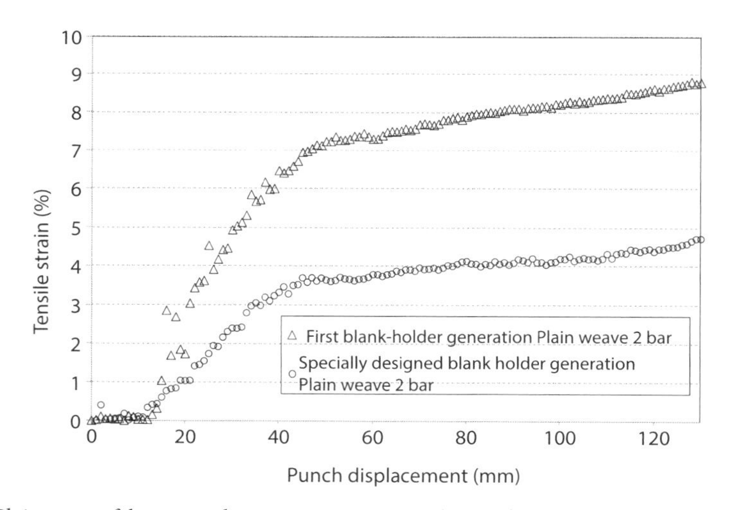

Figure 5.14 Plain weave fabric: tensile strains comparison during forming.

tensile strain reduction on the tows passing by the triple point in the case of the plain weave fabric. This corresponds to a decrease of about 47%.

However, the value of the maximum strain recorded at the end of the forming process on the tightest tow needs to be compared to the limit of the fabric determined by performing biaxial tensile test on the same fabric. Figure 5.11 shows the result of the

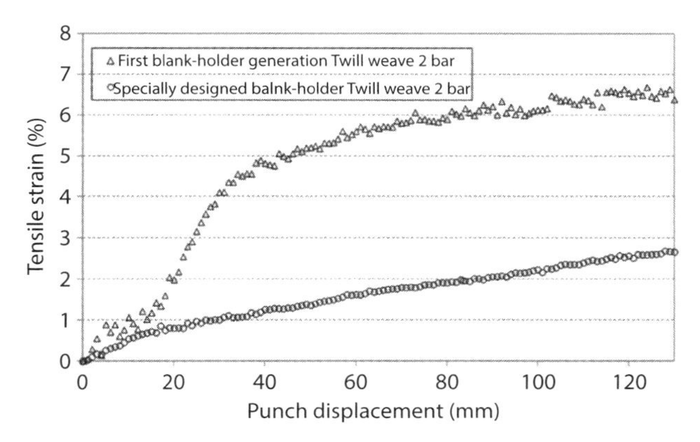

Figure 5.15 Twill weave fabric: tensile strains comparison during forming.

biaxial test in the case of equal deformation in the warp and weft directions and in the case where the warp tows are left free. The last case would be the one to compare with the tensile strain measured during forming as the warp tows exhibiting the tow buckles are not tight and can be considered as almost free. In this case, the maximum limit above which sliding movements within the tow leading to losses of fiber density may happen, is about 4%. This value is lower than the one recorded at the end of the forming process in our forming test (4.7%) suggesting that sliding of group of fiber may have taken place.

Figure 5.15 compares the tensile strains of the same tow measured at the same place close to the triple point in the case of the use of both blank holder sets for the twill weave fabric. Figure 5.15 shows that a large decrease in tensile strain is observed when using the new blank holder set with the same pressure at the end of the forming test. The measured tensile strains are respectively 6.5% and 2.6% for the original and new designs respectively. A reduction of about 60% can be measured in this case. This reduction is in the case of the twill weave more important than in the case of the plain weave fabric but the final tensile strains values need to be compared to biaxial tensile test in order to determine the maximum strain that should not be overcome. Figure 5.12 indicates that the tensile strain limit in the case of the warp tows left free is about 2%. Again, as for the plain weave fabric, the tensile strain recorded during the complex shape forming of the twill weave is higher than the expected limit above which defects due to sliding of fiber packs may happen.

5.4 Discussions

The new blank holders were also designed to release a part of the tension in the vertical tows situated in the zone showing the tow buckles. This zone is also the place where the tightest tows are observed. The new design therefore gives the possibility to help suppressing the tow buckles but also to decrease the tension in the tightest tows. For both fabrics, the reduction in tensile strain for the plain weave and the twill weaves fabrics are respectively 47% and 60%. This means that even if the decrease in tensile strain is consequent, the values of the maximum tensile strain recorded are higher than the values at which internal defect in the tows, in the form of sliding of fiber packs, may happen. However, the maximum tensile strains recorded are not very much higher than the strains at which the defects may start to be generated. As a consequence, it is expectable that low displacement between the fibers took place as described by Moothoo *et al.* [45], but to a low extent. So the fiber density does not, to our opinion, change. In those conditions, the too high strains in the tow should not therefore be a problem concerning the manufacturing of a composite part either by film stacking process [6] or by the RTM process.

5.5 Conclusions

Several defects such as tow buckles and too high strains in tows already described in previous studies need to be avoided and prevented. For this reason a specially designed blank holder set was used to form complex tetrahedron shapes. The defects previously encountered using a non-optimized blank holder set have been suppressed in a great extent. The tow buckles were suppressed by reducing the tension in the vertical tows of the shape and by increasing the tension of the tows exhibiting the buckles. For the plain weave fabric, the buckles were totally suppressed at the end of the forming process, whereas an additional compression stage would probably be necessary in the case of the twill weave to get rid of the tow buckles.

By reducing the tension in the vertical tows of the shape in the zone where tow buckles may take place, the too high strain defect was also reduced to such an extent that it is not a problem anymore.

This study therefore shows that it is possible to form complex shapes using untwisted flax commercial reinforcements not particularly optimized for complex shape forming by well designing the geometry of blank holders. This result is particularly interesting because it is possible to use fabric that do not necessitate as much energy for their production as twisted yarn reinforcement and because the composite parts manufactured using such reinforcements show higher performance because higher fiber volume fraction can be obtained.

References

1. F.P. La Mantia, and M. Morreale,Green composites: A brief review. *Composites Part A*, 42, 579–588 (2011).
2. K.G. Satyanarayana,G.G.C. Arizaga , and F. Wypych, Biodegradable composites based on lignocellulosic fibers—An overview. *Prog. Polym. Sci.* 34, 982–1021 (2009)
3. J. Biagiotti, D. Puglia, and J.M. Kenny, A review on natural fibre-based composites-Part I. *J. Nat. Fibers 1*(2), 37–68 (2004).
4. D.Puglia, J. Biagiotti, and J.M Kenny, A review on natural fibre-based composites—Part II. *J. Nat. Fibers* 1(3), 23–65 (2005).
5. J.K. Pandey, S.H. Ahn, C.S. Lee, A.K. Mohanty, and M. Misra, Recent advances in the application of natural fiber based composites. *Macromol. Mater. Eng.* 295, 975–989 (2010).

6. P. Ouagne, L. Bizet, C. Baley, and J. Bréard, Analysis of the film-stacking processing parameters for PLLA/Flax fiber biocomposites". *J. Compos. Mater.* 44, 1201–1215 (2010).
7. C. Baley, Analysis of the flax fibres tensile behaviour and analysis of the tensile increase. *Compos. A* 33, 2143–2145 (2002).
8. E. Bodros, and C. Baley, Study of the tensile properties of stinging nettle fibres. *Mater. Lett.* 62, 2143–2145 (2008).
9. A. Alawar, A.M. Hamed, and K. Al-Kaabi, Characterization of treated data palm tree fiber as composite reinforcement. *Compos. B* 40, 601–606 (2009).
10. J.T. Kim, and A.N. Netravili, Mercerization of sisal fibers: Effects of tension on mechanical properties of sisal fibers and fiber-reinforced composites. *Compos. A* 41, 1245–1252 (2010).
11. P.K.D. Lehtiniemi, T. Berg, M. Skrifvars, and P. Järvelä, Natural fiber-based reinforcements in epoxy composites processed by filament winding. *J. Reinf. Plast. Compos.* 30 (23), 1947–1955 (2011).
12. I. Angelov, S. Wiedmer, M. Evstatiev, K. Friedrich, and G. Mennig. Pultrusion of a flax/polypropylene yarn. *Compos. A Appl. Sci. Manuf.* 38, 1431–1438 (2007).
13. J. Summerscales, N. Dissanayake, A. Virk, and W.Hall, A review of bast fibres and their composites. Part 2 Composite. *Compos. A 4*1, 1336–1344 (2010).
14. P. Ouagne, L. Bizet, C. Baley , and J.Bréard, Analysis of the Film-stacking Processing Parameters for PLLA/Flax Fiber Biocomposites. *J. Compos. Mater.* 44, 1201–1215 (2010).
15. A. Alawar, A.M. Hamed, and K. Al-Kaabi, Characterization of treated data palm tree fiber as composite reinforcement. *Compos. B* 40, 601–606 (2009).
16. A. Le Duigou, P. Davies, and C. Baley, Environmental impact analysis of the production of flax fibres to be used as composite material reinforcement. *J. Biobased Mater. Bioenerg.* 5, 153–165 (2011).
17. S. Kim, B.E. Dale , L.T. Drzal, and M. Misra, Life cycle assessment of Kenaf fiber reinforced biocomposite. *J. Biobased Mater. Bioenerg.* 2, 85–93 (2008).
18. N. Dissanayake, J. Summerscales, S. Grove, and M. Singh, Life Cycle Impact Assessment of Flax Fibre for the Reinforcement of Composites. *J. Biobased Mater. Bioenerg.* 3, 245–248 (2009).
19. N. Dissanayake, J. Summerscales, S. Grove, and M. Singh, Energy use in the production flax fiber for the reinforcement of composites". *J. Nat. Fibers* 6, 331–346 (2009).
20. S. Goutianos, T. Peijs, B. Nystrom, and M. Skrifvars, Textile reinforcements based on aligned flax fibres for structural composites. *Appl. Compos. Mater.* 13, 199–215 (2006).
21. Madsen,P. Hoffmeyer, A.B. Thomsen, and H. Lilholt, Hemp yarn reinforced composites – I. Yarn characteristics. *Compos.A Appl. Sci. Manuf.* 38, 2194–2103 (2007) .
22. E.A. Subaida, S. Chandrakaran, and N. Sankar, Experimental investigations on tensile and pullout behaviour of woven coir geotextiles. *Geotext. Geomembr.* 26, 384–392 (2008).
23. Bickerton, P. Simacek, S.E. Guglielm, and S.G. Advani, Investigation of draping and its effects on the mold filling process during manufacturing of a compound curved composite part. *Compos. A* 28(9–10), 801–816, (1997).
24. N. Hamila, and P. Boisse, Simulations of textile composite reinforcement draping using a new semi-discrete three node finite element. *Compos. B* 39, 999–1010, (2008).
25. P. Ouagne, D. Soula, G. Hivet, S. Allaoui, and D. Duriatti, Analysis of defects during the performing of a woven flax. *Adv. Compos. Lett.* 20, 105–108 (2011).
26. P. Boisse, N. Hamila, F. Helenon, B. Hagege, and J. Cao, Different approaches for woven composite reinforcement forming simulation. *Int. J. Mater. Form. 1,* 21–29, (2008), DOI: 10.1007/s12289-008-0002-7.
27. M.J. Buntain, and S. Bickerton, Modeling forces generated within rigid liquid composite moldings tools. Part A: Experimental study. *Compos. A* 38, 1729–1741 (2007).

28. C. Lekakou, M.A.K. Johari, Norman, and G. Bader. Measurement techniques and effects on in-plane permeability of woven cloths in resin transfer molding. *Compos. A* 27, 401–408 (1996).
29. P. Ouagne, and J. Bréard, Continuous transverse permeability of fibrous media" *Compos. A* 41, 22–28 (2010).
30. A. Hammani, F. Trochu, R. Gauvin , and S. Wirth, Directional permeability measurement of deformed reinforcement. *J. Reinf. Plast. Compos.* 15, 552–562 (1996).
31. F. Loix, P. Badel, L. Orgeas, C. Geindreau, and P. Boisse, Woven fabric permeability: from textile deformation to fluid flow mesoscale simulations. *Compos. Sci. Technol.* 68, 1624–1630 (2008).
32. S.B. Sharma, M.P.F. Sutcliffe, and S.H. Chang, Characterisation of material properties for draping of dry woven composite material. *Compos. A* 34, 1167–1175 (2003).
33. P. Ouagne, D. Soulat, S. Allaoui, and G. Hivet, Mechanical properties and forming possibilities of a new generation of flax woven fabrics, *Proceeding of the 10th International Conference on Textile Composite (Texcomp).* 26–28 October, Lille, France (2010).
34. E. Capelle, P. Ouagne, Tephany, D. Soulat, D. Duriatti, and S. Gueret. Analysis of the tow buckling defect during the complex shape forming of a flax woven fabrics. *Proceedings of the 15TH European Conference on Composite Materials*, Venice, Italy, 24–28 June (2012).
35. P. Molnar, A. Ogale, R. Lahr, and P. Mitschang, Influence of drapability by using thermoforming. *Compos. Sci. Technol.* 67, 3386–3393 (2007).
36. X. Li, and S. Bai, Sheet forming of the multy-layered biaxial weft knitted fabric reinforcement. Part 1: On hemispherical surfaces. *Compos. A* 40, 766–777 (2009).
37. K. Vanclooster, V.S.Lomov , and I. Verpoest, Experimental validation of forming simulations on fabric reinforced polymers using an unsymmetrical mould configuration. *Compos. A* 40, 530–539 (2009).
38. http://www.groupedepestele.com/pro_recherche.html
39. N. Bretagne, V. Valle, and J.C. Dupré, Development of the marks tracking technique for strain field and volume variation measurements. *NDT & E Int.* 38, 290–8 (2005).
40. P. Boisse, A. Gasser, and G. Hivet, Analyses of fabric tensile behaviour : determination of the biaxial tension-strain surfaces and their use in forming simulation. *Compos. A* 32, 1395–1414 (2001).
41. D. Soulat, S. Allaoui, and S. Chatel, Experimental device for the optimization step of the RTM process. *Int. J. Mater. Form* 2, 181–184 (2009).
42. P. Ouagne, D. Soulat, C. Tephany, J. Moothoo, S. Allaoui , G. Hivet, and D. Duriatti, Complex shape forming of flax based woven fabrics. Analysis of the yarn tensile strain during the process. *Key Eng. Mater. 504*–506, 231–236 (2012).
43. P. Ouagne, D. Soulat, J. Moothoo, E. Capelle, and S.Gueret, Complex shape forming of a flax woven fabric; analysis of the tow buckling and misalignment defect. *Compos. A 51,* 1–10 (2013).
44. C. Baley, Analysis of the flax fibres tensile behaviour and analysis of the tensile stiffness increase. *Compos. A Appl. Sci. Manuf.* 33, 939–948 (2002)
45. G. Romhany, J. Karger-Kocsis, and T. Czigany, Tensile fracture and failure behavior of technical flax fibers. *J. Appl. Polym. Sci.* 90, 1–8 (2003).
46. A. Stamboulis, C.A. Baillie, S.K. Garkhail, H.G.H.V. Melick, Peijs, Environmental durability of flax fibres and their composites based on polypropylene matrix. *Appl. Compos. Mater.* 7, 273–294 (2000).
47. A.S. Virk, W. Hall, and J. Summerscales, Failure strain as the key design criterion for fracture of natural fibre composites. *Compos. Sci. Technol.* 70, 995–999 (2010).

48. G. Coroller, A. Lefeuvre, A.L. Duigou, A. Bourmaud, G. Ausias, T. Gaudry, and C. Baley, Effect of flax fibres individualisation on tensile failure of flax/epoxy unidirectional composite. *Compos. A Appl. Sci. Manuf.* 51, 62–70 (2013).
49. R. Chudoba, M. Vořechovský, and M. Konrad. Stochastic modeling of multi-filament yarns. I. Random properties within the cross-section and size effect. *Int. J. Solids Struct.* 43, 413–434 (2006).
50. J. Moothoo, S. Allaoui, P. Ouagne, and D. Soulat, A study of the tensile behaviour of flax tows and their potential for composite processing. *Mater. Des.* 55, 764–772 (2014).

6

Typical Brazilian Lignocellulosic Natural Fibers as Reinforcement of Thermosetting and Thermoplastics Matrices

Patrícia C. Miléo*[,1], Rosineide M. Leão[2], Sandra M. Luz[3], George J. M. Rocha[4] and Adilson R. Gonçalves[4]

[1] Engineering School of Lorena, University of São Paulo, São Paulo, Brazil
[2] College of Technology, University of Brasília, Brasília, Brazil
[3] University of Brasília at Gama, Brasília, Brazil
[4] Bioethanol Science and Technology Laboratory (CTBE), Campinas, São Paulo, Brazil

Abstract

The use of lignocellulosic fibers and their components as raw material in the production of composites is seen as technological progress in the context of sustainable development. Polymer composites reinforced with natural fibers have become an attractive research field in recent years. These composites are mainly driven by the low cost of natural fibers and also their other unique advantages, such as lower environmental pollution due to their biodegradability, low density, reduced tool wear and acceptable specific strength. In Brazil, there is a wide range of natural fibers which can be applied as reinforcement or fillers. Natural fibers are in most cases cheaper than synthetic fibers and cause less health and environmental hazard problems for people producing the composites as compared to glass fibers. The interaction between the fiber and matrix is one of the determining factors for the formation of composites. However, poor fiber–matrix interfacial adhesion may affect the physical and thermal properties of the resulting composites due to the surface incompatibility between hydrophilic natural fibers and non-polar polymers. Different strategies have been applied to mitigate this deficiency in compatibility, including surface modification techniques. The surface modification of the fibers can be done by chemical treatments. In this chapter, a review of the literature will be presented, focusing attention on the properties in terms of thermal and chemical structure of composites reinforced with different types of natural fibers, processing behavior and final properties of these fibers with thermoplastic and thermosetting matrices, paying particular attention to the use of physical and chemical treatments for the improvement of fiber-matrix adhesion. We have concluded that different types of reinforcement, such as cellulose and lignin from sugarcane bagasse or coconut fibers, are composite materials that may be useful in engineering applications.

Keywords: Composites, natural fibers, lignocellulosic materials, sugarcane bagasse, coconut fiber, polyurethane, polypropylene, lignin

**Corresponding author*: patriciamileo@debiq.eel.usp.br

Vijay Kumar Thakur (Ed.), Lignocellulosic Polymer Composites, (103–124) 2015 © Scrivener Publishing LLC

6.1 Introduction

There is a growing trend to use natural fibers as fillers and/or reinforces in plastic composites as thermosetting and thermoplastic materials. The use of lignocellulosic materials as reinforcements has received increasing attention due to the improvements that natural fibers can provide such as low density, biodegradability and highly specific stiffness, as well as the fact that these materials are derived from renewable and less expensive sources [1].

Despite the variability in their mechanical properties, natural fibers offer many advantages when compounded with plastic materials. Their low density and abrasiveness, high specific properties, and their ability to be incorporated in composites at high levels make the incorporation of lignocellulosic fibers into thermoplastics and thermosetting matrices highly desirable. Their environmentally-friendly character and technical and economic advantages make these fibers attractive for an increasing number of industrial sectors, including the automotive industry, as replacements for glass fibers [2].

Technological developments connected with consumer demands continues to increase the demand for natural global resources, driving the search for materials with wide availability and environmental sustainability [3]. In Brazil, sugarcane bagasse is a coproduct generated in large volumes by the agricultural industry. A fibrous residue of cane stalks is left over after the sugarcane is crushed to extract the juice. This material is a lignocellulosic by-product of the sugar and alcohol industries and is almost completely used by sugarcane factories as fuel for the boilers [4]. The bagasse is a vegetable fiber mainly composed of cellulose, a glucose polymer with a relatively high modulus that is a fibrillar component of many naturally occurring composites (wood, sugarcane straw and bagasse) in association with lignin [1].

Cellulose, the major constituent of all plant materials, is a linear natural polymer of anhydroglucose units with a reducing end group (right side), linked at the one and four carbon atoms by β-glycosidic bonds [5]. The opening of the chain results in the formation of an aldehyde group and a nonreducing end group (left side) in the form of a secondary alcohol [6].

Besides that, coconut broadly grows in tropical and subtropical regions. Annually, approximately 33 billion coconuts are harvest worldwide with only 15% of these coconuts being utilized for fibers and chips [7,8]. This suggests that there is a considerable availability of this material; also, coconut fibers are quite strong and have high heat resistance [9], which are very interesting characteristics for the fiber used in composite materials.

Composites are highly attractive materials due to their ability to combine physical and mechanical properties that do not naturally occur simultaneously in the same material. Thus, composites are obtained by mixing immiscible materials to generate a final material with combined and/or improved properties. Many of these composites combine low density with high mechanical and chemical resistance, desirable properties for a variety of applications.

Biopolymers such as polyurethane derived from castor oil can serve as matrices for composites reinforced with vegetable fibers. In this study, composites were produced from this polyurethane reinforced with lignin and cellulose fibers, both obtained from sugarcane bagasse. Castor oil is readily available as a major product from castor seeds. Castor oil-based polyurethane is a useful, versatile material and is widely used as an individual polymer possessing a network structure because of its good flexibility and elasticity [10]. This oil is extracted by compressing the seeds or by solvent extraction, and it contains approximately 90% (w/w) ricinoleic acid. This triglyceride contains hydroxyl groups that react with isocyanate groups to form urethane links. In addition, castor oil-based polyurethane always shows good miscibility with natural polymers and their derivatives [10].

Among different thermoplastics, polypropylene (PP) presents outstanding properties such as low density, good flex life, good surface hardness, very good abrasion resistance, and excellent electrical properties. The combination of lignocellulosic material with thermoplastic matrix can present a considerable problem: incompatibility between the polar and hygroscopic fiber and non-polar and hydrophobic matrix. The relationship between the fiber and matrix is one of the determining factors of composites final properties. For example, polypropylene can be used as a matrix and a natural fiber (a recyclable material) can be used for reinforcement. The use of natural reinforcement fibers in polymer matrices has been widely applied, resulting in materials with excellent mechanical and thermal properties [11]. Polypropylene (PP) is a semicrystalline polymer obtained by polyaddition, and it is widely used as a thermoplastic in engineering applications. The addition of fibers in polymer matrices, such as PP, is known to modify the mechanical properties of the resulting composite [11]. The main problem with fiber/matrix composites is the lack of interfacial interaction between the hydrophobic matrix and the hydrophilic fibers [12]. A possible solution to improve the fiber polymer interaction is using compatibilizers and adhesion promoters or submitting the natural fibers to a surface treatment for natural fibers [13].

In this chapter, some results will be presented, focusing attention on the properties in terms of physical and chemical structure of the coconut and sugarcane bagasse fibers, processing behavior and final thermal properties of these fibers with thermoplastics or thermosetting matrices, paying particular attention to the use of physical and chemical treatments for the improvement of fiber-matrix interaction.

6.2 Experimental

The experimental part encompasses how to prepare the fillers and reinforcements from sugarcane bagasse and coconut fibers. First of all, the lignin and cellulose (pulp) were extracted from sugarcane bagasse through chemical treatments as we showed below; the coconut fiber was processed by chemical treatments. The methodology also brought aspects as chemical characterization of fibers, as components content, FTIR and thermal characterization before and after chemical treatments. Finally, we presented the preparation of thermoplastic and thermosetting composites reinforced with different types of fibers and their influence in thermal properties.

6.2.1 Preparation of cellulose and lignin from sugarcane bagasse

Cellulose and lignin were obtained from sugarcane bagasse by pretreatment of the bagasse with sulfuric acid, delignification and separation of the fractions.

Pretreatment of sugarcane straw with sulfuric acid solution: *In natura* sugarcane bagasse with a moisture content of approximately 50% (w/w) was pretreated with 10% (w/v) sulfuric acid solution in a stainless steel reactor at 120°C for 10 min.

Delignification: Pretreated sugarcane bagasse was placed in alkaline medium in a 350 L stainless steel reactor with 150 L of distilled water, 10 kg pretreated sugarcane straw and 30 L of an aqueous solution containing 3 kg NaOH. The reaction was performed at 98-100°C for 1 h under stirring at 100 rpm. The final concentration of the mixture was 1.5% (w/v) NaOH, and the solid/liquid ratio was 1:20 (w/v). After delignification, the mixture was centrifuged at 1700 rpm. Cellulose was washed and dried at room temperature. The lignin, present in black liquor, was precipitated with sulfuric acid (pH 2.0). After precipitation, the lignin was filtered, washed (until neutral pH) and dried at 70°C to obtain dry lignin. Cellulose (pulp) and lignin were applied in composites as reinforcement of thermosetting matrices.

6.2.2 Surface Treatment for Coconut Fibers

Three different procedures for surface modification were applied to the coconut fiber: 1) hot water at 80°C for 2 h; 2) aqueous NaOH 2% (w/v) at 80°C for 2 h; and 3) a sequence of chemical treatments in hot water at 80°C for 2 h, extran 20% (v/v) at 80°C for 2 h, acetone/water 1:1 (v/v) at room temperature and aqueous NaOH 10% (w/v) at room temperature. The yield of treatment was calculated in all cases considering the equation 1. Where $w_{initial}$ is the weight of the fibers before the treatment and $w_{modified}$ is the weight of the fibers after the treatment. The treated and untreated coconut fibers were applied as reinforcement of PP composites.

$$\text{Yield}\ (\%) = \left(\frac{w_{initial} - w_{modified}}{w_{initial}}\right) \times 100 \tag{6.1}$$

6.2.3 Chemical Characterization of Fibers and Lignin

6.2.3.1 Carbohydrates and Lignin Determination

Approximately 1 to 2 g of fibers or lignin (sugarcane bagasse or coconut fibers) of known moisture content was put in contact with 10 mL 72% w/w sulfuric acid and stirred constantly in a thermostatic bath at 45 ± 0.5°C for 7 min. The reaction was interrupted by the addition of 50 mL of distilled water. The sample was then transferred to an Erlenmeyer flask for quantitative analysis following the addition of an additional 225 mL of distilled water. To complete the hydrolysis of the remaining oligomers, the Erlenmeyer flask was covered with aluminum foil and placed in an autoclave at 121°C for 30 min. The resulting suspension was cooled to room temperature and filtered through a filter funnel fitted with rapid filtration paper. The solid was dried to constant weight at 105°C and determined gravimetrically as insoluble lignin (Klason lignin). The resulting filtrate was placed in a volumetric flask and distilled water was added to 500 mL. Samples of the filtrate were used to determine the quantities of acid soluble lignin and carbohydrates (glucans and xylans) in the samples. The pH of the hydrolysate was adjusted to 1-3 with 6.5 mol. L^{-1} NaOH, filtered in a Sep-Pak C18 cartridge, and analyzed by high-performance liquid chromatography (HPLC) in a Shimadzu LC10 chromatograph with an Aminex HPX-87H column at 45°C. The mobile phase was 0.005 mol. L^{-1} H_2SO_4 flowing at 0.6 $mL.min^{-1}$.

6.2.3.2 *Determination of Ashes Content in Lignin*

Approximately 2 g of lignin of known moisture content was weighed in a porcelain crucible that had been previously calcined at 800°C and weighed. The lignin was incinerated at 300°C for 1 h and at 800°C for 2 h. The crucible was then cooled and weighed.

6.2.3.3 *Elemental Analysis of Lignin*

Elemental analysis was carried out at the Institute of Chemistry of São Paulo University at São Carlos. Analysis was performed in a Perkin Elmer device (Elemental Analyser 2400 CHN), and the results were used to calculate the lignin C_9 formulae.

6.2.3.4 *Total Acid Determination in Lignin*

The total acid groups, comprising both phenolic and carboxyl groups, were determined using a titration method. Lignin (0.3 g) was mixed with 30 mL 0.1 mol. L^{-1} sodium bicarbonate for 30 min with constant stirring. The mixture was then filtered, and residual sodium bicarbonate was titrated potentiometrically with 0.1 mol. L^{-1} HCl. The equivalent volume was determined by plotting pH against titration volume. The procedure was repeated in the absence of lignin as a control [2].

6.2.3.5 *Total Hydroxyls in Lignin*

In a stoppered test tube, 0.03 g dry lignin was mixed with 0.24 mL reagent (pyridine/acetic anhydride 10:3); the reagent had previously been bubbled with nitrogen for 10 min [2]. The mixture was placed in a heater overnight at 65°C, after which time 15 mL acetone and 15 mL distilled water were added and the mixture was transferred to an Erlenmeyer flask. The mixture was then left for 1 h to ensure the destruction of any residual acetic anhydride. The acetic acid formed in the reaction was titrated against a standard solution of 0.1 mol. L^{-1} NaOH with phenolphthalein indicator. The procedure was repeated in the absence of lignin as a control.

6.2.3.6 *Phenolic Hydroxyls in Lignin*

Hydroxyl groups were measured using the spectroscopic method [14] and the conductimetric method [15] as described below.

Conductimetric method - Between 0.2 and 0.3 g lignin were suspended in 10 mL ethanol and submitted to magnetic agitation. Then, 5 mL acetone and 15 mL distilled water were added, producing a fine lignin suspension. The suspension was bubbled with nitrogen for 5 minutes and titrated with 0.1 mol. L^{-1} LiOH. The equivalent volume was determined by plotting the conductivity against the titrated volume.

Spectroscopic method - Approximately 0.2 g lignin of known moisture content was dissolved in 50 mL dioxane (96%) and the solution was diluted 1:10 in dioxane and then 1:1 in water and the pH was adjusted to 13 by addition of 1 mol. L^{-1} NaOH. A control was prepared with the same dilution but with the pH adjusted to 1 by addition of 1 mol. L^{-1} HCl. The average lignin value was used to calculate the percent phenolic OH.

6.2.3.7 *Determination of Carbonyl Groups in Lignin*

Carbonyl groups were determined using a spectroscopic method [16]. This method takes advantage of the differential absorption that occurs when carbonyl groups are reduced at the benzylic alcohol corresponding with sodium borohydride. A mixture was prepared from 2 mL of 0.2 mg/mL lignin solution in 96% dioxane and 1 mL 0.05 mol. L^{-1} sodium borohydride solution (190 mg in 100 mL 0.03 mol. L^{-1} NaOH). The mixture was left in darkness for 45 h at room temperature, and its absorbance in the 200-400 nm region was determined and compared to an unreduced lignin solution at the same concentration. The purpose of this method was to distinguish between the carbonyl groups in two model lignin structures: coniferyl aldehyde (4-hydroxy-3-methoxycinnamaldehyde) and acetoguaicon (4-hydroxy-3-methoxyacetophenone); the carbonyl groups are located in the aliphatic chain.

6.2.3.8 *Analysis of the Molecular Weight Distribution of Lignin*

The molecular weight distribution was determined using a chromatographic system with a 57 x 1.8 cm Sephadex G-50 column in 0.5 mol. L^{-1} NaOH. The mobile phase was 0.5 mol. L^{-1} NaOH, and the flow rate was 0.4 mL.min^{-1}. Fractions of 4 mL were collected, and the absorbance of each fraction was read at 280 nm. Sample volumes of 0.4 mL at 2.0 mg/mL were injected. The chromatographic column was previously calibrated with proteins of known molecular weight.

6.2.4 Infrared Spectroscopy (FTIR) Applied to Fibers and Lignin

The samples were dried and analyzed by FTIR from 4000 to 400 cm^{-1}, with 128 scans collected at intervals of 4 cm^{-1}. Then, FTIR analyses were carried out in a Spectrophotometer Thermo Scientific Nicolet 6700 equipped with DTGS detector with diffuse reflectance accessory (DRIFT).

6.2.5 Preparation of Thermosetting and Thermoplastic Composites Reinforced with Natural Fibers

Polyurethane and composites were prepared using the following reagents: polyol from castor oil (*Ricinus communis*), polyurethane prepolymer based on MDI (diphenylmethane diisocyanate), and lignin and cellulose fibers extracted from sugarcane bagasse. To maintain a uniform particle size, the lignin was ground with a mortar and pestle, and the cellulose fibers were crushed. Lignin and cellulose with dimensions smaller than 80 mesh and 60 mesh, respectively, were used.

Prepared samples were placed in a glass mold with three independent rectangular plates: two whole boards (upper and lower) measuring 140 x 100 x 5 mm and a plate of the same dimensions with a hollow center of 100 x 60 x 5 mm. Before the materials were distributed in the mold, the mold was covered with adhesive to facilitate the removal of the composites.

Preparation of polyurethane matrix: In a Petri dish, 9.0 g of the polyol of castor oil and 6.0 g of the prepolymer based on MDI were combined to obtain a polyol (castor oil): diisocyanate mass ratio of 1.5:1.0 for all composites. The homogenized mixture was distributed in the mold to fill the entire space of the intermediate plate. Then, the upper plate of the mold was closed and sealed with adhesive tape. After the material was cured for approximately 24 h, the mold was opened.

Preparation of the composites reinforced with cellulose: the same mass ratio between polyol and diisocyanate was maintained in all composites. After weighing the polyol, the desired mass of cellulose was added, and this mixture was homogenized. The prepolymer was then added and the mixture was homogenized again. Finally, the mixture was distributed into the mold as described previously. Composites were prepared with varying amounts of cellulose from 10 to 40 wt%.

Preparation of the composites reinforced with lignin: Composites reinforced with lignin were prepared using the same procedure as for composites reinforced with cellulose. Composites were prepared with varying amounts of lignin from 10 to 60 wt%.

Preparation of the composites reinforced with 10 wt% and 20 wt% of different types of coconut fiber (treated and untreated) as reinforcement of PP: The composites were prepared in batches of 50 g in a thermokinetic mixer (model MH - 50H) rotating at 5250 rpm.

6.2.6 Scanning Electron Microscopy (SEM)

The samples, fibers and fractured composites surfaces were placed in a holder with the aid of carbon tape and subjected to metallic coating by gold to a thickness of 8 nm under an argon atmosphere using the Bal-Tec MED 020 metal coating equipment. The metallic samples were subjected to microscopic analysis in a LEO 440 SEM operating at 20 kW and using a secondary electron detector.

6.2.7 Thermogravimetric Analysis (TGA)

TGA was performed using an SDT 2960 TGA-DTA instrument, in a N_2 atmosphere, at a heating rate of 10°C.min^{-1} from 30 to 600°C to obtain derivative curves (DTG) for composites and fibers.

6.2.8 Differential Scanning Calorimetry (DSC) Characterization

DSC analyses were carried out using a Perkin Elmer Pyris 1, in a N_2 atmosphere, at a heating rate of 10°C.min^{-1} from room temperature to 200°C. After 3 min isotherm, the samples were cooled to 30°C using liquid N_2. This analysis was applied to both fibers and composites.

6.3 Results and Discussion

6.3.1 Chemical Composition and Characterization of Sugarcane Bagasse and Coconut Fibers

Table 6.1 reports the chemical characteristics of the fibers and lignin. In the chemical hydrolysis, the acid pretreatment solubilized an extended part of hemicellulose and a small amount of lignin was removed during this process. Acid pretreatment, mainly using sulfuric acid and hydrothermal methods based on the autocatalytic action of acetic acid released by hydrolytic cleavage of acetyl groups [17], have been shown to be effective in removing part of the hemicellulose present in *in natura* bagasse. Various effective dilute prehydrolysis pretreatment processes have been developed. Candido [18], reported the pretreatment of sugarcane bagasse with 10% H_2SO_4 at 100°C for 1 h and obtained a pretreated material with less lignin and hemicellulose content, but a significant amount of cellulose was lost.

Due to the pretreatment of sugarcane bagasse, the material obtained in this step was much more susceptible to alkaline delignification. The chemical composition of the delignified pulp is shown in Table 6.1. In this process, a significant fraction of lignin was solubilized in the alkaline medium, and also a small portion of the hemicellulose was degraded. Silva [19], studied the hydrothermal pretreatment of sugarcane bagasse at four different temperatures, followed by delignification with 1.0% NaOH (w/v) at 100°C for 1 h. The process of alkaline delignification removed almost 81% of the lignin present in *in natura* sugarcane bagasse. A small portion of cellulose was also degraded.

According to Gouveia *et al.* [20], the chemical composition of the untreated coconut fiber is also shown in Table 6.1. This table shows the amount of each component present in the coconut fiber, which varies according to the type of fiber and its origin. The amount of each component was determined separately. Sulfuric acid was used to hydrolyze and depolymerize the coconut fiber polysaccharides for chemical characterization as described in experimental part.

Table 6.1 Chemical composition of sugarcane bagasse, pretreated sugarcane bagasse, pulp from bagasse and coconut fibers.

Components (%)	Sugarcane Bagasse	Pretreated Bagasse	Pulp from Bagasse	Coconut Fibers
Process yield (%)	100	65	47	100
Cellulose	47.4 ± 0.8	59.6 ± 1.2	83.5 ± 0.2	29.0 ± 1.7
Polyoses	25.1 ± 0.6	15.0 ± 0.1	5.6 ± 0.1	19.8 ± 1.3
Lignin	23.4 ± 0.2	21.9 ± 0.4	7.5 ± 0.1	41.7 ± 0.2
Extractives	–	–	–	8.6 ± 0.18
Ash	3.3 ± 0.1	2.0 ± 0.1	1.3 ± 0.0	1.3 ± 0.25

According to Luz *et al.* [21], acid hydrolysis can be used for the treatment of fibers. The main effect of hydrolysis on lignocellulosic materials is breakage of the linkages between lignin, hemicellulose and cellulose. This results in dissolution of hemicellulose, structural modification of lignin and a reduction in the particle size of the material.

By analyzing the fractions collected after hydrolysis, one may obtain the total amount of cellulose, hemicellulose and lignin present in the fiber. Thus, the chemical composition (by weight) for coconut fiber was 28.0% cellulose, 19.8% hemicellulose and 41.1% lignin. According to Luz *et al.* [21], the cellulose, hemicellulose and the lignin components are responsible for the thermal and mechanical behavior of the material. Thus, it is extremely important to determine the quantity of each of these components in the fiber.

6.3.2 Chemical Characterization of Lignin Extracted from Sugarcane Bagasse

Table 6.2 reports the physico-chemical analysis of lignin obtained from sugarcane bagasse. Based on the elemental analysis of lignin from sugarcane bagasse pretreated with sulfuric acid and considering a methoxyl content of 14%, as has been previously observed for lignin from sugarcane bagasse [22], the formulae $C_9H_{8.34}O_{3.43}N_{0.184}$ $(OCH_3)_{0.91}$ was calculated, with a C_9 molecular weight of 185.11 $g.mol^{-1}$. The presence of nitrogenous compounds in the lignin is somewhat uncommon. In this case, nitrogen accounts for 1.3% according to the elemental analysis, and this could be related to the incorporation of nitrogenous plant compounds such as amino acids or ureas into the lignin during the pulping process [16].

UV-OH is a measure of the hydrolysis of some ether bonds and cannot be considered in the total OH content. The OH and carbonyl values are in agreement with

Table 6.2 Chemical characterization of lignin.

Components	Content (%)
Moisture	6.92 ± 0.06
Ash	1.72 ± 0.08
Klason Lignin	94.30 ± 0.57
Soluble Lignin	1.90 ± 0.03
Glucans	0.69 ± 0.04
Xylans	0.36 ± 0.03
Total acids	0.17 ± 0.01
Cond. phenolic OH	0.12 ± 0.01
UV phenolic OH	1.79 ± 0.17
Total OH	1.38 ± 0.48
Aliphatic OH	1.26 ± 0.48
Aldehyde carbonyls*	0.53 ± 0.15
Acetoguaiacon carbonyls	4.99 ± 0.305

* Conyferyl aldehyde structures.

published results [16] and were not introduced into the C_9 formulae. These values correspond to 0.55 oxygen atoms or 17% of the total oxygen content in the C_9 formulae. This finding demonstrates that this lignin is not oxidized and that most of the oxygen remains as ether linkages between units.

6.3.3 Modification of Coconut Fibers by Chemical Treatment

As mentioned previously, three different procedures for surface modification were applied to coconut fiber. According to Mohanty [23], chemical modification of coconut fiber can increase the compatibility (adhesion) between the fibers and the PP. By optimization of the interface between the fibers and the matrix, composites with good properties can be obtained.

After treatment with water, there was a weight loss of 9.27% for the non-treated fiber, indicating that a small amount of soluble extractives is removed with water. The soluble extractives removed by water are typically carbohydrates, gums, proteins and inorganic salts. Santiago [24] shows that hot water treatment removes part of the waste material in the coconut fiber, without removing the internal components of the fiber, thus avoiding alterations of the fiber properties.

After treatment with NaOH 2% (w/v), a weight loss of 17.93% was observed with respect to the initial weight of the fiber. Lignin and hemicellulose were partially removed from the fiber, promoting better packing of the cellulose chains. This leads to increased crystallinity of the fiber [1].

With a blend of solvents (hot water, extran 20% (v/v), acetone/water 1:1 (v/v) and aqueous NaOH 10% (w/v)), a weight loss of 41.49% was observed with respect to the initial weight of the fiber. Different types of extractives were removed with each treatment step according to their solubility. According to Iozzi *et al.* [25], extran removes fats, oils, saponins and other similar compounds. Acetone and water remove excessive impurities that may be present at the fiber surface.

Chemical treatment with sodium hydroxide and with the chemical treatment sequence resulted in a change in the color of the fiber (Figures 6.1A and 6.1B). This coloration change was probably associated with the breakage of chemical bonds between the lignin and the extractives. As a result, the compatibility between the fibers and PP matrix increases, favoring the interaction between the fibers and the PP [11].

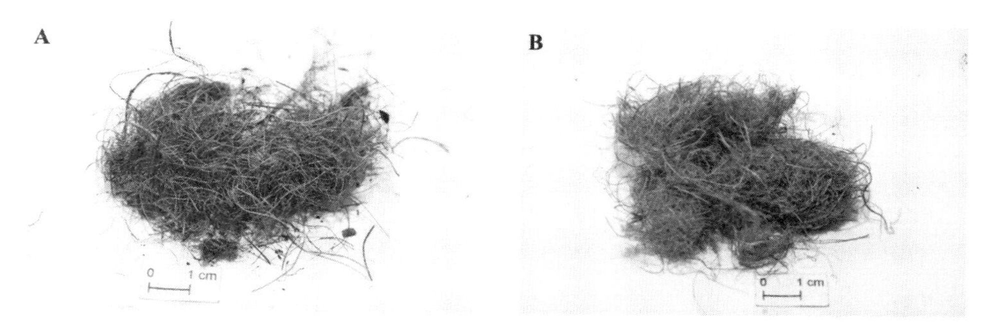

Figure 6.1 Photos of the coconut fibers treated: fiber treated with aqueous NaOH 2% (w/v) (A) sequence of treatment (B).

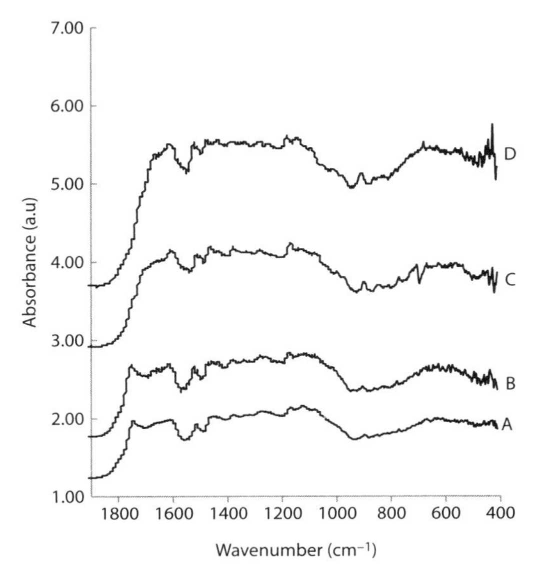

Figure 6.2 FTIR spectra of the treated and untreated coconut fibers: untreated fibers (A), fibers treated with hot water at 80°C (B) fibers treated with aqueous NaOH 2% (w/v) (C) and fibers treated with the chemical sequence (D).

6.3.4 Fourier Transform Infrared Spectrometry Applied to Coconut Fibers

Before and after treatment, the coconut fibers were characterized by FTIR analysis. This investigation uncovered fundamental differences between the untreated and treated fibers. The peak at 1700 cm^{-1} is typical of the carbonyl group found in coconut fiber. This peak was attributed to C = O stretching of the ester linkage between the carboxylic groups of lignin and/or hemicellulose. It was observed a decrease in the C = O peaks (Figures 6.2 C and 2 D) following the chemical treatment sequences. This decrease was likely due to the partial removal of hemicellulose and lignin, confirming the efficacy of the treatment [26, 27]. Each spectra, corresponding to the different treatment types, exhibited a peak at 1515 cm^{-1} that was attributed to the carbon-carbon C = C bond from the aromatic skeletal vibration associated with lignin and hemicellulose [27]. The peak at 1170 cm^{-1} corresponded an asymmetrical stretching of the C-O-C bond in the cellulose, hemicellulose and lignin components [28]. Finally, the vibration peak at 897 cm^{-1} was due to symmetric C-H glycosidic linkages between hemicellulose and cellulose. It was observed an increase in the intensity of the peaks in spectra C and D, indicating a greater exposure of the cellulose due to removal of amorphous constituents during the chemical treatments [21].

6.3.5 Composites with Thermoplastic and Thermosetting as Matrices

6.3.5.1 *Coconut Fibers*

Shortly after drying, the untreated or modified coconut fibers were mixed with the PP matrix to obtain the composites shown in Figures 6.3A, 6.3B and 6.3C. Leão [11], the

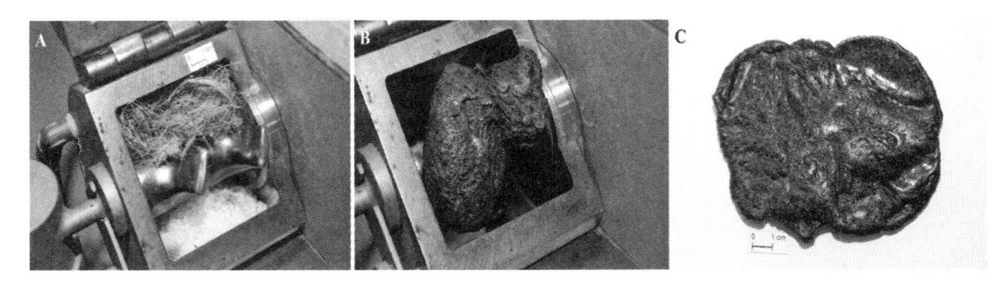

Figure 6.3 Processing of composites in the mixer thermokinetics (A and B) and composites of PP reinforced with coconut fibers (C).

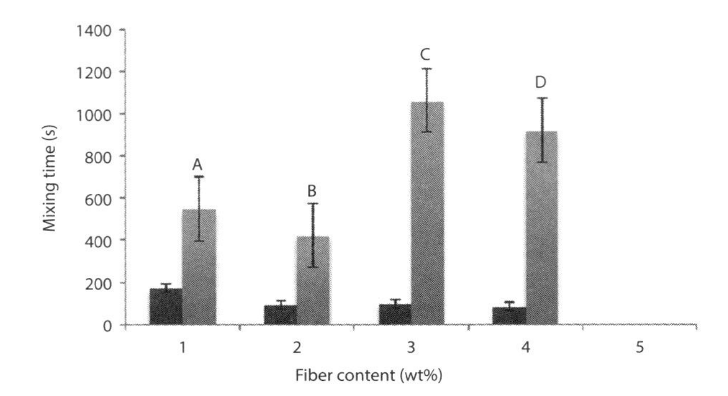

Figure 6.4 Mixing time production as a function of coconut fiber content for: A) *In nature* coconut fiber 10 and 20%/ PP; B) Coconut fiber treated with water 10 and 20%/ PP; C) Coconut fiber treated with NaOH 10 and 20%/ PP; D) Coconut fiber treated with sequence* 10 and 20%/ PP.

thermokinetic mixer works by turning at high speeds to process fiber and PP batches. Mixing between the polymer matrix and the fiber is promoted by melting due to the heating of the system.

Mixing time production for the fiber and the PP matrix varied according to fiber type and content. Figure 6.4 shows the composite mixing time preparation (in seconds) as a function of fiber content (wt%).

The mixing times for composites with 10% and 20% (by weight) fibers were 171 s and 1065 s, respectively. According to Luz *et al.* [21], an increase in the mixing time for composites with 20 wt%/ PP is expected to increase the contact between fibers. Thus, the matrix does not easily contact the wall of the bipartite capsule, complicating matrix melting and fiber incorporation. Furthermore, long mixing times may cause excessive breakdown and thermal degradation of fibers that negatively affect the composite properties [1-11].

6.3.6 Morphological Characterization for Composites Reinforced with Cellulose and Lignin from Sugarcane Bagasse and Coconut Fibers

The obtained composites and SEM micrographs of fractured surfaces are shown in Figure 6.5. No agglomerated lignin could be observed in the pictures, suggesting a well-blended material. In contrast, fibers can be clearly observed in the images of the composites reinforced with cellulose, as well as a poor dispersion of the fibers in the matrix

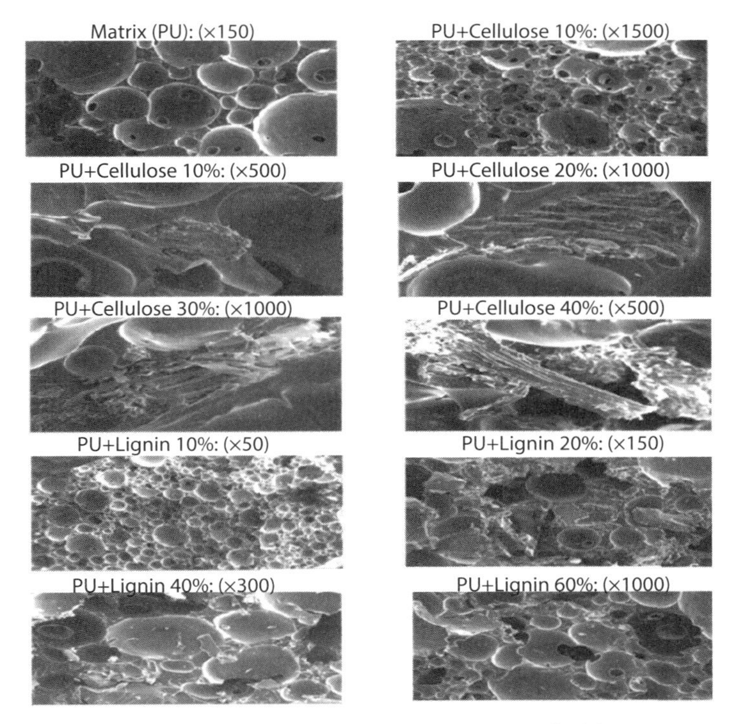

Figure 6.5 SEM micrographs of fractured surfaces of the PU matrix and of the composites reinforced with cellulose and lignin.

and bubbles. The bubbles are mainly formed due to the hydrophilic character of the natural fibers and the high reactivity of the polyurethane with water or moisture [29].

Mileo *et al.* [30] studied the use of cellulose obtained from sugarcane straw as reinforcement for castor oil polyurethane composites. The fractographic analysis showed a poor dispersion of the fibers in the matrix, fractured fibers and good fiber-matrix adhesion. Silva [19] investigated the fracture toughness of sisal and coconut fiber/castor oil polyurethane, and they observed the presence of bubbles in the obtained composites, despite taking great care during the composite processing. SEM analysis of fiber morphology verified our FTIR results.

Figure 6.6 shows representative micrographs for untreated coconut fibers, fibers treated with hot water at 80°C, fibers treated with aqueous NaOH 2% (w/v) and fibers treated with the chemical sequence.

The micrographs of the untreated fibers (Figures 6.6 A and 6.6 A1) showed a rough fiber surface covered by layers of wax and extractives [21]. The micrographs of the fibers treated with hot water (Figures 6.6 B and 6.6 B1) showed that some impurities were removed from the surface layer, thus increasing the contact area by exposing fibrils and globular marks [24]. Ferraz [31] shows that a decrease in extractives present in the fibers after hot water treatment improves the compatibility between the fibers and the matrix.

The micrographs of the fibers treated with aqueous NaOH 2% (w/v) (Figures 6.6C and 6.6C1) showed a rough fiber surface and disaggregation of the fibers into microfibrils. This treatment breaks the hydrogen bonds that connect the cellulose chains, conferring a rough surface that improves mechanical anchoring. The appearance of pores or orifices was also observed all over the fiber surface. This may increase the effective surface for contact with the polymeric matrix [32].

In the treatment with the chemical sequence, the appearance of pores or orifices was observed throughout the fiber surface (Figures 6.6 D and 6.6 D1). As with the

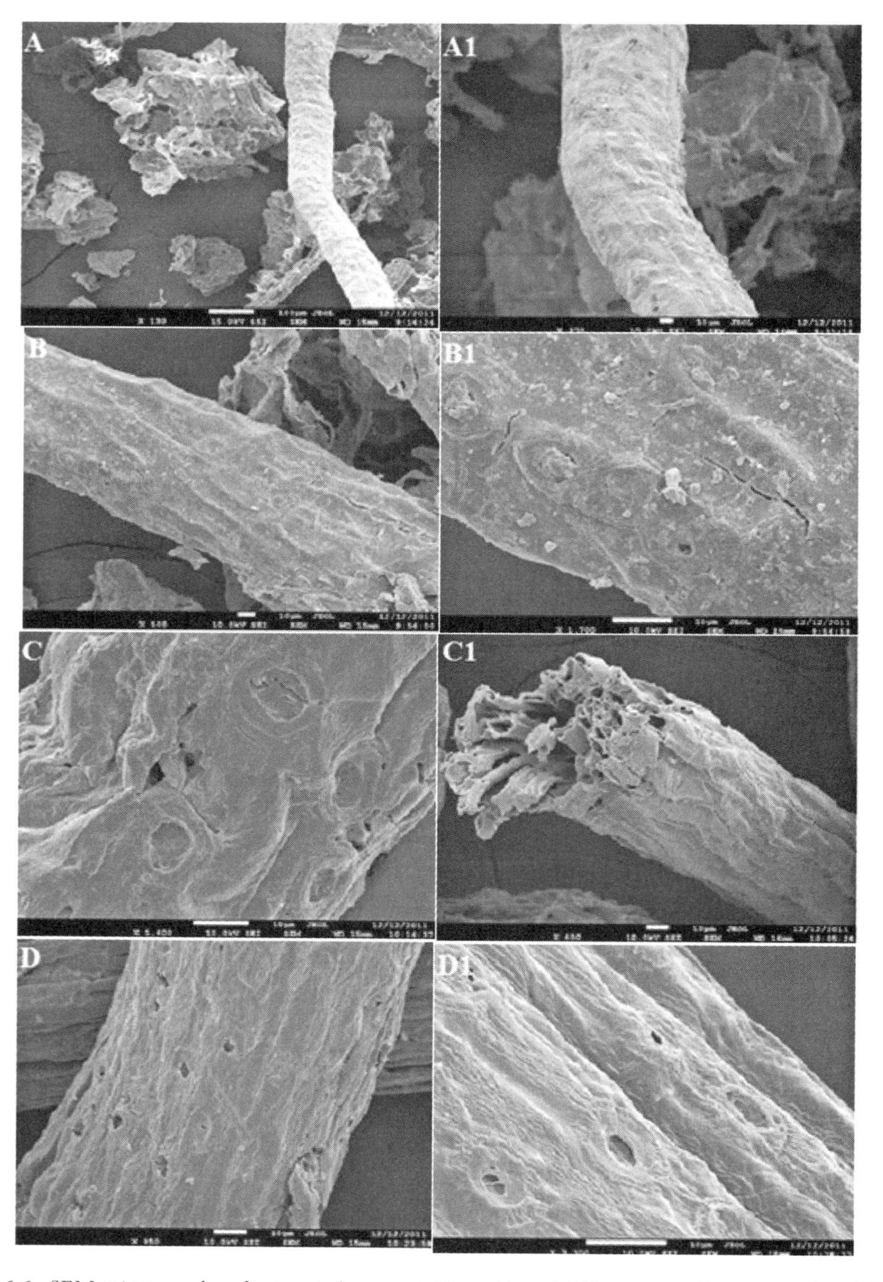

Figure 6.6 SEM micrographs of untreated coconut fibers (A and A1), coconut fibers treated with hot water 80°C (B and B1), coconut fibers treated with NaOH 2% (C and C1) and coconut fibers treated with the chemical sequence (D and D1).

NaOH treatment, this may also increase the effective surface for contact with the polymer matrix [11]. According to Ferraz [31], the treatment sequence removes the surface layer of the fiber, resulting in partial loss of lignin, hemicellulose and extractives. The orifices present on the fiber surfaces were caused by fiber degradation (Figures 6.6 D and 6.6 D1).

6.3.7 Thermogravimetric Analysis for Composites and Fibers

The addition of cellulose fibers (30%) and lignin (40%) to the polyurethane matrix increased the stiffness in the resulting composites when compared with the pure matrix. Cellulose and lignin percentages above these rates lead to a decrease in stiffness. Table 6.3 shows the weight losses of the composites, matrix and fibers. The polyurethane hardly loses weight up to 300°C, but above this temperature the weight loss is rapid and severe. This table also shows that polyurethane obtained from castor oil and lignocellulosic fibers do not lose much mass at low temperatures. Cellulose and lignin undergo significant mass loss above 300°C.

Figure 6.7 presents TGA curves of polyurethane, cellulose, lignin and composites reinforced with cellulose and lignin. The decomposition of the polyurethane takes place in one step, while the composites decompose in two steps. In the first stage, decomposition occurs after the decomposition of the fiber and matrix alone. The small weight loss up to 100°C is related to the moisture content of the material. Above this temperature, some water molecules that are tightly bound to the fibers become volatile [21]. Thus, reinforcement decreases the thermal stability of the composite.

After mixing, part of composite was analyzed by TGA and DTG analyses. Figure 6.8 shows the DTG curves for untreated and treated coconut fibers. A three stage weight

Table 6.3 Weight loss of the fibers, matrix and composites.

Samples	Weight loss (%)				
	100°C	200°C	300°C	400°C	500°C
Lignin	4.0	5.0	13.4	32.6	47.1
Cellulose	5.2	5.5	12.0	77.0	93.5
Polyurethane (PU)	0.3	0.6	4.9	31.0	73.6
Lignin 10 wt% / PU composite	0.8	1.3	5.7	34.7	72.3
Lignin 20 wt% / PU composite	0.8	1.6	6.8	39.4	72.4
Lignin 30 wt% / PU composite	1.1	1.8	6.7	38.2	70.2
Lignin 40 wt% / PU composite	1.4	2.5	8.4	38.4	67.8
Lignin 50 wt% / PU composite	1.6	2.6	8.4	40.1	66.2
Lignin 60 wt% / PU composite	2.0	3.0	8.7	39.5	64.1
Cellulose 10 wt% / PU composite	0.6	1.1	4.5	34.2	73.7
Cellulose 20 wt% / PU composite	0.9	1.7	5.2	38.3	74.1
Cellulose 30 wt% / PU composite	0.9	1.8	5.2	44.7	77.6
Cellulose 40 wt% / PU composite	1.3	2.6	5.7	45.3	74.1

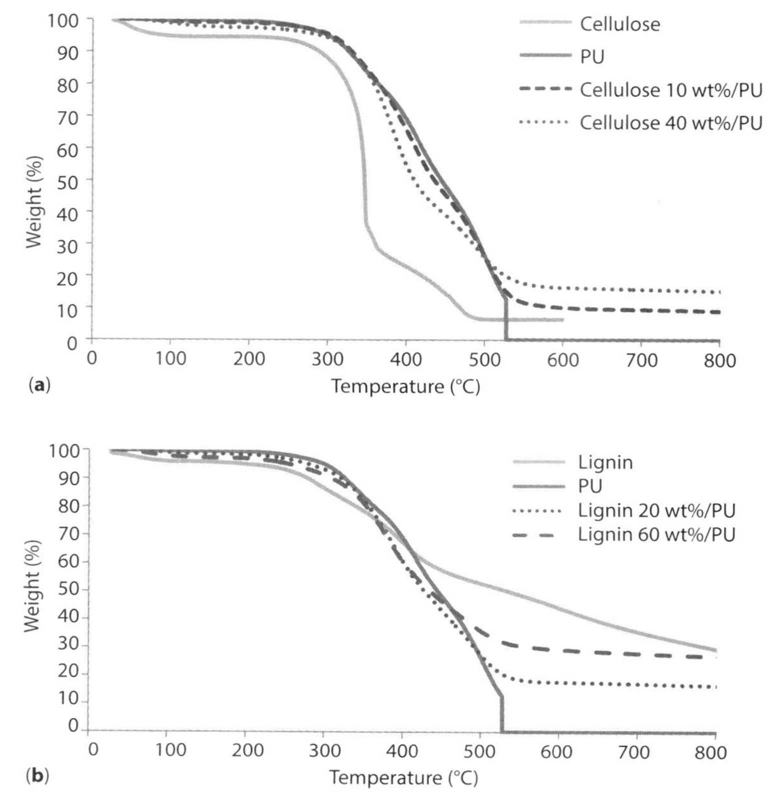

Figure 6.7 TGA curves of polyurethane, fibers and fiber-reinforced composites: (a) cellulose; (b) lignin.

loss was observed for the fibers. The first weight loss occurred between 50 and 100°C, in proportion to the water heat vaporization of the sample. The second stage of weight loss, which was observed between 265 and 282°C, was attributed to lignin degradation, due to the breakdown of ether and carbon–carbon linkages. The third decay stage presented thermal degradation peaks around 327°C [21]. This weight loss was attributed to thermal depolymerization of hemicellulose and cleavage of the glucosidic linkages of cellulose. Ash produced by coconut fibers consists mainly of inorganic compounds. All fibers had a residual weight of approximately 27% at 600°C [11].

The composites reinforced with treated fibers were evaluated by thermogravimetric analysis (TGA) for verification of their thermal characteristics. Figure 6.9 shows the TGA curves for composites reinforced with coconut fiber 20 wt%/ PP compared to modified coconut fibers alone and PP alone.

The composite composed of coconut fibers with 20 wt%/ PP was less stable than PP alone. The thermal stability curves for the coconut fiber/ PP composites fell between the fiber and matrix TGA curves. Figure 6.9 also shows that PP alone exhibited a one-step decomposition process, presenting higher thermal stability than the composites. The thermal stability of PP alone was approximately 337°C. In contrast, the composites clearly showed a two-step decomposition process. The first peak corresponds to lignin decomposition, and the second peak (PP) corresponds to the highest rate of decomposition at 455°C [32].

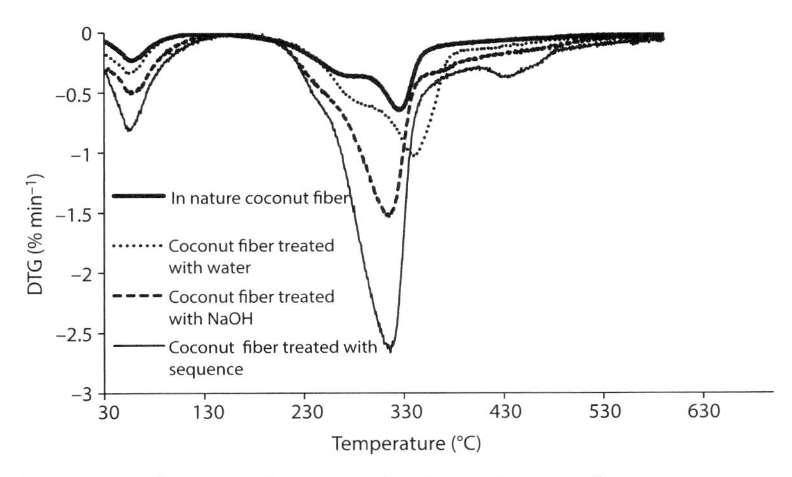

Figure 6.8 Comparative DTG curves for untreated and treated coconut fibers.

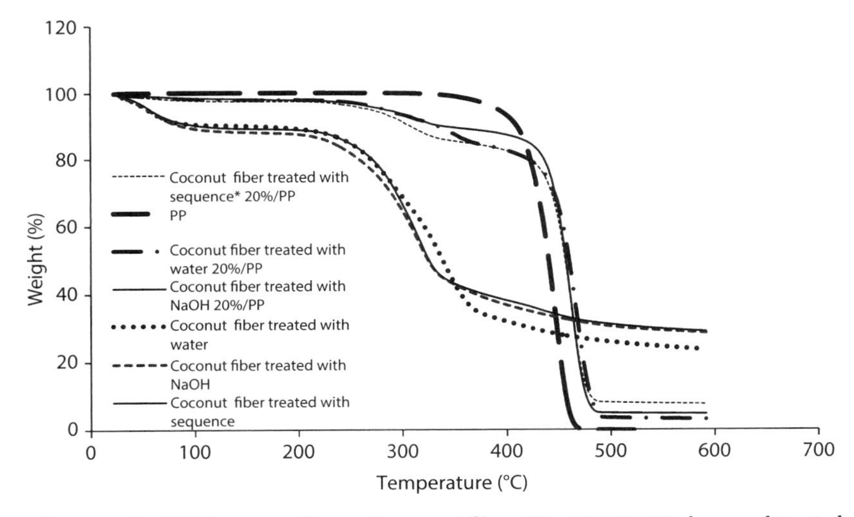

Figure 6.9 Comparative TGA curves of treated coconut fibers 20 wt%/ PP, PP alone and treated coconut fibers alone.

The weight loss for the composite material started at 100°C, and for polypropylene started at 300°C. At 400°C, the weight loss increased with fiber content. Table 6.4 shows that the addition of fiber decreased the ultimate thermal stability in comparison to polypropylene alone.

The peak temperatures for thermal degradation of the composites and the PP matrix occurred at approximately 458°C and 450°C, respectively. Table 6.4 shows that the addition of fibers caused a significant increase in the peak temperature for composite degradation, reaching 466°C for the composites reinforced with 20 wt% fibers treated with hot water.

Table 6.4 Weight loss at different temperatures and degradation temperature peak of polypropylene and fibers/PP composites.

Sample	Weight loss (%)					Degradation temperature peak (°C)
	100°C	200°C	300°C	400°C	500°C	
Polypropylene	0.0	0.0	0.4	6.5	99.7	450.5
Coconut fiber *in nature* 10%/PP	0.5	1.0	3.9	8.9	97.7	459.4
Coconut fiber *in nature* 20%/PP	1.3	1.9	6.7	6.7	93.8	463.1
Coconut fiber treated with water 10%/PP	0.5	1.0	3.3	9.1	98.6	459.3
Coconut fiber treated with water 20%/PP	1.6	1.9	6.6	17.4	96.4	466.8
Coconut fiber treated with NaOH 10%/PP	0.6	0.7	3.2	7.6	95.8	458.2
Coconut fiber treated with NaOH 20%/PP	1.3	2.0	6.3	12.8	95.0	460.6
Coconut fiber treated with sequence* 10%/PP	0.8	1.0	3.6	7.8	96.4	459.3
Coconut fiber treated with sequence* 20%/PP	2.2	2.4	8.9	17.4	91.9	461.7

6.3.8 Differential Scanning Calorimetry Studies for Composites and Fibers

The treated and untreated coconut fibers were also analyzed by DSC. There was a significant change in the crystallization temperature of PP after the addition of fibers, as shown in Figure 6.10. With the addition of 20 wt% fibers treated with hot water at 80°C, there was a significant change in the crystallization temperature profile, as can be observed in Figure 6.10. In this curve, there was an appearance of two crystallization peaks that were coupled, with the first crystallization peak appearing at approximately 109°C and the second one appearing at approximately 124°C.

According to Luz *et al.* [21], this can be explained by interactions between the fibers and the matrix. It is possible that establishment of a layer with different crystallinity, known as a transcrystalline layer, occurs along the fiber where it is in contact with the matrix. This layer provides points for crystallization upon addition of fibers. Another explanation is chemical modification of the lignocellulosic materials. If lignin and hemicellulose were still present on the fibers, they may have contributed to the formation of a transcrystalline layer that differentiated from the rest of the material. Leão [11] evaluated the effectiveness of the chemical treatments by thermal analysis. Treated fibers had higher tensile strength than untreated fibers, a phenomenon attributed to the removal of hemicellulose and lignin. Chemical treatment increases the degree of

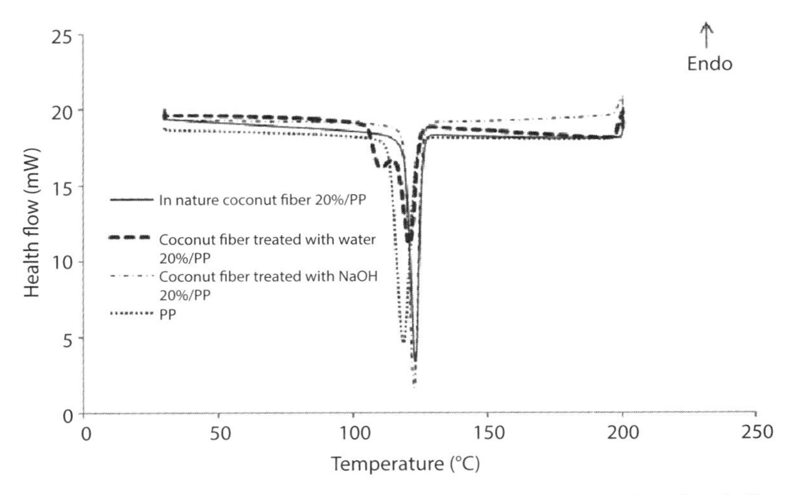

Figure 6.10 DSC curve for (A) melting of the polypropylene composite reinforced with fibers treated with hot water and (B) crystallization.

Table 6.5 DSC properties of PP alone and coconut fiber/ PP composites.

PP and coconut fiber/PP composites	**$\Delta H_c (J/g^{-1})$**	**$\Delta H_m (J.g^{-1})$**	**X_c(%)**
Polypropylene	- 98.3	89.9	65.1
Coconut fiber *in nature* 10%/PP	- 86.4	68.9	55.4
Coconut fiber *in nature* 20%/PP	- 73.6	60.4	54.7
Coconut fiber treated with water 10%/PP	- 87.0	77.7	62.5
Coconut fiber treated with water 20%/PP	-10.5/-76.7	78.1	70.7
Coconut fiber treated with NaOH 10%/PP	- 94.5	77.9	62.7
Coconut fiber treated with NaOH 20%/PP	- 81.2	74.3	67.3
Coconut fiber treated with sequence* 10%/PP	- 82.5	70.7	56.9
Coconut fiber treated with sequence* 20%/PP	- 83.5	74.1	67.1

crystallinity of the fibers and facilitates the separation of the cellulose microfibrils. In addition, the mechanical properties can be improved, which is related to the stiffness, increasing with crystallinity.

The results from the DSC analysis are given in Table 6.5, where the thermal properties such as crystallization heat (ΔH_c), melting heat (ΔH_m) and percentage of crystallinity (X_c) are listed. The crystallinity of the PP component was determined using the equation 2, below:

$$X_c \text{ (\% crystallinity)} = \frac{\Delta H_m \times 100}{\Delta H_m^0 \times w} \tag{6.2}$$

A value of ΔH^o_m 138 J/g was used for 100% of crystalline isotactic PP. The mass fraction of PP in the composite is represented by *w* [1-21].

The crystallinity of PP increased with the addition of coconut fibers. However, the crystallinity also increased because the fiber surface acts as a nucleation site for the crystallization of the polymer, promoting the development and formation of transcrystalline regions around the fiber.

The composite reinforced with coconut fibers presented an increase in the enthalpy of crystallization (ΔH_c) of the PP phase with coconut fiber, indicating that fibers promoted the crystallization process [34] (Table 6.5). This result may be explained by considering the ability of the fibers for nucleating the crystallization of polypropylene. We also observed a decrease in the enthalpy of melting (ΔH_m) for the composites with respect to PP alone. It is clear from Table 6.5 that the addition of coconut fiber to PP results in an increase in X_c of the PP matrix. The (X_c) for composites reinforced with modified fibers remained independent of the treatment. Therefore, the composites reinforced with coconut fibers treated with hot water 20%/ PP showed the best results. This treatment works quickly and is lower in cost than the other treatments.

6.4 Conclusions

Research on biodegradable polymer composites containing lignocellulosic fibers is receiving increasing attention due to dwindling petroleum resources, the low costs of lignocellulosic reinforcements, and their wide variety of properties and increasing ecological concern.

This study showed that the pretreatment step for sugarcane bagasse solubilized a great portion of polyose and a small amount of lignin. The process of delignification solubilized a great fraction of lignin and a small portion of polyose. The presence of lignin or cellulose fibers as reinforcements in a matrix of polyurethane obtained from castor oil seems to improve the thermal properties of the material. Samples of composites reinforced with lignin showed a more homogenous visual appearance by SEM than those reinforced with cellulose. Composite materials might also improve other properties of the final composite, as the composites exhibit a greater thermal stability than the fiber and matrix alone. However, the addition of reinforcement decreases the thermal stability of the composite.

For the coconut fiber, the chemical treatments resulted in discoloration of the fibers, indicating that impurities such as lignin, hemicellulose and extractives were removed from the fibers. Chemical modification was necessary to minimize the hydrophilicity of the fibers and to increase the interfacial adhesion of polar fibers with the nonpolar matrix. FTIR analysis verified that the chemical modification of the coconut fibers was successful, as evidenced by a reduction in the intensity of the characteristic bands for lignin and hemicellulose. SEM allowed us to evaluate the morphology and dimensions of modified and unmodified fibers and indicated that there were considerable morphology changes after the chemical treatments. Thermogravimetric analysis showed that the composites presented higher degradation temperatures than fibers. The DSC results demonstrated that the incorporation of 20% by weight of coconut fibers treated with hot water into PP resulted in an increase in T_c, ΔH_c and X_c. This finding proved that interactions were taking place between the fibers and the PP matrix. Changes in the fiber surface provided by these treatments caused a greater interaction between the fibers and the matrix. This was confirmed by observations of increased crystallinity for the composites. Thus, chemical modification of coconut fibers for use in composite materials may be useful in engineering applications.

Acknowledgements

The authors would like to thank FAPESP, CAPES, CNPq and DPP/UnB for the financial support.

References

1. S.M. Luz, A.R. Gonçalves, and A.P. Del'Arco Jr, Mechanical behavior and microstructural analysis of sugarcane bagasse fibers reinforced polypropylene composites. *Compos. A* 38,1455–1461 (2007).
2. L. Averous, and N. Boquillon, Biocomposites based on plasticized starch: thermal and mechanical behaviours. *Carbohydr. Polym*, 56, 111–122 (2004).
3. O. Faruk, A.K. Bledzki, H.P. Fink, and M. Sain, *Biocomposites reinforced with natural fibers: 2000–2010. Prog. Polym. Sci. 37,* 1552–1596 (2012).
4. M.G. Rasul, V. Rudolph, and M. Carsky, *Physical properties of bagasse. Fuel* 78, 905–910 (1999).
5. R.P. Wool, and X.S. Sun, *Bio-based Polymers and Composites*, Elsevier Academic Press, Burlington, MA, (2005).
6. D. Liu, H. Tian, L. Zhang, and P.R. Chang, Structure and properties of blend films prepared from castor oil-based polyurethane/soy protein derivative. *Ind. Eng. Chem. Res.* 47, 9330–9336 (2008).
7. S.N. Monteiro, L.A.H. Terrones, and *J.R.M. D'Almeida, Mechanical performance of coir fiber/polyester composites. Polym. Test.* 27, 591–595 (2008).
8. W. Wang, and G. Huang, Characterisation and utilization of natural coconut fibres composites. *Mater. Des.* 30, 2741–2744 (2009).
9. J.Y. Jang, T.K. Jeong, H.J. Ogh, J.R. Youn, and Y.S. Song, Thermal stability and flammability of coconut fiber reinforced poly (lactic acid) composites. *Compos. B* 43, 2434–2438 (2012).
10. G. Toriz, P. Gatenholm, B.D. Seiler, and D. Tindall, Cellulose fiber-reinforced cellulose Esters: Biocomposites for the future. *Nat. Fibers Biopolym. Biocompos. (*2005).
11. R.M. Leão, Engenharia Mecânica, Universidade de Brasília, Master dissertation, (2012).
12. S.J. Kim, J.B. Moon, G.H. Kim, and C.S. Ha, Mechanical properties of polypropylene/natural fiber composites: Comparison of wood fiber and cotton fiber. *Polym. Test.* 27, 801–806 (2008).
13. S. Kalia, B.S. Kaith, and I. Kaur, Pretreatments of natural fibers and their application as reinforcing material in polymer composites—A review. *Polym. Eng. Sci.* 49, 1253–1272 (2009).
14. K.V. Sarkanen, and C.H. Ludwig, *Lignins: Ocurrence, Formation, Structure and Reactions,* John Wiley, New York, (1971).
15. G.C. Quintana, G.J.M. Rocha, A.R. Gonçalves, and J.A. Velásques, Lignins for heavy metal removal. *Bioresources* 3, 1092–1102 (2008).
16. G.J.M. Rocha, C. Martin, I.B. Soares, A.M. Souto-Maior, H.M. Baudel, and C.A.M. Abreu, *Biomater. Bioeng,* 35, 663–670 (2011).
17. R.A.A. Nascimento, Engineering College at Lorena, USP, Master's these, (2007).
18. R.G. Candido, Engineering College at Lorena, USP, Master's these, (2011).
19. V.F.N. Silva, Engineering College at Lorena, USP, Master's these, (2009).
20. E.R. Gouveia, R.T. Nascimento, A.M. Souto-Maior, and G.J.M. Rocha, Validação de metodologia para a caracterização química de bagaço de cana-de-açúcar. *Química Nova,* 32, 1500–1503 (2009).

21. S.M. Luz J. Del Tio, G.J.M. Rocha, A.R. Gonçalves, and A.P. Del'Arco. Cellulose and cellulignin from sugarcane bagasse reinforced polypropylenecomposites: Effect of acetylation on mechanical and thermal properties. *Compos. A* 39, 1362–1369 (2008).
22. R.V. Silva, D. Spinelli, W.W. Bose Filho, S. Claro Neto, G.O. Chierice, and J.R. Tarpani, Fracture toughness of natural fibers/castor oil polyurethane composites. *Compos. Sci. Technol,* 66, 1328–1335 (2006).
23. S. Mohanty, and S. K. Nayak, Dynamic and steady state viscoelastic behavior and morphology of MAPP treated PP/sisal composites. *Mater. Sci. Eng.* 443, 202–208 (2007).
24. B.H. Santiago, and P.V.P. Selvam, Tratamento superficial da fibra de coco: estudo de caso baseado numa alternativa econômica para fabricação de materiais compósitos. *Revista Analytica,* 26, **42–45** (2007).
25. M.A. Iozzi et al, *Ciência e Tecnologia*, 20, 25–32 (2010).
26. P.J. Herrera-Franco, A. Valadez-González, A study of the mechanical properties of short natural-fiber reinforcedcomposites. *Compos. B* 36, 597–608, (2006).
27. M. Jonoobi, J. Harun, A. Shakeri, M. Misra, and K. Oksmand Chemical composition, crystallinity, and thermal degradation of bleached and unbleached kenaf bast (Hibiscus cannabinus) pulp and nanofibers. *BioResources* 4, 626–639 (2009).
28. M.F. Rosa, E.S. Medeiros, J.A. Malmonge, K.S. Gregorski, D.F. Wood, L.H.C.Mattoso, G.Glenn, W.J. Orts, and S.H. Ima, Cellulose nanowhiskers from coconut husk fibers: Effect of preparation conditions on their thermal and morphological behavior. *Carbohydr. Polym.* 8, 83–92 (2010).
29. P.C. Mileo, D.R. Mulinari, C.A.R.P. Baptista, G.J.M. Rocha, and A.R. Gonçalves, Mechanical behaviour of polyurethane from castor oil reinforced sugarcane straw cellulose composites. *Proceed. Eng.* 10, 2068–2073 (2011).
30. P.C. Mileo, Engineering College at Lorena, USP, Master's these, (2011).
31. J.M. Ferraz, Engineering Florestal, Universidade de Brasília, Master dissertation, (2011).
32. J.M. Rezende, F.L. Oliveira, and D.R. Mulinari, Hybrid composites evaluation to be applied in engineering. *Cadernos UniFOA* 15 (2011).
33. C.G. Mothé, and A.D. Azevedo, *Análise térmica de materiais*. São Paulo (Ed.), pp. 324, Artliber. (2009).
34. P.V. Joseph, K. Joseph, S. Thomas, C.K.S. Pillai, V.S. Prasad, G. Groeninckx, and M. Sarkissova, The thermal and crystallisation studies of short sisal fibre reinforced polypropylene composites. *Compos. A Appli. Sci. Manuf.* 34, 253–266 (2003).

7

Cellulose-Based Starch Composites: Structure and Properties

Carmen-Alice Teacă*, Ruxanda Bodîrlău and Iuliana Spiridon

"Petru Poni" Institute of Macromolecular Chemistry, Iași, Romania

Abstract
Starch is one of the most promising renewable biopolymers due to its versatility, low cost and applicability to the development of new biomaterials. Cellulose is the most abundant biopolymer on earth and is present in a wide variety of forms. The addition of cellulose fibers is an effective way to stabilize starch-based films, which are known to be very sensitive to air humidity. Cellulose fibers strongly interact with plasticized starch matrix and increase the crystalline feature of composite films, having a positive effect on their mechanical properties. Starch microparticles (StM) from corn starch can be modified by reaction with organic acid to obtain modified starch microparticles (CMSt) by the dry preparation technique. Starch composites may be prepared using CMSt as the filler within a glycerol plasticized corn starch matrix by the casting/solvent evaporation process. Composite films can be obtained by the addition of cellulose originating from different sources as filler within the CMSt/St matrix.

Keywords: Starch, cellulose, composite films, properties

7.1 Introduction

In recent years, an increased interest has been noticed for new composite materials obtained at relatively low cost and presenting significant performance properties, the research being mainly focused on environmentally-friendly materials. Some comprehensive overviews have presented many aspects related to the development of composite materials from renewable resources with special attention on the biodegradable polymers, processing methods, structure, morphology and properties [1-5].

Biodegradable polymers have received much more attention in the last decades [6-9] due their potential applications in the areas related to environmental protection (e.g., packaging, agriculture, etc.) and the maintenance of physical health (e.g., medicine - biomaterials for regenerative therapies; drug release).

Interest in research and development issues related to biopolymers has increased in view of the fact that they play a significant role in global environmental awareness

**Corresponding author*: cateaca14@yahoo.com

Vijay Kumar Thakur (Ed.), Lignocellulosic Polymer Composites, (125–146) 2015 © Scrivener Publishing LLC

because they reduce carbon dioxide emissions and the dependence on fossil fuels. Thus, there is a worldwide increasing demand for the production of polymers derived from renewable resources.

7.2 Starch and Cellulose Biobased Polymers for Composite Formulations

Biosourced products are experiencing renewed interest with diminishing fossil fuels and as researchers work on developing a biorefinery-based economy. Biopolymers are polymers formed in nature during the growth cycles of all living organisms, and are referred to as *natural polymers*. These are synthesized through enzyme-catalyzed polymerization reactions of monomers in living cells by complex metabolic processes.

Generally, polymers from renewable resources have different origins such as: *natural* (e.g., polysaccharides – namely cellulose and starch, which are produced in large amounts; protein; gums), *synthetic* (e.g., polylactic acid, PLA) derived from natural monomers, and *microbial* (e.g., polyhydroxybutyrate, PHB) [1, 5] The main components of biomass are cellulose, lignin, hemicelluloses and extractives and, as a non-wood structural component, starch.

For composite materials applications, the main useful polysaccharides are cellulose and starch, but more complex carbohydrate polymers originated from bacteria and fungi (e.g., exo-polysaccharides such as xanthan, curdlan, pullulan, levan and hyaluronic acid) have attracted increased interest in the last years due to their outstanding potential for various industrial areas [10-14].

Cellulose is a linear structural polysaccharide and the most abundant biorenewable natural polymer produced by plants. Cellulose can be obtained from a variety of sources including seed fibers (cotton), wood fibers (from hardwood and softwood species), plant fibers (flax, hemp, jute, ramie), grasses (bagasse, bamboo), algae (*Valonia ventricosa*), and bacteria (*Acetobacter xylinum*) [15, 16]. As a versatile biopolymer, it represents a valuable raw material with fascinating structure and properties. It is a hydrophilic, biodegradable, highly crystalline polymer with high molecular mass. Cellulose has no thermally processing ability because it degrades before it melts. It is insoluble in water and most organic solvents, but is soluble in ionic liquids [17].

In its natural state, cellulose is highly crystalline in structure. It is a polydisperse linear stiff-chain homopolymer composed of the glucose building blocks which form hydrogen-bonded supramolecular structures. These strong hydrogen bonds are responsible for the stiff, linear shape of the cellulose polymer chains [18].

The elementary cellulose component of wood cell walls is represented by microfibrils which form lamellas or bundles, namely macrofibrils. Under the degradation process, wood first generates macrofibrils, then well-defined, homogeneous microfibrils (MFC) and finally, fibrils [19-21]. The defibrillation of the wood cell wall is obtained using a homogenous mechanical treatment in combination with some appropriate pretreatment, such as enzymatic treatment [22, 23], carboxymethylation [21], or hydrolysis [24].

Microfibrillated cellulose (MFC) originates from many various botanical sources, as well as algae and tunicate animals. It is also well-known that a certain type of bacterium

(*Gluconacetobacter xylinus*) produces a three-dimensional network of bundles of cellulose fibrils. Pure sheets of bacterial cellulose (BC) can be used in composites without any further disintegration [25].

At nanoscale thickness, fibrils are named as nanofibrillar cellulose (NFC) or nanocellulose [26-28]. Highly crystalline cellulose nanowhiskers, also called cellulose micelles, cellulose nanorods, cellulose nanocrystals, or nanocrystalline cellulose are produced under strong acid hydrolysis (e.g., mineral acids such as hydrochloric acid or sulphuric acid) combined with mechanical shearing [22, 29].

The current knowledge related to the structure and chemistry of cellulose, and the development of innovative cellulose derivatives for different applications (coatings, films, membranes, building materials, pharmaceuticals, foodstuffs), as well as the new perspectives, including environmentally-friendly cellulose fiber technologies, bacterial cellulose biomaterials, in-vitro syntheses of cellulose, and cellulose-based biocomposites were highlighted in several important works [30-34].

Starch represents an inexpensive and natural renewable polysaccharide, which was widely investigated as a substitute for petroleum-derived plastics mainly as thermoplastic starch (TPS) [5, 35- 43].

Starch is produced during the photosynthesis and functions as the main polysaccharide reserve source in plants. Starch is deposited in the form of complex structures termed granules, with varied shapes and average sizes in roots, seeds, tubers, stems, leaves and fruits of plants, depending on the botanical origin. It is commercially produced worldwide from corn, wheat, potato, rice and tapioca, these plants producing a large amount of starch. Starch occurs as semi-crystalline small particles, which are insoluble in water at room temperature. Starch is a homopolymer of α-D-glucose, consisting in fact of two polysaccharides, namely amylopectin, and amylose [44, 45]. In its native form, starch is biodegradable in water and is relatively stiff and brittle. However, when starch is heated in the presence of water, the polymer chains are forced apart and starch-starch interactions are replaced by hydrogen bonds between starch and water molecules. This is an irreversible transition, resulting in the decrease of the crystallinity, and the swelling of the starch granules [46]. As a consequence, the viscosity gradually increases, the process being referred to as *gelatinization*. By further subjecting the gelatinized starch to mild shearing forces, it is possible to obtain a "dissolved" starch and a consequent reduced viscosity. In polymer compositions, the native starch granules are no suitable, instead gelatinized or dissolved starch is preferred [47]. The development of new starch-based materials has gained much interest in recent years, this being related to their biodegradability, low cost and wide availability. Thermoplastic starch is one of the starch-based polymers that have been widely investigated [35, 37-43, 48-50].

7.3 Chemical Modification of Starch

Starch is an attractive biosynthesized and biodegradable alternative suitable for film preparation and foaming. Unfortunately, starch presents some disadvantages. It is highly hygroscopic, brittle without plasticizer and its mechanical properties are very sensitive to moisture content [51]. Still, these drawbacks can be avoided by, for example, blending the starch with an appropriate biodegradable polymer. On the market today, there already exist starch-based materials. One example is Mater-Bi materials from Novamont, Italy. These are based on thermoplastic starch and different types of synthetic components [52].

Starch is inherently non-suitable for most applications and, therefore, must be modified chemically and/or physically to enhance its positive properties and/or to minimize its drawbacks [53]. Flexibility in adjusting the properties to the needs of the specific application by appropriately modifying the composition, low-cost blending as opposed to innovative synthetic material development, and biodegradability are some of the main advantages which strongly motivate the development of starch-based materials [9].

Chemical modification of starch makes it suitable for many applications in the food products (e.g., as gelling agents, encapsulating agents, thickeners) and the non-food industry (e.g., as wet-end additives, sizing agents, coating binders, and adhesives in paper industry; as textile sizes; in cosmetic formulations) [46].

Efficient reactions for chemical modification of starch include esterification, etherification, and oxidation of the available hydroxyl groups on the glucose units. Among these reactions, one can mention the following: acetylation [54], succinylation [55], and maleination [56], thus increasing starch functional value and broadening its properties. Crosslinking of the starch granules is an effective physical modification route in order to improve the properties of starch-based materials.

Water is usually involved as reaction media for chemical modification of starch being an environmentally-friendly solvent ("green solvent"), even its use means a relatively low reactivity, or a reduced selectivity due to side reactions. As result of the chemical modification reactions, the crosslinking process of starch occurred, this way being an efficient approach to improve the performance of starch for different applications [57, 58]. Chemically modified starches were investigated by determination of the modification degree [55], and by infrared spectroscopy method [59, 60].

Due to the brittleness of starch materials, plasticizers are commonly used. A frequently utilized low weight hydroxyl compound is glycerol. Another effective plasticizer is water, although not the best because it evaporates easily. Still, starch-based materials readily absorb water and this may result in significant changes in the mechanical properties. Different routes have been explored in order to improve the mechanical properties and water resistance of starch materials. These are chemical modifications to the starch molecule, blends with polymers such as polycaprolactone [61], or reinforcement with different types of cellulose-based fillers, such as ramie crystallites [62], and tunicin whiskers [63], or montmorillonite clay particles [64].

Starch behaves as a thermoplastic polymer in the presence of a plasticizer (water, glycerol, sorbitol, etc.), high temperature values (90-180°C) and shearing, when it melts and fluidizes, enabling its processing as that for synthetic polymers [40]. During this process, hydrogen bonds are formed between plasticizer and starch, the latter one becoming plasticized. The structure of native starch granule is disrupted in the presence of water through breaking the chains hydrogen bonded, but at the same time, water acts as a plasticizer [65]. However, an additional plasticizer besides water (e.g., polyol) is needed in order to allow a melting phase at a lower temperature value than that of the starch degradation process [66]. Starch modification through reaction with mineral or organic acid is an efficient method to influence starch properties. Starch reacts with carboxylic acids at room temperature, in the presence of water (e.g., formylic acid), or does not react in an aqueous medium and requires heating to initiate reaction (e.g., acetic acid, citric acid). Carboxylic acids decreased the viscosity of thermoplastic starch by controlling the macromolecules of starch in the presence of glycerol and water [67].

Derivative resulted from reaction of starch with carboxylic acid is not destroyed through further heating, due to additional dehydration process which determines the crosslinking of starch. The carboxylic acid modified starch is not gelatinized during processing of the films by comparison with starch crystals [68-70], which can be destroyed through gelatinization process at the high processing temperature value. Citric acid can increase the thermal stability of glycerol-plasticized thermoplastic starch, and may decrease the shear viscosity with positive effect on fluidity [71]. Some starch properties were improved in the presence of citric acid, namely elongation and water resistance, while the tensile stress decreased.

7.4 Cellulose-Based Starch Composites

High performance composite materials can be obtained with a good level of dispersion, mainly when the hierarchical structure of cellulose and use of a water soluble polymer to form the matrix are considered. For most materials applications, the main biopolymers of interest are cellulose and starch. The ease of adhesion that occurs in cellulose has contributed to its use in paper and other fiber-based composite materials.

7.4.1 Obtainment

Starch-based composite films reinforced with chemically modified starch have a stable structure under ambient conditions, good resistance to water uptake, and better mechanical properties than those without modified starch. Starch microparticles (StM) were prepared by delivering ethanol into corn starch (St) solution. Chemically modified starch microparticles (CMSt) by reaction with organic acid were obtained by the dry preparation technique [72] and incorporated within glycerol plasticized-corn starch (GCSt) matrix, composite materials being further prepared by the casting/solvent evaporation process [73, 74]. Cellulose fillers (CF), namely beech wood flour, cellulose separated from beech wood [75], birch cellulose [76], spruce cellulose (pulp) [77], *Asclepias syriaca* seed floss ASF, poplar seed floss PSF [78] - were incorporated within the CMSt-GCSt matrix. The surface properties (opacity) and water sorption, as well as mechanical and thermal properties of chemically modified starch/plasticized starch/cellulose filler (CMSt/St/cellulose filler) composite films were investigated.

7.4.1.1 Preparation of Starch Microparticles (StM) and Chemically Modified Starch Microparticles (CStM)

A commercially corn starch (St) was used as the polymer continuous matrix of the composite films. Glycerol was used as plasticizer (30% amount based on starch). Organic acids (malic, adipic, and tartaric ones) were used for chemical modification of starch (as shown in Scheme 7.1, Scheme 7.2 and Scheme 7.3).

A mixture of corn starch (St - 10g) and 200 ml of distilled water was heated at 90°C for 1h in order to result the complete gelatinization of starch under constant stirring. Then, to the solution of gelatinized starch was added dropwise ethanol (200 ml) with constant stirring. The resulted suspensions of starch microparticles (StM) were further cooled at the room temperature, and another 200 ml of ethanol was added dropwise for about 50 min under constant stirring. The suspensions were centrifuged for 20 min, at 8000 rpm, and the settled StM was washed using ethanol for removing water. After complete washing, the StM was dried at 50°C to remove ethanol (Scheme 7.1).

In Scheme 7.2. is represented the chemical reaction between starch microparticles StM and organic acid (adipic, malic and tartaric). Organic acid (20 g) was dissolved in 100 ml of ethanol. StM (3.5 g) was mixed with 15 ml of acid solution in a glass tray and conditioned for 12h at room temperature to allow the absorption of acid solution by StM. The tray was dried in vacuum oven at about 2 mmHg and 50°C for 6h in order to remove ethanol. The resulted mixture was further ground and dried in a forced air oven for 1.5h at 130°C. The dry mixture was then washed three times with water to remove non-reacted organic acid. CMSt mixtures were finally washed with ethanol to remove water, dried at room temperature, and ground (Scheme 7.3). The dried StM and CMSt were used for obtainment of composite films with cellulose fillers. By comparison with starch nanocrystals [79], CMSt is not gelatinized during the processing of the composites. It can be used as a reinforcement for glycerol plasticized starch.

As it is shown in Scheme 7.2, when organic acid was heated, it dehydrated to yield an anhydride, which could react with starch to form a starch-organic acid derivative. Further heating resulted in additional dehydration with crosslinking [70]. Thus, substitution of organic acid groups on starch macromolecular chains could limit their mobility by forming a highly crosslinked starch.

7.4.1.2 Determination of the Molar Degree of Substitution of CMSt

The molar degree of substitution (DS) is the number of organic acid per glucose unit in corn starch and was determined according to the method described in literature [80]. Approximately 0.5 g of CMSt was accurately weighed and placed into a 150 mL conical flask containing 30 mL of 75% ethanol solution, this being further stirred and slowly heated at 50°C for 30 min and then cooled to room temperature. Standard 10 ml of 0.500 M aqueous sodium hydroxide solution was added; the conical flask was tightly stoppered and stirred by means of a magnetic stirrer for 24 h. The excess alkali was back-titrated with a standard 0.200 M aqueous hydrochloric acid solution and re-titrated 2 h later to account for any further alkali that may have leached from the corn

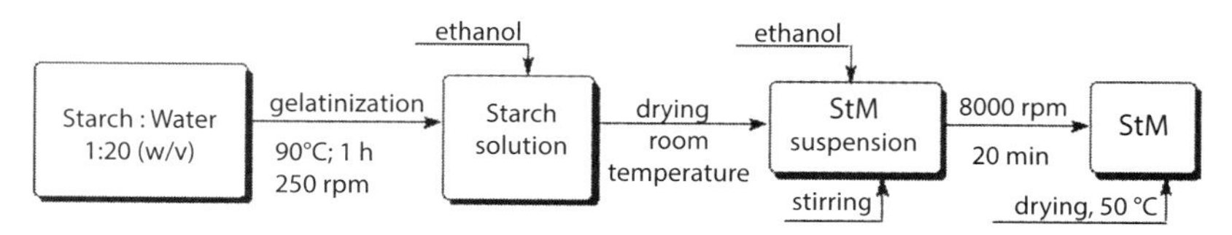

Scheme 7.1 Preparation of starch microparticles (StM).

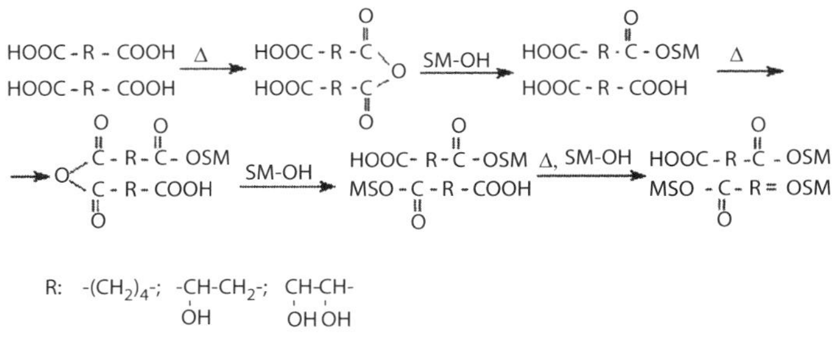

HOOC - R - COOH / HOOC - R - COOH $\xrightarrow{\Delta}$ HOOC - R - C(=O) / HOOC - R - C(=O) (cyclic anhydride, -O-) $\xrightarrow{\text{SM-OH}}$ HOOC- R - C(=O) - OSM / HOOC - R - COOH $\xrightarrow{\Delta}$

$\rightarrow$ O< C(=O) - R - C(=O) - OSM / C(=O) - R - COOH $\xrightarrow{\text{SM-OH}}$ HOOC- R - C(=O) - OSM / MSO - C(=O) - R - COOH $\xrightarrow{\Delta,\ \text{SM-OH}}$ HOOC - R - C(=O) - OSM / MSO - C(=O) - R = OSM

R: $-(CH_2)_4-$; $-CH(OH)-CH_2-$; $CH(OH)-CH(OH)-$

Scheme 7.2 Chemical reaction of organic acid with StM.

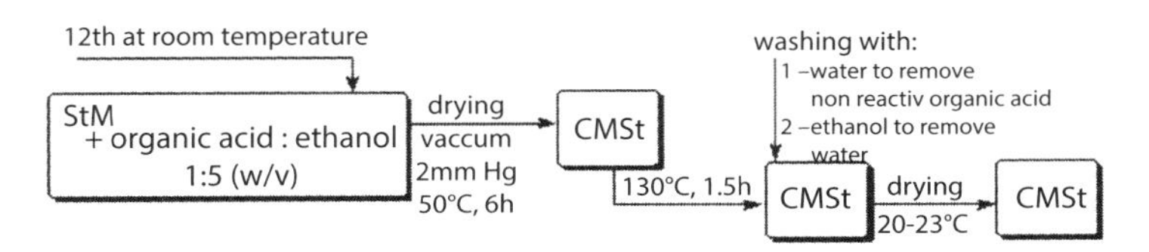

Scheme 7.3 Preparation of chemically modified starch microparticles (CMSt).

starch. Degree of substitution by organic acid was measured in triplicate and had a value of around 0.090 (± 0.005) with some differences given by the organic acid used for starch chemical modification.

7.4.1.3 Preparation of CMSt/St/cellulose Filler Composite Films

Chemically modified starch microparticles CMSt were dispersed in 100 ml solution of distilled water and 1.5 g glycerol for 1 h, then 5 g corn starch and 0.2 g cellulose-based filler (C) were added resulting chemically modified starch/plasticized starch/cellulose filler films (CMSt/St/C). CMSt and cellulose filler loading level (4 wt%) was based on the amount of corn starch. In our recent works [75, 78], different cellulose fillers (cellulose separated from beech wood by TAPPI standard method BWC, respectively beech wood sawdust BWS, spruce cellulose, *Asclepias syriaca* L. seed floss ASF, *Populus alba* L. seed floss PSF; microcrystalline cellulose MC as a reference sample) were incorporated within the starch polymer matrix. The code of samples are presented in Table 7.1. The mixture was heated at 90°C for 0.5 h with constant stirring in order to plasticize the corn starch. To obtain the CMSt/St/C films, the mixtures were cast using a fast coating technique namely the doctor blade technique. Films were obtained by dropping and spreading the different mixtures on a glass plate using a blade with a slit width of 0.8 mm. After degassing in a vacuum oven at 50°C for 24 h up to constant weight, the films were air cooled and detached from the glass surface to be investigated. Films with a thickness of 0.2 mm were obtained, this value resulting from measurements by means of a digital micrometer. The starch-based films were pre-conditioned in a climate chamber at 25°C and 50%RH for at least 48 h prior to the mechanical testing. Water content was around 9 wt%.

The milkweed (*Asclepias syriaca* L.) is a tall plant containing a milky juice in all its parts, native to eastern parts of North America, and naturalized in various parts of Europe. It is a perennial plant that can adapt to adverse soil conditions (e.g., dry and

Table 7.1 Code of starch-cellulose composite samples.

Sample	Composition
StM/St	starch microparticles/starch
AAMSt/St	starch microparticles modified with adipic acid/starch
AAMSt /St/MC	starch microparticles modified with adipic acid/starch/ microcrystalline cellulose
AAMSt /St/BWC	starch microparticles modified with adipic acid/starch/ beech cellulose
AAMSt /St/SC	starch microparticles modified with adipic acid/starch/ spruce cellulose
AAMSt /St/ASF	starch microparticles modified with adipic acid/starch/ asclepias seed floss
AAMSt /St/PSF	starch microparticles modified with adipic acid/starch/ poplar seed floss
AAMSt/St/BWS	starch microparticles modified with adipic acid/starch/ beech wood sawdust

arid, requiring less water). Natural fibers with higher cellulose content (around 75%), better strength and higher elongation than those of milkweed floss have been extracted from the milkweed stems [81], making them suitable for different fibrous applications (e.g., textile, composite, etc.). The seeds present lignified, almost colorless silky hairs (floss), enclosed by a pod. Floss have low density (0.9 g/cm^3) unlike any other natural cellulose fibers, have short lengths and low elongation, thus its applications are limited. Currently, floss is used for comforters.

There are many efforts made for investigation upon additional types of fibers in order to improve performance or reduce costs for cellulose-based materials, A relevant example is represented by poplar seed hairs derived from *Populus* wood species, which represent a cellulose-enriched material. This can be considered a valuable resource derived from the biomass waste to be capitalized under environmentally-friendly conditions through conversion into useful chemical derivatives or use for different fibrous applications. Usually, poplar seed fibers are treated as waste or used as fertilizer, but these can be also efficiently used for oil absorbent applications [82] due to their strong hydrophobic character (fibers are covered by a waxy coating) and specific morphology (hollow microtube type).

The main objective of our studies was to obtain green composites from corn starch matrix and various conventional [73, 76, 77], and non-conventional cellulose sources [78]. Previously, corn starch (St) was converted to starch microparticles (StM). Further, different organic acids (adipic, malic, tartaric) were used for treatment of StM in order to obtain chemically modified starch microparticles (CMSt) according to literature data [72]. After casting and water evaporation, the starch-based films were investigated by means of X-ray diffraction and FTIR spectroscopy methods. Opacity and water uptake of starch-based films were also evaluated.

CMSt were further added to a glycerol plasticized corn starch thermoplastic matrix. In order to obtain starch–cellulose composite materials, different cellulose fillers were added within CMSt/St plasticized starch polymer matrix. The influence of the fillers addition on the properties of composite materials was further investigated.

7.4.2 Characterization of Starch Polymer Matrix

Evidence of starch chemical modification was confirmed by FTIR spectroscopy and X-ray diffraction methods. Compared to starch crystals [68-70], which can be destroyed by using high processing temperature value, CStM is not gelatinized during the processing of the composite film, therefore organic acid modified corn starch can be conveniently used as a filler for plasticized corn starch polymer matrix.

7.4.2.1 FTIR Spectroscopy Investigation

FTIR-ATR spectra were recorded using a spectrophotometer Vertex 70 (Bruker-Germany) in the range of 4000-400 cm^{-1} with a 4 cm^{-1} resolution and a scan rate 32. The spectrophotometer is equipped with MIRacle^{TM} ATR accessory designed for single or multi-reflection attenuated total reflectance (ATR). The ATR crystal plate is from diamond (1.8 mm diameter), and solid materials can be put into intimate physical contact with the sampling area through high-pressure clamping, yielding high-quality, reproducible spectra.

Corn starch (St) and StM exhibited similar FTIR spectra as evidenced in Figure 7.1.

Some differences such as band shape and intensity can be observed in the fingerprint of starch-based films in the spectra, as a result of organic acid used for chemical modification. The analysis of FTIR spectra of the films (Figure 7.2) enabled the hydrogen bond interaction to be identified [80]. A specific peak occurred at around 1650 cm^{-1}, which can be considered a feature of tightly bound water present in the corn starch [83]. The characteristic peaks observed at 1081 cm^{-1} and 1157 cm^{-1}, respectively can be assigned to the C-O bond stretching of C-O-H group. Another two peaks at 1013 cm^{-1} and 1020 cm^{-1} attributed to C-O bond stretching of C-O-C group in the glucose ring shifted from 1013 cm^{-1} (St) at 1018-1022 cm^{-1} (StM). These peaks can be associated with both vibration / solvation of C-OH bond, and transition from an amorphous state to a semi-crystalline one [84, 85]. For CMSt powders, a new peak at 1744 cm^{-1} (MAMSt), 1712 cm^{-1} (AAMSt) and 1708 cm^{-1} (TAMSt - spectrum not represented here) respectively, is characteristic to an ester group.

In corn starch and StM, the oxygen of the C-O-C group could form the hydrogen-bond interaction with the hydrogen of hydroxyl groups, while the ester bonds in CMSt sterically hindered this hydrogen-bond interaction. The shift of the peak observed at 996 cm^{-1} for StM/St to 998 cm^{-1} for CMSt/St film can be associated to the amorphous-crystalline transition in these films. The absorbance values at 995-1014 cm^{-1} for CMSt/St films were different as a function of the organic acid used for chemical modification and higher comparatively with those recorded for StM/St film.

In Table 7.2, are presented the main FTIR bands assigned for components of starch/cellulose filler films.

The main absorption bands, specific to spruce cellulose (which is in fact an industrial bleached pulp), are in the range of intra- and intermolecular hydrogen bonded -OH groups between 3500 and 3200 cm^{-1}, in the fingerprint region such as those at 1337 and

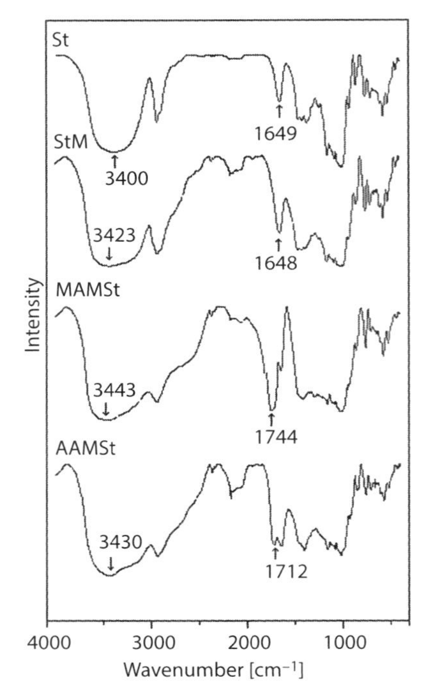

Figure 7.1 FTIR spectra recorded for St, StM and CMSt.

1164 cm^{-1}, and in the range of C–O valence vibrations from 1060 to 1015 cm^{-1} [88]. The peak at 3430 cm^{-1} is specific to amorphous cellulose, being attributed to the OH stretching being related to the intra-molecular hydrogen bonds at the C3 position, but also possible intra-molecular hydrogen bonds between the functional groups at the C2 and C6 positions in cellulose [89]. Technical celluloses, such as bleached pulp (namely, spruce cellulose), contain additional carbonyl and carboxyl groups as a result of the isolation and purification processes involved in the processing of cellulose, these being accompanied usually by residual hemicelluloses [32, 90]. The absence of the band at 1520 cm^{-1} in the FTIR spectrum can evidence that it is no residual lignin in composition of the cellulose sample.

7.4.2.2 X-ray Diffraction Analysis

Corn starch St, StM, and CMSt powders were tightly packed into the sample holder. X-ray diffraction patterns were recorded in the reflection mode in angular range 3°-30° (2θ) at a speed of 2° min^{-1} and at ambient temperature by means of a Bruker AD8 ADVANCE X-ray diffractometer equipment with Cu Kα radiation operating at 40 kV and 35mA.

The X-ray diffraction analysis indicated that during chemical modification process, several crystalline structures of native starch were destroyed, and a new structure of organic acid-modified starch was formed (Figure 7.3). The diffraction peaks for StM and CMSt look like a *V*-type crystalline structure. According to the literature data [92], there can be observed a typical *A*-style crystallinity in the native corn starch. This

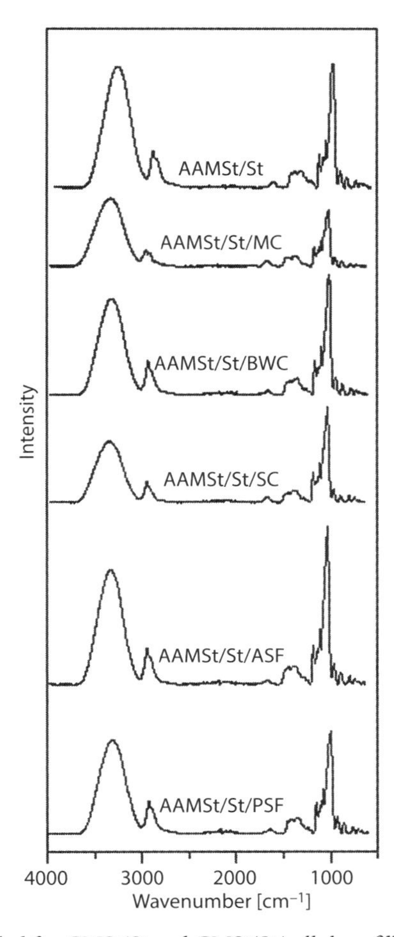

Figure 7.2 FTIR spectra recorded for CMSt/St and CMSt/St/cellulose filler.

Table 7.2 Assignments of FTIR bands to components of starch/cellulose filler films [86–89].

Wavenumber (cm^{-1})	Assignment
3310-3330	free O–H valence vibration
1675–1655	CO stretching
1432	CH_2 symmetric bending
1362	C–H deformation vibration
1337	O–H in plane bending of alcohol groups
1164–1139	C–O–C antisymmetric valence stretching
1080-995	C(3)–O(3)H valence vibration
896	glycosidic C(1)–H deformation with ring valence vibration; OH bending specific to β-glycosidic linkages between glucose units

crystalline state was replaced by the *V*-type crystallinity which was formed in the presence of plasticizer by the inductive effect of the thermal process.

During the thermoplasticization process, the strong interactions between hydroxyl groups of starch macromolecules were substituted by hydrogen bonds formed between plasticizer and starch. Through heating, the starch gelatinization occurred by disruption of its double helix conformations. When organic acid penetrated the StM granules, it could disrupt the *V*-type crystalline structure of starch due to the concentrated solution of organic acid. The reaction should occur mainly in the amorphous phase of starch [93].

According to the X-ray diffraction analysis, organic acid modified starch microparticles CMSt presented an amorphous characteristic, but some crystalline peaks at around $2\theta \cong 20°$ were also evidenced.

7.4.3 Properties Investigation

7.4.3.1 Opacity Measurements

The opacity for cellulose-starch composite films was measured by using a JENWAY 6405 UV–VIS spectrophotometer and defined as the area under the absorbance spectrum between 400 and 800 nm according to the ASTM D 1003-00 method (ASTM D

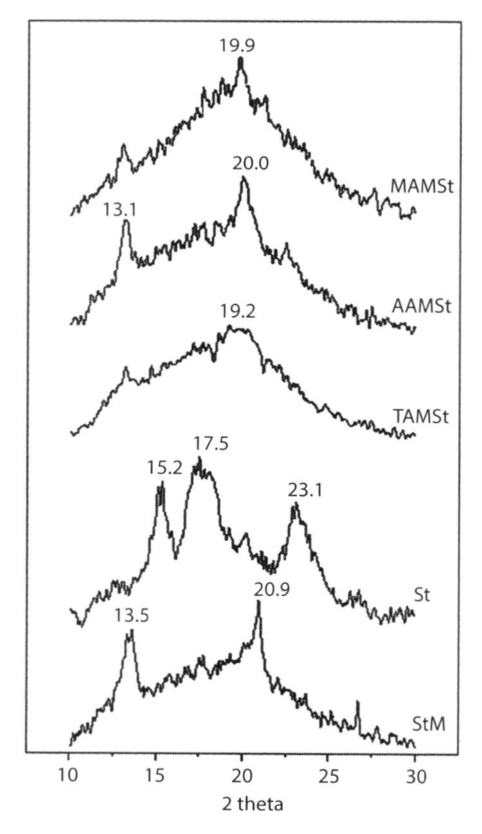

Figure 7.3 X-ray diffraction curves recorded for St, StM and CMSt.

1003-00 Standard Test Method for Haze and Luminous Transmittance of Transparent Plastics). The film samples were cut into a rectangular piece (1×2.5 cm), fixed on the inner side of a 1cm of spectrophotometer cell and the absorbance spectrum recorded. The film opacity measurements were repeated three times.

The opacity is a critical property if the starch-based films are used for food coatings applications. Transparent films are characterized by low values of the area below the absorption curve. The obtained values showed that films with addition of organic acid modified starch microparticles CMSt were more transparent than films with no chemically modified starch [74]. Besides, the opacity of films decreased due to addition of CMSt, this tendency being more noticeable in sample modified with tartaric acid. Film opacity decreased in order: TAMSt/St<MAMSt/St<AAMSt/St<StM/St. In Figure 7.4, opacity recorded for AAMSt/St and AAMSt/St/cellulose filler composite samples is represented [75].

Composite sample comprising lignocellulosic filler-beech wood sawdust, namely AAMSt/St/BWS, is more less transparent comparatively with sample comprising beech cellulose. Film sample based only the plasticized starch matrix with chemically modified starch microparticles in composition maintains relative transparency.

7.4.3.2 Water Sorption Properties

Water uptake is usually used to study the humidity diffusion through the composite films. To determine the water uptake [92, 94, 95], the composite films used were thin rectangular strips with dimensions of 10 mm × 10 mm × 0.2 mm. The films were supposed to be thin enough so that the molecular diffusion was considered to be one-dimensional and were vacuum-dried at 90°C overnight. After weighing, they were conditioned at 25°C in a desiccator containing sodium sulfate in order to ensure a relative humidity (RH) of 95%. They were then removed at specific intervals and gently blotted with tissue paper to remove the excess of water on the surface, and the water uptake was calculated with Eq. 1, as follows:

$$water\ uptake\ (\%) = [(Wt - W_0)/W_0]\ x\ 100 \qquad (7.1)$$

where: Wt and W_0 represents the weight at time *t* and before exposure to 95% RH, respectively. The determinations were performed in triplicate.

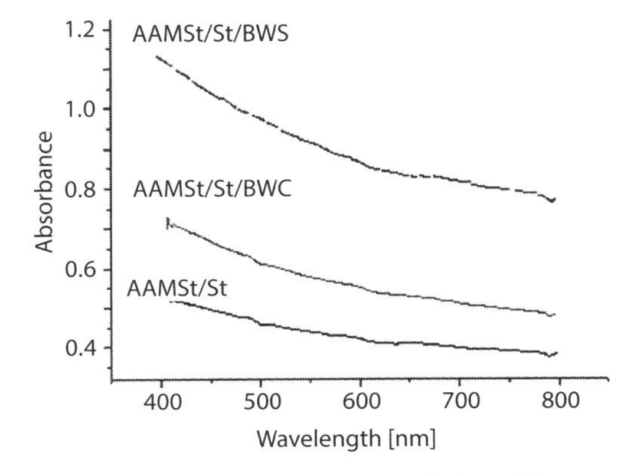

Figure 7.4 Opacity recorded for AAMSt/St and AAMSt/St/cellulose filler composite samples.

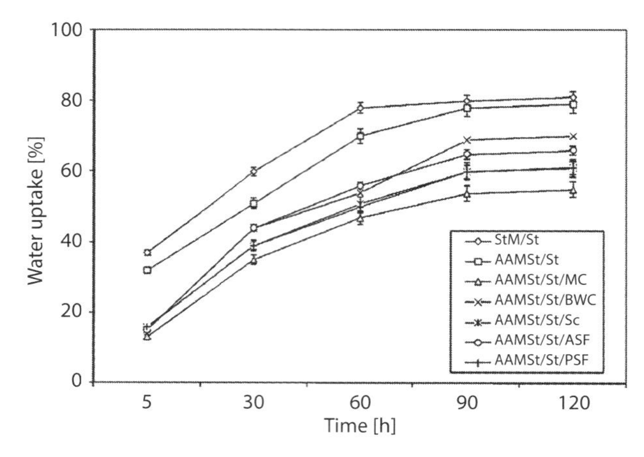

Figure 7.5 Water sorption behavior for different starch-cellulose composite films.

In a multi-component biopolymer system (like starch-cellulose films), each biopolymer engages in biopolymer-biopolymer interactions contributing each one and both to the system properties. Occasionally, these interactions are more important than individual actions. A neat starch film will always display high water absorbency properties, this fact impeding the its application for most water resistance applications.

After drying, the starch films with no CMSt in composition exhibited the higher value for equilibrium moisture content while the lower values were observed by addition of CMSt within plasticized starch polymer matrix due to their hydrophobic properties (Figure 7.5).

This behavior can be explained by the formation of a more tight structure after starch crosslinking which prevents the swelling of starch and also restricts the movement of macromolecular chains, leading to a decrease of the amount of absorbed water in agreement with mechanical properties.

7.4.3.3 Mechanical Properties

Tensile stress and strain at break were evaluated according to ASTM D882-00 (ASTM D882-00 Standard Test Method for Tensile Properties of Thin Plastic Sheeting) using an Instron 3345 with a 5-kN load cell. Before testing, the films were cut into strips and conditioned at 50% RH and 25°C for 48 hours. Testing was done on composite films each measuring 10 cm x 1 cm randomly cut from the cast films. The thickness of each specimen was measured at four points along its length with digital micrometer. The crosshead speed was 10 mm/min. A minimum of 10 replications of each test sample were run.

Table 7.3 summarizes the mechanical properties of some of the starch-cellulose fillers composite films [75].

The tensile strength of the films was influenced by the polymer components. By addition of cellulose fillers, the tensile strength values were improved. Elongation at break for CMSt/St/C film samples was reduced than that for CMSt/St sample. However, as a result of more compact structure due to high intermolecular hydrogen bond

Table 7.3 Mechanical properties recorded for starch–cellulose composite films.

Composite sample	Tensile strength (MPa)	Young's modulus (MPa)	Elongation at break (%)
AAMSt/St	11.5	1162	7.8
AAMSt/St/BWC	17.1	1061	7.4
AAMSt/St/BWS	11.7	722	7.1

Table 7.4 Thermal characteristics of some starch-cellulose composite films [75].

Compositesample	$T_{10\%}$(°C)	T_{peak}(°C)	T_{end}(°C)	T_{end}-$T_{10\%}$(°C)	$T_{50\%}$ (°C)	W_T(%)
AAMSt/St	273	321	336	63	319	88.02
AAMSt/St/BWC	256	322	338	82	319	88.84
AAMSt/St/BWS	265	320	339	74	320	88.41

where: $T_{10\%}$ – temperature corresponding to 10% weight loss; T_{peak} – maximum decomposition temperature; T_{end} – final decomposition temperature; T_{end}–$T_{10\%}$ – decomposition temperature interval; $T_{50\%}$ – temperature corresponding to 50% weight loss; W_T – weight loss on the temperature range (0–600°C).

interactions between starch and cellulose polymers, the films containing cellulose were more brittle than the composite with no filler.

7.4.3.3 Thermal Properties

Thermal analysis of starch-cellulose films was performed with a thermal analyzer STA 449 F1 Netzsch (Germany). Samples (» 5 mg) were placed in Al_2O_3 crucibles and heated under nitrogen from 50°C up to 700°C with 10°C /min heating rate. TG and DTG curves recorded with a ± 0.5°C precision were analyzed with Netzsch Proteus analysis software.

TG/DTG/DSC investigation of the starch-cellulose composite films was performed in order to study their thermal decomposition behavior. Some thermal characteristics of composite materials which contain different cellulose fillers (beech cellulose, beech wood sawdust) were evidenced from thermograms (data presented in Table 7.4).

The degradation occurring around 320°C can be attributed to the starch degradation and is similar in all composite samples, CMSt/St and CMSt/St/cellulose filler. The peak at around 320°C can be also assigned to the decomposition of cellulose, and beech wood sawdust.

7.5 Conclusions/Perspectives

Starch-based composite films reinforced with chemically modified starch have a stable structure under ambient conditions, good resistance to water uptake, and better mechanical properties than those without modified starch.

Starch-cellulose filler composite materials were prepared from modified starch microparticles originating from corn starch previously modified by reaction with organic acid as filler within a glycerol plasticized corn starch matrix by the casting process. X-ray diffraction and FTIR spectroscopy methods evidenced the changes in the structure of StM and CMSt. The absorption peaks corresponding to the hydroxyl, carbonyl and ether groups in starch have a tendency to shift to higher wavenumber values for the starch-cellulose filler films, indicating the occurrence of a hydrogen-bonding interaction between polymer components.

Starch-cellulose filler composite materials presented improved tensile strength values and were more rigid, but had lower elongation capacity. The obtained composite films became more hydrophobic, a decrease of the water absorption in a high humidity atmosphere being observed. Films transparency was also seen to decrease with addition of cellulose filler.

Our work provides a simple and cheap way to prepare fully biodegradable modified starch/cellulose filler films, which may have applications for packaging, or agricultural mulching purposes.

References

1. L. Yu, K. Dean, and L. Li, Polymer blends and composites from renewable resources. *Prog. Polym. Sci.* 31, 576–602 (2006).
2. A.K. Mohanty, M. Misra, and G. Hinrichsen, Biofibres, biodegradable polymers and biocomposites: An overview. *Macromol. Mater. Eng.* 276–277, 1–24 (2000).
3. D. Nabi Saheb, and J.P. Jog, Natural fiber polymer composites: A review. *Adv. Polym. Technol.* 18(4), 351–363 (1999).
4. K.G. Satyanarayana, Biodegradable Composites Based on Lignocellulosic Fibers, *Proceedings of the International Conference on Advanced Materials and Composites ICAMC-2007*, October 24–26, 12–24 (2007).
5. A.K. Mohanty, M. Misra, and L.T. Drzal, Sustainable bio-composites from renewable resources: Opportunities and challenges in the green materials world. *J. Polym. Environ.* 10, 19–26 (2002).
6. R.P. Babu, K. O'Connor, and R. Seeram, Current progress on bio-based polymers and their future trends. *Prog. Biomater.* 2(8), 1–16 (2013).
7. I. Vroman, and L. Tighzert, Biodegradable Polymers. *Materials* 2, 307–344 (2009).
8. R. Chandra, and R. Rustgi, Biodegradable polymers. *Prog. Polym. Sci.* 23, 1273–1335 (1998).
9. G. Scott, and D. Gilead, *Degradable Polymers: Principles and Applications, London*, Chapman & Hall, London (1995).
10. F. Donot, A. Fontana, J.C. Baccou, and S. Schorr-Galindo, Microbial exopolysaccharides: Main examples of synthesis, excretion, genetics and extraction. *Carbohydr. Polym.* 87, 951–962 (2012).
11. R.S. Singh, G.K. Saini, and J.F. Kennedy, Pullulan: microbial sources, production and applications. *Carbohydr. Polym.* 73(4), 515–531 (2008).
12. I.-L. Shih, L.-D. Chen, and J.-Y. Wu, Levan production using Bacillus subtilis natto cells immobilized on alginate. *Carbohydr. Polym.* 82(1), 111–117 (2010).
13. I.-L. Shih, J.-Y. Yu, C. Hsieh, and J.-Y. Wu, Production and characterization of curdlan by Agrobacterium sp. *Biochem. Eng. J.* 43(1), 33–40 (2009).

14. K.I. Shingel, Current knowledge on biosynthesis, biological activity, and chemical modification of the exopolysaccharide, pullulan. *Carbohydr. Res.* 339(3), 447–460 (2004).
15. A. Demirbas, Bioethanol from cellulosic materials: a renewable motor fuel from biomass. *Energy Sources* 27, 327–337 (2005).
16. T.P. Nevell, and S.H. Zeronian, Cellulose chemistry fundamentals, in *Cellulose Chemistry and Its Applications*, T.P. Nevell and S.H. Zeronian, (Eds.), pp. 15–29, Ellis Horwood, Chichester (1985).
17. R.P. Swatloski, S.K.H. Spear, and R.D. Rogers, Dissolution of cellose with ionic liquids. *J. Am. Chem. Soc.* 124, 4974–4975 (2002).
18. *A.C. O'Sullivan, Cellulose: The structure slowly unravels. Cellulose* 4, 173–207 (1997).
19. A.F. Turbak, F.W. Snyder, and K.R. Sandberg, Microfibrillated cellulose, a new cellulose product: properties, uses, and commercial potential. *J. Appl. Polym. Sci.* 37, 815–827 (1983).
20. A. Chakraborty, M. Sain, and M. Kortschot, Cellulose microfibrils: a novel method of preparation using high shear refining and cryocrushing. *Holzforschung* 59, 102–107 (2005).
21. L. Wågberg, G. Decher, M. Norgren, T. Lindström, M. Ankerfors, and K. Axnäs, The build-up of polyelectrolyte multilayers of microfibrillated cellulose and cationic polyelectrolytes. *Langmuir* 24 (3), 784–795 (2008).
22. M. Pääkkö, M. Ankerfors, H. Kosonen, A. Nykänen, S. Ahola, M. Österberg, J. Ruokolainen, J. Laine, P.T. Larsson, O. Ikkala, and T. Lindström, Enzymatic hydrolysis combined with mechanical shearing and high-pressure homogenization for nanoscale cellulose fibrils and strong gels. *Biomacromolecules* 8, 1934–1941 (2007).
23. M. Henriksson, G. Henriksson, L.A. Berglund, and T. Lindström, An environmentally friendly method for enzyme-assisted preparation of microfibrillated cellulose (MFC) nanofibers. *Eur. Polym. J.* 43, 3434–3441 (2007).
24. A. Boldizar, C. Klason, and J. Kubát, Prehydrolyzed cellulose as reinforcing filler for thermoplastics. *Int. J. Polym. Mater.* 11, 229–262 (1987).
25. H. Yano, J. Sugiyama, A.N. Nakagaito, M. Nogi, T. Matsuura, M. Hikita, and K. Handa, Optically transparent composites reinforced with networks of bacterial nanofibers. *Adv. Mater.* 17, 153–155 (2005).
26. S.J. Eichhorn, A. Dufresne, M. Aranguren, N.E. Marcovich, J.R. Capadona, S.J. Rowan, C. Weder, W. Thielemans, M. Toman, S. Renneckar, W. Gindl, S. Veigel, J. Keckes, H. Yano, K. Abe, M. Nogi, A.N. Nakagaito, A. Mangalam, J. Simonsen, A.S. Benight, A. Bismarck, L.A. Berglund, and T. Peijs, Review: current international research into cellulose nanofibres and nanocomposites. *J. Mater. Sci.* 45(1), 1–33 (2001).
27. E.D. Cranston, and D.G. Gray, Morphological and optical characterization of polyelectrolyte multilayers incorporating nanocrystalline cellulos. *Biomacromolecules* 7(9), 2522–2530, (2006).
28. M.A. Hubbe, O.J. Rojas, L.A. Lucia, and M. Sain, Cellulosic nanocomposites: a review. *BioResources* 3(3), 929–980 (2008).
29. I. Siró, and D. Plackett, Microfibrillated cellulose and new nanocomposite materials: A review. *Cellulose* 17(3), 459–494 (2010).
30. M. Gama, P. Gatenholm and D. Klemm, *Bacterial Nanocellulose: A Sophisticated Multifunctional Material*, CRC Press Taylor&Francis Group, Boca Raton, (2013).
31. M.J. John, and S. Thomas, Biofibres and biocomposites. *Carbohydr. Polym.* 71, 343–364 (2008).
32. D. Klemm, B. Heublein, H.-P. Fink, and A. Bohn, Cellulose: Fascinating biopolymer and sustainable raw material. *Angew. Chem. Int. Ed.* 44, 3358–3393 (2005).

33. D.J. Gardner, G.S. Oporto, R. Mills, and A.S.A. Samir, Adhesion and surface issues in cellulose and nanocellulose. *J. Adhes. Sci. Technol.* 22(5–6), 545–567 (2008).
34. A. K. Bledzki, and J. Gassan, Composites reinforced with cellulose based fibres. *Prog. Polym. Sci.* 24, 221–274 (1999).
35. L. Averous, and P.J. Halley, Biocomposites based on plasticized starch. *Biofuels Bioprod. Biorefin.* 3(3), 329–343 (2009).
36. L. Averous, Biodegradable multiphase systems based on plasticized starch: a review. *J. Macromol. Sci. C Polym. Rev.* 44, 231–274 (2004).
37. J. Prachayawarakorn, P. Sangnitidej, and P. Boonpasith, Properties of thermoplastic rice starch composites reinforced by cotton fiber or low-density polyethylene. *Carbohydr. Polym.* 81, 425–433 (2010).
38. X.F. Ma, J. Yu, and J.F. Kennedy, Studies on the properties of natural fibers-reinforced thermoplastic starch composites. *Carbohydr. Polym.* 62, 19–24 (2005).
39. H. Dai, P.R. Chang, F. Geng, J. Yu, and X. Ma, Preparation and properties of thermoplastic starch/montmorillonite nanocomposite using N-(2-hydroxyethyl)formamide as a new additive. *J. Polym. Environ.* 17, 225–232 (2009).
40. A.A.S. Curvelo, A.J.F. Carvalho, and J.A.M. Agnelli, Thermoplastic starch–cellulosic fibers composites: preliminary results. *Carbohydr. Polym.* 45, 183–188 (2001).
41. H.A. Pushpadass, A. Kumar, D.S. Jackson, R.L. Wehling, J.J. Dumais, and M.A. Hanna, Macromolecular changes in extruded starch-films plasticized with glycerol, water and stearic acid. *Starch-Stärke* 61, 256–266 (2009).
42. A. Kaushik, M. Singh, and G. Verma, Green nanocomposites based on thermoplastic starch and steam exploded cellulose nanofibrils from wheat straw. *Carbohydr. Polym.* 82(2), 337–345 (2010).
43. P.R. Chang, R. Jian, P. Zheng, J. Yu, and X. Ma, Preparation and properties of glycerol plasticized-starch (GPS)/cellulose nanoparticle (CN) composites. *Carbohydr. Polym.* 79, 301–305 (2010).
44. J.N. BeMiller, and R.L. Whistler, *Starch: Chemistry and Technology*, 3rd Ed., Academic Press, New York (2009).
45. A.J.F. Carvalho, Starch: major sources, properties and applications as thermoplastic materials, in *Monomers, Polymers and Composites from Renewable Resources,* M. Belgacem and A. Gandini, (Eds.), pp. 321–342, Elsevier, Amsterdam (2008).
46. C.-W. Chiu, and D. Solarek, Modification of starches, in *Starch: Chemistry and Technology*, 3rd Ed., J.N. BeMiller and R. Whistler, (Eds.), pp. 629–656, *Academic Press*, New york (2009).
47. J.L. Willett, Starch in polymer compositions & quot, in *Starch: Chemistry and Technology*, 3rd Ed., J.N. BeMiller and R. Whistler, (Eds.), Academic Press, New York, pp. 715–743, (2009).
48. X. Ma, P.R. Chang, and J. Yu, Properties of biodegradable thermoplastic pea starch/carboxymethyl cellulose and pea starch/microcrystalline cellulose composites. *Carbohydr. Polym.* 72, 369–375, (2008).
49. I.M.G. Martins, S.P. Magina, L. Oliveira, S.R. Freire, A.J.D. Silvestre, C.P. Neto, and A. Gandini, New biocomposites based on thermoplastic starch and bacterial cellulose. *Compos. Sci. Technol.* 69, 2163–2168, (2009).
50. J. Prachayawarakorn, N. Limsiriwong, R. Kongjindamunee, and S. Surakit, Effect of agar and cotton fiber on properties of thermoplastic waxy rice starch composites. *J. Polym. Environ.* 20(1), 88–95 (2012).

51. A. Dufresne, D. Dupeyre, and M.R. Vignon, Cellulose microfibrils from potato tuber cells: Processing and characterization of starch-cellulose microfibril composites. *J. Appl. Polym. Sci.* 76, 2080–2092, (2000).
52. A. Bastioli, Properties and applications of Mater-Bi starch-based materials. *Polym. Degrad. Stabil.* 59(1–3), 263–272, (1998).
53. D. Schwartz, and R.L. Whistler, History and future of starch, in *Starch: Chemistry and Technology*, 3rd Ed., J.N. BeMiller and R. Whistler, (Eds.), pp. 1–10, *Academic Press*, New York, (2009).
54. D.L. Phillips, H. Lui, D. Pan, and H. Corke, General application of Raman spectroscopy for the determination of level of acetylation in modified starches *Cereal Chem.* 76, 439–443 (1999).
55. D.L. Phillips, J. Xing, C.K. Chong, H. Lui, and H. Corke, Determination of the degree of succinylation in diverse modified starches by Raman Spectroscopy. *J. Agric. Food Chem.* 48, 5105–5108 (2000).
56. C.K. Chong, J. Xing, D.L. Phillips, and H. Corke, Development of NMR and raman spectroscopic methods for the determination of the degree of substitution of maleate in modified starches. *J. Agric. Food Chem.* 49, 2702–2708 (2001).
57. J.R. Huang, H.A. Schols, R. Klaver, Z.Y. Jin, and A.G.J. Voragen, Acetyl substitution patterns of amylose and amylopectin populations in cowpea starch modified with acetic anhydride and vinyl acetate *Carbohydr. Polym.* 67, 542–550, (2007).
58. L.F. Wang, S.Y. Pan, H. Hu, W.H. Miao, and X.Y. Xu, Synthesis and properties of carboxymethyl kudzu root starch. *Carbohydr. Polym.* 80, 174–179 (2010).
59. L. Dolmatova, C. Ruckebusch, N. Dupuy, J.-P. Huvenne, and P. Legrand, Identification of modified starches using infrared spectroscopy and artificial neural network processing. *Appl. Spectrosc.* 52, 329–338 (1998).
60. N. Dupuy, C. Wojciechowski, C.D. Ta, J.P. Huvenne, and P. Legrand, Mid-infrared spectroscopy and chemometrics in corn starch classification. *J. Mol. Struct.* 410, 551–554 (1997).
61. P. Matzinos, V. Tserki, A. Kontoyiannis, and C. Panayiotou, Processing and characterization of starch/polycaprolactone products. *Polym. Degrad. Stabil.* 77, 17–24 (2002).
62. Y. Lu, L. Weng, and X. Cao, Morphological, thermal and mechanical properties of ramie crystallites—Reinforced plasticized starch biocomposites. *Carbohydr. Polym. 63,* 198–204 (2006).
63. M.N. Anglés, and A. Dufresne, Plasticized starch/tunicin whiskers nanocomposite materials. 2. Mechanical behavior. *Macromolecules* 34, 2921–2931 (2001).
64. M. Avella, J.J. De Vlieger, M.E. Errico, S. Fischer, P. Vacca, and M. Grazia Volpe, Biodegradable starch/clay nanocomposite films for food packaging applications. *Food Chem.* 93, 467–474 (2005).
65. J.J. G. van Soest, and N. Knooren, Influence of glycerol and water content on the structure and properties of extruded starch plastic sheets during aging. *J. Appl. Polym. Sci.* 64(7), 1411–1422 (1997).
66. L. Averous, and N. Boquillon, Biocomposites based on plasticized starch: thermal and mechanical behaviours. *Carbohydr. Polym.* 56, 111–122 2004.
67. A.J.F. Carvalho, M.D. Zambon, A.A. da Silva Curvelo, and A. Gandini, Thermoplastic starch modification during melt processing: hydrolysis catalyzed by carboxylic acids. *Carbohydr. Polym.* 62, 387–390 (2005).
68. H. Angellier, S. Molina-Boisseau, P. Dole, and A. Dufresne, Thermoplastic starch–waxy maize starch nanocrystals nanocomposites. *Biomacromolecules* 7, 531–539 (2006).
69. H. Angellier, L. Choisnard, S. Molina-Boisseau, P. Ozil, and A. Dufresne, Optimization of the preparation of aqueous suspensions of waxy maize starch nanocrystals using a response surface methodology. *Biomacromolecules* 5, 1545–1551 (2004).

70. J.L. Putaux, S. Molina-Boisseau, T. Momaur, and A. Dufresne, Platelet nanocrystals resulting from the disruption of waxy maize starch granules by acid hydrolysis. *Biomacromolecules*, 4, 1198–1202 (2003).
71. J. Yu, N. Wang, and X. Ma, The effects of citric acid on the properties of thermoplastic starch plasticized by glycerol. *Starch/Stärke* 57, 494–504, (2005).
72. X. Ma, R. J. Jian, P. R. Chang, and J. G. Yu, Fabrication and characterization of citric acid-modified starch nanoparticles/plasticized-starch composites. *Biomacromolecules* 9(11), 3314–3320 (2008).
73. I. Spiridon, C.-A. Teacă, and R. Bodîrlău, Preparation and characterization of adipic acid-modified starch microparticles/plasticized starch composite films reinforced by lignin. *J. Mater. Sci.* 46, 3241–3251 (2011).
74. R. Bodîrlău, C.-A. Teacă, I. Spiridon, and N. Tudorachi, Effects of chemical modification on the structure and mechanical properties of starch-based biofilms. *Monatshefte fur Chemie (Chemical Monthly)* 143, 335–343 (2012).
75. C.A. Teacă, R. Bodîrlău, I. Spiridon, and N. Tudorachi, Multi-component polymer systems comprising modified starch microparticles and different natural fillers, in *Program of the European Polymer Federation Congress (EPF 2013)*, June 16–21, Pisa, Italy, P2–160, p. 42 (2013).
76. C.A. Teacă, R. Bodîrlău, and I. Spiridon, Effect of cellulose reinforcement on the properties of organic acid modified starch microparticles/plasticized starch bio-composite films. *Carbohydr. Polym.* 93, 307–315 (2013).
77. R. Bodîrlău, C.A. Teacă, and I. Spiridon, Influence of natural fillers on the properties of starch-based biocomposite films. *Compos. B* 44, 575–583 (2013).
78. R. Bodîrlău, C.A. Teacă, and I. Spiridon, Green composites comprising thermoplastic corn starch and various cellulose-based fillers. *BioResources* 9(1), 39–53 (2014).
79. D. Le Corre, J. Bras, and A. Dufresne, Starch nanoparticles: A review. *Biomacromolecules* 11, 1139–1153 (2010).
80. G.X. Xing, S.F. Zhang, B.Z. Ju, and J.-Z. Yang, Microwave-assisted synthesis of starch maleate by dry method. *Starch/Stärke* 58, 464–467 (2006).
81. N. Reddy, and Y. Yang, Extraction and characterization of natural cellulose fibers from common milkweed stems. *Polym. Eng. Sci.* 49(11), 2212–2217 (2009).
82. M. Likon, M. Remskar, V. Ducman, and F. Svegl, Populus seed fibers as a natural source for production of oil super absorbents. *J. Environ. Manag.* 114, 158–167 (2013).
83. K. Aoi, A. Takasu, M. Tsuchiya, and M. Okada, New chitin-based polymer hybrids, 3. Miscibility of chitin-graft-poly(2-ethyl-2-oxazoline) with poly(vinyl alcohol). *Macromol. Chem. Phys.* 199, 2805–2811 (1998).
84. J.M. Fang, P.A. Fowler, J. Tomkinson, and C.A.S. Hill, The preparation and characterisation of a series of chemically modified potato starches. *Carbohydr. Polym.* 47, 245–252 (2002).
85. J.J.G. Van Soest, H. Tournois, D. de Wit, and J.F.G. Vliegenthart, Short-range structure in (partially) crystalline potato starch determined with attenuated total reflectance Fourier-transform IR spectroscopy. *Carbohydr. Res.* 279, 201–214 (1995).
86. N.M. Vicentini, N. Dupuy, M. Leitzelman, M.P. Cereda, and P.J.A. Sobral, Prediction of cassava starch edible film properties by chemometric analysis of infrared spectra. *Spectrosc. Lett.* 38, 749–767 (2005).
87. Y. Maréchal and H. Chanzy, The hydrogen bond network in Iβ cellulose as observed by infrared spectrometry. *J. Mol. Struct.* 523(1–3), 183–196 (2000).
88. R.H. Marchessault, Application of infra-red spectroscopy to cellulose and wood polysaccharides. *Pure Appl. Chem.* 5(1–2), 107–130 (1962).

89. B. Hinterstoisser, and L. Salmén, Two-dimensional step-scan FTIR: A tool to unravel the OH-valency-range of the spectrum of Cellulose I. *Cellulose* 6(3), 251–263 (1999).
90. T. Kondo, and C. A. Sawatari, A fourier transform infra-red spectroscopic analysis of the character of hydrogen bonds in amorphous cellulose. *Polymer* 37(3), 393–399 (1996).
91. J. Rőhrling, A. Potthast, T. Rosenau, H. Sixta, and P. Kosma, Determination of carbonyl functions in cellulosic substrates, *Lenzinger Berichte* 81, 89–97 (2002).
92. J.J.G. Van Soest, and J.F.G. Vliegenthart, Crystallinity in starch plastics: consequences for material properties. *Trends Biotechnol.* 15, 208–213 (1997).
93. X.J. Xie, Q. Liu and S.W. Cui, Studies on the granular structure of resistant starches (type 4) from normal, high amylose and waxy corn starch citrates. *Food Res. Int.* 39, 332–341 (2006).
94. A. Dufresne, and M.R. Vignon, Improvement of starch film performances using cellulose microfibrils. *Macromolecules* 31, 2693–2696 (1998).
95. M.A. Bertuzzi, M. Armada, and J.C. Gottifredi, Physicochemical characterization of starch based films. *J. Food Eng.* 82, 17–25 (2007).

8

Spectroscopy Analysis and Applications of Rice Husk and Gluten Husk Using Computational Chemistry

Norma-Aurea Rangel-Vazquez*, Virginia Hernandez-Montoya and Adrian Bonilla-Petriciolet

Division of Graduate Studies and Research, Aguascalientes, México

Abstract

Computational chemistry is a branch of chemistry that uses principles of computer science to assist in solving chemical problems. It uses the results of theoretical chemistry, incorporated into efficient computer programs, to calculate the structures and properties of molecules and solids. The lignocellulosic materials are mainly made up of a complex network of three polymers: cellulose, hemicellulose, and lignin. Due to their hydrophilicity, biodegradability, biocompatibility and low toxicity, hemicelluloses have been studied by numerous research groups with respect to their use as composites in biomedical applications.

In this research, rice husk and gluten husk were analyzed. Rice husk (RH) is a fibrous material, composed mainly of cellulose, lignin and inorganic and organic compounds. Rice husk ash (RHA) is a light material, which is bulky and porous; it amounts to about 20% of the burnt husk. Gluten is the composite of a gliadin and a glutenin, which is conjoined with starch in the endosperm of various grass-related grains. The prolamin and glutelin from wheat (gliadin, which is alcohol-soluble, and glutenin, which is only soluble in dilute acids or alkalis) constitute about 80% of the protein contained in wheat fruit.

The analysis techniques used were: FTIR to study this effect and the optional use of theoretical calculations to justify the obtained results by means of computational chemistry tools. Using QSAR properties, we can obtain an estimate of the activity of a chemical from its molecular structure only. The QSARs have been successfully applied to predict soil sorption coefficients of non-polar and nonionizable organic compounds, including many pesticides. Sorption of organic chemicals in soils or sediments is usually described by sorption coefficients. The molecular electrostatic potential (MESP) was calculated using the AMBER/AM1 method. These methods give information about the proper region by which compounds have intermolecular interactions between their units.

Keywords: Lignocellulosic materials, absorption process, computational chemistry, geometry optimization, QSAR, FTIR, potential electrostatic

**Corresponding author*: normarangelvazquez201301@gmail.com

Vijay Kumar Thakur (Ed.), Lignocellulosic Polymer Composites, (147–174) 2015 © Scrivener Publishing LLC

8.1 Introduction

8.1.1 Computational Chemistry

Computational chemistry is a branch of chemistry that uses principles of computer science to assist in solving chemical problems. It uses the results of theoretical chemistry, incorporated into efficient computer programs, to calculate the structures and properties of molecules and solids. Its necessity arises from the well-known fact that apart from relatively recent results concerning the hydrogen molecular ion (see references therein for more details), the quantum many-body problem cannot be solved analytically, much less in closed form. While its results normally complement the information obtained by chemical experiments, it can in some cases predict hitherto unobserved chemical phenomena. It is widely used in the design of new drugs and materials. Examples of such properties are structure (i.e., the expected positions of the constituent atoms), absolute and relative (interaction) energies, electronic charge distributions, dipoles and higher multipole moments, vibrational frequencies, reactivity or other spectroscopic quantities, and cross-sections for collision with other particles.

In all cases the computer time and other resources (such as memory and disk space) increase rapidly with the size of the system being studied. That system can be a single molecule, a group of molecules, or a solid. Computational chemistry methods range from highly accurate to very approximate; highly accurate methods are typically feasible only for small systems [1].

A single molecular formula can represent a number of molecular isomers. Each isomer is a local minimum on the energy surface (called the potential energy surface) created from the total energy (i.e., the electronic energy, plus the repulsion energy between the nuclei) as a function of the coordinates of all the nuclei. A stationary point is geometry such that the derivative of the energy with respect to all displacements of the nuclei is zero. A local (energy) minimum is a stationary point where all such displacements lead to an increase in energy. The local minimum that is lowest is called the global minimum and corresponds to the most stable isomer. If there is one particular coordinate change that leads to a decrease in the total energy in both directions, the stationary point is a transition structure and the coordinate is the reaction coordinate. This process of determining stationary points is called geometry optimization.

The determination of molecular structure by geometry optimization became routine only after efficient methods for calculating the first derivatives of the energy with respect to all atomic coordinates became available. Evaluation of the related second derivatives allows the prediction of vibrational frequencies if harmonic motion is estimated. More importantly, it allows for the characterization of stationary points. The frequencies are related to the Eigenvalues of the Hessian matrix, which contains second derivatives. If the Eigenvalues are all positive, then the frequencies are all real and the stationary point is a local minimum. If one Eigenvalue is negative (i.e., an imaginary frequency), then the stationary point is a transition structure. If more than one Eigenvalue is negative, then the stationary point is a more complex one, and is usually of little interest.

When one of these is found, it is necessary to move the search away from it if the experimenter is looking solely for local minima and transition structures. The total energy is determined by approximate solutions of the time-dependent Schrödinger equation, usually with no relativistic terms included, and by making use of the Born–Oppenheimer approximation, which allows for the separation of electronic and nuclear motions, thereby simplifying the Schrödinger equation. This leads to the evaluation of the total energy as a sum of the electronic energy at fixed nuclei positions and the repulsion energy of the nuclei. A notable exception is certain approaches called direct quantum chemistry, which treat electrons and nuclei on a common footing. Density functional methods and semi-empirical methods are variants on the major theme. For very large systems, the relative total energies can be compared using molecular mechanics [2,3].

8.1.1.1 Molecular Mechanics Methods

Molecular mechanics uses classical mechanics to model molecular systems. The potential energy of all systems in molecular mechanics is calculated using force fields. Molecular mechanics can be used to study small molecules as well as large biological systems or material assemblies with many thousands to millions of atoms. All-atomistic molecular mechanics methods have the following properties:

- Each atom is simulated as a single particle.
- Each particle is assigned a radius (typically the van der Waals radius), polarizability, and a constant net charge (generally derived from quantum calculations and/or experiment).
- Bonded interactions are treated as "springs" with an equilibrium distance equal to the experimental or calculated bond length.

Molecular mechanics potential energy functions have been used to calculate binding constants, protein folding kinetics, protonation equilibria, active site coordinates, and to design binding sites [4,5].

8.1.1.1.1 AMBER Method

The term "AMBER force field" generally refers to the functional form used by the family of AMBER force fields. This form includes a number of parameters; each member of the family of AMBER force fields provides values for these parameters and has its own name. The functional form of the AMBER force field is (equation 8.1).

$$V(\tau^N) = \sum_{bonds} k_b(l - l_o)^2 + \sum_{angles} k_a(\Theta - \Theta_o)^2 + \sum_{angles} 0.5V_N[1 + \cos(nw - \gamma)] \\ + \sum_{j=1}^{N-1} \sum_{i=j+1}^{N} \left\{ o_{ij} \left(\left(\frac{ro_{ij}}{r_{ij}} \right)^{12} - 2\left(\frac{ro_{ij}}{r_{ij}} \right)^{6} \right) + \frac{q_i q_j}{4\Pi\varepsilon_o ro_{ij}} \right\} \quad (8.1)$$

The meanings of right hand side terms are:

1. First term (summing over bonds): represents the energy between covalently bonded atoms. This harmonic (ideal spring) force is a good

approximation near the equilibrium bond length, but becomes increasingly poor as atoms separate.

2. Second term (summing over angles): represents the energy due to the geometry of electron orbitals involved in covalent bonding.
3. Third term (summing over torsions): represents the energy for twisting a bond due to bond order (e.g., double bonds) and neighboring bonds or lone pairs of electrons. Note that a single bond may have more than one of these terms, such that the total torsional energy is expressed as a Fourier series.
4. Fourth term (double summation over i and j): represents the non-bonded energy between all atom pairs, which can be decomposed into van der Waals (first term of summation) and electrostatic (second term of summation) energies.

The form of the van der Waals energy is calculated using the equilibrium distance (ro_{ij}) and well depth (ε). The factor of 2 ensures that the equilibrium distance is ro_{ij}. The energy is sometimes reformulated in terms of o, where $ro_{ij} = 2^{1/6}$ (o), as used, e.g., in the implementation of the soft core potentials. The form of the electrostatic energy used here assumes that the charges due to the protons and electrons in an atom can be represented by a single point charge (or in the case of parameter sets that employ lone pairs, a small number of point charges) [6].

8.1.1.2 Semi-Empirical Methods

Semi-empirical quantum chemistry methods are based on the Hartree–Fock formalism, but make many approximations and obtain some parameters from empirical data. They are very important in computational chemistry for treating large molecules where the full Hartree–Fock method without the approximations is too expensive. The use of empirical parameters appears to allow some inclusion of electron correlation effects into the methods. Within the framework of Hartree–Fock calculations, some pieces of information (such as two-electron integrals) are sometimes approximated or completely omitted.

In order to correct for this loss, semi-empirical methods are parametrized, that is their results are fitted by a set of parameters, normally in such a way as to produce results that best agree with experimental data, but sometimes to agree with *ab initio* results. Semi-empirical methods follow what are often called empirical methods where the two-electron part of the Hamiltonian is not explicitly included.

For π-electron systems, this was the Hückel method proposed by Erich Hückel. For all valence electron systems, the extended Hückel method was proposed by Roald Hoffmann. Semi-empirical calculations are much faster than their *ab initio* counterparts. Their results, however, can be very wrong if the molecule being computed is not similar enough to the molecules in the database used to parametrize the method. Semi-empirical calculations have been most successful in the description of organic chemistry, where only a few elements are used extensively and molecules are of moderate size. However, semi-empirical methods were also applied to solids and nanostructures but with different parameterization. As with empirical methods, we can distinguish if:

- Restricted to π-electrons. These methods exist for the calculation of electronically excited states of polyenes, both cyclic and linear. These methods, such as the Pariser–Parr–Pople method (PPP), can provide good estimates of the π-electronic excited states, when parameterized well. Indeed, for many years, the PPP method outperformed ab initio excited state calculations [6].

8.1.1.2.1 AM1 Method

AM1 is basically a modification to and a reparameterization of the general theoretical model found in MNDO. Its major difference is the addition of Gaussian functions to the description of core repulsion function to overcome MNDO's hydrogen bond problem. Additionally, since the computer resources were limited in 1970s, in MNDO parameterization methodology, the overlap terms, βs and βp, and Slater orbital exponent's ζs and ζp for *s*- and *p*- atomic orbitals were fixed. That means they are not parameterized separately just considered as βs = βp, and ζs = ζp in MNDO. Due to the greatly increasing computer resources in 1985 comparing to 1970s, these inflexible conditions were relaxed in AM1 and then likely better parameters were obtained.

The addition of Gaussian functions significantly increased the numbers of parameters to be parameterized from 7 (in MNDO) to 13-19, but AM1 represents a very real improvement over MNDO, with no increase in the computing time needed. Dewar also concluded that the main gains of AM1 were its ability to reproduce hydrogen bonds and the promise of better estimation of activation energies for reactions. However, AM1 has some limitations. Although hypervalent molecules are improved over MNDO, they still give larger errors than the other compounds, alkyl groups are too stable, nitro compounds are too unstable, peroxide bond are too short. AM1 has been used very widely because of its performance and robustness compared to previous methods. This method has retained its popularity for modeling organic compounds and results from AM1 calculations continue to be reported in the chemical literature for many different applications.

AM1 is currently one of the most commonly used of the Dewar-type methods. It was the next semiempirical method introduced by Dewar and coworkers in 1985 following MNDO. It is simply an extension, a modification to and also a reparameterization of the MNDO method. AM1 differs from MNDO by mainly two ways. The first difference is the modification of the core repulsion function. The second one is the parameterization of the overlap terms βs and βp, and Slater-type orbital exponents ζs and ζp on the same atom independently, instead of setting them equal as in MNDO. MNDO had a very strong tendency to overestimate repulsions between atoms when they are at approximately their van der Waals distance apart. To overcome this hydrogen bond problem, the net electrostatic repulsion term of MNDO, *f(RAH)* given by equation (8.2), was modified in MNDO/H to be

$$f\left(R_{AH}\right) = Z_A Z_B \left(S_A S_A | S_H S_H\right) [e_{AH}^{-\alpha R^2}] \tag{8.2}$$

Where α was proposed to be equal to 2.0 Å^{-2} for all A-H pairs. On the other hand, the original core repulsion function of MNDO was modified in AM1 by adding Gaussian functions to provide a weak attractive force. The core-core repulsion energy term in AM1 is given by equation 8.3.

$$E_{AB}^{AM1} = Z_A Z_B (S_A\ S_A \mid S_H S_H) [1 + e^{\alpha R}_{A\,AB}] + \frac{Z_A Z_B}{R_{AB}} [F(A) + F(B)] \qquad (8.3)$$

The Gaussian functions F(A) and F(B) are expressed by equation 8.4.

$$\begin{aligned} F(A) &= \sum\nolimits_i K_A ie^{-L_{A,i}} (R_{AB} - M_{A,i})^2 \\ F(B) &= \sum\nolimits_i K_B ie^{-L_{B,i}} (R_{AB} - M_{B,i})^2 \end{aligned} \qquad (8.4)$$

And finally AM1 core-repulsion function becomes (equation 8.5).

$$E_{AB}^{AM1} = E_{AB}^{MNDO} + \frac{Z_A Z_B}{R_{AB}} \left(\sum\nolimits_i K_{A,i} e^{[-L_{A,i}(R_{AB} - M_{A,i})^2]} + \sum\nolimits_j K_{B,j} e^{[-L_{B,j}(R_{AB} - M_{B,j})^2]} \right) \qquad (8.5)$$

In this equation 8.5, K, L and M are the Gaussian parameters. The remaining parameters have the same meaning as in the previous section. L parameters determine the widths of the Gaussians and were not found to be critical by Dewar. Therefore, a common value was used for many of the L parameters. On the other hand, all K and M parameters were optimized. Each atom has up to four of the Gaussian parameters, i.e., K1, …, K4, L1, …, L4, M1, …, M4.

Carbon has four terms in its Gaussian expansion whereas hydrogen and nitrogen have three and oxygen has two terms (only K1, K2, L1, L2, M1, M2). Because in AM1 for carbon, hydrogen and nitrogen both attractive and repulsive Gaussians were used whereas for oxygen only repulsive ones considered, addition of Gaussian functions into the core-repulsion function significantly increased the number of parameters to be optimized and made the parameterization process more difficult.

As for original MNDO, one-center two-electron repulsion integrals *gss*, *gpp*, *gdd*, *gsp*, *hsp* are assigned to atomic spectral values and not optimized. In contrast to MNDO, in which parameters were first optimized for carbon and hydrogen together and then other elements added one at a time, by increased computer resources and improved optimization procedure a larger reference parameterization dataset was used in the parameterization of AM1. All the parameters for H, C, N and O were optimized at once in a single parameterization procedure.

Optimization of the original AM1 elements was performed manually by Dewar using chemical knowledge and intuition. He also kept the size of the reference parameterization data at a minimum by very carefully selecting necessary data to be used as reference. Over the following years many of the main-group elements have been parameterized keeping the original AM1 parameters for H, C, N and O unchanged. Of course, a sequential parameterization scheme caused every new parameterization to depend on previous ones, which directly affects the quality of the results.

AM1 represented a very considerable improvement over MNDO without any increase in the computing time needed. AM1 has been parameterized for many of the main-group elements and is very widely used, keeping its popularity in organic compounds' modeling due to its good performance and robustness. Although many of the deficiencies in MNDO were corrected in AM1, it still has some important limitations as outlined in the historical development section [6].

8.1.2 Lignocellulosic Materials

Lignocellulosic materials comprising forestry, agricultural and agro-industrial wastes are abundant, renewable and inexpensive energy sources. Such wastes include a variety of materials such as sawdust, poplar trees, sugarcane bagasse, waste paper, brewer's spent grains, switch grass, and straws, stems, stalks, leaves, husks, shells and peels from cereals like rice, wheat, corn, sorghum and barley, among others [7,8].

Lignocellulose wastes are accumulated every year in large quantities, causing environmental problems. However, due to their chemical composition based on sugars and other compounds of interest, they could be utilized for the production of a number of value added products, such as ethanol, food additives, organic acids, enzymes, and others. Therefore, besides the environmental problems caused by their accumulation in the nature, the non-use of these materials constitutes a loss of potentially valuable sources. The major constituents of lignocellulose are cellulose, hemicellulose, and lignin, polymers that are closely associated with each other constituting the cellular complex of the vegetal biomass. Basically, cellulose forms a skeleton which is surrounded by hemicellulose and lignin (Figure 8.1).

Cellulose is a high molecular weight linear homopolymer of repeated units of cellobiose (two anhydrous glucose rings joined via a β-1,4 glycosidic linkage). The long-chain cellulose polymers are linked together by hydrogen and van der Walls bonds, which cause the cellulose to be packed into microfibrils. By forming these hydrogen bounds, the chains tend to arrange in parallel and form a crystalline structure. Therefore, cellulose microfibrils have both highly crystalline regions (around 2/3 of the total cellulose) and less-ordered amorphous regions. More ordered or crystalline cellulose is less soluble and less degradable.

Hemicellulose is a linear and branched heterogeneous polymer typically made up of five different sugars – L-arabinose, D-galactose, D-glucose, D-mannose, and D-xylose - as well as other components such as acetic, glucuronic, and ferulic acids. The backbone of the chains of hemicelluloses can be a homopolymer (generally consisting of single sugar repeat unit) or a heteropolymer (mixture of different sugars). According to the main sugar residue in the backbone, hemicellulose has different classifications, e.g., xylans, mannans, glucans, glucuronoxylans, arabinoxylans, glucomannans, galactomannans, galactoglucomannans, β-glucans, and xyloglucans. When compared to cellulose, hemicelluloses differ thus by composition of sugar units, by presence of shorter chains, by a branching of main chain molecules, and to be amorphous, which made its structure easier to hydrolyze than cellulose.

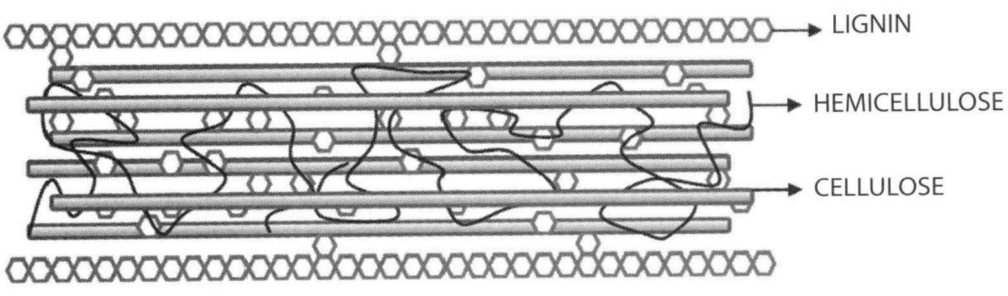

Figure 8.1 Representation of lignocellulosic materials structure showing cellulose, hemicellulose and lignin fractions.

Table 8.1 Main components of lignocellulose wastes.

Lignocellulosic materials waste	Cellulose (Wt%)	Hemicellulose (Wt%)	Lignin (Wt%)
Barley straw	33.8	21.9	13.8
Corn cobs	33.7	31.9	06.1
Corn stalks	35.0	16.8	07.0
Cotton stalks	58.5	14.4	21.5
Oat straw	39.4	27.1	17.5
Rice straw	36.2	19.0	09.9
Rye straw	37.6	30.5	19.0
Soya stalks	34.5	24.8	19.8
Sugarcane bagasse	40.0	27.0	10.0
Sunflower stalks	42.1	29.7	13.4
Wheat straw	32.9	24.0	08.9

Lignin is a very complex molecule constructed of phenylpropane units linked in a large three-dimensional structure. Three phenyl propionic alcohols exist as monomers of lignin: p-coumaryl alcohol, coniferyl alcohol and sinapyl alcohol. Lignin is closely bound to cellulose and hemicellulose and its function is to provide rigidity and cohesion to the material cell wall, to confer water impermeability to xylem vessels, and to form a physic–chemical barrier against microbial attack. Due to its molecular configuration, lignins are extremely resistant to chemical and enzymatic degradation.

The amounts of carbohydrate polymers and lignin vary from one plant species to another. In addition, the ratios between various constituents in a single plant may also vary with age, stage of growth, and other conditions. However, cellulose is usually the dominant structural polysaccharide of plant cell walls (35–50%), followed by hemicellulose (20–35%) and lignin (10–25%). Average values of the main components in some lignocellulose wastes are shown in Table 8.1 [8].

8.1.2.1 Rice Husk

Rice husk ask (Figure 8.2) is one of the most widely available agricultural wastes in many rice producing countries around the world. In majority of rice producing countries much of the husk produced from processing of rice is either burnt or dumped as waste. Burning of rice husk ask in ambient atmosphere leaves a residue, called rice husk ash. For every 1000 kgs of paddy milled, about 220 kgs (22%) of husk is produced, and when this husk is burnt in the boilers, about 55 kgs (25%) of rice husk ask is generated.

The chemical composition of rice husk ask is similar to that of many common organic fibers and it contains of cellulose 40-50%, lignin 25-30%, ash 15-20% and moisture 8-15%. Typical analyses of rice husk ask is shown in Table 8.2. The content of each of them depends on rice variety, soil chemistry, climatic conditions, and even the geographic localization of the culture.

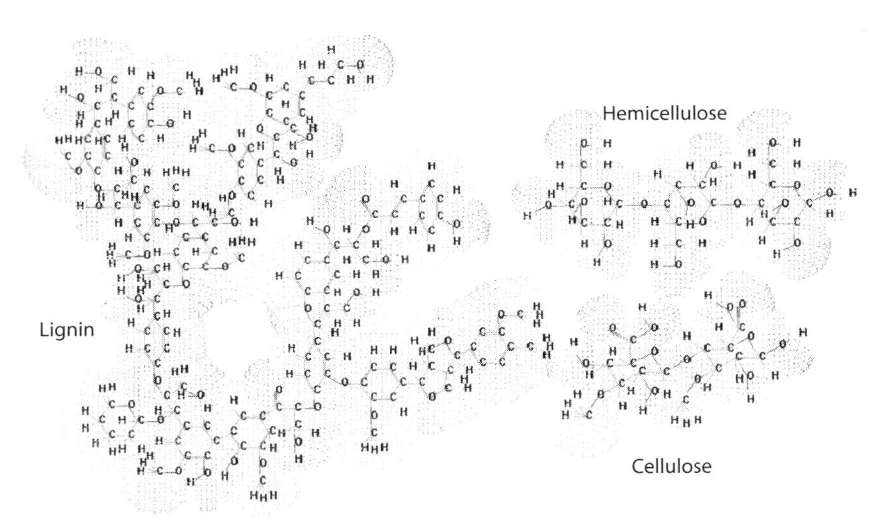

Figure 8.2 Component principal of rice husk ask.

Table 8.2 Typical analysis of rice husk ask.

Property	Range
Bulk density (kg/m³)	96-160
Hardness (Mohr's scale)	5-6
Ash (%)	22-29
Carbon (%)	≈ 35
Hydrogen (%)	4-5
Oxygen (%)	31-37
Nitrogen (%)	0.23-0.32
Sulphur (%)	0.04-0.08
Moisture	8-9

The exterior of rice husk ask are composed of dentate rectangular elements, which themselves are composed mostly of silica coated with a thick cuticle and surface hairs. The mid region and inner epidermis contain little silica. Jauberthie *et al.*, confirmed that the presence of amorphous silica is concentrated at the surfaces of the rice husk and not within the husk itself [9]. The properties of rice husk ash and its main composition are presented in Table 8.3. The organic materials consist of cellulose and lignin which turn to CO_2 and CO when rice husk ask burns in air. The ash contains mainly silica (90%), and a small portion of metal oxides (~5%) and residual carbon obtained from open burning [10].

8.1.2.2 Wheat Gluten Husk

Wheat gluten is a protein composite found in foods processed from wheat and related grain species, including barley and rye. WG gives elasticity to dough, helping it rise and keep its shape and often gives the final product a chewy texture. WG may also

Table 8.3 Chemical composition of Rice husk ask.

Component	%
SiO_2	92.498
Fe_2O_3	0.136
Al_2O_3	0.249
CaO	0.622
MgO	0.442
K_2O	2.490
LOI	3.520
Na_2O	0.023

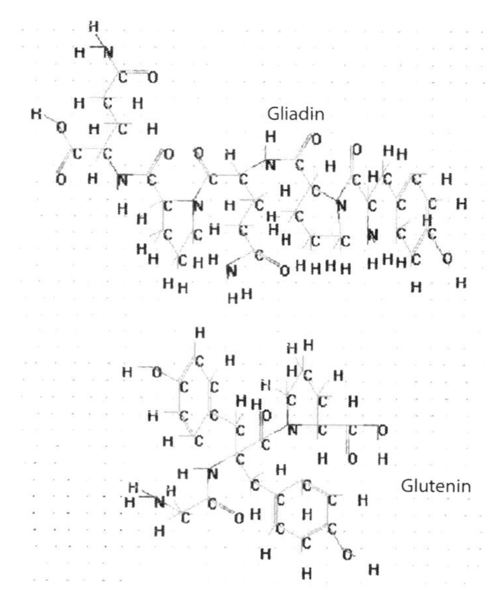

Figure 8.3 Structure of wheat gluten husk.

be found in some cosmetics, hair products, and other dermatological preparations [11,12]. Commercial WG has a mean composition of 72.5% protein (77.5% on dry basis), 5.7% total fat, 6.4% moisture and 0.7% ash; carbohydrates, mainly starches, are the other major component [13]. WG husk is the composite of a *gliadin* and a *glutenin* (Figure 8.3), which is conjoined with starch in the endosperm of various grass-related grains. The prolamin and glutelin from wheat (gliadin, which is alcohol-soluble, and glutenin, which is only soluble in dilute acids or alkalis) constitute about 80% of the protein contained in wheat fruit. Being insoluble in water, they can be purified by washing away the associated starch. Worldwide, gluten is a source of protein, both in foods prepared directly from sources containing it, and as an additive to foods otherwise low in protein [11,12].

Gliadins are monomeric proteins that can be separated into four groups, alpha-, beta-, gamma- and omega-gliadins. Glutenins occur as multimeric aggregates of high molecular weight (HMW) and low-molecular-weight (LMW) subunits held together by disulphide bonds. In wheat, omega- and gamma-gliadins are encoded by genes at the Gli-1 loci located on the short arms of group 1 chromosomes, while alpha and beta-gliadin-encoding genes are located on the short arms of group 6 chromosomes. LMW glutenins are encoded by genes at the Glu-3 loci that are closely linked to the Gli-1 loci. HMW glutenins are encoded by genes at the Glu-1 loci found on the long arms of group 1 chromosomes. Each Glu-1 locus consists of two tightly linked genes encoding one 'x'-type and one 'y'-type HMW glutenin, with polymorphism giving rise to a number of different alleles at each locus.

The y-type genes at the Glu-A1 locus are not expressed in hexaploid wheat. Due to the very close linkage between the x and y type genes, HMW glutenins are classified into alleles according to the x and y type subunits expressed. Considerable efforts have been made to understand the relationship between gliadin and glutenin composition and rheological properties of wheat dough. It is now well understood that the properties of various wheat storage proteins have a major effect on dough rheological properties.

The gliadin and glutenin components contribute to dough quality either in an independent manner (additive genetic effects) or in interactive manner (epistatic effects). It was suggested that the apparent effects of gliadins on dough quality should be attributed to the LMW glutenins due to the close linkage of the Gli-1 and Glu-3 loci. Generally, HMW glutenins have been found to be more important than gliadins and LMW glutenins for dough rheological properties [14].

8.1.2.2.1 Composition and Properties

a) Film Forming

The film forming property of hydrated wheat gluten is a direct outcome of its viscoelasticity. Whenever carbon dioxide or water vapor forms internally in a gluten mass with sufficient pressure to partially overcome the elasticity, the gluten expands to a spongy cellular structure. In such structures, pockets or voids are created which are surrounded by a continuous protein phase to entrap and contain the gas or vapor. This new shape and structure can then be rendered dimensionally stable by applying sufficient heat to cause the protein to denature or devitalize and set up irreversibly into a fixed moist gel structure or to a crisp fragile state, depending on final moisture content.

The open texture of leavened breads; the suspension of solid particles such as fruit pieces or grains; and high fiber bread are examples of success due to the continuous phase of hydrated protein. Where the loading of added solid particles is greater than the strength possible from the flour used, "cripples" result. This is easily correctable by addition of wheat gluten to the flour base. In addition to its film forming potential in food systems, cast or floated films of wheat gluten can be made. Glazing of meat patties is possible, and wheat gluten films in the form of sausage casing, tubes or shreds are recorded in the patent literature as the product of gluten "hot melt" techniques.

b) Flavor

Properly produced and given reasonable care in storage, wheat gluten exhibits a flavor note variously described as "bland" or "slight cereal." Wheat gluten flavors enjoy wide acceptance and wheat gluten merges perfectly into all cereal-based products. Blending with meats in various binding, adhesive and extension roles need not result in off-flavor notes, even at high percentage use levels. Blending of wheat gluten with other food proteins which do possess characteristic flavor notes can result in imporved total flavor as, for example, when soy/wheat gluten blends are used for textured vegetable protein manufacture. Low and acceptable flavor levels of wheat gluten are the result of careful selection of flours, good manufacturing procedures and proper storage at normal ambient temperatures.

c) pH Effects
Since wheat gluten is a complex of proteins it has no sharp isoelectric (minimum solubility and dissociation) point. There is thus no readily discernible point at which the positive and negative charges exactly balance. Because glutenin is essentially insoluble in water over normal pH ranges, wheat gluten tends to reflect the isoelectric behavior of gliadin in pH/solubility properties. When gliadin is separately examined for pH/solubility criteria, it displays minimum solubility over the pH range 6–9. It is in this range that the cohesive, extensible network of wheat gluten is strongest. It is important to note that wheat gluten becomes more soluble in acid or alkaline dispersions (Some manufacturers utilize this effect to produce spray dried wheat gluten.

The aqueous acetic acid or ammonia used is flashed off during the drying step, and the powdered material retains typical vital gluten characteristics). pH manipulation may thus provide interesting property variations in wheat gluten containing foods [15].

8.1.3 Benzophenone

Benzophenone is the organic compound with the formula $(C_6H_5)_2CO$, generally abbreviated Ph_2CO (see Figure 8.4). Benzophenone is a widely used building block in organic chemistry, being the parent diarylketone.

Benzophenone is used as a flavour ingredient, a fragrance enhancer, a perfume fixative and an additive for plastics, coatings and adhesive formulations; it is also used in the manufacture of insecticides, agricultural chemicals, hypnotic drugs, antihistamines and other pharmaceuticals [16].

Benzophenone is used as an ultraviolet (UV)-curing agent in sunglasses, and to prevent UV light from damaging scents and colours in products such as perfumes and soaps. Moreover, it can be added to plastic packaging as a UV blocker, which

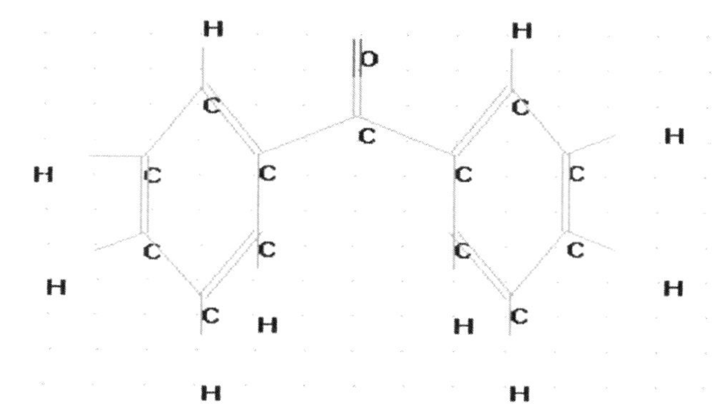

Figure 8.4 Benzophenone structure.

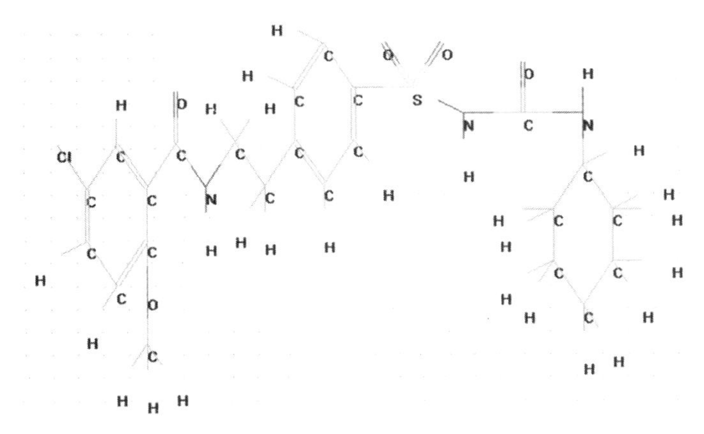

Figure 8.5 Glibenclamide structure.

allows manufacturers to package their products in clear glass or plastic rather than opaque or dark packaging. It is also used in laundry and household cleaning products. Benzophenone is widely used as a photoinitiator for inks and varnishes that are cured with UV light. In addition to being a drying catalyst, benzophenone is an excellent wetting agent for pigments; it can also be used in printing to improve the rheological properties and increase the flow of inks by acting as a reactive solvent [16,17].

8.1.4 Glibenclamide

Glibenclamide (Figure 8.5) is chemically known as 5-chloro-N-[2-[4[(cyclohexylamino) carbonyl] amino] sulfonyl] phenyl] ethyl]-2-methoxy benzamide is second generation sulphonyl ureas drug widely used in treatment of type 2 diabetic patients. It acts by inhibiting ATP-sensitive potassium channels in pancreatic beta cells causing cell membrane depolarization (increasing intracellular calcium in the beta cell) which stimulates the insulin release [18]. It was developed in 1966 in a cooperative study between Boehringer Mannheim (now part of Roche) and Hoechst (now part of Sanofi-Aventis).

8.1.4.1 Mechanism of Action

The drug works by inhibiting the sulfonylurea receptor 1 (SUR1), the regulatory subunit of the ATP-sensitive potassium channels (K_{ATP}) in pancreatic beta cells. This inhibition causes cell membrane depolarization opening voltage-dependent calcium channel. This results in an increase in intracellular calcium in the beta cell and subsequent stimulation of insulin release.

After a cerebral ischemic insult the blood brain barrier is broken and glibenclamide can reach the central nervous system. Glibenclamide has been shown to bind more efficiently to the ischemic hemisphere. Moreover, under ischemic conditions SUR1, the regulatory subunit of the K_{ATP-} and the $NC_{Ca\text{-}ATP}$-channels, is expressed in neurons, astrocytes, oligodendrocytes, endothelial cells and by reactive microglia.

8.1.4.2 Medical Uses

It is used in the treatment of type 2 diabetes. As of 2011, it is one of only two oral antidiabetics in the World Health Organization Model List of Essential Medicines (the other being metformin). As of 2003, in the United States, it was the most popular sulfonylurea. Additionally, recent research shows that glibenclamide improves outcome in animal stroke models by preventing brain swelling and enhancing neuroprotection. A retrospective study showed that in type 2 diabetic patients already taking glyburide, NIH stroke scale scores on were improved on discharge compared to diabetic patients not taking glyburide [19-21].

8.2 Methodology

8.2.1 Geometry Optimization

n this study the semi-empirical methods were used for describing the potential energy function of the system. Next a minimization algorithm is chosen to find the potential energy minimum corresponding to the lower-energy structure.

Iterations number and convergence level lead optimal structure. The optimizing process of structures used in this work was started using the AMBER/AM1 methods, because it generates a lower-energy structure even when the initial structure is far away from the minimum structure. The Polak-Ribiere algorithm was used for mapping the energy barriers of the conformational transitions. For each structure, 1350 iterations, a level convergence of 0.001 kcal/mol/Å and a line search of 0.1 were carried out [22].

8.2.2 FTIR

The infrared spectrum is commonly obtained by passing infrared electromagnetic radiation through a sample that possesses a permanent or induced dipole moment and determining what fraction of the incident radiation is absorbed at a particular energy [23]. The energy of each peak in an absorption spectrum corresponds to the frequency of the vibration of a molecule part, thus allowing qualitative identification of certain bond types in the sample. The FTIR was obtained by first selecting menu Compute, vibrational, rotational option, once completed this analysis, using the option vibrational spectrum of FTIR spectrum pattern is obtained for two methods of analysis.

8.2.3 Electrostatic Potential

After obtaining a free energy of Gibbs or optimization geometry using AMBER/AM1 methods, we can plot two-dimensional contour diagrams of the electrostatic potential surrounding a molecule, the total electronic density, the spin density, one or more molecular orbitals, and the electron densities of individual orbitals. HyperChem software displays the electrostatic potential as a contour plot when you select the appropriate option in the Contour Plot dialog box. Choose the values for the starting contour and the contour increment so that you can observe the minimum (typically about –0.5 for polar organic molecules) and so that the zero potential line appears.

A menu plot molecular graph, the electrostatic potential property is selected and then the 3D representation mapped isosurface for both methods of analysis. Atomic charges indicate where large negative values (sites for electrophilic attack) are likely to occur. However, the largest negative value of the electrostatic potential is not necessarily adjacent to the atom with the largest negative charge [24].

8.3 Results and Discussions

8.3.1 Geometry Optimization

The values of different thermodynamic parameters of both lignocellulosic materials are described in Table 8.4. The negative value of ΔG (Gibbs free energy) reflects the spontaneity of materials [25]. Attractive interactions between π systems are one of the principal non-covalent forces governing molecular recognition and play important roles in many chemical systems. Attractive interaction between π systems is the interaction between two or more molecules leading to self-organization by formation of a complex structure which has lower conformation equilibrium than of the separate components and shows different geometrical arrangement with high percentage of yield (Figures 8.6–8.7).

The difference in the energy values are attributed to the constituents of the lignocellulosic material [6]. Log P negative shows that these lignocellulosic materials can absorb polar solvents because of its hydrophilic character characteristic of cellulose [26,27].

8.3.2 FTIR Analysis

Table 8.5 shows the FTIR bands of rice husk where the characteristic peaks associated with organic components are observed. CH asymmetric (5743 cm^{-1}), C = O stretching hemicelluloses (1745 cm^{-1}) [28]. The absorption band at 3387 cm^{-1} corresponds to the combined bands of the NH_2 and OH group stretching vibration to chitosan [29]. The vibrations of the aromatic rings can be observed at 1846, 1718, 1413 and 1077 cm^{-1}, respectively. At 3125 and 2849 cm^{-1} were attributed at CH stretching in cellulose-rich material. Both cellulose/hemicelluloses -and lignin- associated bands are present in the

Table 8.4 Properties of lignocellulosic materials.

Properties	Rice husk	Wheat gluten husk
ΔG (Kcal/mol)	- 258.07	- 195.25
Surface area ($Å^2$)	2672.40	1479.34
Volumen ($Å^3$)	6966.35	2772.21
Mass (amu)	3468.45	1074.18
Log P	- 22.24	- 15.61

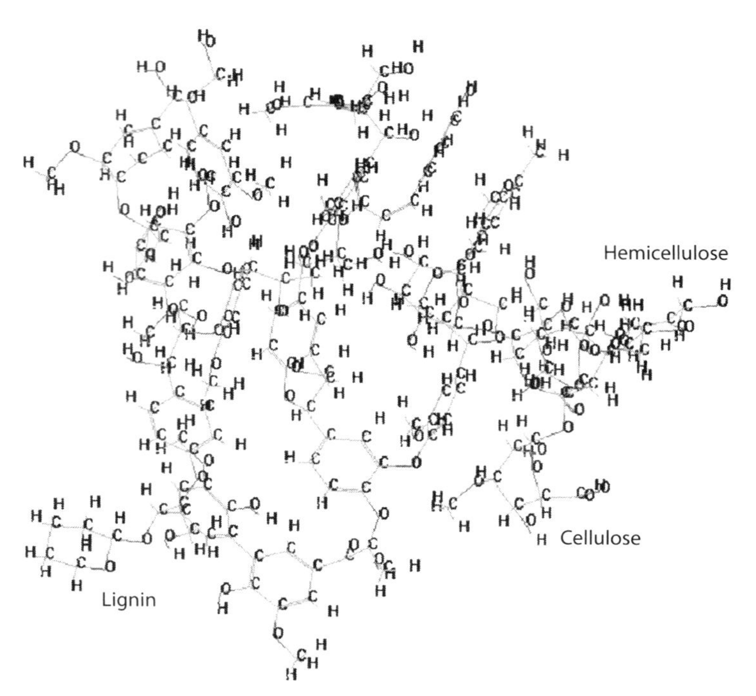

Figure 8.6 Rice husk optimum.

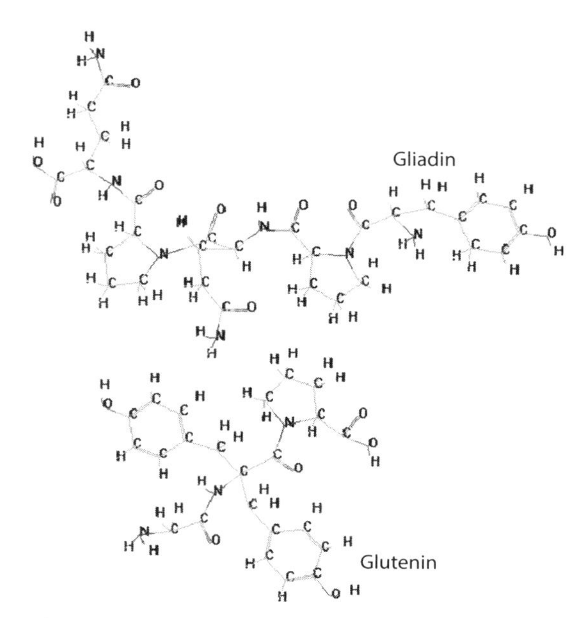

Figure 8.7 Wheat gluten husk optimum.

rice husk, and this suggests the presence of lignin-carbohydrate matrix in rice husk. In fact, the band at 815 cm^{-1} shows the strongest absorption. This band corresponds to the CO stretching vibration in both cellulose/hemicellulose and lignin, and it explains the lignocellulosic nature of rice husk [30].

Wheat gluten husk was analyzed using FTIR to know the various chemical constituents present (see Table 8.6). At 5741 cm^{-1} was assigned to NH stretching, from 4279 at 4029 cm^{-1} was attributed to CH stretching, the vibrations of CH_2 and NH_2 corresponding to 3695 cm^{-1}. The band at 3319 cm^{-1} is assignment to hydrogen bonded OH stretching. The hydrophilic tendency of wheat husk was reflected at 3695 cm^{-1} which is related to the OH groups present in aliphatic in this material. The bands in the range 1450–1370 cm^{-1} were assigned from the CH symmetric and asymmetric deformations. The region of 1200–1000 cm^{-1} represents the CH, C–N and C–O stretching and deformation bands in cellulose and lignin [31]. The lignocellulosic material peak corresponds to C = O and C = C bond appears to 3555, 3366 and 3024 cm^{-1} [32]. Finally, the C–C, C–N and C–O bond were observed to 2619 ad 1096 cm^{-1}.

Table 8.5 FTIR assignments of rice husk. Table 8.6 FTIR assignments of wheat gluten husk.

8.3.3 Electrostatic Potential

Molecular electrostatic potential (MESP), which is related to the electronegativity and the partial charge changes on the different atoms of the molecule, when plotted on the isodensity surface of the molecule MESP mapping is very useful in the investigation of the molecular structure with its physiochemical property relationships. Red and blue areas in the MESP refer to the regions of negative and positive and correspond to the electron-rich and electron-poor regions, respectively, whereas the green color signifies the neutral electrostatic potential [33]. The MESP in case of Figure 8.8(a) clearly

Table 8.5 FTIR assignments of rice husk.

Assignments	Wavenumber (cm^{-1})
CH asymmetric	5743
C=C	3765
C=C	3508
NH^2, OH	3387
C=O	3289, 1745
CH stretching	3125,2849
C-C, C-O	2549, 2127
C-C, C-O ring	1846, 1413
CH=CH	1718
C-C ring	1077
C-O	815
C-C, C-O, C-H	656

Table 8.6 FTIR assignments of wheat gluten husk.

Assignments	Wavenumber (cm^{-1})
NH	5741
CH^2 stretching	4279-4029
OH groups	3695
OH stretching	3319
C=O, C=C	3555, 3366, 3024
C-C, C-N, C-O	2619
C-C, C-N	2226
NH	1454
C-C, C-N, C-O	1289, 1096

suggest that each C–OH, C–O–C bonds represent the most negative potential region of rice husk with a 0.986 at 0.066 eV. Figure 8.8(b) shows that NH and CH bond present neutral potential electrostatic region, glutenine structure represent the most negative potential region and finally the CH_2 and CH represent the most positive potential region. This potential has a value of 1.013 at 0.067 eV.

8.3.4 Absorption of Benzophenone

8.3.4.1 Geometry Optimization

Table 8.7 shows that negative values of ΔH suggested that the exothermic nature of the adsorption. Negative values of ΔG indicated the spontaneous nature of the adsorption process of benzophenone [34]. One can see that the properties change with the addition of benzophenone, for example the Log P values in both cases determined that the materials tend to be more negative which found that tend to absorb water (polar solvents) because of its hydrophilic character characteristic of cellulose [26,27]. Figures 8.9–8.10 show the absorption of benzophenone, in where the formation of hydrogen bonds can be seen after calculating Gibbs free energy, the negative regions were located in the C = C and C–OH bonds respectively.

8.3.4.2 FTIR

Table 8.8 shows the FTIR bands of rice husk/benzophenone where the characteristic peaks associated with absorption process are observed. Comparing the results of Tables 8.5 and 8.8, can be seen that there are shifts in the peaks of rice husk attributed to the absorption of benzophenone, so the existence of one or more aromatic rings in a structure is normally readily determined from the CH and C = C–C ring related vibrations.

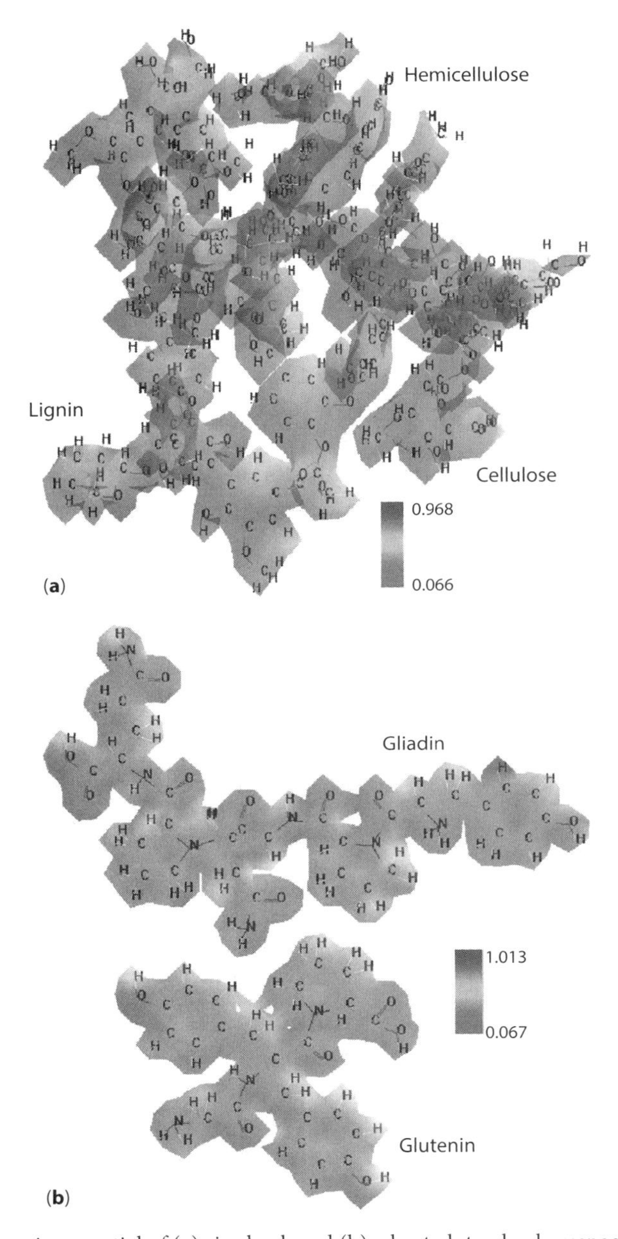

Figure 8.8 Electrostatic potential of (a) rice husk and (b) wheat gluten husk, respectively.

The CH stretching occurs above 2893 cm^{-1} and is typically exhibited as a multiplicity of weak to moderate bands, compared with the aliphatic CH stretch [35–37]. The characteristic infrared absorption frequencies of carbonyl group in cyclic ketones are normally strong in intensity and found in the region 1585–1621 cm^{-1}. The interaction of carbonyl group with other groups present in the system did not produce such a drastic and characteristic change in the frequency of C = O stretch as did by interaction of NH stretch. The carbon–oxygen double bond is formed by pπ–pπ between carbon and oxygen. Because of the different electronegativities of carbon and oxygen

Table 8.7 Properties of lignocellulosic materials with benzophenone.

Properties	Rice husk/ benzophenone	Wheat gluten husk/benzophenone
ΔG (Kcal/mol)	-190	- 55.68
Surface area (Å^2)	2952.04	2395.89
Volumen (Å^3)	7973.28	4409.02
Mass (amu)	4014.10	1620.84
Log P	-27.18	-24.25

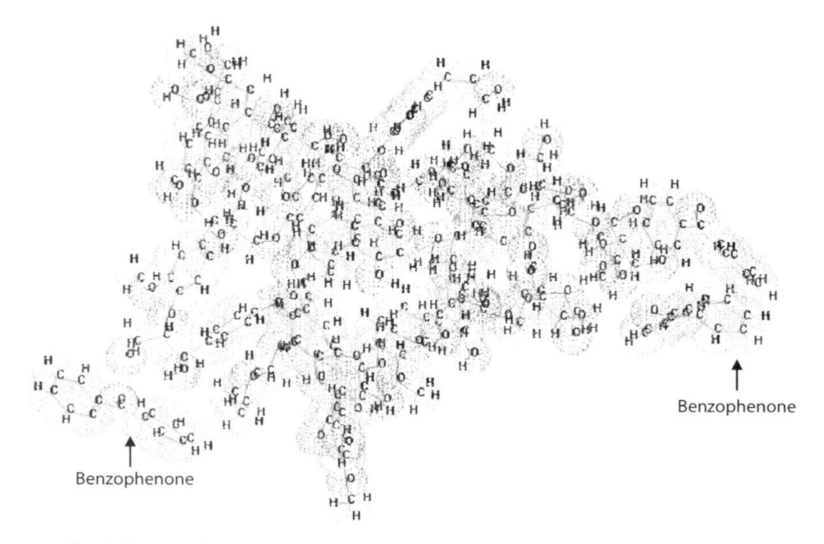

Figure 8.9 Rice husk/benzophenone structure after Gibbs free energy.

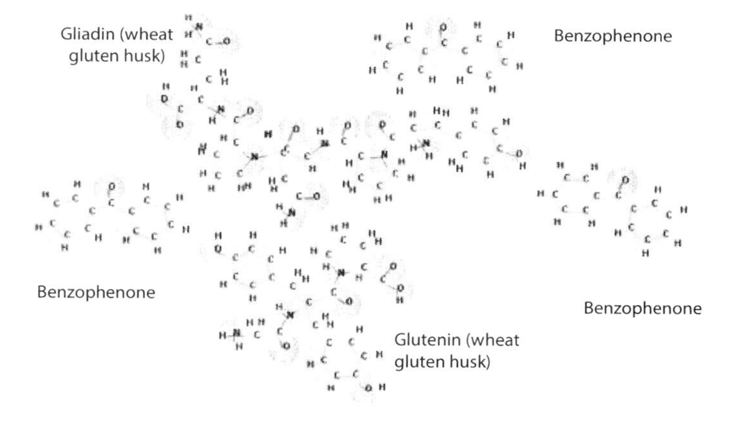

Figure 8.10 Wheat gluten husk/benzophenone structure after Gibbs free energy.

atoms, the bonding electrons are not equally distributed between the two atoms. The lone pair of electrons on oxygen also determines the nature of the carbonyl group. The position of the C = O stretching vibration is very sensitive to various factors such as the physical state, electronic effects by substituents, ring strains. Normally carbonyl group

Table 8.8 FTIR of rice husk/benzophenone.

Assignments	Wavenumber (cm^{-1})
C–H asymmetric stretching (rice husk)	5662
CH_2 asymmetric stretching (rice husk)	5492
C = C (rice husk)	3810, 3724, 3653
CH = CH (rice husk)	3260
C–H (rice husk)	3038
C–H (benzophenone)	2893
C = O (rice husk)	2858
C–H	2678, 2367
C–C scissoring	2026, 869
C–C, C–O (rice husk)	1843, 1008
C = O (benzophenone)	1770, 1685
Ketone	1585, 1621
C–O, C–H (rice husk)	1610
C–C stretching (benzophenone)	1482
C–H (benzophenone)	1416
C–H (benzophenone), C–C and C–O (rice husk)	1326

Table 8.9 FTIR of wheat gluten husk/benzophenone.

Assignments	Wavenumber (cm^{-1})
NH_2	6111
N–H and C–H (wheat gluten husk)	5749, 3502
CH_2 asymmetric stretching (wheat gluten husk)	5411, 4876, 4319
C–H (wheat gluten husk)	4564, 691
CH_2scissoring (wheat gluten husk)	3920
O–H stretching	3394
C = C, C = O (benzophenone)	3715, 3677, 3234
C = O stretching (wheat gluten husk)	3258, 2853
C–C, C–N, C–O (wheat gluten husk)	2619
C–C (benzophenone)	2229
C–H (benzophenone)	1792
C–C, C–N (benzophenone)	1487
C–O, C–C (wheat gluten husk)	992

vibrations occur in the region 1770 and 1685 cm^{-1}. The band at 1482 cm^{-1} is due to C–C stretching frequency of aromatic ring [37].

While that the Table 8.9 shows the FTIR bands of wheat gluten husk/benzophenone where the characteristic peaks associated with absorption process are observed. The NH and CH stretching modes arising from amino groups appear around 5749 and 3502 cm^{-1}. The stretching modes of NH_2 group were assigned to the bands at 6111cm^{-1}[31, 36]. The aromatic CH stretching vibrations appear weak just above 4564 cm^{-1}. A highly intense and well defined peak observed at 3258, 2853 cm^{-1} is due to the C = O stretching vibration of carbonyl group (benzophenone).

The C = C vibrations of aromatic ring are confirmed at 3725, 3677 and 3234 cm^{-1}. The group of bands at 1792 and 691 cm^{-1} is due to aromatic CH bends [38].

8.3.4.3 Electrostatic Potential

The electrostatic potential of rice husk and wheat gluten husk with benzophenone can be observed in Figure 8.11. The results show that the negative (red) regions of MESP were related to electrophilic reactivity and the positive (blue) regions to nucleophilic reactivity. The negative regions are mainly localized on the C = O bond. Also, a negative electrostatic potential region is observed around the OH groups (oxygen atom) [35]. The values of the Figures 8.8–8.9 show that the electronegativity of the benzofenone produces a decrease in nucleophilic areas (blue) of both lignocellulosic materials.

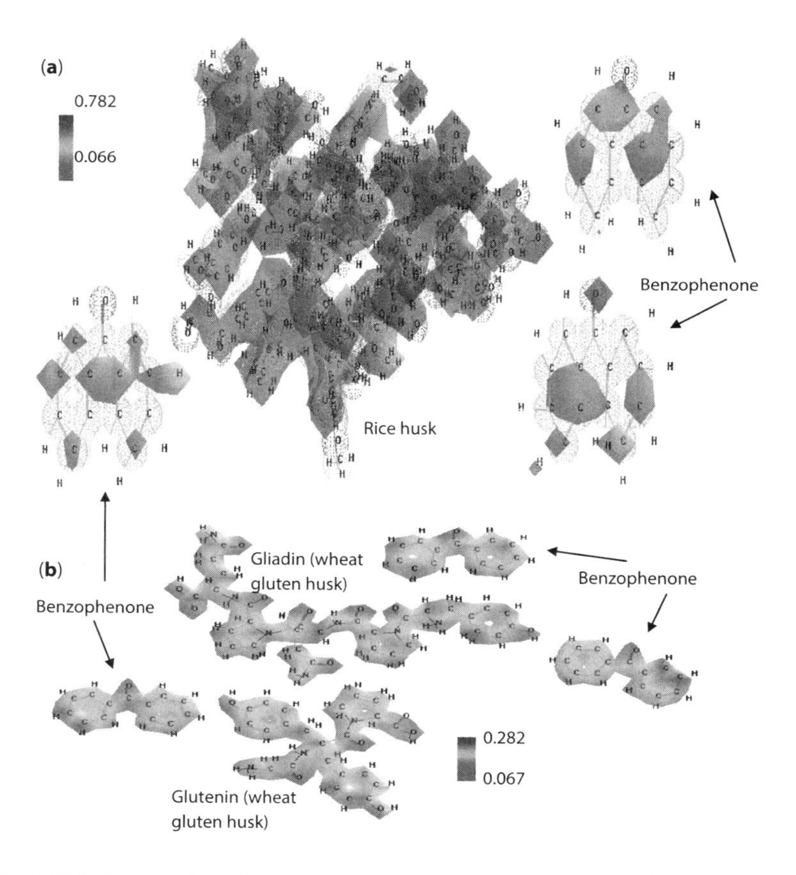

Figure 8.11 MESP of (a) rice husk/benzophenone and (b) wheat gluten husk/ benzophenone, respectively.

8.3.5 Absorption of Glibenclamide

8.3.5.1 Geometry Optimization

Structural properties of rice husk and wheat gluten husk with glibenclamide are listed in Table 8.10 where the Gibbs free energy is spontaneous for both lignocellulosic materials. The solubility of glibenclamide is an important factor in determining the rate and extent of its absorption process [39]. The computed Log P values (P is the partition coefficient of the molecule in the water–octanol system), show that the absorption is effected due to the hydrophilic character and additionally the absorption plays an important role in both partition and receptor binding processes of drug action. Drug design is an iterative process which begins with a compound that displays an interesting biological profile and ends with optimizing both the activity profile for the molecule and its chemical synthesis. It is therefore, important to know if drug molecules exist predominantly in the basic or protonated forms [40]. Rice husk/glibenclamide and wheat gluten husk/glibenclamide structure are appreciated in Figure 8.12, where it

Figure 8.12 Glibenclamide absorption structure, (a) Rice husk (b) Wheat gluten husk structure, respectively.

Table 8.11 FTIR results of Rice husk/glibenclamide.

Assignments	**Wavenumber (cm^{-1})**
CH_3 stretching (O-CH_3 glibenclamide)	5505
CH, O–H (rice husk)	3754
C = C (rice husk)	3640, 3506
NH stretching (glibenclamide)	3368
CH scissoring (rice husk)	3166
C = C (rice husk)	2986
C–C (rice husk)	2709
C–C, C–H and C–O (rice husk)C–C, C–H (glibenclamide)	2412
C–C, C–O (rice husk)	1955, 1334, 1090, 924
C–C, C–N, C–H (rice husk)	1758
C = O, S = O (glibenclamide)	1549
O–H, C–H and C–O (rice husk)C–C, C–H (glibenclamide)	793
O–H, C–H (rice husk)	695

Table 8.12 FTIR results of Wheat gluten husk/glibenclamide.

Assignments	**Wavenumber (cm^{-1})**
CH, O–H asymmetric stretching (glutenin: wheat gluten husk)	5879
NH asymmetric stretching (glibenclamide)	5583
CH asymmetric stretching (glibenclamide)	5007
CH_2 stretching (glibenclamide)	4511
CH_3 (O-CH_3: glibenclamide)	4028
CH = CH (glibenclamide)	3529
CH (gliadin: wheat gluten husk)	3392
C = C (ring: glibenclamide)	2830
C = C, C = O, S = O	2474
C–C, C–H, C–O (glutenin: wheat gluten husk)	2022
CH deformation (glutenin: wheat gluten husk)	1739
C–C, C–N, C–O (glibenclamide)	1332, 1124, 1028, 569
C–C, C–N, C–O (gliadin: wheat gluten husk)	819

observed that the glibenclamide is absorbed through the formation of hydrogen bonds between C = O, NH, C–O and CH bonds.

8.3.5.2 FTIR

In the FTIR results of rice husk/glibenclamide can be seen in Table 8.11, the principal absorption peaks appeared at 5505 cm^{-1} was attributed to CH_3 stretching, at 3368 cm^{-1}

due to the NH stretching, the absorption of C = O and S = O were observed at 1549 cm^{-1}[41]. In the rice huks, the band at 3754 cm^{-1} is representative of the CH and OH bonds. Both bands are ascribed to the stretching of hydrogen bonds and bending of hydroxyl (OH) groups bound to the cellulose structure. The absorption phenomenon was observed at 2412 and 793 cm^{-1} refers to the bending frequency of C–C, CH, C–O and OH respectively, while the absorption around 1955, 1334, 1090 and 924 cm^{-1} were refers to the C–C and C–O of the cellulose component. The results indicate that the glibenclamide was absorbed by rice husk [42]. Table 8.12 shows FTIR results of wheat gluten husk/glibenclamide where, the first characteristic peak appears at 5879 cm^{-1} as a result of C–H and O–H stretching, indicating the presence of bonded hydroxyl groups in the molecular structure of glutenin (wheat husk) [43]. Between 5007 and 5583 cm^{-1} are principally attributed to NH and CH asymmetric stretching vibrations of glibenclamide. The peaks characteristics of glibenclamide were observed to 4511, 4028 and

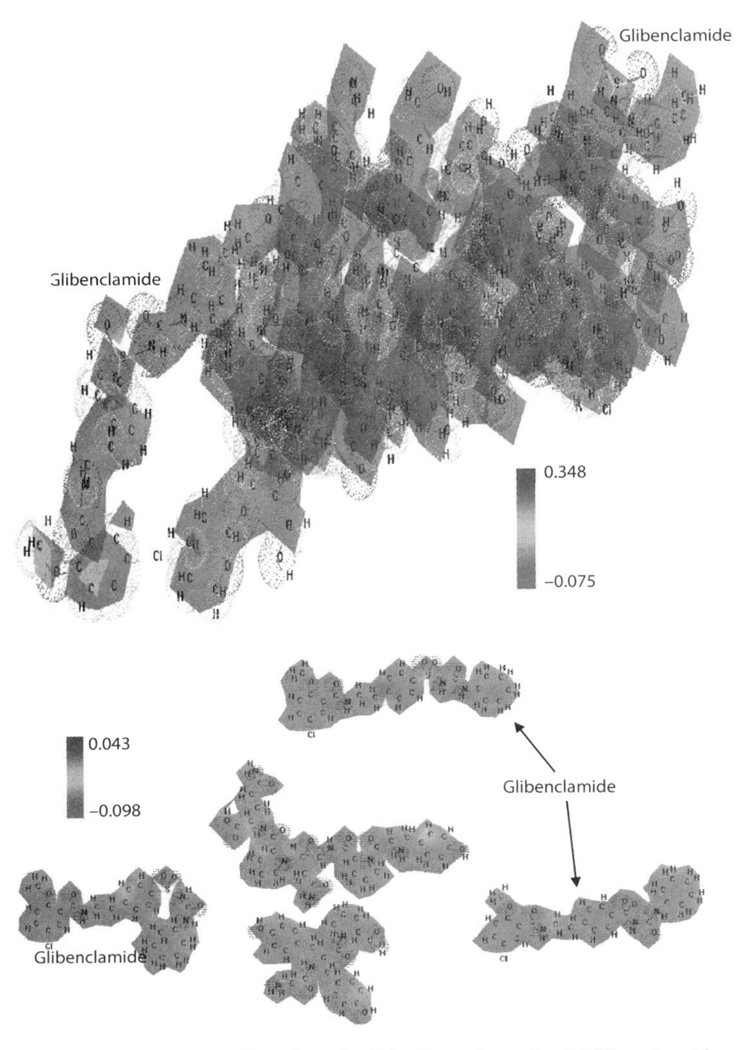

Figure 8.13 MESP of (a) Rice husk/glibenclamide, (b) wheat gluten husk/glibenclamide.

3529 cm^{-1} and attributed to CH_2 stretching, CH_3 (O-CH_3) and CH = CH, respectively. The wheat gluten husk components were assigned to 3392, 2022, 1739 and 819 cm^{-1}.

8.3.5.3 Electrostatic Potential

Figure 8.13 shows the rice husk/glibenclamide and wheat gluten husk/glibenclamide MESP, where can be observed that the values are very similar at lignocellulosic materials (see Figure 8.8). The negative regions were appreciated at OH groups (C–OH bonds). The absorption of the glibenclamide can be seen mainly in to the rice husk by the formation of hydrogen bonds between the OH and NH, respectively.

8.4 Conclusions

The rice husks and wheat gluten husk on an individual basis and such as benzophenone and glibenclamide absorption systems were analyzed to determine the applications of lignocellulosic materials. It was determined that the negative value of the ΔG verifies that the absorption process is carried out in a way spontaneous. The negative values of Log P show that the absorption is affected due to the hydrophilic character and additionally the absorption plays an important role in both partition and receptor binding processes of absorption (benzophenone and glibenclamide). FTIR results show that there are shifts in the peaks of rice husk and wheat gluten husk attributed to the absorption of benzophenone and glibenclamide. The MESP values indicated the nucleophilic and electrophilic regions mainly in the NH, C-O and C = O bonds respectively.

References

1. A.K. Hartmann, *Computer Simulations*, World Scientific, Singapore, (2009).
2. C.J. Cramer, *Essentials of Computational Chemistry*, Wiley, Chichester, (2002).
3. S.J. Smith, and B.T. Sutcliffe, The development of computational chemistry in the United Kingdom. *Rev. Comput. Chem.* 70, 271–316 (1997).
4. D.L. Mobley, A.P. Graves, J.D. Chodera, A.C. Mcreynolds, B.K. Shoichet, and K.A. Dill, Predicting absolute ligand binding free energies to a simple model site. *J. Molecul. Biol.* 37 1(4), 1118–1134 (2007).
5. S. Huo, I. Massova, and P.A. Kollman, Computational alanine scanning of the 1:1 human growth hormone-receptor complex. *J. Comput. Chem.* 23(1), 15–27 (2002).
6. N.A. Rangel-Vázquez, and F. Rodríguez-Felix, *Computational chemistry applied in the analyses of chitosan/polyvinylpyrrolidone/mimosa tenuiflora*, Ed. Science Group Publishing, Hong Kong, (2013).
7. R. Rowell, *ACS Symposium Series*, 476, 26–31 (1992).
8. S.I. Mussatto and J.A. Teixeira, *Current Research, Technology and Education Topics in Applied Microbiology and Microbial Biotechnology*, pp. 897–907, (2010).
9. A. Kumar, K. Mohanta, D. Kumar and O. Parkash, Properties and Industrial Applications of Rice husk:A review. *Int. J. Emerg. Technol. Adv. Eng.* 2(10), 86–90 (2012).
10. S. Balakrishnan, Rice husk ash silica as a support material for iron and ruthenium based heterogeneous catalyst. Malasya, Universiti Sains Malaysia, M. Sc. Thesis (2006).

11. N.M. Edwards, S.J. Mulvaney, M.G. Scanlon, and J.E. Dexter, Role of gluten and its components in determining durum semolina dough viscoelastic properties. *Cereal Chem.* 80(6), 755–763 (2003).
12. P. Tosi, A. Giovangrossi, R. D'Ovidio, F. Bekes, O. Larroque, J. Napier, and P. Shewry, Modification of the low molecular weight (LMW) glutenin composition of transgenic durum wheat: effects on glutenin polymer size and gluten functionality. *Mol. Breed.* 16(2), 113–126 (2005).
13. C. Maningat, S. Bassi, and J.M. Hesser, Wheat gluten in food and non-food systems. *Tech. Bull. 16*(6), 19941–19948 (1995).
14. H.D. Belitz, W. Grosch and P. Schieberle, Food Chemistry, in *Analytical & Bioanalytical Chemistry*, E. Anklam (Ed.), pp. 10–11, New York, Springer (2004).
15. International Wheat Gluten Association, http://www.iwga.net/what_is_wg.htm (2013).
16. HSDB Benzophenone, Hazardous substance database, http://toxnet.nlm.nih.gov/ (2013).
17. National Toxicology Program, NTP technical report on the toxicology and carcinogenesis studies of 2,3,7,8-tetrachlorodibenzo-p-dioxin (TCDD) (CAS No. 1746-01-6) in female Harlan Sprague-Dawley rats (Gavage Studies).*Natl. Toxicol. Prog. Tech. Rep. Ser.* (533), 1–264 (2006).
18. K. Parameswararao, M.V. Satynarayana, T. Naga Raju and G.V. Ramana, Novel spectrophotometric methods for the assay of glibenclamide in pure and dosage forms. *Der Pharma Chemica*, 4(6), 2449–2452 (2012).
19. X, Serrano-Martín, G. Payares, and A. Mendoza-León, Glibenclamide, a blocker of K+ATP channels, shows antileishmanial activity in experimental murine cutaneous leishmaniasis. *Antimicrob. Agents Chemother.* 50(12), 4214–4216 (2006).
20. F.J. Ortega, J. Gimeno-Bayon, J.F. Espinosa-Parrilla, J.L. Carrasco, M. Batlle, M. Pugliese, N. Mahy, and M.J. Rodríguez, ATP-dependent potassium channel blockade strengthens microglial neuroprotection after hypoxia–ischemia in rats. *Exp. Neurol.* 235(1), 282–296 (2012).
21. J.M. Simard, S.K. Woo, G.T. Schwartzbauer, and V. Gerzanich, Sulfonylurea receptor 1 in central nervous system injury: A focused review. *J. Cereb. Blood Flow Metab.* 32(9), 1699–1717 (2012).
22. E. López-Chávez, R. Oviedo-Roa, G. Contreras-Pérez, J.M. Martínez-Magadán, and F.L. Castillo-Alvarado, Theoretical studies of ionic conductivity of crosslinked chitosan membranes. *Int. J. Hydrogen Energy*, 35(21), 12141–12146 (2010).
23. J. Kumirska, M. Czerwicka, Z. Kaczyński, A. Bychowska, K. Brzozowski, J. Thöming, P. Stepnowski, Application of spectroscopic methods for structural analysis of chitin and chitosan. *Mar. Drugs* 8(5), 1567–1636 (2010).
24. Hyperchem® Computational Chemistry, Canada, 1996
25. Y.N. Nkolo Mezeˈe, J. Noah Ngamveng and S. Bardet, Effect of enthalpy–entropy compensation during sorption of water vapour intropical woods: The case of Bubinga (Guibourtia Tessmanii J. Léonard; G. Pellegriniana J.L.). *Thermochim. Acta*, 468, 1–5 (2008).
26. S.M. Leal Rosa, E. Fonseca-Santos, C.A. Ferreira, and S.M. Bohrz Nachtigall, Studies on the properties of rice-husk-filled-PP composites: Effect of maleated PP. *Mater. Res.* 12(3), 333–338 (2009).
27. K. Hardinnawirda and I. Sitirabiatull Aisha, *J. Mech. Eng. Sci.* 2, 181–186, (2012).
28. L. Ludueña, D. Fasce, V. Alvarez and P. Stefani, Nanocellulose from rice husk following alkaline treatment to remove silica. *Bioresources* 6(2), 1440–1453 (2011).
29. M.M. Abdelhady, Preparation and characterization of chitosan/zinc oxide nanoparticles for imparting antimicrobial and UV protection to cotton fabric. *Int. J. Carbohydr. Chem.* 2012, 1–6 (2012).

30. T.N. Ang, G.C. Ngoh, A.S. May Chua and M.G. Lee, Elucidation of the effect of ionic liquid pretreatment on rice husk via structural analyses. *Biotechnol. Biofuels* 5, 67–77 (2012).
31. A.K. Bledzki, A.A. Mamun, and J. Volk, Physical, chemical and surface properties of wheat husk, rye husk and soft wood and their polypropylene composites. *Compos. A* 41, 480–488 (2010).
32. Y.M. Rodríguez, L.P. Salinas, C.A. Ríos, and L.Y. Vargas, Rice husk-based adsorbents in the removal of chromium from tanning industry efluents. *Biotecnología en el sector agropecuario y agroindustrial*, 10(1), 146–156 (2012).
33. R. Gayathri, and M. Arivazhagan, Experimental (FT-IR and FT-Raman) and theoretical (HF and DFT) investigation, NMR, NBO, electronic properties and frequency estimation analyses on 2,4,5-trichlorobenzene sulfonyl chloride. Spectrochim. Acta A Mol. Biomol. Spectrosc. 97, 311–325 (2012).
34. Z. Xi-Ming, and F. Rong-Yu, Solid flux of pulverized coal of high-pressure and dense-phase pneumatic conveying and ANN simulation. *CIESC J.* 64(5), 0–0 (2013).
35. M. Karabacak, M. Kurt, M. Cinar, S. Ayyappan, S. Sudha, and N. Sundaraganesan, The spectroscopic (FT-IR, FT-Raman, UV) and first order hyperpolarizability, HOMO and LUMO analysis of 3-aminobenzophenone by density functional method. *Spectrochim. Acta A Mol. Biomol. Spectrosc.* 92, 365–376 (2012).
36. V. Krishnakumar, S. Muthunatesan, G. Keresztury, and T. Sundius, Scaled quantum chemical calculations and FTIR, FT-Raman spectral analysis of 3,4-diamino benzophenone. *Spectrochim. Acta A Mol. Biomol. Spectrosc.* 62, 1081–1088 (2005).
37. M. Rajasekar, K. Muthu, V. Meenatchi, G. Bhagavannarayana, C. K. Mahadevan, and S.P. Meenakshisundaram, *Spectrochim. Acta A Mol. Biomol. Spectrosc*, 92, 207–211 (2012).
38. M.G. Mohamed, K. Rajarajan, M. Vimalan, J. Madhavan and P. Sagayaraj, Growth and characterization of pure, benzophenone and paratoluidine doped 2A-5CB crystals. *Appl. Sci. Res.* 2(3), 81–93 (2010).
39. A. Ali-Elkordy, A. Jatto, and E. Essa, In situ controlled crystallization as a tool to improve the dissolution of Glibenclamide. *Int. J. Pharm.* 428, 118–120 (2012).
40. M. Remko, Theoretical study of molecular structure, pKa, lipophilicity, solubility, absorption, and polar surface area of some hypoglycemic agents. *J. Mol. Struct. THEOCHEM,* 897, 73–82 (2009).
41. A. Kumar-Nayak, B. Daso, and R. Maji, Calcium alginate/gum Arabic beads containing glibenclamide: Development and in vitro characterization. *Int. J. Biol. Macromol.* 51, 1070–1078 (2012).
42. N. Johar, I. Ahmada, and A. Dufresne, Extraction, preparation and characterization of cellulose fibres and nanocrystals from rice husk. *Ind. Crops Prod.* 37, 93–99 (2012).
43. B. Zhang, X. Li, J. Liu, F. Xie, and L. Chen, Supramolecular structure of A- and B-type granules of wheat starch. *Food Hydrocoll.* 31(1), 68–73 (2013).

9

Oil Palm Fiber Polymer Composites: Processing, Characterization and Properties

S. Shinoj[1] and R. Visvanathan[*,2]

[1]*Krishi Vigyan Kendra, Central Marine Fisheries Research Institute, Narakkal, Kochi, Kerala, India*

[2]*Post Harvest Technology Centre, Tamil Nadu Agricultural University, Coimbatore, Tamil Nadu, India*

Abstract

Oil palm is an edible oil-yielding crop native to Africa, grown in Africa, Malaysia, India, Thailand and other Southeast Asian countries. The fiber extracted from the empty fruit bunches which is left over after oil extraction has been proven to be a good raw material for biocomposites. The cellulose content of oil palm fiber (OPF) is in the range of 43% to 65% and lignin content in the range of 13% to 25%. A compilation of the morphology, chemical constituents and properties of OPF as reported by various researchers have been collected and are presented in this chapter. The suitability of OPF in various polymeric matrices such as natural rubber, polypropylene, polyvinyl chloride, phenol formaldehyde, polyurethane, epoxy, polyester, etc., to form biocomposites as reported by various researchers in the recent past are compiled. The properties of these composites *viz.*, physical, mechanical, water sorption, thermal, degradation, electrical properties, etc., are summarized. Oil palm fiber loading in some polymeric matrices improved the strength of the resulting composites, whereas less strength was observed in some cases. The composites became more hydrophilic upon addition of OPF. However, treatments on the fiber surface improved the composite properties. Alkali treatment on OPF is preferred for improving the fiber-matrix adhesion compared to other treatments. The effects of various treatments on the properties of OPF and that of resulting composites reported by various researchers are compiled in this chapter. The thermal stability, dielectric constant, electrical conductivity, etc., of the composites improved upon incorporation of OPF. The strength properties were reduced upon weathering/degradation. Sisal fiber was reported as a good combination with OPF in hybrid composites.

Keywords: Oil palm fiber, polypropylene composites, polyurethane composites, polyester composites, polyvinyl chloride composites, phenol-formaldehyde, composites, epoxy composites, polystyrene composites, LLDPE composites

**Corresponding author*: drrviswanathan@gmail.com

Vijay Kumar Thakur (Ed.), Lignocellulosic Polymer Composites, (175–212) 2015 © Scrivener Publishing LLC

9.1 Introduction

Oil palm (*Elaeis guineensis* Jacq.) is the highest edible oil-yielding crop in the world. It is produced in 42 countries worldwide on about 11 m ha [1]. West Africa, Southeast Asian countries like Malaysia and Indonesia, Latin American countries and India are the major oil palm cultivating countries [2]. An oil palm plantation produces about 55 tonnes/ha/year of total dry matter in the form of fibrous biomass, while yielding 5.5 tonnes/ha/year of oil [3]. The trunk, frond, empty fruit bunch (the fibrous bundle left behind after stripping of the fruits) and mesocarp waste (the fibrous residue left after extraction of the oil from fruit) are the sources of lignocellulosic fibers from oil palm. Among these, empty fruit bunch (EFB) is preferable for fiber in terms of availability and cost [4]. It was reported that EFB has the potential to yield 73% fibers [5]. The palm oil industry has to dispose of about 1.1 tonne of EFB per every tonne of oil produced [6]. The only current uses of this highly cellulosic material are as boiler fuel [7] in the preparation of fertilizers or as mulching material [8]. When left on the plantation floor, these waste materials create great environmental problems [7, 9]. A view of EFB wastes piled up for disposal in one of the palm oil mills in India is shown in Figure 9.1. Thus there is need to find utilization for the EFB. Lignocellulosic fiber information has been generated on its characterization and utilization, particularly in biocomposite applications.

Biocomposites are defined as the materials made from natural/bio fiber and petroleum-derived nonbiodegradable or biodegradable polymers. The latter category, i.e., biocomposites derived from plant-derived fiber (natural/biofiber) and crop/bioderived plastic (biopolymer/bioplastic), are likely to be more eco-friendly and such composites are termed as green composites [10]. Use of natural fiber as filler in polymeric matrix offers several advantages over conventional inorganic fillers with regard to their lower density, less abrasiveness to processing equipment, environmentally-friendly nature, lower cost [11], greater deformability, biodegradability [12], renewable nature, non-toxicity, flexible usage, high specific strength [13], low energy cost, positive contribution to global carbon budget [14], combustibility, ease of recyclability [15], good

Figure 9.1 View of EFB wastes piled up on the premises of a palm oil mill.

thermal and insulating properties [16, 17], good electrical resistance, good acoustic insulating property, worldwide availability, etc. Replacing man-made fibers with natural fibers in polymeric materials has an additional advantage of composting or recovering the calorific value at the end of their cycle, which is not possible with glass fibers [18]. Considering these advantages, there is a growing interest in natural fiber composites for various applications. Automotive giants such as Daimler Chrysler use flax–sisal fiber mat embedded in an epoxy matrix for the door panels of the Mercedes Benz E-class model. Coconut fibers bonded with natural rubber latex are being used in seats of the Mercedes Benz A-Class model. The Cambridge Industry (an automotive industry in Michigan, USA) is making flax fiber-reinforced polypropylene for Freightliner Century COE C-2 heavy trucks and also the rear shelf trim panels of the 2000 model Chevrolet Impala [10]. Besides the automotive industry, lignocellulosic fiber composites have also found application in the building and construction industries such as for panels, ceilings, and partition boards [16]. Nowadays fiber-reinforced plastic composites are used in many structural applications such as in the aerospace industry, automotive parts, sports, recreation equipment, boats, office products, machinery, etc. [19].

In this chapter, the developments on characterization of oil palm empty fruit bunch fiber and its utilization in biocomposites are compiled. Various methods to improve the compatability of the fiber and polymeric matrix and its effect on the properties of fibers and resulting composites are also discussed. For the purpose of convenience , the term *oil palm fiber* in the subsequent sections refers to *oil palm empty fruit bunch fiber.*

9.2 Oil Palm Fiber

9.2.1 Extraction

It is important to note that oil palm fresh fruit bunches (FFBs) receives a high pressure (3 kg/cm^2) –temperature (130°C) steam treatment for a duration of 1 h before stripping off the fruits and EFB being available for fiber extraction, as steam treatment (sterilization) is an essential step in palm oil milling sequence. A diagram of the fresh fruit bunch (FFB) and cross-section of EFB showing fiber arrangement is presented in Figure 9.2. It is reported that the fiber from bunches can be extracted by retting process. The available retting processes are as follows: mechanical retting (hammering), chemical retting (boiling and applying chemicals), steam/vapor/dew retting and water or microbial retting. Among these, the water retting is the most popular process in extracting fibers from empty fruit bunches [11]. Many researchers in their studies on oil palm fibers followed retting method for fiber extraction [2, 7, 20]. Mechanical extraction would be environmental friendly comparing to other methods as the later pollute water bodies. A machine for the extraction of fiber from oil palm empty fruit bunches was developed by [21]. It decorticates the bunches and separate the pith materials and also grade fiber into different fractions.

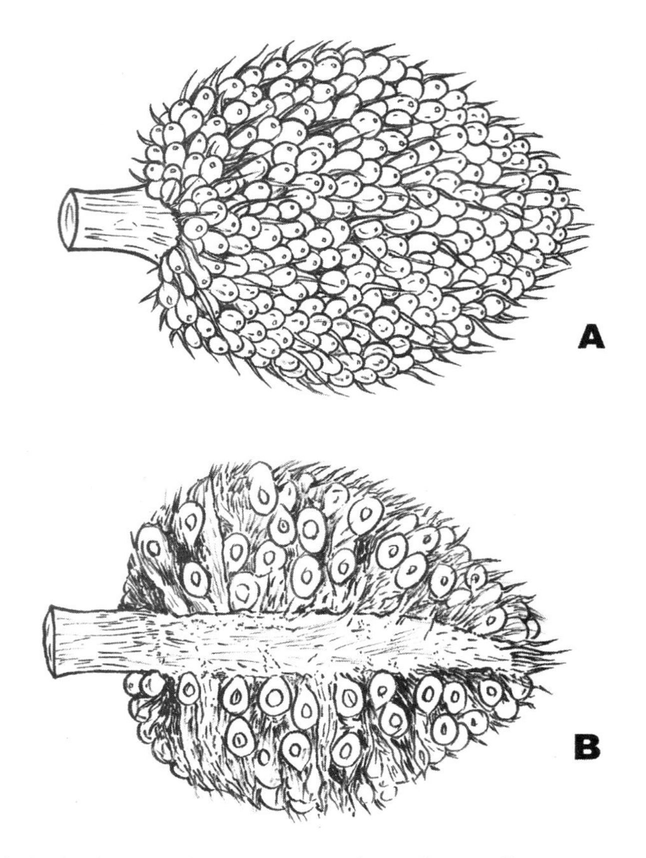

Figure 9.2 Sketch of A) oil palm FFB; B) Cross-section of EFB showing fiber arrangement.

9.2.2 Morphology and Properties

Oil palm fiber is hard and tough, and show similarity to coir fibers [7, 22]. The scanning electron microscopy (SEM) image [1] of the transverse section of oil palm fiber (Figure 9.3) shows a lacuna like portion in the middle surrounded by porous tubular structures [23]. The pores on the surface have an average diameter of 0.07μm. This porous surface morphology is useful for better mechanical interlocking with the matrix resin for composite fabrication [7]. However porous surface structure of oil palm fiber also facilitate the penetration of water into the fiber by capillary action especially when the fiber is exposed to water [14]. The vessel elements in the cross-sectional view of the fibers can be clearely seen in the longitudinal view also (Figure 9.4). However some researchers are of opinion that each fiber is a bundle of many fibers [13, 24].

Granules of starch are found in the interior of the vascular bundle (Figure 9.4). Silica bodies are also found in great number on the fiber strand. They attach themselves to circular craters which are spread relatively uniformely over the strand's surface. The silica bodies are hard, but can be dislodged mechanically. The removal of silica bodies leaves behind perforated silica-crater, which would enhance penetration of matrix in composite fabrication leading to interfacial adhesion stronger [25]. A compilation of the chemical composition and physico-mechanical properties of oil palm fiber

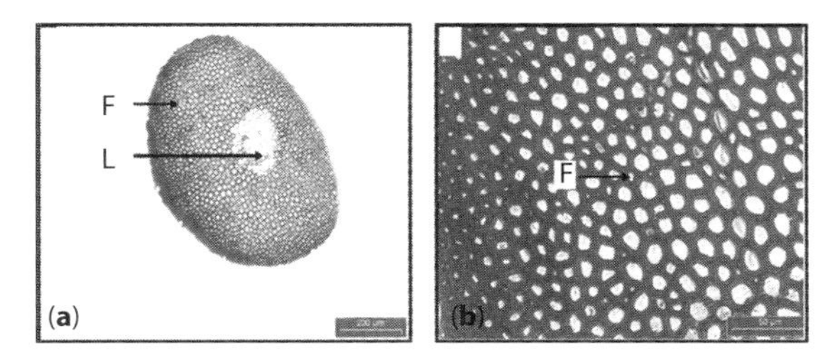

Figure 9.3 SEM images of transverse sections of oil palm fiber (4×). F: fiber, L: lacuna.

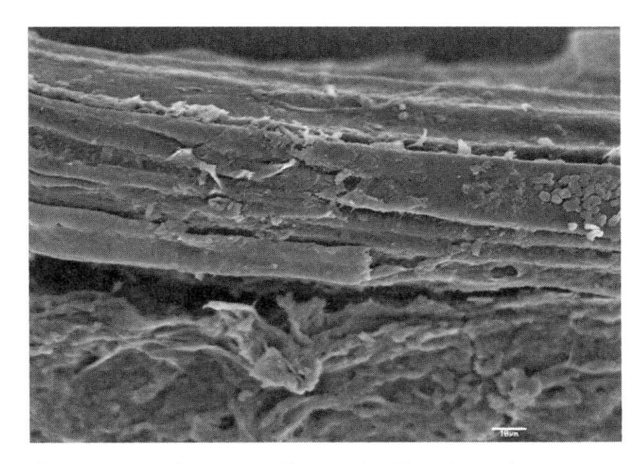

Figure 9.4 SEM image of longitudinal section of oil palm fiber (750×).

reported by various researchers are presented in Tables 9.1 and 9.2, respectively. As differences were observed between the values reported by different researchers, the ranges of values are compiled.

The physical properties and dielectric constant of OPFs [39] are given in Table 9.3. The true density of OPF is 1503 kg/m^3. This value is in confirmation with earlier reported values for most plant fibers in the range of 1400-1500 kg/m^3[40]. The OPFs obtained from oil mill contains nearly one fourth impurities, which necessitates hygienic disposal of EFB in palm oil mills and also ensuring efficient washing and drying system in the fiber processing line towards industrial utilization of OPFs. The values of equilibrium moisture content (EMC) of fibers indicate that increase in relative humidity by 25% nearly doubles the EMC. The dielectric constant value of OPF is in agreement with the values reported earlier for other natural fibers. Size reduction caused increase in dielectric constant, while alkali treatment on fibers caused a decrease in dielectric constant.

The thermal degradation behavior of OPF in comparison with flax and hemp fibers was studied [39]. They reported that OPF is thermally more stable comparing to hemp and flax fibers as its first decomposition peak occurs at 301.71°C, whereas for flax it was 286.66°C and for hemp, it was only 243.18°C. They have also reported the specific heat capacity of OPF at varying temperatures in the range of 20°C to 150°C in comparison to that of flax fiber and hemp fiber. The specific heat capacity of OPF varied from 1.083 J/g°C to 3.317 J/g°C when the temperature increased from 20°C to 150°C. This trend was similar to that exhibited by flax fiber and hemp fiber. However the specific heat capacity values of OPF is less than the other two fibers at all temperatures, whereas

Table 9.1 Chemical composition of oil palm fibers.

Constituents	Range	References
Cellulose,%	42.7-65	[1, 7, 14 , 25, 26]
Lignin,%	13.2-25.31	[5, 7, 14, 27-30]
Ash content,%	1.3-6.04	[1, 5, 7, 25, 27-29, 31, 32]
Extractives in hot water, (100°C),%	2.8-14.79	[5, 7, 25, 26, 28, 32, 33]
Solubles in cold water (30°C),%	8-11.46	[5, 7]
Alkali soluble,%	14.5-31.17	[5, 7, 25, 26, 28, 32]
Alfa-cellulose,%	41.9-60.6	[27, 28, 33]
Hemicellulose,%	17.1-33.5	[14, 25, 30, 33]
Alcohol-benzene solubility,%	2.7-12	[5, 7, 28, 33]
Pentosan,%	17.8-20.3	[28, 30]
Holocellulose,%	68.3-86.3	[1,25, 26-29, 32, 33]
Arabinose,%	2.5	[25]
Xylose,%	33.1	[25]
Mannose,%	1.3	[25]
Galactose,%	1.0	[25]
Glucose,%	66.4	[25]
Silica (EDAX),%	1.8	[25]
Copper, g/g	0.8	[25]
Calsium, g/g	2.8	[25]
Manganese, g/g	7.4	[25]
Iron, g/g	10.0	[25]
Sodium, g/g	11.0	[25]

specific heat capacity of flax fiber and hemp fiber were on par except at higher temperatures. This may be one of the reasons why OPF is thermally more stable than flax and hemp fibers.

The high toughness value [42] and high cellulose content [15] of oil palm fibers make it suitable for composite applications. However the presence of hydroxyl group makes it hydrophilic, which cause poor interfacial adhesion with hydrophobic polymer matrices and lead to low compatibility, causing poor mechanical and physical properties to the composite [11]. Oil palm fibers contain residual oil of the order 4.5% [27]. The fiber-matrix compatibility is adversely affected due to the presence of oil residues on the fiber, the ester components of which may affect the coupling efficiency between the fiber and polymer matrix as well as the interaction between fiber and coupling agents [43]. By the implementation of surface modification of the fibers, the properties can be improved substantially. Chemical treatments would not only decrease the moisture absorption of the fibers but also would significantly increase the wettability of the fibers with the polymer matrix and the interfacial bond strength [44]. There are a

Table 9.2 Physico-mechanical properties of oil palm fiber.

Property	Range	References
Diameter, μm	150-500	[32, 34, 35]
Microfibrillar angle, °	46	[32]
Density, g/cm^3	0.7–1.55	[30, 34. 36, 37]
Average fiber length, mm	1.20	[36]
Tensile strength, MPa	50–400	[13, 27, 32, 36, 38]
Young's modulus, GPa	0.57–9	[7, 24, 27, 30, 32, 35]
Elongation at break,%	4–18	[24, 30, 27, 30, 32, 35, 36]
Tensile strain,%	13.71	[37]
Length-weighted fiber length, mm	0.99	[24]
Cell-wall thickness, μm	3.38	[25]
Fiber coarseness, mg/m	1.37	[25]
Fines (<0.2 mm),%	27.6	[25]
Rigidity index, $(T/D)^3 \times 10^{-4}$	55.43	[25]

Table 9.3 Physical properties and dielectric constant of OPF.

Property	Untreated OPF	Alkali treated OPF
Density, kg m^{-3}	1503	1425
Impurity content,%	23.18	–
EMC,% (30°C, 25% Rh)	3.21	–
EMC,% (30°C, 50% Rh)	6.61	–
EMC,% (30°C, 75% Rh)	12.97	–
Color	L* = 47.83, a* = 5.56, b* = 18.44	L* = 46.22, a* = 5.99, b* = 19.64
Dielectric constant, 425μ-840μ size range	7.94	7.76
Dielectric constant, 177μ-425μ size range	8.12	7.89
Dielectric constant, 75μ-177μ size range	8.31	8.05

number of treatment methods for oil palm fibers to improve their properties and make them compatable with the matrix.

9.2.3 Surface Treatments

The treatments to improve fiber-matrix adhesion includes chemical modification of the lignocellulosic (anhydrides, epoxies, isocyanates, etc.), grafting of polymers onto the lignocellulosic and use of compatabilizers and coupling agents [38].

Various treatment methods onto oil palm fibers are already discussed in detail [22, 23, 45-47] including a detailed discussion on various treatments on lignocellulosic fibers in general to improve their properties [48]. Reviews on the developments in chemical modification and characterization of natural fiber-reinforced composites

Table 9.4 Effect of surface treatments on properties of oil palm fiber.

Treatment	Effect on oil palm fiber
Mercerization	Amorphous waxy cuticle layer leaches out.
Latex coating	Partially masks the pores on the fiber surface.
γ irradiation	Partially eliminates the porous structure of the fiber and causes microlevel disintegration. It degrades mechanical properties considerably.
Silane treatment	Imparts a coating on fiber surface
Toluene diisocyanate (TDIC) treatment	Makes fiber surface irregular as particles are adhered to surface.
Acetylation	Removes waxy layer from the surface and makes the fiber hydrophobic.
Peroxide treatment	Fibrillation is observed due to leaching out of waxes, gums and pectic substances.
Permanganate treatment	Changes the color and makes fibers soft. Porous structure is observed after treatment.
Acrylation	Imparts a coating on fiber surface and removes pits containing silica bodies and keep surface irregular. It improves mechanical properties of fibers.
Silane treatment	Keep the fiber surface undulating and improves mechanical properties
Titanate treatment	Smoothens fiber surface.
Alkali treatment	It makes the surface pores wider and fiber become thinner due to dissolution of natural and artificial impurities
Benzoylation	Imparts a rough surface to the fibers and makes pores prominent, which helps improving the mechanical interlocking with matrix resin.
Oil extraction	Imparts bright color to the fiber. Removal of oil layer exposes surface pits and makes surface coarse.

indicates that alkali treatment of fibers is the most common and efficient method of chemical modification and reported being used to treat almost all the natural fibers with successful results [49]. It is worth referring to [50] the processes involved in the chemical treatment of natural fibers and the effect of treatments.

Different treatments will have different effects on the fiber surface and selection of the best treatment strategy is based on the desired end properties of composites. Effect of some of the treatments on the surface properties of oil palm fibers as reported by various researchers [7, 14, 27, 43, 50-52] are summerized in Table 9.4. Fiber surface properties are most important in the determination of fiber-matrix bonding in composite fabrication.

The surface pits in oil palm fiber became clearer upon alkali treatment as a result of the removal of silica bodies [53]. The microstructure (SEM images) of the alkali-treated oil palm fiber surface in comparison with that of the untreated one is presented in Figure 5. This enhanced the fiber matrix bonding and thereby reduced the water absorption capacity of the composites. The alkali treatment reduces fiber weight by 22%, while silane treatment reduces weight by 6%. The fiber diameter also reduced by chemical treatment [7]. Analysis of the thermal degradation in the same study revealed that initial degradation temperature was higher for alkali-treated fibers (350°C), whereas untreated and acetylated fibers degraded at 325°C and silane treatment increases the degradation temperature to 365°C. Treatment reduces mechanical strength of fibers. Strain to break of the fibers was considerably increased upon treatments except silane. Young's modulus enhanced on mercerization and silane treatments [34]. In an innovative study on the treatment of oil palm fibers with bulk monomer allyl methacrylate (AMA) and curing under ultraviolet (UV) radiation (photocuring), improvement in the physicomechanical properties to a large extent was observed [54]. It was also observed that urea treatment in combination with AMA improved both soil and water weathering characteristics also. The response of individual fibers to applied stress is important as the load applied to composite material is transferred from matrix to fiber. The stress-deformation behavior of treated oil palm fibers shows intermediate behavior between brittle and amorphous materials [50]. The authors are of opinion that the modifications lead to major changes on the fibrillar structure and it removes the amorphous components. This changes the deformation behaviorr of the fibers. The brittleness of the fiber is substantially reduced upon treatments. Thermo gravimetric analysis of the grafted oil palm fiber indicated that grafted fiber is thermally stable than the untreated one [46]. Water sorption characteristics of treated oil palm fibers shown [34] that the equilibrium mole percent uptake of water at 30°C was 13.37% for untreated fibers, which reduced to 7.26% (mercerization), 7.65% (latex coating), 7.36% (γ irradiation), 8.51% (silane treatment), 6.94% (toluene diisocyanate treatment), 7.48% (acetylation) and 7.44% (peroxide treatment). Accordingly, values of diffusion coefficient, sorption coefficient and permeability coefficient show major decrease upon fiber treatment. The toluene diisocyanate (TDIC)-treated fibers exhibited the lowest water absorption. This decrease in uptake value for treated fibers

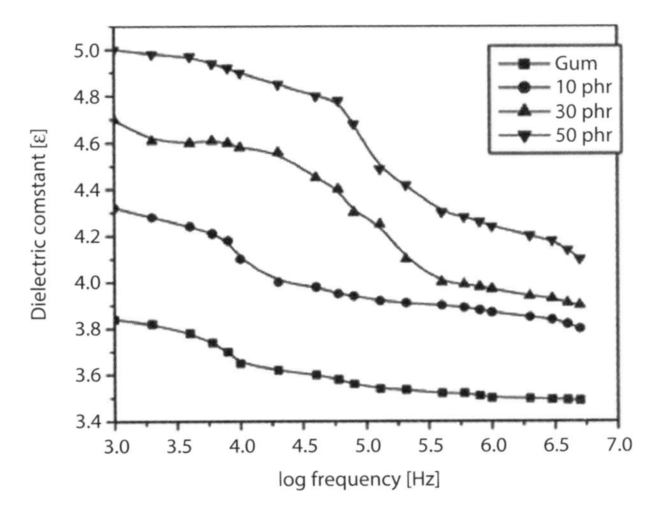

Figure 9.5 Microstructure of the oil palm fiber surfaces: (A) untreated fiber (750×) and (B) alkali-treated fiber (500×).

was attributed to the physical and chemical changes occurred to the fibers upon different treatments.

Interfacial shear strength (ISS) is one of the important parameters controlling the toughness and strength of composites. It depends on fiber surface treatment, modification of matrix and other factors affecting properties of fiber-matrix interface. The ISS of OPF were 1.15 MPa, 1.5 MPa, 1.8 MPa, 1.8 MPa and 1.5 MPa, respectively for polystyrene, metset (unsaturated polyester with 33% styrene content), west system (epoxy), epiglass and crystic (unsaturated polyester with more than 45% styrene content) matrix systems [38]. The acetylation treatment improved ISS of OPFs in all these matrices. Improvement in ISS of polystyrene on acetylation was more prominent than that with other matrices. Chemical modification of fibers create more transactive surface molecules that would readily form bonds with matrix. These results indicate that ISS of OPFs in thermoset matrix was higher than that in the thermoplastic matrix due to higher wettability of thermoset system than thermoplastic system. The lignin content and other main polymerics (hemicellulose, cellulose, etc.) are reactive to thermoset than thermoplastic, thereby improving 'wetting' between the fiber and matrix leading to higher ISS values with thermoset matrix.

9.3 Oil Palm Fiber Composites

A number of studies have been conducted in the recent past on the development and characterization of oil palm fiber filled both thermoset and thermoplastic composites. Mechanical, physical, electrical, thermal properties and biodegradation studies have been carried out. Most of the studies were on the behavior of composites to different mechanical loading, *viz.*, tensile, flexural, impact and static, which is of great importance in the structural and load bearing applications. Studies on stress relaxation and creep behavior for predicting the long-term mechanical performance, dimensional stability of load-bearing structures and retention of clamping force were also conducted on composites [45]. Reliable data on the thermal behavior such as crystallization kinetics of some of developed composites have been generated for understanding them as building materials and using them for commercial applications. Water absorption characteristics and swelling behavior of composites for applications in packaging, building industry, waste water treatment, etc., were also taken care. Electrical properties of composites for applications in electrical insulators, electronic and electrical components, etc., were also attempted. Biodegradation studies on oil palm fiber composites were carried out to some extent. Biodegradation studies are important as the composites must exhibit resistance to fungal attack for use in outdoor conditions and at the end of the product life cycle it may be necessary to compost the product rather than burning it [56].

Parameters like the type of binding agent, process, fiber loading, fiber orientation, size and treatments, affect the fiber-matrix bonding and thereby on the end properties of composites. The effect of these parameters on various properties of oil palm fiber composites is compiled and given in the following sections.

9.3.1 Oil Palm Fiber-Natural Rubber Composites

Natural rubber is derived from latex obtained from the sap of rubber trees. Its use ranges from household articles to industrial products. Tire and tube industries are the largest consumers of rubber and the remaining are taken up by general rubber goods (GRG) sector. A number of research studies have been carried out in the recent past on fabrication and characterization of OPF-NR composites.

9.3.1.1 Mechanical Properties

Effect of fiber loading: Incorporation of OPFs in NR matrix decreased its tensile strength and elongation at break. Tensile strength decreased from 19.2 MPa to 7.28 MPa and elongation at break decreased from 1082% to 496% when the fiber content increased from 5 parts per hundred rubber (phr) to 50 phr for longitudinally oriented fibers [2]. Natural rubber inherently possesses high strength due to strain induced crystallisation. When fibers are incorporated into NR, regular arrangement of rubber molecules was disrupted and hence the crystallization ability decreased causing lower tensile properties. The tensile strength of sisal fiber-OPF-NR hybrid composite was less than that of pure gum [42]. However tensile strength of the composite improved when fiber content was increased in the composite. The tensile strength of hybrid composites from OPF-sisal fiber combination in NR matrix increased from 1.75 MPa to 7.5 MPa when fiber content (OPF and sisal fiber in equal proportions) increased from 10 phr to 30 phr [56]. However, further fiber loading to 50 phr decreased the tensile strength to 3.25 MPa. At intermediate levels of fiber loading (30 phr), the population of the fibers was just right for maximum orientation to cause effective stress transfer. The elongation at break of NR matrix reduced from 875% to 650% upon 10 phr fiber loading (both sisal fiber and OPF in equal proportions). However at 30 phr, it was 800%. The tear strength of NR increased from 20 MPa to 30 MPa when 50 phr fiber was added. However, the trend was not consistent at incremental fiber loadings upto 50 phr. Similarly the abrasion resistance of OPF-sisal fiber-NR bio composites increased with fiber content, as seen from 0.33 cm^3 for the composites containing 50% fibers while pure rubber matrix had an abrasion loss of 0.45 cm^3.

The stress relaxation rate of OPF-sisal fiber-NR hybrid composites decreased with increase in fiber content [35]. The relaxation in the gum compound was due to rubber molecules alone and the relaxation got hindered due to fiber-rubber interface when fibers were added. The dynamic mechanical analysis of OPF-sisal fiber-NR composites indicated that the storage modulus increased with increase in fiber content at all temperatures [57]. The gum compound comprising of rubber phase gives the material more flexibility resulting in low stiffness and hence low storage modulus. The composite stiffness increases upon fiber addition and fibers allows greater stress transfer at the interface resulting in high storage modulus. The loss modulus also increased with fiber loading, reaching a maximum of 755.54 MPa at 50 phr fiber loading, whereas gum has loss modulus of 415 MPa. The damping parameter decreased with fiber loading due to lower flexibility and lower degrees of molecular motion caused by incorporation of fibers in rubber matrix.

Effect of fiber size: The concept of critical fiber length in transmitting load from matrix to fiber was introduced in 2006 [2]. At critical fiber length (*lc*), the load transmittance from matrix to fiber is maximum. If *lc* is greater than length of the fiber, stressed fiber debonds from the matrix and the composite fails at a low load. The tensile properties of OPF-NR composites were maximum when fiber length was 6 mm. Decrease in the properties was observed at higher fiber lengths. This was due to the fiber entanglements prevalent at longer fiber length. The longitudinal orientation of fibers in NR matrix resulted in higher tensile strength (19.2 MPa) than transverse orientation (17.5 MPa). When longitudinally oriented, the fibers were aligned in the direction of strain causing uniform transmission of stress. When transversely oriented, the fibers were aligned perpendicular to the direction of the load and they could not take part in stress transfer. Elongation at break was also less for the transversely oriented (940%) fibers comparing to longitudinally oriented fibers (1082%).

Effect of fiber treatment: There are a number of treatments on fibers to improve the interfacial adhesion and composite mechanical properties. Aqueous alkali treatment at elevated temperatures and various bonding agents improved the physical and adhesion properties of OPF-NR composites. Resorsinol- formaldehyde: precipitated silica: hexamethylenetetramine (5:2:5) was found as the best combination of bonding agents for OPF reinforced in NR matrix [28]. Alkali treatment increased the tensile modulus, tensile strength and tear strength of OPF-NR composites and the properties were observed to be higher when NaOH of 5% concentration was used [2]. However further increase in concentration caused decrease in properties. At higher alkali concentrations, excessive removal of binding materials such as lignin, hemicellulose, etc., happens causing degradation of the fiber properties and obviously the composite quality. Alkali and fluro silane treatments on fibers improved the fiber-matrix interfacial adhesion of OPF-sisal fiber-NR hybrid composites [35]. In this case, increase in alkali concentration further improved the adhesion. Between these treatments, alkali treatment resulted in higher crosslinking [42]. The tensile strength of OPF-sisal fiber-NR hybrid composites was in the range of 6 to 11 MPa for various treatments while that of pure NR was 14 MPa. The shore A hardness of NR was 33 MPa whereas that for hybrid composite was in the range of 40 MPa to 82 MPa.

Fiber treatments also influence the dynamic mechanical properties of composites. The storage modulus (E′) of OPF-sisal fiber-NR hybrid composites increased when chemical treatments were done on fibers [57]. Higher values of E′ was exhibited by the composite prepared from fibers treated with 4% NaOH. Improved interfacial adhesion and increased surface area of fibers on alkali treatment leads to more crosslinks within rubber matrix–fiber network and thus the storage modulus. The loss modulus increased from 634.2 MPa for untreated fiber composites to 655MPa for 0.5% alkali-treated fiber composites (both contain 30 phr fibers). The loss modulus further increased to 800.5 MPa when fibers were treated with 4% alkali. However, treated fiber composites exhibited low mechanical damping parameter ($tan\ \delta_{max}$). The strong and rigid fiber-matrix interface due to improved adhesion reduces molecular mobility in the interfacial zone causing a decrease in $tan\ \delta$. The presence of bonding agents results in more number of crosslinks being formed and alkali treatment on fibers lead to strengthening of these crosslinks. As a result, molecular motion along the rubber macromolecular chain was severely hindered, leading to low damping characteristics.

9.3.1.2 *Water Absorption Characteristics*

Effect of fiber loading: The water absorption of OPF-NR composites increased with fiber loading due to the hydrophilicity of the fibers [58]. The water absorption behavior of NR changed from fickian to non-fickian upon addition of OPF due to the presence of microcracks as well as the viscoelastic nature of the polymer. The water uptake of OPF-NR composite was lower than that of OPF-sisal fiber-NR hybrid biocomposite [58]. The incoroporation of sisal fiber containing comparatively more holocellulose (23%), which is highly hydrophilic caused more water uptake. More over, the lignin content of OPF (19%) was higher than sisal fiber (9%). Lignin being hydrophobic prevents absorption of water.

Effect of fiber treatment: The water uptake by OPF-sisal fiber-NR hybrid composites reduced upon mercirization and alkali treatment of fibers [59]. Mercerization improves the fiber surface adhesive characteristics and alkali treatment leads to fibrillation providing large surface area. This results better mechanical interlocking between fiber and matrix causing less water absorption. Besides, the treatment with NaOH solution promotes activation of hydroxyl groups of the cellulose by breaking hydrogen bond. Among three types of silane treatments given to the fibers *viz.*, fluro silane, amino silane and vinyl silane, the maximum water uptake was exhibited upon vinyl silane treatment and less by flurosilane treatment. The water uptake of OPF-sisal fiber-NR hybrid composites was less at higher alkali concentration due to the increased bonding between fiber and matrix. The hybrid composites approached fickian behavior of water absorption at higher alkali concentration.

9.3.1.3 *Thermal Properties*

Generally, incorporation of plant fibers into polymeric matrix increases the thermal stability of the system. In the thermal analysis of OPF-sisal fiber-NR hybrid composites also, it was found that addition of more fibers resulted in increase of thermal stability as indicated by the higher peak temperatures [35]. Increased thermal stability was also confirmed by the decrease in the activation energy of the composites. The thermal stability of NR-sisal fiber-OPF hybrid composites improved upon chemical modification [35]. However chemical modification resulted in lower activation energy (for decomposition) values compared to the untreated fiber composite.

9.3.1.4 *Electrical Properties*

Effect of fiber content: The dielectric constant of fiber-reinforced composite system is higher due to polarization exerted by the incorporation of lignocellulosic fibers. The dielectric constant of OPF-sisal fiber-NR hybrid composites at all the frequencies increased with increase in fiber content [59] as shown in Figure 9.6. Natural rubber is a non-polar material and has only instantaneous atomic and electronic polarization to account. The presence of two lignocellulosic fibers in NR leads to presence of polar groups giving rise to dipole or orientation polarizability. The overall polarizability of a composite is the sum of electronic, atomic and orientation polarization resulting higher dielectric constant. Hence the dielectric constant increases with increase in fiber loading at all frequencies.

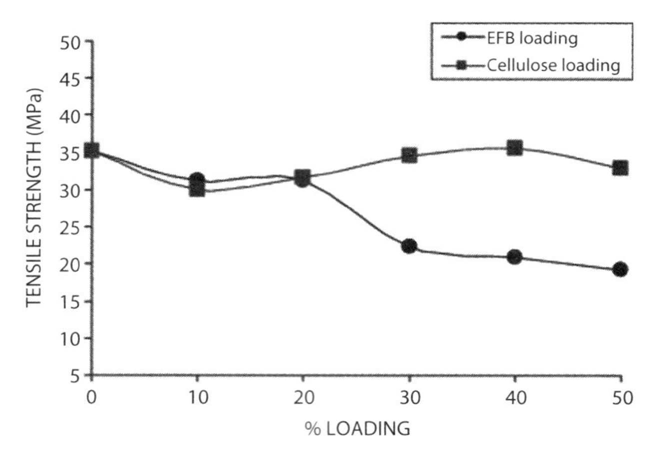

Figure 9.6 Variation of dielectric constant with frequency as a function of oil palm fiber loading.

Volume resistivity is an important property of an insulating material as the most desirable property of an insulator is its ability to resist leakage of electric current. The volume resistivity of OPF-sisal fiber-NR hybrid composites decreased at higher fiber contents indicating increase in the electrical conductivity due to the presence of polar groups [59]. In polymers, most of the current flows through the crystalline regions and passage of current in the amorphous regions is due to the presence of moisture. The presence of two lignocellulosic fibers increases the moisture content and hence increases the conductivity of the system. The observation on the volume resistivity was confirmed by the direct measurement of electrical conductivity, which also increased with fiber loading. At a frequency of 10 KHz, the volume resistivity of pure gum reduced from 650×10^4 Ωm to 150×10^4 Ωm upon addition of 50 phr OPF-sisal fiber combination. Conductivity is also dependent on the dispersion of fibers. At low volume fractions, there was a chaotic dispersion of fibers. The orientation of fibers was too random to promote the flow of current. At high volume fractions of fibers, the population was just right to bring about uniform dispersion facilitating flow of current.

Dissipation factor or loss tangent is defined as the ratio of electrical power dissipated in a material to total power circulating in a circuit. The dissipation factor (*tan* δ) of NR matrix at a frequency of 10 kHz increased from 0.002 to 0.035 upon addition of 50 phr OPF-sisal fiber combination [59]. The polar groups present in the composite increases with fiber content leading to an increment in orientation polarization. The incorporation of fibers in NR matrix disrupts regular arrangement of rubber molecules leading to loss of crystallization ability. Furthermore, addition of fibers enhances flow of current through amorphous region due to their ability to absorb moisture. Therefore addition of fibers resulted in higher loss or higher amount of dissipation that are associated with amorphous phase relaxations.

Effect of fiber treatment: Chemical modification of fibers decreased the dielectric constant of OPF-sisal fiber-NR hybrid composites [59]. This was due to the decrease in orientation polarization of the composites upon treatment. Chemical treatment results in reduction of hydrophilicity of the fibers leading to lowering of orientation polarization and subsequently dielectric constant. Alkali treatment yielded higher dielectric constant comparing to silane treatment. However, higher concentration of alkali reduced the dielectric constant. Alkali treatment resulted in unlocking of the hydrogen bonds making them more reactive. In untreated state, the cellulosic –OH groups are relatively unreactive as they form strong hydrogen bonds. In addition to this, alkali treatment can lead to fibrillation, i.e., breaking down of fibers into smaller ones. All these factors provide a large surface area and give a better mechanical interlocking between the fiber and matrix and thus reduce water absorption. This was attributed as the reason in lowering the overall polarity and hydrophilicity of the system leading to reduction of orientation polarization and consequently dielectric constant of the composites.

9.3.2 Oil Palm Fiber-Polypropylene Composites

Polypropylene (PP) is an addition polymer made from the monomer propylene. It is a thermoplastic polymer mainly used in packaging, stationery, laboratory equipments, automotive components, etc. Resistance to many chemical solvents, bases and acids makes it suitable for a variety of applications. Melt processing of polypropylene can be achieved through extrusion and moulding. However the most common shaping technique followed is injection moulding.

9.3.2.1 Mechanical Properties

Effect of fiber loading: Loading OPFs in PP matrix reduced its tensile strength [33] as shown in Figure 9.7. The reduction in tensile strength upon filler loading was due to interruption caused by the filler in transferring stress along applied force and this problem was intensified by lack of interfacial adhesion of the filler and the matrix. It is interesting to note that addition of OPF-derived cellulose increased the tensile strength of composites eventhough a lowering trend was observed at lower cellulose fractions. The tensile strength was equal to that of pure PP at 40% cellulose loading. An interaction with fairly concentrated aqueous solution of strong base during production of cellulose to produce great swelling was the reason explained for the increase in strength. The flexural modulus increased with increase in filler loading and cellulose composites exhibited a higher modulus than fiber composites. Addition of 50% cellulose to virgin PP increased flexural modulus from 1750 MPa to 4250 MPa, whereas OPF composite of 50% fiber content increased the modulus up to 2750 MPa only. However, increasing percentage of OPFs reduced the flexural strength. In an attempt to make OPF-filled PP composites by reactive processing, maximum tensile strength (21.4 MPa) of the composites was obtained at 20% weight fraction of filler [5]. However the elongation at break decreased with increase in filler loading. The flexural strength, flexural modulus and flexural toughness of OPF-PP composites at 40% fiber loading were 37 MPa, 2.13 GPa and 47 kPa, respectively [4]. The impact strength also increased with filler loading, however OPF composites exhibited a less impact strength than cellulose-filled composites. The izod impact strength of pure PP was 32 J/mm, while 10% fiber and cellulose loadings reduced it to 25 J/mm thereafter steadily increased upto 37 J/mm when 50% cellulose was added [33]. Oil palm fiber loading of 50% resulted same impact strength as that of virgin PP. The irregularity in shape and low aspect ratio of OPF may affect

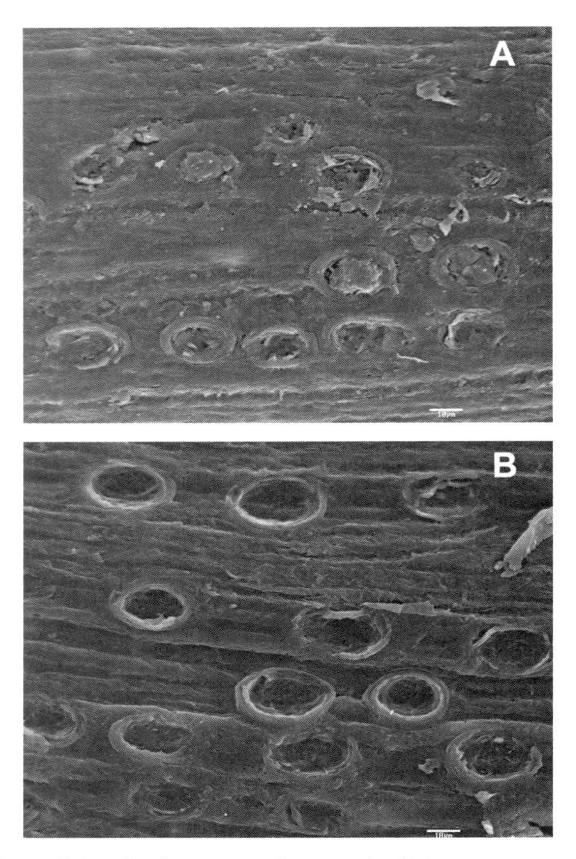

Figure 9.7 Effect of fiber/cellulose loading on tensile strength of PP composites.

their capabilities to support stress transmitted from the PP matrix [4, 33]. The effect would be amplified if the proportion of OPF is increased.

The flexural properties of OPF-glass fiber-PP hybrid composites reduced upon increasing the proportion of OPFs [60]. When the whole glass fibers were replaced with OPFs (30% fiber content), the flexural strength, flexural modulus and flexural toughness values reduced from 37MPa to 20 MPa, 3.75 GPa to 2.75 GPa and 25 kPa to 13kPa, respectively. The reduction in flexural properties were due to irregular shape and low aspect ratio of OPF. Similarly, the tensile properties were also reduced when more OPFs were incorporated. The tensile strength, tensile modulus and tensile toughness of OPF-glass fiber-PP composites containing 30% fiber reduced respectively from 25MPa to 13MPa, 850 MPa to 610 MPa and 180 kPa to 105 kPa when the whole glass fiber was replaced with OPFs.

Effect of fiber treatments: The maleic anhydride grafted polypropylene (MAPP) has significant influence on the surface properties of OPF [44]. The OPF-PP composite exhibited good tensile strength (36 MPa) and impact strength (38 J/m) upon incorporation of 2% MAPP. However there was a substantial decrease in mechanical strength with increase in concentration of MAPP above 2% in the PP matrix. Trimethylolpropane triacrylate (TMPTA) also significantly influenced the mechanical properties of PP-cellulose composite. The toughness of both the composites was significantly improved by addition of TMPTA. However, PP-cellulose composite remained the better choice than PP- OPF composite as the tensile strength (43 MPa), flexural modulus (3300 MPa) and impact strength (43 J/m) were better with the addition of 2% TMPTA. It has been reported [12] that at Maleic anhydride (MAH) chemical loading of 15%, a flexural strength of 40 MPa, flexural modulus of 4 GPa, flexural toughness of 7 kPa and impact strength of 80 J/m was exhibited by the OPF-PP composites prepared with 40 percent of 60 mesh size fibers. The MAH chemical loading increased the flexural strength and flexural modulus of OPF-PP composites which is due to the increased interfacial adhesion caused by the covalent bonding between MAH and PP. The maleic anhydride treatment has enabled the compatibility between polar functional groups of OPF and non-polar PP. The coupling agent E-43 contributed towards higher tensile strength, flexural toughness, flexural modulus and impact strength to OPF-PP composites [4].

The tensile strength and toughness of OPF-glass fiber-PP hybrid composites were significantly improved upon removal of residual oil from OPFs [43]. The tensile strength increased from 7MPa to 13 MPa and toughness increased from 50 kPa to 130 kPa upon oil removal. Oil removal reduced the stress concentrations created by incompatible layer of esters of the oil. However, the tensile modulus (700 MPa) and elongation at break (2.3%) did not vary significantly upon residual oil removal, whereas the flexural properties increased significantly. The flexural strength, modulus and toughness of unextracted fiber composite were 13 MPa, 1.9 GPa and 7.5 kPa, respectively, which increased to 27 MPa, 2.3 GPa and 26 kPa upon residual oil removal. Treatment of OPF with coupling agent- polymethylene (polyphenyl isocyanate) (PMPPIC) further increased both tensile and flexural properties. This is due to the inhibition of the isocyanate activity with hydroxyl group of OPF by the oil. Fibers treated with other coupling agents *viz.*, maleic anhydride–modified PP (commercial name Epolene, E-43) and 3-(trimethoxysilyl)-propylmethacrylate (TPM) also increased the strength properties of the composites and the contribution by E-43 was higher [60]. The greater reinforcement by OPF in the presence of E-43 was the reason for better dispersion of the filler in PP and better bonding between the constituent materials. A good filler-matrix interaction could be derived from the formation of an ester bond between the anhydride groups of E-43 and the hydroxyl groups at the surfaces of OPF fillers and / or hydrogen bonding between the hydroxyl groups and oxygen from carboxylic groups so produced from the former reaction. Being a derivative of PP, E-43 is more compatible with PP.

9.3.2.2 Water Absorption Characteristics

Effect of fiber loading: The water absorption of OPF-PP composites of varying fiber fractions at the end of a six day immersion was in the range of 4.5% to 7.5% [4]. Composites with a higher proportion of OPF absorb more water due to cellulose, lignin and hemicellulose, possessing polar hydroxyl groups leading to formation of hydrogen bonds with water. Absorption of water by the cell wall of lignocellulosic materials causes swelling of the cell wall and thus increases in thickness of the composite. The percent thickness swelling of OPF-PP composites after a six day immersion varied from 2.3% to 7% at different fiber loading range.

Effect of fiber treatments: The hydrophobicity of OPF-PP composites enhanced upon E-43 treatment to the fiber and the effect of treatment was further enhanced by chemical loading [4]. The ability of maleic anhydride residue of E-43 to interact with hydroxyl groups of OPF, blocking the hydroxyl group-water hydrogen bonding and the hydrophobicity imparted by the PP chain of E-43 are the reasons for enhanced hydrophobicity.

9.3.2.3 *Degradation/weathering*

Weathering in general decreases strength of the composites. The flexural stress and flexural modulus of fresh OPF-PP composites were 41.6 MPa and 3.85 GPa, respectively, which reduced to 27.4 MPa and 2.69 MPa after soil exposure for 12 months [61].

9.3.3 Oil Palm Fiber-Polyurethane Composites

Polyurethane polymers are formed through step-growth polymerization by reacting a monomer containing at least two isocyanate functional groups with another monomer containing at least two hydroxyl groups in the presence of a catalyst. Polyurethanes are widely used in high resiliency flexible foam seating, rigid foam insulation panels, microcellular foam seals and gaskets, automotive suspension bushings, carpet underlay, hard plastic parts for electronic instruments, etc. Polyurethane is also used for moldings which include door frames, columns, window headers, etc. Oil palm fiber acts as a good reinforcement in PU matrix.

9.3.3.1 *Mechanical Properties*

Effect of fiber loading: The tensile strength of OPF-PU composites increased from 21 MPa to 30 MPa when fiber content increased from 30% to 45% [29]. Increase in strength was due to increase in reaction between lignin and isocyanates in forming a three dimensional network of crosslinkings when the fiber functioned as both filler and reactive component. Modulus also increased with percent loading of OPFs, obviously because of its inherent stiffness. The tensile strength of OPF-PU composites increased from 10 MPa to 35 MPa when fiber level increased from 30% to 60% [62]. However further addition of fibers lowered the tensile strength. Similar trend was observed for tensile modulus and tensile toughness also. The modulus was 3.9 MPa and 9.5 MPa at 30% and 60% fiber loading, respectively and the corresponding values of tensile toughness were 13 MPa and 60 MPa. However the strength of OPF composites decreased with increase in fiber loading when palm kernel oil (PKO)-based PU was used [63]. However appropriate addition of OPF improved stiffness of PU composites. Hence it is more suitable in structures where stiffness and dimensional stability are important than structural strength. The flexural and impact strength properties of OPF-PU composites were maximum at a fiber loading of 40-50%. Oil palm fiber-PU composites exhibited a flexural strength of 75 MPa, flexural modulus of 2.25 GPa and flexural toughness of 6.5 MPa and impact strength of 100 J/m at 50% loading of 35-60 mesh size fibers [64]. However addition of more fibers decreased hardness of composites. When 65% of fibers were filled in the matrix, the Shore D hardness value was 73, which reduced to 55 when fiber concentration increased to 75% [65]. At higher fiber loadings, the matrix was not able to cover the fibers, which resulted in physical bonding between the fibers where as interfacial adhesion weakened and agglomeration occurred. This resulted discontinuous matrix, producing stress concentration points in the composite. Increased hardness at low fiber content was explained by means of 'polymer–filler' interactions. The hydroxyl (OH) groups present on the fiber surfaces act as electron donors, while the PU carbonyl groups act as electron acceptors. These interactions can therefore be considered as additional physical crosslinks within the polymer network, so increasing the crosslinking capacity and consequently, hardness of the composite. In the same study, it was also reported that impact strength of a material decreases if more OPFs are added to the matrix. The impact strength of PKO-based polyeurathane was 4000 J/m^2 at a fiber content of 65%, which decreased to 2225 J/m^2 when fiber content increased to 75%. The theory explained for decrease in hardness with increasing fiber content hold good for impact strength also. The flexural strength also decreased as the amount of OPFs increased in the matrix. When OPF content increased from 65% to 75% in the OPF-PU-filled composites, the flexural strength decreased from 11 MPa to 4.5 MPa and the flexural modulus decreased from 1450 MPa to 1100 MPa.

The effect of hybridization of kaolinite (a type of clay) with OPFs and PU synthesized from PKO to form composites was examined by [66]. The kaolinite content of 15% was found beneficial and the resulting composites exhibited flexural strength of 11.2 MPa, flexural modulus of 2550 MPa and impact strength of 4600 MPa. The increase in strength of hybrid composites was due to the encapsulation of OPFs by PU in combination with kaolinite, filling the voids between OPFs. Dynamic mechanical analysis revealed that addition of kaolinite increased the storage modulus of the composites at an optimum level of 15% loading (2150 MPa), indicating an increase in the ability to store energy. The width of $tan\ \delta$ peak of 15% kaolinite loaded hybrid composite was much broader compared to others due to better molecular relaxations.

Effect of fiber size: The effect of fiber size on the mechanical properties of OPF filled PKO-based PU foam was studied by [67]. Higher compressive stress was observed for 45-56 μm fiber particulate than higher sizes. This was due to the higher surface area of the fiber in powder form, which may produce better hindrance to stress-impact propagation. Scanning electron microscopic images indicate that small size fibers got embedded in the matrix well comparing to large size fibers. However the flexural strength of OPF-filled PU composites decreased with decrease in filler size [64]. Similar trend was observed for tensile properties also [62].

Effect of fiber treatments: Both TDI treatment and hexamethylene diisocyanate (HMDI) treatment improved the tensile strength of OPF-PU composites [29]. At a fiber loading of 40% and NCO/OH ratio of 1.1, tensile strengths of 26 MPa, 30 MPa, 28 MPa were exhibited by untreated fiber composite, TDI-treated fiber composites and HMDI-treated fiber composites, respectively. The introduction of isocyanates from TDI or HMDI has enhanced the interaction between OPFs and PU matrix.

9.3.3.2 Water Absorption Characteristics

Effect of fiber loading: The water absorption by OPF-PU composites increased sharply on the day of immersion and remained constant during the following 5 days [65]. Composites with 75% fiber content absorbed 55% water and composites with 65% fiber absorbed 48% water after 1 day of immersion. Weak bonding between the matrix and fiber, agglomeration of the fibers and incomplete encapsulation of matrix over OPFs are the factors leading to poor water resistivity of OPF-PU composites.

9.3.3.3 *Degradation/weathering*

The dimensional stability of OPF-filled PU composites at high and low temperature environments were studied by [67]. The dimensional stability of composites prepared with 45-56 μm fibers was within the allowable limits. However higher sizes caused increased deformation due to the deterioration of cellular structure of the PU foam at higher fiber sizes.

9.3.4 Oil Palm Fiber-Polyvinyl Chloride Composites

Polyvinyl chloride (PVC) is a thermoplastic polymer. It is a vinyl polymer constructed of repeating vinyl groups having one of their hydrogens replaced with a chloride group. Polyvinyl chloride is the third most widely produced plastic after polyethylene and PP. It is used for sewege pipe lines and other pipe line applications as it is biologically and chemically resistant. It is also used for windows, door frames and such other building materials by adding impact modifiers and stabilizers. It becomes flexible on addition of plasticizers and can be used in cabling applications as wire insulator. Oil palm fiber is found to be in good combination with PVC to form composites.

9.3.4.1 *Mechanical Properties*

Effect of fiber loading: The tensile strength of OPF-PVC composite decreased with fiber loading [68]. Tensile strength of pure PVC was 60 MPa, which reduced to 34.71 MPa when 40% of OPF was added. The broken ends of short fibers formed during tensile deformation were believed to induce cracks in the matrix and lead to a reduction in tensile strength. The inability of OPFs to support stress transmitted from the matrix was due to its irregular shape and dispersion problems (agglomerate formation). The impact strength also reduced with fiber loading. Pure resin had impact strength of 20 kJ/m^2, which reduced to 17.41 kJ/m^2 at 40% fiber content. The possible reason was the detrimental effect of fibers caused by the volume they take. Oil palm fibers are incapable of dissipating stress unlike the matrix through shear yielding prior to fracture. They also hinder the local chain motions of PVC molecules to shear yield. Alternatively, the tendency of OPFs to cling together in bundles and resist dispersion at higher fiber loadings cause decrease in the ability of the composites to absorb energy during fracture propagation. The flexural strength of pure acrylic impact modified unplasticized polyvinyl chloride (PVC-U) decreased from 80 MPa to 70 MPa when 40% OPF was added, while the flexural modulus increased from 3500 MPa to 4500 MPa [69]. The reduction in flexural strength was attributed to agglomeration of the filler and its inability to support stresses transferred from the PVC-U matrix.

It has been reported that ultimate tensile strength (UTS) of pure PVC- epoxidized natural rubber (ENR) blend reduced from 9 MPa to 6 MPa upon addition of 30% OPFs [11]. The reduction in tensile strength of composites at higher fiber loading was due to the agglomeration of the filler particles to form a domain that acts like a foreign body.

The filler particles or agglomerates were no longer as equally separated or wetted by the polymer matrix above a particular fiber loading. Other mechanical properties *viz.*, tensile modulus and flexural modulus increased with fiber loading, while elongation at break and impact strength reduced with fiber loading. There was only a marginal increase in hardness with fiber loading beyond 5% level. The elasticity or flexibility of the polymer chain was reduced with addition of fiber, resulting in more rigid composites. However the blend was believed to reach hardness of fiber beyond 5% fiber loading, leading to less marginal increase in the hardness. The reduction in impact strength of OPF-PVC-ENR composites was due to the increase in interfacial regions with more fiber content, which is comparatively weaker than the matrix in resisting crack propagation on impact.

Effect of fiber treatments: It has also been reported that [27] the residual oil on fibers has less effect on fiber-matrix interaction and resulting mechanical properties of OPF-PVC composites. Composites with unextracted fiber of 30 phr exhibited impact strength of 6.11 kJ/m^2. Oil extraction caused only a marginal increase of 0.3% in the impact strength. Similarly the flexural strength of OPF-PVC composites with 30 phr unextracted fiber was 72.4 MPa, which increased to 74.5 MPa when oil was removed indicating only a marginal 3% difference. However the flexural modulus of 4142 MPa for unextracted fiber composites reduced to 3900 MPa when extracted fibers were incorporated. Benzoylation treatment to the fibers improved the tensile strength, impact strength and stiffness of OPF-PVC composite due to improved fiber-matrix adhesion [68]. However addition of acrylic impact modifier caused a negative effect on both modulus and strength [69]. Oil palm fibers when grafted with Methyl Acrylate (MA) improved the ultimate tensile strength (UTS) of the PVC-ENR blend composites [11]. At 10% fiber loading, untreated fiber composites exhibited UTS of 8 MPa, which increased to 10 MPa upon grafting. Other mechanical properties *viz.*, tensile modulus (stiffness), flexural modulus, hardness and impact strength also reduced upon grafting, while elongation at break increased. The crystalline structure of OPF was destroyed by grafting with MA and resulting amorphous fiber could be easily deformed. Thus the decline in flexural modulus upon grafting with MA is associated with the reduction in crystallinity of the fiber. It is interesting to note that impact strength of the grafted fiber composites was less eventhough the treatment improved interfacial adhesion as better bonding leads to catastrophic brittle failure.

9.3.4.2 *Thermal Properties*

The glass transition temperature (T_g) of OPF-PVC-ENR composite increased with fiber content [70] However the thermal stability was not affected significantly with addition of OPF. Benzoylation treatment on fiber reduced the glass transition temperature (Tg) of OPF-PVC composites (64°C) than that of pure PVC (79.16°C) whereas Tg of untreated OPF composite (80.29°C) and pure PVC were not significantly different 67 (Bakar and Baharulrazi, 2008). Decrease in Tg was due to plasticization effect of fibers that diffused or dissolved into the PVC matrix.

9.3.5 Oil Palm Fiber-Polyester Composites

Polyester is a category of polymers which contain the ester functional group in their main chain. It is used mainly in textiles and packaging industry. It is also used to manufacture high strength ropes, thread, hoses, sails, floppy disk liners, power belting, etc. Oil palm fiber is a better reinforcement in polyester matrix.

9.3.5.1 Physical Properties

Effect of fiber loading: One of the desirable functions of natural fibers in polymeric matrix is to reduce the mass on account of the inherent low density of the fibers. The density of OPF powder is 1.138 g/cm^3 and that of polyester is 1.202 g/cm^3. Addition of 55% OPF to polyester matrix reduced composite density to 1.17 g/cm^3[14]. Similarly, the density of glass fiber-polyester composites can be reduced by substituting glass fiber with OPF. The density of glass fiber is 2.6 g/cm^3 and that of polyester resin is 1.41 g/cm^3, which obviously results in higher composite density. Oil palm fiber of 1.14 g/cm^3 density used in place of glass fiber resulted in low density composites [7]. Apart from the lower density of the constituent fiber, the reason for reduction in density of composites at higher fiber loading may also be due to the void formation as a result of poor wetting of the fibers with matrix. In that case, strength and performance will be affected. Hence reduction in density should be accounted with strength and other performance parameters. Incorporation of 40% and more OPF in OPF-polyester composites caused reduction in density from 1.15 to 1.11 g/cm^3 [14].

9.3.5.2 Mechanical Properties

Effect of fiber loading: The impact strength of polyester resin increased from 6 kJ/m^2 to 18 kJ/m^2 when 55% OPFs were incorporated [71]. Similarly the elongation at break also increased from 3% to 3.80% upon addition of 55% OPF. However, the flexural strength of polyester resin reduced from 50 MPa to 32 MPa upon addition of 15% fibers and further increased upto 43 MPa upon addition of 55% fibers. The flexural modulus of the pure resin marginally ncreased from 3.2GPa to 3.7 GPa when 55% fiber were incorporated in the matrix [14]. The abrasion characteristics of OPF-filled polyester composites was better than pure resin [14]. Oil palm fiber reinforcement reduced the weight loss on abrasion of pure polyester resin by 50-60%. In addition, the friction coefficient of OPF-polyester composite was less by about 23% comparing to that of the neat polyester [24].

The flexural strength of pure polyester resin increased from 43.3 MPa to 165.4 MPa upon incorporation of 12% glass fiber and the flexural strength remained same when 40% of this fiber fraction was replaced with OPF [7]. However when 70% of the fiber fraction was replaced with OPF, the flexural strength reduced to 143 MPa. The OPFs effectively transferred stress from glass fibers at 40% volume fraction. However, the flexural strength drastically reduced below that of pure resin (36.8 MPa) when whole fiber fraction was replaced with OPF. Hence OPF can be a substitute for glass fiber in potential applications where it do not require high load bearing capabilities. The tensile strength of OPF-glass fiber-reinforced polyester composites increased upto total fiber content of 45% [72]. The OPF and glass fiber when incorporated in the ratio of

70:30 in polyester caused an increase in its tensile strength from 25MPa to 48 MPa, whereas 45% of OPF alone resulted in tensile strength of 31MPa. The elongation at break was high at higher OPF content as it has high strain to failure characteristic (8–18%) compared to low extensibility of glass fiber (3%). The flexural strength and flexural modulus of OPF-glass fiber-polyester hydrid composites were maximum at a total fiber content of 35%. The flexural strength of pure polyester increased from 50 MPa to 75 MPa when 35% fibers were incorporated (OPF: glass fiber ratio of 70:30). The flexural modulus of the hybrid composite was more dependent on the amount of glass fiber rather than OPF due to the high modulus of glass fiber (66–72 GPa). The impact strength also increased with increase in fiber content up to 35%. The fiber plays an important role in the impact resistance of composites as they interact with crack formation in the matrix and act as stress transferring medium. The impact strength of pure polyester increased from 7 kJ/m^2 to 14.7 kJ/m^2 when 35% fibers were incorporated (OPF: glass fiber ratio of 70:30). The Rockwell hardness of pure polyester resin decreased from 113 HRB to 105 HRB when 55% OPF was incorporated. The water molecules in fiber cell wall act as plasticizer which decreases hardness of composites.

Effect of fiber treatments: Alkali treatment on OPFs significantly improved its interfacial shear strength in polyester matrix [14]. Alkali treatment washed out the outer skin, better exposing fiber to the polyester matrix, leading to proper interaction between their surfaces. In addition, the fine holes created on alkali treatment allowed the polyester to penetrate into the fiber bundles in a better way. Acetylation treatment to the fibers improved impact strength of OPF-polyester composites due to improved fiber wettability and resulting fewer void spaces [71]. The tensile stress of OPF-polyester composites increased slightly upon both acetylation and silane treatments on fibers and decreased upon titanate treatment [14]. The flexural modulus of OPF-PP composites also increased considerably upon acetylation treatment on fibers. Similarly the abrasion resistance of OPF-polyester composites was enhanced upon alkali treatment to fibers [13]. Treated fibers enhanced the adhesion resistance of polyester resin by 75-85%, while untreated fibers enhanced the abrasion resistance only by 50–60%.

9.3.5.3 *Water Absorption Characteristics*

Effect of fiber loading: The water absorption pattern of unsaturated OPF-polyester composite followed typical fickian behavior, where the mass of water absorbed increased linearly with a function of square root of time and then gradually decreased until equilibrium plateau or complete saturation is reached [72]. Increased water absorption of the composites is due to swelling of the fibers leading to crack formation in the polyester matrix, which act as pathways for the water molecules to diffuse into the composite material. Temperature facilitate water absorption. Hence the temperature dependence of water absorption of OPF-polyester composites was established by [71]. The water absorption characteristics of OPF-glass fiber-polyester hybrid composites was studied by [30]. Pure polyester resin absorbs 1% water, which increased to 9% when 45% OPFs were incorporated. The absorption percentage reduced to 6% when hybridized with glass fibers in the ratio of 70:30. A from the hydrophilic nature of OPFs, increased water absorption at higher volume fraction of OPF could be attributed to the poor compatibility between OPF and glass fiber and also between OPF and polyester matrix

[23]. Higher volume fractions of OPF in polyester matrix may lead to fiber layering out, which creates micro-void and cracks within the composites. Thus water can easily penetrate and diffuse through the porous structure. This mechanism involves flow of water molecules along fiber-matrix interface followed by diffusion from the interface into matrix and fibers. Absorption of water by the composites obviously cause swelling as mentioned earlier. The percentage thickness swelling of pure polyester resin when submerged in boiling water for 2 hours increased from 1.04% to 1.2% when 12% glass fibers were incorporated [7]. Thickness swelling further increased to 2.03% when 70% of fiber fraction was replaced by OPFs and a maximum thickness swelling of 2.46% was observed when 100% of fiber fraction was replaced with OPFs.

Effect of fiber treatments: Acetylation treatment on fibers reduced the water absorption (at 100°C) of untreated OPF-polyester composites from 15.8% to 5.7% . Good interfacial contact between fiber and matrix and increased hydrophobicity caused by the treatment lead to reduced water absorption.

9.3.5.4 Degradation/weathering

Effect of fiber loading: The loss in tensile properties of OPF-polyester composites upon degradation in soil was quantified by . Loss in tensile strength by 8%, 17% and 35% were observed, respectively after exposure of 3, 6 and 12 months. Similarly tensile modulus, elongation at break and impact strength also reduced upon soil exposure. A loss of impact strength by 6%, 18% and 43% were observed, respectively after 3, 6 and 12 months. Similarly the tensile stress, tensile modulus and elongation at break decreased from 35.1 MPa, 3.29 GPa and 3.75%, 34.6 MPa, 2.32 MPa and 2.48%, respectively upon soil burial for 12 months [61].

Effect of fiber treatment: Chemical treatment to the fiber considerably reduced mass loss of OPF-polyester composites upon weathering [61]. The magnitude of mass loss decreased in the order: unmodified fiber>titanate treated >silane treated > acetylated. Chemical treatments also could conserve the mechanical properties of OPF-polyester composites upon ageing [71]. The loss of tensile stress, tensile modulus and elongation at break of OPF-polyester composites during ageing in deionized water was more prominent for untreated fiber composites. Though further reduction was observed in the subsequent months, an increase in tensile stress and modulus was observed upon acetylation treatment on fiber in the initial 3 month period. The improvement in composite properties at short time period was due to stress relaxation of fiber embedded in composites and also due to less debonding by fiber swelling as fiber is already swollen upon treatment. The treatments such as acetylation, silane and titanate on fibers conserve the tensile properties of OPF-polyester composites even after exposure in soil for many months [56]. The loss in strength decreased in the order: unmodified>titanate> silane>acetylated.

9.3.6 Oil Palm Fiber-Phenol Formaldehyde Composites

Phenol formaldehyde, a synthetic thermosetting resin obtained by the reaction of phenols with aldehyde, is used for making plywood, particle board, medium density fiber boards and other wood- and lignocellulose-based panel and wood joinery. Phenolic laminates are made by impregnating one or more layers of a base material such as paper, fiber or cotton with phenolic resin and laminating the resin-saturated base material under heat and pressure. The resin fully polymerizes (cures) during this process. A number of research reports are available on the properties of OPF-PF composites.

9.3.6.1 Physical Properties

The density of pure PF increased from 1.33 g/cm^3 to 1.48 g/cm^3 when 40% glass fibers were incorporated. This could be reduced to 1.45 g/cm^3 when 24% of fiber fraction was replaced with OPF and further reduction to 1.2 g/cm^3 could be achieved when 96% of the fiber fraction was replaced with OPF [19].

9.3.6.2 Mechanical Properties

Effect of fiber loading: The tensile and flexural strength properties of OPF-glass fiber-PF hybrid composite decreased with increase in fraction of OPF [20]. In case of composite with 40% fiber content (weight fraction), tensile strength reduced from 80 MPa to 50 MPa, tensile modulus reduced from 2500 MPa to 250 MPa, flexural strength reduced from 87 MPa to 50 MPa and elongation at break reduced from 6.5% to 4.75%, when 25% (volume fraction) of the glass fibers were replaced with OPF. However, the impact strength remained the same (230 kJ/m^2) even with OPF substitution. It was also observed that further fiber loadings improved the impact strength.

The stress relaxation in OPF-PF composites was studied by [45]. Higher relaxation was observed for composites with 30% fiber content among the 20, 30 and 40% fiber loadings studied. Stress relaxation mechanism in the OPF-reinforced PF composite exhibited a two step process. Fiber-matrix bond failure contributes to the first step relaxation and the second step relaxation was pre-dominated by the matrix phase relaxation. Hybridization of OPF with glass fiber in PF matrix reduced the relaxation rate than both pure glass fiber-PF composites and OPF-PF composite. This indicates that OPF with glass fiber results in composites of long term higher mechanical performance.

Dynamic mechanical analysis of OPF-PF composites indicated that incorporation of OPF increased the modulus and damping characteristics of the neat sample [20]. This was due to the increase in interface area and more energy loss at interfaces at higher fiber loading. The impact performance of PF resin also largely improved upon reinforcement with OPF [50]. Incorporation of OPF with glass fiber increased the damping value of the OPF-PF composite and notably damping increased with relative volume fraction of OPF [20]. Both the storage modulus and loss modulus values decreased after the relaxation with increase in relative OPF volume fraction in hybrid composites.

Effect of fiber treatments: Oil palm fiber-PF composites exhibited maximum tensile strength of 40 MPa, tensile modulus of 1300 MPa and elongation at break of 9% when the fibers were treated respectively with permanganate, mercerization and latex coating. The changes in tensile strength and tensile modulus of OPF-PF composites followed the order: acetylated (highest) > propionylated > extracted > nonextracted (lowest) [73]. The tensile strength and tensile modulus of acetylated fiber composites were 13 MPa and 1.6 GPa, respectively. The composites with acetylated and propionylated fibers exhibited higher flexural strength than unmodified fiber composites. Acetylated fiber composites exhibited flexural strength of 23 N/mm^2 whereas raw fiber composites exhibited only 9 N/mm^2 strength. The effect of fiber modification on flexural modulus was more significant than flexural strength as the incorporation of modified fiber was still able to impart stiffness to the composites. The stiffness of acetylated and propionylated fiber composites was higher than unmodified fiber composites. This was due to the increased compatibility between the resin and fiber. The increased fiber-matrix compatibility upon modification results in formation of a continuous interfacial region, causing better and efficient stress transfer. The impact properties of OPF-PF composites varied in the following order: non-extracted (lowest) < extracted < propionylated < acetyl (highest). The impact strength of OPF-PF composites increased from 6kJ/m^2 to 6.8 kJ/m^2 upon acetylation treatment on fiber. The impact strength of pure resin increased from 20 kJ/m^2 to 80 kJ/m^2 upon addition of 40% untreated OPF and latex coating imparted impact resistance of 190 kJ/m^2 followed by acetylated (180 kJ/m^2), silane (165 kJ/m^2) and TDI (155 kJ/m^2) [50]. Increased hydrophobicity of the fibers upon these treatments lead to weak interfacial linkage thereby facilitating debonding process on stressed condition. Alkali treatment decreased the relaxation rate of OPF-PF composites due to strong interfacial interlocking between fiber and matrix [45]. Net relaxation in isocyanate-treated sample was less than that of untreated samples, whereas latex treated fiber composite exhibited maximum stress relaxation. The faster relaxation in latex treated fiber composites was due to low interfacial bonding between fiber and matrix upon latex treatment causing decrease in strength and stiffness to the composite.

9.3.6.3 Water Absorption Characteristics

Effect of fiber loading: Oil palm fiber-PF composites exhibited lowest sorption at a fiber content of 40% among various fiber contents in the range of 10% to 50% [23]. It is interesting to note that maximum mechanical properties were also exhibited at this fiber content. The OPF- glass fiber-PF hybrid composites were found more hydrophilic than unhybridized composites due to the poor compatability between the OPF and glass fiber. It was also observed that the water absorption of OPF-PF composites are diffusion controlled and follow fickian behavior, whereas hybridization with glass fibers caused a deviation from fickian behavior.

Effect of fiber treatments: The rate of water uptake by OPF-PF composites upon different fiber treatments was in the order: extracted (highest) > non-extracted > propionylated > acetylated (lowest) [73]. The variation between the lowest and highest was 30%. It is also observed that most of the fiber treatments increased water absorption of the composites except alkali treatment, however the treatments reduced water absoption of the fibers [45]. Alkali treatment removes the amorphous waxy cuticle layer of the fiber and activates hydroxyl groups leading to chemical interaction between the fiber and matrix. In case of PF, the trend was different as it is hydrophilic whereas most polymers used for composite fabrication is hydrophobic. Therefore the more hydrophobic the fiber in OPF-PF composites, less the extent of fiber-matrix interaction, which facilitate sorption process [23]. For example, latex coating make the fibers most hydrophobic and the OPF-PF composites prepared from latex coated fibers exhibit maximum water absorption.

9.3.6.4 Thermal Properties

Effect of fiber loading: The thermal conductivity and thermal diffusivity of OPF-reinforced PF composites was less than that of pure resin. The thermal conductivity (λ) and thermal diffusivity (χ) of untreated OPF-PF composite of 40% fiber loading was 0.29 W/mK and 0.16 mm^2/s, respectively where as the thermal conductivity and thermal diffusivity of pure PF were 0.348 W/m K and 0.167 mm^2/s, respectively [74]. The effective thermal conductivity of composite can also be predicted from individual thermal conductivities of fibers and matrix by employing different models [75]. The incorporation of OPF in PF matrix resulted in decrease of glass transition temperature [20]. This was due to higher void formation at higher fiber content as voids facilitate the chain mobility at lower temperatures resulting decrease in T_g value.

Effect of fiber treatments: The thermal conductivity of fillers in OPF-PF composites increased after chemical treatments such as $KMnO_4$, peroxide treatment, etc., obviously causing increase in conductivity of the resulting OPF-PF composite [52]. The cellulose radicals formed during these treatments enhanced the chemical interlocking at the interface. However silane-treated fibers exhibited lower thermal conductivity and thermal diffusivity compared to alkali-treated fiber composites . Silane treatment made the fibers less hydrophilic leading to less adhesion between fibers and hydrophilic phenolic resin. In addition to this, decrease in pore diameter upon treatment weakened the interlocking with the resin, resulting low effective thermal conductivity and thermal diffusivity. Removal of waxy cuticle layer of the fiber surface and increased polarity of OPFs on reaction of acetic acid with cellulosic OH- groups of the fibers also caused lowering of both thermal conductivity and diffusivity. Increase in thermal stability of OPF-PF composites upon fiber treatments has been repoted [76]. The lignin–cellulose complex formed during treatments imparted more stability to the fiber. Activation energy for crystallization and thermal stability of OPF-PF composites quantified using differential scanning calorimetry indicated that the crystallization energies of treated and untreated samples were not significantly different. Alkali-treated samples had the slowest rate of crystallization and maximum stability.

9.3.6.5 Degradation/weathering

Effect of fiber loading: The tensile strength of OPF-PF composite reduced from 37 MPa to 16 MPa upon thermal ageing and boiling water ageing reduced the tensile strength upto 28 MPa [17]. The reduction in tensile strength when the specimens are aged in thermal environment was due to the decrease in fiber-matrix adhesion owing to shrinkage of the fiber in thermal environment. However tensile strength of cold water aged samples was 38 MPa and biodegraded samples (8 months) was 35 MPa. Increase in strength upon cold water ageing was due to decrease in void size at the fiber–matrix interphase during swelling of the fiber. This could exert a radial pressure leading to higher tensile strength. The flexural strength of OPF-PF composites also decreased on ageing. It was also found that the Izod impact strength decreased to a very low value upon gamma irradiation due to the bond scission and disintegration at the fiber-matrix interface. However, an increased stress relaxation of water aged OPF-PF composites was observed [45]. This was due to the changes in interface properties attained during ageing.

Effect of fiber treatments: The modulus of OPF-PF composites enhanced when fibers were given acetylation, isocyanate, acrylate and silane treatments. Peroxide treatment on resin also caused increase in modulus upon thermal ageing [16]. Similarly, the flexural properties of peroxide-treated, latex-modified and acrylated fiber composites increased upon thermal ageing. However, untreated and most of the treated fiber composites exhibited decrease in flexural performance on water ageing. On contrary, the impact strength of OPF-PF composites (41 KJ/m^2) increased upon water ageing in untreated, acrylonitrile, peroxide and isocyanate treatement systems. However the impact strength of the composite decreased upon thermal ageing, biodegradation and also upon gamma irradiation.

9.3.7 Oil Palm Fiber-Polystyrene Composites

Polystyrene is an aromatic polymer made from styrene, an aromatic monomer which is commercially manufactured from petroleum. Polystyrene is commonly injection moulded or extruded while expanded polystyrene is either extruded or moulded in a special process. Solid polystyrene is used in disposable cutlery, plastic models, CD and DVD cases, etc. Foamed polystyrene is mainly used for packing materials, insulation, foam drink cups, etc. Polystyrene foams are good thermal insulators and therefore used as building insulation materials such as in structural insulated panel building systems. They are also used for non-weight-bearing architectural structures. The information on OPF-polystyrene composites is limited.

9.3.7.1 Mechanical Properties

The modulus of OPF-polystyrene composites increased with fiber loading up to 30%, however the maximum strain and flexural strength decreased [7]. The drop in strain was suspected due to irregular shape of the fibers, which caused inability of transferring stress from the matrix. The flexural properties *viz.*, maximum stress, maximum strain, modulus of elasticity and Young's modulus of OPF-polystyrene composites at 10% fiber content (300-500μm size fibers) were reported as 46.43 MPa, 0.027 mm, 1665.35 MPa and 2685.84 MPa, respectively. The flexural properties were not affected by fiber size when the size was below 300 μm. The flexural properties of polystyrene composites improved upon benzoylation due to better interfacial adhesion and hydrophobicity of the fibers.

9.3.8 Oil Palm Fiber-Epoxy Composites

Epoxy or polyepoxide is a thermosetting polymer formed from reaction of an epoxide resin with polyamine hardener. Epoxy is used in coatings, adhesives and composite materials. They have excellent adhesion, chemical and heat resistance, good mechanical properties and electrical insulating properties. Epoxies with high thermal insulation, thermal conductivity combined with high electrical resistance are used for electronics applications. There are a few studies conducted on OPF-epoxy composites.

9.3.8.1 *Mechanical Properties*

Effect of fiber loading: The ultimate tensile strength of OPF-epoxy composite decreased from 47.78 MPa to 46.13 MPa when fiber content increased from 35% to 55%, whereas the ultimate tensile strength of carbon fiber-epoxy composite and pure epoxy resin are 246.99 MPa and 62.49 MPa, respectively [36]. This indicate that OPF failed to act as a reinforcement in comparison to carbon fiber in epoxy resin. Oil palm fiber do not contribute towards the fatigue strength of epoxy as increase in fiber volume ratio resulted in lowering the fatigue resistance of OPFs. Oil palm fiber-epoxy composite with 10% fiber would be able to support about 200 kg load with a maximum deflection of 0.2 mm for a 7 m span using the T-beam configuration [77]. This outcome indicates the potential use of OPF to build moderate load supporting structures thereby reducing the cost of short span bridge.

The tensile strength of glass fiber-epoxy composite decreased from 111 MPa to 24 MPa when glass fiber was replaced with OPF [17]. This indicates that the hybrid composite has intermediate strength characteristics. Oil palm fibers composed of individual fibers bonded by strong pectin interface, which cause individual fibers not loaded uniformely or in some case no loading at all. Hence the OPF is unable to support the stress transferred from the epoxy matrix successfully. Poor adhesion between epoxy matrix and OPF also lead to a weak interfacial bond, resulting in inefficient stress transfer between epoxy matrix and OPF. The elongation at break of OPF-epoxy composite (4%) was slightly lower than that of glass fiber-reinforced composite (4.75%), however the hybrid composite exhibited higher elongation at break. This exceptional behavior of the hybrid composites was due to the existence of a load sharing mechanism between the plies of glass fiber and OPF. The failed glass fiber plies were able to continue to carry the load while redistributing the remaining load to OPF ply. The incorporation of glass fibers into OPF composite increased the stiffness of the hybrid composites. Increase in stiffness of the hybrid composites with addition of glass fibers was due to the higher tensile modulus of glass fibers (66–72 GPa) than that of OPFs (1–9 GPa). The OPF-epoxy composite exhibited a lower impact strength (18 kJ/m^2) than glass fiber composite (107 kJ/m^2). As OPF composite was subjected to a high speed impact load, the sudden stress transferred from the matrix to the fiber exceeded the fiber strength, resulting in fracture of OPFs at the crack plane without any fiber pullout.

9.3.9 Oil Palm Fiber-LLDPE Composites

Linear low density polyethylene (LLDPE) is significantly stronger than LDPE and has better heat sealing properties. However it has higher melt viscosities and is more difficult to process. Linear low density polyethylene is used mainly for packaging applications, toys, covers, lids, pipes, buckets, containers, covering of cables, flexible tubing, etc. The low processing temperature (less than 130°C) of LLDPE makes biocomposite fabrication possible without partial melting or annealing of the fibers. The high toughness of LLDPE imparts a good impact-resistant to composites. Short processing time, unlimited storage time and solvent free processing are the other advantages of natural fiber composites based on LLDPE.

9.3.9.1 Physical Properties

Effect of fiber loading: Addition of oil palm fibers increased density of OPF-LLDPE composites as reported by [39]. Density of composite is expected to decrease at higher fiber loading due to lack of wetting and resulting void formation. The reverse trend in case of OPF-LLDPE composites was due to much higher density of oil palm fiber (1503 kg/m^3) than that of LLDPE (925 kg/m^3). The maximum values of true density, bulk density and porosity of the oil palm fiber-LLDPE composites were 1177 kg/m^3, 1122 kg/m^3 and 11.3 percent, respectively.

Addition of fiber in LLDPE in the same study reduced the L^* value at lower fiber contents and increased at higher fiber contents. The a^* value of the polymer originally in the negative region turned out to positive (reddish) upon addition of oil palm fiber and continue to be constant upon further fiber loadings. Similarly the b^* value of the polymer in the negative region indicating the color towards blue turned positive (yellow) at higher fiber contents. The L^*, $a^*$and b^*values of the oil palm fiber-LLDPE composites varied in the range, 38.1-59.7, 5.8-7.4 and 15.7-22.2, respectively. Opacity of LLDPE specimen of 3 mm thickness was 63.58 percent. Addition of 10 percent oil palm fiber itself increased opacity to 99 percent and further addition of oil palm fiber resulted 100 percent opaque specimen.

The percent water absorption of oil palm fiber-LLDPE composites initially increased linearly with time of immersion and thereafter maintained constant, which is typical of Fickian behavior [53]. During 8 days period of immersion, the composite specimen of 50 percent fiber content absorbed maximum upto 25 percent water. The rate of water absorption of oil palm fiber-LLDPE composites increased with increase in fiber loading. The values of permeability and thermodynamic solubility also increased with fiber loading. Increase in fiber content caused increase in diffusion coefficient upto 40 percent and reduced thereafter. The thickness swelling increased with fiber loading. In addition to thickness swelling, composites expanded linearly also during water absorption, however linear expansion was considerably less than thickness swelling. Similar to thickness swelling, linear expansion was sharp in initial days of sorption, which leveled off in subsequent days. Higher fiber loading and alkali treatment caused more linear expansion.

Effect of fiber size: Lowering the fiber size caused increase in composite density due to less volume being occupied by the fibers at lower sizes, whereas variation in fiber size did not make much change in porosity [39]. However the composites with higher fiber size exhibited higher water absorption [53].

Effect of fiber treatments: Alkali treatment of fibers slightly increased the density of composites as a result of low void content indicating better interfacial adhesion between matrix and fiber [39]. Alkali treatment considerably reduced the porosity, indicating better fiber-matrix adhesion. The principal cause of increase in porosity in composites is the presence of voids in the fiber-matrix interface due to lack of compatibility. Study on water absorption characteristics of OPF-LLDPE composites [53] indicated alkali treatment on fibers reduced water absorption at higher fiber loadings of 40 and 50 percent. Composites made from alkali-treated fibers exhibited more swelling compared to untreated fiber composites except at 50% fiber loading. Alkali treatment caused more linear expansion.

9.3.9.2 *Electrical Properties*

Effect of fiber loading: From the studies on the dielectric constant of OPF-LLDPE composites, the dielectric constant is found to increase with fiber loading [39]. This trend of increase in effective dipole moment of the composites was due to the polar groups in the filler material [78]. Similar trend for coconut fiber-polypropylene composites were also observed [79]. It is interesting to note that the increase in dielectric constant with OPF loading was more prominent in case of alkali-treated fiber composites.

The volume resistivity of pure LLDPE decreased from 2.96×10^{6} MΩ-m to 2.16×10^{4} MΩ-m when 50 percent fibers were incorporated [17]. The resistance decreased gradually with increase in fiber content. Only slight variation in the breakdown was observed between specimens. The breakdown voltage of the composites was in the range of 5.6 to 6.2 kV/mm.

Effect of fiber size: The dielectric constant of alkali-treated fiber composites changed slightly with fiber size, however untreated fiber composites did not show variation with fiber size [39]. The effective dielectric constant decreases with increasing filler size due to increased interface volume when filler of less particle size was used for a given volume fraction of filler [31]. Also at a given volume fraction of filler, the smaller particle size has more polarization in the interface surface as a result of increased moisture absorption for small size fillers due to increased surface area [78]. Water has unfavorable dielectric properties, which increases the dielectric constant.

Effect of fiber treatments: Alkali-treated fiber composites exhibited a lower dielectric constant, particularly at high fiber loading [39] and similar trend was also observed in case of sisal-LDPE composites [80]. They have explained this trend as reduction in the water absorption capacity of the sisal fiber with alkali treatment. Alkali treatment resulted in unlocking of the hydrogen bonds making them more reactive. In untreated state, the cellulosic–OH groups are relatively unreactive as they form strong hydrogen bonds. In addition to this, alkali treatment can lead to fibrillation, i.e., breaking down of fibers into smaller ones. All these factors provide a large surface area and give a better mechanical interlocking between the fiber and matrix and thus reduce water absorption. This results in lowering the overall polarity and hydrophilicity of the system leading to reduction of orientation polarization and consequently dielectric constant of the composites.

9.3.9.3 *Mechanical Properties*

Effect of fiber loading: The tensile strength of oil palm fiber-LLDPE composites decreased with increase in fiber content [17]. The composites exhibited a tensile strength of less than 17.7 MPa, which is the tensile strength of pure LLDPE. Addition of 10 percent fiber reduced tensile strength to a range of 15-16.57 MPa. At a fiber content of 50 percent, the tensile strength was as low as 8.64-10.44 MPa. The tensile modulus of pure LLDPE was 250.7 MPa, which increased up to 707 MPa upon addition of 50 percent fiber. The elongation at break reduced with fiber content. The elongation at break of pure LLDPE of 18.7 mm decreased upto 1.4 mm for composites with 50 percent fiber content. The flexural strength of oil palm fiber-LLDPE composites decreased steadily with increase in fiber content. The flexural strength of pure LLDPE increased from

20.9 MPa to 23.9 MPa with addition of 20 percent fiber, whereas further fiber loading reduced the flexural strength. The impact strength of pure LLDPE (211 J/m) decreased to 90 J/m when 50 percent fiber was incorporated. The hardness of pure LLDPE was 57.4, which increased to 59.7 when 30 percent fibers were incorporated. However the composites with 50 percent fiber content exhibited hardness of 58.8.

In the stress relaxation behavior study of OPF-LLDPE composites, it was observed that irrespective of the fiber content and fiber size, the stress decayed with time and the decay (relaxation) was faster in the beginning [17]. The stress in the composite decayed to 68-88 percent within 2.5 min of withdrawal of the load. The stress relaxation in the composite was a stepped process in comparison to the uniform stress decay of pure LLDPE. Faster relaxation was observed for pure LLDPE, compared to the fiber-reinforced composites and the rate reduced successively as the fiber content increased in the composite. The same trend was observed for composites with different sized fibers also.

Dynamic mechanical analysis (DMA) of OPF-LLDPE biocomposites showed that as the frequency increased, storage modulus values increased at all temperatures [81]. The modulus values decreased with increase in temperature. The reduction in E' is associated with softening of the matrix at higher temperatures. It was observed that E' increased with increase in fiber loading except for composite with 10% fiber loading. The pure LLDPE is more flexible resulting in low stiffness and low E'. Addition of fibers increases the stiffness of the material causing increase in storage modulus. The increase in E' with fiber addition was also attributed to the decrease in molecular mobility of LLDPE contributed by the stiff fibers. Higher values of E' is an indication of increased ability to store energy. It is also interesting to note that the increase in modulus with fiber loading was in the same order in both glassy and rubbery regions. The loss modulus (E'') values increased upon fiber loading, however, at temperatures lower than 50°C, the composite with 10% fiber content exhibited E'' values lower than that of matrix. It is observed that the loss modulus values decreased sharply between respective glass transition temperatures and 80°C, indicating sharp decrease in the viscosity in this region. This decrease in modulus was less abrupt at higher fiber loadings. Higher $\tan \delta$ was observed for unfilled LLDPE except between a temperature range of -100°C and 25°C.

Effect of fiber size: The composite tensile modulus decreased with increase in fiber size and also upon alkali treatment. It was interesting to note that fiber size did not make any difference in E' values at temperatures above -90°C, whereas the difference was prominent at temperatures below -90°C. Higher modulus was exhibited by fiber in the size range of 425μ-840μ and the modulus successively reduced with fiber size. The high stiffness offered by the large size fibers and resulting decreased molecular mobility is the reason for increased E' for composites with 425μ-840μ size fibers. However, the effect was not prominent at higher temperatures as the system is more flexible with increased molecular mobility at high temperatures, wherein the difference in fiber size could not make much difference. Lowest $\tan \delta$ values were observed for lower size fibers, however only between a temperature range of -40°C to 60°C. A consistent trend could not be observed at other temperatures scanned.

Effect of fiber treatment: The flexural strength of oil palm fiber-LLDPE composites increased with fiber content up to 20 percent and thereafter decreased in case of alkali-treated fiber composites [17]. Alkali-treated fiber caused increase in impact strength whereas reduction in fiber size decreased the impact strength. It was found that E' increased with alkali treatment on fiber indicating better fiber matrix adhesion. The increase in modulus upon alkali treatment is due to greater interfacial adhesion and bond strength between matrix and fiber. It was also observed that the E'' values increased upon alkali treatment at all temperatures scanned. While comparing the composites fabricated with different size fibers, it was found that the lowest values of E'' was exhibited by the lowest size fraction, *viz.*, 75μ-177μ. Lower size fibers get mixed well with the matrix offering a comparatively more homogeneous system. However a consistent trend could not be observed between the other two size fractions throughout the temperature range. Alkali-treated composite exhibited higher *tan* δ values at temperature below 25°C and interestingly, the trend reversed at higher temperatures.

9.3.9.4 Thermal Properties

Effect of fiber content: The degree of crystallinity of OPF-LLDPE composite samples was in the range of 39.3 percent to 53.8 percent when the fiber content varied from 10 to 50 percent [17]. Addition of fibers reduced degree of crystallinity of oil palm fiber-LLDPE composites comparing to that of pure LLDPE and the reduction in degree of crystallinity was proportional to the fiber content. Comparison of melting endotherms of pure LLDPE and composite indicates that each curve is characterized by single peak and a narrow melting region. Higher enthalpy of melting was exhibited by pure LLDPE samples (156 kJ/kg) and as the fibers were incorporated, it reduced successively (80-125 kJ/kg). The pure LLDPE started melting at 124.75°C and its peak melting temperature was less than that of composites. Oil palm fibers do not essentially contribute to the melting endotherm as it does not present any transitions within the melting temperature range of LLDPE. The specific heat capacity (C_p) of the composites in the temperature range 20°C to 100°C in the range of 0.9-2.3 kJ/kg K. The heat deflection temperature (HDT) of the composites increased with an increase in fiber content. Addition of 10 percent fiber in pure LLDPE increased HDT from 76.4 to 99.7°C. Marginal increase (2.5°C) was observed when fiber content was increased from 40 percent to 50 percent. When the fiber content increased from 10 percent to 50 percent, the Coefficient of thermal expansion (CTE) decreased steadily from 9.6×10^{-5}/°C to 1.5×10^{-5} /°C, reflecting the thermal restraint of the LLDPE matrix by the reinforcing oil palm fibers in the oil palm fiber-LLDPE biocomposite. From the TGA thermograms of pure LLDPE and selected combination of oil palm fiber-LLDPE composites, no considerable weight loss was observed in the temperature range 50°C to 300°C. The degradation temperature of pure LLDPE was 443°C whereas the second degradation peak corresponding to matrix degradation in all the composite samples ranged between 454°C and 459°C. The first exothermic peak in the thermogram of alkali-treated oil palm fiber was 297.1°C and the first degradation temperature of composites with 50 percent fiber was 297.9°C. However at 10 percent fiber content, the first degradation temperature was higher (316.2°C).

Effect of fiber size: Fiber in the size range of 425μ-840μ imparted maximum degree of crystallinity to the composite and minimum crystallinity was caused by 177μ-425μ sized fibers [17]. The HDT increased gradually when fiber size was reduced and alkali-treated fiber composites exhibited less HDT comparing to untreated fiber composites. The CTE reduced steadily with reduction in fiber size.

Effect of fiber treatment: Alkali-treated fiber caused higher crystallinity to the OPF-LLDPE composites [16]. Higher CTE was observed for untreated fiber composites than alkali-treated ones.

9.4 Conclusions

Oil palm, the highest oil-yielding crop is popular in many parts of the world as a cheap source of edible oil. The chemical and physical characteristics of the oil palm fiber, a byproduct in palm oil milling, makes it compatable with a wide spectrum of polymer materials for making biocomposites. Easy extraction procedure and availability of this promising fiber in bulk in palm oil mills makes it further suitable for industrial manufacture of composites. Nowadays, the word 'biocomposites' became synonium for 'conservation of nature' itself. Biocomposites are getting wide popularity in number of applications ranging from packaging marterials to automotive industry and aerospace applications. Hybrid composites can also be fabricated from oil palm fibers in combination with other natural and artificial fibers. Completely biodegradable or green composites from oil palm fibers need to be developed in future. The biodegradability of the composites and effect of degradation on various properties also need to be evaluated.

References

1. H.P.S.A. Khalil, M.A. Siti, R. Ridzuan, H. Kamarudin, and A. Khairul, Chemical composition, morphological characteristics, and cell wall structure of Malaysian oil palm fibers. *Polym. Plast. Technol. Eng.* 47, 273–280 (2008).
2. S. Joseph, K. Joseph, and S. Thomas, Green composites from natural rubber and oil palm fiber: Physical and mechanical properties. *Int. J. Polym. Mater.* 55, 925–945 (2006).
3. W. Hasamudin, and R.M. Soom, Road making using oil palm fiber. *Malaysi. Palm Oil Board Inf. Ser.* 171 (2002).
4. H.D. Rozman, C.Y. Lai, H. Ismail, and Z.A.M. Ishak, The effect of coupling agents on the mechanical and physical properties of oil palm empty fruit bunch-polypropylene composites. *Polym. Int.* 49, 1273–1278 (2000).
5. B. Wirjosentono, P. Guritno, and H. Ismail, Oil palm empty fruit bunch filled polypropylene composites. *Int. J. Polym. Mater.* 53, 295–306 (2004).
6. M. Karina, H. Onggo, A.H.D. Abdullah, and A. Syampurwadi, Effect of oil palm empty fruit bunch fiber on the physical and mechanical properties of fiber glass reinforced polyester resin. *J. Biol. Sci.* 8, 100–106 (2008).
7. M.S. Sreekala, M.G. Kumaran, and S. Thomas, Oil palm fibers: Morphology, chemical composition, surface modification, and mechanical properties. *J. Appl. Polym. Sci.* 66, 821–835 (1997).
8. G. Singh, S. Manohan, and K. Kanopathy, In *Proceedings of 1981 In International Oil Palm Conference*, E. Pushparajah, and P.S. Chew, (Eds.), pp. 367–377, Kulalampur, Malaysia (1982).

9. K.N. Law, and X. Jiang, Comparative papermaking properties of oil-palm empty fruit bunch. *Tappi J.* 84, 1–13 (2001).
10. M.J. John, and S. Thomas, Biofibres and biocomposites. *Carbohydr. Polym.* 71, 343–364 (2008).
11. G. Raju, C.T. Ratnam, N.A. Ibrahim, M.Z.A. Rahman, and W.M.Z.W. Yunus, Enhancement of PVC/ENR blend properties by poly(methyl acrylate) grafted oil palm empty fruit bunch fiber. *J. Appl. Poly. Sci.* 110, 368–375 (2008).
12. H.D. Rozman, G.S. Tay, and A. Abusamah, The effect of glycol type, glycol mixture, and isocyanate/glycol ratio on flexural properties of oil palm empty fruit bunch-polyurethane composites. *J. Wood Chem. Technol.* 23, 249–260 (2003).
13. B.F. Yousif, and E.N.S.M. Tayeb, High-stress three-body abrasive wear of treated and untreated oil palm fibre-reinforced polyester composites. *Proc. Inst. Mech. Eng. J. Eng. Tribol.* 222, 637–646 (2008).
14. C.A.S. Hill, and H.P.S.A. Khalil, Effect of fiber treatments on mechanical properties of coir or oil palm fiber reinforced polyester composites. *J. Appl. Polym. Scie.* 78, 1685–1697 (2000).
15. M.S. Sreekala, M.G. Kumaran, M.L. Geethakumariamma, and S. Thomas, Environmental effects in oil palm fiber reinforced phenol formaldehyde composites: Studies on thermal, biological, moisture and high energy radiation effects. *Adv. Compos. Mater.* 13, 171–197 (2004).
16. A.B.A. Hariharan, and K.P.S.A. Khalil, Lignocellulose-based hybrid bilayer laminate composite: Part I - Studies on tensile and impact behavior of oil palm fiber-glass fiber-reinforced epoxy resin. *J. Compos. Mater.* 39, 663–684 (2005).
17. S. Shinoj, Tamil Nadu Agricultural University, Coimbatore, India, *Unpublished Ph.D. thesis* (2010).
18. C.A.S. Hill, H.P.S.A. Khalil, and M.D. Hale, A study of the potential of acetylation to improve the properties of plant fibres. *Ind. Crops Prod.* 8, 53–63 (1998).
19. M.S. Sreekala, J. George, M.G. Kumaran, and S. Thomas, The mechanical performance of hybrid phenol-formaldehyde-based compositesreinforced with glass and oil palm fibres. *Compos. Sci. Technol.* 62, 339–353 (2002).
20. M.S. Sreekala, S. Thomas, and G. Groeninckx, Dynamic mechanical properties of oil palm fiber/phenol formaldehyde and oil palm fiber/glass hybrid phenol formaldehyde composites. *Polym. Compos.* 26, 388–400 (2005).
21. E. Jayashree, P.K. Mandal, M. Madhava, A. Kamaraj, and K. Sireesha, In *Proceedings of the 15th Plantation Crops Symposium- PLACROSYM XV*, K. Sreedharan, P.K.V. Kumar, and C.B.M. Jayarama, (Eds.), pp. 10–13, Chikmagalur, India, (2002).
22. N.A. Ibrahim, F.A. Ilaiwi, M.Z.A. Rahman, M.B. Ahmad, K.Z.M. Dahlan, and W.M.Z.W. Yunus, Graft copolymerization of acrylamide onto Oil Palm Empty Fruit Bunch (OPEFB) Fiber. *J. Polym. Res.* 12, 173–179 (2005).
23. M.S. Sreekala, K.G. Kumaran, and S. Thomas, Water sorption in oil palm fiber reinforced phenol formaldehyde composites. *Compos. A* 33, 763–777 (2002).
24. B.F. Yousif, and E.N.S.M. Tayeb, The effect of oil palm fibers as reinforcement on tribological performance of polyester composite. *Surface Rev. Lett.* 14, 1095–1102 (2007).
25. K.N. Law, W.R.W. Daud, and A. Ghazali, Morphological and chemical nature of fiber strands of oil palm empty-fruit-bunch (OPEFB). *BioResources* 2, 351–362 (2007).
26. K.C. Khoo, and T.W. Lee, Pulp and paper from the oil palm. *Appita J.* 44, 385–388 (1991).
27. A. AbuBakar, A. Hassan, and A.F.M. Yusof, The effect of oil extraction of the oil palm empty fruit bunch on the processability, impact, and flexural properties of PVC-U composites. *Int. J. Polym. Mater.* 55, 627–641 (2006).

28. H. Ismail, N. Rosnah, and H.D. Rozman, Effects of various bonding systems on mechanical properties of oil palm fibre reinforced rubber composites. *Eur. Polym. J. 33,* 1231–1238 (1997).
29. H.D. Rozman, K.R.A. Hilme, and A. Abubakar, Polyurethane composites based on oil palm empty fruit bunches: Effect of isocyanate/hydroxyl ratio and chemical modification of empty fruit bunches with toluene diisocyanate and hexamethylene diisocyanate on mechanical properties. *J. Appl. Polym. Sci.* 106, 2290–2297 (2007).
30. H.P.S.A. Khalil, S. Hanida, C. W. Kang, and N.A.N. Fuaad, Agro-hybrid composite: The effects on mechanical and physical properties of oil palm fiber (EFB)/glass hybrid reinforced polyester composites. *J. Reinf. Plast. Compos.* 26, 203–217 (2007).
31. Hung, T. V., and G. S. Frank. Towards model-based engineering of optoelectronic packaging materials: dielectric constant modeling. *Microelectron. J. 33*, 409–415 (2002).
32. A. Bismarck, S. Mishra, and T. Lampke, Plant fibers as reinforcement for green composites, In *Natural Fibres Biopolymers and Biocomposites,* A.K. Mohanty, M. Misra, and L.T. Drzal, (Eds.), pp. 36–112, CRC Press, Taylor & Francis Group, USA, (2005).
33. M. Khalid, C.T. Ratnam, T.G. Chuah, S. Ali, and T.S.Y. Choong, Comparative study of polypropylene composites reinforced with oil palm empty fruit bunch fiber and oil palm derived cellulose. *Mater. Des. 29,* 173–178 (2008).
34. M.S. Sreekala, and S. Thomas, Effect of fibre surface modification on water-sorption characteristics of oil palm fibres. *Compos. Sci. Technol.* 63, 861–869 (2003).
35. M. Jacob, S. Jose, S. Thomas, and K.T. Varughese, Stress relaxation and thermal analysis of hybrid biofiber reinforced rubber biocomposites. *J. Reinf. Plast. Compos.* 25, 1903–1917 (2006).
36. A. Kalam, B.B. Sahari, Y.A. Khalid, and S.V. Wong, Fatigue behaviour of oil palm fruit bunch fibre/epoxy and carbon fibre/epoxy composites. *Compos. Struct.* 71, 34 –44 (2005).
37. K.M.M. Rao, and K.M. Rao, Extraction and tensile properties of natural fibers: Vakka, date and bamboo. *Compos. Struct.* 77, 288–295 (2007).
38. H.P.S.A. Khalil, H. Ismail, H.D. Rozman, and M.N. Ahamad, The effect of acetylation on interfacial shear strength between plant fibres and various matrices. *Eur. Polym. J.* 37, 1037–1045 (2001).
39. Shinoj, S., R. Visvanathan, and S. Panigrahi, Towards industrial utilization of oil palm fibre: Physical and dielectric characterization of linear low density polyethylene composites and comparison with other fibre sources. *Biosyst. Eng.*106, 378–388 (2010).
40. L.Y. Mwaikambo, and M.P. Ansell, The determination of porosity and cellulose content of plant fibers by density methods. *J. Mater. Sci. Lett.* 20, 2095 - 2096 (2001).
41. N. Chand, Electrical characteristics of sunhemp fibre. *J. Mater. Sci. Lett.* 11, 138–139 (1992).
42. M.J. John, B. Francis, K.T. Varughese, and S. Thomas, Effect of chemical modification on properties of hybrid fiber biocomposites. *Compos. A* 39, 352–363 (2008)
43. H.D. Rozman, G.S. Tay, R.N. Kumar, A. Abusamah, H. Ismail, and Z.A.M. Ishak, The effect of oil extraction of the oil palm empty fruit bunch on the mechanical properties of polypropylene–oil palm empty fruit bunch–glass fibre hybrid composites. *Polym. Plast. Technol. Eng.* 40, 103–115 (2001).
44. M. Khalid, A. Salmiaton, T.G. Chuah, C.T. Ratnam, and S.Y.T. Choong, Effect of MAPP and TMPTA as compatibilizer on the mechanical properties of cellulose and oil palm fiber empty fruit bunch–polypropylene biocomposites. *Compos. Interfaces* 15, 251–262 (2008)
45. M.S. Sreekala, M.G. Kumaran, R. Joseph, and S. Thomas, Stress-relaxation behaviour in composites based on short oil-palm fibres and phenol formaldehyde resin. *Compos. Sci. Technol.* 61, 1175–1188 (2001).
46. G. Raju, C.T. Ratnam, N.A. Ibrahim, M.Z.A. Rahman, and W.M.Z.W. Yunus, Graft copolymerization of methyl acrylate onto oil palm empty fruit bunch (OPEFB) fiber. *Polym. Plast. Technol. Eng.* 46, 949–955 (2007).

47. M.N. Belgacem, and A. Gandini, The surface modification of cellulose fibres for use as reinforcing elements in composite materials. *Compos. Interfaces*, 12, 41–75 (2005).
48. R. Agarwal, N.S. Saxena, K.B. Sharma, S. Thomas, and M.S. Sreekala, Activation energy and crystallization kinetics of untreated and treated oil palm fibre reinforced phenol formaldehyde composites. *Mater. Sci. Eng.* 277, 77–82 (2000).
49. M.J. John, and R.D. Anandjiwala, Recent developments in chemical modification and characterization of natural fiber-reinforced composites. *Polym. Compos.* 29, 187–207 (2008).
50. M.S. Sreekala, , M.G. Kumaran, S. Joseph, M. Jacob, and S. Thomas, Oil palm fibre reinforced phenol formaldehyde composites: influence of fibre surface modifications on the mechanical performance. *Appl. Compos. Materi.* 7, 295–329 (2000).
51. S. Zakaria, and L.K. Poh, Polystyrene-benzoylated EFB reinforced composites. *Polym. Plast. Technol. Eng.* 41, 951–962 (2002).
52. R. Agarwal, N.S. Saxena, K.B. Sharma, S. Thomas, and M.S. Sreekala, Temperature dependence of effective thermal conductivity and thermal diffusivity of treated and untreated polymer composites. *J. Appl. Polym. Sci.* 89, 1708–1714 (2003).
53. S. Shinoj, S. Panigrahi, and R. Visvanathan, Water absorption pattern and dimensional stability of oil palm fiber-linear low density polyethylene composites. *J. Appl. Polym. Sci.* 117, 1064–1075 (2010).
54. K.M. Ashraf, T.J. Ferdous, A.I. Mustafa, and M.A. Khan, Photocuring of empty fruit bunches of oil palm (elaeis guineensis) fibers with allyl methacrylate (AMA): Effect of additives on mechanical and degradable properties. *Polym. Plast. Technol. Eng.* 47,558–566 (2008).
55. M. Jacob, K.T. Varughese, and S. Thomas, Natural rubber composites reinforced with sisal/oil palm hybrid fibers: Tensile and cure characteristics. *J. Appl. Polym. Sci.* 93, 2305–2312 (2004).
56. H.P.S.A. Khalil, and H. Ismail, Effect of acetylation and coupling agent treatments upon biological degradation of plant fibre reinforced polyester composites. *Polym. Test.* 20, 65–75 (2001).
57. M. Jacob, B. Francis, S. Thomas, and K.T. Varughese, Dynamical mechanical analysis of sisal/oil palm hybrid fiber-reinforced natural rubber composites. *Polym. Compos.* 27, 671–680 (2006).
58. M. Jacob, K.T. Varughese, and S. Thomas, Water sorption studies of hybrid biofiber-reinforced natural rubber biocomposites. *Biomacromolecule* 6, 2969–2679 (2005).
59. M. Jacob, K.T. Varughese, and S. Thomas, Dielectric characteristics of sisal–oil palm hybrid biofibre reinforced natural rubber biocomposites. *J. Mater. Sci.* 41, 5538–5547 (2006).
60. H.D. Rozman, G.S. Tay, R.N. Kumar, A. Abusamah, H. Ismail, and Z.A.M. Ishak, Polypropylene–oil palm empty fruit bunch–glass fibre hybrid composites: a preliminary study on the flexural and tensile properties. *Eur. Polym. J.* 37, 1283–1291 (2001).
61. C.A.S. Hill, and H.P.S.A. Khalil, The effect of environmental exposure upon the mechanical properties of coir or oil palm fiber reinforced composites. *J. Appl. Polym. Sci.* 77, 1322–1330 (2000).
62. H.D. Rozman, G.S. Tay, A. Abubakar, and R.N. Kumar, Tensile properties of oil palm empty fruit bunch–polyurethane composites. *Eur. Polym. J.* 37, 1759–1765 (2001).
63. K.H. Badri, Z. Othman, S. H. Ahmad, Rigid polyurethane foams from oil palm resources. *J. Mater. Sci.* 39, 5541–5542 (2004).
64. H.D. Rozman, G.S. Tay, A. Abusamah, and R.N. Kumar, A preliminary study on the oil palm empty fruit bunch-polyurethane (EFB-PU) composites. *Int. J. Polym. Mater.* 51, 1087–1094 (2002).

65. K.H. Badri, K.A.M. Amin, Z. Othman, H.A. Manaf, and N.K. Khalid, Effect of filler-to-matrix blending ratio on the mechanical strength of palm-based biocomposite boards. *Polym. Int.* 55, 190–195 (2006).
66. K.A.M. Amin, and K.H. Badri, Palm-based bio-composites hybridized with kaolinite. *J. Appl. Polym. Sci.* 105, 2488–2496 (2007).
67. K.H. Badri, Z.B. Othman, and I.M. Razali, Mechanical properties of poyurethane composites from oil palm resources. *Iran. Polym. J.* 14, 441–448 (2005).
68. A.A. Bakar, and N. Baharulrazi, Mechanical properties of benzoylated oil palm empty fruit bunch short fiber reinforced poly(vinyl chloride) composites. *Polym. Plast. Technol. Eng.* 47, 1072–1079 (2008).
69. A. AbuBakar, A. Hassan, and A.F.M. Yusof, Effect of oil palm empty fruit bunch and acrylic impact modifier on mechanical properties and processability of unplasticized poly(vinyl chloride) composites. *Polym. Plast. Technol. Eng.* 44, 1125–1137 (2005).
70. C.T. Ratnam, G. Raju, N.A. Ibrahim, M.Z.A. Rahman, and W.M.Z.W. Yunus, Influence of wastewater characteristics on methane potential in food-processing industry wastewaters. *J. Compos. Mater.* 42, 2195–2203 (2008).
71. H.S.A. Khalil, H.D. Rozman, M.N. Ahmad, and H. Ismail, Acetylated plant-fiber-reinforced polyester composites: A study of mechanical, hygrothermal, and aging characteristics. *Polym. Plast. Technol. Eng.* 39, 757–781 (2000).
72. H.P.S.A. Khalil, M.N. Azura, A.M. Issam, M.R., Said, and T.O.M. Adawi, Oil palm empty fruit bunches (OPEFB) reinforced in new unsaturated polyester composites. *J. Reinf. Plast. Compos.* 27, 1817–1826 (2008).
73. H.P.S.A. Khalil, A.M. Issam, M.T.A. Shakri, R. Suriani, and A.Y. Awang, Conventional agro-composites from chemically modified fibres. *Ind. Crops Prod.* 26, 315–323 (2007).
74. K. Singh, N.S. Saxena, M.S. Sreekala, and S. Thomas, Temperature dependence of the thermal conductivity and thermal diffusivity of treated oil-palm-fiber-reinforced phenolformaldehyde composites. *J. Appl. Polym. Sci.* 89, 3458–3463 (2003).
75. R. Agarwal, N.S. Saxena, K.B. Sharma, M.S. Sreekala, and S. Thomas, Thermal conductivity and thermal diffusivity of palm fiber reinforced binary phenolformaldehyde composites. *Indian J. Pure Appl. Phys.* 37, 865–869 (1999).
76. R. Agarwal, N.S. Saxena, K.B. Sharma, S. Thomas, and M.S. Sreekala, Effect of different treatments on the thermal behavior of reinforced phenol-formaldehyde polymer composites. *J. Appl. Polym. Sci.* 78, 603–608 (2000).
77. M.A.A. Bakar, V.D. Natarajan, A. Kalam, and N.H. Kudiran, In *Proceedings of the 13th International Conference on Experimental Mechanics*, E.E. Gdoutos, (Ed.), July 1–6, Alexandroupolis, Greece, 97–98 (2007).
78. Y. Chen, H. ChenLin, and Y. Lee, The effects of filler content and size on the properties of PTFE/SiO_2 composites. *J. Polym. Res.* 10, 247–258 (2003).
79. C.Y. Lai, S.M. Sapuan, M. Ahmad, N. Yahya, Mechanical and electrical properties of coconut coir fiber-reinforced polypropylene composites. *Polym. Plast. Technol. Eng.* 44, 619–632 (2005).
80. A. Paul, K. Joseph, and S. Thomas, Effect of surface treatments on the electrical properties of low-density polyethylene composites reinforced with short sisal fibers. *Compos. Sci. Technol.* 57, 67–79 (1997).
81. S. Shinoj, R. Visvanathan, S. Panigrahi, and N. Varadharaju, Dynamic mechanical properties of oil palm fibre (OPF)-linear low density polyethylene (LLDPE) biocomposites and study of fibre-matrix interactions. *Biosyst. Eng.* 109, 99–107 (2011).

10

Lignocellulosic Polymer Composites: Processing, Characterization and Properties

Bryan L. S. Sipião[1], Lais Souza Reis[1], Rayane de Lima Moura Paiva[1], Maria Rosa Capri[2] and Daniella R. Mulinari[*,1,3]

[1]*Department of Engineering – UniFOA, Volta Redonda/RJ, Brazil*
[2]*Engineering College at Lorena, São Paulo University, Lorena/SP, Brazil*
[3]*Technology College, State University of Rio de Janeiro, Resende/RJ, Brazil*

Abstract

Natural fibers-reinforced polymer matrixes provide more alternatives in the materials market due to their unique advantages. Poor fiber–matrix interfacial adhesion may affect the physical and mechanical properties of the resulting composites due to the surface incompatibility between hydrophilic natural fibers and non-polar polymers. The results presented in this chapter focus on the properties of palm and pineapple fibers in terms of their physical and chemical structure, mechanical properties and processing behavior. The final properties of these fibers with thermoplastics matrixes are also presented, paying particular attention to the use of physical and chemical treatments for the improvement of fiber-matrix interaction.

Keywords: Palm fibers, pineapple fibers, polypropylene, high-density polyethylene, adhesion, mechanical properties

10.1 Introduction

Natural fiber-reinforced polymer composites represent one of today's fastest growing industries. Possessing mechanical properties comparable to those of manmade fibers such as carbon, glass or aramid, natural fibers are a potential alternative in reinforced composites because of growing environmental awareness and legislated requirements [1–3]. These fibers have gained significant importance in technical applications, such as in the automotive industry. Additionally, they are obtainable from renewable sources, are biodegradable, low cost, and have low specific density [4-6]. Natural fibers are used to reinforce the polymer and improve mechanical properties such as stiffness and strength. The interest in using natural fibers such as different non-wood (plant)

*Corresponding author: daniella.mulinari@foa.org.br

Vijay Kumar Thakur (Ed.), Lignocellulosic Polymer Composites, (213–230) 2015 © Scrivener Publishing LLC

and wood fibers as reinforcement in plastics has increased dramatically. Many studies have examined the development of natural fibers such as sugarcane bagasse [7-10] and straw [6], sisal [11], hemp and jute, as green alternatives to conventional materials [12-16]. The potential of these fibers to be incorporated into materials with engineered properties is enormous. Availability, price and performance are some of the factors that have catalyzed the surge in using lignocellulosic fibers, first as fillers and more recently as reinforcements, in polymeric materials [17]. Despite the variability of the mechanical fibers, natural fibers offer many advantages when compounded with, for instance, thermoplastics.

Composites based on thermoplastic resins are now becoming popular due to their processing advantages [18-20]. Among different thermoplastics, polypropylene (PP) processes and high-density polyethylene (HDPE) have outstanding properties such as low density, good flex life, good surface hardness, very good abrasion resistance, and excellent electrical properties. The combination of lignocellulosic material with thermoplastic matrix can present a considerable problem: incompatibility between the polar and hygroscopic fiber and non-polar and hydrophobic matrix. Because of this, treatment of natural fibers or the use of coupling agent and other methods are beneficial in order to improve interfacial adhesion [21-25].

The results presented in this chapterfocus on properties in terms of the physical and chemical structure of palm and pineapple fibers, mechanical properties, processing behavior and final properties of these fibers with thermoplastics matrixes, paying particular attention to the use of physical and chemical treatments for the improvement of fiber-matrix interaction.

10.2 Palm Fibers

Palm trees are considered among one of the oldest plants on the earth, and their records date back 120 million years. They are the most characteristic components of tropical forests that have important features that ensure the sustainable development of agricultural and horticultural systems. These features are due to the variability of shapes, structures in communities of palm trees and various products they offer.

The greatest diversity of these plants are in the Tropics and Subtropics, and interest in the cultivation of this species of plants has increased significantly due to their indisputable landscape, where palms are of immense ecological and economic importance [26].

Archontophoenix alexandrae, commonly known as King Palm, is a species of the family Aracaceae originally from Queensland, Australia, a tropical region with altitude below 1100 m. The climate required to cultivate this species can be hot and humid. This plant adapts to various soil types, very sandy soil or soil with high clay content, and they tolerate low pH [27].

The plants of this species form a very dense root system, which makes them very important in preventing the erosion of river banks. The heart-of-palm, also known as palmito, can be extracted from various species of palms. The *A. alexandrae* produce heart-of-palm of noble type, with higher quality and superior flavor compared to other species of palm. The harvesting of palm heart takes place after a period of 4 years [28].

However, a lot of residue is generated from this cultivation [28]. For each extracted palm harvested there are approximately 400 g of commercial palm heart. The residue constitutes 80-90% of the total palm weight, with some variation depending on species [27]. The residues from king palm are constituted mainly of leaves and leaf sheathes (Figure 10.1). Some quantity of this highly cellulosic material is currently used as boiler fuel, in the preparation of fertilizers or as mulching material, whereas a major portion is left on the mill premises itself. When left in the field, these waste materials create great environmental problems [29].

Several studies are seeking to add value to this raw material produced from the extraction of palm heart, applying it in other ways.

The stem of some species of palm tree is composed of plastic material such as fiber, protein and polysaccharides (cellulose), which give them their shape, and nutritional material that fills the interior of cells such as sugar and starch. Previous studies used the leaves and leaf sheaths in the production of flour and its characterization showed interesting results, especially from the point of view related to the content of dietary fiber and minerals [28].

Other studies attest to more applications of these fibers, for example, filtration with palm fibers could be a potential technology for tertiary wastewater treatment as it provides a "green engineering solution" [30]. Kriker *et al.* examined four types of palm surface fibers and determined their mechanical and physical properties for application of this raw material in concrete structures [31].

It should be mentioned here that included in theresearch into the application of these wastes in a sustainable way, is their increasing use in polymeric composites, particularly in the automobile industry, as they demonstrate increased mechanical performance along with low cost and weight reduction.

10.2.1 Effect of Modification on Mechanical Properties of Palm Fiber Composites

The mechanical properties of the natural fiber-reinforced composites are dependent on some parameters such as volume fraction of the fibers, fiber aspect ratio, fiber–matrix adhesion, stress transfer at the interface, and fiber orientation [32]. Several studies on natural fiber-reinforced composites involve mechanical properties characterization as

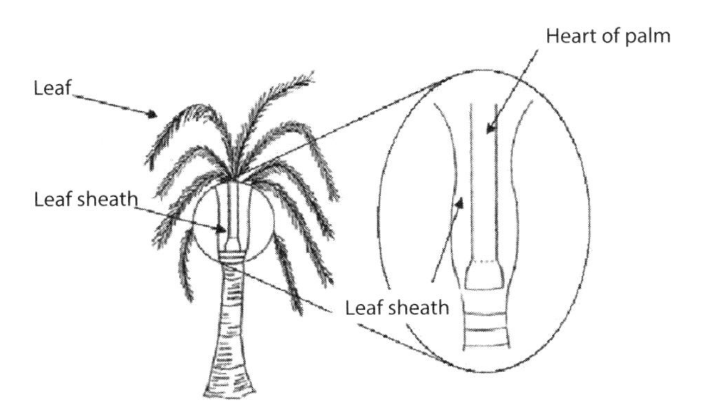

Figure 10.1 Schematic of the residues from palms: leaves and leaf sheath [28].

a function of the fibers content, effect of the fibers treatments, and the use of coupling agents.

Both the matrix and the fibers properties are important to improve the mechanical properties of the composites. The tensile strength is more sensitive to matrix properties, whereas the modulus is dependent on the fibers properties. To improve the tensile strength, a strong interface is required, whereas fiber concentration, fiber wetting in the matrix phase, and high fiber aspect ratio determine tensile modulus. The aspect ratio is very important to determine the fracture properties. In short fiber-reinforced composites, there is a critical fiber length that is required to develop its full stress condition in the polymer matrix [33-35]. For impact strength, an optimum bonding level is necessary. The degree of adhesion, fiber pullout, and a mechanism to absorb energy, are some of the parameters that can influence the impact strength of a short fiber-filled composite.

Below is a brief description of modifications done on surface palm fibers to improve mechanical properties.

10.2.2 Alkali Treatment and Coupling Agent

In this treatment, parameters such as type and concentration of the alkali solution, operational temperature, temperature treatment time, material strength, as well as the applied additives are considered. Optimal conditions of mercerization ensure the improvement of tensile properties and absorption characteristics, which are important in the process [36].

Alkali treatment improves the adhesive characteristics of the fibers surface by removing natural and artificial impurities, thereby producing a rough surface topography. Moreover, alkali treatment provides fiber fibrillation, i.e., breaking down of the composite fiber bundles into smaller fibers. In others words, alkali treatment reduces the fiber diameter and thereby increases the aspect ratio. Therefore, development of a rough surface topography and enhancement in aspect ratio offer higher fiber-matrix interface adhesion with a consequent increase in mechanical properties [36].

The alkaline treatment promotes the removal of partially amorphous constituents such as hemicelluloses, lignin, waxes and oils soluble in alkaline solution, and therefore reduces the level of fiber aggregation, making a surface rougher.

During the alkaline treatment, the OH groups present in the fibers react with sodium hydroxide according to Equation 10.1.

$$\text{Fibre-OH} + \text{NaOH} \rightarrow \text{Fibre-O-Na} + H_2O \tag{10.1}$$

Sipião *et al.* [37] studied the mechanical properties of high-density polyethylene (HDPE) composites reinforced with palm fiber-treated alkali solution.

Firstly, fibers were dried at 50°C for an hour, and after being ground in a mill, finally sieved to obtain a sample that passed through a 45 mesh. To remove the soluble extractives and to facilitate adhesion between fibers and matrix, palm fibers were pretreated with sodium hydroxide solution (1% w/w) for an hour under constant stirring at room temperature. Once the time of treatment was reached, the solution was filtered in a vacuum filter and fibers were washed with distilled water until neutral pH was attained. Then, fibers were dried in an oven at 50°C for 24 hours.

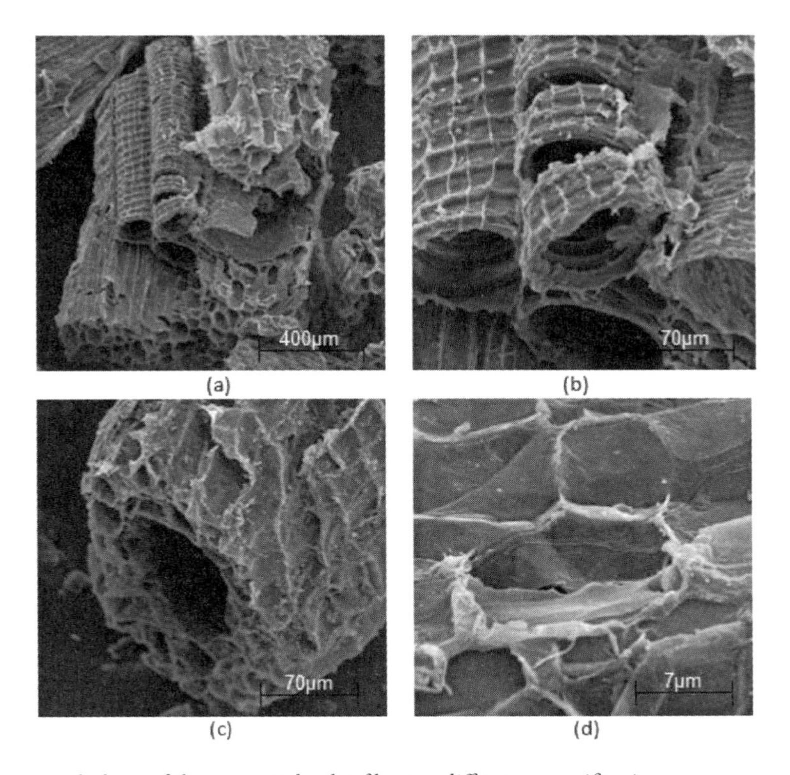

Figure 10.2 Morphology of the untreated palm fibers at different magnifications.

The pretreated palm fibers were mixed with the polymeric matrix (HDPE) in a thermokinetic mixer model MH-50H, with the speed rate kept at 5250 rpm, in which fibers were responsible for 5 and 20 wt% of the composition.

Scanning electron microscopy (SEM) is an excellent technique for examining the surface morphology of the fibers. Figure 10.2 shows SEM micrographs of pretreated and untreated palm fibers.

Examination of the untreated fibers shows a large amount of debris adhering to the surface of the fiber bundles. After the treatment on palm fibers, the removal of ashes on the fibers surface was observed (Figure 10.3). The elimination of superficial layer was also verified, increasing the contact area of exposition of fibrils (reentrance) and globular marks (salience). As a consequence, an increase in the roughness of fibers was observed, which can increase the adhesion between fibers and matrix.

Figure 10.4 shows the infrared spectra of palm fibers. The most visible differences between the spectra of in-nature and pretreated palm fibers are the modifications of the signal at 2885 cm^{-1} and 1732 cm^{-1}, characteristics of the stretching of symmetrical CH groups and stretching of unconjugated CO groups present in polysaccharides and xylans. Considering the first region, the ratio between intensity of the C-H stretching band (~2900 cm^{-1}) is lower in the spectrum of the pretreated palm fibers than observed for the in-nature palm fibers On the other hand, at the second region modifications may be observed, especially in the ratio between the intensities of the $C = O$ stretching band (~1730 cm^{-1}).

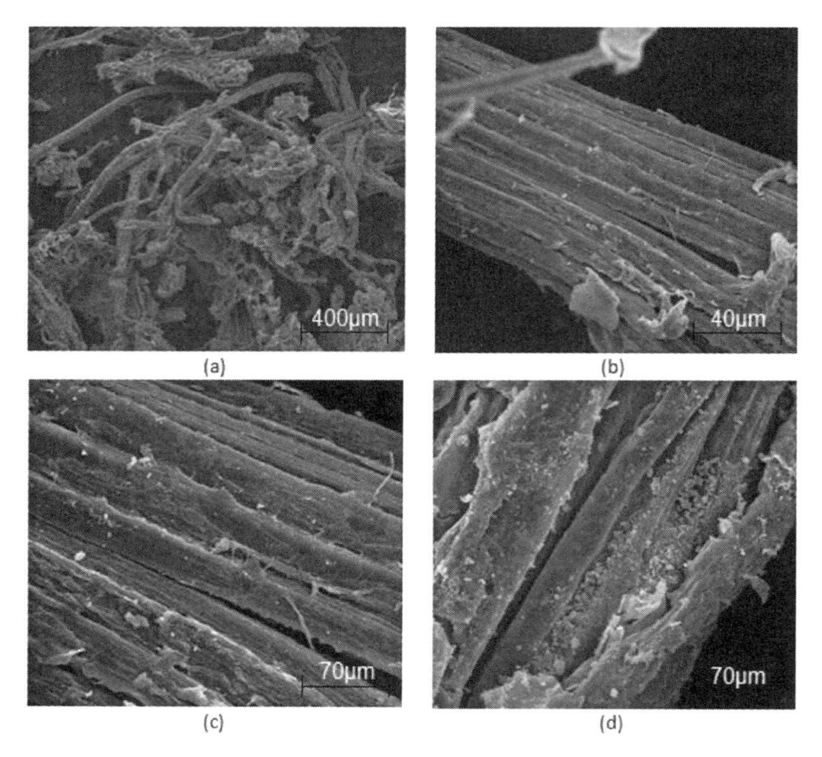

Figure 10.3 Morphology of the untreated palm fibers at different magnifications (a, b, c and d).

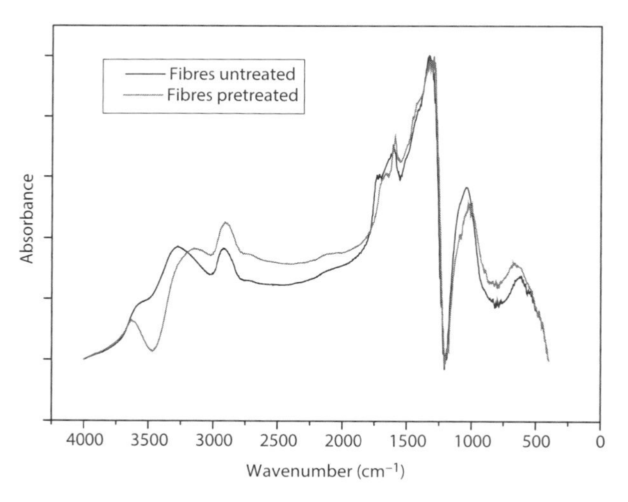

Figure 10.4 FTIR spectra of pretreated and untreated palm fibers.

Mechanical properties of studied composites reinforced with pretreated fibers are summarized in Table 10.1. Composites showed distinct mechanical properties, indicating that the pretreatment affects the fibers-matrix interaction compared to the pure polymer (HDPE). However, the amount of added reinforcement contributes to variation of the tensile modulus.

Table 10.1 Mechanical properties of Pretreated Palm Fibers/High-Density Polyethylene.

Samples	Elongation (%)	Tensile Strenght (MPa)	Tensile Modulus (MPa)
HDPE	8.9 ± 0.8	15.7 ± 1.1	732.5 ± 91
CP5%	8.1 ± 0.6	15.8 ± 1.7	733.7 ± 65
CP10%	7.4 ± 0.3	18.6 ± 0.8	862.2 ± 28
CP20%	6.5 ± 0.6	19.9 ± 0.3	928.4 ± 45
CPT5%	7.4 ± 0.2	18.2 ± 0.7	942.5 ± 99
CPT10%	6.5 ± 0.2	23.5 ± 0.1	979.0 ± 38
CPT20%	5.7 ± 0.3	25.8 ± 0.4	1229.3 ± 35

CP (composites reinforced with in-nature palm fibers); CPT (composites reinforced with modified palm fibers).

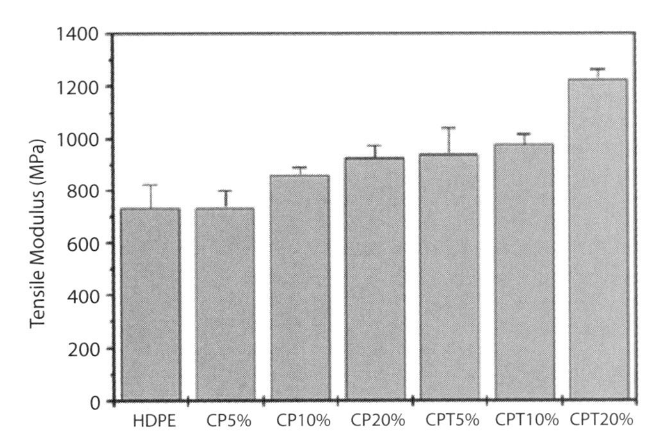

Figure 10.5 Tensile modulus of the composites.

Upon analyzing the data of Table 10.1, it was observed that the tensile strength of composites reinforced with fibers modified by mercerization increased compared to pure HDPE and the other composites. However, an increase in the strength of composites occurred when compared to the pure matrix.

Furthermore, the stiffness of the composites increased by inserting higher fiber content. This difference in composite stiffness can be explained by the chemical modification performed on the fibers. Figure 10.5 evidences tensile modulus obtained in composites reinforced with in-nature and modified fibers; and with different amounts of fibers reinforced in the matrix.

The elongation at maximum tensile of the composites decreases compared to pure HDPE. This difference in elongation occurred due to the amount of reinforcement added in the matrix and the interaction between fiber/matrix, which contributes to increase of the strength.

On the other hand, it was observed that upon using coupling agent in obtaining the composite, a greater decrease in elongation occurred (Table 10.2).

Upon analyzing the composite compatibilized with PP-g-MAH, a significant increase in stiffness was noted when compared to composites without the use of coupling agent.

Table 10.2 Mechanical properties of Pretreated Palm Fibers/High-Density Polyethylene with coupling agent.

Samples	Elongation (%)	Tensile Strenght (MPa)	Tensile Modulus (MPa)
CP5%	8.1 ± 0.6	15.8 ± 1.7	733.7 ± 65
CPT5%	7.4 ± 0.2	18.2 ± 0.7	942.5 ± 99
CPC5%	6.2 ± 0.5	21.3 ± 0.3	1359.4 ± 66

CP (composites reinforced with in-nature palm fibers); CPT (composites reinforced with modified palm fibers); CPC (composites reinforced with fibers and coupling agent (Epolene E-43 Wax)).

This occurred because the coupling agent in contact with the surface fibers interacted strongly by covalent bonds or hydrogen bonds, causing better interaction between fiber and matrix.

10.3 Pineapple Fibers

The pineapple originated in the Americas, and is grown in Asia, Africa and the Americas (North, Central and South). Thailand, the Philippines, Brazil, China and India stand out as the main pineapple producing countries.

The pineapple plant has been an inspiration for painting, architecture and sculpture; and is used on masonry pillars at the entrance of houses, villas and gardens. The stem is used by the food industry asraw material to obtain ethyl alcohol and gums. The rest of the pineapple can be used in animal feed such as silage or fresh material. The fruit is consumed fresh or in the form of ice cream, candy, popsicles, soft drinks and homemade juices. When industrialized, the results can be presented as pulp, syrup, jelly, sweets in syrup or bottled juice. In some countries, in hot and dry regions, wine is obtained with the sweet fruit.

Pineapple fiber is rich in cellulose, has a relatively low cost and is abundantly available. Because of these characteristics numerous research studies are being carried out [38-44]. The fibers from pineapple have been applied as reinforcement in several polymeric matrices due to the advantages they present such as low cost, low density and high specific properties.

Threepopnatkul *et al.* [38] studied the properties of pineapple leaf fiber-reinforced polycarbonate composites. The surface of pineapple leaf fiber (PALF) was pretreated with sodium hydroxide (PALF/NaOH) and modified with two different functionalities such as c-aminopropyl trimethoxy silane (PALF/Z-6011) and c-methacryloxy propyl trimethoxy silane (PALF/Z-6030). The effects of PALF content and chemical treatment were investigated by Fourier transform infrared spectroscopy, scanning electron microscopy and mechanical testing. The modified pineapple leaf fibers composite also produces enhanced mechanical properties. Young's modulus is the highest in the case of the PALF/NaOH composites. The PALF/Z-6011 composites showed the highest tensile strength and impact strength. In thermal property, the results from thermogravimetric analysis showed that thermal stability of the composites is lower than that of neat polycarbonate resin, and thermal stability decreased with increasing pineapple leaf fiber content.

Arib *et al.* [40] investigated the tensile and flexural behaviors of pineapple leaf fiber-reinforced polypropylene composites as a function of volume fraction. The tensile modulus and strength of the composites were found to increase with fiber content in accordance with the rule of mixtures. The flexural modulus gives higher value at 2.7% volume fraction. Scanning electron microscopic studies were carried out to understand the fibers-matrix adhesion and fibers breakage.

George *et al.* [42] evaluated the effects of fiber orientation, fiber loading and fiber length on the viscoelastic properties of pineapple fiber-LDPE composites. Longitudinally oriented composites showed maximum value of the storage modulus. Dynamic storage and loss modulus increased with fiber loading. From the dynamic viscoelastic properties it was found that 2 mm is the optimum fiber length for reinforcement.

Liu *et al.* [43] studied soy-based bioplastic and fiber from pineapple leaf green composites. These composites were manufactured using twin-screw extrusion and injection molding. Thermal, mechanical and morphological properties of the green composites were evaluated with Dynamic Mechanical Analyzer (DMA), United Testing System (UTS) and Environmental Scanning Electron Microscopy (ESEM). The effects of fiber loading and polyester amide grafted glycidyl methacrylate (PEA-*g*-GMA) as compatibilizer on morphological and physical properties of pineapple leaf fiber-reinforced soy-based biocomposites were investigated. The mechanical properties including tensile properties, flexural properties and impact strength of the biocomposites increased with increasing fiber content and the presence of the compatibilizer. The ESEM studies reveal that the dispersion of fiber in the matrix became worse with increasing fiber content but improved with addition of compatibilizer. The addition of the compatibilizer also decreased the water absorption. The corresponding improved mechanical properties of the composites in the presence of the compatibilizer, is attributed to interactions between hydroxyl groups in the pineapple leaf and epoxy groups in PEA-*g*-GMA.

Mangal *et al.* [45] studied the simultaneous measurement of effective thermal conductivity (λ) and effective thermal diffusivity (κ) of pineapple leaf fiber-reinforced phenolformaldehyde (PF) composites by transient plane source (TPS) technique in different weight percentage (15, 20, 30, 40 and 50%). It was found that effective thermal conductivity and effective thermal diffusivity of the composites decreased, as compared with pure PF, as the fraction of fiber loading increased.

Below is a brief description of modifications done on the surface of pineapple fibers to improve mechanical properties.

10.3.1 Alkali Treatment

Sipião *et al.* [22] studied the mechanical properties of polypropylene (PP) composites reinforced with pineapple fiber-treated alkali solution.

Pineapple fibers were extracted from the crown and dried at 80°C for 24 h. Afterwards they were ground in a mill and sieved. To remove the soluble extractives and to facilitate adhesion between fibers and matrix, the in-nature pineapple crown fibers were modified by pretreatment with alkaline solution 1% (w/v). Next the fibers were filtered in a vacuum filter and were washed with distilled water until neutral pH. Then, the fibers were dried in an oven at 100°C for 24 h.

The physical structures of the pineapple crown fibers were evaluated by X-ray diffraction technique. X-ray diffractograms were obtained in a Shimadzu diffractrometer model XRD6000. Conditions used were: radiation CuKα, tension of 30 kV, current of 40 mA and 0.05 (2θ/ 5 s) scanning from values of 2θ it enters 10 to 70° (2θ).

A JEOL JSM5310 model scanning electron microscope (SEM) was used to observe pretreated and untreated pineapple crown fibers. The samples to be observed under the SEM were mounted on conductive adhesive tape, sputter-coated with gold and observed in the SEM using a voltage of 15 kV.

The pretreated pineapple fibers were mixed with the PP in a thermokinetic mixer with speed rate maintained at 5250 rpm, in which fibers were responsible for 5 wt% in the composition. After the mixture, composites were dried and ground in a mill. The composites were placed in an injector camera at 165°C and 2°C min^{-1} heating rate in a required dimension pre-warm mold with specific dimensions for impact specimens.

The mechanical properties of pretreated pineapple fiber-reinforced polypropylene (PP) composites were determined. Five specimens were analyzed, with dimensions in agreement with the ASTM D 6110, ASTM D638 and ASTM D790 standards.

An X-ray diffractogram of pineapple crown fibers is shown in Figure 10.6 . It shows two peaks, which are well defined. The presence of these diffraction peaks indicates that the fiber is semicrystalline. According to several authors [30] the two peaks situated at 2θ = 15.4° and 2θ = 22.5° can be attributed to cellulose I and IV. These two peaks are attributed to the (2 0 0) and (1 1 0) crystallographic planes, respectively. Crystallinity index (CI) is estimated using the Equation 9.2:

$$CI = H_{22.5} - H_{16.5} / H_{22.5} \quad (9.2)$$

where $H_{22.5}$ is the height of the peak at 2θ = 22.5° and $H_{16.5}$ is the diffracted intensity at 2θ = 18.5°. According to this expression 1, treated and untreated pineapple fibers presented 42% and 38% of crystallinity, respectively. These values can be attributed to the fibers modification.

The change in surface morphology of the pretreated pineapple fiberswas studied by scanning electron microscopy. Figures 10.7 and 10.8 show SEM micrographs of pretreated and untreated pineapple fibers. Examination of the untreated fibers shows a large amount of extractives (Figure 10.7). It was observed that after pretreatment of

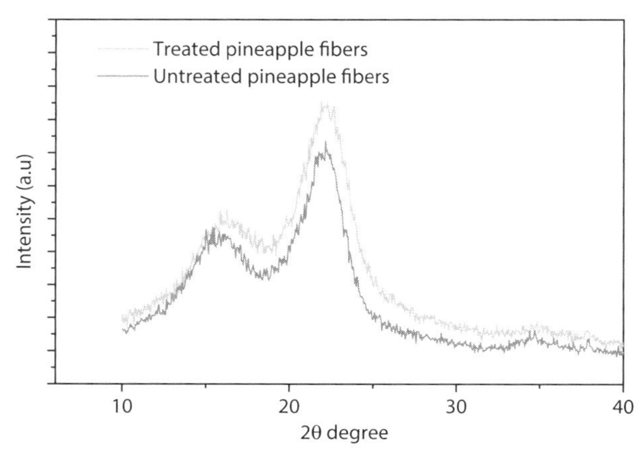

Figure 10.6 X-ray diffractogram of pretreated and untreated pineapple crown fibers.

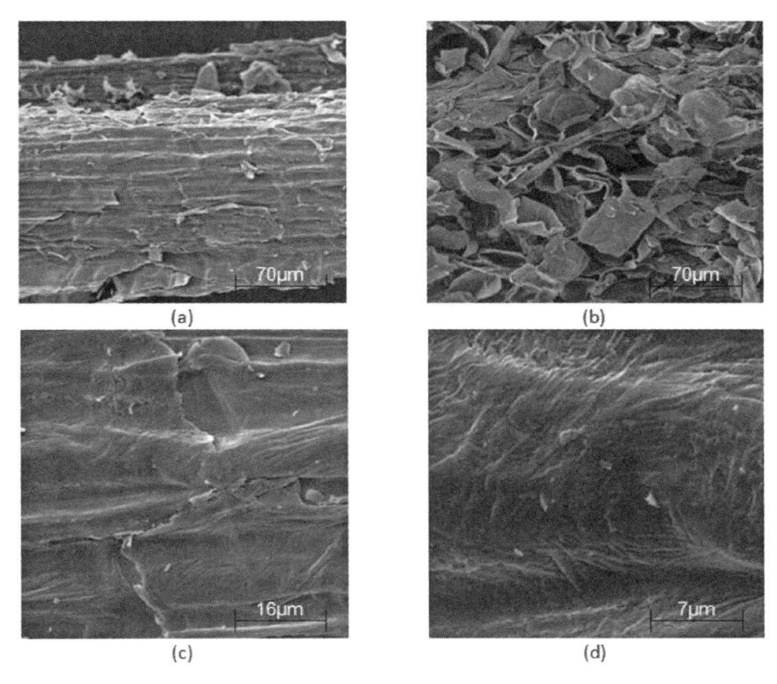

Figure 10.7 Morphology of the untreated pineapple crown fibers at different magnifications (a, b, c and d).

the pineapple fibers, these extractives had been removed from the surface fibers. It was also verified that with the elimination of superficial layer the contact area for exposition of fibrils (reentrance) and globular marks (salience) increased. As a consequence, an increase in the roughness of fibers was observed, which contributes to the increase of the adhesion between fibers and matrix (Figure 10.8).

Table 10.3 presents the interactions between fibers and matrix obtained by mechanical properties during the mixing process, which depends on fiber/matrix interface.

Composites presented better mechanical properties when compared to the pure polypropylene. This fact can be explained by interfacial bonding between fibers–matrix. This interfacial bonding was obtained from examining composite fracture composites (impact tests). Figure 10.9 shows the fractured region after impact t bonding tests, which verified fiber distribution in the matrix, fibers fractured in the matrix and pull-out fibers, characterizing mechanism of fragile fracture. Energy dissipation was also observed during the frictional process mechanics.

10.3.2 Acid Hydrolysis

Sousa *et al.* [46] studied the mechanical properties of polypropylene (PP) composites reinforced with pretreated pineapple fibers with sulfuric solution. Pineapple fibers were extracted from residue SuFresh and dried at 80°C for 24 h. After being ground in a mill and sieved, the fibers were pretreated in a 350 L stainless steel reactor, under these conditions: 1.0% (w/v) H_2SO_4 solution in a 1:10 solid:liquid ratio, 120°C for 10 min.

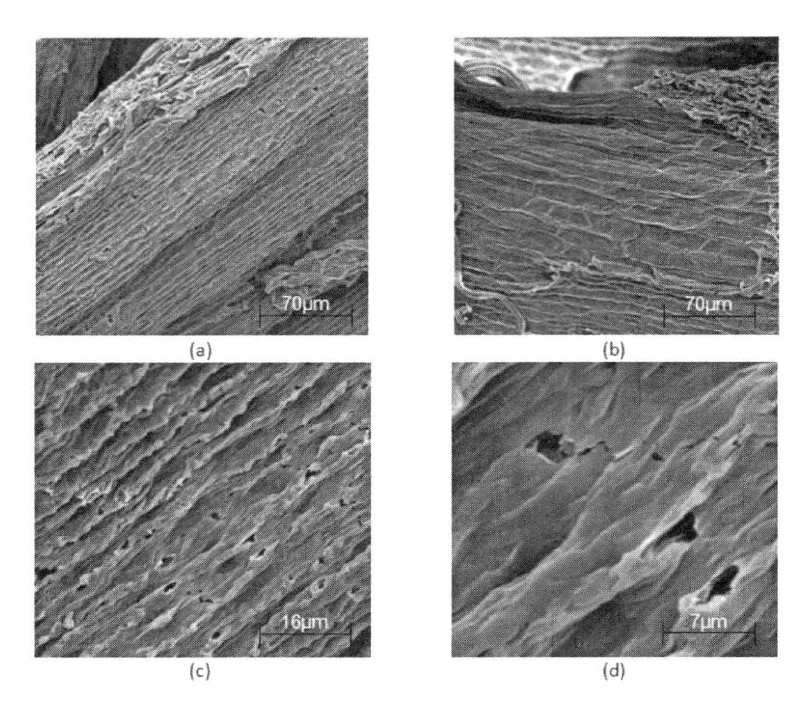

Figure 10.8 Morphology of the pretreated pineapple crown fibers at different magnifications (a, b, c and d).

Table 10.3 Results of materials obtained by impact test.

Samples	Tensile Strength (MPa)	Tensile Modulus (MPa)	Flexural Strength (MPa)	Flexural Modulus (MPa)	Impact Strength ($J.m^{-2}$)
PP	22 ± 0.04	991 ± 71	25 ± 0.9	744 ± 53.5	36 ± 1.1
PP/FA 5%	22 ± 0.1	2509 ± 59	29 ± 0.9	926 ± 28.4	41 ± 0.8

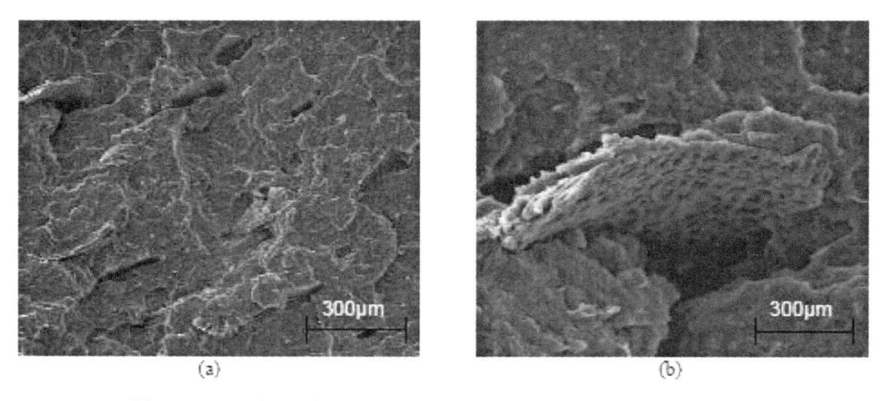

Figure 10.9 SEM of fracture surface of PP/FSB20% composites.

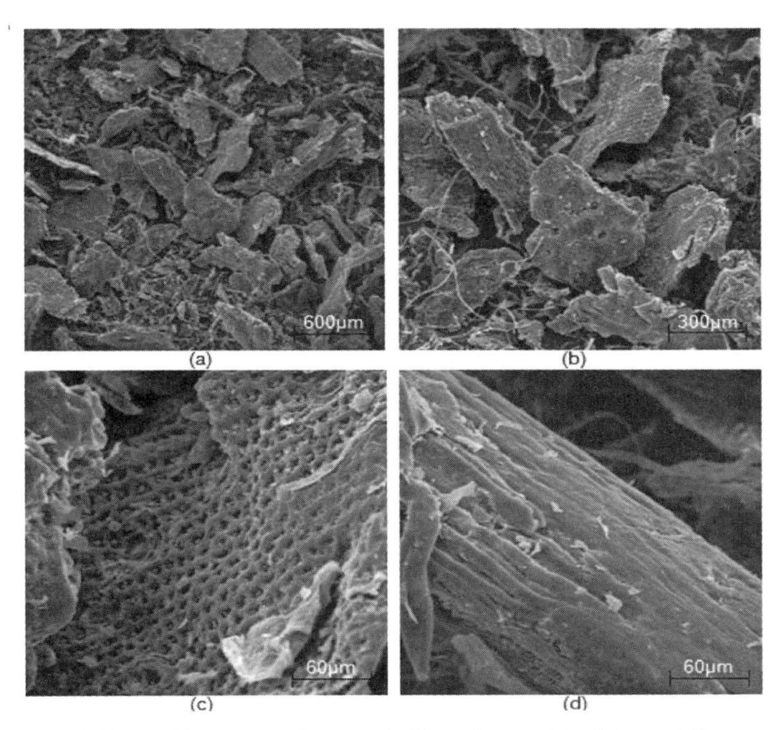

Figure 10.10 Morphology of the untreated pineapple fibers from residue juice at different magnifications (a, b, c and d).

After reaction, the resulting solid material (cellulignin) was separated by centrifugation, washed with water until neutral pH, and dried at 50 ± 5°C to attain 50% moisture content. The modification of the fibers surface was analyzed by scanning electron microscopy (SEM) technique. The pretreated pineapple fibers were mixed with the PP in a thermokinetic mixer with speed rate maintained at 5250 rpm, in which fibers were responsible for 10 and 20 wt% in the composition. After the mixture, composites were dried and ground in a mill. The composites were placed in an injector camera at 165°C and 2°C min^{-1} heating rate in a required dimension pre-warm mold with specific dimensions for impact specimens. The impact tests of pretreated pineapple fiber-reinforced PP composites were determined using a Pantec machine (model PS30). Five specimens were analyzed, with dimensions in agreement with the ASTM D 6110 standard: 12 mm with 63.5 mm length and 12 mm thickness. The absorbed energy and impact strength were evaluated. Figures 10.10 and 10.11 show SEM micrographs of pretreated and untreated pineapple fibers from juice residue. Upon analyzation of the morphology of the untreated fibers (Figure 10.10), an amount of extractives was observed.

The pretreatment of pineapple fibers evidenced the removal of these extractives, causing the elimination of the superficial layer of the contact area. As a consequence, an increase in the roughness of fibers was observed, which contributed to the increase of the adhesion between fibers and matrix (Figure 10.11).

Because of this fact, results obtained in the impact test showed composites have more strength when compared to pure polypropylene. This increase occurred due to

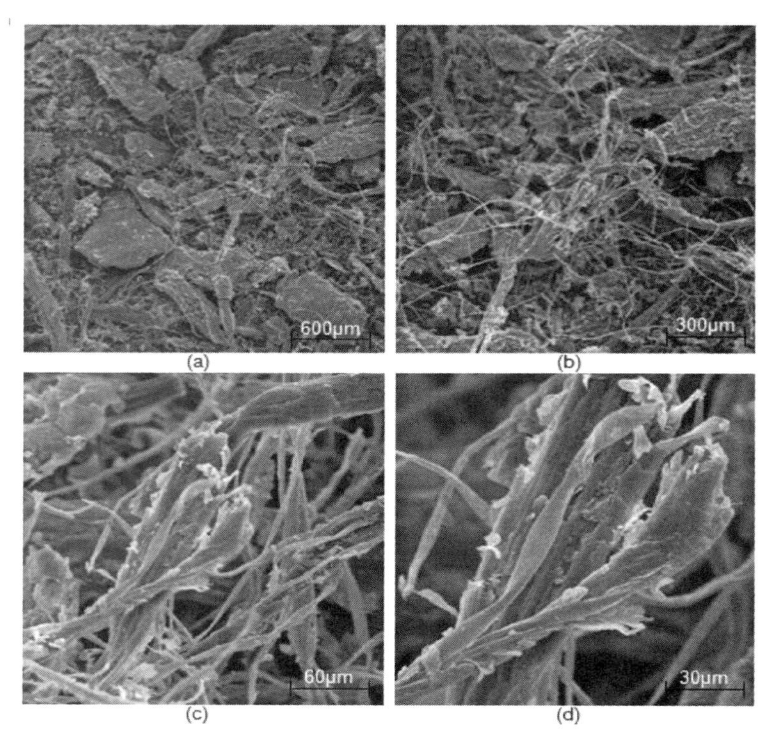

Figure 10.11 . Morphology of the pretreated pineapple fibers from residue juice at different magnifications (a, b, c and d).

Table 10.4 Impact strength of composites.

Samples	Impact Strength ($KJ.m^{-2}$)
PP	36.1 ± 0.3
PP/Fibers in nature 10% m/m	58.7 ± 17.8
PP/Fibers in nature 20% m/m	92.9 ± 16.5
PP/Modified fibers 10% m/m	38.2 ± 12.6
PP/Modified fibers 20% m/m	42.2 ± 18.4

insertion of fibers in matrix, causing an increase in absorbed energy, and consequently in strength. Table 10.4 shows the results obtained in the impact tests.

The pineapple fibers from the juice residue reinforced with polypropylene composites presented higher strength compared to modified fibers from the juice residue reinforced with polypropylene composites.

The amount of reinforcement in the matrix also contributes to the increase in strength. The modification of the pineapple fibers from the juice residue improved the adhesion between fiber/matrix, facilitating the energy transfer of impact, which is one of the influencing factors of this property. However, the unmodified fibers also facilitated the adhesion; this occurred because the acidity in the juice production favored treatment in the fibers from the residue.

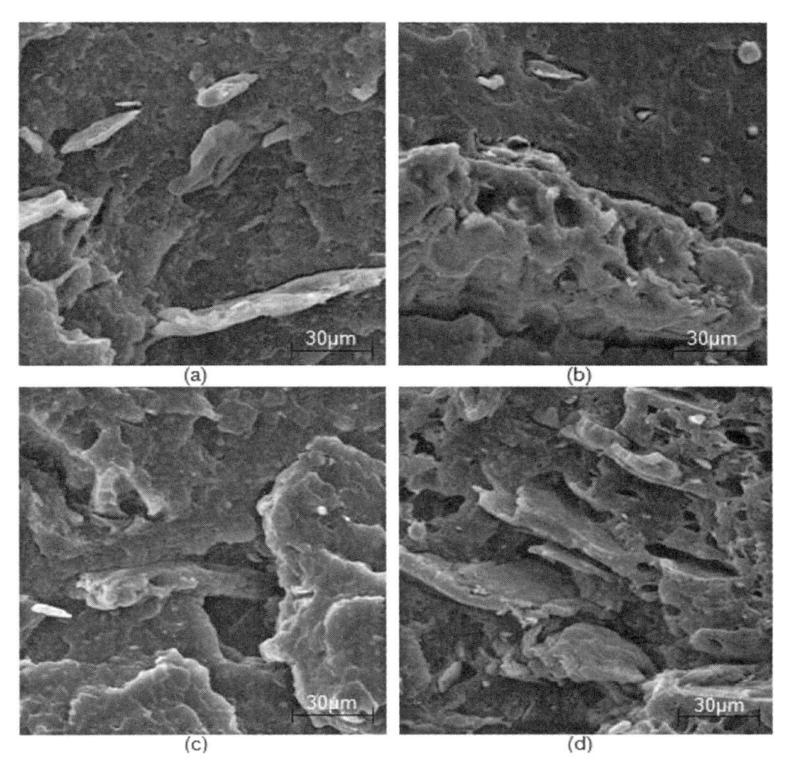

Figure 10.12 SEM of fracture surface of composites at different magnifications (a, b, c and d).

The insertion of fibers in the matrix also increased the impact strength, due to the mechanism of energy dissipation. The fibers were pulled-out of matrix and energy dissipated during the mechanical friction.

Figure 10.12 shows the region of the fracture after the test, in which uniformity of fiber distribution by matrix and fractured fibers were observed.

Thus, it was found that the fibers facilitated the diffusion of matrix inside the fiber, increasing the impact strength.

Acknowledgements

The authors are grateful for the research support of FAPERJ.

References

1. A.C. Ferrão, P.M.C. Silva, A.J. Reis, L.G. Freitas, M. Rodrigues and L.B. Alves, Ecodesign of automotive components making use of natural jute fiber composites. *J. Clean. Prod.* 18, 313–327 (2010).

2. J.R. Araujo, B. Mana, G.M. Teixeira, M.A.S. Spinacé and M.A. De Paoli, Biomicrofibrilar composites of high density polyethylene reinforced with curauá fibers: Mechanical, interfacial and morphological properties. *Compos. Sci. Technol.* 70, 1637–1644 (2010).
3. D.R. Mulinari, H.J.C. Voorwald, M.O.H. Cioffi, M.L.C.P. Da Silva and S.M. Luz, Preparation and properties of HDPE/sugarcane bagasse cellulose composites obtained for thermokinetic mixer. *Carbohydr. Polym.* 75, 317–320 (2009).
4. H.N. Yu, S.S. Kim, I.U. Hwang and D.G. Lee, Application of natural fiber reinforced composites to trenchless rehabilitation of underground pipes. *Compos. Struct.* 86, 285–290 (2008).
5. S.M. Luz, A.R. Gonçalves and A.P. Del'Arco Jr, Mechanical behavior and microstructural analysis of sugarcane bagasse fibers reinforced polypropylene composites. *Compos. A*, 38, 1455–1461 (2007).
6. P.C. Mileo, M.F. Oliveira, S.M. Luz, G.J.M. Rocha and A.R. Gonçalves, Evaluation of castor oil polyurethane reinforced with lignin and cellulose from sugarcane straw. *Adv. Mater. Res.* 123–125, 1143–1146 (2010).
7. E.F. Cerqueira, C.A.R.P. Baptista and D.R. Mulinari, Mechanical behaviour of polypropylene reinforced sugarcane bagasse fibers composites. *Proced. Eng.* 10, 2046–2205 (2011).
8. S.M. Luz, A.C. Pires and P.M.C. Ferrão, Environmental benefits of substituting talc by sugarcane bagasse fibers as reinforcement in polypropylene composites: Ecodesign and LCA as strategy for automotive components. *Resour. Conserv. Recy.* 54(12), 1135–1144 (2010).
9. S.M. Luz, J. Del Tio, G.J.M. Rocha, A.R.G. Gonçalves and A.P. Del'Arco, Cellulose and cellulignin from sugarcane bagasse reinforced polypropylenecomposites: Effect of acetylation on mechanical and thermal properties. *Compos. A* 39, 1362–1369 (2008).
10. W. Han, K. Chen, R. Yang, F. Yang, C. Zhao and W. Gao, Pancreatic cancer. *Bioresouces* 5(3), 1605–1617 (2010).
11. Y. Li, C. Hu and Y. Yu, Interfacial studies of sisal fiber reinforced high density polyethylene (HDPE) composites. *Compos. A* 39, 570–579 (2008).
12. S.K. Acharya, P. Mishra and S.K. Mehar, Effect of surface treatment on the mechanical properties of bagasse fiber reinforced polymer composite. *Bioresource* 6(3), 3155–3165 (2011).
13. M.M. Ibrahim, A. Dufresne, W.K. El-Zawawy and F.A. Agblevor, Banana fibers and microfibrils as lignocellulosic reinforcements in polymer composites. *Carbohydr. Polym.* 81, 811–819 (2010).
14. D.R. Mulinari, C.A.R.P. Baptista, J.V.C. Souza and H.J.C. Voorwald, Mechanical properties of coconut fibers reinforced polyester composites. *Proced. Eng.* 10, 2074–2079 (2011).
15. P.A. Souza, E.F. Rodrigues, J.M.C. Prêta, S.A.A. Goulart and D. R. Mulinari, Mechanical properties of HDPE/textile fibers composites. *Proced. Eng.* 10, 2040–2045 (2011).
16. T.A. Oliveira, A. Teixeira, D.R. Mulinari and S.A.S. Goulart, *Cadernos UniFOA*, 14, 11–17 (2010).
17. H. Anuar and A. Zuraida, Improvement in mechanical properties of reinforced thermoplastic elastomer composite with kenaf bast fibre. *Compos. B* 42, 462–465 (2011).
18. A.C. Karmaker and J.A. Youngquist, Injection molding of polypropylene reinforced with short jute fibers. *J. Appl. Polym. Sci.* 62, 1147–1151 (1996).
19. S. Wong, R. Shanks and A. Hodzic, Properties of poly(3-hydroxybutyric acid) composites with flax fibres modified by plasticiser absorption. *Macromol. Mater. Eng.* 287(10), 647–655 (2002).
20. M.A. Kabir, M.M. Huque, M.R. Islam and A. Bledzki, *Bioresources* 5(2), 854–869 (2010).
21. R.J. Brugnago, K.G. Satyanarayana, F. Wypych and L.P. Ramos, The effect of steam explosion on the production of sugarcane bagasse/polyester composites. *Compos. A*42, 364–370 (2011).

22. B.L.S. Sipião, R.L.M. Paiva, S.A.S. Goulart and D.R. Mulinari, Effect of chemical modification on mechanical behaviour of polypropylene reinforced pineapple crown fibers composites. *Proced. Eng.* 10, 2028–2033 (2011).
23. E.M. Teixeira, T.J. Bondancia, K.B.R., Teodoro, A.C. Corrêa, J.M. Marconcini and L.H.C. Mattoso, Sugarcane bagasse whiskers: extraction and characterizations. *Ind. Crops Prod.* 33, 63–66 (2011).
24. J.R. Monte, M. Brienzo and A.M.F. Milagres, Utilization of pineapple stem juice to enhance enzyme-hydrolytic efficiency for sugarcane bagasse after an optimized pre-treatment with alkaline peroxide. *Appl. Energ.* 88, 403–408 (2011).
25. P.C. Miléo, D.R. Mulinari, C.A.R.P. Baptista, G.J.M. Rocha and A.R. Gonçalves, Mechanical behaviour of polyurethane from castor oil reinforced sugarcane straw cellulose composites. *Proced. Eng.* 10, 2068–2073 (2011).
26. C.M. Belini, I.F. Bonafim and R.F.G. Junior, Crescimento de palmeira real australiana emsubstratos formulados a partir de composto de poda de árvore. *Revista Biologia Fafibe* 1, 1–10 (2011).
27. J.H. Ribeiro, S.O.S. palmito. *Revista Globo Rural,* 3, 24–26 (1996).
28. K.N. Simas, L.N. Vieira, R. Podestá, M.A. Vieira, I.I. Rockenbach, L.O. Petkowicz, J.D. Medeiros, A. Francisco, E.R. Amante and R.D.M. C. Amboni, *Microstructure, nutrient composition and antioxidant capacity of king palm flour: A new potential source of dietary fibre. Bioresour. Technol.* 101, 5701–5707 (2010).
29. S. Shinoja, R. Visvanathanb, S. Panigrahic, and M. Kochubabua, *Ind. Crops Prod.* 33, 7–22 (2011).
30. K. Riahi, A. B. Mammoub and B. B. Thayer, Date-palm fibers media filters as a potential technology for tertiary domestic wastewater treatment. *J. Hazard. Mater.* 161, 608–613 (2009).
31. A. Kriker, G. Debicki, A. Bali, M.N. Khenfer and M. Chabannet, Mechanical properties of date palm fibres and concrete reinforced with date palm fibres in hot-dry climate. *Cem. Concr. Compos.* 27, 554–564 (2005).
32. D.N. Saheb and J.P. Jog, Natural fiber polymer composites: a review. *Adv. Polym. Technol.* 18, 351–363 (1999).
33. M. Krouit, M.N. Belgacem and J. Bras, Chemical versus solvent extraction treatment: Comparison and influence on polyester based bio-composite mechanical properties. *Compos. A* 41, 703–708 (2010).
34. A. Bessadok, S. Roudesli, S. Marais, N. Follain and L. Lebrun, Alfa fibres for unsaturated polyester composites reinforcement: Effects of chemical treatments on mechanical and permeation properties. *Compos. A* 40, 184–195 (2009).
35. J.R. Megiatto Jr, E.C. Ramires and E. Frollini, Phenolic matrices and sisal fibers modified with hydroxy terminated polybutadiene rubber: Impact strength, water absorption, and morphological aspects of thermosets and composites. *Ind. Crops Prod.* 31, 178–184 (2010).
36. H. Gu, Tensile behaviours of the coir fibre and related composites after NaOH treatment. *Mater. Des.* 30, 3931–3934 (2009).
37. B.L.S. Sipião, T. Sousa, S.A.S. and D.R. Mulinari, Analysis of mechanical behavior of polymeric composites reinforced with palm fibers, In *Frontiers Polymer Science, Elsevier, Lyon-France* (2011).
38. P. Threepopnatkul, N. Kaerkitcha and N. Athipongarporn, Effect of surface treatment on performance of pineapple leaf fiber–polycarbonate composites. *Compos. B* 40, 628–632 (2009).

39. B.M. Cherian, A.L. Leão, S.F. Souza, L.M.M.C. Costa, G.M. Olyveira, M. Kottaisamy, E.R. Nagarajan and S. Thomas, Cellulose nanocomposites with nanofibres isolated from pineapple leaf fibers for medical applications. *Carbohydr. Polym.* 86, 1790–1798 (2011).
40. R.M.N. Arib, S.M. Sapuan, M.M.H.M. Ahmad, M.T. Paridah, and H.M.D.K. Zaman, Mechanical properties of pineapple leaf fibre reinforced polypropylene composites. *Mater. Des.* 27, 391–396 (2006).
41. C.H. Weng, Y.T. Lin and T.W. Tzeng, Removal of methylene blue from aqueous solution by adsorption onto pineapple leaf powder. *J. Hazard. Mater.* 170, 417–424, (2009).
42. J. George, S.S. Bhagawan, S. Thomas, Effects of environment on the properties of low-density polyethylene composites reinforced with pineapple-leaf fibre. *Compos. Sci. Technol.* 58, 1471–1485 (1997).
43. W. Liu, M. Misra, P. Askeland, L. T. Drzal and A. K. Mohanty, 'Green' composites from soy based plastic and pineapple leaf fiber: Fabrication and properties evaluation. *Polymer* 46, 2710–2721 (2005).
44. A.V. Tran, Chemical analysis and pulping study of pineapple crown leaves. *Ind. Crops Prod.* 24, 66–74 (2006).
45. R. Mangal, N.S. Saxena, N.S. Sreekala, S. Thomas and K. Singh, Thermal properties of pineapple leaf fiber reinforced composites. *Mater. Sci. Eng.* A339, 281–285 (2002).
46. T.A. Sousa, G.J.M. Rocha and D.R. Mulinari, *Cadernos UniFOA*, 3, 21–29 (2012).

Part II

CHEMICAL MODIFICATION OF CELLULOSIC MATERIALS FOR ADVANCED COMPOSITES

11

Agro-Residual Fibers as Potential Reinforcement Elements for Biocomposites

Nazire Deniz Yılmaz

Department of Textile Engineering, Pamukkale University, Denizli, Turkey

Abstract

Over the last two decades, lignocellulosic fibers have started to be considered as alternatives to conventional manmade fibers, in the academic as well as commercial arena, for a number of areas including transportation, construction, and packaging applications. Some species of plants such as hemp, flax, kenaf, jute, and sisal, which have been utilized as fiber sources since historical times, once again have become the focus of research attention. Nevertheless, agricultural plants have solely been utilized as a food source. Most recently, the possibility of making use of agricultural byproducts has gained interest. Some of these agricultural byproducts are rice and wheat straws; corn husks; corn, okra, cotton, and reed stalk fibers; and banana bunch fibers. Utilization of these byproducts would benefit rural development by providing additional value to the agricultural activities; the environment by saving the field leftovers from burning, which is unfortunately a common practice; and the growing world population by preserving the land for the production of edible plant species. This review investigates agro-based fibers as potential reinforcement elements in biocomposites. Methods of fiber extraction from agricultural residues are explained, characteristics of fibers with resources of different agricultural byproducts are compared, fiber modification techniques are presented and research efforts devoted to utilize these agro-based fibers are listed.

Keywords: Agro-residues, banana bunch fibers, biocomposites, biodegradable, corn husk fibers, natural fibers, okra bast fibers, renewable

11.1 Introduction

Increasing public awareness of the negative environmental effects of synthetic materials together with ever-stricter regulations, have boosted the growth of environmentally-friendly industries which produce "green" products [1].

These environmentally-friendly products include biodegradable and biobased materials based on annually renewable agricultural and biomass feedstock [2], which in turn would not contribute to the shortage of petroleum sources [3]. Biocomposites, which

Corresponding author: ndyilmaz@pau.edu.tr

Vijay Kumar Thakur (Ed.), Lignocellulosic Polymer Composites, (233–270) 2015 © Scrivener Publishing LLC

represent a group of biodegradable biobased products, are produced by embedding natural fibers into polymer matrices [4].

Biocomposites have some distinctive advantages as compared to petro-based non-biodegradable composites. Namely, the plant fibers have lower density, lower cost [5], better crash absorbance, and thermal [6] and sound absorption properties [7] compared to glass fibers. Furthermore, plant fibers cause less tool wear, skin and respiratory irritation than glass fibers [8]. Besides, plant fibers are renewable and biodegradable [5] and achieve good energy recovery if incinerated at the end of the service life [9].

Over the last two decades, lignocellulosic fibers have started to be considered as alternatives to conventional manmade fibers, in the academic as well as commercial arena, for a number of areas including transportation, construction, and packaging applications. Besides wood fibers, some species of plants such as hemp, flax, kenaf, jute, and sisal, which have been utilized as fiber sources since historical times, once again have become the focus of research attention. This has left another fiber source underexploited: agricultural residues [10, 11].

Most recently, the possibility of making use of agricultural byproducts or field crop residues as fiber sources has gained interest. Utilization of agro-based byproducts as fiber sources would extend the level of sustainability several steps further. Besides offering biodegradable fibers based on renewable sources and conserving the non-renewable petroleum sources [12], as in the case for conventional plant fibers, this practice would also preserve the land for the growth of edible agro-porducts [13]. As the world's population is growing, more efficient use of land is necessary in order to feed the people [12]. Furthermore, the field leftovers will be saved from burning. This would also save the environment along with enhancing the economic income of the rural society [13].

Another effect of this practice would be reducing the environmental impact of transportation of plant fibers, as the number of plant fibers utilized will be increased and byproducts of local agricultural economies might be brought into use [14,15].

Exploitation of agricultural residues or byproducts in composites as fillers or reinforcement elements would also diminish wood consumption, especially in regions where there are not sufficient wood resources left [10].

Searching for a use for lignocellulosic agro-residual materials in industry started as an alternative to wood in particleboards. More sophisticated research efforts have taken place in the polymer composite production area [10].

Some of these agricultural byproducts which have been the subject of research efforts as fiber sources are corn husks; rice and wheat straws; corn, okra, cotton and reed stalks; empty banana bunches; pineapple and oil palm leaves; and sugarcanes [5,7,10,13,16].

Exploiting natural fibers from new sources, however, presents the new question of whether these fibers' mechanical, morphological, and thermal characteristics would let them act as effective reinforcement elements for utilization in composites [15]. This necessitates the close study of agro-residual fibers.

This chapter has been written in order to investigate agro-residual fibers in terms of their performance as reinforcing elements for polymer composites. This study excludes agro-residual fibers which replace wood fibers as fillers or are suitable for pulping applications rather than reinforcing polymer composites due to their small dimensions. Nevertheless, several representative examples are given. The chapter is more focused on fibers with high aspect ratio (length\width), which is very important in terms of mechanical strength of the biocomposites. Fibers from agro-residuals which have a long history of utilization as fiber source are also only briefly mentioned, as the main scope of this chapter is focused on novel agro-residual fibers, on which limited research endeavor has been devoted in the open academic literature.

This chapter includes introduction of fibers from different agro-residual sources, extraction methods, chemical constituents and properties of fibers and physical and chemical treatments applied to fibers to provide higher quality reinforcement elements of biocomposites. Due to the broadness of the scope, this chapter will undoubtedly be unfinished, but it will hopefully provide guidance to researchers who want to contribute to the environment and to the rural society, together with the composite industry.

11.2 Fiber Sources

Faruk *et al.* [3] classify the plants, which produce natural fibers, into two groups according to their utilization: primary and secondary. Primary plants are grown for their fiber, while secondary plants are grown for other causes where the fiber is a by-product. Primary plants include jute, hemp, kenaf, and sisal, etc. Some conventional examples of secondary plants are pineapple, oil palm and coir. Some novel secondary plant examples include corn, okra, nettle, etc. This chapter is focused on these novel secondary plants, which give fiber as a by-product and have been the subject of limited research endeavor in the literature so far.

11.2.1 Wheat Straw

Cereal straw is an annually renewable fiber source and is available in abundance throughout the globe. These straws are traditionally used for animal bedding and livestock feeding. The unused straw may be incinerated, incorporated in the field or removed from the field. However, incineration of field residues is forbidden in some countries [10]. Wheat straw, which has already found commercial use as a 20% (w/w) wheat straw-reinforced polypropylene biocomposite, is used in the storage bins in the 2010 Ford Flex [8].

Panthapulakkal and Sain [10] reported better interfacial bonding between wheat straw fibers to HDPE compared to corncob and corn stalk fibers. They attributed this to more-lignin type and carbon rich hydrophobic surface of wheat straws compared to the other agro-based fillers. They observed higher percentage of silica on the wheat straw fiber surface than the other fibers.

11.2.2 Corn Stalk, Cob and Husks

In the past five decades corn production quadrupled to give an all-time high 868 million tons for the year 2011. Corn is the most commonly produced grain in the world surpassing wheat and rice [17], and it is grown in wide regions of the world [13]. Corn yield is also higher than that of wheat and rice giving 5 tonnes per hectare compared to 3 tonnes for hectare for the latter two grains [18]. Similar to wheat straw, cornstalks and corn husks are used as fodder, animal bedding or left on the field. Corncobs are utilized in fuel production [10,11]. Reddy and Yang [12] projects that more than 9 million tons of cellulose fibers with a potential sale value of $19 billion can be produced from the corn husks annually. Yılmaz [11] stated 10 to 33 g of fresh corn husks is needed to produce a gram of fiber. Fibers as long of 20 cm can be extracted from corn husks. The length of corn stalk fibers is limited by the length of stalk segments. There have been early studies related to utilization of corn husks in particle boards as an alternative to hardboards [19].

Corn stover is the residue which is left on the field after corn grain harvest. Corn stover includes approximately 50% stalks, 22% leaves, 15% cob and 13% husk. Stalks of genetically transformed corn may include 33-97% higher lignin compared to the native form [8].

Among several studies related to corn husk fibers, Reddy and Yang [12] extracted fibers from corn husks by applying alkalization and enzymatic treatment. Reddy [20] searched the influence of lignin contstituent of corn husk fibers on the tensile properties and yellowness of samples due to heat and light exposure. Yang and Reddy [21] have a patent on producing corn husks via alkalization and enzymatic treatment.

Yılmaz [11] investigated the effects of alkali extraction treatment parameters on properties of corn husk fibers. Yılmaz [13] compared corn husk fibers produced by water retting, alkalization and enzyme treatment. Huda and Yang [7] produced composites from corn husk fibers and polypropylene (PP) and investigated their flexural, impact resistance, tensile and sound absorption properties.

11.2.3 Okra Stem

Okra plant, which is also known as Lady's Finger and botanically named as *Abelmoschus esculentus*, is a member of the *Malvaceae* family. The plant is well diffused in North-East India [14]. Fibers, seen in Figure 11.1, extracted from the bark of its stem may be more than one meter long.

The thermal stability, chemical constitution and tensile properties of water retted okra fibers were studied by De Rosa *et al.* [14], whereas De Rosa *et al.* [15] investigated the effects of certain chemical treatments including bleaching, alkalization, acetylation on the thermal and mechanical behavior of okra bast fibers. Alam *et al.* [22] studied the UV resistance and the color fastness of dyed bleached and raw okra bast fibers. The effects of bleaching, alkalization and acrylonitrile monomer graft polymerization of okra bast fibers were investigated in terms of morphology, tensile strength and moisture absorption in Khan *et al.* [5,23]. Saikia [24] studied the moisture absorption of okra fibers at different temperatures for different durations.

11.2.4 Banana Stem, Leaf, Bunch

Banana is considered as the fourth most important crop in the developing world [25]. Whereas only 11.6 wt% of the total plant corresponds to fruit [26], 54.3 wt% of the plant total weight is consituted by fibrous by-product [27]. The fibrous by-product

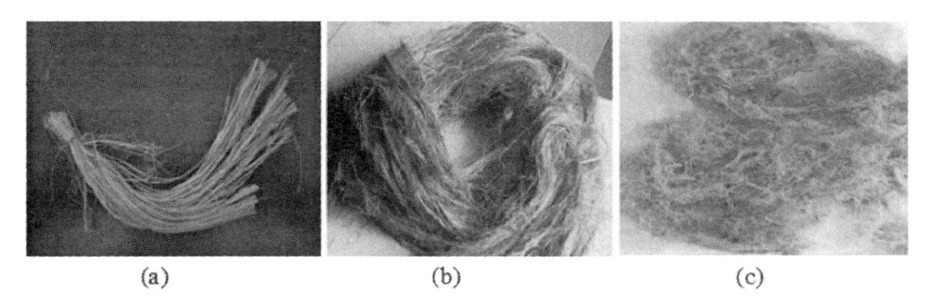

Figure 11.1 Examples of some agro-residual fibers. (a) okra fiber (From I. M. De Rosa *et al.*, *Composites Science and Technology*, 2011 [14]. With Permission from Elsevier); (b) banana fiber (From A. V. R. Prasad, K. M. Rao, G. Ragavinirasulu, *Indian Journal of Fibre and Textile Research*, 2009 [30]. With permission from *Indian Journal of Fibre and Textile Research*); (c) corn husk fiber [31].

finds use in pulping, composite production, or as fodder [25]. Fibers may be extracted primarily from the bunch, stem, and leaf. Banana stems are usually left on the field after cultivation [28]. Fibers are traditional extracted by mechanical means [26].

Ganan *et al.* [26] investigated the effects of alkaline and silane treatments on the moisture absorption behavior of banana bunch and stem fibers. Ganan *et al.* [27] compared the effects of mechanical and biological extraction methods on the chemical and morphological structure and stiffness of banana bunch fibers. Aseer *et al.* [28] investigated the effects of sodium hypochlorite (NaClO) bleaching on chemical structure, surface morphology, thermal stability and water absorption of banana stem fibers. Zaman *et al.* [29] investigated the effects of methyl acrylate grafting on the tensile performance of coir fibers. Prasad *et al.* [30] produced biocomposites from banana bunch fibers and unsaturated polyester and studied the effect of fiber volume fraction on tensile, flexural and impact properties of the composites. The relationship between steam explosion parameters, chemical content, mechanical properties, water absorption and diamensional change of banana bunch fiber board were studied by Quintana *et al.* [25].

11.2.5 Reed Stalk

Fibers from reeds (*Miscanthus*) are generally produced by mechanical separation techniques. The stiffness of reed fiber is between that of sisal and hemp [32]. There are several research studies about incorporation of reed fibers in composites as reinforcement elements in biocomposites [33]. Johnson *et al.* [34] produced fully biodegradable reed fiber-reinforced starch-based biopolymer Novamont Mater-Bi® composites. Kraiem *et al.* [35] investigated the effect of reed fiber content on the mechanical behavior of recycled HDPE composites. Bourmaud and Pimbert [32] studied the mechanical properties of reed fiber-reinforced polypropylene and poly lactic acid composites. Nagarajan *et al.* [36] produced composites by incorporating agro-residual fibers into a biodegradable polymer matrix and the best mechanical performance was obtained from reed fiber-reinforced composite.

11.2.6 Nettle

Nettle (*Urtica dioica L.* or *Girardinia diversifolia*) is a perennial plant which is cultivated for food, fodder, cosmetics and medicine as well as for its fiber [37]. Although nettle fiber had been used for a long time in Europe until World War II [38], there is a very limited research effort on its use in biocomposites compared to flax, hemp, kenaf and so on.

The advantages of nettle crop are the necessity of low agronomic inputs, suitability for organic farming, reduction of soil erosion and recovery of over-fertilized soils due to being a nitrophilous species [38]. Bacci *et al.* [39] obtained around 1700 kg ha^{-1} fiber yield of nettle crop.

Bodros and Baley [37] studied the tensile properties of water retted nettle fibers. Bacci *et al.* [39] investigated fiber yield of nettle cultivation and quality of fibers extracted by alkalization. Bacci *et al.*[38] compared the effects of different fiber extraction method combinations including water retting, mechanical decortication and enzymatic treatments on resultant fibers chemical composition and tensile properties. Bajpai *et al.* [40,41] prepared nettle fiber-reinforced poly lactic acid (PLA) and PP composites by compression molding. Bajpai *et al.* [42] studied the effects of various environments on the tensile strength of nettle fiber-reinforced PP composites.

11.2.7 Pineapple Leaf

Pineapple (*Ananas comosus*), which is native to Brazil, is a tropical plant with leaves rich in cellulose. Being relatively inexpensive and abundantly available, pineapple fiber may be considered for polymer composite reinforcement. Today, pineapple leaves are a by-product of pineapple cultivation [3]. Devi *et al.* [43] investigated the dynamic mechanical behavior of pineapple leaf fiber-reinforced polyester composites. Threepopnatkul *et al.* [44] studied the effects of fiber surface treatments on the performance of pineapple leaf fiber-carbonate composite.

11.2.8 Sugarcane

The fibrous residue that is the remainder when sugarcane stalks are crushed in order to extract their juice is called bagasse. Annual bagasse production is about 54 million dry tons worldwide [45]. As an agro-residue, bagasse has been extensively studied in terms of its use in biocomposites [3]. Mothe and Miranda [45] studied the thermal stability and chemical constituent analyses of bagasse fibers. Acharya *et al.* [46] reported that acetone treatment of bagasse fibers resulted in an increase in strength of bagasse fiber-reinforced epoxy composites. Youssef *et al.* [47] investigated effects of several coupling agents on the physical properties of baggase fiber-reinforced various thermoplastic composites.

11.2.9 Oil Palm Bunch

Only the oil palm industry in Malaysia produces more than a million tones of empty fruit bunches annually as a by-product [48]. Oil palm fiber is anticipated to be a promising raw material due to its toughness [49]. Hasibuan and Daud [48] investigated the effect of super heated steam drying on the surface morphology, color and tensile properties of oil palm empty fruit bunch fibers. John *et al.* [49] developed oil palm fiber-reinforced natural rubber based composites and studied the effect of fiber content on the mechanical performance. Jawaid *et al.* [50] produced hybrid oil palm empty fruit bunch fiber-jute-reinforced epoxy composites and investigated their mechanical properties, Abdul Khalil *et al.* [51] studied the water absorption and swelling of these composites.

11.2.10 Coconut Husk

Coir fibers are obtained from the husk of the coconut. Coir fiber is generally extracted by mechanical means from the plant [52]. Manilal *et al.* [52] extracted coir fibers by a closed retting process in an aerobic retting reactor. Bakri and Eichhorn [53] mechanically extracted coir and celery fibers and studied their tensile behaviors in terms of micromechanics. Mothe and Miranda [45] studied the thermal stability and chemical constituent analyses of coir fibers. Khan and Alam [54] investigated the effects of several chemical treatments on the thermal and meachanical properties of coir fibers. Mahato *et al.* [55] studied the effect of alkalization on the thermal degradation of coir fibers.

11.3 Fiber Extraction Methods

It is theoretically expected that lignocellulosic fibers have better interaction with the hydrophobic polymer matrix compared to cellulosic fibers. This is due to the fact that cellulosic fibers are more polar than the lignocellulosic fibers which contain the aromatic rings of lignin. Nevertheless, non-cellulosic materials have to be removed to some extent to separate fibers from the agro-residue [7]. The reason to this is the fact that the non-cellulose components, such as lignin and hemicellulose decrease the strength of the fiber and accelerate biological, ultraviolet and thermal degradation [8].

The procedures which are applied during the production of conventional lignocellulosic fibers can also be used to extract fibers from agro-based byproducts. Fiber extraction techniques include biological, chemical and physical separation methods [27]. These can be listed as

- Dew retting (biological)
- Cold water retting (biological)
- Hot water retting (biological)
- Mechanical separation (physical)
- Chemical extraction (chemical)
- Enzymatical extraction (chemical)
- Ultrasonic separation (physical)
- Steam-explosion (physical) [56]

In the course of these procedures, the non-cellulosic parts are removed in specific proportions and the cellulosic fibers are exposed and seperated [56]. Each method has its advantages and disadvantages in terms of the yield and quality of the extracted fibers

[27]. One of these processes can be selected, or several processes can be combined to achieve a desired quality [13,56]. Each added step will increase the cellulose content while decreasing variability and fiber diameter [8]. Storing the agro-residue for a certain period such as one year before fiber extraction may also lead to a more efficient fiber seperation than new harvested fiber source [38].

11.3.1 Biological Fiber Extraction Methods

Fermentation of extra-cellulosic materials can be performed by anaerobic or aerobic microorganisms, producing soluble byproducts and gases like methane [27]. During biological fiber extraction methods, the fiber source is subjected to a medium containing bacteria or fungi. The fiber bundles are set apart from the core, epidermis and cuticle of the plant, and the separated fiber bundles are splitted into smaller bundles. These are achieved by the biological activity of bacteria, such as *Bacillus* sp. and *Clostridium* sp., and fungi, such as *Rhizomucor pusillus* and *Fusarium lateritium* [8]. The biological activity can be aerobic or anaerobic [27].

Biological fiber extraction methods include dew retting and water retting techniques. During dew retting, agricultural residues such as plant stems are left out in the field where biological organisms partly degrade some polysaccharides exposing the fiber bundles. The quality of the produced fibers is dependent on many factors including the plant cultivar, soil composition, agricultural management and climate conditions. The downsides of dew retting include the necessity of appropriate climate, variability in fiber quality, and the risk of overretting which degrades fiber cell walls [8]. The advantage of it is lower cost; thus, it is widely used [38]. Sain and Panthapulakkal [57] carried out dew retting in laboratory conditions on wheat straws for three weeks. When compared the retted wheat straw fibers with mechanically split wheat fibers, they found better thermal resistance and lower amount of extra-cellulosic materials for retted fibers.

Water retting includes keeping the fiber source in water tanks for a certain period [27]. The microorganisms proliferating in water lead to fermentation of polysaccharides present in the plant structure exposing cellulose fiber bundles. The drawbacks of this method include high costs of labor, drying and waste water management, high oxygen demand together with unpleasant odor generation [8] caused by methane and hydrogen sulphide production during fermentation [27,52]. The waste water can be reused as it is reported to have affirmative effects on plant growth and insect control. If the water retting is carried out in plant facilities rather than in water ponds, the retting period may be shortened and the fiber yield and quality may be increased. Methane gas may be recovered and the water may be recycled by using a closed-loop system. Accordingly, Manilal *et al.* [52] retted coir fibers in an anaerobic reactor where the retting duration was decreased to one month from 6-12 months in the case of retting in water ponds. The quality of extracted fibers may be enhanced by use of controlled pure bacterial cultures rather than indigenous microorganisms. Bacci *et al.* [38] used pure clones of aerobic ROO40B and anaerobic L 1/6 bacteria for retting of nettle stalks. Although they did not achieve a faster retting, the resultant fibers had lower diameters compared to the ones retted in natural microflora proliferated during immersion of stalks in a water tank.

As an alternative to conventional methods of dew and water retting, Murdy [8] investigated the feasibility of silage retting. By applying solid state retting in an enclosed facility, she aimed to recover fermented polysaccharides.

Retting time depends on the plant species. While okra bast fibers can be obtained after a duration of 15 – 30 days [14,22] it takes several months to extract fibers from corn husk by water retting [13]. The author has not been able to extract fibers from reed leaves and stalks even after two years of immersion in water tanks.

11.3.2 Chemical Fiber Separation Methods

Chemical treatments, such as alkalization, bleaching with sodium hypochlorite (NaClO) or chlorite (NaClO2), are applied on the fiber bundles to extract the technical fiber that can stand the loads of tension and torsion. Chemical treatments are effective in removing the extra celulosic materials and exposing the cellulose. However, if the alkalization conditions are too strong, this may cause damage in the fiber [15].

Bacci *et al.* [39] treated nettle stalks with boiling soda solution until the bark, the source of fiber, was easily removed from the core of the stalk than performed alkalization treatment on the bark. Enzymatical fiber retting may be faster and more reproducible than the traditional biological fiber retting methods and is extensively studied for flax and hemp extraction enhancement [39].

Yılmaz [11] extracted fibers from corn husk by alkalizing them at varying concentrations and for differing durations. She reported that the length, diameter and moisture content of fibers decreased with increasing alkali concentration and duration as given in Figure 11.2. Yılmaz [11] found the effect of alkalization duration to be greater than that of concentration on fiber parameters. Yılmaz [11] reported that the optimum alkalization conditions to achieve the highest performance fibers were a concentration in the range of of 5-10 g/L and a duration between of 60 – 90 minutes for boiling temperature.

Yılmaz [13] reported that corn husk extracted by alkalization had lower lignin and hemicellulose content and showed higher elongation and lower stiffness compared to fibers extracted by water retting. She explained the lower tensile performance of alkalized fibers with the loss of cellulosic crystalline orientation during alkalization treatment carried out under no tension. Alkalization might have caused the native cellulose (cellulose I) to transform to cellulose II, which generally has a lower chain modulus. Another cause maybe that alkalization might have degraded the cellulosic chains.

11.3.3 Mechanical Fiber Separation Methods

Mechanical fiber separation may be carried out in different ways such as beating the fiber source with metal blades (scutching) or combing (hackling) [56]. When combined with other separation techniques, mechanical decortication, which is also known as scutching, leads to production of lighter, softer and finer fibers [8]. Although

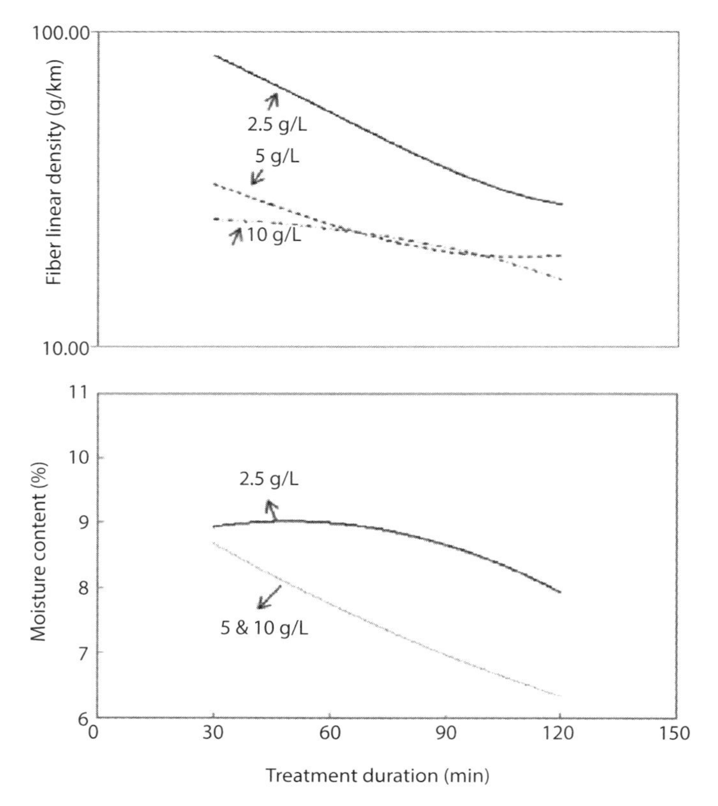

Figure 11.2 Effect of alkalization parameters on corn husk fiber linear density and moisture contents for 2.5, 5 and 10 g/L alkali concentrations. (From N. D. Yılmaz, *Indian Journal of Fibre and Textile Research*, 2013 [11]. With permission from *Indian Journal of Fibre and Textile Research*).

mechanical separation methods traditionally applied following retting, the sequence can be reversed and different combinations may be tried [38].

Due to its economy and high production rate, the mechanical method is commonly employed. However, its efficiency to remove extracellulosic components is low and it negatively affects the fiber bundle length [27], as found by Bacci *et al.* [38], and may cause color darkening which is unappealing in commercial point of view [52]. Ganan *et al.* [26] extracted fibers from stems and bunches of banana plants by mechanical fiber separation. Bacci *et al.* [38] achieved higher tenacity from fibers extracted by only mechanical separation compared to biological or chemical retting methods or combination. In his study, the other retting methods and enzymatic treatments reduced fiber strength. Prasad *et al.* [30] extracted fibers from empty banana bunches mechanically using a decorticator shown in Figure 11.3 and a simplified scheme of a laboratory scutcher designed and built by Bacci *et al.* [38] for nettle stalk decortication is shown in Figure 11.4.

Another way of obtaining fibers from agricultural residues is pulping. However, using this method, the obtained fibers are 0.5 to 1.5 mm in length, which is too short to be processed in textiles, nonwoven production and some other industrial applications [7,12]. As known, aspect ratio is a major factor influencing the mechanical properties of a composite. Higher aspect ratio results in better tensile, flexural and impact

Figure 11.3 Decorticating machine used for mechanical fiber extraction. (From A. V. R. Prasad, K. M. Rao, G. Ragavinirasulu, *Indian Journal of Fibre and Textile Research*, 2009 [30]. With permission from *Indian Journal of Fibre and Textile Research*).

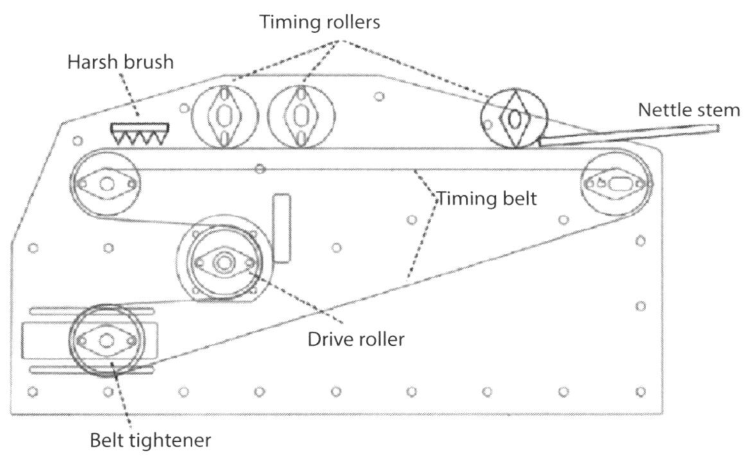

Figure 11.4 Scheme of a scutcher designed and built for mechanical extraction of nettle fibers. (From L. Bacci *et al.*, *Textile Research Journal*, 2011 [38]. With permission from SAGE Publications).

performance, as it provides more space for interaction with the matrix. Having lower aspect ratios, fibers in short lengths tend to act as fillers rather than reinforcing elements in composites [7]. Except for a few representative examples, agricultural waste fibers produced by pulping have not been included into the scope of this chapter.

Mechanical pulping includes subjecting agricultural residue to steam under high temperature and high pressure. To give an example, Ashori and Nourbakhsh [16] obtained fibers of around 1-1.5 mm length by treating sunflower stalk, corn stalk and bagasse with steam at 175°C and 7 MPa for 15 minutes.

Storage of field residuals may have a direct effect on the final quality of the fibers. The residuals are prone to rotting in damp state. Drying may be a practical solution to

Table 11.1 List of Academic Literature on Agro-Residual Fibers, Extraction Methods, Studied Characterics.

Fiber source	Extraction method	Treatment	Chemicals used	Enzymes used	Studied characteristics	Reference
Corn husk	Alkalization Enzymatic treatment	Bleaching		Pulpzyme ® Cellulase	Spectrophotometry	12
Corn husk	Alkalization Enzymatic treatment	Sulfonation	Sodium sulfite Ethylenediamine	Pulpzyme ® Cellulase	Tensile strength, color, light and heat resistance	20
Okra bast	Water retting scouring	Exposure to sun light Bleaching Dyeing			Chemical constituents, Molecular weight of cellulose, Tensile strength, dye asorption	22
Okra bast	Water retting	Bleaching Alkalization Graft copolymarization	Acrilonitril monomer		Water absorption tensile properties	5
Okra bast	Water retting scouring	Bleaching Alkalization Graft copolymarization Dyeing	Acrilonitril monomer, potassium persulphate, ferrous sulphate		UV resistance Color fastness, tensile properties	23
Okra bast	Water retting				Chemical,tensile, thermal properties	14
Okra bast	Water retting	Scouring Bleaching Acetylation	Sulphuric acid Potassium permanganate Dodecyl sulphate		Chemical,tensile, thermal properties, water absorption	15
Banana stem and bunch	Mechanical separation	Silanization, Alkalization	α-glycidoxypropyltrimethoxy-silane		Chemical, thermal analysis, contact angle, surface free energy	26
Banana stem and bunch	Mechanical and water retting				Chemical, morphological, tensile properties	27
Banana bunch	Mechanical				Tensile properties	30
Banana stem	N/S	bleaching	NaClO		Chemical,tensile, thermal properties water absorption	28

Table 1.1 (continued)

Fiber source	Extraction method	Treatment	Chemicals used	Enzymes used	Studied characteristics	Reference
Nettle stalk	Chemical retting, water retting mechanical separation Vat enzyme treatment, spray enzyme treatment		EDTA	Viscozyme_ L, Pectinex_ Ultra SP-L, ROO40B*, L 1/6*	Chemical, tensile, morphological properties	38
Nettle stalk	Chemical extraction					39
Nettle stalk	Water retting				Morphology, tensile testing	37
Coconut husk	Mechanical extraction	Bleaching			Chemical, tensile properties	53
Coconut husk	Water retting mechanical separation	Bleaching, acidation	H_2O_2, HNO_3		Chemical, tensile properties	52
Coconut husk	Water retting	Chemical treatment, grafting	$NaClO_2$, NaOH, acrylamide		Tensile properties, thermal stability	54
Celery	Mechanical extraction	bleaching			Chemical, tensile properties	53
Wheat straw	Mechanical extraction, microbial retting				Physical and tensile properties, chemical, thermal characterization	56
Corn husk	Alkalization	Enzyme treatment		Pentopan ® Mono BG	Tensile properties, moisture regain,	58
Corn husk	Alkalization				Chemical,tensile, thermal properties, moisture regain,	11
Corn husk	Alkalization Water retting	Enzyme treatment		Pentopan ® Mono BG,Celluclast ®	Chemical,tensile, thermal properties	13

*: bacteria

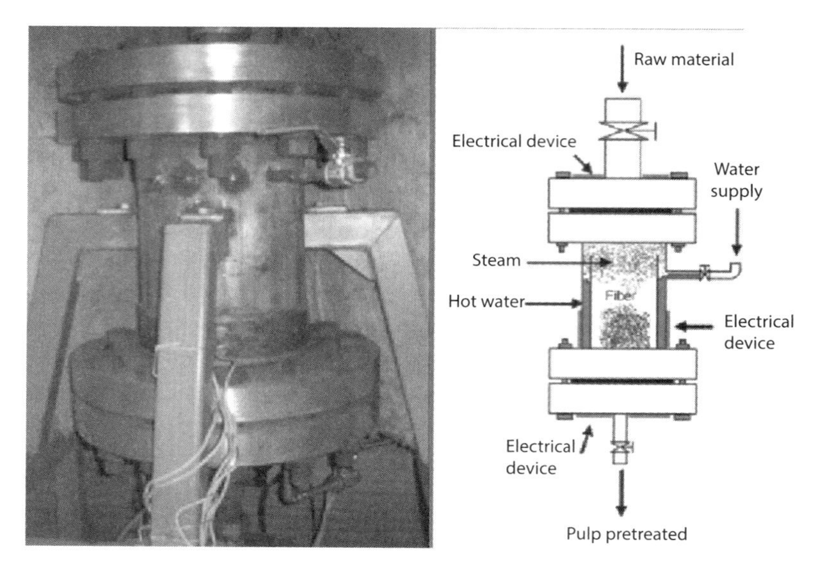

Figure 11.5 Image and scheme of steam explosion reactor. (From G. Quintana, J. Vasqueza, S. Betancourt, P. Ganan, *Industrial Crops and Products*, 2009 [25]. With permission from Elsevier).

avoid agricultural residue degradation. Drying may also have a second effect which is not trivial: Drying reduces the mass of residues in substantial proportions. This might reduce the use of energy and emission of greenhouse gas and other pollutants during transportation and processing. In search of the influence of agricultural residue drying on fiber properties, Yılmaz *et al.* [58] found no adverse effects of drying of corn husks on the characteristics of corn husk fibers. They reported that the weight loss due to drying was as high as 72%, this led to the fiber yield in the case of undried husks to increase to 25% for dried corn husks. Table 11.1 presents a list of research efforts devoted to obtain fibers from agricultural residues by applying various extraction methods.

11.4 Classification of Plant Fibers

There are several different classifications in terms of plant fibers. While the classification of Nishino [59] includes seven groups as bast (soft) fibers (flax, hemp, jute), leaf (hard) fibers (sisal, abaca, pineapple, etc.), stem fibers (bamboo, banana stalk, corn stalk), fruit fibers (coconut), seed fibers (cotton, baobab, kapok), straw fibers (rice, wheat, corn), and others (seaweeds, palm), that of Faruk *et al.* [3] has six groups: bast fibers (jute, flax, hemp), leaf fibers (abaca, sisal and pineapple), seed fibers (coir, cotton and kapok), core fibers (kenaf, hemp and jute), grass and reed fibers (wheat, corn and rice) and all other types (wood and roots).

The traditional plant fiber classification groups are bast, leaf and seed/fruit fibers, other than wood fibers.

The fibers which are found in the stem or stalk of bast plants and mechanically support the plants are called bast fibers [60]. Bast fibers have drawn quite an amount of interest in terms of composite reinforcement purposes [14] due to their superior mechanical properties [9].

Leaf fibers have lower elasticity moduli and higher elongation, and better toughness, and consequently, higher impact resistance than bast fibers. This is caused by their greater microfibril angle [56].

Other than cotton, seed fibers generally exhibit the lowest densities and the highest microfibril angle among plant fibers. They have high failure elongation and low elasticity modulus [61].

11.5 Properties of Plant Fibers

Exploiting natural fibers from agro-residues brings a new question that whether these fibers' characteristics would let them to act as effective reinforcement elements for utilization in composites [15]. This necessitates close study of the characteristics of agro-residual fibers. Accordingly, Table 11.5 lists characterization methods applied on agro-residual fibers.

11.5.1 Chemical Properties of Plant Fibers

Lignocellulosic fibers constitute several chemical materials such as cellulose, hemicellulose and lignin as major constituents [28]. Waxes, fats and pectins also take place in the outer surface of plant fiber bundles [62]. Similar to other lignocellulosics, agro-based materials generally present a composite structure. In this composite structure, the organic matrix which is comprised of lignin, hemicelluloses, pectin, waxes, and proteins, is reinforced by bundles of cellulose microfibers [13]. The proportions of the constituent materials depend on fiber source, growth conditions, and fiber extraction method [12]. The chemical components and their proportions have a substantial influence on the resulting mechanical properties of the fibers [16], so it is of importance to have a thorough knowledge about these components.

11.5.1.1 Cellulose

Cellulose is a linear polymer of D-anhydroglucose units, which contain three hydroxyl groups, and are bonded by β-1,4-glycosidic linkages. The degree of polymerization of cellulose, which is the highest in bast fibers, and the microfibrillar angle which is lower in bast fibers affects fiber mechanical properties [8,56]. Having the highest contribution to the mechanical load carrying, cellulose is the main structural component of plant fibers [61]. The elasticity modulus of cellulose is in the order of 130 GPA [62]. While being resistant to strong alkali (17.5 wt%) and oxidizing agents, cellulose is easily damaged by acids [56]. Figure 11.6 gives the chemical structure of cellulose.

Cellulose includes crystalline and amorphous regions [63]. The degree of crystallinity affects physical and chemical properties of the fiber. While increasing crystallinity means greater strength, decreasing crystallinity means increasing elongation, higher water intake and more sites available for chemical reactions. Reddy and Yang [12] reported a crystallinity degree of 48-50% and crystal size of 3.2 nm for corn husk fibers compared to those of flax and jute which are 65-70% and 2.8, for degree of crstallinity and crystal size, respectively. They also found poorer crystal orientation.

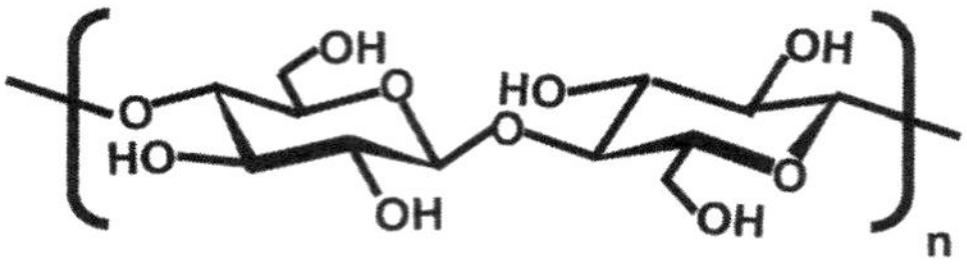

Figure 11.6 Chemical structure of cellulose. (Adapted from W. Kasai, T. Morooka, M. Ek, *Cellulose*, 2013 [64]. With Permission from Springer).

The tensile strength of a fiber is directly proportionate to cellulose content. When comparing tensile and flexural properties of corn stalk, sunflower stalk and bagasse fiber-reinforced polypropylene composites, Ashori and Nourbaksh [16] obtained the highest results from bagasse fiber-reinforced composites, where bagasse fibers has higher cellulose content (53%) compared to those of other fibers (around 37%). All fibers had been extracted by the same method and had similar length, diameter and aspect ratios.

11.5.1.2 Hemicellulose

Hemicellulose is a highly branched polysaccharide attached to the cellulose [8]. It is not a form of cellulose but a group of polysaccharides. In contrast to cellulose, which only includes a 1,4-β-glucopyranose ring, hemicelluloses consist of different sugar units. The polymer chain of hemicellulose is highly branched unlike linear cellulose. Moreover, degree of polymerization of hemicellulose is 10 to 1000 times lower than cellulose. Due to the branched chain structure and the low degree of polymerization [63], hemicelluloses present an amorphous structure [62]. Thus, cellulose is highly hydrophilic owing to the presence of a great number of available sites [6]. Hemicellulose, which is soluble in alkali, easily hydrolyzes in acids. It is the one among major fiber constituents which is the most sensitive to thermal and microbial effects [56].

11.5.1.3 Lignin

Lignin is a phenolic compound [8] that acts as the major binding material that holds the single fiber cells together. It affects the structure, properties, processability and end use applications of lignocellulosic fibers [20]. Lignin is a hydrocarbon polymer [63] that includes phenolic and alcoholic hydroxyl groups which are sensitive to UV rays [22]. It acts as a protective sheath for cellulose and provides structural support to the plant [8]. Lignin gives a stiff handle to the fiber [20], and together with hemicellulose, it increases toughness, i.e., the mechanical energy absorption capability of the fiber [61]. Among the major fiber constituents, lignin is the most hydrophobic one. Lignin, which has an amorphous structure, is a high-molecular-weight three-dimensional thermoplastic polymer that starts to soften at 90°C and starts to melt at 170°C [6]. In contrast to cellulose, lignin is not hydrolyzed by acids, but it is soluble in hot alkali. It condenses easily with phenols and is readily oxidized [56].

Lignin which undergoes photo degradation upon exposure to sunlight imparts a yellowish color to fibers which makes it difficult to bleach the fibers to an acceptable whiteness degree [20]. Presence of lignin also results in greater tenacity and elongation loss and faster degradation upon exposure to sunlight and heat [20,22]. Alam *et al.* [22]

reported greater strength loss in okra bast fibers including greater lignin amounts. Reddy *et al.* [20] found higher resistance to light and heat exposure in delignified corn husk fibers. They used sodium sulfite and ethylenediamine to delignify the corn husk fibers. They did not detect any effect of delignification process on crystal orientation of the fibers.

The lignin constituent can be utilized to bond agrowaste fibers in fiberboard production without a further need for synthetic glues, as Quintana *et al.* [25] produced fiberboards from banana bunch fibers without use of any binders. The performed a thermomechanical pretreatment hydrolyzing many of the hemicelluloses and redistribute the lignin polymer to expose it in the surface of the fiber.

11.5.1.4 Pectin

Pectins which consist of α-1,4-linked galacturonic acid units are heteropolysaccharides. Pectin is hydrophilic [12] and water soluble [63]. It is necessary to remove pectins in order to separate individual fibers of fiber bundles [61]. It is possible to partially neutralize pectin with alkali or ammonium hydroxide [63].

11.5.1.5 Waxes

Waxes which constitute the secondary components together with pectin consist of different types of alcohols [8] that are water soluble and dissolve in acids [63]. Different kinds of agro residual fibers include varying amounts of these chemical consitutents. Table 11.2 gives the chemical composition of agro residual fibers and compared it with conventional plant fibers.

11.6. Properties of Agro-Based Fibers

It is essential to know the fiber characteristics in order to expand the use of agro-residual fibers in biocomposites and to improve their performance [3].

11.6.1 Physical Properties

Except cotton and kapok, most of the natural cellulosic fibers are multicellular. They are usually used as groups of individual cells or as bundles of fiber in industrial applications. As a term, a "fiber", or a technical fiber [14], refers to a bundle of individual cells bound together by hemicellulose, lignin and other non-cellulosic materials [12]. However, the individual fiber cell is drastically stronger than the fiber bundle [56]. For example, the individual fiber of flax is as stiff as aramid [65]. The individual fiber cell has a lumen inside which imparts a hollow structure to the fiber as seen in Figure 11.7. As an example, okra fibers have a void content of 18-32% [15]. The interface between two cells is called middle lamella [14].

The cross-section and longitudinal views of "technical" okra fibers are shown in Figure 11.8.

Table 11.2 Chemical Composition of Agro-Residual Fibers Compared to Some Selected Conventional Fibers.

Fiber	Cellulose	Hemicellulose	Lignin	Pectin	Waxes	Ash	Reference
Okra	60-76	15-20	5-10	3-5	3.9	*	5,22, 24
Nettle	65-86	2.3-12.5	1.6-5.2	*	*	*	38
Sunflower stalk	38.1	*	17.3	*	*	7.0	16
Corn stalk	36.6	*	16.7	*	*	6.0	16
Bagasse	52.7	*	20.6	*	*	1.3	16
Wheat straw	33-38	26-32	17-19	*	*	*	8
Rice straw	28-36	23-28	12-14	*	*	*	8
Oil palm fiber	50-65	*	19-21	*	*	2	49,50
Corn husk	64-84	*	5.7-6.8	*	*	0.3-0.8	7
Banana	60-72	6-10	5-10	*	*	1.16	8,28
Flax	68-85	18-21	2-3	2-3	1-2	*	*
Hemp	70-74	18-23	3.5-6	0.9	0.8	*	*
Jute	61-75	12-21	10-15	0.2-1	0.5	*	*
Kenaf	45-57	21.5	8-13	3-5	*	*	*
Ramie	68-77	13-17	0.5-1	1.9	0.3	*	*
Pineapple	70-82	*	5-13	*	*	*	*
Sisal	66-78	10-14	10-14	10	2	*	*
Abaca	56-63	*	12-13	1	*	*	*
Cotton	85-90	5-6	*	0-1	0.6	*	*
Coir	32-43	0.1-0.3	40-45	3.4	*	*	*

*: Data from Olesen, and Plackett [6] and Bismarck *et al.* [56].

Plant fibers are bundles of elongated dead plant cells with thick walls. The individual fibers look like microscopic tubes of a cell wall surrounding the lumen. The presence of lumen increases moisture absorption. The cell wall is composed of several concentric layers. Whereas the primary (P) cell wall is formed first during cell growth, the secondary cell wall (S) has three layers: S_1, S_2, and S_3. The porous outer cell wall, forming the surface of the fiber, includes the most of the non-cellulosic components, and is considered as the cause of poor absorbency and poor wettability [56]. Table 11.3 compares the physical properties of agrowaste fibers with conventional plant fibers and E-glass fibers.

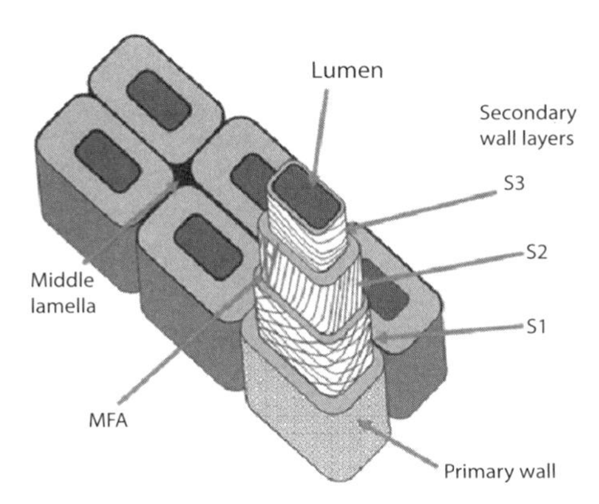

Figure 11.7 Structure of a Plant Fiber. (From D. Kretschmann, Nature Materials, 2003 [66]. With permission from Nature Publishing Group).

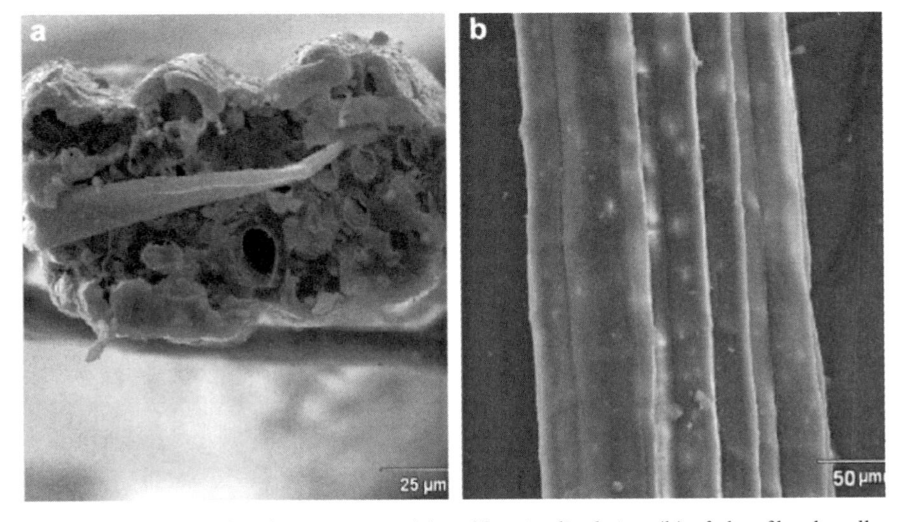

Figure 11.8 SEM micrographs of cross-section (a) and longitudinal view (b) of okra fiber bundle. Source: De Rosa [14] (From I. M. De Rosa *et al.*, *Composites Science and Technology*, 2010 [14]. With permission from Elsevier).

11.6.2 Mechanical Properties

Microfibrillar angle, MFA, (θ), is the name given to the angle between the microfibrils and the longitudinal cell axis in cellulosic fibers. The microfibril angle is inversely proportional to fiber stiffness. The microfibrillar angle of bast fibers is generally lower than that of leaf and seed fibers making the bast fibers stiffest of all [56].

While bast fibers provide higher tensile and flexural strength, leaf finers may present better impact performance [70]. Accordingly, Reddy and Yang [12] reported that the work of rupture value of corn husk fibers, which have low stiffness and tenacity values and high elongation ratios, is greater than that of cotton, flax and jute fibers, the stress-strain curves of which are presented in Figure 11.11. Due to their high stiffness, bast fibers constitute the most researched fibers in the composite production area [56].

Mechanical properties of some agro residual fibers are compared with important bast, leaf and seed fibers in Table 11.4.

Similar to conventional plant fibers, tenacity and stiffness of plant fibers are lower than that of E-glass fiber, the commonly used reinforcement element. Nevertheless, if the fact that the densities of plant fibers (~1.4 g/cm^3) are lower than E-glass (~2.5 g/cm^3) is considered, it will be obvious that the specific tenacity and the specific stiffness of plant fibers becomes comparable to that of E-glass [70].

The mechanical performance of plant fibers is affected negatively from higher temperatures in the presence of oxygen. This should also be taken into account in designing the manufacturing and the use of biocomposites. However, the matrix which surrounds fibers may shield the plant fibers from the adverse effects to some extent [60].

De Rosa *et al.* [14] reported that the tensile tenacity and modulus of okra fibers presented a two-parameter Weibul distribution. They also reported a decrease in tensile tenacity and modulus with increase in okra fiber diameter, as Bodros and Baley [37] found for nettle stalk fibers. Having brittle structures, okra and nettle stalk fibers present straight stress-strain curves [14,37]. Mechanical properties of agro-residual fibers are listed in Table 11.3.

Corn husk fibers the tenacity of which is around 25% lower than that of jute can produce more durable and tough composites than jute due to the elongtaion ratio and work of rupture which are 13 times and 7 times greater than that of jute, respectively [7].

11.6.3 Some Important Features of Plant Fibers

11.6.3.1 Insulation

Partly due to their porous structures, plant fibers present high thermal, electrical and acoustical insulation performance and they can be incorporated in biocomposites produced for these causes [49]. Huda and Yang [7] measured the sound absorption capacity of corn husk and jute reinforced PP composites. They found their noise reduction is comparable, and higher sound absorption was achieved when the corn husks underwent enzyme treatment before composite manufacturing process. It should be considered that moisture absortion substantially deteriorates heat and electrical insulation performance [71].

11.6.3.2 Moisture Absorption

Agro-based fiber-reinforced composites can absorb a significant amount of moisture, as the water retains in the inter-fibrillar spaces of these fibers [10]. The content of voids and the non-crystalline parts determine moisture absorption [3]. When these fibers are used in composites, the moisture can occupy the spaces in the flaws of the interface between the fiber and the matrix and the micro voids in the composites in addition to the inter-fibrillar spaces of the fibers [10]. The moisture absorption behavior of plant fibers is one of the main concerns related to the utilization in composite applications [24]. The moisture absorption that takes place as a result of the hydrophilic character of the fiber affects the performance of the composite negatively. Moisture uptake results in fiber swelling and this consequently changes the dimensional stability of the

Table 11.3 Mechanical Properties of Agro-Residual Fibers Compared to E-Glass and Some Important Conventional Plant Fibers.

Fiber Type	Tensile Strength (MPa)	Specific Strength (MPa.cm^3/g)	Elastic Modulus (GPa)	Specific modulus (GPa. cm^3/g)	Failure Strain (%)	Reference
E-glass	*	*	70-73	29	2.5	*
Oil palm	50-400	32-259	1-9	0.6-5.8	14	49,50
Okra	184-557	127-384	8.8-25	6.1-17	1.9	14,24
Nettle	300-1800	417-2500	87	120	0.7-5.6	38,
Banana bunch	62-160	107-275	2-7	3.4-12	3.10	27,30
Corn husk	283-307[a]	224-242	7.7-8.6	6.1-.68	13-17	12
Celery	33-36	*	2	*	5.7	53
Wheat straw	58-146	*	2.8-7.9	*	*	67
Coir	77-234	92-152	0.5-6	5-6.5	15-43	52,53,68,*
Pineapple	338-1627	235-1130	4-82	3-57	1.6	68
Flax	345-1100	345-620	28-80	34-57	2.7-3.2	*
Hemp	310-750	210-510	30-70	20-47	1.6	*
Jute	200-773	140-320	10-55	7-39	1.16-1.5	*
Kenaf	295-1191	246-993	22-60	18-50	1.6	*
Ramie	400-938	*	61-128	*	1.2-3.8	*
Banana	529-914	392-677	27-32	20-24	*	*
Sisal	468-840	55-580	9-38	6-29	3-7	*
Abaca	430-760	*	*	*		*
Cotton	300-700	194-452	6-12	4-8	7-8	*

Note: *: Data obtained from Drzal *et al.* [69] and Aziz and Ansell [62].

[a] : calculated according to MPa value = 10*ρ*cN/tex = 10*9*. ρ*g/den Taking the density of corn husk fiber to be 1.266 g/cm^2, and assuming that of nettle fiber to be 1.5 g/cm^2 Tex is gram weight of fiber of 1 km length. Denier (den) is gram weight of fiber of 9 km length.

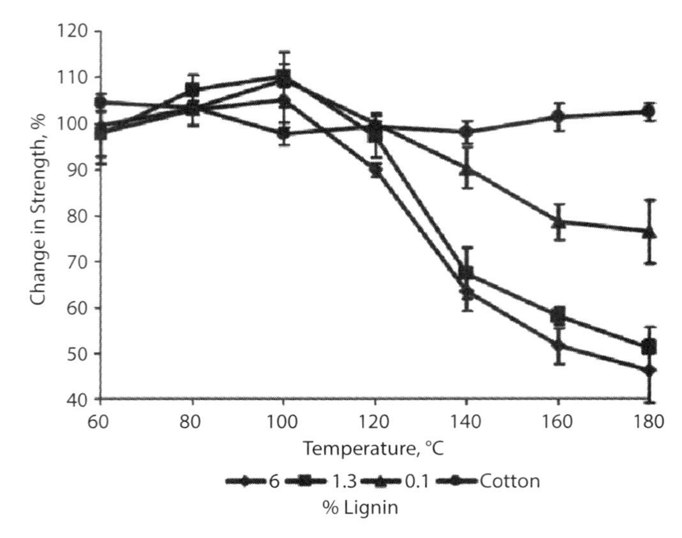

Figure 11.9 Effect of lignin on the breaking tenacity of corn husk fibers with three different lignin contents at various temperatures in comparison to cotton fibers. (From N. Reddy *et al.*, *Macromolecular Materials and Engineering*, 2007 [20]. With permission from John Wiley and Sons).

biocomposite together with its physical and mechanical properties [56]. Therefore, it is very important to understand of the hygroscopic properties of natural fibers in order to improve their long-term performance as reinforcement elements of composites [24].

The mechanisms that take place during water absorption and deteriorate the mechanical performance of the biocomposite may be (a) development of internal stress inside the bicomposite due to swelling of agro-residue fibers, (b) damage of interfacial bonding between agro-residual fibers and polymer matrix and (c) degradation of fibers due to long-time exposure to water [10].

Panthapulakkal and Sain [10] reported that the water absorption of wheat straw-filled high-density polyethylene (HDPE) composite to be lower than corncob-filled HDPE and greater than corn straw-filled composite. The greater amount of water absorption recorded by corncob-filled composites was attributed to the greater amount of hemicellulose present in corncob. When used in composites, incorporation of compatibilizer may decrease water uptake; this may consequently enhance the mechanical properties of the composites upon being subjected to water.

When determining water uptake behavior of composites, short time as well as long time absorption should be recorded [10]. De Rosa *et al.* [15] reported varying moisture absorption ratios of untreated and treated okra bast fibers upon immersion in water for different durations.

De Rosa *et al.* [15] reported a decrease in the moisture content of okra bast fibers after alkalization, bleaching, acetylation treatments. They found the water absorption ability of okra fibers increased after mentioned treatments.

11.6.3.3 Dimensional Stability

Plant fibers are not dimensionally stable under changing moisture conditions as a consequence of the hygroscopicity of the fibers. However, dimensional stability can be

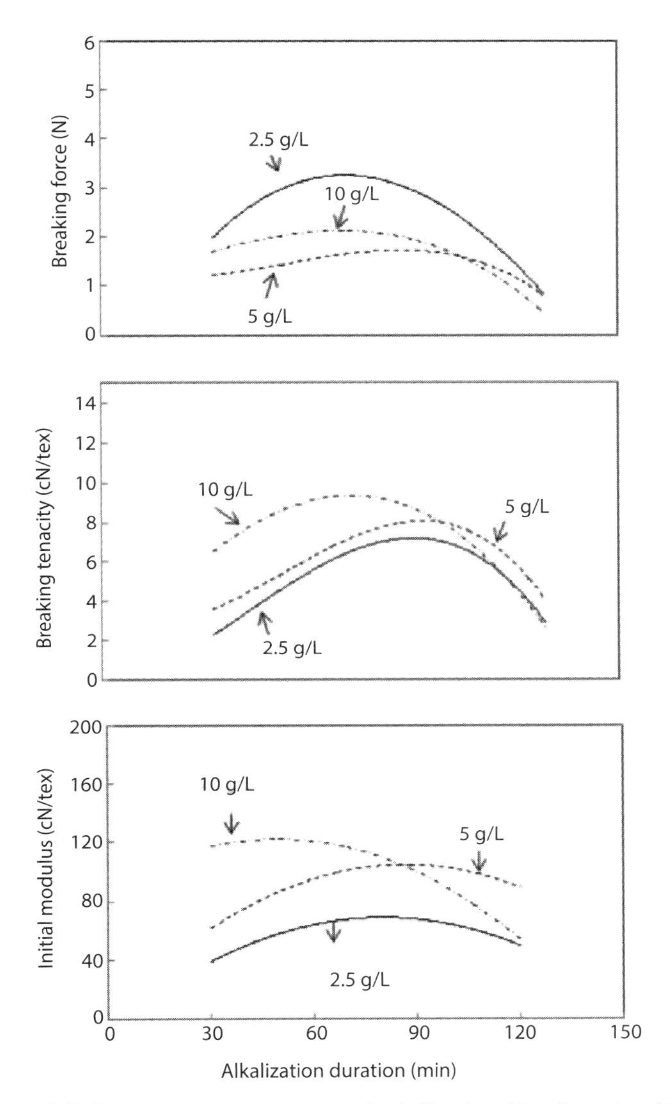

Figure 11.10 Effect of alkalization parameters on corn husk fiber breaking force, breaking tenacity and initial modulus for 2.5, 5 and 10 g/L alkali concentrations. (From N. D. Yılmaz, *Indian Journal of Fibre and Textile Research*, 2013 [11]. With permission from *Indian Journal of Fibre and Textile Research*).

controlled by a number of known treatments such as heat treatments or chemical treatments which at the same time reduce hydrophilicity [6].

11.6.3.4 Thermal Stability

Degradation of plant fibers at elevated temperatures limits their use as reinforcement in composites [56]. Thermal stability of the agro-residual fibers determines the suitability of processing methods and the selection of thermoplastic matrix polymers [10]. As the agro-residual fibers start to decompose at around 200°C, the processing temperature should be lower than 200°C, i.e., the melting point of the thermoplastic polymer should not exceed 200°C such as polyolefins, [10]. Moreover, prolonged processing durations

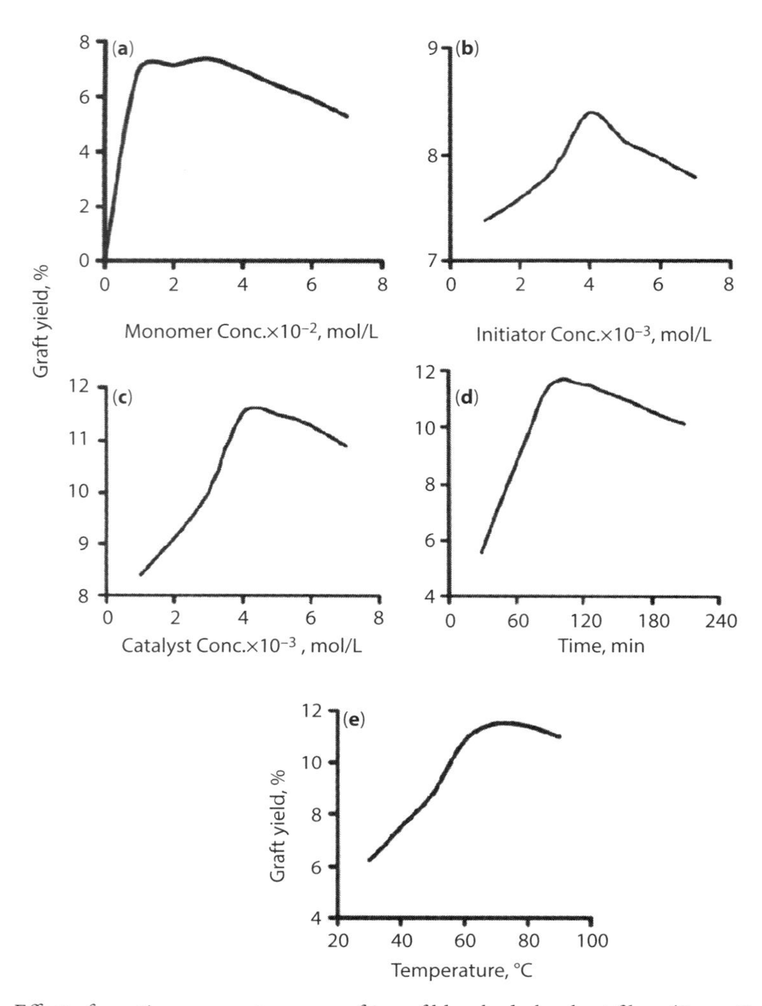

Figure 11.11 Effect of reaction parameters on grafting of bleached okra bast fiber. (From G. M. A. Khan *et al., Indian Journal of Fibre and Textile Research*, 2009 [23]. With permission from *Indian Journal of Fibre and Textile Research*).

should be avoided to prevent degradation [56]. The presence of extracellulosic materials negatively affects thermal stability. Moreover, higher temperatures impart brittleness to fibers [20].

Reddy *et al.* [20] reported the tensile tenacity of corn husk fibers increased by 10% when treated at 100°C for 90 minutes. Over this temperature the fibers continuously lose strength as shown in Figure 11.9. Upon subjecting to 180°C for 90 minutes, untreated corn husk fibers loses 50% of their tenacity, while delignified corn husk fibers loses around 20 to 30%. Lignin has a negative effect on heat resistance of fibers. Heat exposure also leads to loss in elongation do to increased brittleness. Reddy *et al.* [20] reported 40% and 30% decrease in the elongation of untreated and delignified corn husk fibers, respectively after 90 minute treatment at 180°C. They also reported that the thermal treatment increased yellowness of corn husk fibers. This effect is more pronounced in case of untreated fibers compared to delignified fibers.

Three weight loss steps were reported for corn husk fibers as in the range of 24°C-120°C, 195°C-340°C and 340°C-600°C. Upon removal of extra-cellulosic materials with alkalization treatment, thermal degradation of corn husk fibers shifted to over 208°C [11]. Similar to corn husk fibers, okra fibers have three thermal weight loss stages: 30°C-110°C, 220°C-310°C and 310°C-340°C. De Rosa *et al.* [14] attributed the first stage to water vaporization, associated the second stage with depolymerization of hemicellulose, pectin and the breaking of glycosidic linkages of cellulose, and the third stage to the degradation of cellulose. They reported that lignin, which consists of branched aromatic rings, degraded slowly during the whole temperature range. Okra fibers are durable until the temperature of 220°C. Decomposition temperatures of some agro residual fibers are compared with conventional plant fibers in Table 11.4.

11.6.3.4 Photo Degradation

Depending on the type and location of light source and the form of the material, cellulosic materials lose strength upon exposure to light. The reason to the decrease in strength should be the damage occurred at the interactions of cellulose, lignin and hemicellulose and to the deterioration of these polymers themselves. The degree of polymerization of cellulose diminishes. Moreover, lignin absorbs light, transfers it to cellulose and consequently accelerates photo degradation. The higher the lignin content, the higher strength loss. When exposed to 160 AATCC Fading Unit (AFU) of light, untreated corn husk fibers lost 65%(40%) of their tenacity (elongation), while the loss was 30%(10%) in the case of delignified fibers [20]. They also reported a decrease in yellowness upon exposure to light in corn husk fibers. They attributed this to photodegradation of some groups in lignin to colorless groups.

11.6.3.5 Microbial Resistance

Agro-residual fibers, similar to other plant fibers have moderate microbial resistance and thus they tend to decay. Microbial resistance is especially important during shipment and long-term storage [56]. Microbial resistance of agro-residual fibers can be enhanced by increasing hydrophobicity [72].

11.6.3.6 Variability

High variability is a major problem related with agro residual fibers similar to other plant fibers. Namely, Yılmaz [13] reported the coefficient of variation of corn husk fibers' properties to range between 19% and 52%. There are numerous factors to influence fiber quality. These include agricultural parameters including fiber crop variety, location, climate, ripeness at the time of harvesting, the method of harvesting, together with fiber processing methods such as extraction, separation methods [14,56], the portion of the plant from where the fibers are obtained [38]. The variation in properties of agro residual fibers might be reduced by subjecting them to chemical treatments [73].

Fiber quality and fiber yield is affected by fiber cultivar. For example the nettle clone bred by Bredeman had higher fiber content of 16% compared to that of wild nettle as 4-5% [39]. The chemical, physical and mechanical characteristics of fibers differ according to the portion of the plant they are extracted from. Bacci *et al.* [39] reported

Table 11.4 Decomposition Temperatures of Some Plant Fibers.

Plant Fiber	T_o(°C)	T_p(°C)	Reference
Okra	220	359	14
Corn Husk	200	340	11,13
Curaua	230	390	14
Kenaf	219	335	14
Jute	205	284	14

T_o(°C):Temperature of initial decomposition, T_p(°C):Maximum decomposition temperature, (Adapted from I. M. De Rosa *et al.*, *Composites Science and Technology*, 2010 [14]. With permission from Elsevier).

the fiber diameter and lignin content of fiber decreased from 47 μm and 4.4% to 19 μm and 3.5%, respectively from the bottom to the upper part of the stalk, while fiber tenacity increased from 24 to 60 cN tex^{-1}. The fiber content of nettle stalk was found to be the highest in the middle portion.

11.6.3.7 Reactivity

The hydroxyl groups present in the cell wall, which are responsible of the moisture absorption problem, renders agro-residual fibers very reactive. This characteristic allows the fibers to undergo chemical modifications to improve durability and dimensional stability performance [6]. Some physical properties of agro-residual fibers are listed in Table 11.4 and characterization methods of agro-residuals fibers are given in Table 11.5.

11.7 Modification of Agro-Based Fibers

For the improvement of the performance of the biocomposite different alternatives exist including fiber modification, matrix modification, and coupling agent addition. Fiber modification may concern alterations on the physical and chemical properties by causing changes in the surface, chemical composition, and moisture uptake behavior of the plant fiber [26]. Research studies on physical and chemical treatments of agro-residual fibers are listed in Table 11.1.

11.7.1 Physical Treatments

While not changing the chemical composition of the fiber extensively, physical treatments cause variations in structural and surface properties of the fiber and consequently affect the mechanical bonding to the polymer matrix. Thermal treatment, corona and plasma treatments can be given as examples to physical treatments applied on plant fibers [3]. Ragoubi *et al.* [33] reported an increase in mechanical and thermal properties of reed fiber-reinforced PLA and PP composites upon corona discharge treatment of fibers.

Table 11.5 Physical Properties of Agro-Residual Fibers Compared to Some Conventional Fibers.

FiberType	Apparent Density (kg/m^3)	Length (mm)	Diameter (mm)	Aspect Ratio (l/d)	Moisture Content (wt%)	Micro-fibril angle (θ)	Chrystallinity	Reference
E-glass	2550	*	*	*	*	*	*	*
Okra	1450	*	0.04-0.18	*	*	*	*	14,15
Nettle stalk	720	43-58	0.019-0.047	*	*	*	*	37,39
Sunflower stalk	*	1.18	0.022	55	*		*	16
Wheat straw		1.9-3	0.08-0.09		*	*		67
Corn stalk	*	1.22	0.024	50	*	*	*	16
Bagasse	*	1.24	0.023	54	*	*	*	16
Banana bunch	581	63.5-160	0.015-0.12	500-10.000	*	*	*	30
Oil palm fiber	700-1550	6	0.15-0.50	12-40	*	46	*	49,50
Banana	700-1360	*	0.15	*	10-12	11-12	60	28
Corn husk	1266	200	*	300	8-12	*	38-47	7,12,13,58
Coir	1150-1460	*	0.027-0.028	*	8	30-49	*	53
Flax	1500	9-70	0.04-0.6	*	8-12	5-10	*	*
Flax (Fiber bundle)	*	250-1200	0.005-0.038	*	*	*	*	*
Hemp	1500	5-55	0.01- 0.05	*	6-12	2-6.2	*	*
Hemp (fiber bundle)	*	1000-4000	0.5-5	*	*	*	*	*
Jute	1300-1500	2-5	0.010-0.025	*	12-14	7-9	68	12
Jute (fiber strand)	*	1500-3600	*	*	*	*	*	*
Kenaf	1200	*	*	*	*	*	*	*
Ramie	1500-1550	*	*	*	8-17	7-8	*	*
Pineapple	1440	*	*	*	11.8	6-14	*	*
Sisal	1450	*	50-200	*	10-22	10-22	*	*
Abaca	1350	*	*	*	5-10	*	*	*
Cotton	1550	10-40	0.012-0.038	*	7.5-8.5	20-34	*	*
Bamboo	*	1.5-4.	0.007-0.027	*	*	*	*	*

*: Data obtained from Drzal *et al.* [69], Aziz and Ansell [62], Olesen, and Plackett [6] and Bismarck *et al.* [56].

11.7.2 Chemical Treatments

Some chemical treatments may act as removing extra-cellulosic materials. Others, like those utilizing silane or maleated agents, provide a surface coating via a chemical bond reducing the hydrophilicity of the fiber and improving the adhesion to the polymer matrix [15]. Chemical treatmens lead to decrease in variation of properties, as Moniruzzaman *et al.* [73] reported for okra fibers. They also reported that different chemical treatments lead to further separation of okra fibrils, which hold loosely together after retting.

11.7.2.1 Alkalization

Alkaline treatment (mercerization) is one of the most common chemical methods applied to plant fibers which are used as reinforcement in biocomposites. Alkalization partially removes lignin, wax and oils covering the surface of fiber cell wall [3]. Thus, alkalization improves the surface adhesion of fiber to the polymeric matrix with the removal impurities which results in a rough surface topography [8].

Alkalization leads to the loss of hemicellulose and lignin resulting in weight loss of fibers and increase in cellulose content. Properly applied alkalization enhances the tensile performance and thermal resistance of agro-residual fibers. Increasing the concentration, duration and temperature of alkalization treatment results in a decrease in fiber diameter. However, if the applied treatment is too strong that leads to strength loss of fibers [11]. Mahato *et al.* [55] reported decrease of crystallinity upon alkalization of coir fibers.

Decrease in water absorption behavior is reported to occur with alkalizing of okra bast and banana stem and bunch fibers [5,26]. Ganan *et al.* [26] reported an increase in contact angle and decrease in surface free energy. Yılmaz [11] also found a decrease in moisture content of corn husk fibers with the increase in concentration and duration of alkalization. The decrease in moisture content might be due to the reduction in hydroxyl groups.

Khan *et al.* [5] found that the breaking load of okra fibers decreased with the applied alkalization treatment. Similarly, Yılmaz [11] found the breaking load to decrease with increasing alkali concentration. On the other hand, with the increase in treatment duration, breaking force, tenacity and the modulus increased up to a point and then decreased with further duration increase as shown in Figure 11.10. The extra-cellulosic materials in the fibers, which are removed during the alkalization process, have lower strength and extensibility compared to cellulose. They do not contribute to load sharing but lead to crack initiation. Removal of these extracellulosic materials might be the reason to enhancement of tensile performance of fibers at moderate alkalization conditions. The deterioration in tensile properties at longer durations might caused by the harm done to the cellulosic bone itself.

Yılmaz [13] reported that alkaline extracted corn husk fibers posed higher thermal resistance with the onset of the major degradation step of 210 – 225°C compared to water retted and enzyme treated corn husk fiber with that of 185°C. This might be due to the stronger effect of alkalization in removing extra-cellulosic material. She also reported lower yellowness of alkalized corn husk fibers compared to water retted ones which also

Table 11.6 List of Characterization Methods Applied on Agro-Residual Fibers.

Characteristic	Test method	Name	Fiber	Reference
Whiteness and Yellowness	E313-98	ASTM standards on color and appearance measurement	Corn husk	12,13
Color	CieLAB		Oil palm empty bunch	48
Chemical constituents	TAPPI	Cellulose, hemicelluloses, lignin, pectin, wax contents	Okra	22
Chemical constituents		Van Soest's Method	Nettle stalk	38
Cellulose content		Norman and Jenkins Method	Corn husk	12
Cellulose content	AOAC 938.18	Glucose in cacao products. Zerban-sattler method	Corn husk	7,20
Cellulose content	TAPPI procedure T203 om-93		Wheat straw	67
Lignin content	ASTM D1106-96	Standard Test Method for Acid Insoluble Lignin in Wood	Corn husk, banana bunch	7,20,25
Lignin content	TAPPI standard T222		Wheat straw	72
Ash content	ASTM E1755-01	Standard Test Method for Ash in Biomass	Corn husk	7,20
Ash content	ASTM D1102	Standard Test Method for Ash in Wood	Banana bunch	25
Hot water extractive content	ASTM D1110	Standard Test Methods for Water Solubility of Wood	Banana bunch	25
Ethanol toluene extractive content	ASTM D1107	Ethanol-Toluene Solubility of Wood	Banana bunch	25
Moisture content	ASTM D 2495-07	Standard Test Method for Moisture in Cotton by Oven-Drying	Corn husk	11,58
Moisture regain	ASTM D 2654	Test Methods for Moisture in Textiles	Corn husk	12
Humidity	ASTM E871	Standard Test Method for Moisture Analysis of Particulate Wood Fuels	Banana bunch	25
Water absorption			Okra bast, banana	15, 28
Gravimetric water absorption			Okra bast	24
Chrystallinity, christal size and orientation		X-ray diffraction (XRD)	Corn husk, coir, banana	12,20,28,55
Thermal stability		Thermogravimetry (TG), differential thermogravimetry (DT)	Corn husk, okra bast, wheat straw, coir, bagasse	11, 13 14, 15, 28, 45, 54, 57, 67
Thermal stability		Differential scanning calorimetry (DSC)	Coir, bagasse	45,55
Chemical structure		Fourier transform infrared spectrometry (FTIR)	Corn huck, okra bast, wheat straw, coir, bagasse	5, 11, 13, 14, 15, 22,23, 26, 27, 45, 55, 57
Tensile properties	ASTM D 3822	Standard Test Method for Single Textile Fibers	Corn husk	11,13,58

Table 11.6 (contnued)

Tensile properties	ASTM D3379-75(1989)e1	Standard Test Method for Tensile Strength and Young's Modulus for High-Modulus Single-Filament Materials	Okra bast, nettle	14,15,37
Tensile properties			Banana bunch, Nettle stalk, coir, celery fiber	27,39,52,53, 54,57,67
Tensile properties	ISO 5081:1977	Textiles – Woven fabrics – Determination of breaking strength and elongation (Strip method)	Okra bast	23
Tensile properties	ISO 5079:1998	Textile fibers – Determination of breaking force and elongation at break of individual fibers	Nettle stalk	38
Tensile properties		Nanoindentation	Reed	32
Morphology		Scanning Electron Microscopy (SEM)	Okra bast, corn husk, oil palm empty bunch fiber, coir, celery fiber	5,12,14,15,23,27,37,48, 52-54
Morphology		Ligth Microscopy (SEM)	Corn husk	12,27
Morphology		Environmental Scanning Electron Microscope(ESEM)	Okra bast	15
Light resistance	AATCC 169	Weather Resistance of Textiles: Xenon Lamp Exposure	Corn husk	20
Linear Density	ASTM D 1577-07		Corn husk	11,13,58
Density		Sink-float method	Corn husk	7
Density		Piconometric method	Banana bunch	28,30
Density		Immersing in water	Nettle stalk, banana	28,39
Degree of polymerization		Viscosity	Coir	54
Dynamic contact angle		Wilhelmy plate technique	Banana stem and bunch	26
Surface free energy		Owens-Wendt method	Banana stem and bunch	26
	ISO 105-B02:2013	Textiles – Tests for color fastness – Part B02: Colour fastness to artificial light: Xenon arc fading lamp test	Okra bast	23
Micromechanical deformation		Raman spectorscopy	Coir, celery fibers	53

proved the strong effect of alkalization in removing materials that impart a yellow color to the fiber. To the contrary, Alam *et al.* [22] reported that alkalization imparted yellowish color and crimp in the okra bast fiber. If the fibers underwent strong alkalization treatments, they become prone to get damaged by subsequent enzyme treatments [13].

11.7.2.2 Acetylation

The acetylation treatment decreases the hydrophilic nature of the plant fibers. During the acetylation treatment the hydroxyl groups in the cellulosic fiber are replaced with acetyl functional groups. The decrease in hydrophilicity results in lower moisture intake and stronger interfacial bonding [8]. De Rosa *et al.* [15] obtained lower water content in acetylated okra fibers compared to untreated ones. Hill and Khalil [72] reported that the acetylation treatment enhanced the bioresistance of coir and oil palm fiber-reinforced polyester composites.

11.7.2.3 Silane Treatment

By forming stable covalent bonds, the silane treatment would improve the contact angle against water by increasing the hydrophobicity. The mechanism of this treatment can be explained as a silane chemical reacts with water and forms a silanol and an alcohol. With the moisture present, the silanol reacts with the hydroxyl groups of the cellulose and bonds itself to the cell wall [8].

Decrease in water absorption behavior is reported to occur with silanizing of banana stem and bunch fibers [26]. Ganan *et al.* [26] reported an increase in contact angle and decrease in surface free energy and polar component. They found silane deposition on fibers surface according to Fourier transform infrared spectrometry (FTIR) analysis.

11.7.2.4 Bleaching

Bleaching causes the fiber parameters to become more uniform [5]. Due to removal of some extra-cellular materials, the degree of chrystallinity increases [28], fibers acquire a whiter color and the inherent yellowness is decreased. Fibers lose some weight and become finer [12]. Increase in water absorption was reported to occur with bleaching of okra bast fibers. This might be due to the removal of hydrophobic substances which exposes hydrophilic cites [5]. Contrarily, the water absorption of banana fibers was reported to decrease from 61% to 45% for 70 hour water immersion upon sodium hypochlorite bleaching [28].

In terms of tensile properties, breaking load of fiber decreases due to loss of some material [5]. Breaking tenacity of the fibers may decrease or increase depending on the strength of bleaching conditions. In the case of a mild treatment, some weak parts of the fiber are removed and thus the load per area may be increased. On the other hand, during a stronger treatment, the cellulosic chain itself may be damaged through oxidation [12]. Khan *et al.* [5] reported the breaking load of fibers to decrease with bleaching, whereas Reddy and Yang [12] found an increase in fiber tenacity after bleaching.

11.7.2.5 Enzyme Treatment

The utilization of enzymes in the natural fiber modification field is rapidly increasing. This trend may be due to the environmentally-friendly nature of enzyme treatments as the catalyzed reactions are very specific and the performance is very focused [3]. Several enzymes have been used in order to enhance the effectiveness of extracting fibers from the agro-residue or fine-tune the properties of extracted fibers such as lowering their diameter. Xylanases, cellulases, pectinases are the enzyme types that have found more use in agro-residual fiber modification [12,13,38].

Xylanases depolymerize hemicellulose and breaks the covalent bond between lignin and cellulose. The depolymerized hemicellulose and separated lignin may be removed exposing the celulosic fiber and increasing cellulose content. Cellulases are enzyme complexes including enzymes that attack the cellulose chains at random, that hydrolyze the cellulose chains from the end and that hydrolyze the cellulobiose into glucose [12]. Yılmaz [13] detected stronger effect of xylanases on fiber properties compared to that of cellulase enzymes.

During fiber extraction, enzymes alone may be unable to break the outside layer of protective material to separate the fibers as they do not act as effectively as alkali chemicals during removal of extra cellulosic materials. Therefore enzymatic treatment may be applied subsequent to other fiber extraction processes [12,13,38]. Yılmaz [13] reported that xylanase enzyme treatment increased the initial moduli and decreased the elongation at break values of fibers extracted by alkalization. Yılmaz [13] attributed this situation to the removal of nonload bearing mass. In the absence of extra-cellulosic materials, the cellulose chains positioned in the loading direction. This resulted in a higher molecular orientation and a packing density of higher order which in turn led to greater stiffness of the fiber. Huda and Yang [7] reported that the crystallinity of alkali-extracted corn husk fibers increased from 39.5% to 46.4% upon enzyme treatment. Reduction in fiber diameter by separating constituents of fibers which are weaker than cellulose might have also played a role in the increase of the initial modulus.

There have been limited amount of research devoted to enzymatic treatment of agro-residual fibers [12,13,58]. When comparing different extraction procedures including water retting, alkalization and enzyme treatment, Yılmaz [13] found the strongest and finest corn husk fibers to be produced by a fiber production route of water retting followed by enzyme treatment.

Yılmaz *et al.* [58] found that when increasing the concentration in xylanase enzyme treatment of corn husk fibers, the breaking tenacity and initial modulus values are increased up to a point and then started to decrease with further increase in concentration. The initial increase in tensile properties might be due to the removal of extra-cellulosic material which do not contribute to load sharing well. Huda and Yang [7] reported that the cellulose content in corn husk which is 42% to increase to 65% with alkalization and to 84% with a subsequent enzyme treatment. Decrease of tensile properties with further increase in enzyme concentration might be because enzyme treatment at higher concentrations starts to damage the cellulosic backbone.

Bacci *et al.* [38] treated water retted nettle stalk fibers with Viscozyme® and Pectinex® Ultra SP-L enzymes. Viscozyme ® is a multienzymatic solution consisting of arabase, cellulase, b-glucanase, hemicellulase and xylase enzymes. Pectinex ® Ultra SP-L is an active pectolytic enzyme preparation. They carried out the enzme treatment with and without EDTA (ethylenediaminetetraacetic acid). They applied the enzyme treatment in vat form and also in spray form. They reported that EDTA has a positive effect in fiber diameter reduction. They attributed this condition to the fact that EDTA, a chelating agent, contributed to the fracture of cell walls by chelating metal ions from the pectin complex and thus increased retting efficiency. The enzymatic treatment resulted in tenacity reduction when applied to mechanically separated fibers. They obtained better mechanical properties from fibers treated with Pectinex® Ultra SP-L compared to those treated with Viscozyme® L in the presence of EDTA. The enzyme treatment carried out in spray form resulted in coarser fibers with lower tenacity and cellulose content compared to vat enzyme treatment method.

11.7.2.6 Sulfonation

Reddy *et al.* [20] used used sodium sulfite and ethylenediamine to delignify the corn husk fibers. The treatment decreased the lignin content from 6% to 0.1% and the diameter of the fibers was reduced by 35% and the modulus of the fiber was increased by 50%.

11.7.2.7 Graft Copolymerization

The compatibility between the hydroxyl group containing polar cellulose fibers and apolar polymer matrices may not allow for production of high mechanical performance biocomposites. In order to overcome this difficulty, a third material, a coupling agent can be introduced that modifies the interface morphology and the acid-base reactions in the interface, the surface energy and wetting efficiency [8].

The use of maleated coupling agents has received intense research interest. To give an example, maleic anhydride-grafted polypropylene (MAPP or PP-g-MA as shown in Faruk *et al.* [3] has the same molecular structure as polypropylene where the maleic anhydride group is attached to the backbone. Maleated coupling agents target the cellulose fiber as well as the polymer matrix resulting in higher mechanical properties by improving the interfacial bonding [8].

Graft copolymerization treatments using silane or maleated agents, form a surface coating via chemical bonding. This improves the interfacial adhesion between the fiber and the matrix polymer also by impeding fiber's hydrophilic character [15].

Khan *et al.* [23] reported that acrylonitrile, as a vinyl monomer, is promising in terms of graft polymerization with plant fibers. Ionic, radical and chemical initiation systems can be utilized. Among these, the chemical initiation of grafting, which involve oxidizing agents, such as KMnO4, Na2S2O3, K2S2O8, Ce (IV) and V(V), is advantageous economically and quite selective in nature.

The weight and the diameter of fibers increased and the color of the fibers turned to cream after graft copolymerization of okra bast fibers with acrylonitrile monomer [5]. Khan *et al.* [5] reported the optimum acrylonitrile monomer grafting conditions on okra fibers to be a monomer concentration of 3×10^{-2} mol/l, an initiator concentration of 5×10^{-3} mol/l, a catalyst concentration of 5×10^{-3} mol/l, a temperature of 70°C temperature and a reaction time of 90 minutes to obtain a graft yield of 11.43% as shown in Figure 11.11. A substantial decrease in water absorption from 62% to 20% is reported to occur with graft copolimerization of okra bast fibers. Furhthermore, acrylonitrile copolymerization resulted in 30% increase in breaking load of okra fiber [5]. Acrylamid copolymerization led to an increase of 20% in tensile strength and 10% in elongation rate of coir fibers [54], whereas Zaman *et al.* [29] obtained a 72% increase in tensile strength of coir fiber due to methyl acrylate grafting.

Ashori and Nourbakhsh [16] obtained tensile, flexural and impact property increases as high as 50% of corn stalk, sunflower stalk and bagasse fiber-reinforced polypropylene composite due to incorporation of maleic anhydride polypropylene compatibilizer.

Panthapulakkal and Sain [10] reported that incorporation of compatibilizer resulted in 15-25% increase in tensile strength of wheat straw, corn stalk or corncob-filled HDPE composites and 20% increase in flexural strength of wheat straw-filled HDPE composites.

11.8 Conclusion

For the last two decades, lignocellulosic fibers have started to be considered as alternatives to conventional manmade fibers in the academic as well as commercial arena, for a number of areas including transportation, construction, and packaging applications. Some species of plants such as hemp, flax, kenaf, jute, and sisal, which have been utilized as fiber sources since historical times, once again has become the focus of research attention. Nevertheless, agricultural plants have solely been utilized as food source. Most recently, the possibility of making use of agricultural byproducts have gained interest. Some of these agricultural byproducts are rice and wheat straws; corn husks; corn, okra, cotton, and reed stalk fibers; banana bunch fibers. Utilization of these byproducts would benefit rural development by providing additional value to the agricultural activities, the environment by saving the field leftovers from burning, which is unfortunately a common practice, and the growing world population by preserving the land for the production of edible plant species.

This review investigated agro-based fibers as potential reinforcement elements in biocomposites. Methods of fiber extraction from agricultural residues have been explained, characteristics of fibers with resources of different agricultural byproducts have been compared, fiber modification techniques have been presented and research efforts devoted to utilize these agro-based fibers have been listed.

References

1. K. Kavelin, Investigation of Natural Fiber Composites Heterogeneity with Respect to Automotive Structures, Delft University of Technology, The Netherlands, PhD Dissertation, (2005).
2. I.I. Negulescu, Y. Chen, O.I. Chiparus, D.V. Parikh, *International Development of Kenaf and Allied Fibers, Part II, p.* 84, (2003).
3. O. Faruk, A.K. Bledzki, H.-P., Fink, and M. Sain, Biocomposites reinforced with natural fibers: 2000–2010. *Prog. Polym. Sci.* 37, 1552–1596 (2012).

4. A.K. Mohanty, M. Misra, and L.T. Drzal, Sustainable bio-composites from renewable resources: opportunities and challenges in the green materials world. *J. Polym. Environ.* 10, 19–25 (2002).
5. G.M.A. Khan, M.D. Shaheruzzaman, M.H. Rahman, S.M. Abdur Razzaque, M.D. Sakinul Islam, and M.D. Shamsul Alam, Surface modification of okra bast fiber and its physico-chemical characteristics. *Fibers Polym.* 10, 65–70 (2009).
6. P.O. Olesen and D.V. Plackett, Perspectives on the performance of natural plant fibres, *Proceedings of the Natural Fibres Performance Forum*, Copenhagen, Denmark (1999).
7. S. Huda and Y. Yang, Chemically extracted corn husk fibers as reinforcement in light-weight poly(propylene) composites. *Macromol. Mater. Eng.* 293, 235–243 (2008).
8. R.G.C. Murdy, Exploring Cornstalk and Corn Biomass Silage Retting as a New Biological Fibre Extraction Technique, University of Waterloo, Ontario, Canada, PhD Dissertation (2013).
9. M. Pervaiz, M. Sain, and A. Ghosh, Evaluation of the influence of fibre length and concentration on mechanical performance of hemp fibre reinforced polypropylene composite. *J. Nat. Fibers* 2, 67–84 (2006).
10. S. Panthapulakkal and M. Sain, Agro-residue reinforced high-density polyethylene composites: Fiber characterization and analysis of composite properties. *Compos. A* 38, 1445–1454 (2007).
11. N.D. Yılmaz, Effect of chemical extraction parameters on corn husk fibers characteristics. *Indian J. Fibre Text. Res.* 38, 29–34 (2013).
12. N. Reddy and Y. Yang, Properties and potential applications of natural cellulose fibers from corn husks. *Green Chem.* 7, 190–195 (2005).
13. N. D. Yılmaz, Effects of enzymatic treatments on the mechanical properties of corn husk fibers. *J. Text. Inst.* 104, 396–406 (2013).
14. I.M. De Rosa, J.M. Kenny, D. Puglia, C. Santulli, and F. Sarasini, Morphological, thermal and mechanical characterization of okra (Abelmoschus esculentus) fibres as potential reinforcement in polymer composites. *Compos. Sci. Technol.* 70, 116–122 (2010).
15. I.M. De Rosa, J.M. Kenny, M.D. Maniruzzaman, M.D. Moniruzzaman, M. Monti, D. Puglia, C. Santulli, and F. Sarasini, Effect of chemical treatments on the mechanical and thermal behaviour of okra (Abelmoschus esculentus) fibres. *Compos. Sci. Technol.* 71, 246–254, (2011).
16. A. Ashori and A. Nourbakhsh, Bio-based composites from waste agricultural residues. *Waste Manage.* 30, 680–684, (2010).
17. J. Larsen, Bumper 2011 Grain Harvest Fails to Rebuild Global Stocks, Eco-Economy Indicators (2012), http://www.earth-policy.org/indicators/C54/grain_2012
18. J. Larsen, Global Grain Stocks Drop Dangerously Low as 2012 Consumption Exceeded Production, Grain Harvest, Eco-Economy Indicators, (January 17, 2013), http://www.earth-policy.org/indicators/C54/grain_2013
19. S.R. White, N.R. Sottos, and T.M. Mackin, Corn-based structural composites, US Patent 5834105, assigned to The Board of Trustees of the University of Illinois, (November 10, 1998).
20. N. Reddy, A. Salam, and Y. Yang, Effect of lignin on the heat and light resistance of lignocellulosic fibers. *Macromol. Mater. Eng.* 292, 458–466, (2007).
21. Y. Yang and N. Reddy, Natural cellulosic fiber bundles from corn husk and the method for making the same, US patent WO/2007/008228, assigned to Y. Yang and N. Reddy, (January 18, 2007).
22. M.S. Alam and G.M.A. Khan, Chemical analysis of okra bast fiber (Abelmoschus esculentus) and its physico-chemical properties. *J. Text. Apparel Technol. Manage.* 5, 1, (2007).

23. G.M.A. Khan, M.D. Shaheruzzaman, S.M. Abdur Razzaque, M.D. Sakinul Islam, and M.D. Shamsul Alam, Grafting of acrylonitrile monomer onto bleached okra bast fibre and its textile properties. *Indian J. Fibre Text. Res.* 34, 321–327 (2009).
24. D. Saikia, Studies of water absorption behavior of plant fibers at different temperatures. *Int. J. Thermophys.* 31, 1020–1026 (2010).
25. G. Quintana, J. Vasqueza, S. Betancourt, and P. Ganan, Binderless fiberboard from steam exploded banana bunch. *Ind. Crops Prod.* 29, 60–66 (2009).
26. P. Ganan, J. Cruz, S. Garbizu, A. Arbelaiz, and I. Mondragon, Stem and bunch banana fibers from cultivation wastes: Effect of treatments on physico-chemical behavior. *J. Appl. Polym. Sci.* 94, 1489–1495 (2004).
27. P. Ganan, R. Zuluaga, J.M. Velez, and I. Mondragon, Biological natural retting for determining the hierarchical structuration of banana fibers. *Macromol. Biosci.* 4, 978–983 (2004).
28. J.R. Aseer, K. Sankaranarayanasamy, P. Jayabalan, R. Natarajan, and K. Priya Dasan, Morphological, physical, and thermal properties of chemically treated banana fiber. *J. Nat. Fibers* 10, 365–380, (2013).
29. H.U. Zaman, M.A. Khan, R.A. Khan, and S. Ghosgal, Effect of ionizing and non-ionizing preirradiations on physico-mechanical properties of coir fiber grafting with methylacrylate. *Fibers Polym.* 13, 593–599, (2012).
30. A.V.R. Prasad, K.M. Rao, and G. Ragavinirasulu, Mechanical properties of banana empty fruit bunch fiber reinforced polyester composites. *Indian J. Fibre Text. Res.* 34, 162–167, (2009).
31. E. Çalışkan, Fiber Extraction from Agro-Residuals: Properties of Fibers Extracted from Undried and Dried Corn Husks by Alkalization and Enzymatic Treatment, Pamukkale University, Turkey, Undergraduate thesis, under supervision of N. D. Yılmaz (2013).
32. A. Bourmaud and S. Pimbert, Investigations on mechanical properties of poly(propylene) and poly(lactic acid) reinforced by miscanthus fibers. *Compos. A* 39, 1444–1454, (2008).
33. M. Ragoubi, B. George, S. Molina, D. Bienaimé, A. Merlin, J.-M. Hiver, and A. Dahoun, Effect of corona discharge treatment on mechanical and thermal properties of composites based on miscanthus fibres and polylactic acid or polypropylene matrix. *Compos. A* 43, 675–685 (2012).
34. R.M. Johnson, N. Tucker, and S. Barnes, Impact performance of Miscanthus/Novamont Mater-Bi® biocomposites. *Polym. Test.* 22, 209–215 (2003).
35. D. Kraiem, S. Pimbert, A. Ayadi, and C. Bradai, Effect of low content reed (Phragmite australis) fibers on the mechanical properties of recycled HDPE composites. *Compos. B Eng.* 4, 368–374 (2013).
36. V. Nagarajan, A.K. Mohanty, and M. Misra, Sustainable green composites: Value addition to agricultural residues and perennial grasses. *ACS Sustainable Chem. Eng.* 1, 325–333, (2013).
37. E. Bodros and C. Baley, Study of the tensile properties of stinging nettle fibres (Urtica dioica). *Mater. Lett.* 62, 2143–2145, (2008).
38. L. Bacci, S. Di Lonardo, L. Albanese, G. Mastromei, and B. Perito, Effect of different extraction methods on fiber quality of nettle (Urtica dioica L.). *Text. Res. J.* 81, 827–837 (2011).
39. L. Bacci, S. Baronti, S. Predieri, and N. di Virgilio, Fiber yield and quality of fiber nettle (Urtica dioicaL.) cultivated in Italy. *Ind. Crops Prod.* 29, 480–484 (2009).
40. P.K. Bajpai, I. Singh, and J. Madaan, Joining of natural fiber reinforced composites using microwave energy: Experimental and finite element study. *Mater. Des.* 35, 596–602 (2012).
41. P. K. Bajpai, I. Singh, and J. Madaan, Comparative studies of mechanical and morphological properties of polylactic acid and polypropylene based natural fiber composites. *Reinf. Plast. Compos.* 31, 1712–1724 (2012).

42. P.K. Bajpai, D. Meena, S. Vatsa, and I. Singh, Tensile behavior of nettle fiber composites exposed to various environments. *J. Nat. Fibers* 10, 244–256 (2013).
43. L.U., Devi, S.S. Bhagawan, and S, Thomas, Dynamic mechanical properties of pineapple leaf fiber polyester composites. *Polym. Compos.* 32, 1741–1750 (2011).
44. P. Threepopnatkul, N. Kaerkitcha, and N. Athipongarporn, Effect of surface treatment on performance of pineapple leaf fiber–polycarbonate composites. *Compos. B Eng.* 40,628–632 (2009).
45. C.G. Mothe, and I.C. de Miranda, Characterization of sugarcane and coconut fibers by thermal analysis and FTIR. *J. Therm. Anal. Calorim.* 97, 661–665 (2009).
46. S.K. Acharya, P.P. Mishra, S.K. Mehar, and V. Dikshit, Weathering behavior of bagasse fiber reinforced polymer composite. *J. Reinf. Plast. Compos.* 27, 1839–1846 (2008).
47. H.A. Youssef, M.R. Ismail, M.A.M. Ali and A.H. Zahran, Effect of the various coupling agents on the mechanical and physical properties of thermoplastic-bagasse fiber composites. *Polym. Compos.* 29, 1057–1065, (2008).
48. R. Hasibuan, W.R.W. Daud, Quality changes of superheated steam–dried fibers from oil palm empty fruit bunches. *Drying Technol.* 27, 194–200 (2009).
49. M.J. John, K.T. Varughese, and S. Thomas, Green composites from natural fibers and natural rubber: Effect of fiber ratio on mechanical and swelling characteristics. *J. Nat. Fibers* 5, 47–60 (2008).
50. M. Jawaid, H.P.S. Abdul Khalil, and A. Abu Bakar, Woven hybrid composites: Tensile and flexural properties of oil palm-woven jute fibres based epoxy composites. *Mater. Sci. Eng. A* 528, 5190–5195 (2011).
51. H.P.S. Abdul Khalil, M. Jawaid, and A. Abu Bakar, Woven hybrid composites: Water absorption and thickness swelling behaviours. *Bioresources,* 6, 1043–1052 (2011).
52. V.B. Manilal, M.S. Ajayan, and S.V. Sreelekshmi, Characterization of surface-treated coir fiber obtained from environmental friendly bioextraction. *J. Nat. Fibers* 4, 324–333, (2010).
53. B. Bakri and S.J. Eichhorn, Elastic coils: Deformation micromechanics of coir and celery fibres. *Cellulose* 17, 1–11, (2010).
54. G.M.A. Khan and M.S. Alam, Thermal characterization of chemically treated coconut husk fibre. *Indian J. Fibre Text. Res.* 37, 20–26, (2012).
55. D.N. Mahato, B.K. Mathur, and S. Bhattacherjee, DSC and IR methods for determination of accessibility of cellulosic coir fibre and thermal degradation under mercerization. *Indian J. Fibre Text. Res.* 38, 96–100 (2013).
56. A. Bismarck, S. Mishra, and T. Lampke, Plant fibers as reinforcement for green composites in *Natural fibers, biopolymers, and biocomposites,* A.K. Mohanty, M. Misra, and L.T. Drzal, (Eds.), pp. 37–108, *Taylor and Francis,* (2005).
57. M. Sain and S. Panthapulakkal, Bioprocess preparation of wheat straw fibers and their characterization. *Ind. Crops Prod. 23,* 1–8 (2006).
58. N. D. Yılmaz, E. Çalışkan, K. Yılmaz, Effect of xylanase enzyme on mechanical properties of fibres extracted from undried and dried corn husks. *Indian J. Fibre Text. Res.* 39, 60–64 (2014).
59. T. Nishino, Natural fibre sources, in *Green Composites: Polymer Composites and the Environment*, C. Baillie, (Ed.), pp. 49, Woodhead Publishing, England (2004).
60. A.S. Hermann, J. Nickel, and U. Riedel, Construction materials based upon biologically renewable resources—from components to finished parts. *Polym. Degrad. Stab.* 59, 251–261, (1998).
61. L.Y. Mwaikambo, Plant-Based Resources for Sustainable Composites. University of Bath, UK, PhD thesis (2002).

62. S.H. Aziz and M.P. Ansell, Optimizing the properties of green composites, in *Green Composites: Polymer Composites and the Environment, C. Baillie, (Ed.),* Woodhead Publishing, England (2004).
63. M. Sain and S. Panthapulakkal, Green fiber thermoplastic composites, in *Green Composites: Polymer Composites and the Environment, C. Baillie, (Ed.),* Woodhead Publishing, England (2004).
64. W. Kasai, T. Morooka, and M. Ek, Mechanical properties of films made from dialcohol cellulose prepared by homogeneous periodate oxidation. *Cellulose*, 21, 769–776 (2014).
65. P.J. Herrera-Franco, and A. Valadez-Gonzales, A study of the mechanical properties of short natural-fiber reinforcedcomposites. *Compos. B* 36, 597–608 (2005).
66. D. Kretschmann, Natural materials: Velcro mechanics in wood. *Nat. Mater.* 2, 775–776 (2003).
67. S. Panthapulakkal, A. Zereshkian and M. Sain, Preparation and characterization of wheat straw fibers for reinforcing application in injection molded thermoplastic composites. *Bioresource Technol.* 97, 265–272 (2006).
68. M.S.A.S. Buana, P.Pasbaskhsh, K.L. Goh, F. Bateni, M.R.H.M. Haris, *Fibers and Polymers*, Vol. 14, P. 623, 2013.
69. L.T. Drzal, A.K. Mohanty, R. Burgueño and M. Misra, Biobased Structural Composite Materials for Housing and Infrastructure Applications: Opportunities and Challenges. Pat hnet (2006), http://www.pathnet.org/si.asp?id=1076
70. A.K. Mohanty, M. Misra, L.T. Drzal, S.E. Selke, B.R. Harte and G. Hinrichsen, Natural fibers, biopolymers, and biocomposites: An introduction, in *Natural Fibers, Biopolymers, and Biocomposites*, A.K. Mohanty, M. Misra, and L.T. Drzal, (Eds.), pp. 1–34, Taylor and Francis Group, Boca Raton, FL (2005).
71. W.E. Morton, and J.W.S. Hearle, *Physical Properties of Textile Fibers*, The Textile Institute, Manchester, UK (1997).
72. C.A.S. Hill and H.P.S Abdul Khalil, Effect of fiber treatments on mechanical properties of coir or oil palm fiber reinforced polyester composites. *J. Appl. Polym. Sci.* 78, 1685–1697 (2000).
73. M. Moniruzzaman, M. Maniruzzaman, M.A. Gafur, and C. Santulli, Lady's finger fibres for possible use as a reinforcement in composite materials. *J. Biobased Mater. Bioenerg.* 3, 286–290 (2009).

12

Surface Modification Strategies for Cellulosic Fibers

Inderdeep Singh*, Pramendra Kumar Bajpai

Department of Mechanical and Industrial Engineering, Indian Institute of Technology Roorkee, India

Abstract
Lignocellulosic polymer composites (biocomposites) are being considered as a replacement for manmade fiber-reinforced composites in various application areas. But the natural fibers have poor adhesion efficiency with polymeric matrix because of their waxy nature and the presence of moisture content in the lignocellulosic fibers. Therefore, to achieve high compatibility between fibers and matrix, it is important to choose a suitable processing technique for modification of the natural fiber surface. This chapter mainly focuses on the different techniques to improve interfacial properties between natural fibers and matrix to develop high performance polymer composites for different application areas. The chapter covers the effect of different processing techniques on fiber/matrix interface and consequently on the properties of the developed composites. Different surface modification techniques (chemical treatment) of natural fibers and their effect on various properties of the developed composites are discussed in detail in the chapter.

Keywords: Lignocellulosic polymer composites, surface modification techniques, chemical treatment of fibers

12.1 Introduction

During the last few years, cellulosic fibers (natural fibers derived from plants) have attracted widespread attention due to ecological and global energy problems. Researchers are involved in investigating the exploitation of cellulosic fibers in diverse fields such as load-bearing constituents in polymer composites (due to their low cost, their ability to recycle, light weight, ease of availability, their abundant natural resources and good structural properties) [1–4]. Wood is an interesting example of a natural fiber composite; the longitudinal hollow cells of wood are made up of layers of spirally wound cellulose fibers with varying spiral angles, bonded together with lignin during the growth of the tree. Wood is the natural cellulosic composite which is used in many applications due to good structural properties.

**Corresponding author*: dr.inderdeep@gmail.com

Vijay Kumar Thakur, Lignocellulosic Polymer Composites, (271–280) 2015 © Scrivener Publishing LLC

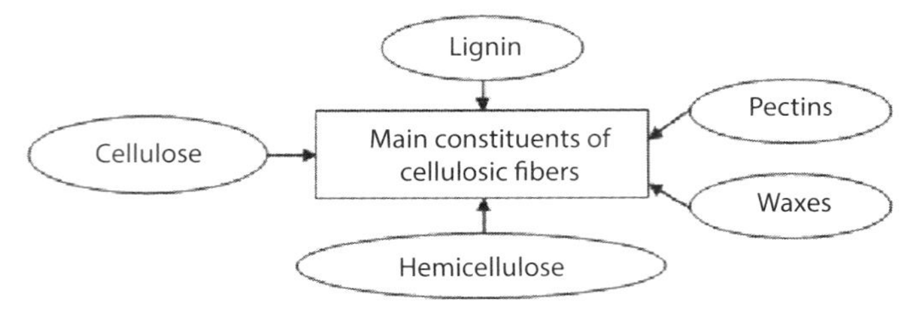

Figure 12.1 Constituents of cellulosic fibers.

The use of cellulosic fibers in polymer composites has increased in the last decade. The reinforcing ability of natural fibers is governed by the nature of cellulose and its crystallinity. Natural fibers are generally lignocellulosic in nature [1]. The main constituents of cellulosic (natural fibers) fibers are shown in Figure 12.1.

Natural fibers are basically derived from three natural resources, which are plants, animals, and minerals. Fibers from plants can be obtained from leaf (sisal fibers), bast (nettle fibers), seed (cotton), fruit (coconut) and wood (hard and soft wood). Silk, wool and feathers are examples of animal fibers. Natural fibers from plants are widely used in fabrication of biocomposites for various applications [5].

Lignocellulosic polymer composites (biocomposites) are of two types: one is partially biodegradable biocomposite and the other is fully biodegradable biocomposite. In partially biodegradable composite, natural fibers are combined with synthetic polymer. An example is sisal/epoxy composite in which sisal fibers are biodegradable but epoxy polymer is non-biodegradable in nature. In the case of fully biodegradable composites, natural fibers are incorporated in biodegradable polymer. An example is sisal/PLA composite in which sisal fibers and PLA polymer are both biodegradable.

The adhesion efficiency between cellulosic fibers and polymer matrix is generally poor because cellulosic fibers are hydrophilic in nature and the surface of these fibers contain waxy impurities which hinders its strong binding with polymer matrix which is generally hydrophobic in nature. The important considerations in developing lignocellulosic polymer composites (natural fiber-reinforced polymer composites) of superior and desired properties are efficient natural fiber treatment, and selection of efficient processing techniques to have strong interfacial adhesion of natural fibers and polymer matrix. There are several surface modification techniques for the treatment of cellulosic fibers to improve the interfacial adhesion between fibers and matrix. These lignocellulosic polymer composites are developed by using any of the existing primary processing techniques such as hand layup, compression molding, extrusion and injection molding, pultrusion, resin transfer molding, sheet molding techniques, etc. The selection of a suitable primary processing technique is the first and foremost step and has a significant effect on the interaction of natural fiber surface and polymer matrix (that is in improving the interfacial characteristics and hence properties of the developed composite).

The same biocomposite can be processed with different processing techniques and the effectiveness and suitability of a process is decided based on the comparison of the properties of developed composites with different processing routes. Flax fiber-reinforced composites based on poly hydroxyl butyrate (PHB) and its copolymer with hydroxyl valerate (HV) has been developed with two different primary processing techniques. The influence of the manufacturing process (compression molding of non-woven mats and injection molding of short-fiber compounds) and processing conditions (cooling temperature and annealing) on the mechanical properties of the composites have shown that injection molding is advantageous over compression molding because of shorter cycle times and higher reproducibility. But, injection-molded composite parts have shown lower impact strength than those developed through compression molding because of the shorter fiber lengths [6]. Chopped sisal fibers in the mat form reinforced the polyester/matrix using resin transfer molding (RTM) and compression molding (CM) techniques. The comparison of mechanical properties showed that the tensile strength, Young's modulus, flexural strength and flexural modulus of RTM samples were higher than CM samples. The void content and water absorption property of composites at different temperatures prepared by RTM were lower as compared to CM due to good fiber/matrix interaction. Resin transfer molding has been found to be a good processing technique for fabrication of polyester/sisal composites having superior mechanical properties [7]. Chopped hemp fiber-reinforced cellulose acetate biocomposites have been developed successfully using powder impregnation processing through compression molding (process I) and conventional extrusion followed by injection molding (process II). Injection-molded biocomposite specimens have exhibited better strength than those fabricated with compression molding due to sufficient shear forces for the intimate mixing of powdered polymer, liquid plasticizer and fibers. It has been confirmed from the ESEM images of fractured surfaces during impact that biocomposites made through process I have poor adhesion and lack of fiber dispersion as compared to process II [8]. Big blue stem grass fiber-reinforced thermoplastic biocomposites were fabricated with both extrusion followed by injection molding and biocomposite sheet-molding compounding panel (BCSMCP) manufacturing process followed by compression molding to evaluate the influence of processing on their performance. Biocomposites fabricated with BCSMCP followed by compression molding resulted in improved physical properties such as modulus at higher temperatures, greater impact strength, and a higher heat deflection temperature as compared to those of injection-molded composites from pelletized extrudate [9]. Kenaf fiber-reinforced soy-based plastic biocomposites have been developed by extrusion with injection molding as well as compression molding techniques. Compression molded samples have a higher heat deflection temperature (HDT) and notched Izod impact strength. In addition, compression molded samples have a modulus similar to injection-molded samples. It has been established that the compression molding process is beneficial for both thermal and mechanical properties of the developed biocomposites [10].

12.2 Special Treatments during Primary Processing

Generally, biocomposites are processed with the techniques discussed in the above section. But, some special treatments in addition to above mentioned primary processing techniques have also been attempted to investigate their effects on performance of biocomposites. Among these treatments, microwave curing (in place of conventional thermal curing) during primary processing has been tried and the results have shown comparable outcomes. The treatment (chemical treatment) of bio-fibers before composite processing has been found effective and convenient way to enhance the interfacial adhesion between fibers and polymers. The following section discusses these two treatments in detail:

12.2.1 Microwave Curing of Biocomposites

All the primary processing techniques employ conventional (thermal) heating for curing of biocomposites. Sometimes, the properties of natural fiber get deteriorated due to thermal heating. Microwave curing is now evolving as an alternate unconventional method of curing for biocomposite processing. Microwave heating is also an energy efficient, cost effective and environment friendly process. The feasibility of microwave curing of randomly oriented short sisal fiber (SF)-reinforced polypropylene (PP) biocomposites was studied. A fixed-frequency microwave with a maximum output power of 900 W was employed during the feasibility study. A mold of vitrified ceramic tiles was fabricated for the microwave curing. The optimal processing parameters for successful curing of PP/SF biocomposite were found at the wattage level of 900W and exposure time of 600 seconds. PP/SF biocomposites were also developed through conventional thermal curing technique. Tensile strength evaluation of both types of biocomposites was found comparable [11]. The effectiveness of microwave curing was studied in a variety of natural fiber-reinforced biocomposites using hemp, flax, kenaf, henequen and glass fibers. The same biocomposites were also processed with convection oven-based cure processing. Fibers (15% by weight) were mixed into the epoxy and the mixture was degassed in a 100 ^{0}C vacuum oven for 20 min and placed in a silicone rubber mold and capped with a Teflon lid. At the start of the microwave curing process, microwave primarily heats the epoxy while at the end fibers are primarily heated. Hemp and kenaf composites reached the same extent of cure in the microwave and convection oven. Glass and flax composites reached a higher final extent of cure with microwave curing than with thermal curing. Glass, flax, and hemp cured faster in the microwave as compared to convection oven [12].

12.2.2 Chemical Treatments of Fibers During Primary Processing of Biocomposites

Sometimes, degree of adhesion between fiber and matrix achieved through various primary processing of biocomposites is weak which results in poor performance of developed composites. Natural fibers comprise of waxy and fatty materials on their surface which causes improper bonding between fiber and matrix. There are various chemical treatment techniques (such as alkali-treatment, silane treatment, Benzoylation, Acetylation, etc.) which can be applied to natural fibers before primary processing to condition the surface and consequently improve the fiber matrix adhesion to develop high performance biocomposites.

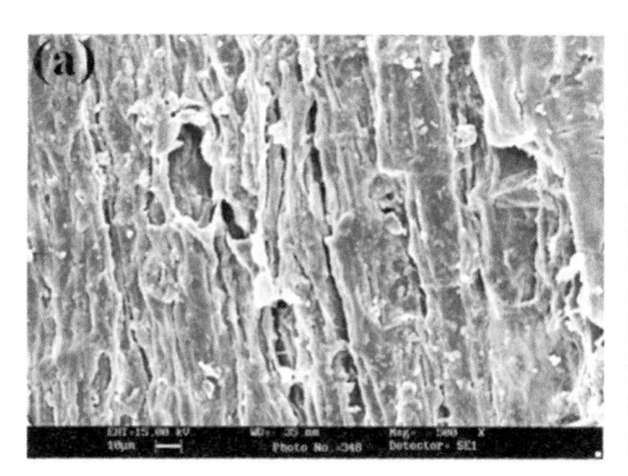

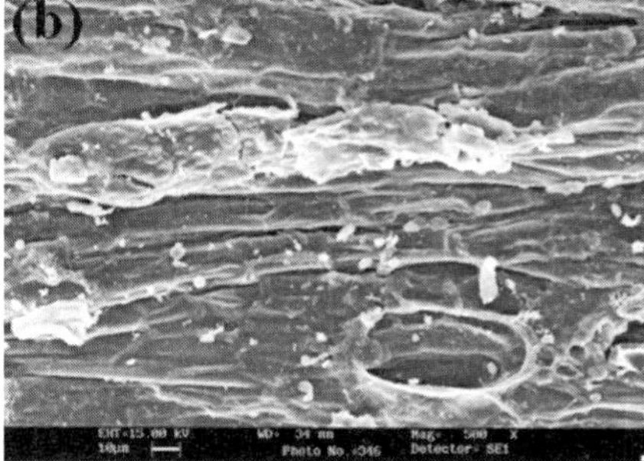

Figure 12.2 (a) Untreated *grewia optiva* fiber, and (b) Alkali-treated *grewia optiva* fiber.

These surface modification techniques have been applied by many researchers on various types of natural fibers during composite processing. The work being carried out on these techniques has been discussed in the following section in detail.

12.2.2.1 Alkaline Treatment

In this treatment, natural fibers are dipped in aqueous sodium hydroxide solution (NaOH) for definite period. After that, fibers are taken out of solution, washed several times with distilled water and sometimes neutralized with HCl solution to remove any alkalinity. Alkaline treatment removes waxy materials and lignin which covers the external surface of the natural fibers [13]. The typical SEM images of alkali-treated and untreated surface of *grewia optiva* fibers are shown for its morphological evaluation (Figure 12.2). It can be seen from images that surface of alkali-treated fibers is rougher than that of untreated fiber surface which confirms the removal of waxy contents from the surface and hence enhanced interfacial adhesion with polymer matrix during composite fabrication.

Gomes *et al.* [14] applied highly concentrated alkali treatment to curaua fibers to improve mechanical properties of green composites developed by reinforcing a corn starch-based biodegradable resin with alkali-treated curaua fibers. Tensile test results showed that alkali-treated fiber composites increased in fracture strain two to three times more than untreated fiber composites, without a considerable decrease in strength. Thus the study proved that appropriate alkali treatment is a key technology for improving mechanical properties of cellulose-based fiber composites. Joseph *et al.* [15] investigated the effect of alkali treatment along with other chemical treatments on the tensile properties of sisal-LDPE composites as a function of fiber loading, fiber length and orientation. Sisal fibers were first dipped in 10% solution of NaOH for 1 hour with subsequent stirring. After that, sisal fibers were washed with water, distilled water containing little acid to remove any trace of NaOH sticking to the fibers and finally the fibers were dried in air. The study concluded that the developed natural fiber composites with alkali-treated sisal fibers showed better performance in terms of tensile properties than untreated sisal fiber-reinforced composites due to increased aspect ratio and rough surface topography of treated fibers. Munawar *et al.* [16] studied the effects of alkali (NaOH (2%)), mild steam (0.1 MPa), and chitosan (4% and 8%) treatments on the surface morphology, fiber texture, and tensile properties of pineapple, ramie, and sansevieria fiber bundles. The study showed that the degree of crystallinity, crystallite orientation factor, and crystallite size of steam-treated fibers were higher than the alkali-treated fibers. Results of mechanical properties (tensile strength, Young's modulus, and toughness) of the steam-treated fibers were similar to those of 4% chitosan-coated fibers, which were higher than those of the alkali-treated fibers.

Goda *et al.* [17] investigated the effect of mercerization (with load application technique) on tensile properties of a ramie fiber. The ramie fibers were alkali-treated by 15% NaOH solution with applied loads of 0.049 and 0.098 N. The results showed that tensile strength of the treated ramie fiber was improved, 4–18% higher than that of the untreated ramie fiber and Young's modulus of the treated fibers decreased but fracture strains of the treated ramie fiber drastically increased which is, twice to three times higher than those of the untreated ramie fiber. The fair improvement in properties upon mercerization was correlated with change of morphological and chemical structures in microfibrils of the fiber. Yuhazri *et al.* [18] studied the effect of different concentrations and soaking time in NaOH solution on kenaf fiber. It has been found that the increasing of soaking time in NaOH showed the damage on fiber surface. The mechanical properties of alkali-treated kenaf fiber-reinforced polyester composites developed by vacuum infusion method have been found to be improved. Kim and Netravali [19] carried out alkali treatment (mercerization) of sisal fibers under tension (to minimize fiber shrinkage and to lower the microfibrillar angle) and no tension, to improve their tensile properties and interfacial adhesion with soy protein resin. It has been found that mercerization has increased the fracture stress and Young's modulus of the sisal fibers with decrease in fracture strain and toughness. Alkali-treated sisal fiber-reinforced composites with soy protein resin showed improvement in both fracture stress and stiffness by 12.2% and 36.2%, respectively, compared to the unmercerized fiber-reinforced composites.

Gassan and Bledzki [20] optimized the mechanical properties of tossa jute fibers by alkali treatment with different alkali (NaOH) concentrations and shrinkages. It has been found that shrinkage of the fibers during treatment has the most significant effect on the fiber structure and, thus on the mechanical properties of fibers such as tensile strength, modulus and toughness. Bisanda [21] investigated the effect of alkali treatment on the wetting ability and coherence of sisal/epoxy composites. It has been shown that treatment of sisal fibers in a 0.5N solution of NaOH, resulted into more rigid composites with lower porosity (increased density). There was an improvement in the adhesion characteristics due to treatment and the developed composites showed improvements in the compressive strength and water resistance. The study has recommended the alkali treatment on plant fibers for removing various waxy materials on the surface and thereby increasing the possibility for mechanical interlocking and chemical bonding with polymer matrix.

12.2.2.2 Silane Treatment

Silane treatment of natural fibers is highly effective and widely used chemical treatment technique. Silanes act as coupling agents during modification of natual fiber surfaces. Silones react with cellulose hydroxyl group of the natural fibers due to which fiber matrix interfacial characteristics are improved and thus improve the composite properties [22]. Herrera-Franco and Valadez-Gonzalez [23] examined the effect of fiber surface modification using an alkaline treatment and a matrix pre impregnation together with a silane coupling agent on the mechanical behavior of high-density polyethylene (HDPE) reinforced with continuous henequen fibers (Agave fourcroydes). It was observed that the resulting strength and stiffness of the composite depended on the amount of silane deposited on the fiber. It was found that the mechanical properties did not improve significantly when high silane concentrations were used to treat the fiber surface. It has been observed that after silane treatment, natural fiber surface has shown a layer of polymer even after the fracture which confirms stronger adhesion of fibers with polymer and the failure mode changed from interfacial failure to matrix failure. Bisanda and Ansell [24] modified the surface of sisal fibers by mercerization and silane (gamma-aminopropyltriethoxy silane) treatment to improve adhesion characteristics and moisture resistance. Sisal/epoxy composites have been developed with and without surface pretreatments. Silane treatment has been found the most effective in reducing moisture uptake of fibers in humid environments. The study shows that mercerization and silane treatment improve the compressive strength but flexural strength and stiffness are not much affected by chemical treatment.

Fully biodegradable kenaf fiber-reinforced poly lactic acid laminated composites have been developed by compression molding using the film stacking technique. Three surface treatment methods have been used to improve the kenaf fiber-matrix interfacial bonding for developing high performance biocomposites. These fiber surface treatments are: (i) kenaf fiber treated with an aqueous alkaline solution (FIBNA), (ii) kenaf fiber treated with silane coupling agent (FIBSIL), and (iii) kenaf fiber treated first with an aqueous alkaline solution and then with silane coupling agent (FIBNASIL). In alkaline treatment, sodium hydroxide solution of 5% w/v was used. 3-Amino propyl triethoxy silane (APS) (5 wt%) was used in a mixture of water–ethanol for silane treatment of the kenaf fibers. For alkaline and silane treatment, firstly, kenaf fibers were treated by NaOH solution and then treated by APS using the same procedure as was used during separate fiber surface treatment. Evaluation of mechanical properties of processed biocomposites showed that both silane-treated fiber-reinforced composites and alkali-treated fiber-reinforced composites offered superior mechanical properties as compared to untreated fiber-reinforced composites. The alkali- followed by silane-treated fiber-reinforced composite also significantly improved the mechanical properties [25]. Chemical treatment of the natural fibers can stop the moisture absorption process, clean the fiber surface, chemically modify the surface or increase the surface roughness. Silane treatment (using amino propyl triethoxy silane) of natural fibers is a promising process for improving physical and chemical properties of fibers such as *grewia optiva* and pinus fibers [26, 27].

12.3 Other Chemical Treatments

The chemical treatments such as benzoylation, polystyrene maleic anhydride (PSMA) coating and acetylation of short sisal fibers have been found effective in improving fiber matrix adhesion of polystyrene/sisal composites. It has been reported that the thermal and dynamic mechanical properties of polystyrene composites reinforced with short sisal fibers treated with different chemical modification techniques have significantly improved. For benzoylation, alkali-treated (with 18% NaOH solution) short sisal fibers were kept in 10% NaOH solution with benzoyl chloride followed by soaking in ethanol to remove the unreacted benzoyl chloride and finally washed with water and dried. In acetylation, alkali-treated (with 18% NaOH solution) short sisal fibers were immersed in glacial acetic acid and finally in acetic anhydride containing two drops of conc. H_2SO_4 followed by filtering and drying. In polystyrene maleic anhydride (PSMA) coating, a fixed amount of sisal fibers were kept in 5% solution of PSMA in toluene and refluxed for 30 min followed by filtering and drying [28]. Admicellar-treatment with a poly (methyl methacrylate) film coating has been found effective for sisal fibers for enhancing the interfacial adhesion of the fiber and matrix and for subsequently improving the mechanical properties of sisal fiber-reinforced unsaturated polyester biocomposites. The three main steps of admicellar polymerization are admicelle formation, adsolubilization, and in-situ polymerization which are carried out in sequence [29].

Rahman and Khan [30] carried out the surface modification of coir fibers to improve their adhesion efficiency with polymer matrix. Coir fibers were treated by ethylene dimethylacrylate (EMA) and cured under UV radiation along with pretreatment with UV radiation and mercerization before grafting to improve their physico-mechanical performance. The study has shown that coir fiber shrinkage is higher at low temperature and 20% alkali-treated coir fibers yielded maximum shrinkage and weight losses which showed enhanced physico-mechanical properties. The grafting of alkali-treated fiber showed an increase of polymer loading (about 56% higher) and tensile strength (about 27%) than 50% EMA grafted fiber. Rao *et al.* [31] developed polyester matrix composites reinforced with treated elephant grass stalk fibers (with $KMnO_4$) solution to improve adhesion with matrix which were extracted using retting and chemical (NaOH) extraction processes. The developed composites comprising a maximum of 31% volume of fibers resulted in a tensile strength of 80.55 MPa and tensile modulus of 1.52 GPa for elephant grass fibers extracted by retting. The tensile strength and the modulus of chemically extracted elephant grass fiber composites increased by approximately 1.45 times to those of elephant grass fiber composites extracted by retting.

12.4 Conclusions

Green composites have generated a great deal of research interest because of their various favorable properties over conventional polymer composites. Inspite of having significantly favorable properties, a number of challenges lie ahead in full commercialization of green composites. These are the economic and the technical challenges. Among the technical challenges, better composite properties should be achievable through improvements in adhesion between biodegradable polymer and natural fibers.

The present chapter highlights the important techniques adopted to improve interfacial adhesion and to achieve high performance green composites. The chapter covers the effect of various processing techniques on performance of green composites and also special processing method such as microwave curing of green composites. Various chemical treatments of natural fibers and their effect on mechanical behavior of developed green composites have been discussed in detail.

References

1. M.J. John, and S. Thomas, Biofibres and biocomposites. *Carbohydr. Polym.* 71, 343–364 (2008).
2. P.K. Bajpai, I. Singh, and J. Madaan, Comparative studies of mechanical and morphological properties of polylactic acid and polypropylene based natural fiber composites. *J. Reinf. Plast. Compos.* 31, 1712–1724 (2012).
3. P.K. Bajpai, I. Singh, and J. Madaan, Joining of natural fiber reinforced composites using microwave energy: Experimental and finite element study. *Mater. Des.*35, 596–602 (2012).
4. P. K. Bajpai, I. Singh, and J. Madaan, Tribological behavior of natural fiber reinforced PLA composites. *Wear* 297, 829–840 (2013).
5. P. K. Bajpai, I. Singh, and J. Madaan, Development and characterization of PLA-based green composites A review. *J. Thermoplast. Compos. Mater.* 27(1), 52–81 (2014).
6. N.M. Barkoulaa, S.K. Garkhaila and T. Peijs, Biodegradable composites based on flax/polyhydroxybutyrate and its copolymer with hydroxyvalerate. *Ind. Crops Prod.* 31, 34–42 (2010).
7. P.A. Sreekumara, K. Joseph, G. Unnikrishnana, and S. Thomas, A comparative study on mechanical properties of sisal-leaf fibre-reinforced polyester composites prepared by resin transfer and compression moulding techniques. *Compos. Sci. Technol.* 67, 453–461 (2007).
8. A.K. Mohanty, A. Wibowo, M. Misra, and L.T. Drzal, Effect of process engineering on the performance of natural fiber reinforced cellulose acetate biocomposites. *Compos. A* 35, 363–370 (2004).
9. W. Liu, K. Thayer, M. Misra, L.T. Drzal, and A.K. Mohanty, Processing and physical properties of native grass-reinforced biocomposites. *Polym. Eng. Sci.* 47, 969–976 (2007).
10. W. Liu, L.T. Drzal, A.K. Mohanty, and M. Misra, Influence of processing methods and fiber length on physical properties of kenaf fiber reinforced soy based biocomposites. *Compos. B* 38, 352–359 (2007).
11. S. Ali, I. Singh, and A.K. Sharma, in *Proceedings of the Third international Multicomponent Polymer Conference* (2012).
12. N. Sgriccia, and M.C. Hawley, Thermal, morphological, and electrical characterization of microwave processed natural fiber composites. *Compos. Sci. Technol.* 67, 1986–1991 (2007).
13. A.K. Mohanty, M. Misra, and L.T. Drzal, Surface modifications of natural fibers and performance of the resulting biocomposites: An overview. *Compos. Interfaces* 8, 313–343 (2001).
14. A. Gomes, T. Matsuo, K. Goda, and J. Ohgi, Development and effect of alkali treatment on tensile properties of curaua fiber green composites. *Compos. A* 38, 1811–1820 (2007).
15. K. Joseph, S. Thomas, and C. Pavithran, Effect of chemical treatment on the tensile properties of short sisal fibre-reinforced polyethylene composites. *Polymer,* 37, 5139–5149 (1996).
16. S.S. Munawar, K. Umemura, F. Tanaka, and S. Kawai, Effects of alkali, mild steam, and chitosan treatments on the properties of pineapple, ramie, and sansevieria fiber bundles. *J. Wood Sci.* 54, 28–35 (2008).
17. K. Goda, M.S. Sreekala, A. Gomes, T. Kaji, and J. Ohgi, Improvement of plant based natural fibers for toughening green composites—Effect of load application during mercerization of ramie fibers. *Compos. A* 37, 2213–2220 (2006).
18. M. Yuhazri, Y. Phongsakorn, P.T.H. Sihombing, A.R. Jeefferie, A.M. Puvanasvaran, P. Kamarul and K. Rassiah, Mechanical properties of kenaf/polyester composites. *Int. J. Eng. Technol.* 11, 127–131 (2011).

19. J.T. Kim, and A.N. Netravali, Mercerization of sisal fibers: Effect of tension on mechanical properties of sisal fiber and fiber-reinforced composites. *Compos. A* 41, 1245–1252 (2010).
20. J. Gassan, and A.K. Bledzki, Possibilities for improving the mechanical properties of jute/epoxy compositesby alkali treatment of fibres. *Compos. Sci. Technol.* 59, 1303–1309 (1999).
21. E.T.N. Bisanda, The effect of alkali treatment on the adhesion characteristics of sisal fibres. *Appl. Compos. Mater.* 7, 331–339 (2000).
22. X. Li, L.G. Tabil, and S. Panigrahi, Chemical treatments of natural fiber for use in natural fiber-reinforced composites: a review. *J. Polym. Environ.* 15, 25–33 (2007).
23. P.J. Herrera-Franco, and A. Valadez-Gonzalez, Mechanical properties of continuous natural fibre-reinforced polymercomposites. *Composites: Part A*, 35, 339–345 (2004).
24. E.T.N. Bisanda, and M.P. Ansell, The effect of silane treatment on the mechanical and physical properties of sisal-epoxy composites. *Compos. Sci. Technol.* 41, 165–178 (1991).
25. M.S. Huda, L.T. Drzala, A.K. Mohanty, and M. Misra, Effect of fiber surface-treatments on the properties of laminated biocomposites from poly(lactic acid) (PLA) and kenaf fibers. *C ompos. Sci. Technol.* 68, 424–432 (2008).
26. A.S. Singha, and V.K. Thakur, Morphological, thermal, and physicochemical characterization of surface modified pinus fibers. *Int. J. Polym. Anal. Charact.* 14, 271–289 (2009).
27. A. S. Singha, and V. K. Thakur, Synthesis and characterizations of silane treated grewia optiva fibers. Int. J. Polym. Anal. Charact. 14, 301–321 (2009).
28. K.C.M. Nair, S. Thomas, and G. Groeninckx, Thermal and dynamic mechanical analysis of polystyrene compositesreinforced with short sisal fibres. *Compos. Sci. Technol.* 61, 2519–2529 (2001).
29. S. Sangthonga, T. Pongprayoona, and N. Yanumet, Mechanical property improvement of unsaturated polyester composite reinforced with admicellar-treated sisal fibers. *Compos. A* 40, 687–689 (2009).
30. M.M. Rahman, and M.A. Khan, Surface treatment of coir (Cocos nucifera) fibers and its influence on the fibers' physico-mechanical properties. *Compos. Sci. Technol.* 67, 2369–2376 (2007).
31. K.M.M. Rao, A.V.R. Prasad, M.N.V.R. Babu, K.M. Rao, and A.V.S.S.K.S. Gupta, Tensile properties of elephant grass fiber reinforced polyester composites. *J. Mater. Sci.* 42, 3266–3272 (2007).

13

Effect of Chemical Functionalization on Functional Properties of Cellulosic Fiber-Reinforced Polymer Composites

Ashvinder Kumar Rana[*,1], Amar Singh Singha[2], Manju Kumari Thakur[3] and Vijay Kumar Thakur[4]

[1]*Department of Chemistry, Sri Sai University, Palampur, H.P., India*
[2]*Department of Chemistry, NIT Hamirpur, H.P., India*
[3]*Division of Chemistry, Government Degree College Sarkaghat, Himachal Pradesh University, Summer Hill, Shimla, India*
[4]*School of Mechanical and Materials Engineering, Washington State University, Washington, U.S.A.*

Abstract

The successful combination of vegetable natural fibers with polymer matrices results in an improvement of the mechanical properties of the composites compared with the matrix material. These fillers are cheap and nontoxic and can be obtained from renewable sources and thus are easily recyclable. Moreover, despite of their low strength, they can lead to composites with high specific strength and low density, and thus can play a major role in the automotive, sports and transportation industries, packaging, leisure-based products, etc. Increased technical advancement, identification of new applications, continued political and stringent environmental norms and government investment in new methods for fiber harvesting have played a key role in the adoption of these novel materials. Natural fibers are undergoing a high-tech revolution that could see them replace synthetic materials in various applications such as boat hulls, archery bows and bathtubs, etc. Recently, there has been an increasing interest in the commercialization of natural fiber-reinforced polymer composites and their use for interior paneling in the automotive industry. Biocomposites will continue to expand their role in various industries, especially in automotive applications, only if technical challenges such as moisture stability and fiber polymer interface compatibility are solved to the satisfaction of users.

Keywords: Grewia optiva, flammability, thermal stability, mechanical properties, UPE, benzoylation, FTIR

**Corresponding author*: ranaashvinder@gmail.com

Vijay Kumar Thakur (Ed.), Lignocellulosic Polymer Composites, (281–300) 2015 © Scrivener Publishing LLC

13.1 Introduction

Within the last two decades, the field of natural fiber-reinforced composites has experienced a great deal of interest from researchers and scientists, particularly with regard to substituting the synthetic fillers such as glass, carbon, steel, etc. The potential of natural fiber-based composites for the automotive industry as well as in the fabrication of boats, skis, agricultural machinery, cars, etc., is becoming widely recognized, with many products now well established in the market [1, 2]. Automobile industrialists are looking for the incorporation of natural fiber-reinforced composites in both interior and exterior car parts, which would serve a two-fold goal; to lower the overall weight of the vehicle, thus increasing fuel efficiency, and to increase the sustainability of their manufacturing process. However, two important factors which are responsible for the lesser use of natural fibers in the preparation of composites are their low strength as compared to synthetic fillers and high water absorption characteristics. The latter drawback is highly responsible for the incompatibility between the hydrophobic matrix and natural fiber. A large number of surface modification techniques have been reported in literature for surface modification of hydrophilic fibers in order to increase their hydrophobicity, which ultimately will enhance the strength of the composites [3-5]. However, poor inherent flammability resistance of natural fiber-reinforced polymer composites is a major drawback for their use in the transportation and aerospace sectors. Different kinds of fire retardant strategies have been used by researchers to improve the fire-retardant behavior of polymer composites [6, 7]. The strategies could be used for a variety of fire-retardant additives and intumescent systems for thermoplastic composites. In the case of thermosets, the use of coatings and intumescent systems, and for reinforcing natural fibers, a flame-retardant treatment of the fibers prior to incorporation in the matrix could be the best option [8].

The flammability of composites depends upon a large number of factors, i.e., type of polymer, type of fiber used and adhesion between matrix and fibers [9, 10]. Polymers undergo thermal and thermal-oxidative decomposition on exposure to heat, which further leads to the production of heat, smoke and volatile components (mixtures of monomers, hydrocarbon, carbon monoxide, and noncombustible gases). However, the type of products to be produced during decomposition of polymers depends upon their chemical composition. Relative flammability properties of different polymers were generally assessed by using a limited oxygen index (LOI) test (which measures the percentage of oxygen that is needed to sustain combustion; normal O_2 in the atmosphere is ~20.95%) in accordance with ASTM D-2863. There are large numbers of flame retardants and mineral fillers which can increase the fire-resistance behavior of polymer composites. For example, phosphorus based [6] and halogen-free flame retardants [7], nanocomposites and mineral fillers [11, 12] and flame retardants for specific polymers such as polyolefins, PS and polyvinylchlorides [13-15]. Rai *et al.* [16] have evaluated the effect of euphorbia coagulum binder on the fire properties of polyester-banana fiber composites and reported an increase in limiting oxygen index value from 18 to 21% with reduction in smoke density. Ayrilmis *et al.* [17] have conducted an LOI test of coir fiber-reinforced PP composites and found an increase in flame retardancy behavior with an increase in coir fiber content. Schartel *et al.* [18] have studied the effects of fire-retardant additives (ammonium polyphosphate (APP) and expandable graphite (EG)) on fire retardancy behavior of flax fiber PP composites.

Polyesters are a very important class of high performance polymers, which find extensive use in a number of diverse applications [19]. Unsaturated polyester resins was chosen in the present study for making fiber-reinforced composites because of their good strength and low cost in comparison to other thermosetting resins.

In the present chapter, we report some of our study on both raw and surface-modified *Grewia optiva* fiber-reinforced UPE matrix-based composites, which possess enhanced mechanical and physico-chemical properties when compared with UPE matrix. In addition to the effect of flame retardants, i.e., magnesium hydroxide and zinc borate, on flame resistance, the behavior of resulted *Grewia optiva* fiber-reinforced composites have also been evaluated and was found to be improved. A significant discussion on the work of other researcher's work has also been added in the chapter.

13.2 Chemical Functionalization of Cellulosic Fibers

Prior to fiber's surface fabrication, raw fibers were cut into 3mm size and were subsequently chemically modified at optimized parameters as per methods reported earlier [20, 21]. The various surface modification processes are presented below.

13.2.1 Alkali Treatment

In the case of alkali treatment, purified raw fibers were soaked in optimized NaOH concentration (10 wt%) at room temperature for 240 min, as per procedure reported earlier [20]. Scheme 1 represents the possible mechanism of alkali treatment.

13.2.2 Benzoylation

Benzoylation of *Grewia optiva* fiber was carried out by using a 5 vol% benzoyl chloride in 10% alkali solution for 10-15 min as per procedure reported earlier [21]. Scheme 2 shows the possible mechanism for grafting of benzoyl chloride onto cellulosic fibers.

13.2.3 Composites Fabrication

For the preparation of polymer composites, unsaturated polyester resin was reinforced with raw and surface-modified particle (90 microns) fibers in four different proportions

$$\text{Fiber—OH} + \text{NaOH} \longrightarrow \text{Fiber—O}^-\text{Na}^+ + H_2O$$

Scheme 13.1 A possible mechanism for alkali treatment of cellulosic fibers.

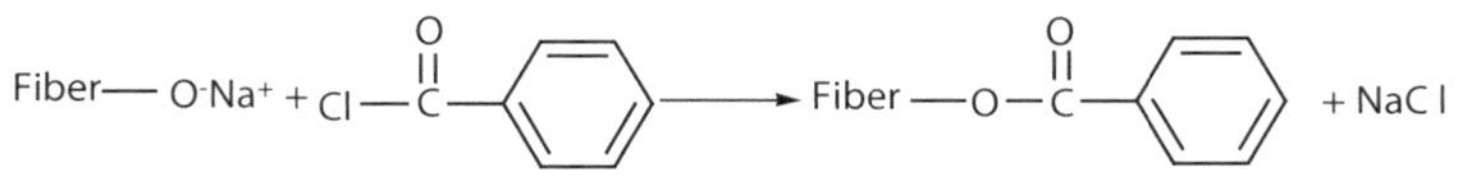

$$\text{Fiber—O}^-\text{Na}^+ + \text{Cl—C(=O)—C}_6\text{H}_5 \longrightarrow \text{Fiber—O—C(=O)—C}_6\text{H}_5 + \text{NaCl}$$

Scheme 13.2 A possible mechanism for grafting of benzoyl chloride onto *Grewia optiva* fiber [22].

(w/w; 10, 20, 30 and 40 wt%). The composite samples were subsequently made by hand lay-up method followed by compression molding method. A stainless steel mold having dimensions of (150 × 150 × 5) mm^3 has been used for the preparation of polymer composites. The curing of the polyester resin was done by incorporation of 2% cobalt naphthenate (as accelerator) mixed thoroughly in the polyester resin followed by addition of 1% methyl-ethyl-ketone-peroxide (MEKP) as a hardener prior to reinforcing with fibers. The fibers were then mixed manually in the mixture in order to disperse the fibers properly in the matrix. However, in the case of fire retardants/*Grewia optiva*/ unsaturated polyester matrix-based composites, the fire retardants were mixed properly before the incorporation of particle fibers. The cast of each composite was cured under a pressure of 100Kg/cm^2 for 12 hrs, followed by post-curing in open air for 24 hrs. Specimens of suitable dimensions were used for mechanical as well as LOI testing. Utmost care has been taken to maintain uniformity and homogeneity of the composite, although reproducibility is somewhat difficult in hand lay-up methods.

13.3 Results and Discussion

13.3.1 Mechanical Properties

The specimens of suitable dimensions were cut from the composite sheets by using a Diamond cutter and were subjected for analysis of tensile, compressive and flexural strength in accordance with ASTM standards on a computerized universal testing machine (Housfield H25KS). Three specimens of each sample were used for the measurement of each mechanical property at ambient laboratory conditions and average results have been reported.

13.3.1.1 Tensile Strength

Tensile strength is the resistance of a material to a force tending to tear it apart, and is measured as the maximum tension that material can withstand without tearing. For tensile strength analysis, specimens of dimensions 100 mm x 10 mm x 5.0 mm were used, and the tensile test was conducted in accordance with ASTM D3039 method. The test was conducted at constant strain rate of 10 mm/min and stress was applied till the failure of sample.

It has been observed that tensile strength of UPE matrix increases after its reinforcement with raw and surface-functionalized *Grewia optiva* particle fibers (Figure 13.1). Neat UPE specimens have been found to bear a maximum load of 18.7 MPa. The raw fibers-reinforced UPE composites have been found to exhibit tensile strength of 23.36, 24.63, 24.78 and 23.73 MPa at 10, 20, 30 and 40% fiber loading, respectively. Further, it has also been observed that the tensile strength of cellulosic fiber-reinforced UPE composites increases upon surface modification of fiber. When reinforced with mercerized and benzoylated fibers, the UPE matrix has been found to exhibit tensile strength of 26.2, 26.84, 27.48 and 24.36 MPa; 26.24, 28.04, 31.52 and 25.98 MPa at 10, 20, 30 and 40% fiber loading, respectively. The results obtained after surface modification of fibers have been found to be consistent with results obtained by Herrera-Franco and

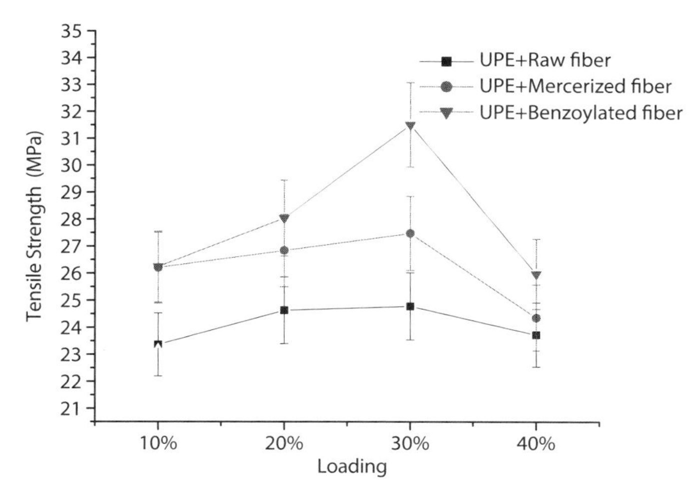

Figure 13.1 Tensile strength of different surface-modified *Grewia optiva* particle fibers-reinforced unsaturated polyester composites.

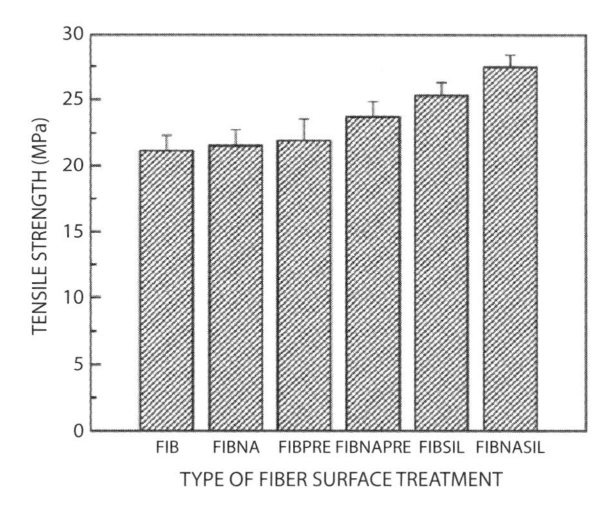

Figure 13.2(a) Tensile strength data of raw (FIB), mercerized (FIBNA), pre-impregnated with dissolved HDPE (FIBPRE), mercerization followed by pre-impregnation with HDPE (FIBNAPRE), silanated (FIBSIL) and mercerized + silane coupling agent treated Henequen fibers-reinforced HDPE (20:80 v/v) composites. Reprinted with permission from [23]. Copyright 2005 Elsevier.

Valadez-Gonzalez [23]. These results have been depicted in Figure 13.2 (a, b). Further, tensile strength of composites was maximum when reinforced with benzoylated fibers followed by mercerized and raw *Grewia optiva* fibers reinforcement. Higher tensile strength of benzoylated fiber-reinforced UPE matrix-based biocomposites may be due to development of better hydrophobic character on the fiber surface [24]. The increase in tensile strength of mercerized fiber-reinforced polymer composites as compared to raw ones may be due to the improvement in adhesive characteristics between fiber and matrix phase, because mercerization removes natural and artificial impurities, thereby producing a rough surface topography [24]. Pothan *et al.* have also found similar

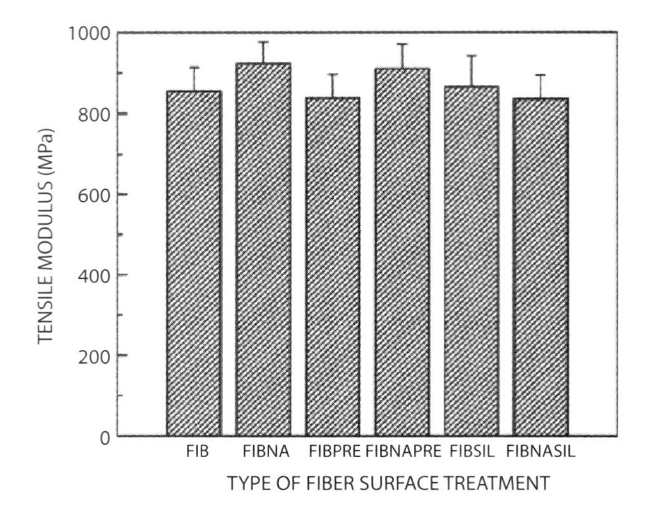

Figure 13.2(b) Elastic modulus of a HDPE/henequen-fiber (80:20 v/v) composite, plotted as a function of fiber surface treatment. Reprinted with permission from [23]. Copyright 2005 Elsevier.

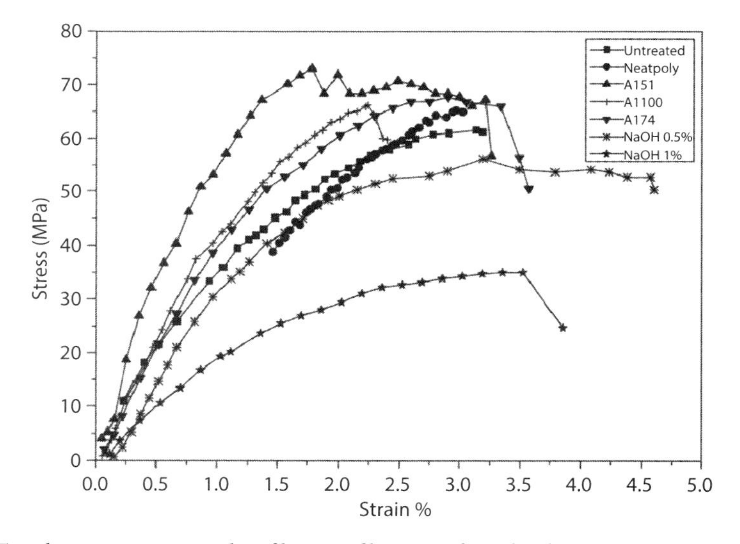

Figure 13.3 Tensile stress-strain results of banana fibers-reinforced polyester composites. Reprinted with permission from [25]. Copyright 2007 Sage Publications.

results after the surface modification of fibers and their results have been represented in Figure 13.3 [25].

13.3.1.2 Compressive Strength

The compressive test was conducted in accordance with ASTM D3410 method on a computerized universal testing machine. Specimens of dimensions 100 mm x 10 mm x5 mm were used for the compressive test. A constant strain rate of 10 mm/min was applied till failure of samples.

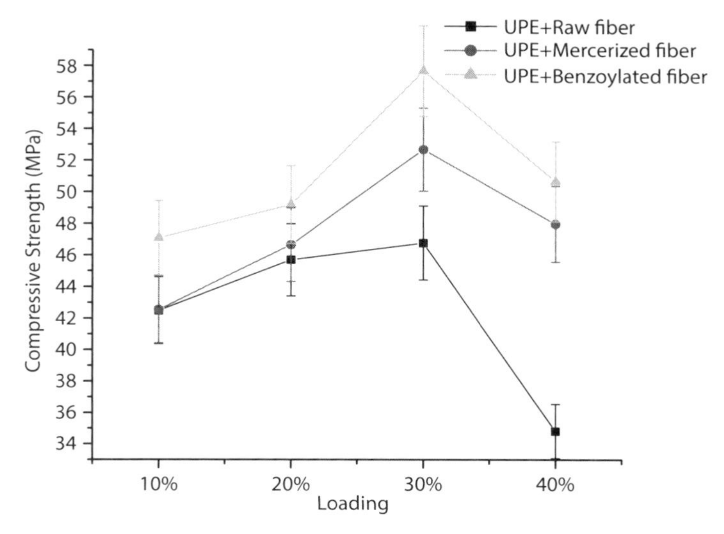

Figure 13.4 Compressive strength of different surface-modified *Grewia optiva* particle fibers-reinforced unsaturated polyester composites.

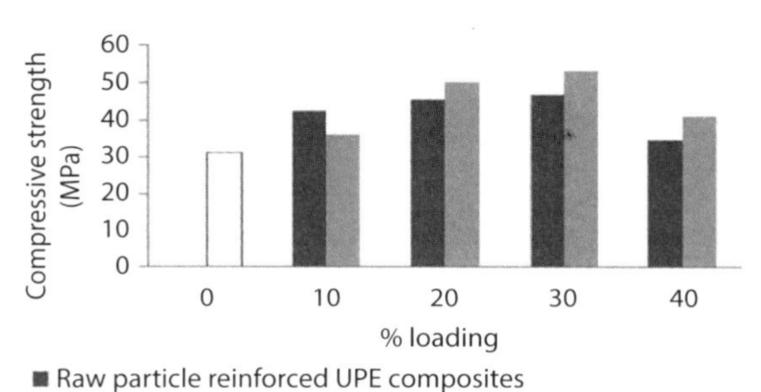

Figure 13.5 Compressive strength of raw and AN-grafted particle fibers-reinforced unsaturated polyester composites [26]. Reprinted with permission from [26]. Copyright 2012 John Wiley and Sons.

Neat UPE specimens have been found to exhibit a maximum compressive strength of 31.25 MPa. However, a considerable increase in compressive strength has been observed after reinforcing UPE matrix fibers (Figure 13.4). Raw particle fibers-reinforced UPE composites have been found to exhibit compressive strength of 42.49, 45.7, 46.78 and 34.82 MPa at 10, 20, 30 and 40% fiber loading, respectively.

When reinforced with mercerized and benzoylated particle fibers, the UPE matrix has been found to exhibit compressive strength of 42.55, 46.66, 52.7 and 47.99 MPa; 47.08, 49.2, 57.7 and 50.68 MPa at 10, 20, 30 and 40% fiber loading, respectively (Figure 13.4). These results were also in conformity with AN graft copolymerized *Grewia optiva* fibers-reinforced composites [26]. Figure 13.5 shows the compressive strength results of the *Grewia optiva* fibers-reinforced composites.

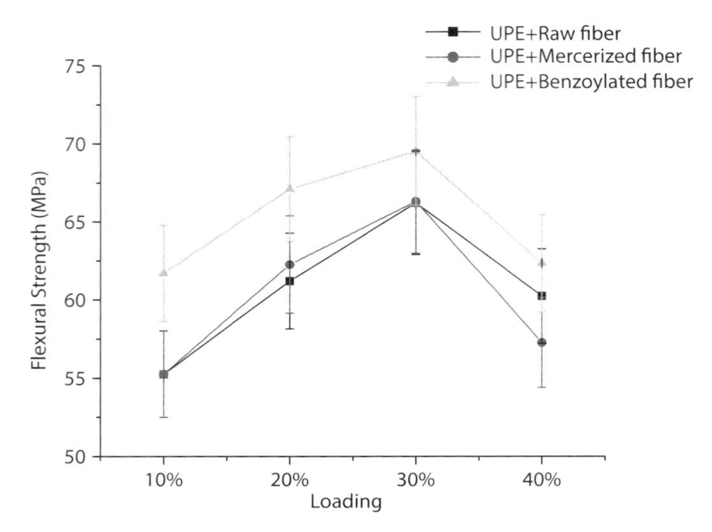

Figure 13.6 Flexural strength of different surface-modified *Grewia optiva* particle fibers-reinforced unsaturated polyester composites.

13.3.1.3 Flexural Strength

Flexural strength is also referred to as transverse beam testing and is used to determine the flexure or bending properties of a material. The value of flexural strength represents the highest stress experienced by a material before rupture. The flexural test was performed on a universal testing machine by using specimens of dimensions 100 mm x 10 mm x 5 mm in accordance with ASTM D790. The results of the flexural test have been found to follow similar trends as obtained above in the cases of the tensile and compressive strength tests (Figure 13.6). From Figure 13.6, it is evident that flexural strength increases with increase in fiber content up to 30% and then decreases with further increase in fiber content. The flexural strength of neat UPE is 51.7 MPa, which increases upon reinforcement with raw, mercerized and benzoylated particle fibers. The UPE composites have been found to show flexural strength of 55.25, 61.2, 66.2 and 60.23 MPa when reinforced with raw particle fibers at 10, 20, 30 and 40% fiber loading, respectively.

When reinforced with mercerized and benzoylated particle fibers, The UPE matrix has been found to exhibit flexural strength of 55.26, 62.26, 66.29 and 57.26 MPa; 61.7, 67.1, 69.52 and 62.32 MPa at 10, 20, 30 and 40% fiber loading, respectively. The results obtained were found to be consistent with results obtained in the case of AN graft copolymerized *Grewia optiva* fibers-reinforced unsaturated polyester composites [26]. Rai *et al.* have also reported similar results during their studies [16].

13.3.2 FTIR Analysis

The FTIR spectra of pure UPE and its composites reinforced with raw, mercerized and benzoylated fibers were recorded on Perkin Elmer infrared spectrophotometer over a range of 400-4000 cm^{-1} [Figure 13.7 (A-D)]. The FTIR spectrum of UPE resin (Figure 13.7A) showed peaks at 3026.87 cm^{-1} (due to unsaturated stretching C-H vibrations),

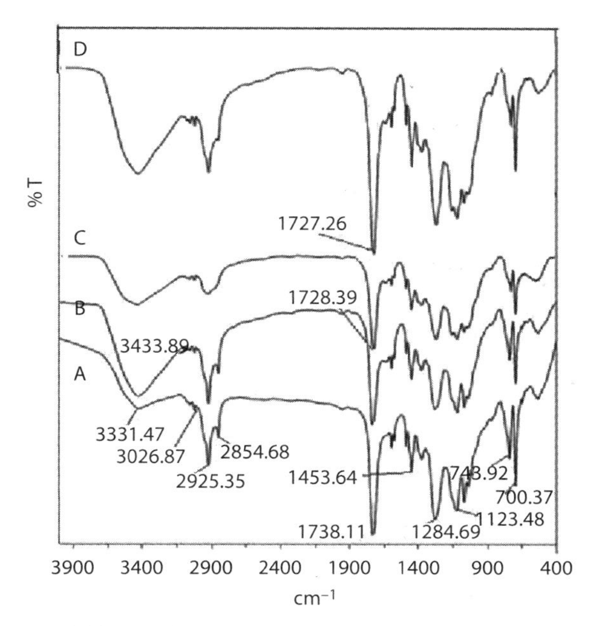

Figure 13.7 FTIR spectra of (A) neat, (B) raw, (C) mercerized and (D) benzoylated particle fibers-reinforced UPE composites.

between 2985-2880 cm^{-1} (due to saturated, aliphatic hydrocarbon C-H stretching), at 1738.11 cm^{-1} (due to ester carbonyl group stretching), at 1600.18 & 1453.64 cm^{-1} (due to aromatic ring), at 1070.79 cm^{-1} (due to unsaturated flow in plane deformation) and at 743.92 & 700.37 cm^{-1} (due to aromatic out-of-plane bending deformations). However, an additional broad peak at 3433.89 cm^{-1} due to O-H stretching of H-bonded groups has been observed after reinforcement of UPE matrix with raw *Grewia optiva* fibers (Figure 13.7B). Further reinforcement of UPE matrix with mercerized fibers (Figure 13.7C) causes a decrease in the intensity of peak at 1728.39 cm^{-1} (due to removal of lignin and hemicellulose contents) and with benzoylated fibers (Figure 13.7D) an increase in intensity of peak at 1727.26 cm^{-1} (due to formation of ester linkage on fiber surface after benzoylation, which causes enhancement in ester group intensity) have been observed.

13.3.3 SEM Analysis

Surface morphology of polymeric UPE resin and UPE-based polymer composites reinforced with particle form of raw, mercerized and benzoylated *Grewia optiva* fibers has been examined by scanning electron microscopy technique. It has been observed from Figure 13.8 (A-D) that the surface morphology of pure matrix is different from its polymer composites reinforced with raw, mercerized and benzoylated fibers in terms of smoothness and roughness. Further micrographs clearly show the weak interfacial adhesion between raw *Grewia optiva* fiber and UPE matrix. However, a proper intimate mixing of fibers and matrix has been observed on reinforcement of UPE matrix with mercerized and benzoylated *Grewia optiva* fiber.

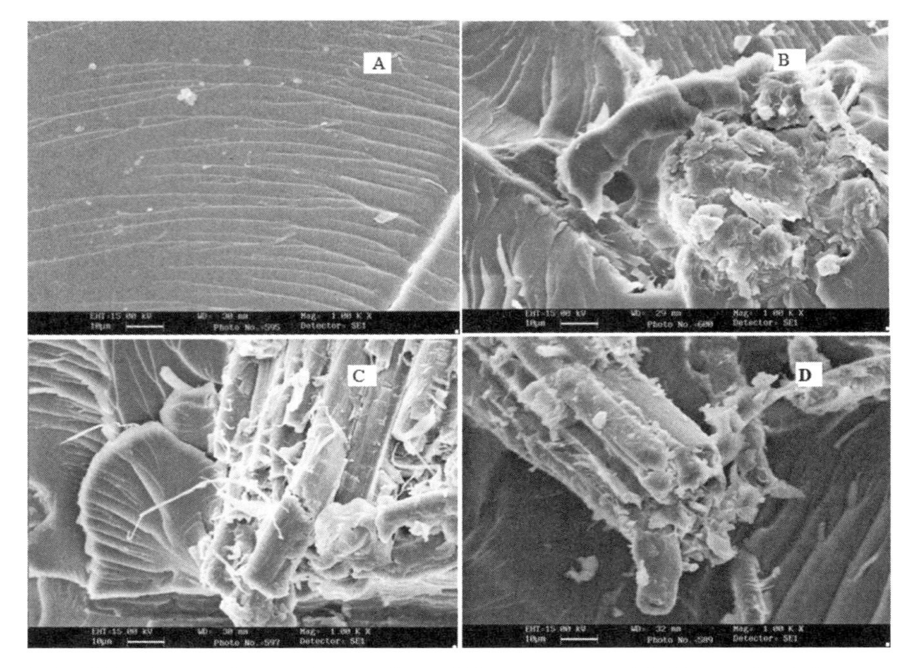

Figure 13.8 SEM micrographs of (A) neat, (B) raw, (C) mercerized and (D) benzoylated fibers-reinforced UPE matrix-based composites.

13.3.4 Thermogravimetric Analysis

Thermogravimetric (TGA) studies of optimized samples of raw and surface-functionalized fibers-reinforced UPE composites were conducted on a TGA with auto sampler (Mettler Toledo) analyzer at a heating rate of 15°C/min between temperature ranges from 20°C to 1000°C in nitrogen atmosphere (Figure 13.9). Neat UPE has been found to decompose in a single step with initial decomposition (IDT) and final decomposition (FDT) temperature values of 325.05°C (12.28% weight loss) and 441.48°C (83.89% weight loss), respectively. On reinforcement of UPE matrix with raw particle fibers, IDT & FDT values have been found to be 316.51°C (21.20% weight loss) and 442.49°C (85.35% weight loss), respectively. However, after fibers surface functionalization, the IDT & FDT values of the finally fabricated composites have been found to have increased. Between mercerized and benzoylated fibers-reinforced composites, the latter one has been found to have higher IDT (325.51°C) & FDT (459.14°C) values and thus higher thermal stability.

13.3.5 Evaluation of Physico-Chemical Properties

13.3.5.1 Water Absorption

A water absorption study was carried out by immersing neat UPE and composite specimens of dimensions 10 mm x 10 mm x 2 mm in distilled water for two months at 35°C. The samples were dried completely in a hot air oven before performing the water absorbance study and then initial weight was recorded. For water absorption measurements,

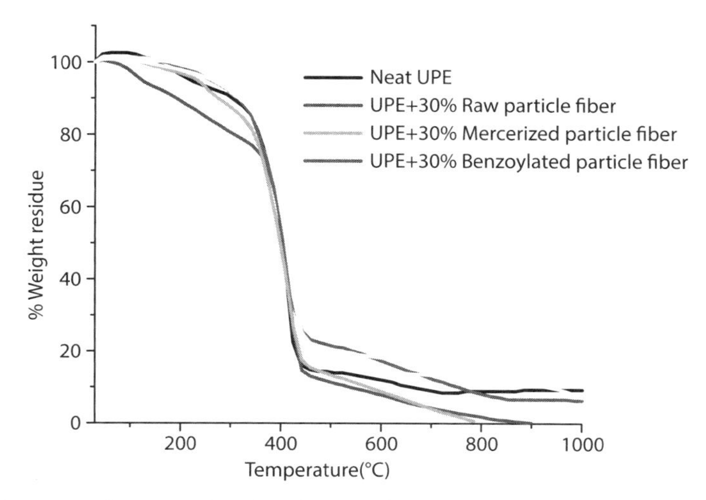

Figure 13.9 TGA curves for raw and surface-functionalized *Grewia optiva* fibers-reinforced unsaturated polyester matrix-based composites.

specimens were withdrawn from water after every 10 days, dried by pressing between a fold of filter paper to remove excess water and then weighed again. The percentage weight gain was calculated using the following equation:

$$W_G = \left(\frac{W_f - W_i}{W_i}\right)100 \tag{13.1}$$

Where W_G, W_i and W_f are percentage weight gain, initial weight and final weight, respectively.

Water absorption results of neat UPE and its composites reinforced with raw, mercerized and benzoylated fibers have been shown in Table 13.1. It can be seen from the table that water absorption characteristics of polymer composites depend upon the content of fiber loadings, water immersion time and surface modification techniques. The water absorption of raw and surface-modified fiber-reinforced UPE composites has been found to increase with the increase in percent loading. Similar results were also reported by Rashdi *et al.* during their studies on the water absorption behavior of kenaf fibers-reinforced polyester composites [27]. This may be due to greater affinity of water for OH groups present on the fiber backbone, whose number increases with the increase in fiber contents. In comparison to raw filler, composites reinforced with surface-modified filler exhibited low water absorption, which may be due to the reduction in the hydrophilic character of cellulosic biofibers after surface modification.

Between mercerized and benzoylated fibers-reinforced UPE composites, benzoyl chloride grafted fibers-reinforced UPE matrix has been found to have better water resistant behavior, which may be due to better compatibility of benzoylated fibers and hydrophobic matrix.

Table 13.1 % Water absorbance by neat UPE and its composites reinforced with particle and short form of raw, mercerized, silanated, benzoylated, *Grewia optiva*-g-poly (AN) and *Grewia optiva*-g-poly-(AAc) fibers.

Sample Designation	10 Days	20 Days	30 Days	40 Days	50 Days
Neat UPE	0.01	0.09	0.19	0.99	1.65
UPE + 10% raw fiber	2.14	2.87	3.25	3.49	4.09
UPE + 20% raw fiber	3.12	4.63	5.44	5.72	6.99
UPE + 30% raw fiber	4.71	5.07	5.88	7.05	8.06
UPE + 40% raw fiber	5.70	6.93	7.17	7.53	9.59
UPE + 10% mercerized fiber	2.01	2.69	3.03	3.49	4.01
UPE + 20% mercerized fiber	3.05	3.81	4.72	5.21	5.65
UPE + 30% mercerized fiber	3.65	5.02	5.81	6.76	7.84
UPE + 40% mercerized fiber	4.33	6.24	7.11	7.55	8.59
UPE + 10% benzoylated fiber	1.26	1.97	2.25	2.31	3.12
UPE + 20% benzoylated fiber	2.76	3.54	3.86	4.13	4.42
UPE + 30% benzoylated fiber	3.46	4.60	5.03	5.25	5.73
UPE + 40% benzoylated fiber	4.18	6.08	6.78	7.02	7.46

13.3.5.2 *Chemical Resistance*

For studying chemical resistance, the dried specimens of neat UPE and its polymer composites were immersed in 100 ml of 1 and 3 N solutions of NaOH and HCl for different time periods ranging from 10 to 30 days. After this, the samples were taken out and washed twice with distilled water, dried and weighed. The percent chemical resistance (P_{cr}) was calculated in terms of weight loss by the following relationship:

$$\text{Perecent chemical resistance}(\text{Pcr}) = \frac{W_i - W_{aci}}{W_i} \times 100 \tag{13.2}$$

Where W_i = initial weight and W_{aci} = weight after a certain interval

It has been observed from Tables 13.2 & 13.3 that chemical resistance of neat UPE and its composites reinforced with raw, mercerized and benzoylated particle fibers towards acid and base decreases with increase in normality, as well as with increase in immersion time. This behavior could be due to the propagation of microcracks which cause more and more internal penetration of acids and bases into the composite samples and hence decreased chemical resistance behavior. Among raw, mercerized and benzoylated particle fibers-reinforced composites, the benzoylated one has been found to have better chemical resistance properties towards acids, which may be due to the better interfacial adhesion between treated fibers and matrix because of the enhanced hydrophobic character of benzoylated fibers. However, the lower chemical resistance behavior of benzoylated fibers-reinforced UPE composites towards NaOH could be

Table 13.2 Data on % chemical resistance shown by neat UPE and its composites reinforced with raw and surface-modified fibers against 1.0 and 3.0 N NaOH.

Sample Designation	1.0 N			3.0N		
	10 Days	20 Days	30 Days	10 Days	20 Days	30 Days
Neat UPE	0.39	2.42	4.76	6.95	9.06	12.25
UPE + 10% raw fiber	2.59	5.20	6.08	7.22	15.70	21.12
UPE + 20% raw fiber	3.88	10.70	12.04	15.69	27.69	28.08
UPE + 30% raw fiber	5.16	11.90	14.05	17.50	29.93	32.65
UPE + 40% raw fiber	8.44	17.32	19.45	24.11	33.78	34.62
UPE + 10% mercerized fiber	1.37	5.00	8.10	6.16	12.83	20.97
UPE + 20% mercerized fiber	3.64	9.44	12.31	13.31	21.89	26.78
UPE + 30% mercerized fiber	4.31	11.75	13.36	16.41	25.90	27.02
UPE + 40% mercerized fiber	7.30	16.42	17.55	19.54	30.68	31.19
UPE + 10% benzoylated fiber	5.36	13.13	15.13	12.56	20.82	27.51
UPE + 20% benzoylated fiber	9.33	17.94	20.72	20.99	29.73	35.00
UPE + 30% benzoylated fiber	11.78	20.66	23.33	27.54	32.63	38.88
UPE + 40% benzoylated fiber	15.41	25.42	27.51	30.63	33.15	40.30

due to the hydrolysis of ester linkage formed during benzoylation, which enhances the penetration of base into the composites and hence increased weight loss.

13.3.5.3 Moisture Absorption

The moisture absorbance study of composite samples was carried out in a humidity chamber at 20, 50 and 80% relative humidity. The results have been reported in Table 13.4. Known weights (W_i) of dried samples were placed in a humidity chamber maintained at a particular humidity level for a time interval of 8 hrs. Final weights (W_f) of samples were taken instantly by taking sample out of the humidity chamber after the desired time period. The percent moisture absorption was calculated by using the following relationship:

$$\%\,\text{Moisture Absorption} = \frac{W_f - W_i}{W_i} \times 100 \tag{13.3}$$

Table 13.3 Data on % chemical resistance shown by neat UPE and its composites reinforced with raw and surface-modified fibers against 1.0 and 3.0 N HCl.

Sample Designation	1.0 N			3.0N		
	10 Days	20 Days	30 Days	10 Days	20 Days	30 Days
Neat UPE	0.0001	0.0412	0.0702	0.1212	0.1895	0.2746
UPE + 10% raw fiber	0.0036	0.7209	1.5695	0.9116	1.4242	2.9499
UPE + 20% raw fiber	0.0075	1.1667	4.5913	3.0057	3.5183	5.5654
UPE + 30% raw fiber	0.0116	1.6472	5.7695	3.1857	3.5999	6.0767
UPE + 40% raw fiber	0.0205	1.9517	6.7085	3.6585	4.1663	7.1737
UPE + 10% mercerized fiber	0.0017	0.4766	1.2531	0.8981	1.4107	2.7708
UPE + 20% mercerized fiber	0.0054	0.5794	1.6803	2.0616	2.5742	4.3235
UPE + 30% mercerized fiber	0.0098	0.8762	2.0309	2.9931	3.5057	5.5824
UPE + 40% mercerized fiber	0.0176	0.9675	2.6556	3.0863	3.5989	5.8855
UPE + 10% benzoylated fiber	0.0009	0.2735	0.9791	0.5836	1.0962	2.0513
UPE + 20% benzoylated fiber iberfiberparticle fiber	0.0027	0.5086	1.4533	1.107	1.6196	3.2091
UPE + 30% benzoylated fiber particle fiber	0.0049	0.6980	1.6192	1.546	2.0586	3.5224
UPE + 40% benzoylated fiber particle fiber	0.0099	0.8134	2.2629	2.7153	3.2279	5.1563

It has been observed from the Table 13.4 that percent moisture absorbance of neat UPE and its composites increases with an increase in percent humidity level from 20-80%. Also, moisture absorbance has been found to increase with increase in percent fiber loading. This could be due to the presence of more –OH groups on the polymeric backbone at high loading, which causes more moisture absorption by UPE matrix-based composite materials. Further, among raw, mercerized and benzoylated fibers-reinforced UPE matrix-based composites, benzoylated fibers-reinforced ones have been found to have the highest moisture resistance, followed by mercerized and raw fiber. This trend matches with the trend obtained in the case of the water absorbance study and can be accounted for by the same explanation as given earlier in the water absorbance section.

Table 13.4 Data on % of moisture absorption shown by neat UPE and its composites reinforced with raw and surface-modified fibers at different humidity levels.

Sample Designation	%Humidity Levels		
	20	50	80
Neat UPE	0.08	0.11	0.25
UPE + 10% raw fiber	0.33	0.54	0.74
UPE + 20% raw fiber	0.39	0.65	0.81
UPE + 30% raw fiber	0.46	0.71	0.88
UPE + 40% raw fiber	0.54	0.85	1.05
UPE + 10% mercerized fiber	0.31	0.52	0.68
UPE + 20% mercerized fiber	0.38	0.63	0.76
UPE + 30% mercerized fiber	0.44	0.69	0.84
UPE + 40% mercerized fiber	0.50	0.78	1.02
UPE + 10% benzoylated fiber	0.19	0.40	0.55
UPE + 20% benzoylated fiber	0.25	0.43	0.61
UPE + 30% benzoylated fiber	0.32	0.47	0.67
UPE + 40% benzoylated fiber	0.37	0.57	0.74

13.3.6 Limiting Oxygen Index (LOI) Test

The LOI test of the neat UPE, raw, mercerized and benzoylated fibers-reinforced UPE composites and composites prepared with different fire-retardant [Magnesium hydroxide $Mg\ (OH)_2$ & Zinc borate $Zn_3(BO_2)_3 \cdot 5H_2O$] fillers were measured in accordance with ASTM D-2863-77 in an Oxygen Index Analyzer (Model No LOI-smoke-230, Dynsco Company, ALPHA Technologies). Sheets having dimensions of (150 × 150 × 5) mm^3have been used for the LOI tests. An oxygen analyzer has been used to determine the concentration of oxygen in the oxygen-nitrogen mixture. The test specimen was clamped in vertical position in a specimen holder with its top edge at least 100 mm below the top of the glass column. The entire top of the specimen was ignited by an ignition flame till it was well-lighted. The LOI was recorded as the lowest percentage oxygen level that continued the samples into their burning position for at least 180 sec duration. During this test, the oxygen levels were adjusted upward or downward by approximately 0.2% oxygen depending on whether the specimen continued to burn or not. Each sample has been characterized three times at different flow rates within 3–5 cm length.

The effect of different fire retardants and the fiber's surface modification on LOI values of polyester-*Grewia optiva* composites has been reported in Table 13.5. It can be observed from the table that when reinforced with raw fibers, the UPE matrix exhibits low LOI value in comparison to neat UPE matrix. Similar results have also been reported by Rai *et al.* during their study on the effect of coagulum on LOI of polyester

Table 13.5 Effect of fire retardants on limiting oxygen index (LOI) of UPE/*Grewia optiva*/Fire-retardant composites.

UPE: *Grewia optiva* Ratio (% by weight)	**Fire Retardant (% by weight of UPE)**	**LOI**
100% UPE		20
70%UPE + 30% raw particle fiber		19
70%UPE + 30% mercerized particle fiber		21
70%UPE + 30% benzoylated particle fiber		21
100% UPE	10% $Mg(OH)_2$	21
100% UPE	15% $Mg(OH)_2$	22
100% UPE	20% $Mg(OH)_2$	22
100% UPE	25% $Mg(OH)_2$	22
100% UPE	30% $Mg(OH)_2$	23
70%UPE + 30% raw particle fiber	30% $Mg(OH)_2$	22
70%UPE + 30% mercerized particle fiber	30% $Mg(OH)_2$	23
70%UPE + 30% benzoylated particle fiber	30% $Mg(OH)_2$	23
100% UPE	10% Zinc borate	20
100% UPE	15% Zinc borate	21
100% UPE	20% Zinc borate	21
100% UPE	25% Zinc borate	22
100% UPE	30% Zinc borate	22
70%UPE + 30% raw particle fiber	30% Zinc borate	21
70%UPE + 30% mercerized particle fiber	30% Zinc borate	22
70%UPE + 30% benzoylated particle fiber	30% Zinc borate	22

banana fiber composites [16]. This lower LOI value of raw particle fibers-reinforced composites may be due to the fact that cellulosic fibers are more flammable. However, on reinforcement of UPE composites with mercerized and benzoylated fibers, an increase in LOI value in comparison to neat and raw fibers-reinforced UPE composites has been observed. This increase in LOI value after surface modification of cellulosic fibers could be due to the higher thermal stability of surface-functionalized fibers, which causes combustion at higher temperature. These results have also been found to be consistent with the TGA results, which also support higher thermal stability of benzoylated fiber-reinforced UPE matrix-based composites.

From the table it has also been observed that the LOI value of neat UPE increases on reinforcement with fire-retardant fillers, and this LOI value increases considerably with a further increase in magnesium hydroxide/zinc borate loading (10 to 30% by weight). A considerable increase in fire retardancy has been found with magnesium hydroxide loading in comparison to zinc borate because of the high thermal stability of magnesium hydroxide [28]. The LOI values of *Grewia optiva* fibers-reinforced UPE composites have been found to increase when fire retardants were used in combination with these surface-modified fibers. In the case of both retardants (Magnesium hydroxide & Zinc Borate), maximum fire retardancy has been found with benzoylated fibers, followed by mercerized and raw particle fibers.

13.4 Conclusion

It has been observed from the above discussion that mechanical, physico-chemical and fire retardancy properties of UPE matrix increases considerably on reinforcement with surface-modified natural cellulosic fibers. The benzoylated fibers-reinforced composite materials have been found to have the best mechanical and physico-chemical properties, followed by mercerized and raw *Grewia optiva* fibers-reinforced composites. From the above data it is also clear that polymer composites reinforced with 30% fibers loading showed the best mechanical properties. Further, benzoylated fibers-reinforced composites were also found to have better fire retardancy properties than mercerized and raw fibers-reinforced polymer composites. Fire retardancy behavior of raw and surface-modified *Grewia optiva*/UPE composites have been found to increase when fire retardants were used in combination with fibers. This increase in fire retardancy behavior of resulted composites was attributed to the higher thermal stability of magnesium hydroxide/zinc borate.

References

1. T. Kikuchi, Woodfiber-plastic composites for furniture applications, in *Seventh international conference on wood fiber-plastic composites proceedings*, pp. 159–161, Forest Products Society, Madison, WI (2004).
2. B.C. Suddell, W.J. Evan , The increasing use and application of natural fiber composite materials within the automotive industry, in *Seventh international conference on wood fiber-plastic composites proceedings*, pp. 7–14, Forest Products Society, Madison, WI (2003).
3. A. Vazguez, J. Riccieri, L. Carvalho, Interfacial properties and initial step of water sorption in unidirectional unsaturated polyester/vegetable fiber composites. *Polym. Compos.* 20, 29–37 (1999).
4. M. Pommet, J. Juntaro, J.Y.Y. Heng, A. Mantalaris, A.F. Lee, K. Wilson, G. Kalinka, S.P. Shaffer Milo, and A. Bismarck, Surface modification of natural fibers using bacteria: depositing bacterial cellulose onto natural fibers to create hierarchical fiber reinforced nanocomposites. *Biomacromolecules* 9, 1643–1651 (2008).
5. S.S. Nair, S. Wang, and C.C. Hurley, Nanoscale characterization of natural fibers and their composites using contact-resonance force microscopy. *Compos. A* 41, 624–631 (2010).
6. S.V. Levchik, and E.D. Weil, A review of recent progress in phosphorus-based flame retardants. *J. Fire Sci.* 24, 345–364 (2006).

7. S.Y. Lu, and I. Hamerton, Recent developments in the chemistry of halogen-free flame retardant polymers. *Prog. Polym. Sci.* 27, 1661–1712 (2002).
8. S. Chapple, and R. Anandjiwala, Flammability of natural fiber-reinforced composites and strategies for fire retardancy: A review. *J. Thermoplast. Compos. Mater.* 23, 871–893 (2010).
9. A.R. Horrocks, and B.K. Kandola, Flammability having improved fire resistance, In *Desigen and Manufacture of Textile Composites*, A.R. Horrocks, and D. Price, (Ed.), pp. 330–363, Woodhead, Cambridge (2005).
10. B.K. Kandola, and E. Kandare, Composites having improved fire resistance, In *Advances in Fire Retardant Materials*, A.R. Horrocks, and D. Price, (Eds), pp. 398–442, Woodhead, Cambridge (2008).
11. M. Le Bras, C.A. Wilkie, and S. Bourbigot, *Fire retardancy of polymers: New applications for mineral fillers, Royal society of chemistry*, Cambridge (2005).
12. S. Bourbigot, and S. Duquesene, Fire retardants polymers: Recent developments and opportunities. *J. Mater. Chem.* 17, 2283–2300 (2007).
13. E.D. Weil, and S.V. Levchik, Flame retardants in commercial use or development for polyolefins. *J. Fire Sci.* 26, 5–43 (2008).
14. E.D. Weil, and S.V. Levchick, Flame retardants for polystyrene in commercial use or development. *J. Fire Sci.* 25, 241–266 (2007).
15. E.D. Weil, S.V. Levchick, and P. Moy, Flame and smoke retardants in vinyl chloride polymers-commercial usage and current developments. *J. Fire Sci.* 24, 211–236 (2006).
16. B. Rai, G. Kumar, R.K. Diwan, and R.K.Khandal, Study on effect of euphorbia coagulum on physic-mechanical and fire retardant properties of polyester-banana fiber composite. *Ind. J. Sci. Technol.* 4, 443 (2011).
17. N. Ayrilmis, S. Jarusombuti, V. Fueangvivat, P. Bauchongkol, and R.H. White, Coir fiber reinforced polypropylene composite panel for automotive interior applications. *Fibers Polym.* 12, 919–926 (2011).
18. B. Schartel, K.H. Pawlowski, and R.E. Lyon, Pyrolysis combustion flow calorimeter: A tool to assess flame retarded PC/ABS Materials. *Thermochim. Acta* 462, 1–14 (2007).
19. E. Braun, and B.C. Levin, Polyesters: A review of the literature on products of combustion and toxicity. *Fire Mater.* 10, 107–123 (1986).
20. A.S. Singha, and A.K. Rana, Effect of aminopropyltriethoxysilane (APS) treatment on properties of mercerized lignocellulosic Grewia optiva fiber. *J. Polym. Environ.* 21, 141–150 (2013).
21. A.S. Singha, and A.K. Rana, Effect of surface modification of Grewia optiva fibers on their physicochemical and thermal properties. *Bull. Mater. Sci.* 35, 1099–1110 (2012).
22. K. Joseph, and S. Thomas, *J. Reinf. Plast. Compos.* 12, 134 (1993).
23. P.J. Herrera-Franco, and A. Valadez-Gonzalez, A study of the mechanical properties of short natural-fiber reinforced composites. *Compos. B* 36, 597–608 (2005).
24. A.K. Mohanty, M.A. Khan, and G. Hinrichsen, Influence of chemical surface modification on the properties of biodegradable jute fabrics-polyester amide composites. *Compos. A* **31**, 143–150 (2000).
25. L.A. Pothan, C.N. George, M. Jacob, and S. Thomas, Effect of chemical modification on the mechanical and electrical properties of banana fiber polyester composites. *Compos. Mater.* 41, 2371–2386 (2007).
26. A.S. Singha, and A.K. Rana, Effect of graft copolymerization on mechanical, thermal and chemical properties of Grewia optiva/Unsaturated polyester biocomposites. *Polym. Compos. 33*, 1403–1414 (2012).

27. A.A.A. Rashdi, M.S. Salit, K. Abdan, and M.M.H. Megat, Water absorption behavior of kenaf reinforced unsaturated polyester composites and its influence on their mechanical properties. *Pertanika J. Sci. Technol.* 18, 433–440 (2010).
28. A. Durin-France, L. Ferry, J-M. Lopez Cuesta and A. Crespy, Magnesium hydroxide/zinc borate/talc compositions as flame-retardants in EVA copolymer. *Polym. Int.* 49, 1101–1105 (2000).

14

Chemical Modification and Properties of Cellulose-Based Polymer Composites

Md. Saiful Islam[*,1], Mahbub Hasan[2] and Mansor B. Ahmad[1]

[1]*Department of Chemistry, Faculty of Science, Universiti Putra Malaysia, Serdang, Selangor, Malaysia*

[2]*Department of Materials and Metallurgical Engineering, Bangladesh University of Engineering and Technology, Dhaka, Bangladesh*

Abstract

Chemical modification is an often followed route to improve specific characteristic properties of cellulose-based polymer composite materials. In this chapter, several types of chemical modifications along with the incorporation of nanocaly into cellulose-based materials are described, and their effect on the structural and mechanical properties of the resulting composites are discussed. All chemical treatments are intended to improve at least one property of the composites. They can, however, have a positive or negative impact on other composite properties. Interaction, adhesion and compatibility between the cellulose fiber and the polymer are the main concerns of chemical treatment. Furthermore, chemical modification reduces -OH groups from the cellulose fiber surface.

Among various chemical treatments, alkali treatment with NaOH solution is often chosen to modify cellulose-based materials. Since NaOH reacts with -OH groups of cellulose and reduces hydrophilicity and impurities from cellulose fiber, it improves the compatibility between the cellulose and the polymer matrix. Consequently, significant improvement in the mechanical and morphological properties of composite such as MOE, MOR, compressive modulus, SEM morphology and XRD pattern are observed for the alkali-treated composite materials. Benzene diazonium salt is also an important chemical, which reacts with cellulose in fiber and produces 2,6-diazo cellulose by a coupling reaction.

Also discussed in this chapter are the influence of other chemical treatments on the physical and mechanical properties of cellulose-based materials andthe effect of nanoclay on the morphological and mechanical properties of polymer composite.

Keywords: Chemical modifications, alkali treatment, nanocaly treatment, wood polymer composites, coir/PP composites, mechanical strength, FTIR, SEM, XRD

**Corresponding author*: msaifuli2007@gmail.com

Vijay Kumar Thakur (Ed.), Lignocellulosic Polymer Composites, (301–324) 2015 © Scrivener Publishing LLC

14.1 Introduction

In the recent years, cellulose-based polymer composites have drawn a lot of interest due to their wide range of applications. Cellulose fibers have many advantages, such as low cost, low density, high specific strength and high elasticity modulus, and at the same time are non-abrasive, nontoxic, can be easily modified by chemical agents, are abundant and come from renewable sources making their use very interesting [1]. Natural fibers are the main source of cellulose and the purpose of reinforcement is to provide strength, rigidity and help in sustaining structural load. Position and orientation of the reinforcement is maintained by the matrix or binder. Using natural fibers with polymer based on renewable resources will allow many environmental issues to be solved. Using biodegradable polymers as matrices, natural fiber-reinforced plastics are becoming the most environmentally-friendly materials; because they can be composed at the end of their life cycle [2].

Nowadays, synthetic polymers are combined with various reinforcing natural fillers in order to improve the physico-mechanical properties and obtain the characteristics demanded in definite applications [3]. Remarkable research is moving towards using lignocellulosic fibers as reinforcing fillers [4]. Compared to synthetic fibers (i.e., talc, silica, glass fiber, carbon fiber, etc.), the lignocellulosic fibers (i.e., cornstalk, coir, bamboo, plam, rice husk, rice straw, jute, abaca, sawdust, wheat straw, sisal, coconut, pineapple and grass) are lightweight, biodegradable, easily available, renewable and inexpensive, which makes them generally more competent than the others in terms of economics and practicality [5].

Compared to other thermoplastics, polypropylene (PP) possesses many outstanding properties such as low density, good flex life, sterilizability, good surface hardness and excellent abrasion resistance [6]. However, it has been established that PP and its matrix do not form strong bonds with cellulose fibers. Poor chemical and physical interfacial interactions (adhesion and compatibility) between the fiber surface and polymer are two of the most important mechanisms of bond failure. Therefore, the polymer and cellulose fibers of the composite simply mix together without making a strong interaction between them. It can therefore be deduced that if bonding were to take place between the polymer and the natural fibers, the overall properties of the composite may be further improved [7]. Chemical treatments with suitable chemicals could be a promising way to improve the adhesion and compatibility of natural fiber-reinforced polymer composite [8]. In order to improve the bonding strength between polymer and cellulose fiber, several types of chemical treatments such as alkali, benzene diazonium salt, *o*-hydroxybenzene diazonium salt, succinic anhydride, maleic acid, acrylonitrile, etc., have been carried out on cellulose fibers. Among them, some are very effective in improving the adhesion and compatibility of cellulose fiber to polymer composite, whereas some have a negative impact [9,10].

In this chapter, various types of chemical treatments including alkali, benzene diazonium salt, *o*-hydroxybenzene diazonium salt, succinic anhydride, maleic acid, acrylonitrile and nanocaly onto cellulose-based materials will be described and their effect on the structural and mechanical properties of the resultant composites will also be discussed.

14.2 Alkali Treatment

Alkali treatment with sodium hydroxide (NaOH) has been widely used to modify natural fibers [2, 11,12]. This treatment has the potential to reduce of hydrophilic hydroxyl groups and impurities from the cellulose fiber and to increase adhesion and compatibility of fiber to polymer matrix. Usually, 2-5% NaOH solution has been used for alkali treatment. Alkaline sodium hydroxide removes natural fats and waxes from the cellulose fiber surface, thus revealing chemically reactive functional groups like hydroxyl groups. The removal of the surface impurities from the cellulose fibers also improves the surface roughness of the fibers or particles, thus opening more hydroxyl groups and other reactive functional groups on the surface. Sodium hydroxide also reacts with accessible –OH groups according to the following proposed chemical reaction [13,14].

$$\text{Cellulose- OH} + \text{NaOH} \;\; \text{Cellulose- O}^{-}\text{Na}^{+} + H_2O + \text{impurities} \tag{14.1}$$

The reactions of NaOH with cellulose in natural fiber yielded cellulose-ONa compound and removed impurities from the fiber surface [11,12]. This is confirmed by the FTIR spectroscopic analysis as shown in Figure 14.1. The FTIR spectrum of the raw wood clearly shows the absorption band in the region of 3407 cm^{-1}, 2917 cm^{-1} and 1736 cm^{-1} due to O-H, C-H and C = O stretching vibration respectively. These absorption bands are due to hydroxyl groups in cellulose, carbonyl groups of acetyl ester in hemicellulose and carbonyl aldehyde in lignin.

After treatment with 5% NaOH, the characteristic peak at 2917 cm^{-1}, 1736 cm^{-1} and 1636 cm^{-1} fully disappeared and there is a slight shift of the –OH peak by about 31 cm^{-1} to the high frequencies obtained. These characteristic peaks may be due to the removal of surface impurities and the formation of cellulose-ONa compound on the fiber surface [11,15].

Islam *et al.* (2012) reported that the mechanical and morphological properties of wood polymer composites (WPCs) were improved significantly after pretreatment with alkali solution. For his study, methyl methacrylate (MMA) and styrene (ST)

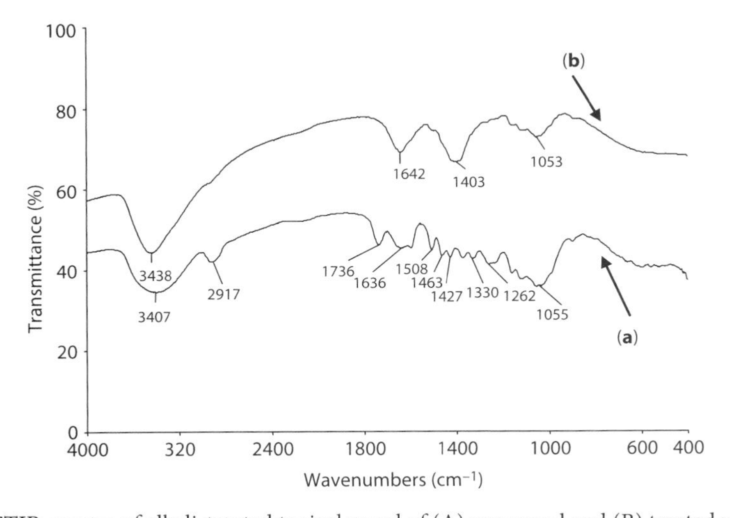

Figure 14.1 FTIR spectra of alkali-treated typical wood of (A) raw wood and (B) treated wood.

vinyl monomer mixture (50:50; volume: volume) was used in preparation of WPCs. Before being impregnated with an MMA/ST monomer mixture, wood species were chemically pretreated with 5% sodium hydroxide (NaOH) solution for the reduction of hydrophilic hydroxyl groups and impurities from the cellulose fiber in wood and to increased adhesion and compatibility of wood fiber to polymer matrix. Monomer mixture (MMA/ST) was impregnated into raw wood and NaOH pretreated wood specimens to manufacture WPC and pretreated wood polymer composite (PWPC).

Figure 14.2 (i) shows a number of void spaces and uneven layers in the raw wood fiber surface, which is removable by the suitable chemical treatment [16]. Figure 14.2 (ii) depicts the micrograph of WPC, while Figure 14.2 (iii) is that of PWPC. Figure 14.2 (ii) shows clean polymer stands throughout the wood fibers with remarkable gaps between this polymer and the cell walls, while Figure 14.2 (iii) shows no noticeable gaps and strong bonds between the polymer and the cell wall. Furthermore, fibrous cellulose material adhered to the surface of the polymer stands. It is thus deduced that the reaction of NaOH with wood cellulose had increased the adhesion and compatibility of the polymer to the cellulose fibers of the wood.

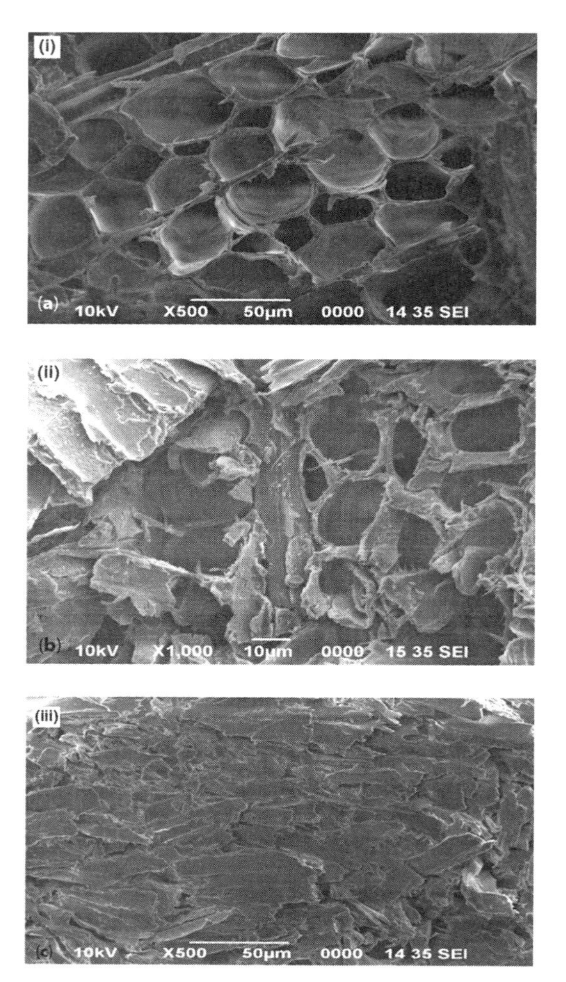

Figure 14.2 SEM photographs of typical wood of (i) raw wood, (ii) WPC and (iii) PWPC.

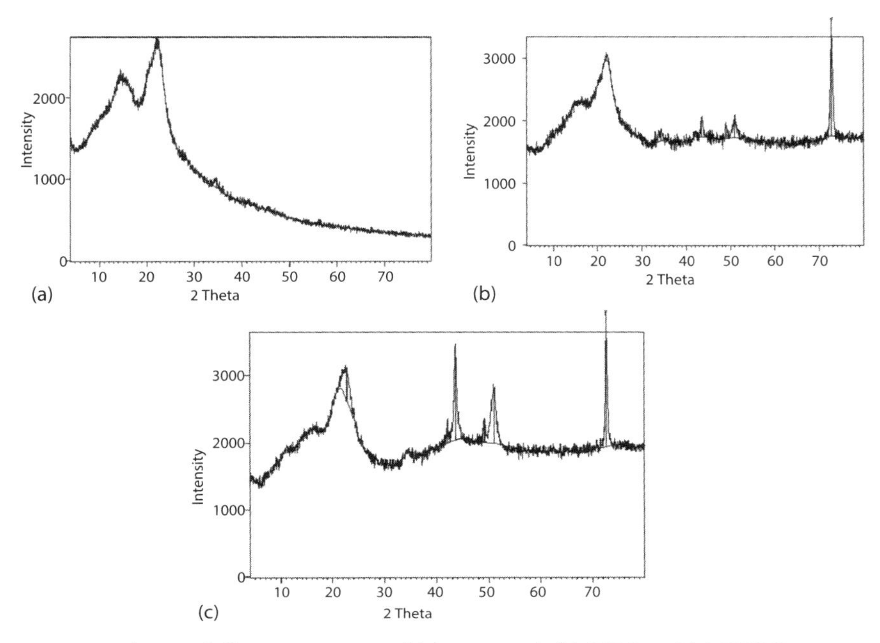

Figure 14.3 Typical x-ray diffraction patterns of (a) raw wood, (b) WPC and (c) PWPC.

The study also reported XRD results for composites sample. The X-ray diffraction patterns of raw wood, WPC and PWPC are given in Figure 14.3(a-c) respectively. As seen in Figure 14.3(a), the patterns of raw wood fibers exhibit three well defined peaks (2θ) at 15.1, 22.8 and 34.7°. The 15.1, 22.8 and 34.7° reflections correspond to the (110), (200) and (023) or (004) crystallographic planes, respectively [17]. On the other hand, comparing all the X-ray diffractograms (Figures 14.3(a), (b) and (c)) it is observed that there are some new peaks (2θ) of various intensities in the amorphous region 40° to 75°. These peaks may be due to the incorporation of MMA/ST inside wood and the formation of wood composites. The diffraction patterns of WPC in (Figure 14.3(b)) exhibits four new diffraction peaks at 43.6 °, 49.1 °, 50.9 °, and 72.7 ° whereas, the PWPC (Figure 14.3(c)) shows five new peaks of high and sharp intensity at 42.1 °, 43.6 °, 49.2 °, 51.0 °, and 72.6 °. This result also indicates that the manufactured WPC and PWPC significantly increased the crystallinity of wood as reported by other researchers [18]. However, PWPC had more crystallinity peaks compared to the raw wood and WPC as shown in (Figure 14.3(b)). Such a result is expected because the alkali pretreatment removed all impurities from the wood fiber surface thus increasing the impregnation of MMA/ST and the degree of polymerization inside the wood fiber [11].

The mechanical property of this study was also reported. The modulus of elasticity (MOE) and modulus of rupture (MOR) of raw wood, WPC and PWPC were measured and results are given in Table 14.1. The MOE of WPC and PWPC were higher than those of their raw ones. It is worth noting that the MOE was more significantly affected by the NaOH pretreatment. This result is expected because NaOH reacts with cellulose in wood and enhances the adhesion and compatibility between wood fibers and the polymer resulting in improved MOE. The higher MOE of both WPC and PWPC

Table 14.1 Static Young's modulus of raw wood, WPC and PWPC.

Wood species	Sample particulars	MOE (GPa)	H. G.	MOR (MPa)	H. G.
Jelutong	Raw	5.31 ± 0.44	A	46 ± 3.35	A
	WPC	7.90 ± 0.39	B	62 ± 3.02	B
	PWPC	8.52 ± 0.49	C	72 ± 5.60	C
Terbulan	Raw	7.39 ± 1.15	D	60.5 ± 4.2	D
	WPC	10.15 ± 0.39	E	77 ± 5.29	E
	PWPC	11.14 ± 0.40	F	88 ± 6.86	F
Batai	Raw	6.51 ± 0.57	G	55.4 ± 2.24	G
	WPC	9.30 ± 0.48	H	73 ± 4.64	H
	PWPC	10.00 ± 0.40	I	83 ± 7.31	I
Rubberwood	Raw	11.62 ± 1.05	J	105.8 ± 5.12	J
	WPC	15.20 ± 0.39	K	130 ± 1.63	K
	PWPC	17.24 ± 0.32	L	150 ± 2.03	L
Pulai	Raw	4.12 ± 1.80	M	37.80 ± 1.5	M
	WPC	6.00 ± 0.26	N	50 ± 7.11	N
	PWPC	6.50 ± 0.53	O	58 ± 6.61	O

Mean value is the average of 10 specimens.

The same letters are not significantly different at $\alpha = 5\%$. Comparisons were done within each wood species group. (H. G. = Homogeneity Group).

compared to the raw wood were due to the chemical modification and impregnation, which is in accordance with previous research [19].

In the wood specimens, NaOH reacted with OH groups of cellulose fiber and yielded cellulose-ONa compound and removed the impurities from the wood fiber surfaces, thus enhancing polymer loading and degree of polymerization, which further increased the MOE of PWPC. It is also apparent from Table 14.1 that the MOE of WPC and PWPC of the Jelutong wood were highest, followed by Pulai, Batai, Terbulan, and Rubberwood respectively. However, for Rubberwood, a small increase was found for its WPC and PWPC due to high density of this species and a little amount of MMA/ST incorporation inside the cell wall, as found by other researchers [20]. On the other hand, the MOR of PWPC was higher than those of WPC and raw wood. As one can see from Table 14.1 there was significant improvement in MOR of PWPC for all wood species. These results suggest that the pretreatment enhanced the interfacial bonding strength between wood fibers and polymer.

14.3 Benzene Diazonium Salt Treatment

Recently, natural fiber has also been treated with benzene diazonium salt to improve its physical, mechanical and thermal properties. Benzene diazonium salt reacted with cellulose in wood and produced 2, 6-diazocellulose by a coupling reaction, as confirmed

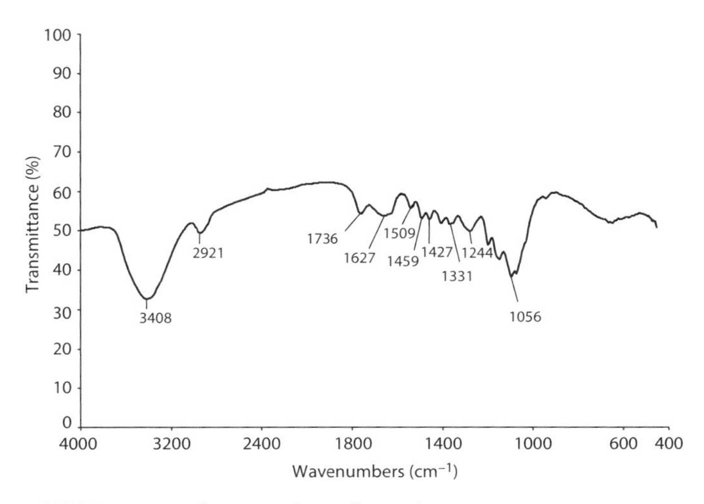

Figure 14.4 Typical FTIR spectra of untreated wood samples.

by FTIR spectrum analysis. The mechanical properties of treated samples in terms of dynamic Young's modulus (E_d),modulus of elasticity (MOE), compressive Young's modulus was seem to be enhanced, whereas modulus of rupture (MOR) shown to drop off on treatment. The morphological properties were studied by FTIR and SEM and found to be changed. A significant improvement was found in the treated woods compared to the untreated ones.

The formation of 2,6-diazo cellulose compound by the coupling reaction with diazonium salt and cellulose fiber was confirmed by the FTIR spectroscopic analysis of the untreated and treated wood as shown in Figures 14.4 and 14.5. The FTIR spectrum of the untreated wood clearly shows the absorption bands in the region of 3408 cm^{-1}, 2921 cm^{-1}, and 1736 cm^{-1} due to O-H stretching vibration, C-H stretching vibration, and C = O stretching vibration respectively. These absorption bands are due to the hydroxyl group in cellulose, carbonyl group of acetyl ester in hemicellulose, and carbonyl aldehyde in lignin [21]. On the other hand, FTIR spectra of treated wood in Figure 14.5 clearly show the presence of the characteristic band of NO group in the region of 1512 cm^{-1} and 1646 cm^{-1}. The peaks at 1403 and 1457 cm^{-1} are due to the -N = N- moiety of the azo compound, and the absorption band at 1309 cm^{-1} may be attributed to the symmetric deformations of NO_2 presence in the cellulose azo compound [1]. The absorption band of O-H group also shifted towards higher wave number (3408 cm^{-1} to 3437 cm^{-1}) with weak band intensity, which gives further evidence of the reaction of cellulose hydroxyl groups with diazonium salt.

Yields of 2,6-diazo cellulose can be explained as being due to the presence of three hydroxyl groups in the cellulose anhydroglucose unit. One is the primary hydroxyl group at C_6 and the other two are secondary hydroxyl groups at C_2 and C_3. The primary hydroxyl group is more reactive than the secondary ones, and the coupling reaction at C_2 and C_6 results in the formation of 2,6-diazo cellulose (Figure 14.6).

The SEM images of untreated and treated wood samples are shown in Figures 14.7 (a) and (b) in transverse section. The untreated wood in Figure 14.7 (a) clearly shows the number of void spaces and agglomeration of fiber in the wood surface, which could

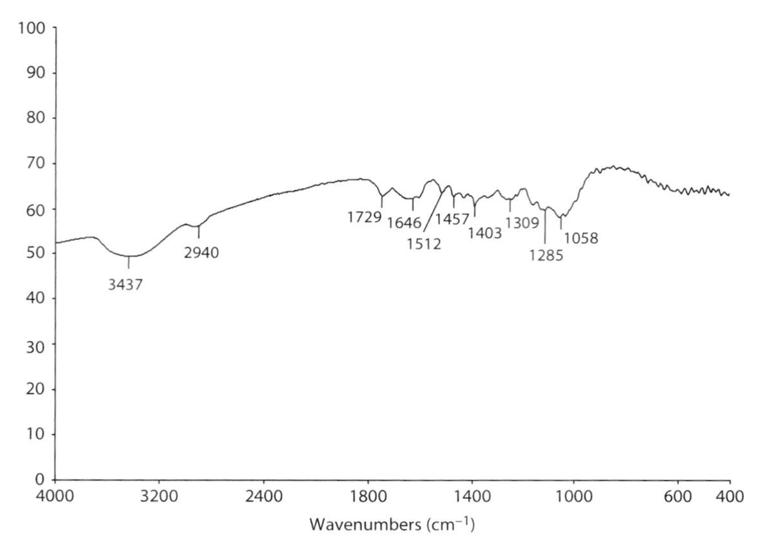

Figure 14.5 FTIR spectra of treated wood samples.

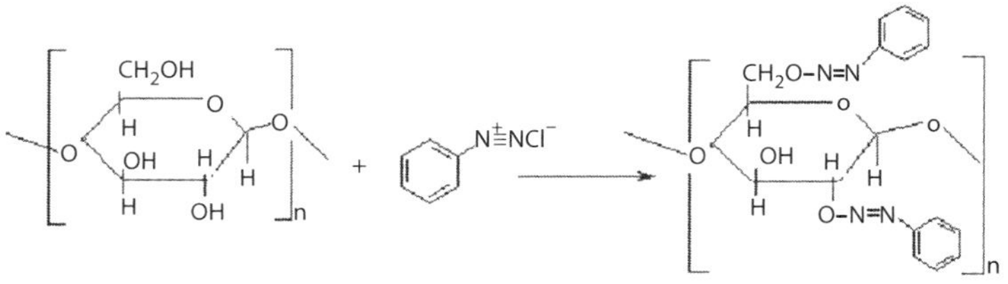

Cellulose in wood

2, 6-diazo cellulose

Figure 14.6 The coupling reaction of benzene diazonium salt with cellulose unit.

be removable by suitable chemical treatment displays the micrograph of a treated wood sample [22].

Comparing the images of untreated wood and treated wood reveals that the diazonium salts in the wood form a rough coating and interaction on the fiber surface, which enhanced the morphological properties. Furthermore, diazonium salts treatment partially filled the void spaces of wood and aligns uneven fiber, which provided a smoother surface texture than raw wood. This result also suggests that the chemical reaction formed a strong interface with wood cell walls, accounting for the observed increase in mechanical and thermal properties.

The Dynamic Young's modulus (E_d) of untreated and treated wood samples is given in Figure 14.8. The E_d for all treated wood samples was found to be higher than for untreated ones. The treated Batai wood samples exhibited the highest percentage increase in E_d by 11.08% followed by Pulai (7.13%), Jelutong (5.90%), Terbulan (5.18%) and Rubberwood (2.02%) respectively. This result is expected, because the coupling reaction with cellulose in wood enhanced the hydrophobic nature of wood and coating among wood fibers, improving Young's modulus. The higher E_d in treated sample seen in all species compared to untreated wood was due to the chemical modification, which is in accordance with other researchers [23]. In the wood specimens, the coupling reagent reacts with OH groups of cellulose fiber to yield 2,6-azo compound as well as cellulose compound, thus enhancing the E_d of the treated wood sample.

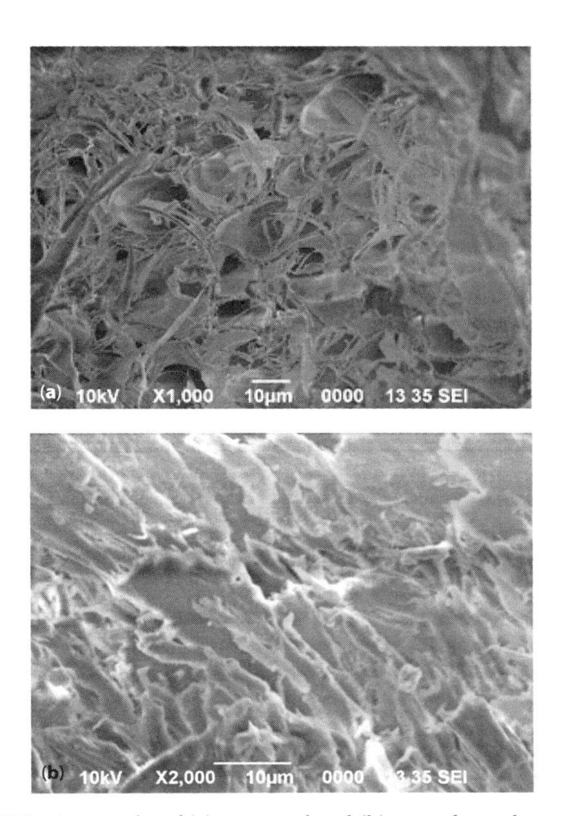

Figure 14.7 Typical SEM micrographs of (a) untreated and (b) treated wood samples.

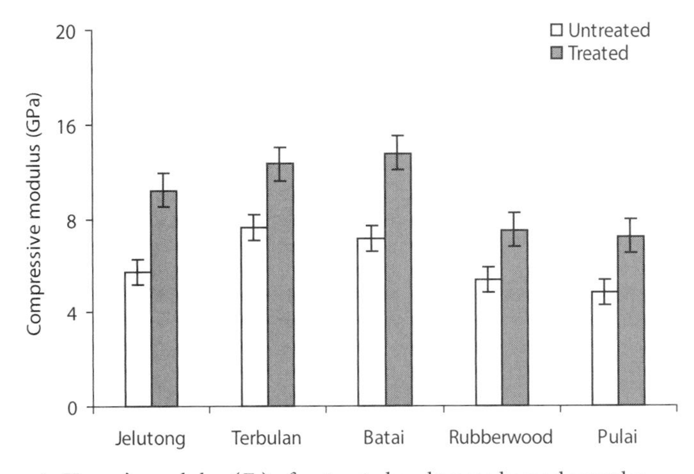

Figure 14.8 Dynamic Young's modulus (E_d) of untreated and treated wood samples.

The modulus of elasticity (MOE) and modulus of rupture (MOR) of untreated and treated wood samples is shown in Table 14.2. A slight increased was observed for MOE of the all treated wood samples. The MOE of Jelutong showed highest increment on treatment, followed by Pulai, Batai, Terbulan and Rubberwood respectively. The higher MOE of treated samples compared to the untreated wood was due to the chemical modification. In the wood specimens, coupling reagent reacted with OH groups of cellulose fiber and yielded 2, 6- azo compound, thus enhancing the MOE of woods. However, for

Table 14.2 Modulus of elasticity (MOE) and modulus of rupture (MOR) of untreated and treated wood. (Std.: standard deviation, HG: Homogeneity group).

Wood species	Sample particulars	MOE (GPa)	Std.	H.G	MOR (MPa)	Std.	HG
Jelutong	Untreated	5.31	0.45	a	46.58	0.47	a
	Treated	5.45	0.52	a	28.17	0.65	b
Terbulan	Untreated	7.39	1.10	b	60.19	0.56	c
	Treated	7.75	0.75	b	55.27	1.10	d
Batai	Untreated	6.51	0.45	c	55.35	0.76	e
	Treated	7.55	1.13	c	51.55	0.77	f
Rubberwood	Untreated	11.62	1.30	d	105.94	0.74	g
	Treated	11.75	0.65	d	39.00	0.54	h
Pulai	Untreated	4.12	1.43	e	37.83	1.30	k
	Treated	4.15	0.65	e	18.60	0.65	i

Mean value is the average of 10 specimens.

The same letters are not significantly different at $\alpha = 5\%$. Comparisons were done within the each wood species group.

Rubberwood, a small difference was found between untreated and its treated samples because of its high density and a little amount of chemical incorporation inside the cell wall [24]. It can be shown from Table 14.2 that the modulus of rupture (MOR) decreased after chemical treatment.

The lower MOR of treated samples compared to the untreated one is due to the reaction of coupling reagent with cellulose unit in wood yielded azo compound as stated earlier which improved hardness and decreased MOR. This indicates that coupling reaction in wood enhanced brittleness of wood.

Figure 14.9 shows the compressive modulus parallel to the grain for treated and untreated wood species. The figure shows that there was significant increase (35-62%) of compressive modulus for all the treated wood species. Untreated wood species failed in compression because of the bulking of relatively thin cell walls due to a long column type of instability. The chemical modification of raw wood puts a coating on the walls which thickens them, thus greatly increase their lateral stability [25].

It is also expected because benzene diazonium salt has the ability to react with the cellulose, thus forming azo compound with improve compressive modulus. This enhanced the lateral stability of the cell wall.

14.4 *o*-hydroxybenzene Diazonium Salt Treatment

The physico-mechanical properties of *o*-hydroxybenzene diazonium salt-treated coir-reinforced polypropylene (PP) composites been investigated by Islam *et al.* [26].

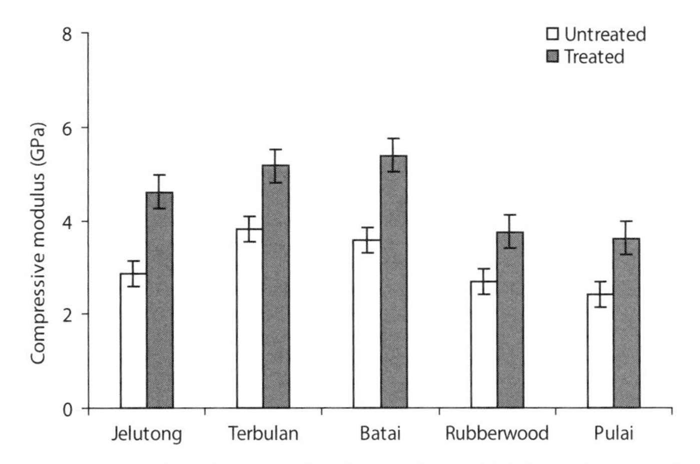

Figure 14.9 Compressive modulus of untreated and treated wood (different letters reflected statistical significance).

Coir fiber was chemically treated with *o*-hydroxybenzene diazonium salt to improve mechanical properties of the composites. The mechanical properties of the composites prepared from chemically treated coir were found to be much better compared to those of untreated ones. The surface morphologies of the fractured surfaces of the composites were also found to be changed.

In the study, surface modification of the coir fiber was carried out to achieve better mechanical properties of composites and the results were compared with those of the untreated ones. *o*-hydroxybenzene diazonium salt breaks the OH groups at C_2 and C_6 positions of the cellulose in the coir, producing 2,6-diazocellulose (Figure 14.10). Maybe due to the steric effect, the hydroxyl group at C_3 did not undergo chemical reaction with the bulky diazo group of the salt. It is clear in the spectrum that untreated coir shows presence of absorption band near 1734 cm^{-1}, which is assigned to CO group of acyl ester of hemicellulose and ahdehyde group in lignin (Figure 14.11). On the other hand, chemically treated coir shows the absorption band in the regions 1700–1600 cm^{-1} and 1300–1000 cm^{-1} for N–O and C–O stretching, respectively. These results suggest that chemical modification of coir has occurred upon treatment with diazonium chloride.

In order to achieve better mechanical properties of the composites, the fiber was chemically treated with *o*-hydroxybenzene diazonium salt synthesized from the reaction of *o*-hydroxyaniline and sodium nitrite dissolved in hydrochloric acid. It is to be noted that phenolic OH present in the aromatic ring of the diazonium salt reacted with the excess sodium hydroxide during treatment of coir producing phenoxide. Therefore, the hydrophilic nature of the phenolic OH group was diminished during the treatment of coir. The hydroxyl groups of cellulose couple with the diazonium salt to produced 2,6-diazocellulose. This reduced the hydrophilic property of cellulose responsible for moisture absorption. Consequently, better interfacial adhesion between the fiber and the matrix occurred, imparting improvement in the mechanical properties of the composites. It is interesting to note here that the tensile strength values of the treated coir PP composites at all mixing ratios were found to be higher than that of the neat PP, suggesting better compatibility between the chemically modified filler and the

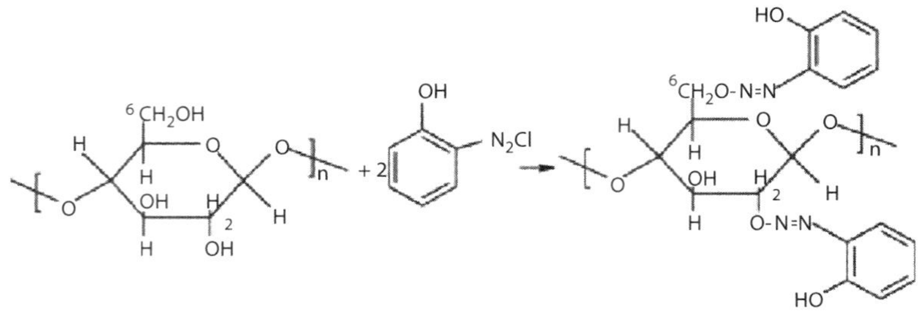

Figure 14.10 Treatment of coir with o-hydroxybenzene diazonium salt.

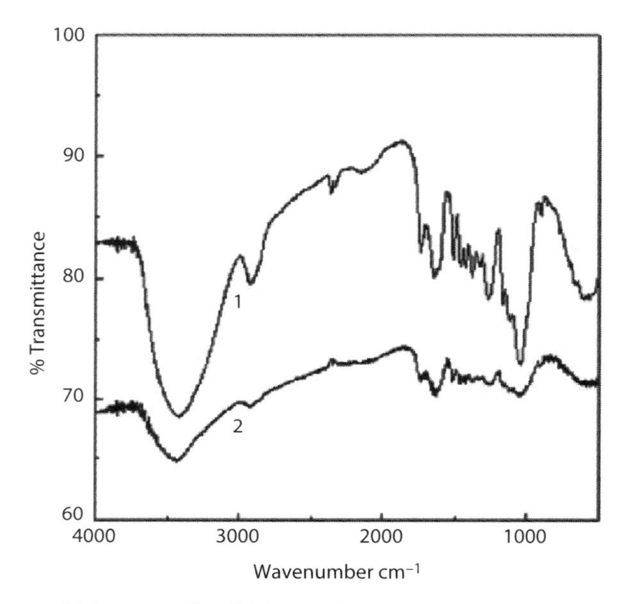

Figure 14.11 IR spectra of (1) untreated and (2) treated coir.

hydrophobic matrix. This indicates that interfacial bonding between the filler and the matrix has significantly improved upon chemical treatment leading to increased stress transfer efficiency from the matrix to the fiber

The morphological features of tensile fractured surfaces of 20% coir-reinforced PP composites were also investigated by SEM in the study. It can be easily traced out that there are a number of agglomerations, resulting from strong fiber–fiber interaction and fiber pullout marks in the SEM images of the untreated coir PP composite. This implies that there is poor interfacial bonding between the matrix and the untreated filler (image A). On the other hand, chemically treated coir-PP composites show almost no signs of fiber agglomeration and micro-voids in the fractured surface of the composite. This result suggests that interfacial adhesion between the chemically treated coir and the PP matrix became much more favorable upon treatment of the fiber with diazonium salt. Furthermore, the pullout traces were found to have substantially decreased for treated coir PP composites, indicating that better filler–matrix interaction has occurred upon chemical treatment of coir.

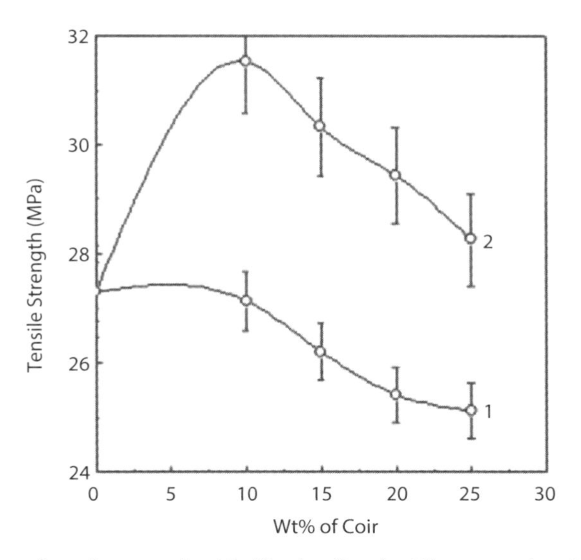

Figure 14.12 Variation of tensile strength with filler loading for (1) untreated and (2) treated coir-reinforced PP composites.

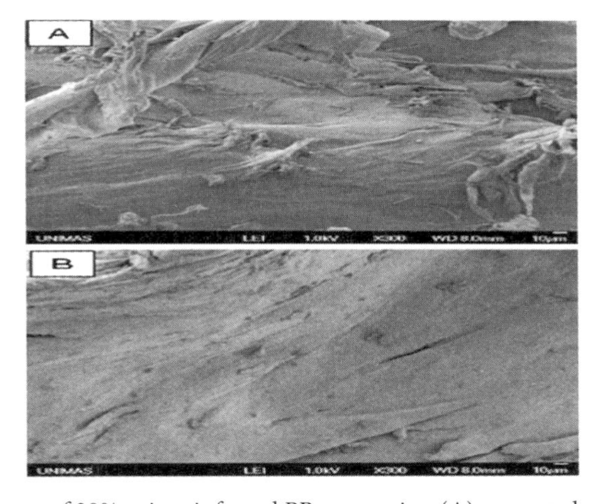

Figure 14.13 SEM images of 20% coir-reinforced PP composites (A) untreated coir- and (B) treated coir PP composites.

14.5 Succinic Anhydride Treatment

The effect of succinic anhydride treatment on physical and mechanical properties of jute fiber-reinforced polypropylene (PP) biocomposites with different fiber loading was investigated by Ahmed *et al.* [27]. Before composite manufacturing, raw jute fiber was chemically treated with succinic anhydride to react with the cellulose hydroxyl group of the fiber and to increase adhesion and compatibility with the polymer matrix. Jute fiber/PP composites were fabricated using high voltage hot compression technique. Fourier transform infrared spectroscopy (FTIR) and scanning electron microscopy (SEM) were employed to evaluate the morphological properties of composite. Succinic anhydride underwent a chemical reaction with raw jute fiber, which was confirmed

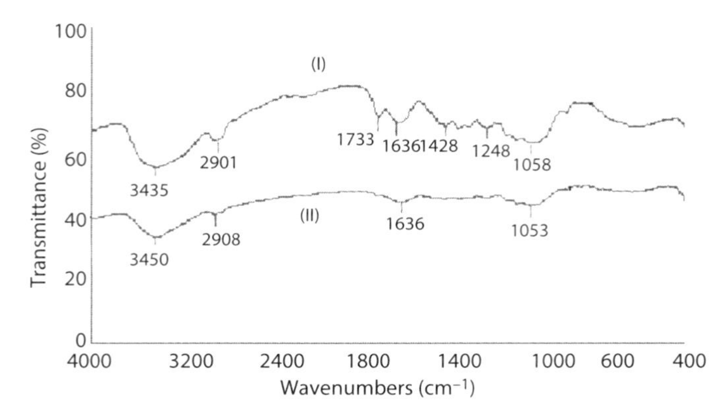

Figure 14.14 FTIR spectra of (I) raw and (II) treated jute fiber.

through FTIR results. SEM micrographs of the fractured surface area were taken to study the fiber/matrix interface adhesion and compatibility. Reduced fiber agglomeration and improved interfacial bonding was observed under SEM in the case of treated jute fiber/PP composites. The mechanical properties of jute/PP composite in terms of the tensile strength and Young's modulus was found to be increased with fiber loading up to 15 wt% and decreased at 20 wt%. Conversely, flexural strength and flexural modulus increased with fiber loading up to 10 wt% and start decreasing at 15 wt%. The treated jute/PP composite samples had higher hardness (Rockwell) and lower water absorption value compared to that of the untreated ones.

Reaction between succinic anhydride and cellulose in jute fibers was confirmed by FTIR spectrum analysis of the raw and treated jute fiber as shown in Figure 14.14. The FTIR spectrum of the raw fiber clearly shows the absorption band in the region of 3435 cm^{-1}, 1733 cm^{-1} and 2901 cm^{-1} due to O-H stretching vibration, C = O stretching vibration and C-H stretching vibration respectively. These absorption bands are due to the hydroxyl group in cellulose, carbonyl group of acetyl ester in hemicellulose and carbonyl aldehyde in lignin [19]. On the other hand, the characteristic peak at 1733 cm^{-1} and other peaks, i.e., 1428 cm^{-1} and 1248 cm^{-1} were completely disappeared upon chemical treatment with succinic anhydride (Figure 14.14 (II)). This observation indicates that succinic anhydride successfully interacted with jute fiber and removed some surface impurities/waxy substances from the fibers surfaces which led to improve the adhesion and compatibility of jute fiber to the PP matrix. It can also be seen from the Figure 14.14 (II), the absorption band of O-H and C-H group shifted towards higher wave numbers with narrow band intensity, which gave further evidence of the reaction between jute fiber and succinic anhydride. All these, as stated above, confirmed the suggested chemical reaction as mentioned in Figure 14.15.

The typical SEM micrographs of the tensile fracture surface of raw and treated jute fiber/PP composites at 20 wt% fiber loadings are shown in Figures 14.16 (a) and (b). The SEM image of raw jute fiber/PP composite shows number of pullout traces of fiber with rough surfaces and micro-voids as well as agglomeration of the fiber in the PP matrix (Figure 14.16 (a)) [28]. In addition, the matrix and fiber are clearly distinguishable in the raw fiber composite compared to the treated one. This feature indicated that there was poor dispersion and weak interfacial bonding between the matrix and the

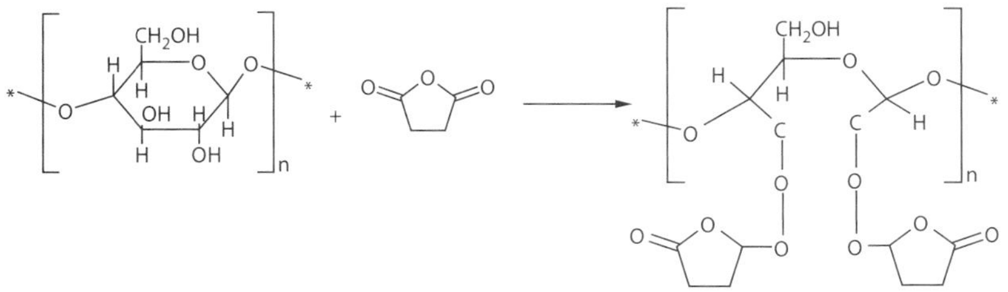

Figure 14.15 The chemical reaction scheme of raw jute fiber with succinic anhydride.

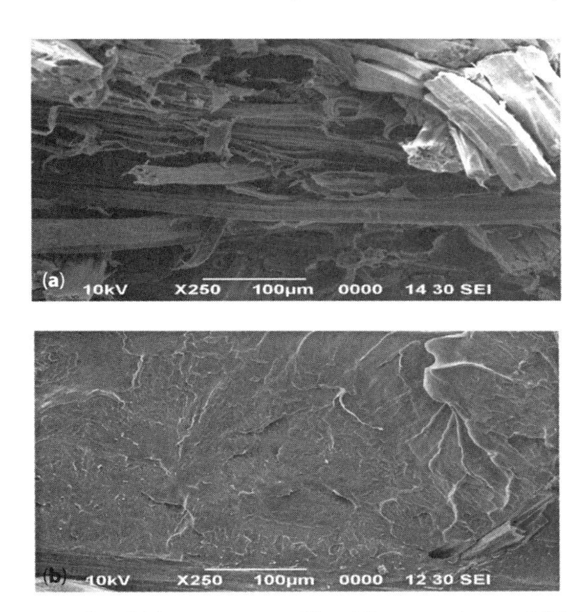

Figure 14.16 SEM micrographs of (a) 20% raw jute fiber-PP composites and (b) 20% treated jute fiber PP composites.

fiber, which confirmed that the interfacial bonding between the filler and the matrix polymer was poor and weak. In contrast, the SEM image for the treated jute/PP composite shows a better dispersion of filler throughout the matrix (Figure 14.16 (b)). This resulted in better interfacial bonding between the fiber and the matrix, as is reflected in the FTIR results. The strong interfacial bonding between the fiber and PP matrix is due to the chemical surface modification of raw jute fiber with the succinic anhydride [29]. Therefore, it can be deduced from the SEM micrographs that the chemical treatment enhanced the phase compatibility and adhesion between the fiber and the PP matrix, which significantly increased the surface morphology of the composite.

The tensile strength of raw and treated jute fiber/PP composites at different fiber loading is shown in Figure 14.17. It is clear from the figure that values of tensile strengths of raw jute fiber/PP composites were increased with an increase in filler content up to 10% and decreased at 15%. In contrast, significant improvement in tensile strengths is found for treated jute fiber/PP composites up to 15% filler content. The decreasing trend of raw composites may be due the weak interfacial area between the filler and matrix which increases with an increase of filler content to the composite [30]. Hydrophilic nature of

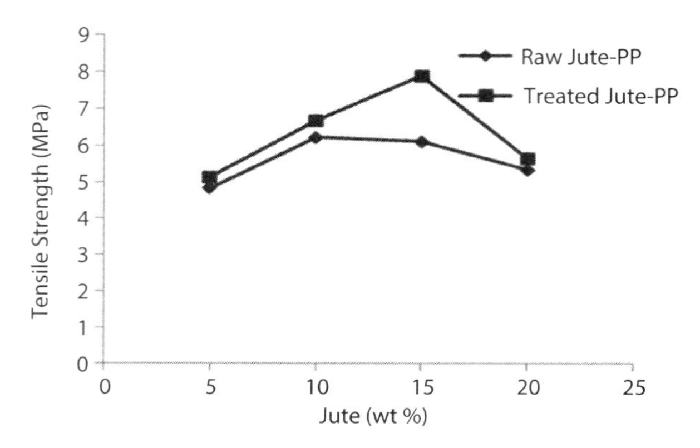

Figure 14.17 Tensile strength of raw and treated jute fiber-reinforced PP composites.

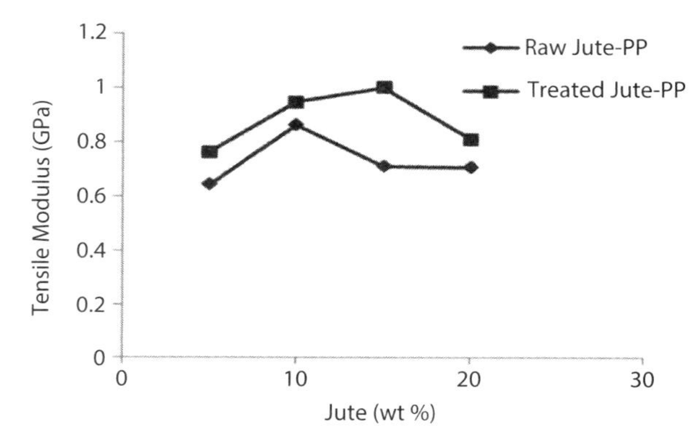

Figure 14.18 Tensile modulus of raw and treated jute fiber reinforced PP composites.

jute fibers were significantly reduced by the succinic anhydride treatment. As a result, interfacial adhesion between the filler and matrix has improved. This in turn improved the tensile strengths of treated jute fiber composites. Significant improvement in tensile strengths is found for treated composites compared to raw composites. It is noted that the tensile strength values of the treated composites increased by approximately 6-29% over the raw composites. This result indicates that the chemical treatment improved the adhesion and compatibility between jute fibers and PP composite resultant composites significantly increased the value of tensile strength. Furthermore, the tensile strength of treated jute fiber/PP composites at 15% loading exhibited higher improvement compared to raw one (i.e., treated composite-7.88MPa and raw composite- 6.1MPa). This result indicates that fiber treatment was able to improve the fiber-matrix interfacial adhesion, leading to better stress transfer efficiency from the matrix to the fiber and consequently improved mechanical properties. It is noted that the tensile strength for jute fiber/PP composites found was 4.82-7.88MPa.

Figure 14.18 shows the Young's modulus of the composites of raw and treated jute fiber at different filler loading. The chemically treated jute fiber/PP composites are found to show higher modulus compared to those of the untreated ones. This indicates that homogeneous dispersion of jute particles and better filler-matrix interaction has occurred upon treatment of jute fiber. During tensile loading, partially separated micro spaces are created, which obstructs stress propagation between the fiber and matrix. As the fiber loading increases, the degree of obstruction increases, which consequently increase the stiffness. However, the treated jute fiber/PP composites are found to show higher modulus compared to raw composites over the filler content.

14.6 Acrylonitrile Treatment

In the study, chemically treated kenaf-reinforced PP composites were fabricated using a single extruder and an injection molding technique [31]. Raw kenaf was chemically treated with acrylonitrile to increase its hydrophobicity and compatibility with polymer matrix (PP). Acrylonitrile reacted with kenaf fiber, which was confirmed through FTIR spectroscopy. Both raw and treated kenaf was utilized and four levels of filler loading (25, 30, 35 and 40 wt%) were used during composite manufacturing. The mechanical properties of the treated kenaf-reinforced PP composites were found to be much improved compared to the corresponding values of the untreated ones.

Water absorption of the composites increased with an increase in fiber content. However, treated kenaf/PP composites showed lower water uptake capacity compared to those prepared from raw kenaf, indicating that upon chemical treatment the number of hydroxyl groups in the cellulose of kenaf was decreased, thus reducing hydrophilic nature of the fiber. The surface morphology of the composites was investigated through scanning electron microscopy (SEM). This result showed that raw kenaf/PP composites possess microvoids, fiber agglomerates and surface roughness with extruded fiber moieties. However, due to satisfactory interaction between the treated kenaf and the PP matrix, agglomerates and micro-voids in the composites was largely minimized showing better dispersion of the fiber in the matrix. Hydrophilic nature of kenaf was significantly minimized upon surface modification, giving better fiber-matrix interfacial adhesion and improved mechanical properties of the composites.

FTIR test were performed in order to confirm the reaction between kenaf and acrylonitrile. The raw kenaf exhibited some characteristic absorption band in the region of 3425 cm^{-1}, 1726 cm^{-1} and 2915 cm^{-1} due to O-H stretching vibration, C = O stretching vibration and C-H stretching vibration respectively. These absorption bands are due to the hydroxyl group in cellulose, carbonyl group of acetyl ester in hemicellulose and carbonyl aldehyde in lignin [1,19]. On the other hand, the carbonyl stretching vibration peak (1726 cm^{-1}) was shifted towards higher wave number with high peak intensity when treated with Acrylonitrile. FTIR spectra of treated kenaf clearly indicated the presence of the characteristic band of cayanoethylation (-CN) bonds in the region of 2895 cm^{-1}. In addition, the absorption band of O-H group also shifted towards higher wave number (3425 cm^{-1} to 3396 cm^{-1}) with narrowed band intensity, which gives further evidence of the reaction of wood hydroxyl groups with ethylene groups of acrylonitrile. Therefore, it can be concluded that the kenaf and acrylonitrile performed the following reaction.

$$\text{Kenef-OH} + CH_2 = \text{C-CN} \rightarrow \text{Kenef-O-}CH_2\text{-}CH_2\text{-CN} \quad (14.2)$$

According to SEM analysis, the raw kenaf fillers were clearly distinguishable due to the weak interfacial bonding between the filler and the matrix. On the other hand, the filler and the matrix were not clearly differentiable due to the improved interfacial bonding between them. This smooth surface is due to the reduction of –OH group from the kenef fiber and strong interaction between fiber and polymer matrix [10].

For raw kenaf composites, the tensile strength decreased with an increase in fiber loading. As the filler load increased, the weak interfacial area between the filler and the matrix increased, which in turn decreased the tensile strength. In order to increase mechanical properties of the composites, kenaf was chemically treated using acrylonitrile. Due to the elimination of most of the hydroxyl groups in the treated kenaf, the interfacial bonding between the kenaf filler and the PP matrix increased in the resultant composites. This in turn increased the tensile strength of the 25% fiber loaded treated composites compared to PP matrix itself.

From the study it is concluded that the mechanical and morphological properties of treated kenaf/PP composites was significantly improved by the acrylonitrile treatment. The tensile strength of the composites decreased with an increase in kenaf filler loading. However there was an increase in the tensile strength of the 25% treated kenaf reinforced composite compared to the untreated one. FTIR and SEM results indicated the changes morphology after treatment. Water absorption was significantly reduced for the treated kenaf/PP composite.

14.7 Maleic Anhydride Treatment

Reaction of acetic anhydride with (typical wood fiber) OH groups forms an ester bond and acetic acid is formed as a by-product. Reaction without catalyst was conducted at 70°C [33]. It is known that the rate of reaction is promoted by wood-swelling agents such as pyridine that can be used only at laboratory scale. The improved dimensional stability of wood as a result of anhydride modification has been found to be a function of weight percentage gain (WPG) only, regardless of the anhydride used for modification.

$$\text{Wood-OH} + (H_3C\text{-}C(=O))_2O \longrightarrow \text{Wood-O-}\underset{\underset{O}{\|}}{C}\text{-}CH_3 + CH_3COOH$$

14.8 Nanoclay Treatment

Cellulose-based polymer nanocomposites preparation by the nanotechnology with silicate nanoclays as *in situ* reinforcement has been intensively investigated in the recent year. Nanotechnological modification has also proven to be very effective means of improving the essential properties of wood [34]. Different types of silicates nanoclay such as montmorillonite, hectorite and saponite have a layered structure that upon exfoliation leads to composites with very high stiffness and strength. Moreover, considerable improvements in physical and mechanical properties, including tensile modulus and strength, flexural modulus and strength, thermal stability, flame retardant and barrier resistance have been observed for various thermoplastic and thermoset nanocomposites at low silicates content [35].

Cai *et al.* reported that wood polymer nanocomposites were prepared by impregnation of solid aspen wood with water-soluble melamine-urea-formaldehyde (MUF) resin in combination with hydrophilic and hydrophobic nanofillers [36]. Significant improvements in physical and mechanical properties (density, surface hardness, MOE) can be obtained by impregnation with MUF resin and nanoclay/MUF resin formulations. The hardness of the wood impregnated with hydrophobic nanoclay/MUF resin was improved from 1.09 to up to 3.25 MPa. Impregnation with MUF resin or nanofiller/ MUF resin caused improvements in MOEs. The average MOE of treated wood was almost twice as much as that of untreated wood. The MOR did not differ significantly between treated and untreated woods.

The water absorption and dimensional stability of wood impregnated with melamine-urea-formaldehyde resin and wood impregnated with different nanofiller/MUF resin formulations were also investigated [36]. Significant improvements in water repellence and better dimensional stability were obtained for the nanofiller/MUF resin-treated wood. The untreated wood absorbed approximately 63% of moisture after 24 hours soaking in water and water uptake was about 125% after 1 week immersion in water. The MUF resin-impregnated wood absorbed about 8.3% and 38.5% of moisture after 24 hours and 1 week immersion in water, respectively. For the organophilic nanoclay/ MUF resin-impregnated wood, even lower water absorption was observed (about 5% water uptake in 24 hours and 22% after 1 week). The anti-swelling efficiency (ASE) was also improved from 63.3% to 125.6% for the nanofiller/MUF-treated wood. The significant improvement in water resistance and dimensional stability of the resulting wood polymer nanocomposites were attributed to the introduction of MUF and nanofillers into the wood. X-ray fluorescence photo showed that some nanoparticles had migrated into the wood cell wall [34]. Wood treatments with MUF resin and nanofiller/MUF resin showed no significant influence on the color of the wood, which is important for practical applications of the treated wood in some specific areas such as flooring. Based on the literature, it is believed that phosphates, ammonium compounds, zinc chloride, sodium tetraborate, boric acid and polymeric materials are commonly used in Europe and the USA as flame retardants for wood.

Wood polymer nanocomposites were prepared through impregnation of wood with PF/Nanoclay by another research Islam *et al.* and their co-researcher [37]. They reported that wood polymer nanocomposites (WPNCs) which were manufactured from five tropical wood species by vacuum-impregnation method have increased dynamic mechanical thermal properties.

In the study, thermo-mechanical properties of wood samples were investigated by the dynamic mechanical thermal analysis (DMTA) over the temperature range of -100°C to 200°C. The intrinsic properties of the components, morphology of the system and the nature of interface between the phases were also determined through DMTA test composite. Storage modulus (*E'*) of WPNC samples exhibited significant improvement over the temperature range, in both glassy region and rubbery plateau in relation to their corresponding raw wood samples and wood polymer composites (WPCs). Furthermore, damping (loss tanδ) peaks of all wood species were lowered by PF-nanoclay system treatment, an indication of improved surface interphase of wood.

Dynamic Young's modulus (E_d) of wood was also calculated using free-free vibration testing. A significant increment was obtained for the PF-Nanoclay impregnated WPNC samples.

They conclude that significant change in dynamic mechanical properties in storage modulus (*E¢*), glass transition (*Tg)* temperature and dynamic Young's modulus (E_d) for WPNC. It is also reported that the nanoclay as a co-reinforcement increased the adhesion and compatibility of wood fiber to polymer matrix, thus enhancing the degree of polymerization and interaction between PF, nanoclay and wood component. This, in turn, significantly increased the dynamic mechanical properties of WPNC.

14.9 Some other Chemical Treatment with Natural Fibers

14.9.1 Epoxides Treatment

The reaction of fiber with an epoxide leads to the formation of an ether linkage and a new OH group. Therefore graft-polymerization reactions are possible [38]

$$\text{Fiber-OH} + \text{R-CH(-O-)CH}_2 \rightarrow \text{Fiber-O-CH}_2\text{-CH(OH)-R} \quad (14.3)$$

Several epoxides have been used the past decades for fiber modification purposes. They include ethylene oxide, propylene oxide and butylene oxide [39]. Usually the reaction is catalyzed under mild basic conditions. In most experiments, triethylamine is used as a catalyst. Decay resistance of wood treated with propylene oxide was ineffective towards *G. trabeum* decay.

14.9.2 Alkyl Halide Treatment

The alkyl halides in the presence of a strong base can be used for fiber etherification.

$$\text{Fiber-OH} + \text{NaOH} \rightarrow \text{Fiber-ONa} + \text{R-X} \rightarrow \text{Fiber-O-R} \quad (14.4)$$

Reactions of fiber with crotyl chloride, methyl iodide and butyl chloride have led to an improvement in dimensional stability. Decay resistance has also been investigated for wood treated with fatty dialkyldimethylammonium chlorides and bromides. Treated wood showed good resistance against brown rot fungi [40].

14.9.3 β- Propiolactone Treatments

The reaction of β-propiolactone with fiber can be catalyzed by acids or bases to yield two different products. Treated wood in acidic conditions with 30% WPG resulted in good decay resistance and anti swelling efficiency (ASE) of 60%. Nevertheless a strong degradation of wood has been observed [40].

Wood - OH + [β-propiolactone] —Acid→ Wood - O - CH_2 - CH_2 - COOH

—Base→ Wood - O - C(=O) - CH_2- CH_2- OH

14.9.4 Cyclic Anhydride Treatments

Cyclic anhydrides do not yield a by-product when reacting with the hydroxyl groups of fiber. The anhydride covalently bonded on wood by an ester function yields a free carboxylic group at its end. The free carboxylic group can in theory react with another OH group to crosslink the cell walls polymers of wood [41].

Wood- OH + [succinic anhydride] → Wood - O-C(=O)-CH₂-CH₂-C(=O)-OH —Wood - OH→ Wood - O-C(=O)-CH₂-CH₂-C(=O)-O - Wood

Cyclic anhydrides have been studied as chemical reagents for wood modification: phtalic anhydride, maleic anhydride, glutaric anhydride, succinic anhydride and alkenyl succinic anhydrides (ASA) [42]. Scots pine-treated samples with petrochemical octenyl succinic anhydrides (ASAs) dissolved in pyridine that after treatment did not present enough resistance against fungi decay but increased dimensional stability.

14.9.5 Oxidation of Natural Fiber

Cellulose-based natural fiber can be oxidized using oxidizing agent with acidic catalyst. Oxidation reactions applied to cellulose in fiber for chemical modifications [43]. Oxidation reactions occur on cellulose selectively at particular position. The reaction of sodium metaperiodate with cellulose in wood fiber in the presence of sulfuric acid catalyst at 120°C and 85 KPa pressure yielded the oxidized product. Sodium metaperiodate reacts with hydroxyl groups of cellulose and produce 2,3-dialdehyde cellulose which improved the physical and mechanical properties of polymer composites [44].

[cellulose repeat unit, CH_2OH]$_n$ + $NaIO_4$ —H^+, 120°C, 2h→ [2,3-dialdehyde cellulose repeat unit, CH_2OH]$_n$

14.10 Conclusions

Chemical modification with suitable chemicals and innovative technology provides a noble and facile route for the development of intrinsic negative properties of cellulose-based composites materials. Although, the chemical treatments sometimes improved the certain properties composites, however, negative effect also observed. Alkali treatment improved adhesion and compatibility considerably among all conventional chemical treatments. The modification with nanotechnology in addition to the conventional techniques becomes effective to improve some negative inherent properties such as physical, mechanical and thermal properties of composites. Finally, it can be concluded that Nanotechnological modification is the most powerful and suitable technique among all the techniques used in the previous research. However, other traditional modification also appears to be effective in partially or fully. More research on certain chemicals is required to investigate their effects on the polymer composites materials.

References

1. M.M. Haque, M. Hasan, M.S. Islam, and M.E. Ali, Physico-mechanical properties of chemically treated palm and coir fiber reinforced polypropylene composites. *Bioresour. Technol.* 100, 4903–4906 (2009).
2. T.H. Nama, S. Ogihara, N.H. Tung, and S. Kobayashi, Effect of alkali treatment on interfacial and mechanical properties of coir fiber reinforced poly(butylene succinate) biodegradable composites. *Compos. B* 42, 1648–1656 (2011).
3. H.U. Zaman, M.A. Khan, R.A. Khan, M.A. Rahman, L.R. Das, and M, Al-Mamun, Role of potassium permanganate and urea on the improvement of the mechanical properties of jute polypropylene composites. *Fibers Polym.* 11(3), 455–463 (2010).
4. M.R. Rahman, M. Hasan, M.M. Huque, and M.N. Islam, Physico-mechanical properties of maleic acid post treated jute fiber reinforced polypropylene composites. *J. Thermoplast. Compos. Mater.* 22(4), 365–381 (2009).
5. P. Wambua, J. Ivens, and I. Verpoest, Natural fibres: can they replace glass in fibre reinforced plastics? *Compos. Sci. Technol.* 63, 1259–1264 (2003).
6. M.A. Kabir, M.M. Huque, M.R. Islam, and A.K. Bledzki, Jute/polypropylene composites I. Effect of matrix modification. *Compos. Sci. Technol.* 66, 952–963 (2010).
7. E. Baysal, S.K. Ozaki, and M.K. Yalinkilic, Dimensional stabilization of wood treated with furfuryl alcohol catalysed by borates. *Wood Sci. Technol.* 38, 405–415 (2004).
8. M.M. Haque, M.S. Islam, M.D.S. Islam, M.N. Islam, M.M. Huque, and M. Hasan, Physicomechanical properties of chemically treated palm fiber reinforced polypropylene composites. *Reinf. Plast. Compos.* 29(11), 1734–1742 (2010).
9. M.S. Islam, S. Hamdan, M.R. Rahman, I. Jusoh, and A.S. Ahmed, The effect of crosslinker on mechanical and morphological properties of tropical wood material composites. *Mater. Des.* 32, 2221–2227 (2011).
10. M.S. Islam, S. Hamdan, I. Jusoh, M.R. Rahman, and Z.A. Talib, Dimensional stability and dynamic young's modulus of tropical light hardwood chemically treated with methyl methacrylate in combination with hexamethylene diisocyanate crosslinker. *Ind. Eng. Chem. Res. 50,* 3900–3906 (2011).
11. B.S. Ndazi, S. Karlsson, J.V. Tesha, and C.W. Nyahumwa, Chemical and physical modifications of rice husks for use as composite panels. *Compos. A* 38(3), 925–935 (2007).
12. M.S. Islam, S. Hamdan, M.R. Rahman, I. Jusoh, and A.S. Ahmed, The effect of alkali pretreatment on mechanical and morphological properties of tropical wood polymer composites. *Mater. Des.* 33, 419–424 (2012).
13. J. George, M.S. Sreekala, and S. Thomas, A review on interface modification and characterization of natural fiber reinforced plastic composites. *Polym. Eng. Sci.* 41(9), 1471–1485 (2001).

14. A.K. Mohanty, M.A. Khan, and G. Hinrichsen, Surface modification of jute and its influence on performance of biodegradable jute-fabric/Biopol composites. *Compos. Sci. Technol.* 60, 1115–1124 (2000).
15. S.L. Fávaro, M.S. Lopes, A.G.V.C. Neto, R.R.R. Santana, E. Radovanovic, Chemical, morphological, and mechanical analysis of rice husk/post-consumer polyethylene composites. *Compos. A* 41(1), 154–160, (2010).
16. N.E. Zafeiropoulos, D.R. Williams, C.A. Baillie, and F.L. Matthews, Engineering and characterisation of the interface in flax fibre/polypropylenecomposite materials. Part I. Development and investigation of surface treatments. *Compos. A* 33(8), 1083–1093 (2002).
17. U.S. Elesini, A.P. Cuden, and A.F. Richards, Study of the green cotton fibres. *Acta Chim. Solv.* 49, 815–833 (2002).
18. S. Borysiak, B. Doczekalska, X-ray diffraction study of pine wood treated with NaOH. *Fibers Text. East Eur.* 13 (5), 87–89 (2005).
19. R.M. Rowell, Taylor and Francis, CRC. pp.381–420 (2005).
20. M. Deka, and C.N. Saikia, Chemical modification of wood with thermosetting resin: effect on dimensional stability and strength property. *Bioresour. Technol.* 73, 179–181 (2000).
21. H. Ismail, M. Edyhan, and B. Wirjosentono, Bamboo fibre filled natural rubber composites: the effects of filler loading and bonding agent. *Polym. Test.* 21(2), 139–144 (2002).
22. R. Rashmi, A. Devi, T.K. Maji, and A.N. Banerjee, Studies on dimensional stability and thermal properties of rubber wood chemically modified with styrene and glycidyl methacrylate. *J. Appl. Polym. Sc.* 93, 1938–1945 (2004).
23. U.C. Yildiz, S. Yildiz, and E.D. Gezer, Mechanical properties and decay resistance of wood–polymer composites prepared from fast growing species in Turkey. *Bioresour. Technol.* 96, 1003–1011 (2005).
24. M. Deka, C.N. Saikia, and K. K. Baruah, Studies on thermal degradation and termite resistant properties of chemically modified wood. *Bioresour. Technol.* 84, 151–157 (2002).
25. H.D. Rozman, R.N. Kumar, H.P.S. Khalil, A. Abusamah, and R. Abu, Fibre activation with glycidyl methacrylate and subsequent copolymerization with diallyl phthalate. *Eur. Polym. J.* 33(8), 1213–1218, (1997).
26. M.N. Islam, M.M. Haque, and M.M. Huque, Mechanical and morphological properties of chemically treated coir-filled polypropylene composites. *Ind. Eng. Chem. Res.* 48 (23), 10491–10497 (2009).
27. A.S. Ahmed, M.S. Islam, A. Hassan, M.K. Mohamad Haafiz, K.N.l Islam, and R. Arjmandi, *Fibers Polym.* 1–8.
28. B. Öztürk, Hybrid effect in the mechanical properties of jute/rockwool hybrid fibres reinforced phenol formaldehyde composites. *Fibers Polym.* 11(3), 464–473 (2010).
29. F. Vilaseca, J.A. Mendez, A. Pelach, M. Llop, N. Canigueral, J. Girones, X. Turon, and P. Mutje, Composite materials derived from biodegradable starch polymer and jute strands. *Process Biochem.* 42, 329–334 (2007).
30. M.N. Islam, M.R. Rahman, M.M. Haque, and M.M. Huque, Carbon nanofiber paper for lightning strike protection of composite materials. *Compos. A* 41, 192–198 (2010).
31. M.A.M, Bin Mohd Idrus and M.S. Islam, *World conference on integration of knowledge*, Malaysia, November, 25–26 (2013).
32. H.G.B. Premlal, H. Ismail, and A.A. Baharin, Comparison of the mechanical properties of rice husk powder filled polypropylene composites with talc filled polypropylene composites. *Polym. Test.* 21(7), 833–839, (2002).
33. H.P.M. Bongers, and E.P.J. Beckers, Mechanical properties of acetylated solid wood treated on pilot plant scale. In *Proceeding of the first european conference on wood modification*, J. Van Acker, and C.A.S. Hill, (Eds.), pp. 341–350, Ghent University, Belgium (2003).

34. X. Cai, B. Riedl, S.Y. Zhang, and H. Wan, The impact of the nature of nanofillers on the performance of wood polymer nanocomposites. *Compos. A* 39, 727–737 (2008).
35. S.S. Ray, and M. Okamoto, Polymer/layered silicate nanocomposites: a review from preparation to processing. *Prog. Polym. Sci.* 28, 1539–1641 (2003).
36. X. Cai, B. Riedl, S.Y. Zhang, and H. Wan, Effects of nanofillers on water resistance and dimensional stability of solid wood modified by melamine-urea-formaldehyde resin. *Wood Fiber Sci.* 39(2), 307–318 (2007).
37. M.S. Islam, S. Hamdan, Z.A. Talib, A. Ahmed, M.R. Rahman, Tropical wood polymer nanocomposite (WPNC): The impact of nanoclay on dynamic mechanical thermal properties. *Compos. Sci. Technol.* 72 (2012), 1995–2001 (2012).
38. S. Kumar, Chemical modification of wood. *Wood Fiber Sci.* 26(2), 270–280 (1994).
39. M. Norimoto, J. Gril, and R.M. Rowell, Rheological properties of chemically modified wood: relationship between dimensional and creep stability. *Wood Fiber Sci.* 24, 25–35 (1992).
40. C.A.S. Hill, pp. 175-190, John Wiley & Sons Ltd., (2006).
41. H. Matsuda, Preparation and utilization of esterified woods bearing carboxyl groups. *Wood Sci. Technol.* 21(1), 75–88 (1987).
42. M. Morard, C. Vaca-Garcia, M. Stevens, J. Van Acker, O. Pignolet, and E. Borredon, Durability improvement of wood by treatment with Methyl Alkenoate Succinic Anhydrides (M-ASA) of vegetable origin. *Int. Biodeterior. Biodegrad.* 59, 103–110, (2007).
43. D.N.-S. Hon, and N. Shiraishi, *Chapter* 14, pp. 599–626, Marcel Dekker, Inc., (2001).
44. M.R. Rahman, M.M. Haque, M.N. Islam, and M. Hasan, Mechanical properties of polypropylene composites reinforced with chemically treated abaca. *Compos. A* 40, 511–517 (2009).

Part III

PHYSICO-CHEMICAL AND MECHANICAL BEHAVIOUR OF CELLULOSE/ POLYMER COMPOSITES

15

Weathering of Lignocellulosic Polymer Composites

Asim Shahzad and D. H. Isaac

[1]*Materials Research Centre, College of Engineering, Swansea University, Swansea, UK*

Abstract

Lignocellulosic polymer composites (LPCs) are finding increasing applications (siding, windows, decks, roofs, etc.) where they are exposed to weathering agents, which can cause changes in the microstructure and the chemical composition of these composites. These changes in turn can modify properties such as strength, modulus, impact resistance and fatigue behavior, which can have adverse effects on the service life of these composites. Therefore for long service life, the weathering phenomenon and how its effects can be mitigated must be understood.

Moisture and sunlight are the two main agents of weathering in LPCs. The ultraviolet (UV) portion of sunlight initiates the process of weathering by photochemical degradation of the surface of composites. Photodegradation leads to changes in wood and plant fiber composites' appearance such as discoloration, roughening and checking of surfaces, and deterioration of mechanical and physical properties. The hydrophilic nature of wood and plant fibers also results in absorption of water which causes them to swell. In composites, the swelling of fibers can cause local yielding of the polymer matrix due to swelling stresses, fracture of fibers due to restrained swelling, and fiber/matrix interfacial breakdown.

The weathering properties of LPCs can be studied by either exposing them to outdoor conditions or by simulating the outdoor conditions in the laboratory in accelerated weathering machines. The use of accelerated weathering machines has gained in popularity because it is less laborious and time consuming than the use of outdoor conditions. Studies on cellulose fibers, cellulose fiber-reinforced thermoplastics and thermosets, and fully degradable "green" composites show that, generally, they all suffer fading of color and deterioration in physical and mechanical properties following exposure to weathering conditions. However, various studies have also demonstrated that the use of coupling agents, fiber surface treatments, ultraviolet absorbers and pigments can improve the weathering resistance of these composites.

Keywords: Cellulose fibers, UV, moisture, discoloration, mechanical properties, accelerated weathering, natural weathering

**Corresponding author*: mr_asim_shahzad@yahoo.com

Vijay Kumar Thakur (Ed.), Lignocellulosic Polymer Composites, (327–368) 2015 © Scrivener Publishing LLC

Table 15.1 Lignocellulosic fibers used as reinforcement in composites.

Bast (stem) fibers: Flax, Hemp (and Sunhemp), Kenaf, Jute, Mesta, Ramie, Urena, Roselle, Papyrus, Cordia, Indian Malow, Nettle
Leaf fibers: Pineapple, Banana, Sisal, Pine, Abaca (Manila hemp), Curaua, Agaves, Cabuja, Henequen, Date-palm, African palm, Raffia, New Zealand flax, Isora
Seed (hairs) fibers: Cotton, Kapok, Coir, Baobab, Milkweed
Stalk fibers: Bamboo, Bagasse, Banana stalk, Cork stalk
Fruit fibers: Coconut, Oil palm
Wood fibers: Hardwood, Softwood
Grasses and Reeds: Wheat, Oat, Barley, Rice, Bamboo, Bagasse, Reed, Corn, Rape, Rye, Esparto, Elephant grass, Canary grass, Seaweeds, Palm, Alpha

15.1 Introduction

Lignocellulosic fibers are being increasingly used as reinforcements in composite materials, particularly in the automotive, building and construction industries. They have many advantages compared with synthetic fibers, including biodegradability, superior environmentally-friendly credentials and lower cost. Lignocellulosic fibers are classified according to the part of the plant from which they are extracted. These sites include: the seeds of the plants, such as cotton; the tissues of the stems of plants, referred to as bast fibers, such as wood, hemp and flax; the leaves of plants, such as sisal and banana; and the fruit of plants, such as coconut. The major lignocellulosic fibers used as reinforcements in composites are shown in Table 15.1 and of these, jute, ramie, flax, coir, kenaf, sisal, wood and hemp are the most widespread.

According to the Food and Agriculture Organization (FAO) of the United Nations, about 35 million tonnes of natural fibers, both animal and plant, are produced annually worldwide [1].

Because of their superior mechanical properties, bast fibers are the most widely used of all the lignocellulosic fibers for composite reinforcement. Bast fibers form the fibrous bundles in the inner bark of the stems, where their high stiffness and strength help to hold the plant erect. Their lower density (1200–1600 kg/m^3) compared with synthetic fibers (e.g., 2500 kg/m^3 for glass) means that they have specific strengths and stiffnesses that are comparable to or even better than those of synthetic fibers.

Lignocellulosic fibers themselves are cellulose fiber-reinforced composite materials as they consist of crystalline cellulose micorfibrils in an amorphous matrix, mainly composed of lignin and/or hemicelluloses, with relatively small amounts of other components. Although most of the cellulose is crystalline, some cellulose surrounding the core of microfibrils tends to be in a non-crystalline state (i.e., amorphous cellulose). With the exception of cotton, lignocellulosic fibers include cellulose, hemicellulose, lignin, pectin, waxes and water soluble substances. Generally the fibers contain 60–80% cellulose, 10–20% hemicelluloses, 5–20% lignin, and up to 20% moisture [2]. Thus cellulose, hemi-cellulose and lignin are the basic components that determine the physical and mechanical properties of the fibers. Of these, cellulose is the stiffest and strongest constituent and so it provides the stiffness and strength to the fibers. The fibers consist of several fibrils that run all along the length of the fiber. The high strength and stiffness of the fibers is provided by the covalent bond linkages within the macromolecular chains that are oriented along the length of the fibrils, together with weaker inter- and intra-chain hydrogen bonds.

15.2 Wood and Plant Fibers

Lignocellulosic fibers are divided into two main classes according to their origin: wood fibers and plant fibers. These two classes are distinguished by the following main features, which affect the fibers and the composites made from them [3].

a) Wood fibers are short (~1–5 mm length) and are typically used for making composites with in-plane isotropic properties, that is, composites with a non-specific (random) fiber orientation. Plant fibers are longer (~5–50 mm length) and are typically used for making composites with anisotropic properties, that is, composites with a specific fiber orientation. However they are also widely used for making isotropic composites.
b) In wood fibers, the luminal area is in the range 20–70% of the fiber cross-sectional area, compared with just 0–5% in plant fibers. This higher luminal area in wood fibers provides a greater capacity for storing moisture.
c) The proportions of hemicellulose and lignin is higher in wood fibers, and this is particularly true for lignin which has a content of about 30% w/w in wood fibers, compared with only about 5%w/w in plant fibers. Nevertheless, in both cases the proportions have a significant effect on weathering properties of their composites because higher lignin and hemicellulose contents mean greater propensity to absorb UV radiation and moisture respectively.
d) Wood fibers exhibit lower cellulose crystallinity than plant fibers, with typical values in the ranges of 55–70% and 90–95%, respectively. The crystallinity also has an effect on moisture absorption of the fibers, since amorphous regions can absorb more water.
e) The microfibril angle (the angle between the majority of cellulose microfibrils and the long cell axis), which dictates overall mechanical performance of fibers, varies in the range 3°–50° in wood fibers depending on the type and location of the fibers in the wood (e.g., late and early wood), whereas the microfibril angle in plant fibers is more constant in the range 6°–10°.

In addition to its use as fibers, wood is also used in particulate form (as woodflour) in polymer composites. When used as woodflour, it acts as a filler rather than a reinforcement. Wood and plant fibers are available as non-woven fiber mats and these are the most widely used preforms for composite reinforcement. Lignocellulosic fibers may also be spun into yarn which can be used to make preforms of woven fabrics and non-crimp fabrics. Woven fabrics are mostly used for textile applications while non-crimp fabrics of flax fibers have recently been introduced specifically for use as reinforcement

Table 15.2 Influence of weathering agents on lignocellulosic fiber constituents in order of importance [4].

UV degradation
Lignin; hemicellulose; non-crystalline cellulose; crystalline cellulose
Moisture absorption
Hemicellulose; non-crystalline cellulose; lignin; crystalline cellulose

in composites [3]. Lignocellulosic fibers are used to reinforce thermosets and thermoplastics, including biodegradable polymers.

The main uses of lignocellulosic polymer composites (LPCs) are in the automotive and construction industries. However, it is their use in the construction industry that gives significant exposure to weathering conditions. Their main growth has been in the USA. The wood-plastic composites (WPCs) market share has been growing rapidly, especially for applications such as decking and railing, in the past few years. Two-thirds of the WPCs produced in the USA are now for decking and railing products. Other applications for WPCs include window and door frames, siding, roofing, residential fencing, picnic tables, benches, landscape timber, patios, gazebos and walkways, and playground equipment.

One of the main drawbacks of LPCs compared with synthetic fiber composites is their poor weathering resistance, a fact that limits their usage in outdoor applications. The two most significant agents of weathering effects are moisture and UV radiation and, therefore, this chapter focuses on these two phenomena. Lignocellulosic fibers are not only good absorbers of moisture but their constituents are also susceptible to UV radiation. Table 15.2 shows the influence of the weathering agents on lignocellulosic fiber constituents in order of importance.

Photochemical degradation by UV light occurs primarily in the lignin component, which is also responsible for color change, while cellulose is less susceptible to UV degradation. Hemicellulose and amorphous cellulose are mainly responsible for the high water uptake of lignocellulosic fibers, since they contain numerous easily accessible hydroxyl groups which give a high level of hydrophilic character to the fibers. The moisture content of fibers can vary between 5 and 20% [5], which can lead to dimensional changes, weakening of the fiber/matrix interfacial bonding strength and reduction in the mechanical properties of composites. It can also cause fibers to swell and ultimately rot through attack by fungi. A possible way to reduce these effects is to improve fiber-matrix interface by using compatibilizers and adhesion promoters since, with better adhesion, the moisture sensitivity is usually reduced. Additionally, surface treatments of fibers with various chemicals can make the fibers more hydrophobic.

15.3 UV Radiation

Figure 15.1 shows the percentage of the total spectral irradiance that falls on earth for UV, visible and infrared radiation, along with their corresponding wavelengths. Although UV light makes up only about 6.8% of sunlight, it is responsible for most of the sunlight damage to materials, especially polymers, exposed outdoors. This is because the energy of a photon and hence the photochemical effectiveness of light increases with decreasing wavelength according to the Planck-Einstein equation,

$$E = hv = hc/\lambda \tag{15.1}$$

where E, h, v, c and λ are energy, Planck's constant, frequency, speed of light and wavelength respectively. Also, as shown in Figure 15.2, the UV radiation spectrum is commonly divided into three ranges: UV-A (400–315nm), UV-B (315–280nm) and UV-C (280–100nm). The upper wavelength value for UV-A radiation (400nm) is the boundary between visible and UV light. The region around the 315 nm boundary between UV-A and UV-B corresponds to the point at which the radiation can start to have adverse effects on polymers. UV-B includes the shortest wavelengths of sunlight incident on the earth's surface (~295 nm) and can cause severe polymer damage [6]. UV-C radiation (<280nm), although of the highest energy, is completely absorbed by the ozone layer and so is not a significant factor in outdoor weathering.

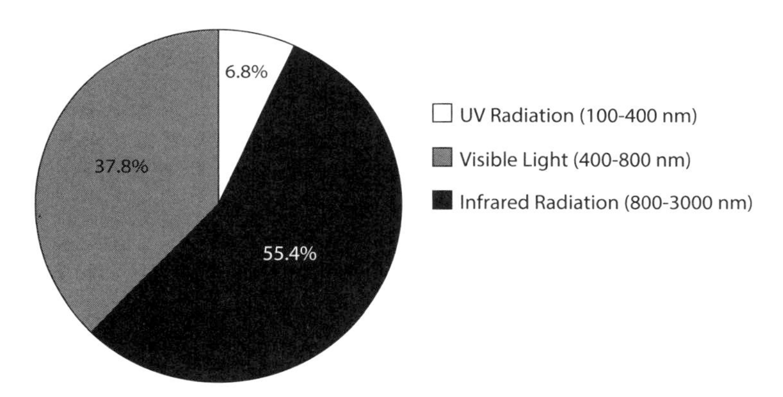

Figure 15.1 Percentage of total solar spectral irradiance for UV, visible and infrared radiation.

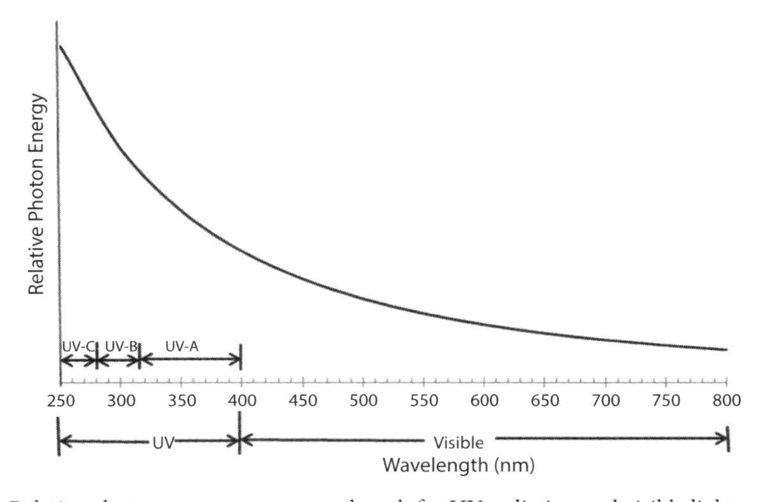

Figure 15.2 Relative photon energy versus wavelength for UV radiation and visible light.

Table 15.3 Some typical bond dissociation energies and corresponding radiation wavelengths [8].

Bond	**Bond dissociation energy (kJ/mol)**	**Wavelength (nm)**
C–C (Aromatic)	519	231
C–H (Aromatic)	431	278
C–H (Methane)	427	280
O–H (Methanol)	419	286
C–O (Ethanol)	385	311
C–O (Methanol)	373	321
CH_3COO–C (Methyl ester)	360	333
C–C (Ethane)	352	340
C–Cl (Methyl chloride)	343	349
C–$COCH_3$ (Acetone)	331	362
C–O (Methyl ether)	318	376

15.3.1 Lignocellulosic Fibers

The energy required to break chemical bonds depends on the type of chemical bond. Typical bond dissociation energies for some combinations of elements found in lignocellulosic fibers are shown in Table 15.3. By comparing the energy available from the photons in the UV range of the spectrum, it is apparent that there is sufficient energy to break bonds in the chemicals that comprise lignocellulosic fibers.

In order for a bond to break, energy must be absorbed by some component of the lignocellulosic fiber. The absorbed energy may not result in a degrading chemical reaction, but the absorption is a necessary condition. The degradation process depends on the surface composition of the fiber. The lignin component is primarily responsible for UV absorption by lignocellulosic fibers. Of the total amount of UV light absorbed by these fibers, lignin absorbs 80–95% [7]. Lignin absorbs UV radiation throughout the UV radiation spectrum and into the visible light spectrum; however it is only the absorption above 295 nm that is significant for weathering of lignocellulosic fibers since lower wavelengths are absorbed by the ozone layer. Chromophoric functional groups present in lignin which interact with UV light include phenolics, hydroxyl groups, double bonds, and carbonyl groups. A free radical is formed, which then reacts with oxygen to form a hydroperoxide. Additional reactions result in the formation of carbonyls. Both the free radical and hydroperoxide can initiate a series of chain scission reactions to degrade the polymeric components of the lignocellulosic fibers. This can considerably reduce the mechanical properties of these fibers.

Exposure to light of wavelengths between 300 and 360nm results in yellowing of the fiber, whereas exposure to wavelengths between 380 and 400nm, on the fringe of the visible spectrum, has a bleaching effect. The final color is the resultant of the two processes. UV light degrades lignin into water-soluble compounds that are washed from the fiber with rain, leaving a cellulose-rich surface with a fibrous appearance [7]. These poorly bonded cellulose-rich fibers erode easily from the surface, which exposes new lignin to further degradative reactions. This process causes the surface of the composite to become rough and can account for a significant loss in surface fibers.

15.3.2 Polymer Matrices

Since the polymer matrix forms the directly exposed outer surface of the composite material, it is crucial also to consider the response of the polymer matrix to UV light. It has been shown that in composite materials, significant sunlight exposure effects are confined to surface layers only down to about 10μm depth, which is the typical thickness of resin at which the intensity of UV light is reduced to one half of its incident value [9]. Since the polymeric matrix materials contain similar elements with bond dissociation energies to those discussed above for lignocellulosic materials, the photon energies of UV light are sufficient to break bonds in the polymer molecules. As discussed, in order to induce such chemical changes, the photon must first be absorbed by the material and a material impervious to the UV range will not exhibit photo-initiated degradation. Many chemical entities found in polymer molecules have characteristic absorption in the UV range. For example, the carbonyl group has an absorption peak at approximately 300 nm because of the excitation of a non-bonding electron into the π^* molecular orbital [10].

UV-absorbing groups may be present naturally in the polymer or they may be introduced adventitiously by several means, including oxidation during fabrication, polymerization anomalies, and introduction of various additives. The energy contained in photon-excited high energy orbitals may be dissipated as heat, but a certain fraction of these states may relax by initiating chemical change. This latter process may produce degradation in the polymer molecule.

The breakage of molecular bonds can cause changes in molecular weight, lead to formation of crosslinks, excited states and radicals, chain scission and reaction with oxygen. These structural changes may lead to gross physical changes such as chalking, cracking, surface embrittlement, discoloring and loss of tensile and impact strength. The broken chain ends are often highly reactive free radicals which undergo further reactions with the polymer, oxygen and water to cause more chain breakage which leads to embrittlement, loss of strength and material removal. Catalyst residues, hydroperoxide groups, and carbonyl groups introduced during polymer manufacturing, processing, and storage absorb UV light above about 290 nm and initiate photochemical reactions. The range of dissociation energies between atoms in polymer molecules is typically about 280–420 kJ/mol [11]. UV light in the wavelength range 290-320 nm has an energy range of 374-413 kJ/mol, but, fortunately, it accounts for only 0.5% of sunlight.

The photodegradation of polyolefins is mainly caused by the introduction of chromophores, such as catalyst residues, hydroperoxide groups, carbonyl groups, and double bonds, during polymer manufacture [12]. In order to understand the degradation processes in polymers, the example of degradation mechanisms in polyethylene (PE) is shown in Figure 15.3. It has been postulated that carbonyl groups are the main light-absorbing species responsible for the photochemical initiation reactions of UV-exposed PE. The degradation reactions proceed from carbonyl group precursors

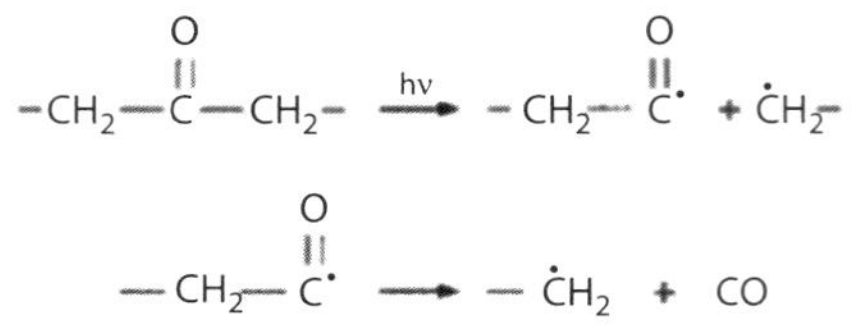

$$-CH_2-\overset{\overset{O}{\|}}{C}-CH_2- \xrightarrow{h\nu} -CH_2-\overset{\overset{O}{\|}}{C}{}^{\bullet} + \dot{C}H_2-$$

$$-CH_2-\overset{\overset{O}{\|}}{C}{}^{\bullet} \longrightarrow -\dot{C}H_2 + CO$$

(**a**) Norrish I

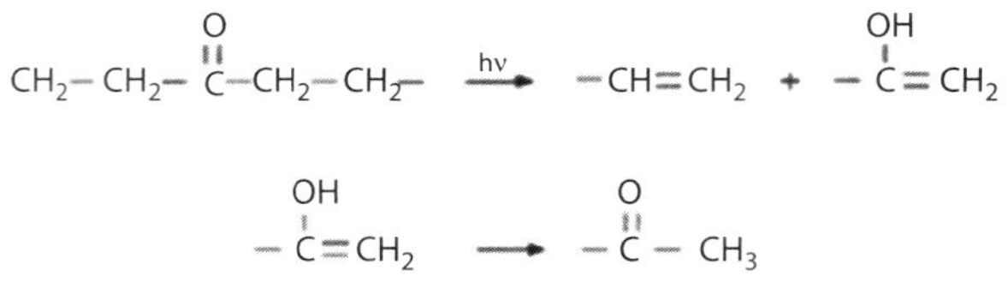

$$CH_2-CH_2-\overset{\overset{O}{\|}}{C}-CH_2-CH_2- \xrightarrow{h\nu} -CH=CH_2 + -\overset{\overset{OH}{|}}{C}=CH_2$$

$$-\overset{\overset{OH}{|}}{C}=CH_2 \longrightarrow -\overset{\overset{O}{\|}}{C}-CH_3$$

(**b**) Norrish II

Figure 15.3 Norrish degradation mechanisms in polyethylene [12].

according to Norrish I and II reactions. If degradation of the carbonyl groups proceeds according to a Norrish I reaction, the resulting free radicals that form can attack the polymer [Figure 15.3 (a)]. Free-radical attack may lead to termination via crosslinking or chain scission. If the degradation proceeds according to a Norrish II reaction, carbonyl groups and terminal vinyl groups are produced [Figure 15.3 (b)], and chain scission occurs. The carbonyl group that forms is also capable of further degradation. During the course of PE photodegradation, the two mechanisms—chain scission and crosslinking—are competing. Although chain scission occurs in the amorphous phase of the polymer, imperfect crystalline regions degrade because of crosslinking. Tie molecules—chains traversing the amorphous phase from one crystalline lamella to another—can also be affected during photodegradation. Decreasing the tie-molecule density can increase environmental stress cracking. The formation of carbonyl groups, vinyl groups and an increase in crystallinity indicate that main chain scission has occurred. The shorter chains produced during chain scission are more mobile and are able to crystallize more readily; this results in increased crystallization and associated embrittlement. Crosslinking, on the other hand, does not affect polymer chemistry. Table 15.4 compares the resistance of the most widely used polymer matrices and fibers to UV radiation.

15.3.3 Methods for Improving UV Resistance of LPCs

The UV resistance of LPCs can be improved by two main methods as outlined in Table 15.5: incorporating additives into the composite or increasing the surface resistance of the composite. Additives may be used to physically absorb or block the absorption of UV light, whereas surface treatments retard the loss of color-imparting extractives from the cellulose fibers. To overcome degradation in polymers, ultraviolet absorbers (UVAs), such as carbon black or aromatic ketones, may be added to them. The damaging effects of UV light can be mitigated by using these photostabilizers. They screen material from UV light by reflecting or absorbing the radiation (and degrading it to heat), or harmlessly mopping up free radicals. They are generally classified according

Table 15.4 Resistance of polymers and fibers to UV radiation [13].

Polymers	
PEEK	Excellent
Conventional Polyesters	Good (some yellowing with time)
Nylons	Moderate
Epoxies	Moderate (need surface coatings for good protection)
Vinyl Esters	Moderate (more degradation due to –OH groups)
PP	Poor (needs stabilizing)
Fibers	
Glass	Good
Carbon	Good, dark color gives UV screening
Aramid (Kevlar)	Moderate / poor
Cellulose	Poor

to the degradation mechanism that they hinder. Commercial UVAs are readily available such as benzophenones and benzotriazoles [9]. A relatively new class of materials, hindered amine light stabilizers (HALS), has also been extensively examined for UV protection [9]. Pigments may be used to physically block light, thereby protecting the composite from photodegradation. Pigments have been shown to mitigate the increase in lightness and significantly increase the flexural property retention of WPCs after accelerated weathering.

However, such addition of UVAs or pigments into the bulk of the composites is not a cost effective method. A better option is to improve the surface resistance of the composites where the weathering occurs primarily. The methods used for improving surface resistance of the composites include coating, lamination and coextrusion. All of these methods entail producing a UV resistant layer on the surface of the composites. Of these, coating is the most widely used method in the wood industry [14].

15.4 Moisture

Polymer composites absorb moisture when exposed to humid environments or when immersed in water, the amount of moisture absorbed depending on various factors such as the matrix and fiber type, exposure time, component geometry, relative humidity, temperature and exposure conditions. Typical consequences of this moisture uptake include matrix and fiber swelling, fiber-resin debonding, matrix cracking, and chain scission.

The major mechanism for moisture penetration into composite materials is diffusion, mainly direct diffusion of water molecules into the matrix and fibers. The other common mechanisms are capillary flow along the fiber/matrix interface, followed by diffusion from the interface into the bulk resin and fibers, and transport by microcracks. Capillary flow is mostly active only after debonding between fiber and matrix has occurred whereas transport of moisture by microcracks involves both flow and

Table 15.5 Methods of improving UV resistance of LPCs [14].

1. Use of additives i) UVAs - absorb UV light. Effective. ii) HALs - scavenge the radical intermediates formed in the photo-oxidation process. High effectiveness in the initial stage due to their fast diffusion to the surface but, low effectiveness with increased exposure time due to their loss by surface evaporation. iii) Pigments – block the penetration of UV light at the surface. Darker color pigments more stable.
2. Increasing surface resistance of composites i) Coating - opaque pigmented coatings block the UV/visible light. Mostly acrylic films used. Issues with the poor bonding and lower color stability of the coating. ii) Lamination - achieved by melt fusion or an adhesive layer. Issue with the bonding of the lamination. iii) Coextrusion - produce a multilayered product with different materials for the outer and inner layers, thus offering controllable and different properties between surface and bulk. iv) Color dye – dying the fibers with stain. Oil-based stain more successful than water-based stain when used for woodflour.

storage of water in micro-damaged regions. The rate at which a composite material exposed to moisture absorbs water reduces with time until eventually the liquid content reaches an equilibrium (saturation) level. The time taken for a composite laminate of a given thickness to attain this equilibrium moisture concentration depends on the temperature and relative humidity of the environment. Typically, synthetic fiber/thermoset matrix composites have saturation limits ranging from 1 to 2%, while synthetic fiber/thermoplastic matrix composites have lower limits of 0.1 to 0.3% [15].

The absorption of liquid molecules into the polymer matrix or fibers results in significant swelling. The degree of swelling is linked to the solubility and molecular volume of the absorbed liquid, and the stiffness of the material also plays a part. Swelling is found most commonly in the polymer matrix, but is also seen in some polymer fibers and can be quite significant in cellulose fibers. This swelling can interact with any internal residual stresses that may already be present in the composite as a result of the production process, such as the shrinkage of the polymer matrix during curing. For example, unsaturated polyester resin is well known for its high shrinkage during curing. If the fibers do not undergo a similar shrinkage, this leads to compressive stresses on the fibers and tensile stresses in the matrix. Subsequent swelling of the fibers then tends to increase the tensile stresses in the matrix and the compressive forces in the fibers. Some of the possible effects of absorbed moisture on polymeric composites have been summarized as follows [16]: plasticization of the matrix, resulting in reduction in the glass transition temperature (T_g) and usable temperature range, dimensional changes due to matrix swelling, increased creep and stress relaxation responses, increased ductility, change in the coefficient of expansion, reduction in ultimate strength and stiffness.

15.4.1 Lignocellulosic Fibers

Moisture may be present in lignocellulosic fibers in the cell voids or lumina (free water) and adsorbed to the cellulose and hemicellulose in the cell wall (bound water).

Table 15.6 Equilibrium moisture content (EMC) of cellulose fibers at 65% RH and 21°C [18].

Fiber	EMC (%)
Wood	12
Jute	12
Flax	7
Hemp	9
Ramie	9
Sisal	11
Pineapple	13

Free water only exists when all sites for the adsorption of bound water in the cell wall are filled. This point is called the fiber saturation point (FSP) and all the water added to fibers after the FSP has been reached exists as free water. Thus the total moisture content is the sum of the bound and free water. When a cellulose fiber remains at a constant relative humidity (RH) and temperature for a long enough period of time, the proportion of water reaches a steady state known as the equilibrium moisture content (EMC). Test results show that, typically, for small pieces of wood at a constant RH, the EMC is reached in about 14 days [17]. The EMC varies significantly with the source of the fibers and values for some of the most widely used fibers (at typical ambient conditions of 65% RH and 21°C) are given in Table 15.6.

Cellulose fibers are hygroscopic in nature since the polymers that make up the hemicellulose and cellulose components are rich in hydroxyl groups that are responsible for moisture sorption through hydrogen bonding. Although the hemicellulose and cellulose components are hydrophilic, lignin is comparatively hydrophobic. This means that the cell walls have a great affinity for water, but the ability of the walls to take up water is limited to some extent by the presence of lignin.

The moisture content of cellulose fibers has a significant effect on their properties and hence the properties of their composites. When cellulose fibers are exposed to moisture, there is the potential for hydrogen bonds to form between the numerous hydroxyl groups of the cellulose molecules and water as shown in Figure 15.4 [19]. The first water molecules may be absorbed directly onto the hydrophilic groups of the fibers and form a relatively strong hydrogen bond. Following this, other water molecules are attracted either to other hydrophilic groups or they may form further layers on top of the water molecules already absorbed by weaker hydrogen bonding. Thus the cellulose fibers absorb and desorb moisture from the atmosphere until they reach the equilibrium stage.

Cellulose fibers swell when moisture is absorbed and shrink when moisture is desorbed, but it is only the water present in cell walls (bound water) that is responsible for the dimensional changes. As water is added to the cell wall, the increase in fiber volume is approximately proportional to the volume of water added. Swelling of the fiber continues until the cell walls reach the FSP, but the water uptake above the FSP is free water in the lumina, which does not contribute to further swelling. This process is reversible, so that the fiber shrinks as it loses moisture below the FSP. The governing factor limiting the swelling of the cell walls occurs when the forces of water

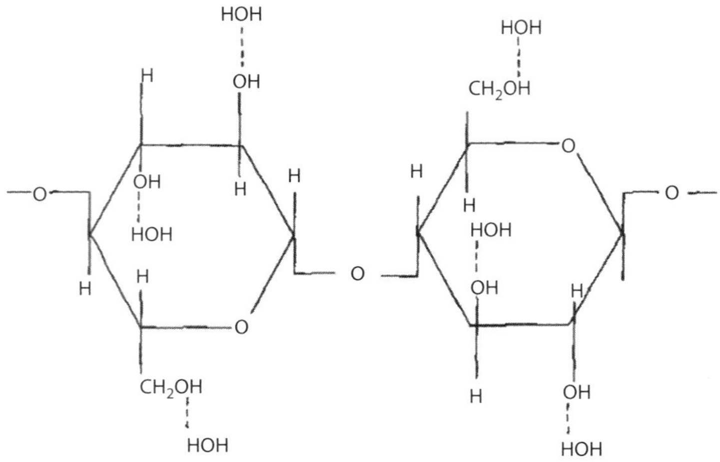

Figure 15.4 Absorption of moisture in cellulose molecules [19].

absorption are counterbalanced by the cohesive forces of the cell walls. The swelling of the fibers is found to be directional with the maximum swelling occurring in the lateral direction and minimum in the longitudinal direction, due to the predominance of molecular chains in this latter direction. The amorphous regions of cellulose, where the hydroxyl groups are more readily accessible to water, contribute more to swelling than the crystalline regions, where only the crystallite surfaces are available for water sorption. Therefore increasing the crystallinity of fibers reduces their swelling capacity. Since the hydrophilic hemicellulose also contributes greatly to fiber swelling, reducing the amount of hemicelluloses also reduces the swelling capacity of fibers. In LPCs, the swelling of fibers, resulting from moisture absorption, affects their dimensional stability and weakens the fiber/matrix interfacial bonding in the composite, leading to adverse effects on the mechanical properties.

Although below the FSP the moisture content has a major effect on the mechanical properties of fibers, increases in water content above the FSP causes very little further change in the mechanical behavior. The most significant effect of sorption of bound water molecules within the cell wall polymers is to plasticize or loosen the cell wall microstructure. The resulting decreases in cohesive forces in the cell wall reduce the stiffness, thus increasing the deformation under stress. Most mechanical properties decrease with increasing moisture content up to the FSP because increased amounts of bound water interfere with and reduce hydrogen bonding between the organic polymers of the cell wall.

The moisture absorption in LPCs has been found to increase with increase in fiber content and this has been demonstrated in a wide range of fiber-matrix combinations, including jute-epoxy [20]; jute-polyester [21]; hemp-polyester [22], pineapple-polyester [23], banana-polyester [24], woodflour-polyester [25], hybrid ramie/cotton-polyester [26], sisal-epoxy [27], sisal-polypropylene [28], rice husk-polypropylene [29], pineapple-polyethylene [30], cellulose-polypropylene [31], and rice hull-HDPE [32].

It has been shown that the water absorption and desorption pattern of LPCs at room temperature follows Fickian behavior where the water uptake process is initially more

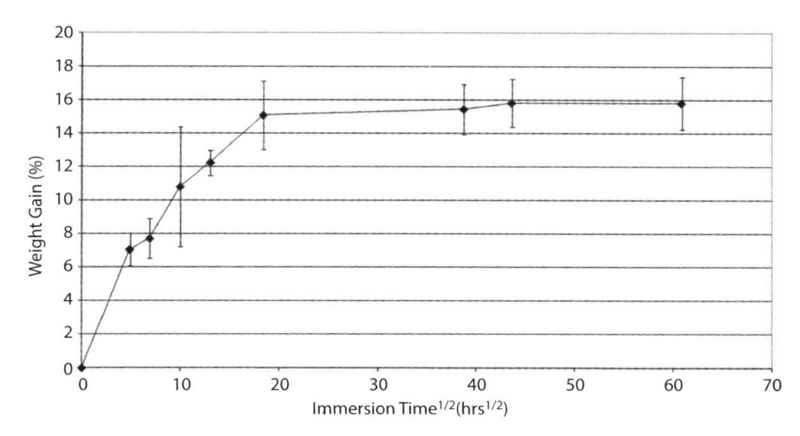

Figure 15.5 Water uptake in hemp fiber reinforced polyester composites against square root of time, showing Fickian behavior [33].

rapid, then slows down until eventually the water content approaches saturation after a prolonged time. A typical Fickian curve for hemp fiber/unsaturated polyester composites is shown in Figure 15.5. For Fickian diffusion into thin laminate samples, the initial mass uptake versus time can be expressed by the following equation:

$$M_t = M_\infty \left(\frac{16Dt}{h^2 \pi} \right)^{1/2} \tag{15.2}$$

where M_t is the mass uptake at time t, M_∞ is the equilibrium mass uptake, D is the diffusion coefficient and h is the sample thickness. The initial mass uptake is therefore proportional to the square root of time, so that a plot of M_t against $t^{1/2}$ can be used to determine whether Fickian behavior is observed or not.

15.4.2 Polymer Matrices

All organic matrices are to some extent permeable to moisture with a consequent reduction in modulus. The most important type of chemical degradation mechanism in the polymer matrix is hydrolysis, in which water or ions (OH^-, H^+, or H_3O^+) attack chemical groups within the matrix. These hydrolysis reactions are more pronounced in acidic or alkaline environments. Polymers containing polar groups, such as esters, amides and carbonates, are most susceptible to hydrolysis so unsaturated polyester resin is particularly liable to undergo hydrolysis reactions, with the ester groups being broken primarily by OH^- ions. The other important type of chemical degradation in the matrix material is chemical oxidation with oxidizing acids such as nitric and sulfuric, or other oxidizing agents such as peroxides and hypochlorites. Attack occurs via active free radicals, like H_2O and HO radicals, which attack main chain bonds in the polymer.

As the key degradation reaction is main chain scission in most polymers, the activation energies for degradation in a particular environment are strongly dependent on the strength of the weakest bonds in the polymer backbone. Thus polymers with simple carbon-carbon bonds in the backbone, which have relatively high activation energies, are less susceptible to degradation than those containing oxygen bridges, with lower activation energies, in the main chain, such as polyesters.

Another important effect of moisture absorption on the polymer matrix is plasticization. The small solute molecules disrupt the intermolecular bonding between polymer chains, allowing easier chain movement, leading to a reduction in the glass transition temperature (T_g) of the polymer. The lowering of T_g can have serious effects on the properties of the composite. This mechanism primarily occurs in amorphous regions, so it is particularly important for glassy polymers. Generally the higher the level of equilibrium solubility, the greater is the degree of plasticization. Plasticization can lead to a significant decrease in stiffness, increase in creep rate, increase in diffusion coefficient, and the potential for environmental stress cracking of the polymer. In polymer-based composites plasticization has most significant effects on matrix dominated properties such as shear, transverse tension and longitudinal compression behavior.

15.4.3 Methods for Improving Moisture Resistance of LPCs

Measures to improve the moisture resistance of LPCs consist primarily of retarding the water absorption of cellulose fibers because it is virtually impossible to completely prevent their moisture absorption. A number of methods have been shown to retard moisture absorption in cellulose fibers, but they fall under two main categories: water repellant treatments and dimensional stability treatments [17], as shown in Table 15.7. Whereas the water repellant treatment slows down the rate of water absorption, the dimensional stability treatment reduces the swelling in the fibers.

Water repellant treatments involve the deposition of a thin layer of a hydrophobic substance onto external surfaces and, to some extent, internal cell lumen surfaces. Generally the equilibrium moisture content is not significantly altered by these water repellent treatments, although the absorption rate is reduced, so that it may take months or even years to reach equilibrium. For short to medium term protection, polymer coatings can be applied to reduce moisture absorption. For example, short jute fiber/PLA composites, coated with 100 μm thick polypropylene plastic adhesive tape, have been shown to be moisture resistant in humid environments, although the effects were not long lasting [34].

Chemical treatments improve moisture resistance by reducing fiber hydrophilicity, improving fiber/matrix bonding and/or plugging water penetration pathways in the

Table 15.7 Methods for retarding moisture absorption of LPCs.

Method	Action	Examples
Water repellent treatment	Slows down the rate of water absorption	Coatings, surface applied oils, chemical treatments
Dimensional stability treatment	Reduces or prevents swelling in the fibers	Bulking the cell wall with polyethylene glycol, penetrating polymers, or bonded cell wall chemicals, or crosslinking cell wall polymers. Mostly used in wood fibers.

fiber. They do so by changing the chemical composition of the fibers. Unlike coatings these treatments are applied to fibers prior to incorporation into the matrix, and so they have an effect throughout the whole composite material and are more effective and longer lasting than surface isolation through coatings [35]. A combination of coating treatment with surface treatment gives even better moisture resistance as coatings provide a front line defense against moisture absorption while fiber treatments ensure longer-term durability. The most widely used chemical treatment methods are mercerization, acetylation, silane treatment, and maleated coupling.

Mercerization is an alkaline treatment method and its efficacy depends on a number of factors, including the type and concentration of the alkaline solution, the temperature and time of treatment, stress levels applied to the fibers as well as additive chemicals that may be used. This process removes most of the non-cellulosic components and part of the amorphous cellulose, and leads to a reduction in the hydrogen bonding potential of the network structure. Such fiber modifications help to reduce the moisture absorption. Potassium hydroxide (KOH) or sodium hydroxide (NaOH) are commonly used to decrease the hydrogen bonding capacity of cellulose and eliminate open hydroxyl groups that would be available to bond with water molecules. The process also dissolves hemicelluloses (the most hydrophilic component of the fibers) thus reducing the ability of the fibers to absorb moisture. In the amorphous regions, cellulose micromolecules are separated at relatively large distances with water molecules in between. Alkali sensitive hydroxyl (OH) groups present among the molecules are broken down, and then react with water molecules (HAOH) and move out from the fiber structure. The remaining reactive molecules form fiber–cell–O–Na groups between the cellulose molecular chains. This leads to a reduction in hydrophilic hydroxyl groups and an increase in the moisture resistance of the fibers.

Acetylation methods have been used for many years to improve some properties of wood cellulose such as moisture repellency, dimensional stability and resistance to environmental degradation. Their use for improving the properties of natural fibers has increased significantly in the last decade or so. The methods are based on the reaction of lignocellulosic material with acetic anhydride at elevated temperature, with or without a catalyst. The acetic anhydride reacts with the more reactive hydroxyl groups according to the equation [36],

$$\text{Fiber-OH} + (CH_3CO)_2O \leftrightarrow \text{Fiber-O-CO-}CH_3 + CH_3COOH \tag{15.3}$$

The acetylation of hydroxyl groups reduces their hydrophilicity, causes bulking of the cell wall and renders the material less susceptible to biological decay. The acetylated fibers show lower ECM, less weight loss and reduced erosion rate compared with non-acetylated fibers.

The use of various coupling agents, which build chemical and hydrogen bonds between the matrix and fibers, has also been shown to reduce the moisture absorption of cellulose fibers in LPCs. The improved fiber-matrix adhesion generated by the coupling agent reduces the moisture absorption that results from fiber-matrix debonding.

Silanes are multifunctional molecules which are used as coupling agents to modify fiber surfaces and create a chemical link between the fiber and matrix through a siloxane bridge. They go through various stages of hydrolysis, condensation and bond formation during the fiber treatment process. This treatment is used to stabilize

LPCs by increasing the resistance to water leaching. Silicon is accumulated in the cell lumina and bordered pits of fibers thus blocking these penetration pathways for water [35]. Treatment in bulk is more effective than surface coating because it may reduce the cell wall nano-pore size and deactivate or mask the hydroxyl functionalities thereby decreasing water sorption.

Maleated coupling agents contain combinations of chemical groups that allow one end of the molecule to be grafted onto the lignocellulosic fibers and the other end to interact favorably with the polymer molecules forming the matrix. The maleic anhydride end of the molecule can react with and bond to the hydroxyl groups in the amorphous regions of the cellulose structure, leaving the other end of the molecule, the structure of which is determined by its compatibility with the matrix, on the surface of the fiber. So, for example, in the case of a polypropylene (PP) matrix, the link molecule commonly used is maleic anhydride-polypropylene, which produces a fiber surface compatible with the hydrophilic PP molecules [37].

Other less frequently used treatments for improving moisture resistance are benzoylation treatment, peroxide treatment, sodium chlorite treatment and isocyanate treatment [37].

15.5 Testing of Weathering Properties

The weathering properties of LPCs can be studied either by exposing the composites to a natural environment for an extended period of time, or by exposing them to an accelerated weathering environment in the laboratory for a shorter period of time. The first method, whilst providing a scenario closer to realistic ambient conditions, is more difficult to control and analyze, laborious and time consuming. The second method, which has gained popularity in recent years, provides an easily controllable and reproducible environment using an accelerated weathering testing machine. In order to simulate the physical damage caused by sunlight it is not necessary to reproduce the entire spectrum of sunlight, since in most cases it is only necessary to simulate the more damaging, higher energy short wavelengths (in the UV range). In typical outdoor weathering conditions materials are exposed to alternating cycles of UV light (daytime), periodic condensation (dew at nighttime), and intermittent periods of rain. Accelerated weathering machines have the facility to simulate these outdoor conditions of sunlight, dew and rain in appropriate sequential cycles at controlled, elevated temperatures. The effects of sunlight are simulated with fluorescent ultraviolet lamps and those of dew and rain are simulated with condensing humidities and water spraying. This helps to reproduce in a few days or weeks the type of damage to the materials that would actually occur over months or years of outdoor exposure, which is the major advantage of this method. Conditions of temperature, moisture and light are controlled independently to enable simulation of the closest possible match to natural weathering.

UVA 340 lamps are normally used for generating UV radiation (as recommended in national standards, such as ISO or ASTM) to simulate the UV part of daylight. These lamps provide the best available simulation of sunlight in the critical short wavelength UV region between 360 nm and the solar cut off of 290 nm. They have a radiant emission below 300 nm of less than 2% of the total light output, an emission peak at 343 nm, and are most commonly used for simulation of daylight from 300 nm to 340 nm. For materials sensitive to higher wavelengths, xenon arc-type lights are also sometimes used for simulating sunlight since they produce a broader spectrum including wavelength up to infrared.

It must be emphasized, however, that it is very difficult to convert hours in the weathering machine into months or years of outdoor exposure. The major reason is the inherent variability and complexity of outdoor exposure conditions. The amount of sunlight, rain and moisture in one year in London will be different to that in Lahore or Los Angeles. Some of the variable factors include: the geographical latitude of the exposure site (closer to equator means more sunlight); altitude (higher means more UV); local geographical features, such as proximity to a body of water to promote dew formation; random year-to-year variations in the weather in the same city; seasonal variations (winter exposure may only be one seventh as severe as summer exposure); sample insulation (outdoor samples with insulated backing often degrade 50% faster than un-insulated samples). It is clearly impossible to set up an operating cycle for the machine that could take into account such diverse variables. Thus the data obtained from accelerated weathering testing machines cannot be used accurately to predict outdoor lifetimes, but can be used to give a reliable indication of the relative durability ranking of different materials.

The main effects of weathering on LPCs are color fading and loss in mechanical and physical properties. Studies have shown that, for both natural and accelerated weathering, longer exposure times increase the degree of color change and lightness, carbonyl concentrations, and fiber loss on weathered LPC surfaces.

The change in surface color of the samples is normally measured with a photometer or chroma meter in which lightness (L^*) and chromaticity coordinates (a^* and b^*)

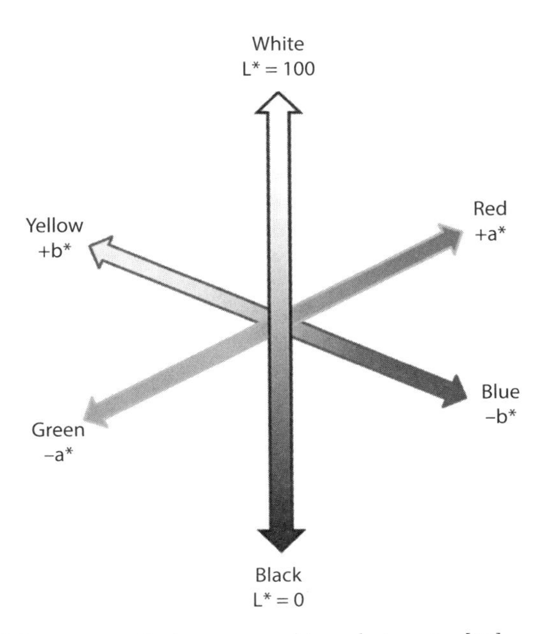

Figure 15.6 Color and lightness coordinates measured in a photometer [35].

are measured. L^* represents the lightness coordinate and varies from 100 (white) to 0 (dark), a^* represents the red ($+a^*$) to green ($-a^*$) coordinate, and b^* represents the yellow ($+b^*$) to blue ($-b^*$) coordinate. These coordinates are represented in Figure 15.6 [35]. An increase in L^* means that the sample has lightened or faded (positive ΔL^* for lightening and negative ΔL^* for darkening). Similarly, a positive Δa^* represents a color shift toward red, and a negative Δa^* represents a color shift toward green; a positive Δb^* represents a color shift toward yellow, and a negative Δb^* represents a color shift toward blue. The total color change (ΔE) is calculated according to ASTM D 2244:15,

$$\Delta E = [(\Delta L^*)^2 + (\Delta a^*)^2 + (\Delta b^*)^2]^{1/2} \tag{15.4}$$

It should be noted that ΔE represents the magnitude of the color difference but does not indicate the direction of this difference.

Whilst fading is more of an aesthetics issue, the loss in many other properties is crucial when these composites are to be used for outdoor applications. Both UV light and moisture degrade LPCs independently, but they can also act synergistically. The effects of weathering are measured in terms of retained physical, chemical and mechanical properties, by using techniques such as mechanical testing, microscopy, FTIR spectroscopy, and DSC analysis.

Figure 15.7 shows the typical effects of weathering on the surface of woodflour-filled high-density polyethylene (HDPE) composites, (a) before weathering, (b) after 1000 hours, (c) after 2000 hours, and (d) after 3000 hours of weathering. The swelling of wood particles at the surface and the matrix cracking due to UV light result in a flaky

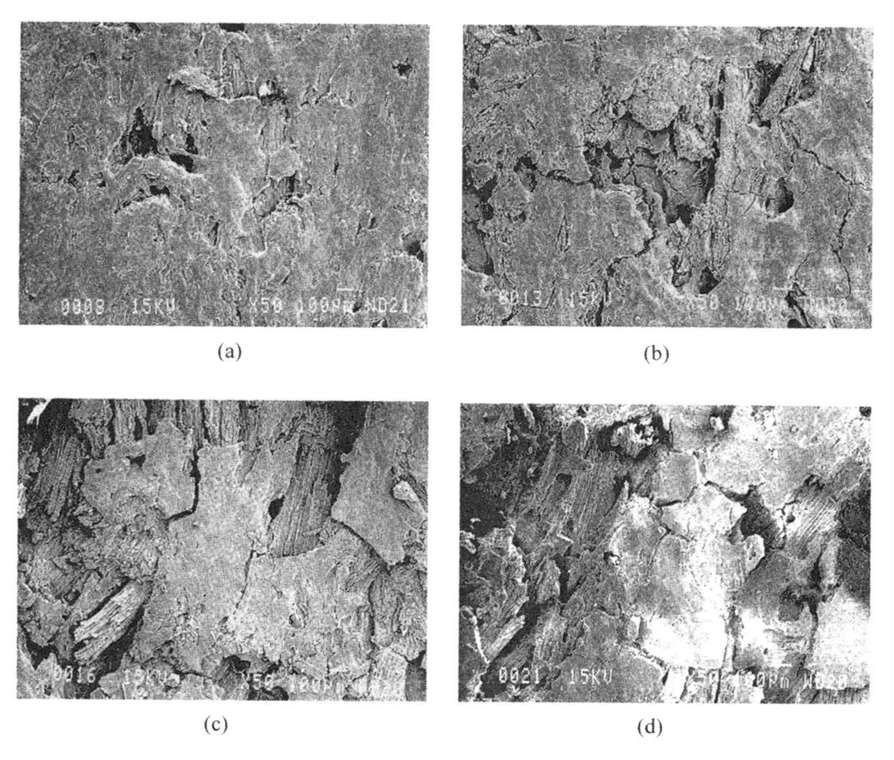

Figure 15.7 Micrographs of woodflour filled HDPE composite demonstrating surface deterioration: (a) before weathering, (b) after 1000 hours, (c) after 2000 hours, (d) after 3000 hours of weathering [7].

cracked surface. In addition, the level of oxidation at the surface, which is also a measure of weathering, can be determined by using FTIR spectroscopy.

15.6 Studies on Weathering of LPCs

A range of studies have explored weathering properties of composites including lignocellulosic fibers, cellulose fiber-reinforced thermosets, cellulose fiber-reinforced thermoplastics, and cellulose fiber-reinforced biodegradable polymers. Most of these studies have also focused on understanding how to mitigate the deleterious effects of weathering.

15.6.1 Lignocellulosic Fibers

Methacnon *et al.* [38] studied the accelerated weathering properties of cellulose fibers from four sources, namely, water hyacinth, reed, sisal, and Thai kenaf for their potential applications as geotextiles. Fibers were exposed to cycles of UV light for 8 hours followed by water spray for 4 hours at 50°C for a total exposure time of 840 hours. All the fibers exhibited about 50% reduction in tensile strength but negligible change in elongation to break following the exposure time. However they cautioned that it was difficult to relate the artificial weathering properties to life expectancy under natural weathering conditions since the rate of fiber degradation in nature depends on various external factors such as intensity of radiation, temperature, humidity, air pollution, etc., as discussed above.

A study providing a comprehensive picture of the color change responses to artificial indoor sunlight of some of the most important wood species in Europe was reported by Otlean *et al.* [39]. Sixteen wood species, including hardwoods and softwoods, were exposed for 600 hours to a xenon-arc lamp via a 3 mm thick window glass filter, in order to simulate natural sunlight indoors. Different wood species showed different color change behaviors, but there were some general patterns. Typically, for softwoods the surfaces discolored particularly rapidly in the first 24 hours of irradiation whereas in the case of hardwood species the more rapid discoloration occurred over a shorter period of about 12 hours. Although there were some exceptions, in most cases the rate of discoloration decreased with time and the colors stabilized, so that changes after about 120 hours were relatively small.

Rahman *et al.* [40] studied the influences of various surface pretreatments of photografted oil palm and henequen fibers on weathering responses by measuring mechanical and other degradable properties. In the case of oil palm fibers, different percentages of alkyl methacrylate (AMA) were grafted onto the fibers in methanol and Darocur 2959 was added to initiate photochemical polymerization. The fiber surfaces underwent a pretreatment with $KMnO_4$, and were dewaxed. Fibers were exposed to alternating cycles of UV light (for 4 hours at 65°C) and dew and condensation (for 2 hours at 45°C) for a total period of 100 hours. The losses in properties of the virgin fibers were significantly higher than those of treated fibers. The minimum losses of fiber weight (2%) and tensile properties (strength 10%, elongation 11%) were observed for the alkali-treated sample grafted with a formulation containing a silane coupling agent. In another study [41], untreated henequen fibers, 3% ethylacrylate (EA) treated fibers, and 3% EA treated fibers grafted with 0.2% silane were subjected to accelerated weathering cycles of UV light, dew and condensation over a period of 300 hours. These weathering conditions produced a tensile strength loss of 13.5% in untreated fibers, compared with just 6.8% for EA-grafted fibers and even less for the fibers treated with the silane additive. It was also noted that the minimum loss of all of the tested properties due to the severe weathering treatment was from the EA treated fibers containing silane.

It is clear from these studies that surface treatments of cellulose fibers can have significant positive effects on their response to weathering.

15.6.2 Lignocellulosic Thermoplastic Composites

The weathering properties of thermoplastic matrix cellulose fiber composites have been studied more extensively than other matrix types. HDPE-based composites have received most attention, followed by polypropylene (PP) matrix composites. The response of these composites to both natural weathering and accelerated weathering conditions has been studied, and Table 15.8 includes a summary of some of the studies into the effects of accelerated weathering on mechanical properties of cellulose fiber-reinforced thermoplastics. It is clear from the table that weathering has adverse effects on the mechanical properties of almost all the composites studied.

LPCs undergo surface chemistry changes when exposed to weathering conditions and these changes can be related to their subsequent performance. Stark and Matuana [60] studied surface chemistry changes of woodflour/HDPE composites using XPS and FTIR spectroscopy, following exposure to UV radiation for 2000 hours. Their results indicated that surface oxidation occurred in both neat HDPE and the composites as soon as samples were exposed. The surfaces of the composites were oxidized to a greater extent than those of the neat HDPE, suggesting that the addition of wood flour to the HDPE matrix resulted in greater weather-related damage. FTIR spectroscopy was used to study carbonyl group formation, vinyl group formation and crystallinity changes during weathering. An increase in the carbonyl index, vinyl index and matrix crystallinity indicates an increase in polymer chain scission. Although initially the carbonyl index was higher for WF/HDPE composites than for neat HDPE, it increased at approximately the same rate for both materials as shown in Figure 15.8 (a). The results indicated that chain scission may have occurred immediately upon exposure and that the number of chain scissions increased with increasing exposure time. The calculated vinyl group index, remained constant through the first 250 hours of exposure time, increased through 1000 hours, and then reached a plateau for both WF/HDPE composites and neat HDPE samples, as shown in Figure 15.8 (b). The increase in concentration of vinyl groups was much larger for neat HDPE than for WF/HDPE composites, suggesting that vinyl group formation takes place mainly in the polymer component of the composite. The plateau is thought to be due to degradation via the Norrish type II mechanism, which causes vinyl groups to be the major products in the first stage followed by a slower conversion to carbonyl groups. Increases in matrix crystallinity can also occur as a result of chain scission, since the shorter, more mobile chains in the

Table 15.8 Effects of accelerated weathering on mechanical properties of lignocellulosic polymer composites.

Matrix	Fiber/filler	Weathering cycle	Duration (h)	Degradation in mechanical properties (%)	Reference
Thermoplastics					
HDPE	Kenaf	UV (102 min) + water spray (18min)	2000	F. S. 24; F. M. 42	[42]
	Woodflour	UV (102 min) + water spray (18min)	2000	F. S. 20; F. M. 33	[42]
	Woodflour	Constant UV	2985	F. S. 2; F. M. 25 *increase*	[43]
		UV (108 min) + water spray & UV (12min)	3000	F. S. 27; F. M. 33	[43]
	Woodflour	UV (108 min) + water spray & UV (12min)	2000	F. S. 22; F. M. 26	[44]
	Wood fiber	UV (8h) + moisture (4h)	2000	F. S. 24; F. M. 21	[45]
	Woodflour	UV (108 min) + water spray &UV (12 min)	1500	F. S. 30; F. M. 30	[46]
PP	Wood fiber	Constant UV	2000	I. S. 18	[47]
	Vetiver	UV + water spray	240	T. S. 13; T. M. 35; I. S.40	[48]
	Woodflour	UV (12h) + moisture + UV (6h)	2000	F. S. 19; F. M. 11	[49]
	Jute	Constant UV	1350	F. S. and F. M. 0; I. S. 30	[50]
	Rice straw/ Seaweed	UV (4h) + moisture (2h)	720	T. S. 5; F. M. 20	[51]
	Sisal	Constant UV	2000	T. S. 22; T. M 35	[52]
	Jute	Constant UV	500	T. S. 25; T. M. 7	[53]
	Wood fiber	UV (4h) + moisture (4h)	1350	F. S. 20; F. M. 20; I. S. 20	[54]
PVC	Woodflour	UV (8h) + moisture (225 min) + water spray (15)	504	T. S. 12; T. M. 10; F. S. 12; F. M. 17	[55]
Thermosets					
Phenolic	Jute	UV (102 min) + UV & water spray (18 min)	750	T. S. 50; F. S. 50	[56]
Phenolic	Jute	UV (102 min) + UV & water spray (18 min)	500	T. S. 5; F. S. 20; F. M. 2	[57]
Biodegradable polymers					
PLA	Hemp	UV (1 h) + water spray (1 min) + moisture (2 h)	1000	T. S. 87; T. M. 88; F. S. 89; F. M. 67	[58]
PHB	Hemp	UV (150 min) + water spray (30 min) + moisture (1440 min)	1998	T. S. 47; T. M. 62	[59]

Abbreviations: T. S.: Tensile Strength; T. M.: Tensile (Young's) Modulus; F. S.: Flexural Strength; F. M.: Flexural Modulus; I. S.: Impact Strength.

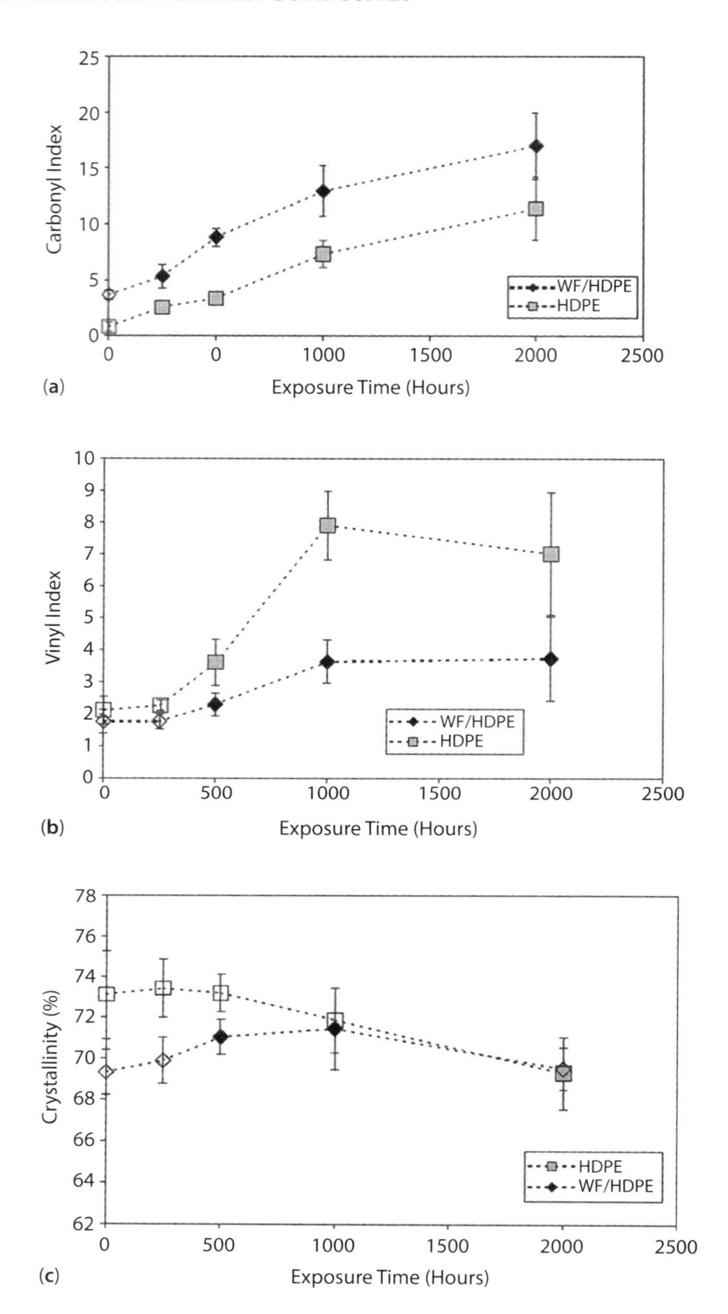

Figure 15.8 Carbonyl group formation (a), vinyl group formation (b), and crystallinity (c), as a function of exposure time for neat HDPE and WF/HDPE samples [60].

amorphous phase of the polymer may crystallize. There were differences between the crystallinity changes for neat HDPE and WF/HDPE composites, as shown in Figure 15.8 (c). There were not significant changes in the crystallinity of neat HDPE samples through the first 500 hours of exposure, but it then decreased steadily during further exposure up to 2000 hours. On the other hand, crystallinity of the HDPE matrix in the WF/HDPE composites started at a lower level due to the WF hindering crystallization, but increased with exposure time so that by 1000 hours it had reached the same level as neat HDPE at 1000 hours exposure. They then both decreased at the same rate between 1000 and 2000 hours. The results suggested that while neat HDPE may undergo crosslinking in the initial stages of accelerated weathering, WF may physically hinder the ability of HDPE to crosslink, resulting in the potential for HDPE chain scission to dominate in the initial weathering stage.

Similar increases in carbonyl index, vinyl index and crystallinity have also been reported for rape straw flour (RSF)/HDPE and nano-SiO_2/RSF/ HDPE composites exposed to natural weathering for 120 days by Zuo *et al.* [61].

Fabiyi *et al.* [62] studied natural and accelerated weathering properties of wood fiber/HDPE and wood fiber/PP composites. Both types of composites were exposed to weathering cycles of UV light and water spray for 2000 hours, and outdoor exposure for 2 years. Figure 15.9 shows scanning electron micrographs of wood fiber/HDPE composites; un-weathered (top left), 400 hours UV weathered (top right), 2000 hours UV weathered (bottom left), and 2 years natural weathered (bottom right). The study showed that for both natural and accelerated weathering, longer exposure times increased the degree of color change (and lightness), carbonyl concentrations, and wood loss on weathered surfaces. Three stages of degradation were evident in this study. First, the surface layer was eroded, creating cavities on the surface of weathered WPC.

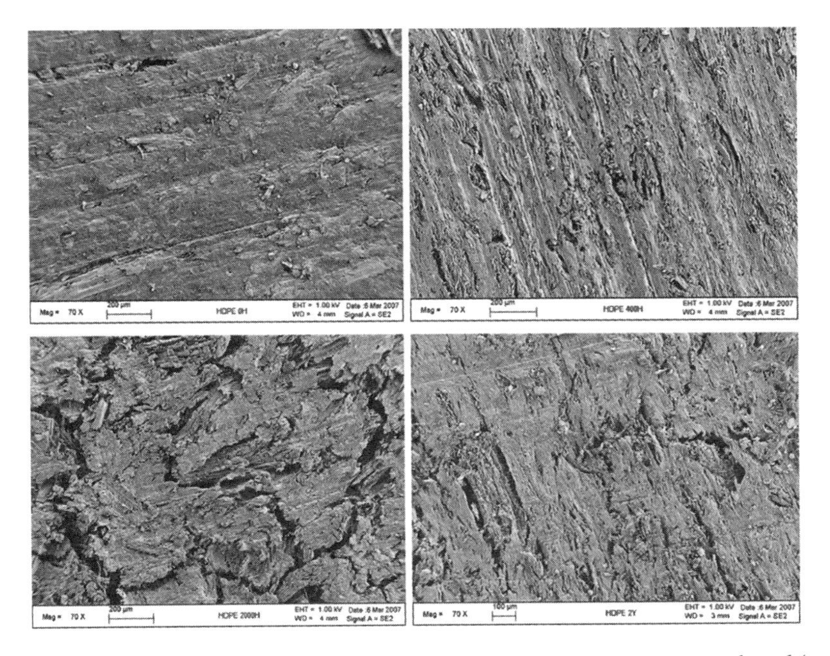

Figure 15.9 Scanning electron micrographs of HDPE/wood fiber composite: un-weathered (top left), 400 hours UV weathered (top right), 2000 hours UV weathered (bottom left), and 2 years natural weathered (bottom right) [62].

Second, the frequency and size of the cavities increased upon extended exposure time during both interior and exterior weathering. Third, the development of small cracks on the weathered surface of WPC followed the second stage. HDPE-based composites exhibited less lightening, carbonyl concentrations, and wood content loss when compared with PP-based composites. Oxidation and degradation of wood lignin were considered to have been the main mechanisms influencing the composite color changes (lightening) during weathering.

Fabiyi and McDonald [63] studied the effects of various wood species on mechanical, thermal and accelerated weathering behavior of wood fiber/HDPE composites. The selected wood species were poplar, Douglas-fir, black locust, white oak, and ponderosa pine. Samples exposed to alternating cycles of UV light and water spray for 2000 hours, exhibited color and lightness changes that increased with exposure time, the degree of increase varying between wood species. Oxidation of the composites during weathering, which was assessed by carbonyl group concentration, was shown to increase with exposure time. Poplar and ponderosa pine were found to have the best color stability among the wood species examined.

Comparison of wood and rice husk composites was carried out by Rahman *et al.* [64] who studied both natural and accelerated weathering properties of Kempas wood, HDPE and rice husk/HDPE composites. Natural weathering was conducted over a four month period. The accelerated weathering testing regime exposed samples to cycles with a periodicity of 12 hours, corresponding to 8 hours of UV exposure at 70°C followed by 4 hours of condensation at 50°C, for a total of 2000 hours. The results showed that natural and accelerated weathering degraded the properties of all three materials studied, that is, the rice husk/HDPE composite, neat HDPE, and Kempas wood. Of these materials, wood samples experienced the most serious degradation, but the color of the composite faded greater than that of the neat HDPE and Kempas wood under both testing conditions. Attacks by rain water and dew contributed to degradation of the lignin portion in rice husk and natural wood, which subsequently led to loss of impact strength. The HDPE matrix protected the composite from substantial degradation from the prolonged accelerated weathering attacks. The rice husk/HDPE composite showed better weathering resistance than wood and thus has the potential to be used as a durable eco-friendly furniture material.

Panthapulakkal *et al.* [65] studied the effects of accelerated UV radiation exposure on the flexural properties of rice husk/HDPE composites for 745 hours. Although the surface of the composites showed considerable discoloration, their flexural strength and modulus were not significantly different from those of the unexposed composite samples. It was suggested that the exposure time was not sufficient to cause any significant deterioration in flexural properties. On the other hand, water immersion for 1600 hours had a significant effect of flexural properties, particularly the flexural stiffness.

Some studies have been carried out to assess the effects of adding fillers on the weathering behavior of composites. Accelerated weathering properties of wood fiber/PP composites containing three different mineral fillers, calcium carbonate, wollastonite and talc, were studied by Butylina *et al.* [66]. Samples were exposed to UV light for up to 2000 hours, and this weathering resulted in significant color fading of all the composites. The composites incorporating mineral fillers exhibited greater changes in color (increased lightness) but lower drop in lignin index than the control composite sample. Scanning electron microscopy analysis revealed deterioration of the polymer surface layer in all the weathered composites. The samples containing talc were found to be more efficient in retaining their Charpy impact strength after weathering and it was concluded that owing to its relatively good mechanical properties and low sensitivity to weathering, the composite containing talc as a filler seemed to be most suitable for outdoor applications.

The properties of wood flour/recycled PP composites have been studied by Adhikary *et al.* [49]. They applied an exposure cycle consisting of five steps: 12 hours of UV light exposure at a temperature of 60°C, maintaining the sample at room temperature for 3 hours, spraying the exposed surface of the sample with water until its surface was saturated, exposing the sample to UV light for 6 hours at a temperature of 50°C, and keeping the sample at room temperature for 3 hours, for a total exposure time of 2000 hours. Significant property and microstructural changes were found to be induced by this accelerated weathering regime. Water absorption and thickness swelling of the composites were increased and the surface of the composites underwent significant color changes and lightening. Young's modulus and flexural strength were reduced and microstructural observations revealed a decrease in interfacial bonding between the wood flour and PP matrix resulting from the weathering. Whilst the crystallinity and melting temperature of the PP–wood flour composites were decreased after weathering, neat PP showed a 6% increase in crystallinity. The addition of MAPP as a coupling agent reduced the degradation in flexural strength and flexural modulus in the composite by about 50 and 27%, respectively.

The positive UV stabilizing effect of jute fibers was demonstrated by Van den Oever and Snijder [50] who studied the effects of UV radiation exposure on jute fiber/PP composites over a period of 56 days duration. They found that although the flexural and impact properties of neat PP were reduced dramatically, the irradiation had virtually no effect on the flexural properties and limited effect on the impact properties of the composites. Although UV irradiation caused bleaching of the very top layer of the compound from dark brown to light brown, the jute fibers clearly exhibited a UV stabilizing effect on the mechanical performance of PP.

However cellulose fibers can also have destabilizing effects on some polymer matrices. Properties of oil palm empty fruit bunch (EFB)-filled unplasticized poly (vinyl chloride) (UPVC) composites were studied by Bakar *et al.* [67]. The samples were alternatively exposed to UV (4 hours at 70°C) and to condensation (4 hours at 50°C) in repetitive cycle for 504 hours. The results showed that the incorporation of EFB fibers into the UPVC matrix accelerated the photo-degradation of the matrix when exposed to UV irradiation. Polymer crosslinking and chain scission were considered to be the main causes for the observed loss of impact strength and flexural modulus following weathering. Similarly Matuana and Kamdem [68] reported that the wood in wood/PVC composites was an effective chromosphore; wood/PVC degraded faster than neat PVC during accelerated weathering with water spray.

The use of hybrid fibers to improve the properties of composites is well established. The positive effect on the durability of hybrid rice straw/seaweed-PP composites was shown by Hasan *et al.* [51]. The weathering test was performed in alternating cycles of UV light of 4 hours (at 65°C) followed by condensation for 2 hours (at 65°C) for up to 720 hours. The results showed that the extent of tensile property loss of hybrid composites was less than that of single fiber (rice straw or sea weed) composites. The composites were also buried in soil (25% water) for a period of 16 weeks to study the effect of environmental condition on the degradation of the samples. Again it was found that the weight loss and tensile property loss was lower for the hybrid composite compared with the single fiber composites.

It has also been shown that the surface composition of LPCs can have a significant effect on their weathering behavior. Rowell *et al.* [60] used xenon-arc/water-spray weathering cycles (over 2000 hours) to evaluate the effects of the fiber content in aspen/PE composites (with fiber loadings of 0, 30–60%, by weight). As the fiber content increased, the weight loss increased, but at the conclusion of the 2000 hours, if the degraded surfaces were scraped off the samples, the weight loss was greater for the polyethylene with no wood fiber. It seems that although the wood degradation products are washed from the surface during weathering, the polyethylene degradation products are not. These studies exemplify the importance of the surface composition on weathering and the significance of water in the process. Composites having a high concentration of wood fibers at the surface can absorb water, whereas surfaces having fibers encapsulated by a polymer layer cannot absorb water until it has degraded.

Similarly, surface textures of LPCs can also affect their weathering behavior, as shown by a study of the weathering properties of rice husk/PE composites by Wang *et al.* [70]. They used weathering cycles consisting of 8 hours of UV irradiation and 4 hours of dew for a total of 2000 hours to assess differences between three surface textures, namely, smooth, lined, and sawn surfaces. Samples with smooth surfaces generally exhibited greater and more rapid fading than those with lined surfaces. However, compared with samples with sawn surfaces, samples with smooth surfaces showed much less change in morphology over longer periods. Also, whereas for samples with sawn surfaces, both flexural modulus of rupture and modulus of elasticity decreased significantly upon weathering, for those samples with smooth surfaces, the flexural properties declined only slightly during the whole weathering period. FTIR analysis indicated that both the cellulosic component and the PE matrix had degraded after weathering and oxidation might have occurred. It was concluded that the smooth surface was more durable in terms of mechanical properties than the other surface textures due to the improved protection of the natural fibers by the plastic matrix.

15.6.2.1 Effects of Photostabilizers and Surface Treatments

UV absorbers (UVAs) and pigments are the two most widely used methods for improving weathering resistance of LPCs and both methods have been reported to result in better weathering behavior. Stark and Matuana [43, 71] studied the influence of photostabilizers (hydroxy phenyl benzotriazole UVA and zinc ferrite pigment) on woodflour/HDPE composites exposed to xenon-arc radiation both with and without water spray. Samples were either exposed to 2 hour cycles consisting of 108 minutes of UV radiation followed by 12 minutes of simultaneous water spray and UV radiation for a total of 3000 hours, or to continuous UV radiation for 2985 hours. Adding the UVA and/or pigment was shown to help protect against increases in lightness and mechanical property loss to varying extents. Adding either UVA or pigment decreased the reduction in flexural properties, with the pigment having a more significant effect than the UVA. Although increasing the concentration of UVA resulted in reduced loss of both strength and modulus, increasing the concentration of pigment decreased the loss in strength but not the loss in modulus. The carrier wax of the pigment was thought to make the woodflour surface more hydrophobic, thus protecting the interface. For both unstabilized and photostabilized composites, the amount of lightening and loss in mechanical properties was less when the composites were exposed to UV light only compared with exposure to UV light and water spray. This is likely to be a result of the water spray washing away the degraded layer and wood extractives during weathering, as well as causing the wood-plastic interface to be affected through dimensional changes in the wood particles.

In another study, Stark and Matuana [44] investigated properties of photostabilized WF/HDPE composites following exposure to UV weathering. A hydroxybenzotriazole UVA, a low-molecular-weight hindered amine light stabilizer (HALS), a high-molecular-weight HALS, and zinc ferrite pigment were added to the composites as photostabilizers. The exposure cycle consisted of 108 min of UV radiation followed by 12 min of simultaneous water spray and UV radiation for a total time of 2000 hours. For both unfilled HDPE blends and the WF/HDPE composites, the addition of UVA and colorant decreased the lightening effect, but the composites experienced more significant lightening than the unfilled HDPE blends as a result of the bleaching of the woodflour. HDPE blends experienced a drop in flexural strength and modulus after 2000 hours of exposure, but it was found that the addition of photostabilizers could moderate the losses and even lead to increases. The flexural properties (both modulus and strength) of unfilled HDPE blends were affected by just 250 hours of cyclic exposure, whereas those of the composites were generally not significantly affected until between 1000 and 2000 hours of exposure. However, by 2000 hours the mechanical properties of the composites had deteriorated much more significantly than the unfilled HDPE blends. The addition of UVA and a colorant to the composite significantly reduced the strength and modulus loss. Although the addition of HALS helped maintain the flexural properties of unfilled HDPE blends during exposure, they did not have a significant effect on retaining the flexural properties of the composite formulations tested. This was attributed to acid–base interactions resulting from the acid-sensitive hindered amines and the acidic characteristics of woodflour.

Lundin [72] also found that HALS was ineffective in reducing the lightness and mechanical property loss in a study of 50% WF/PE composites weathered for 1500 hours. It was reported that the addition of HALS (either 0.25% or 0.5% by weight) to the composites did not affect color change caused by accelerated weathering and that the modulus of the composites decreased by between 26% and 30% regardless of the addition of HALS. On the other hand, the addition of 0.5% HALS reduced loss in flexural strength by about 3%, though the statistical significance was not reported.

The structural changes in photostabilized HDPE and WF/HDPE composites, resulting from accelerated weathering, were also studied by Stark and Matuana [12]. A hydroxyphenyl benzotriazole UVA, and a zinc ferrite pigment in a carrier wax were used as photostabilizers. The weathering cycle period of 120 min consisted of 108 min of UV exposure only followed by 12 min of combined UV light exposure and water spray for a total of 2000 hours. For the unfilled HDPE during the initial stages of exposure, the main structural change was crosslinking, leading to an increase in modulus. On further exposure, chain scission became more significant, and photodegradation of the polymer continued to the point at which the tie molecules were affected. This led to environmental stress cracking and a decrease in modulus. The addition of photostabilizers such as UVAs or pigments helped to reduce the loss in crystallinity during 2000 hours of exposure, and this in turn delayed the appearance of surface cracking and the consequent loss in modulus. On the other hand, for the WF/HDPE composites during the initial stages of exposure, crosslinking in the matrix did not predominate. Instead, the crystallinity of the polymer matrix increased as chain scission occurred. However, the increased HDPE crystallinity did not translate into modulus increase for the composites. It was suspected that the loss in modulus was due mainly to the effects of moisture. The modulus decreased during the second 1000 hours of weathering, possibly because of a combination of chain scission of the tie molecules in the matrix, photodegradation of the woodflour, and degradation of the interface due to moisture. The addition of photostabilizers prevented a significant loss in crystallinity between 1000 and 2000 hours of exposure. However, the modulus decreased during this time for the photostabilized composites, and so this further suggested that the modulus loss was due to the degrading effect of moisture on the composite interface.

The micrograph in Figure 15.10(a) of the unstabilized WF/HDPE composite before weathering shows a relatively smooth surface. However, following 2000 hours of weathering, two distinct phenomena were observed [Figure 15.10(b)]: the woodflour

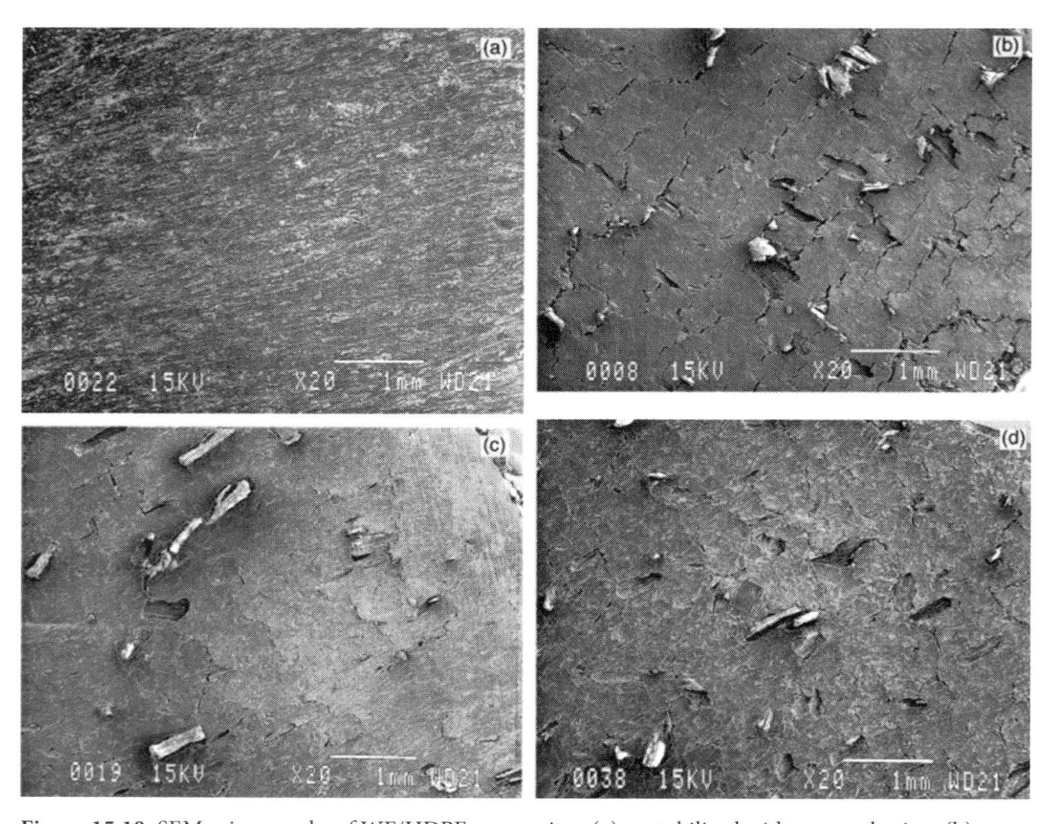

Figure 15.10 SEM micrographs of WF/HDPE composites: (a) unstabilized without weathering, (b) unstabilized and exposed to 2000 h of weathering, (c) stabilized with UVA and exposed to 2000 h of weathering, and (d) stabilized with pigment and exposed to 2000 h of weathering [12].

particles appeared to have risen from the surface, and surface cracks appeared in the HDPE matrix. The addition of UVA or pigment did not prevent the woodflour particles from protruding through the sample surface after weathering [Figures 15.10(c & d)]. However, the cracks in the HDPE matrix appeared to be less severe for the composites with UVA and pigment added than for the unstabilized composites [Figure 15.10(b)].

The effects of two different UVAs, UV-326 (benzotriazoles) and UV-531 (substituted o-hydroxybenzophenones), on the durability of rice hull/HDPE composites were studied by Du *et al.* [73]. These UVAs prevent excitation of the polymer by absorbing UV light and converting it into heat. Samples were subjected to accelerated weathering cycles consisting of 8 hours of dry UV exposure at 60°C, followed by 4 hours of exposure to condensation without UV at 50°C for up to 2000 hours. The composites without UVAs showed surface discoloration post-weathering as a consequence of extensive chemical degradation involving chain-scission reactions and the formation of carboxyl groups. The use of either of the UVAs resulted in less discoloration and smaller changes in the carbonyl index. However on the basis of less surface discoloration, a smaller change in the carbonyl index, and a higher fiber index with increasing exposure time, the sample with UV-326 was considered to have shown the least photodegradation. Although surface cracks were apparent in all of the composites, cracks in those containing UVAs appeared to be less severe than in those without UVAs. It was concluded that the UVAs protected the composites from UV degradation to a certain extent, with UV- 326 being more effective than UV-531.

Zhang *et al.* [74] studied the effects of four kinds of iron oxide pigments (red, yellow, blue and black) on the properties of WF/HDPE composites caused by accelerated UV weathering. These pigments were chosen because of their wide applications in the plastic industry in China. The weathering schedule involved 8 hours UV irradiation at 60ºC and 4 hours condensation at 50ºC for 2000 hours. They found that these iron oxide pigments provided very good protection against discoloration of the composites. However, apart from the black iron oxide that contained carbon black, the pigments used in this study did not appear to help maintain the mechanical properties of the composites. Black and red iron oxide proved to be more beneficial than the other two kinds of iron oxide pigments.

The effects of different color pigments on the durability of WF/HDPE were also evaluated by Du *et al.* [75]. The composites were dyed using inorganic pigments of three different colors (ferric oxide, carbon black, and titanium dioxide). The weathering cycle consisted of 8 hours of dry UV exposure at 60°C, followed by 4 hours of condensation exposure at 50°C for a total of 1500 hours. All the samples showed significant fading and color changes in exposed areas. Surface oxidation occurred within the first 250 hours of exposure for all samples, evidenced by an increase in carbonyl index, as shown in Figure 15.11. The surfaces of the control composites were oxidized to a greater extent than those of the dyed composites. Samples containing pigments exhibited less deterioration, less surface discoloration and showed a lower change in carbonyl index than the control with increasing exposure time. Cracks in the HDPE matrix also appeared to be less severe for composites containing pigment than for those without pigment. This suggests that the addition of pigments to the composites resulted in less weather-related damage. Carbon black was found to have a more positive effect on color stability than the other pigments.

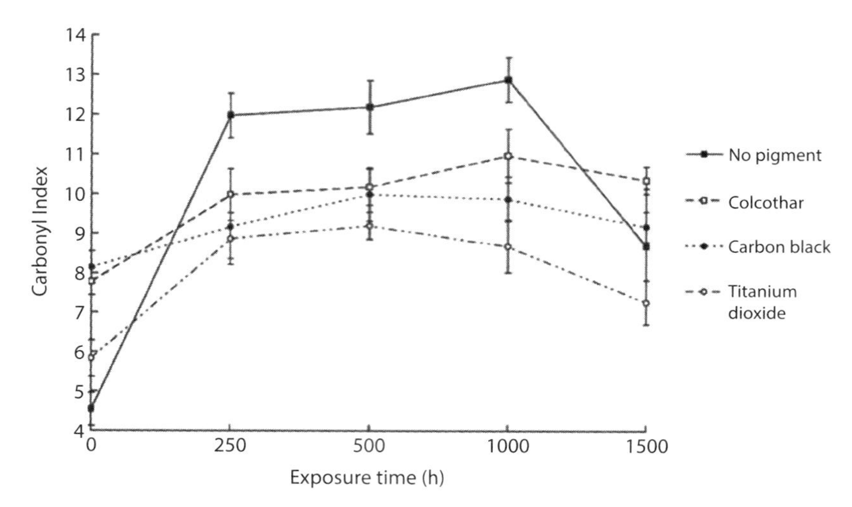

Figure 15.11 Change in the carbonyl index of control and pigmented WF/HDPE composites after accelerated UV weathering [75].

Matuana *et al.*. [76] also reported that rutile titanium dioxide was an effective UV stabilizer for wood/PVC composites.

Falk *et al.* [46] studied the effects of red and black pigments on properties of WF/HDPE and WF/PP composites following exposure to UV light. Samples were exposed to cycles of 102 min of UV irradiation followed by 18 min of UV and water spray for a total exposure time of 1500 hours. Nearly all exposed specimens experienced fading after exposure. The PP-based composites faded at a more rapid rate than the HDPE-based composites. In addition, the specimens with higher woodflour content faded more than specimens with lower woodflour content. The addition of either red or black pigment as UV absorbers was effective in reducing the fading effect for both HDPE and PP composites. Flexural testing of samples showed that although the addition of pigments decreased both flexural strength and stiffness of unexposed specimens, samples containing pigments had better flexural properties after accelerated weathering exposure than control samples.

Kiguchi *et al.* [77] studied the surface deterioration of WF/PP composites following exposure to natural and accelerated weathering. They found that the addition of darker color pigment as a stabilizer reduced the discoloration caused by weathering. Natural weathering was carried out for 2 years and accelerated weathering cycles consisting of continuous exposure to UV light and 18 min of deionized water spray every 2 hours for a total of 2000 hours. Although the darker colored pigmented composites showed less color change than non-pigmented composites following natural weathering, chalking on the surfaces still occurred.

Similar results were reported by Butylina *et al.* [78] who carried out a more extensive study of the natural weathering properties of wood-PP composites with and without pigments. Samples were exposed to natural weathering for one year in a Finnish climate and their color changes monitored. Darker pigments (such as brown and grey) were found to provide the best resistance to help reduce the discoloration of the composites. Although these darker pigments were able to reduce the color changes caused by photo-oxidation, they did not prevent the chemical changes in the lignin and polypropylene which occurred for all the composites whether pigments had been added or not. Degradation only occurred in the surface layers, never below a depth of 0.5mm. It was also found that samples having a higher polypropylene content and containing grey pigment and lubricant provided the composites with the best weathering resistance. Their results from Charpy impact tests showed that impact resistance was particularly reduced for the composites which absorbed higher quantities of water during weathering, a similar effect to that noted by Huang *et al.* [79].

Abu-Sharkh and Hamid [80] studied the effects of the compatibilizer maleated polypropylene and stabilizers Irgastab FS410FF and Tinuvin-783 on the natural and artificial weathering properties of date palm fiber/PP composites. Samples were exposed to natural weathering for nine months or to constant UV irradiation at 50% relative humidity for 2000 hours. The composites were found to be much more resistant than neat PP to losses in mechanical properties and melting point in both the natural and accelerated weathering trials. It was thought that the retention of mechanical strength in the composites during weathering resulted from the enhanced stability imparted by the presence of the fibers and enhanced interfacial adhesion resulting from oxidation of the polymer matrix. It was also found that compatibilized samples were less resistant to weathering than non-compatibilized ones due to the lower stability of the maleated PP compatibilizer.

However the positive effects of the compatibilizer maleated polypropylene have been reported for wood fiber/PP composites containing 3% UV stabilizer based on HALS technology by Seldén *et al.* [54]. Accelerated weathering consisted of 4 hours of dry UV irradiation at 60°C followed by 4 hours of condensation without radiation at 50°C, for a total of 8 weeks. Whereas this weathering caused a 10% reduction in flexural strength and a 50% reduction in impact strength in neat PP, the composites displayed a maximum 20% reduction in flexural strength but almost retained their original impact strength. The degradation in the composites was again found to be restricted to a thin surface layer, and this was put down to the screening effect of the wood-fibers. The degraded layer had a chalky appearance, due to degradation of the PP matrix, leading to chemicrystallization and extensive surface cracking. The rate of degradation of the PP matrix was approximately twice as high in samples with 50 wt% wood fibers, compared with samples containing 25 wt% fibers, owing to the higher number of chromophores in the former.

Surface treatments of cellulose fibers have been extensively used to try to improve their weathering properties as well as their bonding to the matrix. Joseph *et al.* [52] studied the effects of UV radiation on the mechanical properties of sisal fiber/PP composites following exposure of up to 12 weeks, with and without fiber surface treatments. The sisal fibers were subjected to chemical treatments such as urethane derivative of polypropylene glycol (PPG), poly [methylene poly (phenyl isocyanate)] (PMPPIC) and maleic anhydride PP (MAPP). It was observed that the tensile properties of composites containing both untreated and chemically treated fibers decreased after exposure to UV radiation, resulting from chain scission caused by photo-oxidation. The chemical treatments did not appear to enhance the resistance to weathering, with a greater decrease in tensile properties for composites containing chemically treated fibers compared with untreated fibers being noted.

Khan *et al.* [53] studied the physico-mechanical properties of starch-treated jute yarn/PP composites. They found that by using starch-treated jute yarns, PP matrix unidirectional composites showed better mechanical properties than those of the untreated jute-based composites. Samples were exposed to alternating cycles of 4 hours of sunlight and 2 hours condensation for a total period of about 500 hours. It was found that initially (after 50 hours) tensile strength and modulus values increased slightly and subsequently decreased steadily. After 500 hours of weathering, almost 75% of the original strength and about 93% of the modulus were retained for composites incorporating fibers treated with 3% starch.

Ke-Cheng Hung *et al.* [81] studied the natural weathering properties of acetylated bamboo fiber/ HDPE composites by exposing samples with varying degrees of acetylation to natural weathering for 1080 days. Acetylation of bamboo fibers was found to be a useful method for improving the weathering resistance of the composites. Composites made from bamboo fibers with a high degree of acetylation exhibited superior strength and modulus retention ratios after natural weathering compared with those containing unmodified fibers, which had the lowest ratios. However, chain scission of the polymer matrix occurred in all the composites during natural weathering, with no significant differences being observed between the composites, indicating that matrix degradation was independent of the acetylation level of reinforcements. Similarly, color fading was observed in all composites, but those composites containing acetylated fibers exhibited better mildew resistance than those with untreated fibers.

Benini *et al.* [82] studied the effects of weathering on neat high impact polystyrene (HIPS) and HIPS composites reinforced with mercerized or bleached sugarcane bagasse fibers. Samples were exposed to 12 hour cycles with UV radiation for 8 hours at 60°C followed by 4 hours of water condensation at 50°C for a total of 900 hours. It was found that the degradation processes were influenced by both the fiber volume fraction and fiber chemical pre-treatment, although SEM analysis indicated that the weathering did not cause degradation of the fiber/matrix interface.. Weathering resistance, as measured by tensile properties, was reduced by the addition of fibers to HIPS, with composites containing higher fiber fractions showing less resistance. The removal of residual lignin by bleaching treatment led to composites reinforced with these fibers being less susceptible to UV radiation and moisture than those reinforced with mercerized fibers.

Kumar *et al.* [83] studied the weathering properties of ethylene–propylene copolymer (EPC) matrix composites with three different reinforcement materials, namely, 3% NaOH-treated jute fibers, 17.5% NaOH-treated jute fibers and commercial microcrystalline cellulose powder, using maleated EPC as a compatibilizer. The samples were subjected to UV radiation at 60°C in air for 150 hours. Again, the neat polymer samples were more resistant to weathering than the composites. The samples reinforced with commercial microcrystalline cellulose were the most stable of the composites and those made with fibers treated with the lower concentration of NaOH were the most susceptible to photo-oxidation. It was concluded that optimizing the durability and mechanical properties of the natural fiber-reinforced composites was closely dependent on selecting the appropriate treatment for the fibers.

Another method for improving the weathering behavior of LPCs that has yielded encouraging results is to use a coating to cover the complete surface of the LPC. Matuana *et al.* [14] have reported investigations into the effects of coextruding a clear hydrophobic HDPE covering layer onto WF/HDPE composites. They measured the discoloration of samples when exposed to an accelerated UV testing regime. Samples were exposed to 12 hour cycles consisting of 8 hours of UV radiation at 60°C, 15 min of water spray, and 3 hours 45 min of condensation at 50°C. Both chemical and physical changes occurred, giving rise to two separate discoloration characteristics in the weathering responses of the composites. In the uncapped control composites, photo-oxidation of the lignin in the wood components occurred but the degraded colored wood components were washed from the surface by the water spray and condensation, which also increased the surface roughness resulting in further lightening of the composite surface. In contrast, for the co-extruded composite samples, the presence of a hydrophobic HDPE coating prevented the loss of the colored components from the surface by protecting it against the water spray and condensation and so initially it darkened due to degradation of the WPC at the interface. However eventually the surface covering layer failed due to erosion and induced thermal stresses from temperature changes, leading to significant discoloration during weathering as the water washed away the water soluble degraded wood components. The initial darkening of the coextruded composites was thought to be mainly due to photo-oxidation of wood components at the interface before the failure of the covering layer. The hydrophobic cap prevented wood fiber access to moisture/water, reducing the negative effect of water on weathering. Furthermore, the covering layer absorbed some UV light and reduced the availability of oxygen at the interface of coextruded composites, thus decreasing the photodegradation rate. The decreased lignin degradation rate and the oxidation rate at the interface of coextruded WPCs were demonstrated by FTIR and XPS spectra, respectively.

Pattamasattayasonthi *et al.* [55] studied the effects of an acrylic coating on the accelerated weathering properties of WF/PVC composites. The acrylic coating contained cerium dioxide (CeO_2) as a UV absorber. The weathering cycle used consisted of 8 hours of UV irradiation at 60°C, followed by 3 hours and 45 min of condensation at temperatures of 40, 50, or 60°C and then spraying with water for 15 min for a total of 504 hours. They found that the tensile and flexural properties of the composites decreased with aging time and the effect was greater at higher condensation temperatures. The hydrophilicity of the composites, as quantified by contact angle measurements, was seen to increase with aging time due to hydrogen bond formation between hydroxyl groups on the wood surfaces and water molecules. It was suggested that the relatively small changes in mechanical properties and hydrophilicity induced by the weathering showed that the protection provided by the acrylic coating was effective, although no weathering tests on uncoated control samples were reported.

15.6.3 Lignocellulosic Thermoset Composites

Lignocellulosic thermosets have also been studied for their weathering properties, although not to the same extent as lignocellulosic thermoplastics.

Singh *et al.* [56] studied the physical and mechanical properties of jute fiber/phenolic composites under natural and accelerated weathering conditions. Samples were exposed to 2 years of natural weathering and accelerated weathering using a cycle of 102 min of UV light at a temperature of 40°C followed by 18 min of UV light and water spray at a temperature of 30°C for a total of 750 hours. Observations in the SEM revealed fiber accentuation as well as fiber breakage/splitting and surface discoloration in samples that had been subjected to both natural and accelerated weathering. The tensile strength decreased by about 50% after 2 years of outdoors exposure and by about 47% in the 750 hours of accelerated weathering. The reduction in flexural strength was more pronounced in the accelerated weathering than in the outdoor exposure. They also noted that coating the jute/phenolic composites with polyurethane significantly reduced the mechanical property losses caused by weathering.

The performance of pultruded jute fiber-reinforced phenolic composites, typically used in the building trade for door frames, in natural and accelerated weathering conditions has been studied by Singh and Gupta [57]. Samples were subjected to natural weathering over 5 years and accelerated weathering for 500 hours with a cycle of 102 min of UV light at 40°C followed by 18 min of UV light and water spray at 30°C. After 5 years, the naturally exposed samples showed surface roughness and discoloration. Mechanical testing revealed reductions in flexural strength of 22% after 2 years natural weathering and 19% after 500 hours of accelerated weathering. It was concluded the properties were satisfactory for jute fiber-reinforced phenolic composite door frames to be considered as viable alternatives to the more conventional wooden door frames.

O'Dell [84] studied the weathering properties of jute fiber/unsaturated polyester composites and compared them with glass fiber/unsaturated polyester composites. Samples were exposed to cycles of continuous UV exposure for 102 min, followed by 18 min of water spray for a total of 1200 hours. It was noted that there were no observable visual differences between weathered glass and jute fiber specimens in color changes or surface erosion effects.

15.6.4 Lignocellulosic Biodegradable Polymer Composites

Cellulose fiber-reinforced biodegradable polymers result in completely biodegradable composites, the so-called 'green composites'. Interest in these composites has increased in recent years, with a corresponding increase in the studies looking into the weathering properties of these composites.

Islam *et al.* [58] studied the weathering properties of untreated and alkali-treated hemp fiber/polylactic acid (PLA) composites. Cycles consisting of 1 hour UV irradiation, followed by 1 min of spray with de-ionized water and a subsequent 2 hour condensation at 50°C were carried out for a total of 1000 hours. Every 250 hours during this accelerated ageing, tensile strength, flexural strength, tensile modulus, flexural modulus and fracture toughness were measured. It was found that all of these properties decreased with weathering, whereas the impact strength was found to increase for both untreated and alkali-treated fiber composites. Although during these accelerated weathering tests the untreated composites were found to be more resistant over the first 250 hours, the alkali-treated composites were found to be more resistant after 500 to 1000 hours. So overall, the untreated composites had a greater total reduction in mechanical properties than the alkali-treated composites. Results obtained from FTIR analysis and crystallinity measurements of the composites were also consistent with the deterioration of mechanical properties following exposure to the accelerated ageing environment.

The positive effects of alkali treatment on weathering properties of unidirectional bamboo strips-novolac resin composites were also reported by Das *et al.* [85]. The composite specimens were subjected to natural sunlight exposure for 60 days. It was observed that with increasing alkali concentration, the UV stability of the composite surfaces improved. This was attributed to the fact that the reaction between free hydroxyl groups of cellulose fibers (which are gradually more and more exposed with higher alkali concentration) and methylol group of novolac resin imparted the chemical bonds with higher UV stability. Although all the samples suffered deterioration in flexural properties during weathering those with fibers treated with higher alkaline concentrations retained higher flexural strengths and stiffnesses.

Yew *et al.* [59] studied the properties of rice starch/PLA composites following exposure to natural weathering for 8 weeks. The tensile properties of the composites dropped drastically upon exposure to natural weathering. Addition of epoxidized natural rubber as a filler further increased the degradation of the composites owing to oxidation and ozone attack on the unsaturated sites of the rubber. The incorporation of a modified rice starch with glycerol and modified rice starch with glycerol and water further decreased the tensile properties of the PLA composites upon exposure to natural weathering.

The accelerated weathering properties of neat poly-hydroxybutyrate (PHB) biopolymer films and hemp fiber-reinforced PHB composites have been studied by Michel and Billington [86]. Simulated weathering conditions included exposure to UV, heat, water spray, and elevated relative humidity. Two distinct weathering procedures were performed, one with elevated relative humidity and one without it. Hemp fiber/PHB composites exhibited mass loss, increased cross-sectional area, and decreased ultimate strength and modulus for both procedures, while ultimate strain increased in the presence of moisture and decreased otherwise. The presence of moisture generally exacerbated polymer and composite mechanical degradation, cracking, and fading. The composites exhibited weathering degradation similar to several synthetic polymer-based composites, but typically experienced either faster degradation rates or greater aggregate deterioration. In addition, changes in physical, mechanical, and visual properties of the polymer and composite specimens did not stabilize at the conclusion of either weathering procedure, indicating that these properties would likely further degrade with continued weathering.

In summary, susceptibility to weathering has clearly been shown to be significant for fully bio-based composites, and improved resistance to environmental attack will be necessary for fully bio-based composites to be considered viable replacements for any long term applications where materials are subject to hostile environmental conditions such as in the building sector.

15.7 Conclusions

LPCs are being increasingly used for applications where they are exposed to weathering conditions in which their performance is still far from satisfactory. Consequently studies in natural and accelerated weathering properties of LPCs have increased in recent years. The lignin and hemicellulose components of lignocellulosic fibers are primarily responsible for their poor weathering resistance. The two major agents of weathering are UV radiation and moisture and their two major effects on LPCs are fading of color and deterioration in mechanical properties. Oxidation and degradation of lignin is the principal cause of color lightening whereas the hygroscopic nature of hemicellulose is the principal contributor towards high moisture absorption of fibers. Various methods have been used to try to improve the weathering resistance of LPCs and these include the addition of UVAs, pigments and coupling agents during manufacture of the composites, incorporation of weathering resistant surface layers on LPCs, and surface treatment of lignocellulosic fibers prior to incorporation. The laborious and time-consuming nature of natural weathering testing has made accelerated weathering testing a popular method for assessing LPCs. These accelerated weathering machines can simulate outdoor conditions of sunlight, dew and rain with UV lamps and condensation.

Composites made from different wood species have been found to exhibit different color fading behaviors during weathering, for example, poplar and ponderosa pine show greater stability than Douglas-fir, black locust and white oak. The use of photografting along with silane coupling agents has been shown to improve the resistance of lignocellulosic fibers to weathering. Lignocellulosic thermoplastics have been the most widely studied group of LPCs and they have been shown to exhibit measurable changes in surface chemistry following exposure to weathering conditions. Exposure to natural or accelerated weathering results in immediate surface oxidation and the effect is exacerbated by the addition of cellulose fiber/filler. Increases in carbonyl index, vinyl index and crystallinity following exposure are indicators of polymer chain scission, which is accompanied by deterioration in mechanical properties of LPCs. Three stages are generally evident during surface degradation: erosion of the surface layer, increase in size and frequency of pits thus formed, and development of small cracks. The use of UV light alone has a less damaging effect on color and mechanical properties of LPCs than the combination of UV light and moisture. The composite performance improves and mechanical property loss is delayed when the composite has fewer fibers at the surface because fibers provide readily accessible pathways for moisture to enter the composite, contributing to the reduction of properties.

A number of methods have been tried in attempts to reduce color fading and deterioration in mechanical properties of LPCs during weathering. Although the use of HALS has not been reported to have any positive effect, the use of UVAs and darker pigments does improve the weathering resistance of LPCs. In particular, the pigment zinc ferrite has been shown to be more effective than UVAs in retaining composite properties. Although maleated PP has been shown to act as a compatibilizer between lignocellulosic fibers and PP matrices, its presence in LPCs reduces their long term weathering resistance. Acetylation of fibers improves the weathering resistance of LPCs but does not reduce the color fading. Removal of lignin from lignocellulosic fibers by mercerization results in improved resistance of LPCs to UV radiation and moisture. The use of starch as a chemical pre-treatment of fibers has also been shown to improve the weathering properties in the case of jute yarn/PP composites. The use of calcium carbonate, wollastonite or talc as a filler increases the color fading of the composite. The use of hybrid fibers also improves weathering resistance of LPCs. Composites made by using injection molding and extrusion processes exhibit similar patterns of color fading and mechanical property loss. Susceptibility to weathering is even more significant for fully biodegradable lignocellulosic composites than cellulose fiber/synthetic polymer matrix composites. Coextruding a clear hydrophobic HDPE coating layer over wood/ HDPE composites significantly decreases the discoloration during the weathering process. The use of acrylic and silicone coatings has also been shown to be effective in improving the weathering resistance and this type of approach seems to have potential for producing LPCs that can withstand long term weathering adequately.

References

1. Food and Agriculture Organization (FAO), Natural Fibers, http://www.naturalfibers2009.org/en/fibers (2009)
2. A.K. Bledzki and J. Gassan, Composites reinforced with cellulose based fibres. *Prog. Polym. Sci.* 24, 221–274 (1999).
3. B. Madsen and E. Kristofer Gamstedt, Wood versus plant fibers: Similarities and differences in composite applications. *Adv. Mater. Sci. Eng.* 2013, 1–14 (2013).
4. R. M. Rowell, Property enhanced natural fiber composite materials based on chemical modification, in *Science and Technology of Polymers and Advanced Materials*, P.N. Prasad, J.E. Mark, S.H. Kandil, and Z.H. Kafafi, (Eds.), pp. 717–732, Plenum Press, New York (1998).
5. D. Nabi Saheb and J.P. Jog, Natural fiber polymer composites: A review. *Adv. Polym. Technol.* 18(4), 351–363 (1999).
6. D. Kockott, Weathering, in *Handbook of Polymer Testing*, R. Brown, (Ed.), pp. 697–734, Marcel Dekker Inc., New York, (1999).
7. N.M. Stark and D.J. Gardner, Outdoor durability of wood-polymer composites, in *Wood-Polymer Composites,* K.O. Niska and M. Sain, (Eds.), pp. 142–165, Woodhead Publishing Limited, Cambridge (2008).
8. R. S. Williams, Weathering of Wood, in *Handbook of Wood Chemistry and Wood Composites*, R.M. Rowell, (Ed.), pp. 139–185, CRC Press, Boca Raton (2005).
9. K.H.G. Ashbee, *Fundamental Principles of Fiber Reinforced Composites*, 2nd Ed., Technomic Publishing Co. Inc, Lancaster (1993).
10. D. Roylance and M. Roylance, Weathering of fiber-reinforced epoxy composites. *Polym. Eng. Sci.* 18 (4), 249–254 (1978).
11. F. Trojan, *Engineering Materials and Their Applications*, 4th Ed., Houghton Mifflin Company, Boston (1990).
12. N. M. Stark and L. M. Matuana, Surface chemistry and mechanical property changes of wood-flour/high-density-polyethylene composites after accelerated weathering. *J. Appl. Polym. Sci.* 94, 2263–2273 (2004).
13. J.C. Arnold, The Durability of Composites for Outdoor Applications, Welsh Composites Consortium, http://www.welshcomposites.co.uk (2012).
14. L.M. Matuana, S. Jin, and N.M. Stark, Ultraviolet weathering of HDPE/wood-flour composites coextruded with a clear HDPE cap layer. *Polym. Degrad. Stabil.* 96, 97–106 (2011).

15. S. Mall, Laminated polymer matrix composites, in *Composites Engineering Handbook,* P.K. Mallick, (Ed.), pp. 811–890, Marcel Dekker Inc., New York (1997).
16. C.A. Harper, *Handbook of Plastics, Elastomers, and Composites*, 3rd Ed., McGraw-Hill, New York (1996).
17. R.M. Rowell, Moisture properties, in *Handbook of Wood Chemistry and Wood Composites,* R. M. Rowell, (Ed.), pp. 77–98, CRC Press, Boca Raton (2005).
18. O. Faruk, A.K. Bledzki, H. Fink and M. Sain, Biocomposites reinforced with natural fibers: 2000–2010. *Prog. Polym. Sci.* 37, 1552–1596 (2012).
19. D. Ray and J. Rout, Thermoset biocomposites, in *Natural Fibres, Biopolymers and Biocomposites,* A.K. Mohanty, M. Misra, and L.T. Drzal, (Eds.), pp. 291-345, Taylor & Francis, New York, (2005).
20. J. Gassan and A.K. Bledzki, Effect of moisture content on the properties of silanized jute-epoxy composites. *Polym. Compos.* 18(2), 179–184 (1997).
21. M.A. Semsarzadeh, Fiber matrix interactions in jute reinforced polyester resin. *Polym. Compos.* 7(2), 23–25 (1984).
22. D. Rouison, M. Couturier, M. Sain, B. MacMillan, and B.J. Balcom, Water absorption of hemp fiber/unsaturated polyester composites. *Polym. Compos.* 26, 509–525 (2005).
23. L.U. Devi, K. Joseph, K.C.M. Nair, and S. Thomas, Ageing studies of pineapple leaf fiber-reinforced polyester composites. *J. Appl. Polym. Sci.* 94, 503–510 (2004).
24. L.A. Pothan and S. Thomas, Effect of hybridization and chemical modification on the water-absorption behavior of banana fiber-reinforced polyester composites. *J. Appl. Polym. Sci.* 91, 3856–3865 (2004).
25. N.E. Marcovich, M.M. Rebordeo and M.I. Aranguren, Moisture diffusion in polyester–woodflour composites. *Polymer* 40, 7313–7320 (1999).
26. C.Z.P. Junior, L.H. de Carvalho, V.M. Fonseca., S.N. Monteiro and J.R.M. d'Almeida, Accelerated aging of hybrid ramie-cotton resin matrix composites in boiling water. *Polym. Plast. Technol. Eng.* 43, 1365–1375 (2004).
27. M.Z. Rong, M.Q. Zhang, Y. Liu, Z.W. Zhang, G.C. Yang and H.M. Zeng, Mechanical properties of sisal reinforced composite in response to water absorption. *Polym. Compos.* 10, 407–426 (2002).
28. P.V. Joseph, M.S. Rabello, L.H.C. Mattoso, K. Joseph and S. Thomas, Environmental effects on the degradation behaviour of sisal fibre reinforced polypropylene composites. *Compos. Sci. Technol.* 62, 1357–1372 (2002).
29. Z.A.M. Ishak, B.N. Yow, B.L. Ng, H.P.S. Abdul-Khalil and H.D. Rozman, Hygrothermal aging and tensile behavior of injection-molded rice husk-filled polypropylene composites. *J. Appl. Polym. Sci.* 81, 742–753 (2001).
30. J. George, S.S. Bhagwan and S. Thomas, Effects of environment on the properties of low-density polyethylene composites reinforced with pineapple-leaf fibre. *Compos. Sci. Technol.* 58, 1471–1485 (1998).
31. A. Espert, F. Vilaplana and S. Karlsson, Comparison of water absorption in natural cellulosic fibres from wood and one-year crops in polypropylene composites and its influence on their mechanical properties. *Compos. A* 35, 1267–1276 (2004).
32. B. Wang, M. Sain and P.A. Cooper, Study of moisture absorption in natural fiber plastic composites. *Compos. Sci. Technol.* 66, 379–386 (2006).
33. A. Shahzad, Impact and Fatigue Properties of Natural Fiber Composites, Swansea University, Swansea, *Ph. D. thesis* (2009).
34. R.H. Hu, M.Y. Sun and J.K. Lim, Moisture absorption, tensile strength and microstructure evolution of short jute fiber/polylactide composite in hygrothermal environment. *Mater. Des.* 31, 3167–3173 (2010).

35. Z.N. Azwa, B.F. Yousif, A.C. Manalo and W. Karunasena, A review on the degradability of polymeric composites based on natural fibres. *Mater. Des.* 47, 424–442 (2013).
36. G.T. Pott, Natural fibres with low moisture sensitivity, in *Natural Fibers, Plastics and Composites,* F.T. Wallenberger and N. Weston, (Eds.), pp. 105–122, Springer (2004).
37. M.M. Kabir, H. Wang, K.T. Lau and F. Cardona, Chemical treatments on plant-based natural fibre reinforced polymer composites: An overview. *Compos. B* 43, 2883–2892 (2012).
38. P. Methacanon, U. Weerawatsophona, N. Sumransina, C. Prahsarna and D.T. Bergadob, Properties and potential application of the selected natural fibers as limited life geotextiles. *Carbohyd. Polym.* 82, 1090–1096 (2010).
39. L. Oltean, A. Teischinger and C. Hansmann, Wood surface discolouration due to simulated indoor sunlight exposure. *Holz. Roh. Werkst.* 66, 51–56 (2008).
40. M.M. Rahman, A.K. Mallik and M.A. Khan, Influences of various surface pretreatments on the mechanical and degradable properties of photografted oil palm fibers. *J. Appl. Polym. Sci.* 105, 3077–3086 (2007).
41. M.M. Rahman, I.A.M.S. Chowdhury, M.J. Islam and M.A. Khan, Surface modification of henequen (Agave fourcroydes) fibers by ultraviolet curing with 2-hydroxyethylacrylate and ethylacrylate: Effect of additives on degradable properties. *J. Appl. Polym. Sci.* 102, 4000–4006 (2006).
42. T. Lundin, R.H. Falk and C. Felton, Accelerated weathering of natural fiber-thermoplastic composites: Effects of ultraviolet exposure on bending strength and stiffness, in *The Sixth International Conference on Woodfiber-Plastic Composites,* The Madison Concourse Hotel Madison, Wisconsin, 2001, pp. 87–93, Forest Products Society (2002).
43. N.M. Stark and L.M. Matuana, Influence of photostabilizers on wood flour–HDPE composites exposed to xenon-arc radiation with and without water spray. *Polym. Degrad. Stabil.* 91, 3048–3056 (2006).
44. N.M. Stark and L.M. Matuana, Ultraviolet weathering of photostabilized wood-flour-filled high-density polyethylene composites. *J. Appl. Polym. Sci.* 90, 2609–2617 (2003).
45. Z. Zhang, H. Du, W. Wang and Q. Wang, Property changes of wood-fiber/HDPE composites colored by iron oxide pigments after accelerated UV weathering. *J. Forest Res.* 21(1), 59–62 (2010).
46. R.H. Falk, T. Lundin and C. Felton, The effects of weathering on wood-thermoplastic composites intended for outdoor applications, in *Proceedings of the 2nd Annual Conference on Durability and Disaster Mitigation in Wood-Frame Housing,* Madison, Wisconsin, pp.175–179, Forest Products Society (2001).
47. S. Butylina, M. Hyvärinen and T. Kärki, A study of surface changes of wood-polypropylene composites as the result of exterior weathering. *Polym. Degrad. Stabil.* 97, 337–345 (2012).
48. Y. Ruksakulpiwat, N. Suppakarn, W. Sutapun and W. Thomthong, Vetiver–polypropylene composites: Physical and mechanical properties. *Compos. A*, 38, 590–601 (2007).
49. K.B. Adhikary, S. Pang, and M.P. Staiger, Accelerated ultraviolet weathering of recycled polypropylene--Sawdust composites. *J. Thermoplast. Compos. Mater.* 22, 661–679 (2012).
50. M.A.J. Van den Oever and M.H.B. Snijder, Jute fiber reinforced polypropylene produced by continuous extrusion compounding, part 1: Processing and ageing properties. *J. Appl. Polym. Sci.*, 110, 1009–1018 (2008).
51. M.M. Hassan, M. Mueller, D.J. Tartakowska, and D.H. Wagner, Mechanical performance of hybrid rice straw/sea weed polypropylene composites. *J. Appl. Polym. Sci.* 120, 1843–1849 (2011).
52. P.V. Joseph, M.S. Rabellob, L.H.C. Mattosoc, K. Joseph and S. Thomas, Environmental effects on the degradation behaviour of sisal fibre reinforced polypropylene composites. *Compos. Sci. Technol.* 62, 1357–1372 (2002).

53. M.A. Khan, R.A. Khan, H. S. Ghoshal, M.N.A. Siddiky, and M. Saha, Study on the physico-mechanical properties of starch-treated jute yarn-reinforced polypropylene composites: effect of gamma radiation. *Polym. Plast. Technol. Eng.* 48, 542–548 (2009).
54. R. Seldén, B. Nyström, and R. Långström, UV aging of poly(propylene)/wood-fiber composites. *Polym. Compos.* 25, 543–553 (2004).
55. N. Pattamasattayasonthi, K. Chaochanchaikul, V. Rosarpitak and N. Sombatsompop, Effects of UV weathering and a CeO_2-based coating layer on the mechanical and structural changes of wood/PVC composites. *J. Vinyl Additive Technol.* 17, 9–16 (2011).
56. B. Singh, M. Gupta and A. Verma, The durability of jute fibre-reinforced phenolic composites. *Compos. Sci. Technol.* 60, 581–589 (2000).
57. B. Singh and M. Gupta, Performance of pultruded jute fibre reinforced phenolic composites as building materials for door frame. *J. Polym. Environ.* 13, 127–137 (2005).
58. M.S. Islam, K.L. Pickering and N.J. Foreman, Influence of accelerated ageing on the physico-mechanical properties of alkali-treated industrial hemp fibre reinforced poly (lactic acid) (PLA) composites. *Polym. Degrad. Stabil.* 95, 59–65 (2010).
59. G.H. Yew, A.M.M. Yusof, Z.A.M. Ishak and U.S. Ishiaku, Natural Weathering Effects on the Mechanical Properties of Polylactic Acid/Rice Starch Composites, in *Proceedings of the 8th Polymers for Advanced Technologies International Symposium*, Budapest, Hungary, Volume 1 (2005).
60. N.M. Stark and L.M. Matuana, Surface chemistry changes of weathered HDPE/wood-flour composites studied by XPS and FTIR spectroscopy. *Polym. Degrad. Stabil.* 86, 1–9 (2004).
61. P. Zou, H. Xiong and S. Tang, Natural weathering of rape straw flour (RSF)/HDPE and nano-SiO_2/RSF/HDPE composites. *Carbohydr. Polym.* 73, 378–383 (2008).
62. J.S. Fabiyi, A.G. McDonald, M.P. Wolcott and P.R. Griffiths, Wood plastic composites weathering: Visual appearance and chemical changes. *Polym. Degrad. Stabil.* 93, 1405–1414 (2008).
63. J.S. Fabiyi and A.G. McDonald, Effect of wood species on property and weathering performance of wood plastic composites. *Compos. A*, 41, 1434–1440 (2010).
64. W.A.W.A. Rahman, L.T. Sin, A.R. Rahmat, N.M. Isa, M.S.N. Salleh and M. Mohktar, Comparison of rice husk-filled polyethylene composites and natural wood under weathering effects. *J. Compos. Mater.* 45, 1403–1410 (2010).
65. S. Panthapulakkal, S. Law and M. Sain, Effect of water absorption, freezing and thawing, and photo-aging on flexural properties of extruded HDPE/rice husk composites. *J. Appl. Polym. Sci.* 100, 3619–3625 (2006).
66. S. Butylina, M. Hyvärinen and T. Kärki, Accelerated weathering of wood–polypropylene composites containing minerals. *Compos. A*, 43, 2087–2094 (2012).
67. A. Abu Bakar, A. Hassan and A.H.M. Yusof, Effect of accelerated weathering on the mechanical properties of oil palm empty fruit bunch filled UPVC composites. *Iran. Polym. J.* 14, 627–635 (2005).
68. L.M. Matuana and D.P. Kamdem, Accelerated ultraviolet weathering of PVC/wood-flour composites. *Polym. Eng. Sci.* 42(8), 1657–1666 (2002).
69. R.M. Rowell, S.E. Lange and R.E. Jacobson, Weathering performance of plant-fiber/thermoplastic composites. *Mol. Cryst. Liquid Cryst.* 353, 85–94 (2000).
70. W. Wang, Q. Wang and W. Dang, Durability of a rice-hull--polyethylene composite property change after exposed to UV weathering. *J. Reinf. Plast. Compos.* 28, 1813–1822 (2009).
71. N.M. Stark, Changes in woodflour/HDPE composites after accelerated weathering with and without water spray, in *Proceedings of 2nd Wood Fiber Polymer Composites Symposium: Applications and Perspectives*, Bordeux, France, pp.1–12 (2005).

72. T. Lundin, Effect of Accelerated Weathering on the Physical and Mechanical Properties of Natural Fiber Thermoplastic Composites, University of Wisconsin, Madison, *M.S. thesis* (2001).
73. H. Du, W. Wang, Q. Wang, S. Sui, and Y. Song, Effects of ultraviolet absorbers on the ultraviolet degradation of rice-hull/high-density polyethylene composites. *J. Appl. Polym. Sci.* 126, 906–915 (2012).
74. Z. Zhang, H. Du, W. Wang, and Q. Wang, Property changes of wood-fiber/HDPE composites colored by iron oxide pigments after accelerated UV weathering. *J. Forest Res.* 21(1), 59–62 (2010).
75. H. Du, W. Wang, Q. Wang, S. Sui, Z. Zhang, and Y. Zhang, Effects of pigments on the UV degradation of wood-flour/HDPE composites. *J. Appl. Polym. Sci.* 118, 1068–1076 (2010).
76. L.M. Matuana, D.P. Kamdem and J. Zhang, Photoaging and stabilization of rigid PVC/wood-fiber composites. *J. Appl. Polym. Sci.* 80(11), 1943–1950 (2001).
77. M. Kiguchi, Y. Kataoka, H. Matsunaga, K. Yamamoto and P.D. Evans, Surface deterioration of wood-flour polypropylene composites by weathering trials. *J. Wood Sci.* 53, 234–238 (2007).
78. S. Butylina, M. Hyvärinen and T. Kärki, A study of surface changes of wood-polypropylene composites as the result of exterior weathering. *Polym. Degrad. Stabil.* 97, 337–345, (2012).
79. S.H. Huang, P. Cortes and W.J. Cantwell, The influence of moisture on the mechanical properties of wood polymer composites. *J. Mater. Sci.* 41, 5386–5390 (2006).
80. B.F. Abu-Sharkh, and H. Hamid, Degradation study of date palm fibre/polypropylene composites in natural and artificial weathering: mechanical and thermal analysis. *Polym. Degrad. Stabil.* 85, 967–973, (2004).
81. K. Hung, Y. Chen, and J. Wu, Natural weathering properties of acetylated bamboo plastic composites. *Polym. Degrad. Stabil.* 97, 1680–1685 (2012).
82. K.C.C.A. Beninia, H.J.C. Voorwald and M.O.H. Cioffi, Mechanical properties of HIPS/sugarcane bagasse fiber composites after accelerated weathering. *Proced. Eng.* 10, 3246–3251 (2011).
83. A.P. Kumar, D. Depan and R.J. Singh, Durability of natural fiber-reinforced composites of ethylene–propylene copolymer under accelerated weathering and composting conditions. *J. Thermoplast. Compos. Mater.* 18, 489–508 (2005).
84. J.L. O'Dell, Natural Fibers in Resin Transfer Molded Composites, in *Proceedings of Fourth International Conference on Woodfiber-Plastic Composites*, pp. 280–285, Madison, Wisconsin (1997).
85. M. Das, V.S. Prasad and D. Chakrabarty, Thermogravimetric and weathering study of novolac resin composites reinforced with mercerized bamboo fiber. *Polym. Compos.* 30, 1408–1416, (2009).
86. A.T. Michel and S.L. Billington, Characterization of poly-hydroxybutyrate films and hemp fiber reinforced composites exposed to accelerated weathering. *Polym. Degrad. Stabil.* 97, 870–878 (2012).

16

Effect of Layering Pattern on the Physical, Mechanical and Acoustic Properties of Luffa/ Coir Fiber-Reinforced Epoxy Novolac Hybrid Composites

Sudhir Kumar Saw*, **Gautam Sarkhel and Arup Choudhury**

Department of Chemical and Polymer Engineering, Birla Institute of Technology, Mesra, Ranchi, India

Abstract

Different layering patterns of native luffa mat and non-woven coir fiber-reinforced epoxy novolac hybrid composites were prepared by keeping the relative weight ratio of luffa and coir of 1:3 and the total fiber loading 0.40 weight fractions. The action of water in composite material was studied so as to produce greater swelling with resultant changes in the physical, mechanical and acoustic properties. The hybrid fiber-reinforced composites exhibited fair water absorption and thickness swelling. Hybridization of coir composites with luffa fibers can improve the dimensional stability, extensibility, and mechanical behavior of pure coir composites. The fracture surface morphology of the tensile samples of the hybrid composites was performed by using scanning electron microscopy. The excellent sound absorption performance and sufficient strength of the hybrid composites made them suitable for use in inner walls as sound-absorbing material or in interior trim parts in automotive applications. The experimental results provided evidence that addition of coir fiber enhanced the performance of hybrid composites.

Keywords: Hybrid composites, lignocelluloses fibers, water absorption, thickness swelling, mechanical properties, acoustic properties, SEM

16.1 Introduction

At the present time, due to lack of wood and petroleum resources, as well as increasing awareness of the environment and energy, the areas of application for new types of natural fiber functional materials are expanding. Natural fibers have the advantages of active surface, low density, low cost, worldwide availability, renewability, biodegradability, ease of preparation, low energy consumption and relative non-abrasiveness over traditional reinforcing synthetic fibers [1-3]. Moreover, natural fibers are

**Corresponding author*: sudhirsawbit@gmail.com; sudhirsaw@yahoo.co.in

Vijay Kumar Thakur (Ed.), Lignocellulosic Polymer Composites, (369–384) 2015 © Scrivener Publishing LLC

environmentally friendly and neutral with respect to carbon dioxide emissions. Making full use of certain characteristics of natural fibers such as protection of environment, lightness of weight, their softness, sound absorption and insulation in order to develop natural fiber-based functional composites has become a hot topic. More attention has been paid to the hybrid composites of natural fibers and processed hybrid composites in the production of wood replacement materials.

In view of this, study has been concentrated on the use of luffa fiber (*Luffa cylindrica*) and coir fiber (*Cocos nucifera*) that is easily available due to its natural distribution around the world. Luffa fiber has some inherent advantages over other fibers for its renewable nature, biodegradability, natural network (special arrangements), high strength, and initial modulus. It is obtained from the fruit of two cultivated species of the genus Luffa in the cucurbitaceous family, *Luffa Cylindrica* (smooth fruit) and *Luffa Acutangula* (angled fruit). Luffa fiber is not a single filament like glass fiber but a bundle of cellulose fibrils making a fibrous vascular system in a hierarchical structure. It is found in plenty of tropical and subtropical countries of Asia (India, China, Pakistan and Indonesia), Africa and South America. The young and greenish fruit is eaten as a vegetable all over the world. The ripe and dried fruit is the source of luffa or vegetable sponge or sponge gourd as an industrial fiber as shown in Figure 16.1a. Parts of the plant are used to create bath or kitchen sponges, a natural jaundice remedy (Juice), furniture and even houses. Luffa cultivation worldwide has been steadily increasing in the past 20 years in response to the rising demand for renewable fiber sources and clean agricultural practices [4,5].

Coir is one of the most common and familiar natural fibers on earth. It is the seed-hair fibers obtained from the outer shell or husk of the coconut, the fruit *Cocos nucifera*, a tropical plant of the Arecaceae (Palmae) family as shown in Figure 16.1b. It grows extensively in tropical countries such as India, Sri Lanka, Thailand, etc. Because of its hardwearing quality, durability and other advantages, it is used for making a wide variety of floor furnishing materials, yarns, heavy cord, coarse nets, mats, mattresses, brushes, sacking, caulking boats, rugs, geotextiles, and insulation panels in traditional industries. Moreover, the production of these traditional coir products reaches about 450 thousand tons annually, which is only a small percentage of the total world production of coconut husk [6]. Several studies have been carried out to understand the structure, properties and the influence of chemical modification on coir fibers. Hence, research and development efforts have been underway to find new application areas for coir fibers, including utilization of coir fibers as reinforcement in polymer composites.

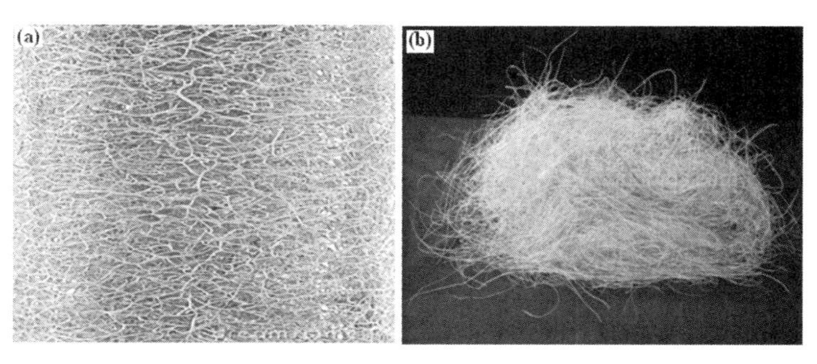

Figure 16.1 A photograph of (a) luffa and (b) coir fiber.

The application areas of these fibers include the automobile industry, the building industry, the packaging and insulation industries [7-9]. Due to the composition and morphological features of the coir fiber, the potential for improving the mechanical properties of the composite by using coir fiber as the reinforcement is not great, but the application prospects look good as the functional architecture and decoration materials have been researched [10, 11].

A large number of polymers that have good performance as matrix materials provide a wide range of properties and, therefore, a large variety of composite materials [12]. Compared to thermoplastic materials such as polyethylene, polypropylene, polystyrene, polyamides, etc., thermoset resins have a superior thermal stability and lower water absorption [13]. Owing to its outstanding advantages, phenol formaldehyde (PF) resin has some limitations of brittleness, evolution of volatiles during processing and shelf life in terrible environment. A strategy to solve these problems is to introduce addition-curable and thermal stable groups, i.e., an epoxy group into phenolic backbone by the process of epoxidation. At this moment, the laboratory prepared epoxy novolac resin (ENR) fulfills the requirements for higher performance applications having some outstanding properties like excellent adhesion to many different material strengths, toughness and resilience, resistance to chemical attack and moisture. Apart from these, it has better electrical insulation properties, absence of volatility on heat curing as well as negligible shrinkage.

Composite materials which are produced today are used for hulls, surfboards, sporting goods, swimming pool linings, building panels and car bodies. There are two types of fibers that are used as reinforcements: natural and manmade or synthetic fibers. Composite materials comprised of two or more fibers in a single matrix are called hybrid composites. Hybridization with more than one fiber type in the same matrix provides another dimension to the potential versatility of fiber-reinforced composite materials. Properties of the hybrid composites may not follow a direct consideration of the independent properties of the individual components. Czigany investigated the comparison carried out between a hybrid of carbon fibers and basalt fibers with polypropylene matrix. In order to achieve sufficient interfacial adhesion, the fibers were treated with the reaction mixture of maleic acid anhydride and sunflower oil. The strength properties of hybrid composites were improved by surface treatment, and this proven by mechanical tests and microscopic analysis. An acoustic emission test was also carried out by using a suitable microphone to find out mode of failure [14]. Padma priya and Rai experimentally analyzed the degree of mechanical reinforcement that can be obtained by the introduction of glass fibers in plant fiber (silk fabric)-reinforced epoxy composites. The addition of a small amount of glass fiber to the silk fabric-reinforced epoxy matrix enhances the mechanical properties of the resulting hybrid composites. It has also been observed that the properties increase with the increase in the weight fraction of reinforcement content to the maximum extent. The water uptakes of hybrid composites are observed to be less than that of unhybridized composites [15]. In an investigation by Jawaid *et al.*, trilayer hybrid composite of oil palm empty fruit bunches (EFB) and jute fiber was prepared by keeping oil palm EFB as skin material and jute as the core material and vice versa. A remarkable reduction in the void content of hybrid composites in different layering patterns was observed; also, the tensile property was slightly higher for the composite having jute as skin and oil palm EFB as core material [16]. Amico *et al.* and Mariatti *et al.* studied the mechanical properties of pure sisal, pure glass, and hybrid sisal/glass compression-molded composites, in which various stacking sequences of fiber mat layers were used. It was shown that hybridization originated a material with general intermediate properties between pure glass and pure sisal. However, the importance of controlling the stacking sequence to enhance properties was evident [17, 18]. A lot of work has been done on the composites based on these fibers.

In the present study, coir fiber is hybridized with small quantities (25%) of stronger luffa fiber mat and reinforced with laboratory prepared epoxy novolac resin (ENR) to develop cost effective and high performance composites. The intrinsic properties of luffa and coir fibers are given in Table 16.1. The luffa fibers have high cellulose content, high aspect ratio due to low value of diameter and low microfibrillar angle, i.e., 11-12°,due to parallel orientation of microfibrils possessing high tensile properties. Hence the inherent physicomechanical properties of luffa fiber are better than coir fiber. The coir fiber is porous in nature and has high value of microfibrillar angle, i.e., 30-39°, due to spiral orientation of microfibrils. Hence, coir fiber has less tensile properties; but the coir fiber has certain unique properties which make it very attractive. Among natural fibers, coir fibers have high lignin content and high value of failure strain, i.e., 40-45% between fiber and matrix. This has the potential to ameliorate the toughness when they are used in composites. The internal porosity and structural characteristics of coir fiber have been used to make good sound absorbing materials. So, native luffa mat and non-woven coir fiber can be selected to hybridize and reinforce with epoxy novolac resin (ENR) in order to study some mechanical and physical properties of these composites in terms of long-term thickness swelling, water absorption and acoustic properties. Microstructures of various formulated composites were examined to understand the mechanisms for the fiber-matrix interaction, which affects dimensional stability and mechanical strength of the hybrid composites.

Table 16.1 Physical properties of luffa and coir fiber.

Parameter	Luffa fiber	Coir fiber
Density (gm/cc)	1.3±0.1	1.1±0.1
Cellulose content (%)	60.42±1.8	33.28±1.21
Hemicelluloses content (%)	20.88±1.4	12.67±1.44
Lignin content (%)	11.69±1.2	46.84±0.8
Moisture content (%)	7±1	9±1
Microfibrillar angle (°)	11±2	39±5
Diameter (mm)	150±20	240±10
Lumen width (mm)	5±2	12±5
Average shape factor (%)	93±2	90±2
Tensile Strength (MPa)	375	144.6
Young's Modulus (MPa)	19500	3101.2
Elongation at break (%)	1.2±0.2	32.3±0.2
Aspect ratio (L/D)	350±10	100±5

16.2 Experimental

16.2.1 Materials

The coir fiber (Bristol fiber, *Cocos nucifera*, diameter = 200-240 μm; density = 1.0-1.2 g/cm^3 and micro-fibril angle = 30-39°) was obtained from Central Coir Research Institute, Alleppy, Kerala, India. Luffa fiber (*Luffa cylindrica*) was obtained from special markets of Jharkhand state, India. To separate the inner fiber core from outer mat core, the as-received luffa-gourd was cut carefully. Only the outer mat core of luffa was used in this study. Phenol and 37% w/w formalin were obtained from E. Merck, India were used as received for novolac preparation. Epichlorohydrin and sodium hydroxide were used for epoxidation of novolac resin and purchased from S. D. Fine Chemicals, India. Di-ethylene triamine (DETA) was used as a hardener for the epoxy novolac resins. The physical properties of luffa and coir fibers are given in Table 1.

16.2.2 Synthesis of Epoxy Novolac Resin (ENR)

The epoxy novolac resin was prepared by two step reactions. The first step was involved in the formation of novolac resin through chemical reaction between phenol and formaldehyde in acid environment. In the second step, the epoxidation of previously prepared novolac resin was carried out by reacting epichlorohydrin with a novolac resin in a specific molar ratio. The detail of the synthesis of novolac resin and its epoxidation was described in our previous report [19, 20]. The typical properties of the synthesized ENR are given in Table 16.2.

16.2.3 Fabrication of Composite Materials via Hot-pressing

A hand lay-up technique followed by compression molding was adopted for composite fabrication. The netted luffa (L) mat and non-woven coir (C) fibers of 1 cm length were used in the preparation of hybrid composites. ENR was impregnated into fiber within a mold cavity (dimensions: 100 mm 100 mm 5 mm). In each formulated composite, the

Table 16.2 Typical properties of synthesized liquid epoxy novolac resin (ENR).

Parameter	Evaluation
Appearance	A clear pale yellow liquid
Specific gravity at 25°C (gm/cc)	1.2
Viscosity (cps)	475
Epoxy equivalent number	191
Gel point in minutes	26
Solid content (%)	84
Tensile strength (MPa)	7.4
Tensile modulus (MPa)	176
Impact strength (kJ/m^2)	1.2

weight ratio of matrix to fiber is 60:40. Composites with different sequence of fiber mat arrangement such as luffa/coir/luffa (L/C/L) and coir/luffa/coir (C/L/C) were prepared by keeping the relative coir/luffa fiber weight ratios constant, i.e., 3:1. The curing of resin was done by the incorporation of epoxy novolac and hardener (DETA) in weight ratio of 80:20 in 100 g of acetone by making homogenous liquid with mechanical stirring. Acetone is used as the distribution medium for the resin to be adhered to the fiber. The composite was allowed to set for 24 hours at room temperature. Thereafter, the composite was compression-molded at constant pressure of 5 MPa and 110°C for half one hour followed by post-curing at 105°C for two hour in an oven. Before making the composite, the press machine was degassed to remove air bubbles and voids and also to make the raw materials maintain a certain shape. The hot-press temperature was then set to 110 °C for 10 minutes. A neat epoxy novolac matrix (unfilled) sample was prepared and epoxy novolac resin with pure coir fiber and pure luffa mat composites were also prepared. A sketch of different configurations of the layering pattern of composites is shown in Figure 16.2. The test specimens for mechanical properties, acoustic properties, water absorption and thickness swelling were cut from the composites according to ASTM standards.

16.3 Characterization of ENR-Based Luffa/Coir Hybrid Composites

16.3.1 Dimensional Stability Test

The dimensional stability tests involved were thickness swelling and water absorption. Thickness swelling and water absorption were conducted as per ASTM D 5229. Before testing, the weight and thickness of each specimen were measured. Three specimens of each formulated composite were immersed in distilled water at room temperature. The specimens were removed from the water after certain period of time and wiped with a cotton cloth before the weight and thickness value was measured. The weight and thickness value of the specimens were taken and then immersed again in water. The dimension stability test was continued for several hours until the constant weight and thickness of specimens obtained.

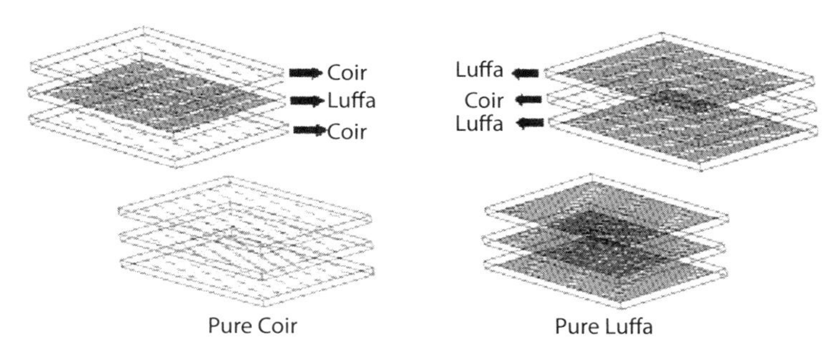

Figure 16.2 A sketch of different configurations of layering pattern of composites.

16.3.2 Mechanical Strength Analysis

The tensile test was carried out with rectangular specimens (width = 10 mm and thickness = 5 mm) using a universal tensile machine (UTM 3366, Instron) according to ASTM D 638. A crosshead speed of 1 mm/min was applied. All tests were conducted under ambient conditions. The data reported were averages of at least six measurements. For cured ENR (The results are reported in Table 16.2) and its composites, six measurements were executed with each sample. The tensile properties of both of these coir and luffa fibers were determined using the same tensile machine at a strain rate of 1 mm/min and a gripping length of 50 mm at 23°C ±1°C and 58% relative humidity (The results are reported in Table 16.1).

The impact strength of the pure ENR and its hybrid composites was measured with a standard Izod impact testing machine (model IT 1.4, Fuel Research Instruments, Maharashtra, India) according to ASTM D 256. The measurements were done on unnotched samples (70×10×4 mm) with an impact speed of 1 m/s and energy of 1.4 joules. For each specimen, six measurements were recorded.

16.3.3 Sound Absorption Test

In order to evaluate the acoustic properties of the various formulated composites, the sound absorption coefficients were determined by the impedance tube method according to ASTM E 1050-12. A sound absorption coefficients and impedance measurements instruments (Figure 16.3) made by Walen Audio Technologies, Maharashtra, India was adopted. The sample diameter was 100 mm, and each value represented the average of six samples. The sound absorption coefficients were measured in six frequencies: 100, 200, 400, 800, 1600 and 3200 Hz.

16.3.4 Scanning Electron Microscopy (SEM)

Before their incorporation in the composites, the cross-section of luffa and coir fibers were analyzed by SEM (Model: Jeol-JSM 6390 LV, Japan) with the following specifications: accelerating voltage, 20 KV; image mode, secondary electron image; working distance, 20 mm. The test samples were mounted on aluminum stub and coated with gold metal to avoid electrical charging during examination. To study the morphological features of fiber-matrix interface in the composite samples, the tensile test samples

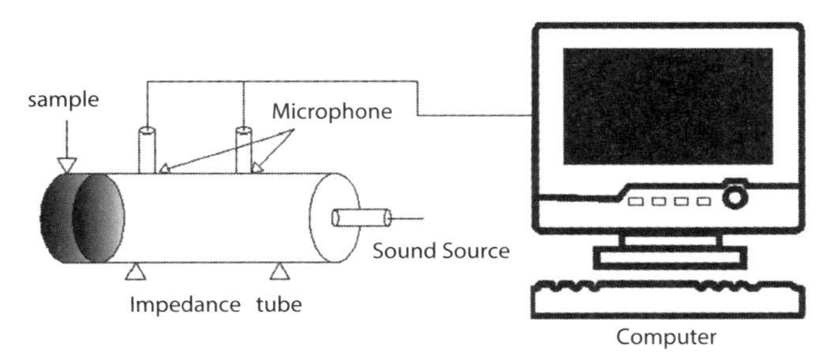

Figure 16.3 Schematic diagram of the impedance tube method.

Table 16.3 Evaluation of water absorption and thickness swelling test of various formulated hybrid composites.

Type of composites	Absorption of water (%)	Thickness swelling (%)
Epoxy novolac	0.22	0
Pure coir	21.87	10.57
Pure luffa	9.35	7.29
Coir/luffa/coir	16.32	9.43
Luffa/coir/luffa	12.73	8.63

were fractured in liquid nitrogen after 15-20 min of freezing in liquid nitrogen. The fractured surfaces were also sputtered with gold and analyzed in same specifications. All the test samples were scanned into 500 times of magnification.

16.4 Results and Discussion

16.4.1 Water Absorption Test

Water absorption test was used to determine the amount of water absorbed under specified conditions. In general, moisture diffusion in a composite depends on factors such as volume fraction of fiber, voids, viscosity of matrix, humidity and temperature. The effect of layering pattern on the water absorption properties of the composites is shown in Table 16.3. From Table 16.3 and Figure 16.4, it is observed that the water absorption for the pure coir composite with the value of 21.87% was the highest among the different formulated composites. This indicated that the high porosity or the presence of void on the surface of pure coir composite. With the presence of voids on the surface of composite, the weight of composite will increase by trapping the water inside the voids. In contrast, the water absorption for the epoxy novolac composite was the lowest with the value of 0.22%. This is because epoxy novolac resin limits the absorption of water into the fiber-mat composite because epoxy novolac resin acts as water resistant matrix. The pure luffa composite shows low absorption of water, i.e., 9.35%; the water absorption behavior of the polymer composites depends on the ability of the fiber to absorb water due to the presence of hydroxyl groups. The hydroxyl groups absorbed moisture or water through the formation of hydrogen bonding. The higher the moisture content of the natural fiber, the higher is the change in the mechanical and physical properties of composite. It leads to poor wetability with matrix, which results in a weak interfacial bonding between the fiber and matrix [21].

The hybrid composite shows relatively fair water absorption, which is 16.32% for hybrid coir/luffa/coir composite and 12.73% for hybrid luffa/coir/luffa composite. This showed that the packing and hybrid arrangement of fiber would limit the absorption of moisture into the composite because the voids have been filled up during the formation of hybrid composite. It is observed that water uptake of all composites increased with immersion time, reaching a steady state value at saturation point (0 to 50 h), at which point no more water was wrapped up and composite water content remained constant.

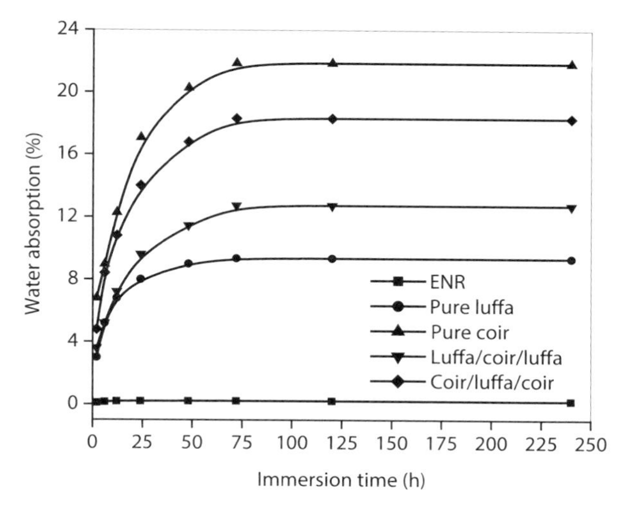

Figure 16.4 Water absorption (%) of various formulated composites.

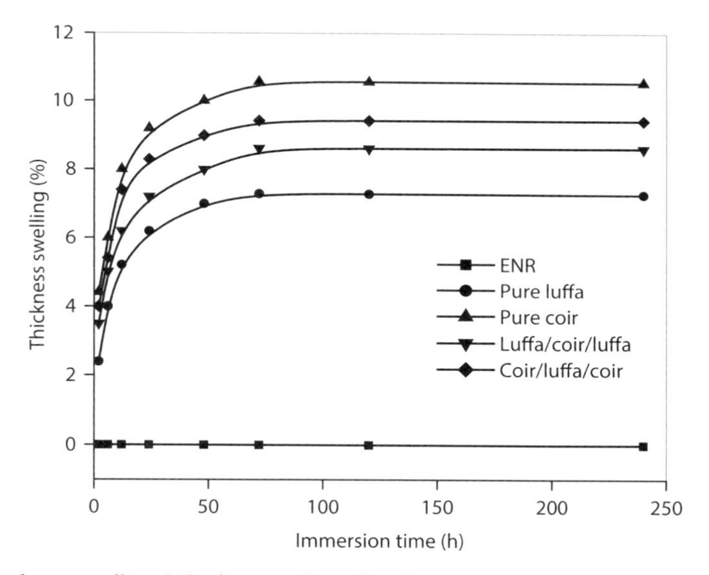

Figure 16.5 Thickness swelling (%) of various formulated composites.

16.4.2 Thickness Swelling Test

The result of thickness swelling test is shown in Table 16.3. The thickness swelling of composite was mainly because of the exposure of the lignocellulosic fiber on the surface of composite. However, dimension of lignocellulosic fiber change with different moisture content. The hydrophilic properties of lignocellulosic materials and the capillary action will cause the intake of water when the samples were soaked into water and thus increase the dimensions of composite. This will cause the swelling of fiber and thus increase the dimensional stability. Figure 16.5 shows the long-term immersion effect on thickness swelling of different composites.

Thickness swelling of composites was carried out for several hours until a constant weight was obtained. From the Table 16.3 and Figure 16.5, it is observed that the thickness swelling for the pure coir composite with the value 10.57% was the highest among the different type of composite. This indicated that the high porosity or the presence of void on the surface of pure coir composite. This is responsible for the changes in the dimension of cellulose-based composites, particularly in the thickness, and the linear expansion due to reversible and irreversible swelling of the composites [21]. In contrast, the dimensional stability for the epoxy novolac composite was the lowest with the value 0%. In other words, there was no thickness swelling in the epoxy novolac composite. This is due to epoxy novolac resin limit the absorption of water into the fiber-mat composite because the nature of epoxy novolac resin as water resistant matrix. The pure luffa composite shows a moderate dimensional stability with moderate thickness swelling, with the value of 7.29%. The dimensional stability of coir/luffa/coir and luffa/coir/luffa hybrid composites shows a moderate value, which are 9.43 and 8.63%. This indicated that layering pattern would affect thickness swelling due to packing and hybrid arrangement of fiber. It will limit the absorption of moisture into the composite, thus limit the swelling of fiber and the percentage of dimensional stability. Thickness swelling of hybrid composites follows a trend similar to the water absorption behavior, increasing with immersion time until an equilibrium condition is attained.

16.4.3 Effect of Different Configurations on Mechanical Properties

The importance of mechanical properties is to quantify the reinforcing potential of the composite system. However, mechanical properties can also give indirect information about interfacial behavior in composite systems, because the interaction between the components has a great effect on the mechanical properties of the composites. The mechanical properties of luffa and coir fibers were reported in Table 16.1. It is observed that the tensile strength and modulus of luffa fiber is higher than that of coir fiber while the diameter and lumen size is lower than that of coir fiber. Their trend of variation in the strength, modulus and elongation of the various formulated composites are shown in Figure 16.6 (a-d), respectively.

The tensile properties of different layering patterns of hybrid composites consisting of trilayer such as coir/luffa/coir and luffa/coir/luffa composites are shown in Figure 16.6. There is significant difference in their properties due to difference in morphological and bonding characteristics. These properties are also affected by relative weight ratios of the two fibers (i.e., coir: luffa = 3:1). The tensile strength (Figure 16.6a) was observed to be higher when luffa was used as the skin material and coir as the core material. The tensile strength will be higher when the high strength material is used as the skin, which is the main load bearing component in tensile measurements. In coir/luffa/coir, the value is slightly lower because the low strength coir fiber is used as the skin material. The diameter and lumen width of luffa fiber is lower than that of coir fiber. The surface area of the luffa fiber per unit area of composite is greater than that of coir fiber and hence, increased the stress transfer maximum. This is the reason for high tensile strength of luffa/coir/luffa composite. The same pattern of analysis was

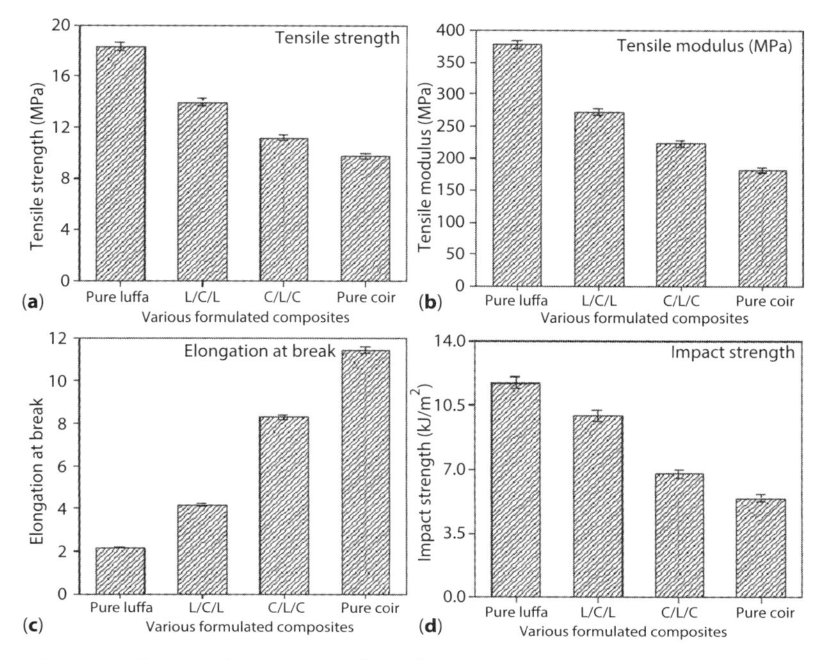

Figure 16.6 Mechanical properties of various formulated composites.

observed in the properties of tensile modulus shown in Figure 16.6b. The reason for these results is the same as discussed in the measured tensile properties.

From the measured percent elongation of various composites in Figure 16.6c, the elongation was found to be higher in the case of pure coir-reinforced composites while this value is lower in the case of pure luffa-reinforced composites. Regarding the elongation values of natural fiber-reinforced composites, structural parameters of fibers such as microbrillar angle, fiber strength and fiber aspect ratio play an important role. The stiffness and ductility of the fiber is also an important factor affecting the elongation of the corresponding composites. These values are dominated by the fiber breakage and matrix cracking. The ductile nature of coir fiber provides strain compatibility between fiber and matrix. The strain compatibility also imparts when luffa fiber is hybridized with coir fiber, elongation is increased. The percentage of elongation is higher in luffa/coir/luffa hybrid composite with respect to pure luffa composite but lower than coir/luffa/coir hybrid composite and pure coir composite as addition of coir increases the ductility of composites.

The impact strength of composite materials was directly related to their overall toughness. The impact performance of fiber-reinforced composites depended on many factors, including the nature of the constituent, fiber/matrix interface, the construction and geometry of the composite, and test conditions. Figure 16.6d shows the variation of impact strength as a function of different layering patterns of luffa-coir hybrid fiber-filled composites. The impact strength of the pure epoxy novolac matrix is recorded as 1.2 kJ/m^2 (Table 16.2). Comparing the impact strength values of pure matrix and its hybrid composites, it can be inferred that for all composites the presence of luffa-coir hybrid fibers increased the impact strength, that is, the composites have better energy absorbing capacity compared to that of pure ENR. The impact strength of luffa fiber-reinforced composites is observed to be higher than that of the coir fiber-reinforced composites. This might be due to the better fiber-matrix adhesion [22]. The coir fiber has a greater difficulty in adhering to the matrix which may generate fissures and decrease impact strength.

16.4.4 Sound Absorption Performances

One increasingly important property for interior components in automotive applications, where natural fiber-reinforced composites have performance advantages, is the acoustic absorption. Natural fiber-reinforced composite materials in an interior component have an open cell structure. It will therefore contribute to sound absorption and may result in a reduced need for absorbers. Multilayer components further offer many possibilities to tailoring the acoustic absorption of the system.

Because of natural growth characteristics, coir fiber has porous cellular structure. It has a greater lumen width, diameter and shape factor within its cellular structure when compared to luffa fibers as mentioned in Table 16.1 [23]. The scanning electron micrographs of cross-section of coir and luffa fiber are seen in Figure 16.7. Figure 16.7a and b represents the cross-section of coir fiber in 200 and 500 times of magnification, respectively and Figure 16.7c represents the cross-section of luffa fiber in 500 times of magnification. It has been observed from Figure 16.7 that coir fiber has a large number of lamina surrounded with lacuna compared to luffa fiber, which promotes the sound absorption capacity. Moreover, the density of coir fiber is less than the density of luffa fiber due to high content of amorphous lignin. The densities of lignin, hemicelluloses, and cellulose vary with values from 1.387 g/cm^3, 1.559 g/cm^3, and 1.552 g/cm^3 respectively and the water density is 1.0 g/cm^3. [24]. According the calculation of the void content fraction of lightweight composites described in the literature [25], the supposed densities of the other constituents are about 1.0 g/cm^3. The inner porosity of coir fiber can be calculated as:

$$V_{cp} = 1 - \sum_{i=1}^{n} v_i / v_{total} = \sum_{i=1}^{n} W_i P_{total} / P_i \tag{16.1}$$

Where, *Vcp* is the inner porosity of coir fiber, *Vi* is the volume of other constituent, *Vtotal* is the volume of coir fiber, *Wi* is the weight fraction of other constituent, *Ptotal* is the density of coir fiber, and *Pi* is the density of other constituent.

The inner porosity of coir fiber was calculated to be 16% to 25% by using Eq. 1. Similarly, for luffa fiber, this value was calculated to be only 7% to 13%. The high lignin contents and the lower density of lignin are two factors that cause the greater inner porosity of coir fiber, which leads to the better sound absorption and heat insulation characteristics. The samples with various configurations were tested for the sound-absorption coefficient using the impedance tube method from 100 Hz to 3200 Hz as shown in Figure 16.8. The sample of pure coir composites had the best sound absorption performance, while the sample of pure luffa composites had the least performance due to poor acoustic property. The greater holes ratio inside the coir fiber was the main

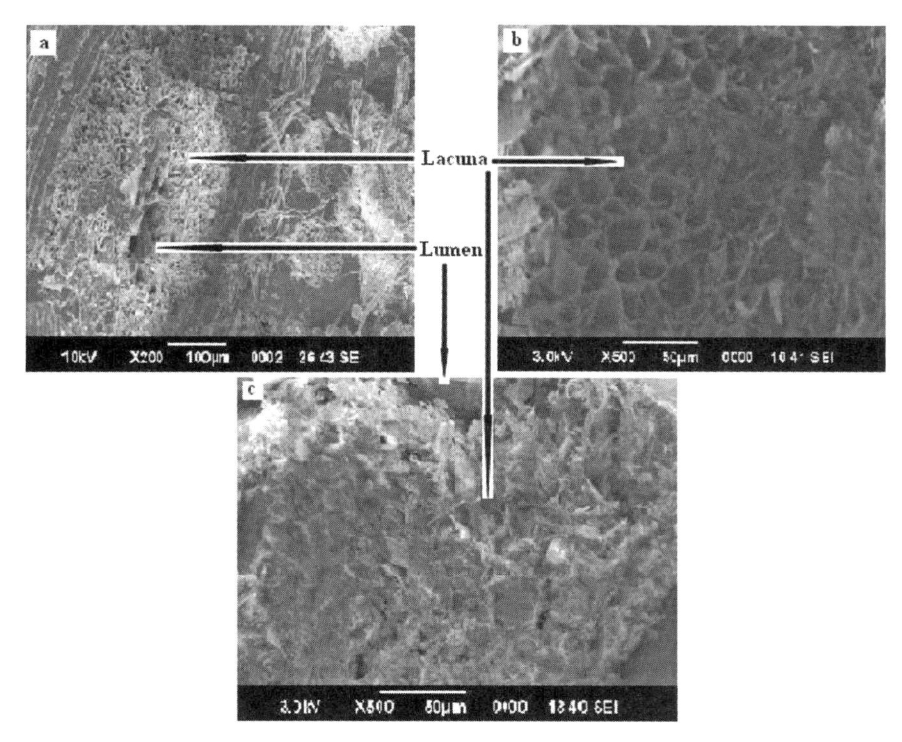

Figure 16.7 SEM micrographs of cross-section of coir fiber (7a. 200x and 7b. 500x magnification) and luffa fiber (7c. 500x magnification).

reason for the better sound absorption ability. Because of the problems with processing technology and the natural growth characteristics of natural fibers, porosity is inevitably generated in natural fiber composites [26], and the porosity, no doubt plays a certain role in improving the performance of sound absorption. In this study, the porosity in non-woven mats of coir was more even and better for sound absorption characteristics. The sound absorption coefficients of pure luffa composites increased with addition of coir fiber in hybrid composites. The acoustical property is more pronounced, when coir is used as skin material as in coir/luffa/coir configuration of hybrid composite and it is less pronounced, when coir is used as core material as in luffa/coir/luffa configuration of hybrid composite.

16.4.5 Study of Hybrid Composite Microstructure

The mechanical properties of composites could be corroborated with the morphological evidences. The fiber-matrix adhesion in the hybrid composites can be understood by examining the SEM micrographs of cryogenically fracture surfaces of tensile test specimens. Figure 16.9 (a and b) shows SEM micrographs of the tensile fracture surfaces of ENR-luffa/coir/luffa and ENR-coir/luffa/coir hybrid composites, respectively. A large number of fiber pull-out, fiber agglomeration, fiber-matrix incompatibility, and matrix cracking were noticed in the ENR-coir/luffa/coir hybrid composites, compared with ENR-luffa/coir/luffa hybrid composite. The observed fiber pull-out phenomena in

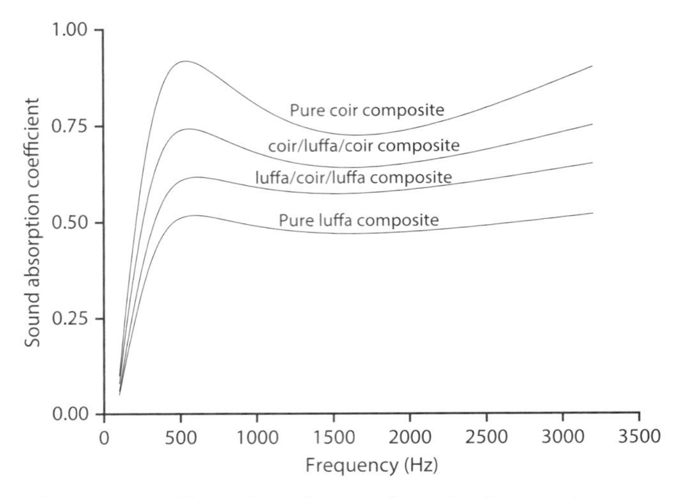

Figure 16.8 Sound absorption coefficient (%) of various formulated composites.

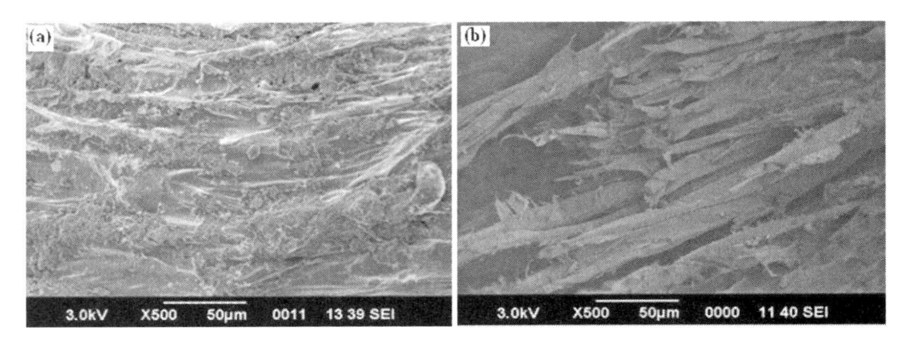

Figure 16.9 SEM micrographs of tensile fracture samples of (a) ENR-luffa/coir/luffa and ENR-coir/luffa/coir hybrid composite.

the fracture surfaces of the composites is a kind of index of the adhesiveness between the fibers and the matrix resin. Fracture of fiber with little or no pull out of luffa fiber is observed in the SEM micrograph of ENR-luffa/coir/luffa hybrid composite (Figure 16.9a). This observation indicates lesser extensibility of luffa fibers leading to little fiber pull-out and matrix failure under tensile loading. It can also be noted for these composites that the fiber failed by tearing, but no complete interfacial failure was observed; indicating that adhesion between the hybrid fibers and ENR matrix was quite good for reinforcing. There is substantial ENR matrix adhering to the fiber surfaces; indicating that the interfacial bond strength is fairly high in configuration of luffa/coir/luffa composites than coir/luffa/coir due to difference in the interfacial characteristics between the fibers and the matrix. The wettability of the luffa fiber with the ENR matrix enhances the fiber-matrix interfacial bond strength by diffusion of resin matrix into the fiber structure network or vice-versa. This interdiffusion depends on the correlation between the cohesive energy of the pre-cured resin matrix and the surface energy of the fiber, among other factors.

16.4 Conclusions

The present investigation deals with a potential opportunity for the development of novel luffa/coir fibers-reinforced epoxy novolac hybrid composites. The effect of hybridization of luffa fiber on the physical, mechanical and acoustic properties was studied. From physical properties testing, it can be concluded that pure coir composite had the highest 21.87% of water absorption and highest 10.57% of thickness swelling among the different type of composite. Pure luffa composites showed the lowest percentage in both water absorption and thickness swelling. The hybridization of coir and luffa fibers substantially improved both the dimensional stability and mechanical properties of the composites. The maximum and minimum values of mechanical and water resistance properties were recorded for the epoxy novolac composites reinforced pure luffa and pure coir fibers, respectively, which is due to the high complex conformity and stability of the luffa fibers. It can be concluded that luffa fibers act as barrier to prevent diffusion of water. The trilayer luffa/coir/luffa composite showed better mechanical and water resistance properties compared to coir/luffa/coir hybrid composites. Inversely, the trilayer coir/luffa/coir hybrid composites showed better sound absorption properties compared to luffa/coir/luffa composite. In this comparative study between different hybrid configurations based on coir fiber and luffa fiber shows a symbiotic relation between them. Lacking of mechanical properties of coir fiber was compensated by the luffa fiber while acoustic properties of luffa fiber were compensated by coir fiber. Both of the fibers are complementary in improving the properties with reduced water absorption character. The morphological features of the composites were well corroborated with the mechanical properties. On the basis of above studies, it can be concluded that an optimal configuration of luffa-coir hybrid fibers could effectively reinforce the epoxy novolac resin and enable to achieve satisfactory properties of the composites for various engineering applications. For future application of coir fiber, the hybrid composites could be studied with other plant fibers that complement them. The thermal insulation performance should be detected further. Also, the optimization design of the coir fiber hybrid composite must be performed.

Acknowledgements

The authors are thankful to Birla Institute of Technology, Mesra, Ranchi, India for providing instrumental facility (CIF) and financial assistance during the research work.

References

1. K.H.P.S. Abdul, A.M. Issam, S.M.T. Ahmad, R. Suriani, and A.Y. Awang, Conventional agro-composite from chemically modified fibres. *Ind. Crops Prod.* 26, 315–323 (2007).
2. A.K. Bledzki, A. Abdullah, and M.J. Volk, *Compos. Part A* 41, 480 (2010).
3. C. Asasutjarit, S. Charoenvai, J. Hirunlabh, and J. Khedari, *Compos. Part B* 40, 633 (2009).

4. Y. Seki, K. Sever, S. Erden, M. Sarikanat, G. Neser, and C. Ozes, *J. App. Polym. Sci.* 123, 2330 (2011).
5. C.A. Boynard, and J.R.M. D'Almeida, *Polym-Plas. Technol. Eng.* 39, 489 (2000).
6. N. Ayrilmis, S. Jarusombuti, V. Fueangvivat, Bauchongkol, and R. White, *Fibers Polym.* 12, 919 (2011).
7. S.N. Monteiro, L.A.H. Terrones, F.P.D. Lopes, and J.R.M. D'Almeida, *Revista Mater.* 10, 571 (2005).
8. S.N. Monteiro, L.A.H. Terrones, F.P.D. Lopes, and J.R.M. D'Almeida, *Polym. Testing* 27, 591 (2008).
9. J. Yao, Y. Hu, and W. Lu, *Bioresour.* 7(1), 663 (2012).
10. W. Wang, and G. Huang, *Mater. Des.* 30, 2741 (2009).
11. C. Asasutjarit, *Constr. Build. Mater.* 21, 277 (2007).
12. W. Hoareau, W.G. Trindade, B. Seigmund, A. Castellan, and E. Frollini, *Polym. Degrad. Stab.* 86, 567 (2004).
13. W.G. Trindade, W. Hoareau, J.D. Megiatto, I.A.T. Razera, A. Castellan, and E. Frollini, *Biomacromol.* 6, 2485 (2005).
14. T. Czigany, *Compos. Sci. Technol.* 66(16), 3210 (2006).
15. P. Padma priya, and S. K. Rai, *J. Indus. Textiles* 35, 217 (2006).
16. M. Jawaid, K.H.P.S. Abdul, A.A. Baker, and K.P. Noorunnisa, *Materials And Design* 32, 1014 (2011).
17. S.C. Amico, C.C. Angrizang, and M.L. Drummond, *J. Rein. Plas. Comp.* 29, 179 (2010).
18. M. Mariatti, M. Nasir, and H. Ismail, *Polym. Plast. Technol. Eng.* 12, 65 (2003).
19. S.K. Saw, G. Sarkhel, and A. Choudhury, *Fibers Polym.* 12, 506 (2011).
20. S.K. Saw, G. Sarkhel, and A. Choudhury, *J. Appl. Polym. Sci.* 125, 3038 (2012).
21. S.K. Saw, G. Sarkhel, and A. Choudhury, *Polym.Compos.* 33, 1824 (2012).
22. N. Venkateshwaran, A. Elayaperumal, A. Alavudeen, and M. Thiruchitrambalam, *Mater. Des.* 32, 4017 (2011).
23. A. Espert, F. Vilaplana, and S. Karlsson, *Compos. Part A* 35, 1247 (2004).
24. E.M.L. Ehrnrooth, *J. wood Chem. Technol.* 4, 91 (1984).
25. S. Bhatnagar, and M.A. Hanna, *Trans. ASAE* 38, 567 (1995).
26. J. Yao, Y. Hu, and W. Lu, *BioResour.* 7, 663 (2012).

17

Fracture Mechanism of Wood-Plastic Composites (WPCS): Observation and Analysis

Fatemeh Alavi, Amir Hossein Behravesh* **and Majid Mirzaei**

Department of Mechanical Engineering, Tarbiat Modares University, Tehran, Iran

Abstract

This chapter focuses on the observation and analysis of fracture mechanism in wood-plastic composites (WPCs). In general, different fracture mechanisms of WPCs can be attributed to the cracking of wood, plastic and the wood-plastic interfaces (also known as debonding). The occurrence of these mechanisms depends on many parameters, including the temperature and the size, shape and orientations of the wood particles. The WPC fracture mechanisms can be observed *in situ* using a high-speed digital microscope or a microtomography instrument to capture the material response to the applied loading. Application of the Digital Image Correlation (DIC) technique to the captured images can provide local displacement or strain fields within the crack domain. The fracture analysis of WPCs is described based on the identification of wood-plastic interfacial properties. Also presented is the usage of finite element methods for simulation of damage initiation and propagation.

Keywords: Wood-plastic composite (WPC), fracture mechanism, observation, analysis

17.1 Introduction

Particulate polymer composites (PPCs) are mainly comprised of micro- or nanoscale particles of various sizes and shapes which are usually distributed randomly in the matrix. Changing the filler dispersion can control several properties such as overall stiffness, strength, and fracture toughness of the produced composites. In particular, the wood-plastic composites (WPCs), consisting of thermoplastic polymers filled with lignocellulosic, can be categorized as green materials with direct applications in various industries such as construction, furniture, and transportation [1, 2]. In general, WPCs possess both the desirable performance and the cost benefits of thermoplastics and wood. In practice, the usage of wood is beneficial over organic fillers (such as kanaf and flax), with considerations like density, thermal insulation, availability, cost, biodegradability, tool wear, strength and stiffness. Thus, wood fillers are overwhelmingly employed in thermoplastic-based composites with polypropylene, polyethylene

**Corresponding author*: amirhb@modares.ac.ir

Vijay Kumar Thakur (Ed.), Lignocellulosic Polymer Composites, (385–416) 2015 © Scrivener Publishing LLC

and PVC as the matrix [3]. On the other hand, the main disadvantages of WPCs are the poor adhesion between the filler particles and the polymer matrix, low impact strength, and thermal decomposition at temperatures over 200°C [4]. Although such materials have found wide applications, the demand for higher quality and cheaper product has invoked many investigations. Also, there has been an increasing tendency towards improving the mechanical properties of WPCs throughout the last decades [5-7]. It is well known that the properties of heterogeneous polymer composites can be determined by factors such as component characteristics, composition, structure and interfacial interactions [8]. Thus, fundamental studies on the strength and fracture mechanisms of WPCs can provide insights into further development of these composites.

In this chapter, the effects of various parameters on the fracture behavior of wood-plastic composites are discussed. Moreover, different methods of composite damage analysis are reviewed and the fracture analysis of WPC is presented.

17.1.1 Fracture Behavior of Particulate Composites

It is known that the essential characteristics of composites (such as energy absorption) depend on key parameters like: shape and dimension of particles, mechanical properties of the filler and the host matrix, filler-matrix interfacial strength, as well as volume fraction and particle dispersion in the matrix [9]. More specifically, the fracture behavior of particulate polymer composites depends on the following factors:

- Particle size
- Volume fraction
- Fillers orientation
- Characteristics of matrix and fillers (stiff or compliant)
- Loading
- Temperature
- Interface

In this section, the effects of the above factors on the facture behavior of particulate polymer composites are discussed.

17.1.1.1 Particle Size, Volume Fraction, and Fillers Orientation

Three main mechanisms are reported for fracture of WPCs: debonding, wood cracking, and plastic cracking. These mechanisms may alternatively occur depending on wood flour size and orientation. A review on the development of polymer composites filled with inorganic particles has been reported by Fu *et al.* [9].

In order to determine fracture toughness, stress intensity approach and energy approach are applied. The stress intensity approach yields fracture toughness (K_{IC}). The energy approach provides a critical energy release rate (G_{IC}). Fracture toughness of particle-filled polymer composites shows a very complex variation with increasing particle fraction. The particle size distribution parameter is the one aspect which plays a decisive role on the structural and mechanical properties of the components and the interface that influence the crack resistance [10]. The effect of size distribution of particles on crack resistance in particle-reinforced polymers cannot be ignored. In order to consider the effects of size-scale and particle volume fraction on the fracture behavior of nano- and micro-particle-filled composites under quasi-static loading conditions, several comparative studies have also been carried out.

It is demonstrated that reducing the particle size can increase the fracture toughness. Many investigations proposed the modeling of crack resistance for composites with particles of the medium sizes [11-13]. The crack resistance, the size of the dissipation zone, and the volume specific debonding energy were calculated as a function of the distance to the crack tip for different debonding stress criteria. It was demonstrated that whenever the energy criterion is used, the fracture toughness is affected by debonding process and seems independent of particle size. However, applying the interfacial stress criterion results in the increase of facture toughness (K_{IC}) by decreasing the particle size [14].

The monotonic increase in K_{IC} with volume fraction for aluminum-reinforced polyster resin with micro- and nanoparticle sizes was reported by Singh *et al.* [10]. They showed higher K_{IC} in the case of nano-filled composites than the micron-sized ones for up to a volume fraction of 2.3%.

The effect of particle size on toughening mechanisms of epoxy–silica nanocomposites was investigated by Liang and Pearson [15]. They also improved $K_{IC}$and fracture energy with the volume fraction. In Addition, Adachi *et al.* [16] observed the improvement in K_{IC} of silica-filled epoxies in the nano-filler cases compared to the micron-size counterparts. However, in the investigation of the effect of filler volume fraction, Hussain *et al.* [17] indicated that TiO_2-filled epoxy composites with micron size particles exhibited higher fracture toughness with increasing volume fraction compared to the nanoparticle counterparts. Moreover, mechanical response of nano- and micron-size TiO_2 particle-reinforced epoxies was compared by Ng *et al.* [18]. They revealed that nanoparticle-filled epoxy in comparison with micron-sized filled one, show higher stiffness, failure strain and toughness.

Hsieh *et al.* [11, 19] reported that in the fracture toughness, the contribution of matrix shear band is more than the plastic void growth. With increasing the particle volume fraction, they observed an increase in the elastic modulus, fracture toughness, and fracture energy and determined the localized shear bands. They realized that the particle debonding as the main toughening mechanism, leads to void growth. The enhanced fracture toughness and energy due to void growth and particle–matrix debonding in nanosilica-filled epoxies were reported by Johnsen *et al.* [20].

Jordan *et al.* [21] reviewed the processing and mechanical responses of nano- vs micron-sized particulate composites in experimental trends. This fact that nanocomposites serve better in some mechanical aspects than the micron-sized counterparts, was one of their conclusions. However, due to physical and chemical differences between the constituent phases of nanocomposites and fabrication challenges, large deviations in properties exist and no general trends were established.

Attractions have been drawn towards the improvement of natural fiber-reinforced composites properties. More extensive results have been published on the effect of fiber characteristics on properties [22, 23]. The debonding of excessive sized particles and their premature failure is reported to be common. In addition, thin fibers easily break

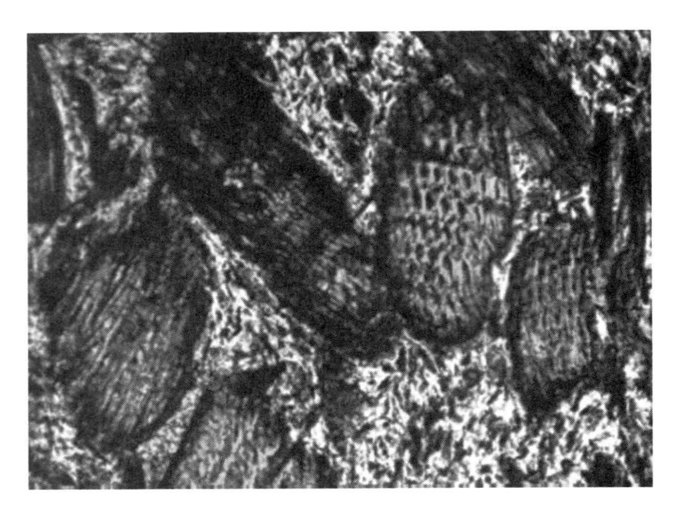

Figure 17.1 The color segmentation of a microscopic image of wood particles embedded in PVC matrix [26].

during the processing. Increasing the aspect ratio of the fibers boosts both stiffness and strength, while the importance of the fiber orientation highlights as the anisotropy increases [24].

Cui *et al.* [25] investigated the effect of wood fiber length, weight fraction and surface treatment on the mechanical properties and fracture mechanism of wood/recycled plastic composite (WRPC) material. They concluded that increasing wood fiber content increased and decreased flexural strengths and impact strengths, respectively. On the other hand, the increase of wood fiber length increased the flexural modulus, while both the flexural strength and the impact strength decreased. Due to the large size and low surface energy of wood particles, aggregation does not play a significant role in the mechanical properties of WPCs. However, for geometrical reasons, wood particles may come into contact [7]. Wang [26] used RGB color segmentation techniques for wood plastic composite specimen to separate and quantify extensive interphase between wood particles and polymer matrix. Figure 17.1 illustrates the wood particle orientation in PVC matrix captured using this technique.

Particle orientation can affect the fracture behavior in particulate composites. The effect of wood particle orientation on wood plastic composites was investigated by Alavi *et al.* [27]. To experimentally identify the effect of particle orientation on the fracture mechanism, three different specimens were prepared, as shown in Figure 17.2. They demonstrated that wood orientation significantly affected the fracture mechanism of WPCs. The major fracture mechanism is debonding for a particle oriented normal to the direction of tension and opening failure mode has occurred (Figures 17.2(a), (d), and(g)). However, Figure 17.2(e) indicates that sliding mode is not the dominant fracture mechanism as particle cracking has taken place instead of debonding. Therefore, it can be stated that interfacial shear strength is stronger than the strength of wood particles even in the direction of the tree growth. Figures 17.2(f) and 17.2(i) clearly show debonding phenomenon for a particle tilted towards the direction of tensile loading. It is apparent that the maximum strength belongs to the sample in which shear strength plays the major role in the interfacial resistance. It is also evident that the weakest

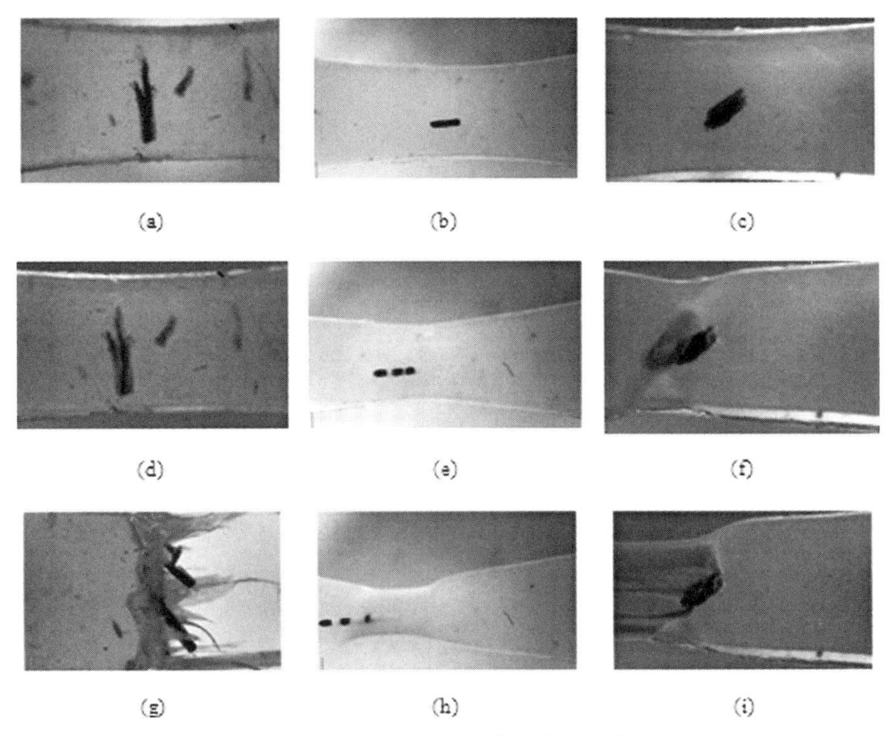

Figure 17.2 The diverse fracture mechanisms which is affected by different wood orientation in HDPE matrix [27].

strength is attributed to the specimen in which normal strength plays the major role in the interfacial resistance. They compared the strength of these specimens and dedicated that the extreme strength values, as the strength of oriented particle is between those of horizontal and vertical ones. This can be attributed to the combined effect of normal and shear strengths on the overall strength of oriented particles and wood orientation angle has the main effect on the amount of its strength.

17.1.1.2 Fillers and Polymers Characteristics

The main contribution of fracture toughness complexity of particulate composites relates to the fracture toughness of the polymeric matrix. So that, the addition of particulate fillers in ductile matrices increase the brittleness of composites while, the brittleness is reduced for brittle matrices. Besides, for brittle thermosetting resins, inorganic particles account for effective toughening agents [28]. On the other hand, significant decrease of fracture toughness compared with the neat polymers occurs in ductile thermoplastics filled with rigid particles. In this case the attempts of Fu and Lauke [29] as well as Levita *et al.* [30] can be mentioned.

Polymer-based composites can be divided into thermoset and thermoplastic composites, which due to their different properties show diverse fracture mechanism. Due to tight three-dimensional molecular network structure of the most of thermoset matrixes such as epoxy resins, they exhibit inherent brittle fracture behavior and poor

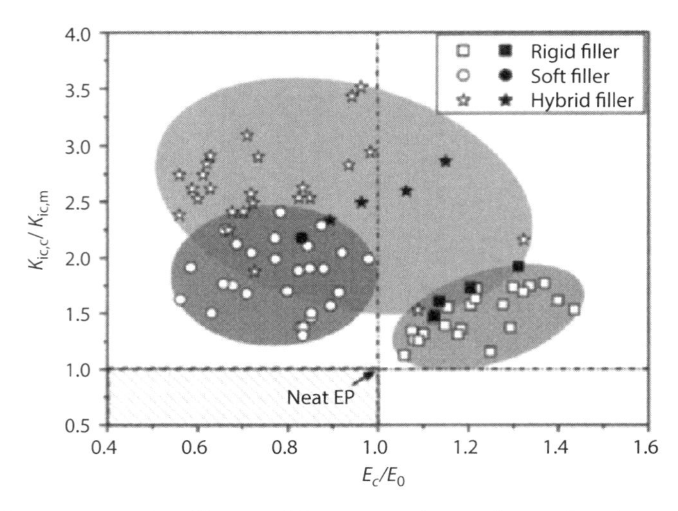

Figure 17.3 The improvements in stiffness and fracture toughness of epoxy based composite with soft, rigid, and rigid-soft fillers [36].

crack growth resistance. A common way in order to toughen these matrices is to introduce soft rubber particles in the resin, which change the fracture mechanism of these matrices. Rubber particle debonding/cavitation, localized shear banding of matrix as well as rubber particle bridging are the major toughening mechanisms. However, adding such filler improve fracture toughness, it may decrease some other properties, such as the elastic modulus, failure strength as well as thermal–mechanical properties [31-33].

In contrast, adding rigid fillers, despite of their size, can improve the fracture toughness, stiffness and even strength of these thermoset resins simultaneously [34, 35]. The researches which carried out the effect of rigid fillers on epoxy reveal that size of fillers affect the fracture mechanism significantly. They demonstrated that micro-fillers toughen epoxy through crack pinning, crack deflection/bifurcation effects, whereas the rigid nano-fillers and the soft rubber particles may toughen epoxy similarly [20]. That is, the toughening mechanism is mostly contributed by the filler debonding, and the resulting void growth as well as the matrix shear band. The toughening efficiency of both rigid micro- and nano-fillers is lower than that of rubber particles. To both simultaneously improve the toughness and stiffness of epoxy resins, the usage of soft-rigid fillers has been proposed. Figure 17.3 illustrates a comparison of the effect of soft and rigid particles addition on some epoxy mechanical properties [20, 33].

Tang *et al.* [36] investigated the mechanical properties of epoxy resins filled with different amounts of soft rubber particles and/or rigid silica nanoparticles, with special focus on the related mechanical and fracture behaviors. Additionally, increases in fracture toughness by the rigid particles addition in polypropylene [37, 38] and polyethylene [38, 39] were addressed in some researches.

Lei and Wu [40] prepared wood plastic composites based on *in-situ*-formed polyethylene terephthalate (PET) sub-micro-fibril (less than 500 nm in diameter) reinforced high-density polyethylene (HDPE) matrices through strand die extrusion and hot strand stretching. The PET fibrils obviously increased mechanical properties of the blend, especially the moduli. They resulted that subsequent addition of wood fibers did not change the size and morphology of PET fibrils. The effect of wood type was investigated by several researchers [41, 42], but it is still undetermined that whether the use of soft or hard wood boosts the properties of WPCs.

17.1.1.3 Loading

In addition to the discussed factors, the type of loading (quasi-static or dynamic) affects the mechanical response and fracture behavior of particulate polymer composites.

In spite of several investigations on the fracture behavior of particulate composites under quasi-static condition, very limited investigations have been reported the dynamic crack growth in rapid loading condition. However, the stress-wave loading appears in wide range of engineering applications like aerospace, automotive, and sporting equipment. The aim of few investigations was towards comparing the fracture response of particulate composites under different types of loading, i.e., quasi-static and dynamic loading conditions [43, 44]. The effects of particle size, filler volume fraction, and loading type on the fracture behavior of particulate composites were considered by Jajam and Tippur [43].

The mode-I fracture behavior of glass-filled epoxy in dynamic loading condition was studied by Kitey and Tippur [45]. Shukla *et al.* [46] compared the fracture toughness of nanoparticle-filled resin and the neat one in the dynamic loading condition. They reported the increase of fracture toughness and crack velocity in nanoparticles- (TiO_2 and Al_2O_3) filled resin. The higher fracture toughness of TiO_2-filled-polyester nanocomposites under dynamic loading compared to the quasi-static case was addressed by Evora and Shukla [47]. Moreover, Evora *et al.* [48] developed a relationship between mode-I dynamic stress intensity factor and the crack velocity in TiO_2-polyester nanocomposite. The higher crack velocity and the toughness of nanocomposite compared to the neat polyester were reported.

The development of the WPCs for load-bearing structural applications necessitates the characterization of their strain rate-dependent mechanical properties. In this regard, the effect of strain rate on flexural properties of WPC was addressed by Tamrakar and Lopez [49]. The strain at failure was not significantly influenced by the strain rate variation. A prediction model for the effects on strain rate on the modulus of elasticity (MOE) of WPC material was demonstrated based on the viscoelastic standard solid model. Yu *et al.* [50] analyzed the variability of the dynamic young's modulus of WPC, which was measured by different non-destructive test (NDT) methods. They also estimated the correlativity between the dynamic Young's modulus and the static MOE of WPC.

17.1.1.4 Temperature

In spite of the several investigations carried out on the mechanical properties and fracture behavior of particulate polymer composites at room temperature, there are limited attempts on micro-fracture mechanisms under high temperatures. The dependence of creep deformation on stress, time, loading speed, and temperature is well known [51].

Temperature effect is usually considered in the study of viscoelastic materials. In order to describe the mechanical behavior of polymer bonded explosives (PBX) under various temperatures and strain rates, a nonlinear constructive relation was proposed by Luo *et al.* [52]. Additionally, a nonlinear viscoelastic constructive relationship with damage, called Norton–Bailey creep law, was proposed which determines strain rate as a function of stress and temperature [53, 54].

Wade and Cantwell [55] worked on plasma-treated glass fiber-reinforced nylon-6,6 composite bonded using a silica-reinforced epoxy adhesive. They studied the effect of test temperature and loading rate on sliding mode of fracture toughness of end notch flexural test specimen. They reported that the temperature increase reduced the mode-II fracture toughness and raised the crosshead displacement rate. They also observed a similar trend in compression tests and demonstrated that mode-II fracture energy of adhesive system is related to the yield stress of adhesive. In order to understand the effect of temperature and loading rate on the failure mechanism, they employed double end flexure geometry. The failure mechanism is caused by changes in matrix shear yielding and particle–matrix debonding.

The effects of matrix weight fraction and interfacial area at ultra-low temperatures were evaluated by Ray [56] with different loading rates on interlaminar shear failure mechanism of glass/epoxy composites. The effect of high temperature exposure on bending properties and damage mechanism of carbon fiber composite sandwich panel with pyramidal truss cores was considered by Liu *et al.* [57]. They found that degradation of matrix and fiber–matrix interface property at high temperatures caused a decrease in residual bending strength of the composite sandwich panels. Interfacial interactions and the strength of adhesion determine the micromechanical deformation processes and the failure mode of the composites. The effect of temperature on the failure mechanism and bending properties of three-dimensional E-glass/epoxy braided composite was reported by Li *et al.* [58]. They demonstrated that bending properties decreased significantly with a temperature increase and failure mechanism changed from fiber fracture and brittle mode at room temperature, to the matrix microcracking and fiber–matrix debonding at the elevated temperatures.

The effect of inorganic fiber on thermal properties of WPC was investigated by Farhadinejad *et al.* [59]. The effect of temperature on the mechanical properties of hemp fiber polypropylene composites at three representative temperatures of 256, 296, and 336 K was studied by Tajvidi *et al.* [60]. They reported that the impact resistance was independent of temperature, while flexural and tensile properties (specially modulus of elasticity) were severely affected. However, they did not consider the effect of temperature on WPCs fracture mechanism. Some researchers [61, 62] characterized the WPC formulation to describe nonlinear response of time and temperature dependence. Bajwa *et al.* [63] evaluated the variation of WPC physical and mechanical properties under accelerated aging process. Azwa *et al.* [64] reviewed the degradability of polymer composites based on natural fibers, including the degradation due to moisture, thermal effects, fire, and ultraviolet rays. The effect of time and temperature on the mechanical properties of an extruded WPC was investigated by Tamrakar *et al.* [65]. Quasi-static and creep tensile tests were carried out at -10°C, 21°C, 30°C, 45°C, and 65°C. The long-term creep performance was modeled by the time–temperature superposition principle. They characterized the decrease in the modulus of elasticity and the modulus of rupture with the increase in temperature.

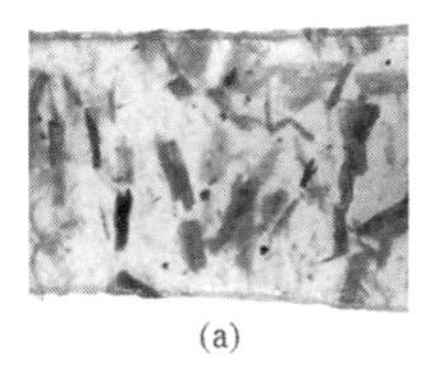
(a)
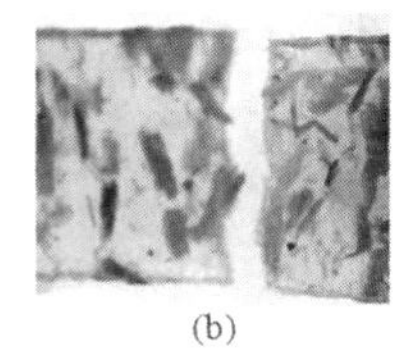
(b)

(c)
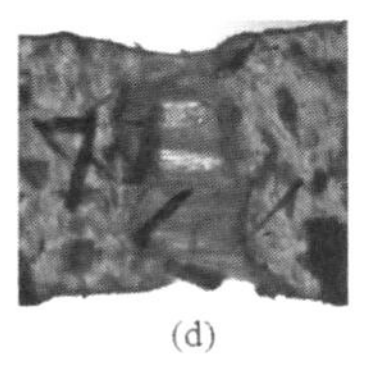
(d)

Figure 17.4 The different fracture mechanisms in wood plastic composite specimen (a) before fracture under -30°C, (b) after fracture (wood cracking) under -30°C, (c) before fracture under 53°C, (d) after fracture (debonding) under 53°C [66].

To investigate the effect of temperature on the mechanical properties and the fracture mechanism of WPCs, an experimental study was carried out by Alavi *et al.* [66]. The deformation and fracture behaviors of the WPC specimens at different temperatures (-30°C, -6°C, 6°C, 25°C, and 53°C) were studied. The results indicated the significant effect of the test temperature on the fracture mechanism of WPC specimens. At the room temperature, the dominant fracture mechanism for specimens without coupling agent (maleic anhydride polyethylene (MAPE)) was debonding, whereas wood cracking was the dominant fracture mechanism in presence of MAPE. At high temperatures, debonding was prominent over wood cracking in all samples (with and without MAPE), whereas at low temperatures (below zero) wood cracking was the dominant fracture mechanism. Figure 17.4 shows the different fracture mechanisms in WPC specimens at different temperatures.

17.1.1.5 Interface

The polarity of adjacent molecules significantly affects the bonding interfaces. Polar (hydrophilic) molecules are those that have a permanent electric dipole moment, which gives the molecule a partial charge enabling hydrogen bond with water. Non-polar (hydrophobic) molecules have no partial negative or positive charge, therefore they repel water. Most polymers, especially thermoplastics, are non-polar substances that are not compatible with polar wood fibers. The result is poor adhesion between polymer and wood filler in WPC [27].

Thus, the role of a coupling agent, by improving compatibility and adhesion in wood plastic composites, become very significant. The following mechanisms in modified WPC specimens with coupling agents can be observed:

- chemical bonding formation between the filler and the matrix,
- reduction of the surface energy of the wood particle which decreases the static interactions between the particles,
- mechanical interlocking caused by the improvement of the physical penetration of the plastic matrix into the microstructure of wood particles.

Maleic anhydride polypropylene (MAPP) and MAPE are overwhelmingly employed as coupling agents in polypropylene- and polyethylene-wood composites, respectively. Their performance for WPCs relays on the content of grafted maleic anhydride groups and on the polymer chain length. Grafted maleic anhydride groups lead to an improvement in the interaction between wood and the matrix, while the co-crystallization of

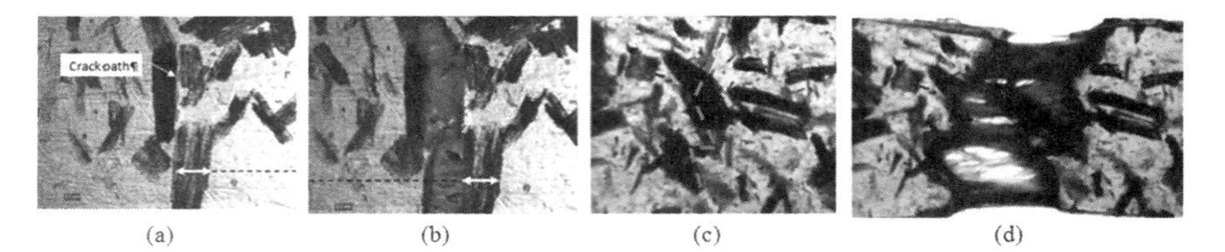

Figure 17.5 The different fracture mechanisms in wood plastic composite specimen (a) before fracture, (b) after fracture (debonding) in WPC specimen and (c) before fracture, (d) after fracture (wood cracking) in WPC specimen in the presence of MAPE [27].

MAPP and MAPE in the PP- and PE- matrix are provided by the high polymer chain length [67].

The mechanism of stress transfer from the plastic matrix to the wood particles and vice versa was investigated by Sretenovic *et al.* [68]. They measured the strain distribution around the particle using electronic speckle pattern interferometry (ESPI). They also conducted an analytical analysis and finite element modelling of strain and stress distribution in the wood plastic composite.

In order to improve the interfacial adhesion, the addition of functionalized polymers, i.e., MAPP or MAPE to the composite is employed. The presence of these polymers slightly modifies stiffness, but significantly increases tensile yield stress, tensile strength, and in some cases deformability [69, 70]. The numerous other surface modification techniques like silane treatment [71, 72], isocyanates [73, 74], surfactants [75], and various monomers as well as chemical modification [76, 77] are also utilized.

In heterogeneous polymer composites, stress concentration develops around the heterogeneities upon loading due to different elastic properties of components. Local deformation processes are induced by local stresses with different mechanisms. The debonding is the dominant micromechanical deformation process in the wood plastic composites without coupling agent [27]. In the weak adhesion condition, besides debonding, fiber pull-out may also occur in fiber-reinforced composites. In contrast, in the strong adhesion using functionalized polymers, fillers fracture is expected to take place. Figure 17.5 illustrates the different fracture mechanisms under the effect of MAPE coupling agent in the wood plastic composite.

As the performance of the composite is profoundly dominated by the micromechanical deformation process, its knowledge and control are critical for the improvement of composite properties. The effect of particle characteristics and interfacial adhesion on the micromechanical deformation processes in PP-wood composites was investigated by Renner *et al.* [7]. They proposed a failure map as well as the practical results and considered the influence of matrix characteristics on deformation and failure in PP-natural fiber composites in other research [24]. Hietala *et al.* [78] studied the effect of chemical pre-treatment and moisture content of wood chips on the wood particle aspect ratio during the processing and mechanical properties of WPCs. The use of pretreated wood chips enhanced the flexural properties of the wood chip-PP composites. Moreover, the use of undried wood chips compared to dried one can improve and reduce the flexural strength and flexural modulus, respectively. On the other hand, they concluded that the use of pretreated and undried wood chips lead to the highest aspect ratio after compounding. The effect of composition and the incorporation of MAPP on the tensile strength and fracture behavior of WPC were investigated by Pérez *et al.* [79]. An increase in the Young's modulus and decrease in tensile strength, strain at break, and fracture toughness were observed with wood content increase, in unmodified composites. In spite of the beneficial effect of MAPP addition on the tensile strength and ductility, it had no significant effect on fracture toughness.

Usage of silane coupling agents for natural fiber polymer composites was reviewed by Xie *et al.* [80]. Most established ones are trialkoxysilanes bearing a nonreactive alkyl or reactive organofunctionality. Silane, as the hydrolyzed forming reactive silanols, is adsorbed and condensed on the fiber surface (sol–gel process) at a specific pH and temperature. Heating the treated fibers at a high temperature converts the hydrogen bonds. These bonds form between the adsorbed silanols and hydroxyl groups of natural fibers/fillers into covalent bonds, although such bonds are susceptible to hydrolysis. The organofunctionality of silane and the matrix characteristics governs the interaction modes of the silane and matrix. A chemical bonding between the organofunctionalities of silanes and the matrices is necessary for the improvement of the interfacial adhesion. In the presence of catalysts or radical initiators, the organofunctionalities of silanes can react with the functional groups of thermoset matrices. Proper treatment of fillers with silane can improve the mechanical and outdoor performance of the resulting particle-filled polymer composites by boosting the interfacial adhesion of the target polymer matrices.

Godara *et al.* [81] investigated the influence of the amount of coupling agent, biopolymers, and the inorganic fillers on the structural integrity of polypropylene matrix composites reinforced by natural wood fibers using digital image correlation (DIC) coupled with tensile tests. They concluded that acetylation process of wood and coupling agent addition improved mechanical strength and reduced failure strains.

There are some interdependent adhesion mechanisms which govern the particle/matrix interfacial strength in particulate composites such as mechanical interlocking, molecular entanglement, secondary force interactions, electrostatic attraction, chemical bonding, and polymer diffusion.

The mechanical interlocking is a topography-based effect and the surface preparation/treatment is an important factor in effectiveness of mechanical interlocking. When the composite and its interface are strained, the shapes of certain surface irregularities may increase the local stress concentrations. Wood is one of the porous materials which have tortuous pore networks. Mechanical interlocking is improved by permeating the adhesive/matrix into these networks [82].

Additionally, molecular entanglements of polymer molecules at interfaces improve the strength of interlocking. Secondary force interactions are perhaps the dominating interactions within adsorption theory of adhesion. Van der Waals and hydrogen bonds also can dominate interface interactions between materials. Electrostatic attraction is a secondary bonding mechanism where oppositely charged surfaces interact. Chemical bonding (primary bonding) may also occur across the interface.

The adhesive failures can be categorized as: adhesive type and cohesive type. In adhesive type, the failure occurs along the interface of the adhesive and the adherent in, while in cohesive type, failure is contained within the adhesive layer. Cohesive failure occurs more commonly in the heterogeneous materials than neat polymer adhesives.

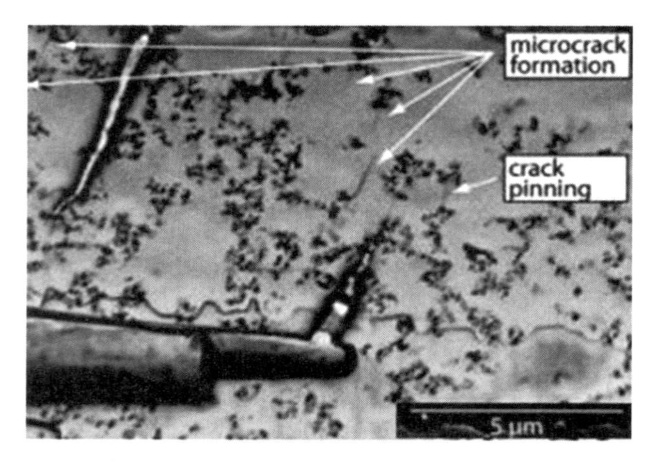

Figure 17.6 The formation of microcracks and crack pinning in metal oxide nanoparticle reinforced epoxy system [89].

From a mechanical design point of view, understanding the link between the failure type of an adhesive and several design parameters is essential [83].

17.2 Fracture Mechanism

In order to evaluate the toughness variation of particulate polymer composites, diverse mechanisms such as crack pinning [84], crack path deflection [85], particle debonding and subsequent void growth [20, 86], and plastic deformation [87] have been distinguished. Moreover, plastic deformation and the plastic void growth were considered as the responsible toughening mechanisms [88]. The fracture testing conducted by Boesl *et al.* [89] provided an insight into the fracture mechanisms on the microscale. They provided validation of the existence of microcracking and crack pinning in the material. The formation of microcracks and crack pinning in metal oxide nanoparticle-reinforced epoxy system is illustrated in Figure 17.6.

Lee and Yee [90, 91] reported that micro-cracking just near the fracture surface is generally due to debonding in particulate polymer composites. However, it does not take place in a large dissipation zone around the propagating crack. On the other hand, Norman and Robertson [92] realized fracture toughness increase in these composites compared to the neat matrix because of the off-plane particle–matrix debonding and local matrix yielding which occur around the particles. Moreover, for such composites, some other toughening mechanism such as diffuse shear yielding and microshear banding are also likely. As a result, the initiation and propagation of the damage process are determined by interfacial debonding.

To investigate the toughening mechanisms of composites, Vickers indentation is used. Porwal *et al.* [93] analyzed the fracture patterns generated from Vickers indentation in silica–graphene oxide nanoplatelets (GONP) composites. The scanning electron microscopy (SEM) images of Vickers indentation fracture surfaces for silica and silica-GONP nanocomposites are illustrated in Figure 17.7. Figure 17.7(a) shows a typical straight crack path for pure silica. Figures 17.7(b) and 17.7(c) indicate the wavy crack path and the deflected crack path in silica-GONP nanocomposites, respectively. The produced necking and joining the crack during indentation of GONP can be observed. Additionally, the crack bridging, crack branching, and GONP pull out from silica matrix are depicted in Figures 17.7(d), 17.7(e), and 17.7(f), respectively.

Initiation and propagation of micro-cracks along the particles/matrix interfaces as well as inside the matrix are typical damage mechanisms in composite materials. Damage mechanism phenomena such as crack deflection, plastic deformation, and crack pinning appear simultaneously. Moreover, localized shear bands are the major toughening mechanisms of nanocompoites, which can initiate by the stress concentrations surrounding the particles, micro-cracks formation inside the composites, delamination or debonding between particles and matrix, the fracture surface area increment, and matrix deformation [94-96].

The potential of nanoparticles for the damage resistance of polymers enhancement depend on several factors such as; aspect ratio [94], hardness [95] and shape [97] as the nanoparticles structural parameters, which should be considered along with the morphological parameters like orientation [94], clustering [95] and weight content [94-97].

It is generally expected that the addition of hard and soft nanoparticles improves and reduces the elastic modulus of nanocomposite, respectively [95]. On the other hand, as an example of shape effect, spherical nanoparticles illustrated a weaker toughening effect than the plate-like nanoparticles [97]. The nanoparticle distribution morphology in the matrix is a significant factor, such that aligned nanoparticle arrangement improves the elastic modulus and reduces fracture toughness [94]. In addition, Fracture toughness and modulus are influenced by the clustering of nanoparticles in matrix, which is described in terms of exfoliation or intercalation [98].

Johnsen *et al.* [20] and Dittant and Pearson [99] researched on the toughening mechanisms of epoxy nanocomposites reinforced with nanoparticles and indicated that

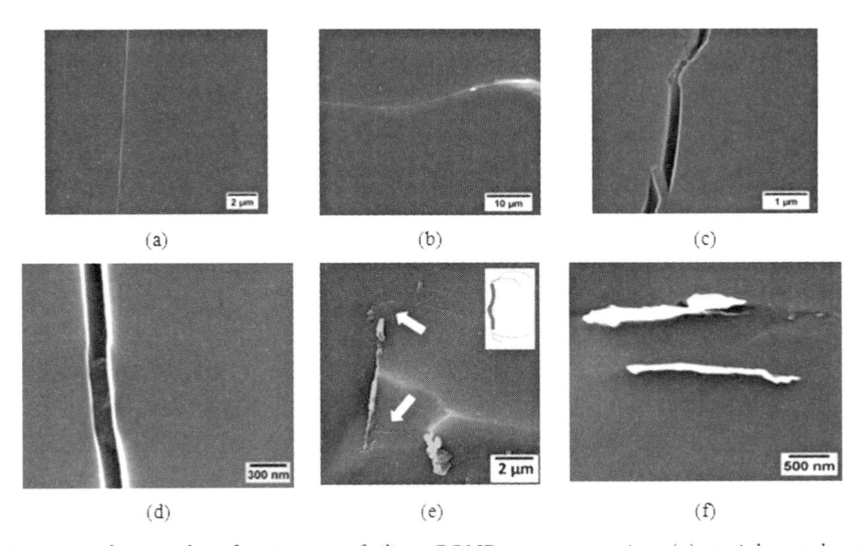

Figure 17.7 SEM fractured surface images of silica–GONP nanocomposites: (a) straight crack path for pure silica; (b) low magnification image showing wavy crack path for silica-GONP nanocomposites; (c) crack deflection and GONP necking toughening mechanisms; (d) high magnification image showing GONP crack brigding; (e) chevron notch fractured surface image showing crack branching; and (f) GONP pull out from silica matrix [93].

localized shear bands, debonding, and subsequent plastic void growth are the prevailing dissipation mechanisms. In order to explain the complex dependence of fracture toughness on particle diameter, as plastic matrix voiding after debonding, they concluded the necessity of consideration of other mechanisms, such as the one proposed by Williams [100]. He assumed the plastic void growth as the dominant toughening mechanism in particle-filled polymers. The yielding energy for the applied stress at particle/matrix debonding was calculated, which determines the energy density. This energy is independent of the position within the dissipation zone. Neither the debonding process with subsequent yielding, nor matrix yielding without debonding can explain the effect of the particle size or size distribution on fracture toughness. Lauke [101] considered the debonding mechanism on the basis of a cubic lattice arrangement of particle and yielding without previous debonding [13]. For the calculation of void growth energy, Zappalorto *et al.* [102] extended the William's approach by considering an interphase zone between the particle and matrix. The fracture toughness enhancement due to plastic void growth was reported by Lauke [103]. The different fracture processes close to the crack tip are caused by high multiaxial stress fields in front of a crack. These processes result in the energy dissipation of the moving crack, and thus increase the crack resistance of the material. When debonding stress is larger than the minimum yielding initiation stress, debonding energy, matrix yield stress, and particle size become significant. The debonding process is demonstrated to be affected by many factors, such as the filler content, filler size, and interaction between the filler particles and the matrix. The debonding process can affect the mechanical properties of particle-filled polymers.

The fracture behavior of WPC and its fracture mechanism were investigated by Jeong [104]. As it was outlined in the previous sections, the interfacial fracture, fiber pull out, and fiber breakage are the main fracture mechanisms of WPC. Additionally, intermolecular cleavage may occur with cracks running across fiber, interfacial area, and polymer. The fracture surface of WPC specimens, captured by SEM are illustrated in Figure 17.8.

17.3 Toughness Characterization

Some investigations [105-107] on micron-size particle-filled composites demonstrated that fracture toughness is dramatically influenced by the particle shape, size, volume fraction, and particle/matrix interfacial strength. Brunner *et al.* [108] reviewed the development of fracture mechanics test methods in determination of fracture toughness and delamination resistance of fiber-reinforced polymer composites.

The polymeric composites can be divided into two different categories of brittle and ductile. The concept of K_{IC} is used to characterize the toughness of brittle polymer composites, which is obtained from well-established linear elastic fracture mechanics [109, 110] by the standard test methods. Several types of standard test are employed for direct determination of the fracture toughness using; single-edge notched tension (SENT), double-edge notched tension (DENT), double cantilever beam (DCB), single-edge notched three point bending (SENB), end notched flexure (ENF), and compact tension (CT) specimens [111,112].

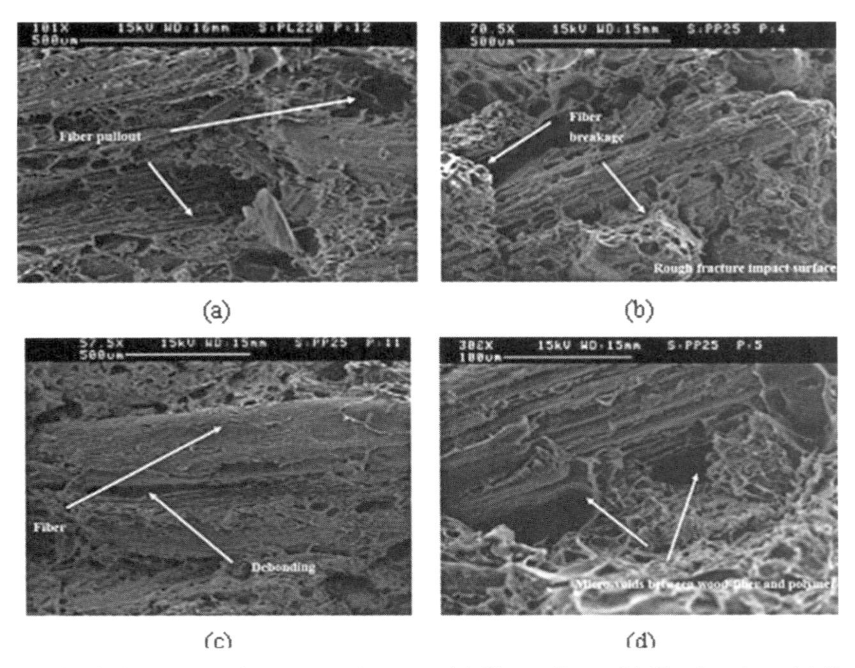

Figure 17.8 The different WPC fracture mechanisms, (a) fiber pull out, (b) fiber breakage (c) fiber/matrix debonding, (d) The formation of micro-voids between wood fiber and polymeric matrix [104].

Although the expressions for K_{IC} and its relation to G_{IC} are originally derived for homogeneous, isotropic materials, it has been successfully utilized for particle-filled composites as follows;

$$G_{IC} = {K_{IC}^2}\Big/{E_c} \tag{17.1}$$

where E_c is effective composite modulus. It was demonstrated that macroscopically measured G_{IC} and K_{IC} are affected by particle size, interfacial adhesion, and particle loading [9]. The following model given by Lange [113] relates the G_{IC} to the line tension T:

$$G_{IC} = G_m + \frac{F\left(d_p\right)T}{L} \tag{17.2}$$

where G_m is the matrix toughness, $0 \leq F(d_p) \leq 1$ is a non-dimensional function, and $L = 2d_p \dfrac{(1-V_p)}{3V_p}$ is the interparticle spacing as a function of the particle diameter d_p in which V_p is the particle volume fraction.

However, the toughness characterization of the more ductile polymer composites is more challenging. Traditionally, toughness of these composites has been characterized by the Izod or Charpy impact energy. The impact energy is a function of plastic and fracture work. Although, the Izod and Charpy tests cannot be directly used in design, they have been used for comparing the toughness of particulate polymer composites. The toughness characterization of ductile polymer composites can also be conducted from fracture mechanics point of view. It is generally believed that fracture initiation at high stress cannot be explained by linear elastic fracture mechanics. In this regard, to characterize the toughness of ductile polymer composites, post yield fracture mechanics J-integral approach [114] is used. The initial value of the J-integral (J_{IC}) depends on the degree of constraint at the crack tip, which is affected by geometry, size, and the thickness of the specimens [115]. Besides the J-integral approach which is a developed form of linear elastic fracture mechanics for less ductile fractures, the essential work of fracture (EWF) approach is proposed for plane stress ductile metal fractures [116] and polymers [117]. The EWF approach can also be applied to plane strain fracture such as slow strain rate tests [118] and impact tests [119]. Lauke and Fu [13] developed the theoretical model for the fracture toughness of particulate polymer composites by considering a simple geometrical model of particle–particle interaction in a regular particle arrangement. They revealed the effect of some structural properties such as particle volume fraction and matrix mechanical properties on fracture toughness.

Jeong [104] evaluated the fracture toughness of WPC as 1.79 $(MPa\sqrt{m})$. He realized that while the stress intensity factor is a function of crack length, the fracture toughness of WPC is independent of notch length. The shear-mode (Mode-II) fracture toughness of wood-wood and wood-fiber-reinforced plastic (FRP) bonded interfaces was evaluated by Wang and Qiao [120] using tapered end notched flexure specimens. The compliance rate variation of these specimens was verified experimentally and numerically by compliance calibration and finite element analysis, respectively. Accordingly, the critical strain energy release rates were evaluated.

Semrick [121] determined the fracture toughness of low- and high-density polyethylene and wood plastic composites using the orthogonal cutting. The tool interface method was proposed to separate tool forces from crack propagation energy. Moreover, the measurement of the toughness was carried out through modeling internal stresses in the chip. Souza [122] carried out shear testing on WPC and concluded that the fracture toughness evaluation of the aspects and parameters of the WPC reinforcing process is preferred, as it is more sensitive to parameter adjustments than shear testing.

Hristov [123] studied the impact fracture behavior of WPC modified by MAPP as compatibilizer and poly (butadiene styrene) rubber as impact modifier using Charpy impact testing. They demonstrated that the crack propagation energy compared to the crack initiation energy is much more influenced by the morphology.

17.4 Fracture Observation

The traditional measuring devices such as strain gauge and extensometer are capable of providing a single point strain and average strain in a specific direction, respectively. It is well known that the damage processes are microstructure dependent and macromechanical behavior of particulate polymer composites mainly depend on their microstructures. For example, consider the material under strain in which microvoid in the matrix, or along the particle/matrix interface, appears as the irreversible microstructural damage. By increasing the applied force, the damage growth occurs by either successive nucleation or coalescence of the microvoids, or polymer binder tearing [124]. Therefore, characterizing the macromechanical properties of particulate polymer composites without studying the microstructural damage mechanisms and their effects on macromechanical properties is not possible.

To study the microscopic fracture behavior, observation of the whole field deformations in real time seems to be necessary. An efficient optical method, called DIC has been developed for measuring inhomogeneous deformation. This method was developed in the early' 80s at the University of South Carolina to measure full-field in plane displacements and displacement gradients of a strained body at the macroscale. It is a non-contact optical and computer-based process to obtain 2D full field information by recording deformation and motion of speckled patterns on a specimen surface before and after deformation of the body. Moreover, it uses image processing to go through several images of material, and calculate the deformation at any point in the field and find deformation and strain values. In recent years this method found its application in several fields of science. As an example it is employed to measure crack initiation mode and stress intensity factor in fracture mechanics. The DIC method provides the whole field contour maps of displacements with high sensitivity. In addition to DIC, several other techniques such as; Moiré interferometry [125], digital speckle photography [126], and *in-situ* tensile tests under SEM [127] have been developed for direct observation of the microstructural damage process. Zheng *et al.* [127] studied the crack initiation and propagation of the plerosphere/PP composites by using the *in-situ* SEM observation. Moreover, DIC method combined with SEM observation was used to investigate the microscopic fracture behavior and the creep properties at different temperatures. Stark and Matuana [128] studied the changes of WPC's surface chemistry after weathering using spectroscopic techniques. X-ray photoelectron spectroscopy (XPS) was employed to characterize the occurrence of surface oxidation.

Alavi *et al.* [66] developed a portable mechanical testing rig equipped with a light microscope to observe the fracture mechanism of WPCs under different temperature as shown in Figure 17.9.

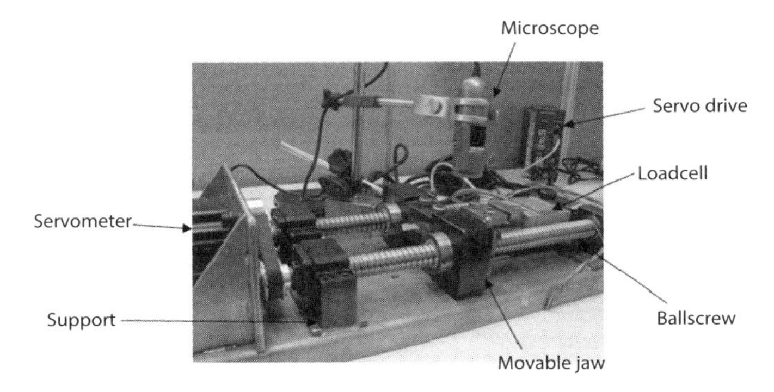

Figure 17.9 Manufactured portable mechanical test setup [66].

17.5 Fracture Analysis

17.5.1 Macroscale Modeling

Determination of the macroscopic overall characteristics of heterogeneous materials is an essential problem in many engineering applications. The investigation of these characteristics could be carried out by either performing the experiments with the material sample or using analytical methods making rather strong assumptions on the mechanical field variables. Due to the expensive and time-consuming nature of the experimental procedures, numerical or computational modeling techniques are preferred. While the analytical and numerical techniques have given valuable insight into composites deformation behavior, they also simplify their heterogeneous microstructure. However, these simplifications typically make modeling and analysis more efficient and straightforward. The macroscopic deformation behavior of composites is severely affected by the microstructural complexities such as the inhomogeneous spatial distribution of particles, irregular morphology of the particles, and anisotropy in particle orientation. However, conventional models cannot precisely represent the effective properties and localized damage mechanisms, which are microstructural dependent.

Modeling both the overall elastic-plastic response and the local damage mechanisms in composite materials is a very complex problem. The values of composite effective properties such as Young's modulus are provided by analytical and empirical models from the known properties of its constituents [129]. Some models like Hashin-Shtrikman (H-S) and Halpin-Tsai (H-T) do not consider the composite's microstructure and its anisotropy. They provide the analytical bound on the composite behavior. H-S model is based on variational principles of the linear elasticity and treats the composite as an isotropic aggregate. However, H-T is a semi-empirical approach modified from continuous fibers to discontinuous reinforcement. Chen *et al.* [130] proposed an effective analytical approach for spherical particle/matrix imperfect interface in particulate composites. When the elastic interphase is thin, the boundary problem of hydrostatic compression of an infinite spherical symmetrical body is obtained by solving the displacement and the stress field in the inclusion and matrix. On the other hand, numerical modeling is often more effective than the analytical one in the case of particulate composites. The advantage of numerical modeling is due to the capability of obtaining the deformation and damage characteristics on a local scale. The behavior of composites is numerically modeled by assuming a single fiber, whisker, or particle with simple geometry in a unit cell model [131]. Typically, modeling of damage in particulate composites must take into consideration several factors such as particle fracture, void nucleation as well as growth, coalescence of voids within the matrix, decohesion, and crack growth along the particle/matrix interface. In order to directly study the effects of particle cracking on composite properties, the unit-cell models have been employed. Although, these models do not invoke particle fracture during loading, they represent an upper bound and lower bound for particle damage and effective composite strength, respectively [132]. Exploring the effects of particle fracture on the plastic response of the composite is conducted by several researchers. Ghosh *et al.* [133] modeled damage in the composite in 2D by assuming the particle morphology as ellipsoids. They also carried out the damage modeling based on a 3D elastic Voronoi cell approach [134]. The effect of spatial distribution of the particles accounts for another aspect of the microstructure in the composite. However, there are few researches about linking between spatial distribution and mechanical behavior [135, 136]. Lloyd [137] and Prangnell [138] demonstrated the randomness of the particles' distribution in particulate composites. Thus, some particle-rich regions usually called clusters were formed. It was revealed that damage evolution prefers to originate in clustered regions [137] and the different shapes of particles show different susceptibility to interfacial decohesion [139]. Boselli *et al.* [135] modeled the effect of crack growth using idealized 2D microstructures composing of circular disks. They revealed that clustering had a significant effect on the local shielding as well as "anti-shielding" effects at the crack tip. The effect of particle clustering on composite damage is also investigated by Segurado *et al.* [136]. The spherical particles with different degrees of clustering were modeled. It was demonstrated that the standard deviation in stress, in spite of the average stress in the particles, vary significantly with clustering. In this regard, in highly clustered composites, which exposed to a given far-field applied stress a given particle is more prone to fracture, because it locally may have much higher stress deviation from the average stress. The mechanical response of the WPC was studied by Hansis *et al.* [140]. Two explicit semi-analytical micromechanical methods, i.e., Mori–Tanaka method (MTM) and generalized method of cells (GMC) were used to model the elasto-plastic response of the WPC. They revealed that while, both methods can accurately predict the response of WPC in the elastic area; Mori–Tanaka method provides more precise results when predicting plastic deformations of WPC. Mohamadzade *et al.* [141] studied the failure modes of the WPC single shear plane joint experimentally and numerically. The failure loads of these joints were determined using Yamada-Sun failure criterion.

A major challenge in understanding the fracture mechanisms of some composites (specially nanocomposites) is the vast range of scale. In such systems, modeling the material behavior is an essential step over the length scales.

17.5.2 Multi-scale Modeling

In the case of particulate composites, matrix cracking, particle–matrix debonding, and particle cracking are three fracture mechanisms. In fiber-reinforced composite, however, failure modes include fiber breakage, fiber pullout, and delamination between plies. The operation of these mechanisms at various length scales consequences the complexity in the modeling of damage process. The fracture at large scale (centimeter and millimeter) is often analyzed with continuum mechanics. However, severe limitations exist in such traditional approaches due to this fact that the macromechanical theories cannot exactly describe the mechanisms involved near the crack tip [142]. On the other hand, the fillers may be modeled individually and microcracks can be introduced at micromechanics scale [143]. Moreover, the failure of composite interface can be modeled at the molecular level by considering the molecular dynamics of multiple polymer chain scissions [144]. Among these various length scales, the bridge is required, such that the properties at larger length scales can be obtained from relevant fundamental properties derived from smaller length scales. Due to the complexity of the fracture mechanisms in composites and the requirement to a multiscale (from microscale to macroscale) approach, the traditional fracture theories cannot always be directly applied [145].

In order to address these particular topics, some numerical solutions such as the damage-based micro–meso–macro-approaches proposed by Ladevèze *et al.* [146] and purely numerical approaches as discussed by Llorca *et al.* have been developed [147]. However, an explicit prediction of the composite fracture behavior using microscale simulations is a challenging procedure. A novel element-failure method (EFM) was developed to model the damage, fracture and delamination in composites. This approach was combined with a new micromechanics-based failure theory called strain invariant failure theory (SIFT) [148] for the modeling of damage initiation and propagation in composite structures [149]. The mechanisms which bridge micro- and macro-length scales were took into account by a simple nodal force modification scheme, which was called SIFT–EFM approach.

17.5.3 Cohesive Zone Model (CZM)

In order to predict a mesoscale fracture criterion from numerical simulations at the microscale, the analysis of microscale deformation mechanisms is conducted using finite elements combined with the cohesive zone method (CZM). This method was pioneered by Barenblatt [150] and Dugdale [151] and accounts for the fracture processes ranging from micro-crack initiation to propagation. The traction between crack lips during the separation process is introduced by the traction separation law (TSL), which models a progressive damage of the material. The fracture energy is contributed by the dissipated energy during this process, which illustrated in the concept of the cohesive laws in Figure 17.10. Two different TSLs exist; the first one is called intrinsic cohesive law which is illustrated in Figure 17.10(b). In this case, the cohesive elements are inserted at the beginning of the simulation and the TSL should model the reversible fracture stage. The cohesive law consist of an initial reversible portion and successive irreversible decreasing law initiated at the onset of material strength σ_c. While unloading during the irreversible stage, the cohesive law becomes reversible again, although the elastic stiffness decrease occurs due to the damage. This type of CZM is attractive (especially in the case of well-known crack path) because it can easily be implemented in the applications such as debonding [152,153], material interface decohesion [154], structure interface decohesion [155], and composite delamination [156-158]. Generally, it is the efficient and accurate tool in order to simulate a crack initiation and propagation by inserting cohesive elements between all bulk elements. Mixed mode loading and the corresponding TSL shape have been investigated [159, 160]. In spite of the merits of intrinsic approach, several drawbacks attracted researchers to develop other type of CZM. The significant dependence of this scheme to mesh can be considered a main drawback [161]. Although increasing the initial slope of the TSL reduces this error [162, 163], it consequences an ill-conditioned stiffness matrix for static simulations or it leads to an unacceptable small values of the critical time step for explicit dynamic simulations [164].

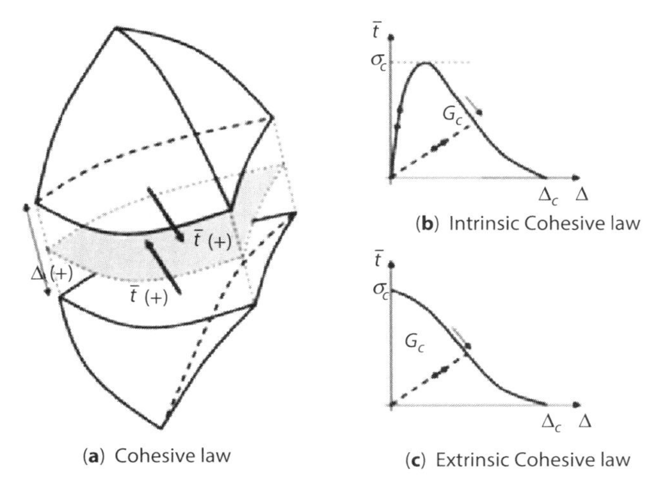

Figure 17.10 Implementation of the Cohesive Zone Model (CZM): (a) the Traction Separation Law (TSL), (b) Intrinsic TSL; and (c) Extrinsic TSL. The cohesive laws are characterized by the strength , the critical opening and the critical energy release rate G_C [145].

The use of an extrinsic cohesive law has been proposed by Camacho and Ortiz [165] and Ortiz and Pandolfi [166]. As it is demonstrated in Figure 17.10(c), this approach models only the irreversible part of the response. Therefore, in simulation the cohesive elements introduced at the onset of fracture. However, the complexity of 3D implementation increases drastically due to the mesh topology changes.

For numerical analysis of bonded structures, at the macro-scale, the cohesive finite element (CFE) method provides a simple approach by collapsing the cohesive layer to a surface in 3D cases or a line in 2D ones. Although this method is attractive due its simplicity, the phenomenological and mathematically convenient cohesive laws proposed in the literature do not directly represent the complex failure processes occurring at the micro-level in heterogeneous adhesives. Bi-linear [156] and exponential cohesive laws [167] are among the two widely employed models, which establish a relation between the tractions acting along the crack faces and the displacement jump. Due to the complexity of the failure process at microscale, the direct determination of the macroscale cohesive law of heterogeneous adhesives directly from microscale simulations is limited. Matouš *et al.* [168] proposed a multi-scale cohesive scheme to relate the microscopic failure details in heterogeneous layers to the macroscopic traction-separation law. Several investigations have been conducted to experimentally measure the interfacial traction–separation laws [169, 170]. On the other hand, various fracture problem solutions have been presented based on CZM numerical simulations [161,165, 171].

Kulkarni *et al.* [83] studied the failure processes occurring at the micro-scale in heterogeneous adhesives using a multi-scale cohesive scheme. They also considered failure effect on the macroscopic cohesive response. Investigating the representative volume element (RVE) size has demonstrated that for the macroscopic response to represent the loading histories, the microscopic domain width needs to be 2 or 3 times the layer thickness. Additionally, they analyzed the effect of particle size, volume fraction and particle–matrix interfacial parameters on the failure response as well as effective

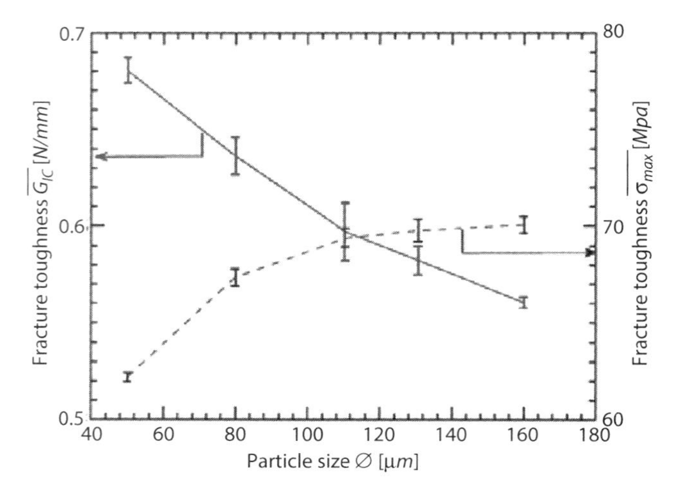

Figure 17.11 Effect of particle size on macro-scale fracture toughness and failure strength under mode I loading [83].

macroscopic properties. Contrary to the perfect bonding between the particle and the matrix, in which the failure is naturally adhesive-cohesive, a weak interface generally results in a cohesive failure. As it is shown in Figure 17.11, the macroscopic failure properties, namely, the mode I fracture toughness G_{IC} and failure strength σ_{max} can be extracted from the traction-separation law.

The hybrid discontinuous Galerkin/cohesive zone model (DG/CZM) has been proposed by Mergheim *et al.* [172] and developed by Radovitzky *et al.* [173] and Prechtel *et al.* [174]. The DG method is applied to the non–linear solid mechanics in which the unknown field between bulk elements is considered. It can be combined with the CZM, therefore on the beginning of simulation, the interface elements are inserted between bulk elements. Wu *et al.* [145] studied composite failures at the microscale using hybrid DG/extrinsic cohesive law framework.

The cohesive zone stress can be extracted from experimental data. To determine the cohesive stresses, the J-integral as a function of crack opening displacement (COD) can be determined experimentally for a DCB specimen using the following relationship [175],

$$J = \frac{2P\theta}{b} \tag{17.3}$$

Where P is the reaction force at the loading pin location, θ is rotation at the loading pin obtained through DIC calculation, and b is the width of the specimen. Then, a smoothing spline fit is obtained to take the first derivative of J-integral with respect to COD (δ) at each data point with following equation:

$$\sigma^{\text{cohesive}} = \frac{\partial J}{\partial \delta} \tag{17.4}$$

The COD can be obtained by measuring the crack opening displacement between two points (one above and one below) at the location of the crack tip in the unloaded state [176]. Fuchs and Major [177] described that cohesive stresses can be evaluated by taking the first derivative of J-integral with respect to COD (δ) for Mode-I type failure.

As long as the discontinuities at the microscale are concerned, the convergence of the extracted response is not easily obtained, which is not the case for homogenized response [178]. However, Verhoosel *et al.* [179] developed an approach to extract a mesoscopic cohesive law based on modeling the fracture of the micro-structures by the cohesive elements. They proposed to subtract the elastic and plastic deformations from the boundary displacement such that the mesoscopic crack opening is extracted. The convergence of the multiscale analysis in the absence of irreversible behaviors prior to the mesoscopic crack is proven to be assured. Extraction of a mesoscale cohesive law takes into account the modification of the elastic properties of the composite material in the method proposed by Verhoosel *et al.* [179]. They demonstrated that the extracted law is independent of the cell size. The predicted behaviors prior to catastrophic failure of the tensile samples are found to be in good agreement. A computational homogenization framework based on the same approach is proposed by Nguyen *et al.* [180].

Tan *et al.* [181] used bilinear cohesive law for particle/matrix interface in high-explosive composites. Their used model consist of three stages, namely, the rising, descending and complete debonding sections which are governed by three parameters consist of interface cohesive strength and the rising and descending modulus obtained by experiments. They investigated the effect of interface debonding on particulate composites during uniaxial tensile test [182]. Some other researches are carried out Adopting the cohesive interface model, such as Inglis and Geubelle [183] which applied micromechanics solution in study of the effect particle/matrix debonding on the constitutive response of particulate composite materials and compared results with the multi-scale finite element analysis.

17.5.4 Other Numerical Methods

Kwon *et al.* [184] investigated the crack behavior based on the damage initiation, growth and the local saturation in notched particulate composite specimens using numerical simulation. The results of experimental study and the simulations were compared and indicated that the crack behavior is inherently affected by an inhomogeneous material property. Wilkins and Shen [185] computationally determined the stress evolution inside the particles using finite element method. They predicted the stress enhancement in the inclusion using simplistic model. To analyze the particulate composite, Okada *et al.* [186] proposed a boundary element method (BEM) formulation. They used homogenization analysis based on RVE considering a unit cell including randomly distributed particles. The predicted effective properties are shown to be accurate for the particle volume fraction less than about 25%. An investigation of the mechanical properties and evolution of the debonding process in particulate composites were developed by Chen and Kulasegaram [187] using smooth particle hydrodynamics (SPH). An extended element free galerkin (EFG) method was proposed by Belytschko *et al.* [188] and was employed by Rabczuk and Zi [189] to deal with cohesive cracks. Moreover, Rabczuk *et al.* [190] applied a three-dimensional extended EFG method for arbitrary crack initiation and propagation that ensured crack path continuity for non-linear material models and the cohesive law. The method can be applied to static, quasi-static and dynamic crack problems and provides precise numerical results compared with other experimental and analytical results.

The extended finite element method (XFEM) for treating fracture in composite materials is proposed by Huynh and Belytschko [191]. This methods work with meshes that are independent of matrix/inclusion interfaces and the discontinuities and near-tip enrichments were modeled. In order to describe the geometry of the interfaces and cracks in this method, level sets were employed, so that there is no need for explicit representation of either the cracks or the material interfaces. The other researchers such as Du *et al.* [192] and Ying *et al.* [193] used XFEM to model material interfaces in particulate composites with more in detail.

On the other hand, some other alternative approaches have been developed to model the fracture as the XFEM [194] or the embedded localization method [195]. These approaches can also be combined with a cohesive zone model [196].

17.5.5 Inverse Method

Besides the direct methods, identification procedures based on an inverse methodology have also been widely used. Kang *et al.* [197] used genetic algorithms to construct an inverse approximation procedure to identify interfacial parameters based on the experimental data. Lee *et al.* [198] proposed a systematic procedure for determination of the mixed-mode cohesive parameters by employing an optimization technique for the co-cured single leg bending (SLB) joints. Identification of CZM parameters for analysis of adhesive joints based on an experimental–numerical methodology was carried out by Fedele *et al.* [199, 200]. Guessasma and Bassir [201] developed a hybrid approach based on finite element calculation and inverse/genetic algorithm to identify the correlation between the interface, intrinsic properties, and the effective properties of the composite. Shen and Paulino [202] addressed the inverse identification of elastic properties and CZM of fiber-reinforced cementitious composites using the full-field displacement through an optimization technique in a finite element framework. Ferreira *et al.* [203] coupled DIC and boundary element method to identify Young's modulus and fracture parameters associated with a cohesive model, and Han *et al.* [204] extracted cohesive parameters of HTPB propellant via an inverse identification method.

In order to analyze the damage of wood plastic composite, the identification of wood/plastic interfacial properties seems to be essential. Therefore, the wood/plastic cohesive parameters ware determined through optimization technique [205]. The Nelder-Mead algorithm was applied to optimize the force-displacement results of the numerical simulation and experimental data using the following equation.

$$\psi\left(\hat{\mathbf{\imath}}\right)=\sum_{i=1}^{2}\left|\mathbf{P}_{FEM}^{i}-\mathbf{P}_{Exp}^{i}\right|^{2} \tag{17.5}$$

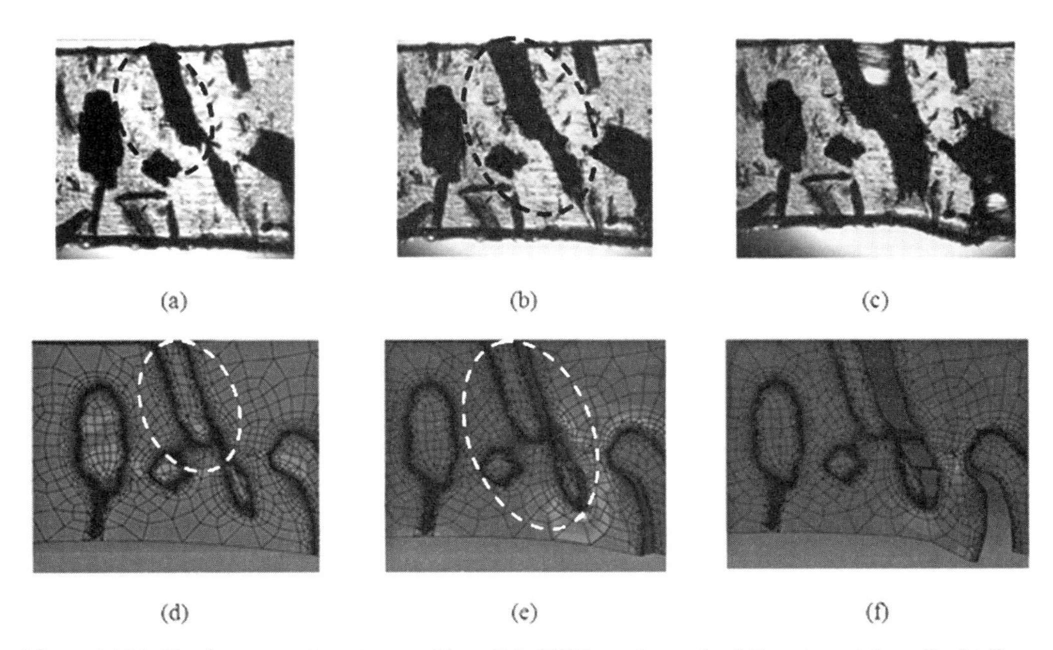

Figure 17.12 The fracture pattern in a multi-particle WPC specimen (a-c) Experimental results, (d-f) simulated results [205].

where ψ is the objective function, ξ is the vector of cohesive zone parameters, and P^i_{Exp} and P^i_{FEM} are points on the experimental and FEM simulated curves, respectively. Accordingly, the obtained cohesive parameters were implemented in the finite element (FE) models for damage prediction of the wood plastic specimens. A multi-particle WPC specimen was analyzed using the determined CZM parameters, for which the simulated and experimental fracture patterns are shown in Figure 17.12. The experimental and simulated fracture initiation and growth patterns are depicted in Figures 17.12(a) to 17.12(c) and 17.12(d) to 17.12(f), respectively. As shown in these figures, debonding is the dominant fracture mechanism in WPC specimen in the absence of any coupling agents.

17.6 Conclusions

In this chapter, the fracture of WPCs as particle-filled polymer composites was elaborated. The characterization of particulate polymer composites' fracture behavior and the influencing factors such as particle size as well as orientation, temperature, and loading were discussed. The fracture observation using special setup was described and the diverse numerical methods to analyze the fracture of such composites were reviewed. Finally the finite element simulation of the fracture for WPC specimen with real geometrical model was conducted and the agreement of results compared to the experimental ones was demonstrated.

References

1. P. Shahi, A.H. Behravesh, S.Y. Daryabari, and M. Lotfi, *Polym. Comp.* 33(5), 753–763 (2012).
2. F. Alavi, A.H. Behravesh, D. Karimi and A. Milani, *Polymer Processing Society Regional Meeting*, Kish, Iran (2011).
3. C. Clemons, *Forest Product. J.* 52(6), 1–20 (2002).
4. A.N. Netravali andS. Chabba, Mater. *Today* 6, 22–26 (2003).
5. A. H. Behravesh, E. Shakouri, A. Zolfaghari, and M. Golzar,J. *Comp. Mater.* 44(11), 1293–1304 (2010).
6. A.H. Behravesh, A. Zohdi, and E. Soury, *J. Rein. Plas. Compos.* 29, 456–465 (2010).
7. A. Zolfaghari, A. H. Behravesh, A. Adli and M. TabkhPazSarabi, *J. Rein. Plas. Compos.* 32(1), 52–60 (2013).
8. K. Renner, C. Kenyó, J. Móczó, and B. Pukánszky, *Compos. Part A* 41, 1653–1661 (2010).
9. S.Y. Fu, X.Q. Feng, B. Lauke, and Y.W. Mai, *Compos. Part B* 39(6), 933–961 (2008).
10. R.P. Singh, M. Zhang, and D. Chan, *J. Mater. Sci.* 37, 781–788 (2002).
11. T.H. Hsieh, A.J. Kinloch, K. Masania, J. Sohn Lee, A.C. Taylor, and S. Sprenger, *J. Mater. Sci.* 45, 1193–1210 (2010).
12. J.G. Williams, *Comput. Sci. Technol.* 70, 885–891 (2010).
13. B. Lauke andS.Y. Fu,*Comp. Part B* 45, 1569–1574 (2013).
14. B. Lauke, *Comput. Sci. Technol.* 68, 3365–3372 (2008).
15. Liang YL, Pearson RA. *Polymer* 50(20), 4895–4905 (2009).
16. T. Adachi, M. Osaki, W. Araki, and S.C. Kwon. *Acta Mater*. 56(9), 2101–2109 (2008).
17. M. Hussain, A. Nakahira, S. Nishijima, and K. Niihara, *Mater. Lett.* 27, 21–25 (1996).
18. C.B. Ng, B.J. Ash, L.S. Schadler, and R.W. Siegel, *Adv. Compos. Lett.* 10(3), 101–111 (2001).
19. T.H. Hsieh, A.J. Kinloch, K. Masania,A.C. Taylor, and S. Sprenger, *Polymer* 51(26), 6284–6294 (2010).
20. B.B. Johnsen, A.J. Kinloch, R.D. Mohammed, A.C. Taylor, and S. Sprenger, *Polymer* 48(2), 530–541 (2007).
21. J. Jordan, K.I. Jacob, R. Tannenbaum, M.A. Sharaf, and I. Jasiuk, *Mater. Sci. Eng. A* 393, 1–11 (2005)
22. A. Bledzki and O Faruk, *Appl. Compos. Mater.* 10, 365–379 (2003).
23. N.M. Stark and R.E. Rowland, *Wood Fiber Sci.* 35, 167 (2003).
24. K. Renner, J. Móczó, Suba, and B. Pukánszky, *Compos. Sci. Technol.* 70, 1141–1147 (2010).
25. Y. Cui, S. Lee, B. Noruziaan, M. Cheung, and J. Tao, *Compos. Part A*39, 655–661 (2008).
26. Y. Wang, Morphological characterization of wood plastic composite (WPC) with advanced imaging tools: Developing methodology for reliable phase and internal damage characterization, *Master Thesis*, Oregon State University (2007).
27. F. Alavi, A.H. Behravesh,M. Mirzaei, *Compos. Interfaces* 20(3), 211–220 (2013).
28. J. Spanoudakis and R. J. Young, *Mater. Sci.* 19, 473–486 (1984).
29. S.Y. Fu and B. Lauke, *Compos. Part A* 29, 631–641 (1998).
30. G. Levita, A. Marchetti, A. Lazzeri, *Polym. Compos.* 10, 39–43 (1989).
31. J. Kong, R.C. Ning, and Y.S. Tang, *J. Mater. Sci.* 41(5), 1639–1641 (2006).
32. T. Raju, Y.M. Ding, Y.L. He, L. Yang, M. Paula, and W.M. Yang, *Polymer* 49(1), 278–294 (2008).
33. Q.H. Le, H.C. Kuan, J.B. Dai, I. Zaman, L. Luong, J. Ma, *Polymer* 51(21), 4867–4879 (2010).
34. J. Ma, M.S. Mo, X.S. Du, Rosso, K. Friedrich, and H.C. Kuan, *Polymer* 49(16), 3510–3523 (2008).

35. H. Zhang, L.C. Tang, Z. Zhang, K. Friedrich, and S. Sprenger, *Polymer* 49(17), 3816–3825 (2008).
36. L.C. Tang, H. Zhang, S. Sprenger, L. Ye, and Z. Zhang, *Compos. Sci. Technol.* 72, 558–565 (2012).
37. B. Pukanszky,*Polypropylene: structure, blends and composites* 3, pp. 1–70, Chapman & Hall, London. (1995).
38. R.A. Baker, L.L. Koller, and P.E. Kummer, *Handbook of fillers for plastics*, pp. 119–42. Van Nostrand Reinhold Co, New York. (1987).
39. Z. Bartczak, A.S. Argon, R.E. Cohen, and M. Weinberg, *Polymer* 40, 2347–2365 (1999).
40. Y. Lei and Q. Wu, *Compos. Part A* 43, 73–78 (2012).
41. G. Cantero, A. Arbelaiz, F. Mugika, A. Valea, and I. Mondragon, *J. Rein. Plas. Compos.* 22(1), 37–50 (2003).
42. M.N. Ichazo, C. Albano, J. Gonzalez, R. Perera, and M.V. Candal, *Compos. Struct.* 54(2), 207–214 (2001).
43. K.C. Jajam and H.V. Tippur, *Compos. Part B* 43, 3467–3481 (2012).
44. L. Sun, R.F. Gibson, F. Gordaninejad, and J. Suhr, *Compos. Sci. Technol.* 69(14), 2392–2409 (2009).
45. R. Kitey and H.V. Tippur, *Acta Mater.* 53(4), 1153–1165 (2005).
46. A. Shukla, V. Parameswaran, Y. Du, and V. Evora, *Adv. Mater. Sci.* 13(1), 47–58 (2006).
47. V.M.F. Evora and A. Shukla, *Mater. Sci. Eng. A* 361, 358–366 (2003).
48. V.M.F. Evora, N. Jain, and A. Shukla, *Exp. Mech.* 45(2), 153–159 (2005).
49. S. Tamrakar and R.A. Lopez-Anido, *Forest Prod. J.* 60(5), 465–472 (2010).
50. G.W. Yu, Y. C. Hu, and J. Y. Gu, *Mater. Sc. Forum* 658, 475 (2010).
51. Z.W. Liu, H.M. Xie, K.X. Li, W. Chen, and F.L. Huang, *Polym. Testing* 28, 627–635 (2009).
52. J.R. Luo, S.Q. Zhang, and F.F. Zhao, *J. Energetic Mater.* 8, 42–45 (2000).
53. H.T. Yao, F.Z. Xuan, Z. Wang, and S.T. Tu, *Nucl. Eng. Des.* 237, 1969–1986 (2007).
54. C. Chandenduang and A.A. Becker, *Comput. Struct.* 81(16), 1611–1618 (2003).
55. G.A. Wade and W.J. Cantwell, *J. Adhesion* 76(3), 245–264 (2001).
56. B.C. Ray, *J. Appl. Polym. Sci.* 100, 2289–2292 (2006).
57. J. Liu, Z. Zhou, L. Wu, L. Ma, and S. Pan, *Appl. Compos. Mater.* 20(4), 1216–1225 (2012).
58. D. Li, D. Fang, G. Zhang, and H. Hu, *Mater. Des.* 41, 167–170 (2012).
59. Z. Farhadinejad, M. Ehsani, B. Khosravian, and G. Ebrahimi, *Eur. J. Wood Wood Prod.* 70, 823–828 (2012).
60. M. Tajvidi, N. Motie, G. Rassam, R.H. Falk, and C. Felton, *J. Rein.Plas. Compos.* 29(5), 664–674 (2010).
61. D.J. Pooler, The temperature dependent non-linear response of a wood plastic composite, Master Thesis, Washington State University (2001).
62. S.E. Hamel, Modeling the time-dependent flexural response of wood-plastic composite materials, *PhD thesis*, University of Wisconsin (2011).
63. S.G. Bajwa, D.S. Bajwa, and A.S. Anthony, *J. Thermoplastic Compos. Mater.* 22(2), 227–243 (2009).
64. Z.N. Azwa, B.F. Yousif, A.C. Manalo, and W. Karunasena, *Mater. Des.* 47, 424–442 (2013).
65. S. Tamrakar, R. A. Lopez-Anido, A. Kiziltas, and D. J. Gardner, *Compos. part A* 42, 834–842 (2011).
66. F. Alavi, A.H. Behravesh, and M. Mirzaei, *Thermo Plas. Compos. Mater.* 27(1) (2014).
67. M. Kazayawoko, J.J. Balatinecz, and L.M. Matuana, *J. Mater. Sci.* 34, 6189 (1999).
68. A. Sretenovic, U. Müller, and W. Gindl, *Compos. A* 37, 1406 (2006).
69. J.Z. Lu, Q. Wu, and I.I. Negulescu, *J. Appl. Polym. Sci.* 96(1), 93–102 (2005).
70. X.Y. Liu and G.C. Dai, *Express Polym. Lett.* 1(5), 299–307 (2007).

71. M. Abdelmouleh, S. Boufi, M.N. Belgacem, A. Dufresne, and A. Gandini, *J. Appl. Polym. Sci.* 98(3), 974–984 (2005).
72. H. Demir, U. Atikler, D. Balköse, and F. Tihminlioglu, *Compos. A* 37(3), 447–456 (2006).
73. C. Zhang, K. Li, and J. Simonsen, *J. Adhesion Sci. Technol.* 18(14), 1603–1612 (2004).
74. C. Zhang, K. Li, J. Simonsen. *Polym. Eng. Sci.* 46(1), 108–113 (2006).
75. R.G. Raj and B.V. Kokta, *J. Appl. Polym. Sci.* 38(11), 1987–1996 (1989).
76. C.S.R. Freire, A.J.D Silvestre, C.P. Neto, M.N. Belgacem, and A. Gandini, *J. Appl. Polym. Sci.* 100(2), 1093–1102 (2006).
77. P. Jandura, B. Riedl, and B.V. Kokta, *Polym. Degrad. Stability* 70(3), 387–394 (2000).
78. M. Hietala, E. Samuelsson, J. Niinimäki, and K. Oksman, *Compos. A* 42, 2110–2116 (2011).
79. E. Pérez, L. Famá, S.G. Pardo, M.J. Abad, and C. Bernal, *Compos. B* 43, 2795–2800 (2012).
80. Y. Xie, C.A.S. Hill, Z. Xiao, H. Militz, and C. Mai, *Compos. A* 41, 806–819 (2010).
81. A. Godara, D. Raabe, I. Bergmann, R. Putz, and U. Müller, *Compos. Sci. Technol.* 69, 139–146 (2009).
82. P. Alam, *Porous Particle-Polymer Composites, Advances in Composite Materials, Centre for Functional Materials*, Abo Akademi University, Finland.
83. M.G. Kulkarni, H. Geubelle, and K. Matouš, *Mech. Mater.* 41, 573–583 (2009).
84. L.F.R. Rose, *Mech. Mater.* 6, 11 (1987).
85. P. Rosso, L. Ye, K. Friedrich, and S. Sprenger, *J. Appl. Polym. Sci.* 101, 1235 (2006).
86. M. Leonard, N. Murphy, and A. Karac, *(6)th International Conference on Fracture of Polymers, Composites and Adhesives* (2011).
87. J. Ma, M.S. Mo, X.S. Du, Rosso, K. Friedrich, and H.C. Kuan, *polymer* 5, 43 (2008).
88. Y.L. Liang andR.A. Pearson, *Polymer* 50(20), 4895–4905 (2009).
89. B.P. Boesl, G.R. Bourne, and B.V. Sankar, *Compos. B* 42, 1157–1163 (2011).
90. J. Lee and A.F. Yee, *Polymer* 42, 577–588 (2001).
91. J. Lee and A.F. Yee, *Polymer* 42, 589–597 (2001).
92. A.D. Norman and R.E. Robertson, *Polymer* 44, 2351–2362 (2003).
93. H. Porwal, Tatarko, S. Grasso, C. Hu, A.B. Boccaccini, I. Dlouhý, and M.J. Reece, *Sci. Technol. Adv. Mater.* 14, 55 (2013).
94. Y.L. Liang and R.A. Pearson, *Polymer* 51, 4880–4890 (2010).
95. H.Y. Liu, G.T. Wang, Y.W. Mai, and Y. Zeng, *Compos. B* 42, 2170–2175 (2011).
96. S.U. Khan, A. Munir, R. Hussain, and J.K. Kim. *Compos. Sci. Technol.* 70, 2077–2085 (2010).
97. B. Cotterell, J.Y.H. Chia, K. Hbaieb, *Eng. Fract. Mech.* 74, 1054–1078 (2007).
98. G. Lai and L. Mishnaevsky Jr, *Compos. Sci. Technol.* 74, 67–77 (2013).
99. P. Dittanet, R.A. Pearson, *Polymer* 53, 1890–1905 (2012).
100. J.G. Williams, *Compos. Sci. Technol.* 70, 885–891 (2010).
101. B. Lauke, *Comput. Mater. Sci.* 77, 60–73 (2013).
102. M. Zappalorto, M. Salviato, M. Quaresimin, *Compos. Sci. Technol.* 72, 49–55 (2011).
103. B.Lauke, *Compos. Sci. Technol.* 86, 135–141 (2013).
104. G.Y. Jeong, Fracture behavoir of wood plastic composite (WPC), Louisiana state university, Master Thesis (2005).
105. Y. Nakamura, M. Yamaguchi, M. Okubo, and T. Matsumoto, *J. Appl. Polym. Sci.* 45(7), 1281–1289 (1992).
106. Y. Nakamura, S. Okabe, and T. Iida, *Polym. PolymPolym. Compos.* 7(3), 177–186 (1999).
107. K.C. Jajam, Fracture Behavior of Particulate Polymer Composites (PPCs) andInterpenetrating Polymer Networks (IPNs): Study of Filler Size, Filler Stiffness and Loading Rate Effects, PhD Thesis, Auburn University (2013).
108. A.J. Brunner, B.R.K. Blackman, and P. Davies, *Eng. Fract. Mech.* 75, 2779–2794 (2008).

109. *Plastics – Determination of Fracture Toughness (G_{IC} and K_{IC}) – Linear Elastic Fracture Mechanics (LEFM) Approach, ISO Standard ISO-13586,* Geneva (2000).
110. J.G. Williams, *Fract. Mech. Testing Polym.*, ESIS pub 28, 11–26 (2001).
111. N. Dourado, S. Morel, M.F.S.F. de Moura, G. Valentin, and J.Morais, *Compos A* 39, 415–427 (2007).
112. M.J. Laffan, S.T. Pinho, Robinson, and A.*J. McMillan, Polym. Testing* 31, 481–489 (2012).
113. F.F. Lange, *J. Am. Ceramic Soc.* 54, 614–620 (1971).
114. G.E. Hale and F. Ramsteiner, *Fract. Mech. Testing Polym.*, ESIS pub 28, 23–158 (2001).
115. R.H. Dodds, C.F. Shih, and T.L. Anderson, *Int. J. Fract.* 64(2), 101–133 (1993).
116. B. Cotterell and J.K. Reddel, *Int. J. Fract.* 13(3), 267–277 (1977).
117. Y.W. Mai and B. Cotterell,*Int. J. Fract.* 32, 05–126 (1986).
118. P. Luna, C. Bernal, A. Cisilino, Frontini, B. Cotterell, and Y.W. Mai, *Polymer* 44(5), 145 (2003).
119. J.S. Wu, Y.W. Mai, and B. Cotterell, *J. Mater. Sci.* 28, 3373–3384 (1993).
120. J. Wang and P. Qiao, *J. Compos. Mater.* 37, 875 (2003).
121. K. Semrick, *Determining Fracture Toughness by Orthogonal Cutting of Polyethylene and Wood-Polyethylene Composites,* Oregon State University, Master Thesis (2012).
122. J. Souza, Fracture Mechanics Characterization of WPC-FRP Composite Materials Fabricated by the Composites Pressure Resin Infusion System (Compris) Process Volume I, Electronic Theses and Dissertations. *Paper* 862 (2005).
123. V.N. Hristov, R. Lach, W. Grellmann, *Polym. Testing* 23, 581–589 (2004).
124. E.E. Gdoutos and G. Papakaliatakis, *Fatigue Fract. Eng. Mater. Struct.* 24(10), 637–642 (2001).
125. P.W. Chen, H.M. Xie, F.L. Huang, T Huang, Y.S. Ding, *Polym. Testing* 25, 333 (2006).
126. L. Larsson, S.O. Mikael and T Fredrik, *Optics Laser Eng.* 41, 767 (2004).
127. Y.H. Zheng, Z.G. Shen, M.Z. Wang, S.L. Ma, and Y.S. Xing, *J. Appl. Polym. Sci.* 106, 3736 (2007).
128. N. M. Stark, L. M. Matuana, *Polym. Degrad. Stability* 92, 1883–1890 (2007).
129. J.C. Halpin and S.W. Tsai, Environmental Factors Estimation in Composite Materials Design, *AFML TR* 67 (1967).
130. J. Chen, J. Tong and Y. Chen, J. Wuhan *Univ. Technol. Mater. Sci.* Ed. 17(2), 78–82 (2002).
131. Y.L. Shen, M. Finot, A. Needleman and S. Suresh, *Acta Metallurgica et Mater.* 42, 77 (1994).
132. N. Chawla and K.K. Chawla, *J. Mater. Sci.* 41, 913–925 (2006).
133. M. Li, S. Ghose, T.N. Rouns, H. Weiland, O. Richmond, and W. Hunt, *Mater. Characteristics* 41, 81 (1998).
134. S. Ghose and S. Moorty, *Comput. Mech.* 34, 510 (2004).
135. J. Boselli, D. Pitcher, J. Gregson and I. Sinclair, *Mater. Sci. Eng. A* 300, 113 (2001).
136. J. Segurado, C. Gonzalez, and J. Llorca, *Acta Mater.* 51, 23–55 (2003).
137. D.J. Lloyd, *TMS annual meeting*, pp. 39–47, Las Vegas, USA. (1995).
138. P.B. Prangnell, S.J. Barnes, S.M. Roberts, and P.J. Withers, *Mater. Sci. Eng. A* 220, 41–56 (1996).
139. A.F. Whitehouse and T.W. Clyne, *Acta Metallurgicaet Mater.* 41, 1701–1711 (1993).
140. A. Hančič, F. Kosel, K. Kuzman, and J. M. Slabe, *Compos. Part B Eng.* 43(3), 1500–1507 (2012).
141. M. Mohamadzadeha, A. RostampourHaftkhania, G. Ebrahimia, and H. Yoshiharab, *Mater. Des.* 35, 404–413 (2012).
142. T.E. Tay, *Appl. Mech. Rev.* 56(1), 1–32 (2003).
143. J.M. Mahishi, D.F. Adams, *J. Compos. Mater.* 16, 457–469 (1982).
144. V.B.C. Tan, M. Deng, and T.E. Tay, *Polymer* 45(18), 6399–6407 (2004).

145. L. Wu, D. Tjahjanto, G. Becker, A. Makradi, A. Jérusalem, and L. Noels, *Eng. Fract. Mech.* 104, 162–183 (2013).
146. P. Ladevèze, G. Lubineau, D. Marsal, *Compos. Sci. Technol.* 66(6), 698–712 (2006).
147. J. Llorca, C. González, J.M. Molina-Aldareguía, J. Segurado, R. Seltzer, and F. Sket, *Adv. Mater.* 23(44), 5130–5147 (2011).
148. T.E. Tay, S.H.N. Tan, V.B.C. Tan, and J.H. Gosse, *Compos. Sci. Technol.* 65(6), 935–944 (2005).
149. T.E. Tay, V.B.C. Tan, and G. Liu, *Mater. Sci. Eng. B* 132, 138–142 (2006).
150. G. Barenblatt, *Adv. Appl. Mech.* 7(08), 55–129 (1962).
151. D.S. Dugdale, *J. Mech. Phys. Solids* 8(2), 100–104 (1960).
152. A. Needleman, *J. Appl. Mech.* 54(3), 525–531 (1987).
153. H.M. Inglis, H. Geubelle, K. Matouš, H. Tan, and Y. Huang, *Mech. Mater.* 39(6), 580–595 (2007).
154. V. Tvergaard and J.W. Hutchinson. *Philosophia Mag.* A 70(4), 641–656 (1994).
155. P.P. Camanho, C.G. Dávila, and S.T. Pinho, Fatigue Fract Eng. *Mater. Struct.* 27(9), 745–757 (2004).
156. P.H. Geubelle and J.S. Baylor, *Compos. B* 29(5), 589–602 (1998).
157. A. Pantano and R.C. Averill. *Int. J. Solids Struct.* 41(14), 3809–3831 (2004).
158. A. Turon, C.G. Dávila, P. Camanho, and J. Costa, *Eng. Fract. Mech.* 74(10), 1665–1682 (2007).
159. V. Tvergaard, *J. Mech. Phys. Solids* 49(9), 2191–2207 (2001).
160. V. Tvergaard and J.W. Hutchinson, *J. Mech. Phys. Solids* 40(6), 1377–1397 (1992).
161. X.P. Xu and A. Needleman, *J. Mech. Phys. Solids* 42(9), 1397–1434 (1994).
162. S.H. Song, G.H. Paulino, and W.G. Buttlar, *Eng. Fract. Mech.* 73(18), 2829–2848 (2006).
163. N. Blal, L. Daridon, Y. Monerie, and S. Pagano. *Comptes Redus Méchan.* 339(12), 789–795 (2011).
164. P.D. Zavattieri and H.D. Espinosa. *Acta Mater.* 49(20), 4291–311 (2001).
165. G.T. Camacho and M. Ortiz, *Int. J. Solids Struct.* 33, 2899–2938 (1996).
166. M. Ortiz and A. Pandolfi. *Int. J. Numerical Methods Eng.* 44(9), 1267–1282 (1999).
167. A. Needleman, *Int. J. Fract.* 42, 21–40 (1990).
168. K. Matouš, M. Kulkarni, Geubelle, *J. Mech. Phys. Solids*56, 1511–1533 (2008).
169. B.F. Sørensen and P. Kirkegaard, *Eng. Fract. Mech.* 73, 2642–2661 (2006).
170. R.B. Sills and M.D. Thouless, *Eng. Fract. Mech.* 109, 353–368 (2013).
171. V Tvergaard and J.W. Hutchinson, *J. Mech. Phys. Solids* 41(6), 1119–1135 (1993).
172. J. Mergheim, E. Kuhl, and P. Steinmann, *Commun. Numerical Methods Eng.* 20(7), 511–519 (2004).
173. R. Radovitzky, A. Seagraves, M. Tupek, and L. Noels, *Computer Methods Appl. Mech. Eng.* 200(14), 326–344 (2011).
174. M. Prechtel, G. Leugering, Steinmann, M. Stingl, *Eng. Fract. Mech.* 78(6), 944–960 (2011).
175. J. Anthony and P. Paris, *Int. J. Fract.* **38**, 19–21 (1988).
176. P. Upadhyaya, S. Roy, M. H. Haque, H. Lu, *Compos. Struct.* 104, 118–124 (2013).
177. P. Fuchs and Z. Major, *Exp. Mech.* 51, 779–786 (2011).
178. V.P. Nguyen, O. Lloberas-Valls, M. Stroeven, and L.J. Sluys, *Comput. Methods Appl. Mech. Eng.* 199, 3028–3038 (2010).
179. C.V. Verhoosel, J.J.C. Remmers, M.A. Gutiérrez, and R. de Borst, *Int. J. Numerical Methods Eng.* 83, 1155–1179 (2010).
180. V.P. Nguyen, M. Stroeven, L.J. Sluys, *Eng. Fract. Mech.* 79, 78–102 (2012).
181. H. Tan, C. Liu, Y. Huang, and P.H. Geubelle, *J. Mech. Phys. Solids* 53, 1892–1917 (2005).

182. H. Tan, Y. Huang, C. Liu, G. Ravichandran, H.M. Inglis, and P.H. Geubelle, *Int. J. Solids Struc.* 44(6), 1809–1822 (2007).
183. H.M. Inglis, H. Geubelle, K. Matous, H. Tan, and Y. Huang, *Mech. Mater.* 39(6), 580–595 (2007).
184. Y.W. Kwon, J.H. Lee, and C.T. Liu, *Compos. B* 29(4), 443–450 (1998).
185. T. Wilkins and Y.L. Shen, *Comput. Mater. Sci.* 22, 291–299 (2001).
186. H. Okada, Y. Fukui, and N. Kumazawa, *Comput. Model. Eng. Sci.* 5(2), 135–149 (2004).
187. Y. Chen, and S. Kulasegaram, *Comput. Mater. Sci.* 47(1), 60–70 (2009).
188. T. Belytschko, Y.Y. Lu, and L. Gu, *Int. J. Numerical Methods Eng.* **37**, 229–256 (1994).
189. T. Rabczuk, and G. Zi, *Comput. Mech.* 39(6), 743–760 (2007).
190. T. Rabczuk, S. Bordas and G. Zi, *Comput. Mech.* 40(3), 473–495 (2007).
191. D.B.P. Huynh and T. Belytschko, *Int. J. Numerical Methods Eng.* 77(2), 214–239 (2008).
192. C. Du, Z. Ying and S. Jiang, *Materials Science and Engineering, IOP Conference Series,* Volume 10, pp. (2010).
193. Z. Ying, C. Du, and Y. Wang, *J. Hydraulic Eng.* 42(2), 198–203 (2011).
194. N. Moës and T. Belytschko, *Eng. Fract. Mech.* 69(7), 813–833 (2002).
195. F. Armero and C. Linder, *Int. J. Fract.* 160, 119–141 (2009).
196. R. de Borst, M.A. Gutiérrez, G.N. Wells, J.J.C. Remmers, H. Askes, *Int. J. Numerical Methods Eng.* 60(1), 289–315 (2004).
197. Y.L. Kang, X.H. Lin, and Q.H. Qin, *Compos. Struct.* 66, 449–458 (2004).
198. M.J. Lee, T.M. Cho, W.S. Kim, B.C. Lee, and J.J. Lee, *J. Adhesion Adhesives*30, 322–328 (2010).
199. R. Fedele, B. Raka, F. Hild, and S. Roux, *J. Mech. Phy. Solids.* 57, 1003–1016 (2009).
200. N. Valoroso and R. Fedele, *Int. J. Solids Struc.* 47, 1666–1677 (2010)
201. S. Guessasma and D.H. Bassir, *Mech. Mater.* 42, 344–353 (2010).
202. B. Shen and G.H. Paulino, *Cement Concrete Compos.* 33, 572–585 (2011).
203. M.D.C. Ferreira, W.S. Venturini, and F. Hild, *Eng. Fract. Mech.* 78, 71–84 (2011).
204. B. Han, Y. Ju, and C. Zhou, *Eng. Fail. Anal.* 26, 304–317 (2012).
205. F. Alavi, Damage prediction of composites filled with irregular particles, *phD Thesis* (2013).

18

Mechanical Behavior of Biocomposites under Different Operating Environments

Inderdeep Singh*, Kishore Debnath and Akshay Dvivedi

Department of Mechanical and Industrial Engineering, Indian Institute of Technology, Roorkee, Uttarakhand India

Abstract

Cellulose-based natural/plant fiber composites, or biocomposites, have received significantly more attention than conventional synthetic fiber composites due to their comparable physical and mechanical properties. Lightweight characteristics, ease of handling and biodegradable nature are a few of the salient characteristics that make biocomposites a unique class of engineering materials. The non-load bearing applications of biocomposites have spread in many engineering fields ranging from automobiles to sports components. The widespread application of these materials necessitates long-term service under different environmental conditions. Therefore, there is an imminent need to explore the mechanical properties of these materials under different environmental conditions to further enhance their application spectrum. In this chapter, the state-of-the-art is presented regarding the mechanical characterization of biocomposites when subjected to different environmental conditions.

Keywords: Natural fiber, biocomposites, mechanical properties, environmental effect

18.1 Introduction

Synthetic fiber (carbon, glass, Kevlar, etc.) fiber-reinforced plastics are extensively used for high-end, sophisticated applications due to the fact that these materials have high strength and stiffness, low density and high corrosion resistance. Fiber-reinforced plastics have already replaced many automobile, aircraft and spacecraft components - which previously were made with metals and alloys [1–4]. Despite having several good properties, these materials are now facing problems due to their non-biodegradable nature. The products engineered with petroleum-based fibers and polymers are found to be severely lacking when their service life meets the end. The non-biodegradable nature of these materials is a threat to the environment where ecological balance is concerned. Depletion of fossil resources, release of toxic gases, and the volume of waste increases with the use of petroleum-based materials. These are some issues which have

**Corresponding author*: dr.inderdeep@gmail.com

Vijay Kumar Thakur (Ed.), Lignocellulosic Polymer Composites, (417–432) 2015 © Scrivener Publishing LLC

led to the reduced utilization of petroleum-based non-biodegradable composites and the development of biocomposites.

Currently, the main focus of researchers and engineers is to develop environmentally-friendly, cost-effective materials and processes. "Go green" has become the only concern in many aspects of the modern manufacturing industries. Development of biocomposites are one of the greatest examples of environmentally-friendly, cost-effective materials. Based upon the type of polymers, biocomposites may be broadly classified into two categories: partially and fully biodegradable biocomposites (Figure 18.1). Reinforcing biodegradable polymers (polylactic acid [PLA], polyhydroxybutyrate [PHB], poly hydroxy alkanoates [PHA], etc.) with natural fiber results in fully biodegradable composites; whereas, reinforcing traditional petroleum-based non-biodegradable polymers (epoxy, polyester, polypropylene [PP], polyethylene [PE], polystyrene [PS], etc.) with natural fiber results in partially biodegradable biocomposites. Partially biodegradable biocomposites have been evaluated to have comparable mechanical properties to the conventional polymer composites. Furthermore, the product based on partially biodegradable biocomposites can be recycled if the product fails during or after the service. But, these types of biocomposites remain non-biodegradable in nature. To fill this gap, completely biodegradable biocomposites were developed in 1989 by DLR Institute of Structural Mechanics [5]. The most common fibers that are used as reinforcement in biocomposites are sisal, hemp, jute kenaf, bamboo, banana, bagasse, etc. But, there are certain issues and disadvantages associated with natural fibers when they are used as reinforcement in biocomposites. For instance, the use of biocomposites is limited to interior and non-structural applications only, because the degradation rate of biocomposites is high as compared to the synthetic fiber composites. Furthermore, the performance characteristic of biocomposites has been found to deteriorate when exposed to a humid environment, because natural fibers have a very high tendency to absorb moisture. The presence of hydroxyl and other polar groups in natural fiber is the main reason for the poor moisture resistance of biocomposites. The hydrophilic nature of the natural fibers results in poor bonding with hydrophobic polymer, which ultimately affects the load carrying and load transfer capacity of the developed biocomposites [6-8]. The other issues related to biocomposites are fiber-matrix interfacial bonding, poor moisture and fire resistance, a tendency to form aggregates during primary processing, low processing temperature, short durability, variations in quality in the final product, etc. [9,10]. These are the issues that need to be elucidated in order to open new application areas for these attractive materials. Many efforts have been made towards thedevelopment of new types of biocomposites to expand their application spectrum. For instance, the use of natural fiber composites were found in Mercedes-Benz E-class, where the door panels of the car were manufactured using biocomposites. The materials used earlier for the door panel were replaced by flax or sisal fiber-reinforced epoxy composites, which resulted in 20% weight reduction of the door panel. Further, the use of these materials improves passengers' protection in case of any accident [11]. For the manufacturing of other parts of automobiles, such as headliners, dashboards, seat backs, and interior parts, natural fibers or natural fiber-reinforced biocomposites are often used [12,13]. There are numerous such examples available where biocomposites replaced conventional composite materials in order to improve the overall performance of the components/parts.

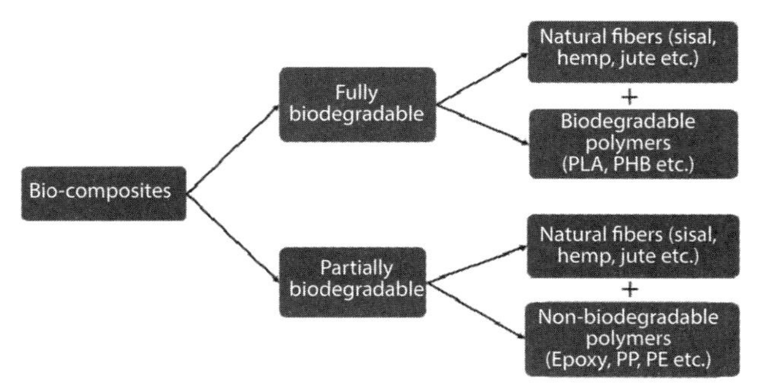

Figure 18.1 Characteristics and components of biocomposites.

The biocomposites used in various applications may be exposed to a variety of environmental conditions such as water, soil, sunlight, oil, etc. The mechanical performance of biocomposites changes with respect to the environmental condition to which they are exposed during the service life. For instance, composite materials are often used for the manufacturing of liquid storage tanks and pipes. If the tank and pipes are used for storing and transporting harmful chemicals, these may attack the constituents of the composites. Another engineering application of composite materials is found in aerospace and marine equipment, where composites are subjected to freezing conditions [14]. From prior discussion, it is clear that biocomposites are exposed to different environmental conditions, where their physical structure and mechanical properties change with respect to the exposure time and corresponding environmental condition. The investigation regarding the mechanical characterization of biocomposites when subjected to different environmental conditions may further expand their usage in various sectors. This chapter has been designed to address the mechanical performance of various developed biocomposites when subjected to different environmental conditions (water, soil, diesel, oil, UV irradiation, temperature, etc.).

18.2 Classification and Structure of Natural Fibers

As fibers are the major load-bearing member/constituent of biocomposites, it becomes imperative to understand the basic structure and characteristics of the fibers. As the constituents of the fibers vary a lot, it is necessary to study the structure and characteristics of general types of natural fibers. Natural fibers can be broadly classified according to the origin from where they are extracted. Depending upon their origin, there are three types of natural fibers which are: plant fibers, animal fibers and mineral fibers. Plant fibers can be further subdivided into leaf, bast, seed and stalk fibers. A broad classification of natural fibers with examples is presented in Figure 18.2. Among all the natural fibers, plant fibers are widely used as reinforcement for the development of biocomposites. The use of mineral fibers is very limited because of their carcinogenic characteristics.

Figure 18.3 shows a complete structure of a plant fiber. The physical structure of a plant fiber typically consists of a primary cell wall and three secondary cell walls. The middle secondary cell wall (secondary wall-II, as shown in Figure 18.2) is thick in

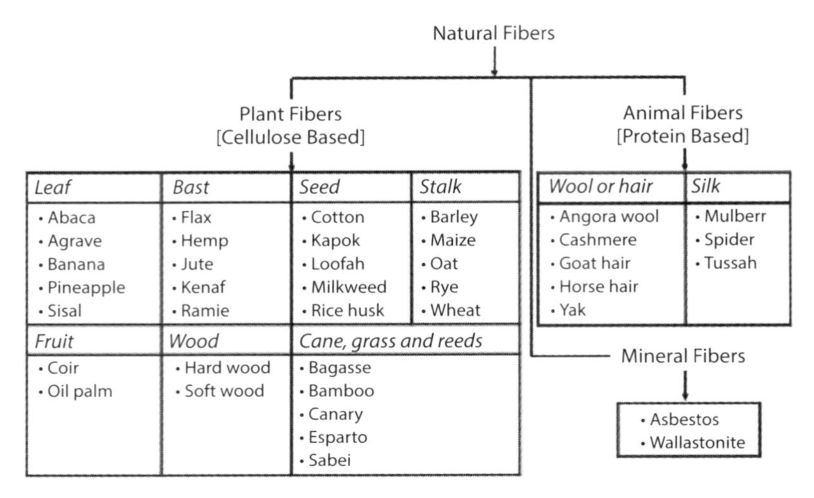

Figure 18.2 Classification of natural fibers [15].

nature as compared to the other two secondary cell walls which determine the overall mechanical properties of a single fiber. The primary and secondary cell walls consist of a number of crystalline cellulose microfibrils. The microfibrils are basically composed of cellulose molecules which act as fibers, whereas lignin and hemicellulose act as matrix [9,16]. Basically, lignin holds the fibrils together and forms a fiber. The other constituents which are found in fibers include pectin, oil, and wax [16,17]. The lumen present in natural fibers makes it a hollow structure, unlike synthetic fibers [18]. The main constituent of plant fibers is cellulose, whereas protein is the main constituent of animal fibers. The three major constituents of plant fibers are cellulose, hemicellulose, and lignin. Cellulose consists of three hydroxyl groups in each repeating unit and is found in the form of slender rod-like crystalline microfibrils. The microfibrils are found to be mostly aligned along the length of the fiber. The major characteristics of cellulose, which is one of the main components of the fiber, is its resistance to hydrolysis, strong alkali and oxidizing agents. But, cellulose may degrade when exposed to chemicals. Hemicellulose is composed of lower molecular weight polysaccharides that act as a matrix to hold the microfibrils together and ultimately form the fiber cells. Hemicelluloses are by nature hydrophilic and can be easily hydrolyzed if exposed to dilute acids and bases. Lignin is a complex hydrocarbon polymer which is insoluble in most solvents and hydrophobic in nature. Lignin provides rigidity to the plant fiber and also helps in the transportation of water. Pectin is basically a heteropolysaccharide that provides flexibility to the fiber [18,19]. The other components like wax and oil present on the fiber surface provide protection to the fiber [17]. Apart from the components discussed above, the other factors that determine the overall properties of the fibers are microfibrillar angle, cell dimensions, defects and structure of the fiber [9,16,17]. Microfibrillar angle is the angle between fiber axis and microfibrils. The diameter of a typical microfibril may vary from 10-30 nm. Higher stiffness and strength of the fiber is obtained when the microfibrillar angle is found to be small, but, higher microfibrillar angles lead to higher ductility in the fiber. The ductility of the fiber also depends upon the orientation of the microfibrils. If the microfibrils are spirally oriented to the fiber

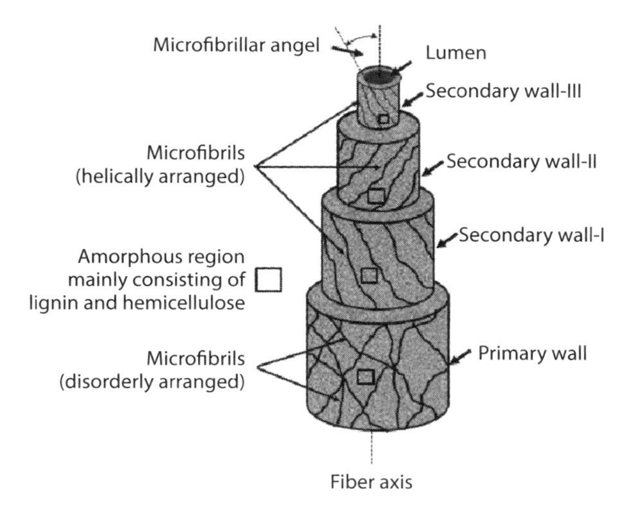

Figure 18.3 Structure of natural fiber.

axis, the ductility of the plant fibers is found to be more. But, if the microfibrils have a parallel orientation to the fiber axis, the fibers characteristics will be rigid and inflexible with high tensile strength. The tensile strength and the Young's modulus of the fibers were found to increase with an increase in the cellulose content and cell length [20]. Generally, plant fibers contains 60-80% cellulose, 5-20% lignin and up to 20% moisture. The percentage composition of each of these constituents varies for different fibers. The chemical compositions and moisture content of some plant fibers are shown in Figure 18.4.

18.3 Moisture Absorption Behavior of Biocomposites

The major disadvantage of using fiber-reinforced composites is that to a certain degree they have a tendency to absorb moisture when exposed to humid atmosphere and when immersed in water. But, if natural fibers are used as reinforcing constituent in composites, then the degree of moisture absorption becomes high. This characteristic of natural fibers leads to poor interfacial bonding between the fiber and matrix, which results in low stress transfer capabilities. Further, moisture absorption leads to changes in the mechanical properties and dimensions of developed biocomposites [23]. The moisture absorption characteristics of polymer matrix composites can be explained by three different mechanisms. The first mechanism is the diffusion of water molecules inside the micro-gaps between polymer chains. The second mechanism involves capillary transport into the gaps and flaws at the interfaces between fiber and matrix, which mainly occurs due to the poor wettability and impregnation. The third mechanism involves propagation of microcracks in the matrix that are initiated due to the swelling of fibers. The third mechanism is more imperative in the case of natural fiber composites [24,25]. Depending upon the mechanisms, the moisture absorption behavior of the fiber-reinforced polymer composites can be categorized into five different types which include: (a) linear Fickian (moisture weight gains gradually and

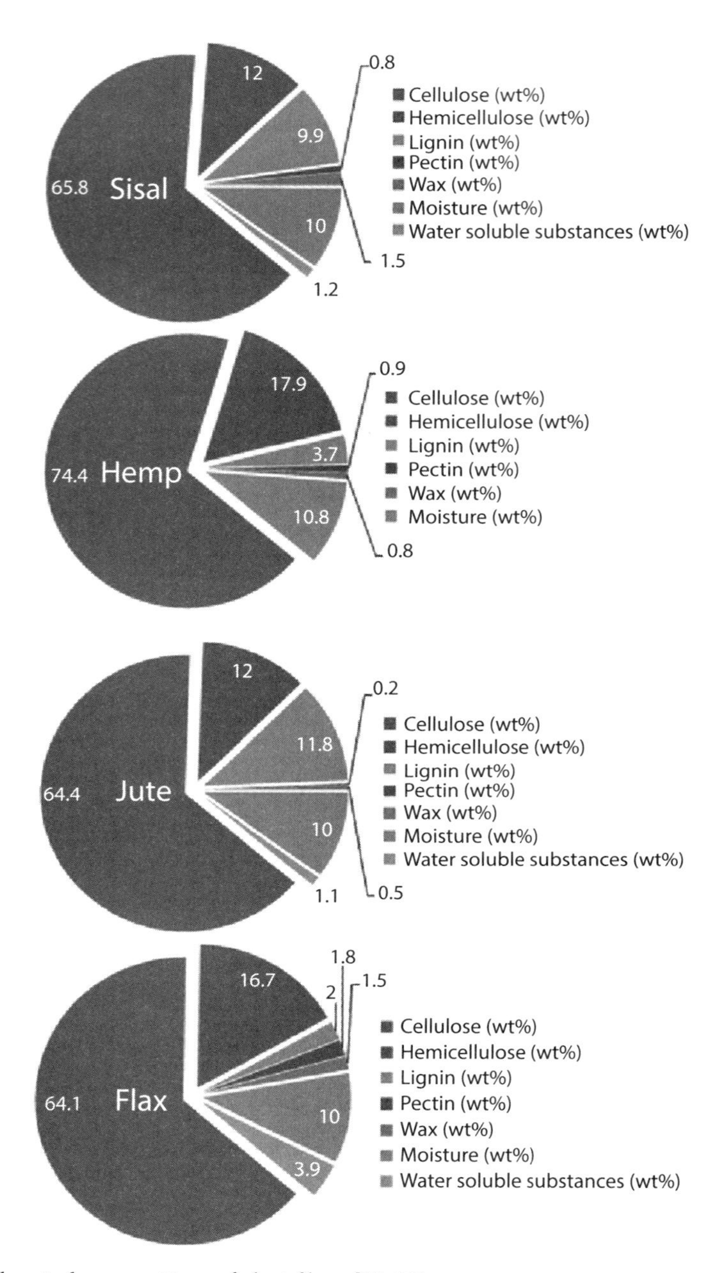

Figure 18.4 Chemical compositions of plant fibers [21, 22].

then reached an equilibrium state after initial take off); (b) pseudo-Fickian (moisture weight gains never reached an equilibrium state after initial take off); (c) two-stage diffusion (moisture weight gains jump abruptly after initial take off); (d) rapid moisture weight gain due to fiber-matrix debonding and matrix debonding; and (e) decreasing trend of moisture weight gains after initial take off that mainly occurs due to the leaching out of constituents from the bulk composites resulting from physical or chemical breakdown [26]. Moisture diffusion in the polymer composites depends upon several factors such as volume fraction of fibers, voids content, viscosity of matrix, humidity and temperature [27].

18.4 Mechanical Characterization of Biocomposites in a Humid Environment

The investigation on water absorption behavior and its influence on the mechanical properties of natural fiber-reinforced composites is a critical concern for further expanding their outdoor application spectrum. The effect of river water on tensile strength of nettle fiber-reinforced PP composites has been investigated at different exposure times [28]. The results showed that the tensile strength of the developed biocomposites is decreased by approximately 20% when exposed to river water for 64 hours as compared to the tensile strength of the unexposed composites. The reason for this can be explained by the fact that as the composite specimens are immersed in river water, the specimens absorb some water. Due to the capillary action, the water molecules enter into the interface of the fiber and matrix, which adversely affects the interfacial bonding properties. The deterioration of the interfacial bonding properties mainly occurs due to the hydrolysis reaction (chemical breakdown of the components due to the reaction with water) of unsaturated polymer groups. It has also been observed that initially the tensile strength of the composite specimens decreases at a slower rate. At the initial stage, the interfacial bonding strength of the developed biocomposites becomes poor, which may be one of the plausible reasons for low tensile strength. As the exposure time reached a higher level, the percolation of water molecules reached the saturation point. At this condition, the voids or cracks present in the composite specimen are completely filled with water, which acts as a plasticizer and may result in improved tensile strength. The effect of water absorption on the tensile and flexural properties of hemp fiber-reinforced unsaturated polyester composites (HFRUPC) has been investigated for fiber volume fraction of 0 to 0.26 [22]. The composite specimens were immersed in deionized water at room temperature (25°C) and boiling temperature (100°C) for different time durations. The results reveal that, as the hemp fiber volume fraction increased, the moisture uptake increases, because with an increase in fiber volume fraction, the cellulose content increases. The other part of the study discloses that the tensile and flexural properties of HFRUPC were found to decrease with an increase in the percentage of moisture uptake. The effect of water absorption on the mechanical properties of recycled cellulose fiber (paper)-reinforced epoxy composites showed that flexural strength, modulus, and fracture toughness decreased due to absorption of moisture. But, it is interesting to note that the impact strength of the developed biocomposites increases when specimens are exposed in wet conditions for two weeks as compared to dry conditions [29]. The impact strength of jute fiber-reinforced PP composites was also found to be enhanced after being exposed to water for two weeks [30]. The un-notched impact properties of recycled cellulose fiber (RCF)-reinforced epoxy composites were significantly improved when exposed to seawater, whereas a detrimental effect on the flexural properties has been observed in the case of long-term exposure to seawater. This dramatic improvement in impact properties may be due to the plasticization effect of epoxy matrix by seawater [31]. The water absorption behavior of pultruded jute fiber-reinforced unsaturated polyester composites has been investigated in order to analyze the effect of different environmental conditions on the mechanical properties of the developed composites. The water absorption characteristic of the jute fiber-reinforced polyester was found to follow pseudo-Fickian behavior. The exposure of natural fiber composites in aqueous environments results in a significant drop in strength and modulus, whereas a significant improvement in the maximum strain was observed, which may be attributed to the breakdown of cellulose structure after immersion, which results in an increase in ductility of natural fibers [32]. An investigation regarding the effect of a resin system on the water absorption characteristics of biocomposites has also been conducted, which showed that kenaf fiber-reinforced vinyl ester laminates absorbed the least amount of water as compared to the kenaf fiber-reinforced polyester and kenaf fiber-reinforced epoxy composites [33]. Water absorption behavior of chemically treated (NaOH) areca fiber-reinforced urea-formaldehyde resin has been investigated [34]. The composite specimens were immersed in seawater, river water, pond water and borewell water at room temperature to investigate the water absorption behavior of the developed composites. From the results and discussion, it has been found that the percentage of water absorption is maximum when the specimens are soaked in the bath of pond water. The minimum percentage of water absorption was found when composite specimens were submerged in borewell water. This means the water absorption behavior of the biocomposites also depends on the type of water. The investigation also reveals that areca fiber-reinforced composites absorb less water as compared to the conventional wood-based particleboard. An adequate number of research initiatives have been performed to highlight the moisture absorption behavior and its effects on the mechanical properties of biocomposites. From the discussion it is clear that there are several factors that control the mechanical properties of biocomposites when they are exposed in a humid environment. Exposure time, types of fibers and resin, fiber concentration, fiber-matrix bonding, voids content, and types of water (seawater, river water, ground water, etc.) are significant factors that determine the overall mechanical properties of biocomposites when they are conditioned in a humid environment.

18.5 Oil Absorption Behavior and Its Effects on Mechanical Properties of Biocomposites

The products made out of polymeric materials may come in contact with different types of liquids during their service. The effect of moisture absorption on the mechanical characteristic of biocomposites has already been discussed. In this section, the effects of other types of liquids like naphthenic oil, diesel, petrol, etc., will be discussed. It is well understood that when composites are used for oil seals, marine articles, etc., the performance of the materials is affected due to the direct attack by the liquid substances. Besides, there is the possibility of accidental splashing of oils and greases that frequently happens in automobiles. The leaking of such substances may adversely affect the various parts of automobiles manufactured from composite materials. Therefore, the study of the effect of various liquids on the performance characteristics of polymer composites is an important consideration. The investigation on mechanical properties of fully biodegradable "green" composites was performed in the presence of three different liquids: water, naphthenic oil, and diesel oil [35]. To develop the fully biodegradable green composites, Ecoflex polymer and woven ramie fibers were used. The study revealed that there is more of an interaction of diesel oil with neat Ecoflex compared to the Ecoflex-ramie composites due to the present of aromatic hydrocarbon in diesel. The hydrocarbons present in diesel easily penetrate the neat Ecoflex. But, in the case of Ecoflex-ramie composites, the hydrocarbon molecules diffuse through the fiber and matrix interface and voids if any are present in the composites. The hindrances exerted by the ramie fibers restrict the movement of diesel and result in decreased liquid diffusion. But, in the case of the diffusion of napthenic oil, there is more oil sorption for Ecoflex-ramie composites compared to the neat Ecoflex, which was found to be just the opposite of the diesel sorption behavior. The study on the effects of fiber (isora) loading on the solvent swelling characteristics of the isora fiber-reinforced natural rubber composites showed that percentage of swelling index and swelling coefficient of the composites decreases with an increase in fiber loading [36]. With an increase in fiber loading, the hindrance exerted by the isora fibers also increases and results in low swelling of fibers. The good interaction between the fiber and rubber may be another reason for obtaining lower swelling of composites. The investigation also disclosed that the maximum uptake of solvent took place in the presence of petrol, followed by diesel and then lubricating oil. Further, it has been found that composites consisting of alkali-treated fibers absorbed less solvent as compared to the composites consisting of untreated isora fibers, which indicates that alkali treatment improves the adhesion between fiber and matrix. The effect of diesel oil on the tensile strength of nettle-PP composites showed that the tensile strength of the respective composite specimen is decreased by 13% as compared to the unexposed specimen after exposing for 64 hours [28]. The tensile strength drops down gradually with exposure time, because diesel oil acts as an organic solvent and enters the voids and pores of biocomposite and then the chemical reaction takes places, which eventually affects the interface of fiber and matrix. As the exposure time increases, the drop in tensile properties of the biocomposites was found to reach to an equilibrium state. With further exposure of diesel oil, the interface of the fiber and matrix achieved stability and the tensile strength became almost constant.

18.6 UV-Irradiation and Its Effects on Mechanical Properties of Biocomposites

Biocomposites are often being used for outdoor applications where they are exposed to sunlight or UV light. Degradation or any change in mechanical properties due to exposure to UV light usually reduces the service life. The effect of UV irradiation on mechanical properties of biocomposites is a significant consideration in regards to their outdoor application spectrum. Basically, when biocomposites are exposed to UV light, both weight and tensile strength of the composites decrease. The tensile strength of nettle-PP composites under UV irradiation was found to decrease rapidly (tensile strength decreased by 11% as compared to the tensile strength of the unexposed sample) after initial exposure for 64 hours [28]. But, after 64 hours, the tensile strength of the composite specimens was found to decrease at a very slow rate. During initial exposure of composite specimen to UV irradiation, the nature of the polymer (PP) changes from ductile to brittle, which results in unfavorable changes in fiber matrix interface, and in turn results in low tensile strength. The influence of UV-C irradiation on the mechanical properties of thermoplastic starch (TPC) and polycarprolactone (PCL) biocomposites with bleached sisal fibers has been investigated [37]. Structural and morphological changes were observed in UV-C irradiated biocomposites when exposed for 142 h at constant temperature and relative humidity of 25°C and 65%, respectively. The tensile properties of the biocomposites decreased due to chain scission. Composites of 40% HDPE and 60% natural flour have been developed in order to measure the changes in mechanical properties due to exposure of composites to UV light [38]. The composite samples were exposed under different combinations of temperature (T) and relative humidity (RH) (23°C-34%, 40°C-34%, 23°C-93%, and 40°C-90%) for 200 hours. The experimental findings reveal that at lower levels of RH, with an increase in T, the modulus of elasticity decreases. On the other hand, at higher levels of RH, with an increase in T the modulus of elasticity increases. The embrittlement of polymer at higher levels of T and RH is the main reason of lower value of modulus of elasticity. Whereas, the modulus of rupture of the composite specimens was found to decrease with an increase in T and RH for both exposed and unexposed specimens. But, the effect of temperature on modulus of rupture is less as compared to the moisture. Further, it has been concluded that the modulus of rupture was not significantly affected by the UV irradiation. The weathering (rain, sunlight, wind, etc.) behavior of neat PP, PP with basic stabilization, and talc-filled PP (10%) was studied for an exposure period of six months [39]. The composite samples were prepared by an injection molding machine and were placed on the roof of the building and collected every two months to study the effect of weathering. The retention of tensile strength of the talc-filled composite specimens was found to be high as compared to the other two types of materials. This means that talc as reinforcing filler may be added to PP when used for engineering applications involving exposure to different types of climate conditions. The hemp fiber-reinforced PP composites exposed to UV radiation show that the impact strength, tensile strength and elongation at break decrease with weathering time. The tensile strength of the composites was found to decrease because of the degradation of polymers that can be attributed to the photo-oxidation of PP and lignin from hemp shives. The elongation at break drops by 20% of the initial value after an exposure of 600 hours to UV light because of extensive crosslinking or chain scission of PP in the composites [40].

18.7. Mechanical Behavior of Biocomposites Subjected to Thermal Loading

Biocomposites often are being used in cold regions where their mechanical property varies with the changing temperature. For instance, the tensile strength of nettle-PP biocomposites was evaluated in freezing conditions, where tensile strength decreased with respect to exposure time [28]. The tests were conducted at a low temperature of −8°C in order to understand their performance at low temperatures. The tensile strength of the biocomposites decreased by 6% after 64 hours and then gradually reached a steady-state condition. After exposure of 512 hours in freezing conditions, the tensile strength of the developed biocomposites decreased by 12%. It was also observed that the percentage reduction in tensile strength is less for nettle-PP biocomposites when exposed to freezing environment as compared to the biocomposites exposed to river water and diesel. This may be due to the fact that in freezing conditions the moisture uptake by the composite specimen is less, which results in less deterioration of interfacial characteristics of fiber and matrix and hence less drop in tensile strength. This behavior of developed biocomposites reveals that these materials are also suitable for low temperature applications. Thermal aging behavior of woven jute fiber-reinforced epoxy composites has been investigated [41]. The composite samples were kept at a temperature of 75°C for 10 hours and at a temperature of −75°C for 6 hours. The tensile strength of the developed composites increased from 189.48 MPa to 197.51 MPa when exposed to high temperature (75°C) and decreased to 168.11 MPa when subjected to low temperature (−75°C). Matrix cracking, delamination growth and fiber fracture are the most common damage modes observed in composites due to thermal aging. At lower temperature, fiber-matrix adhesion properties deteriorate, which results in a decrease in tensile strength, but at higher temperature, fiber crosslinking is highly probable, which results in superior adhesion properties and in turn results in an increase in tensile strength. The effects of freezing and thawing on flexural properties of rice husk-filled HDPE composites have been investigated [42]. Two different compatibilizers, terpolymer based on ethylene-(acrylic ester)-(maleic anhydride) and a polymer based on ethylene and glycidylmethacrylate were mixed during the fabrication of composites in order to evaluate the mechanical performance of the developed composites after freeze-thaw cyclic tests. A complete freeze-thaw cycle consists of freezing at −21°C for 24 hours, thawing at room temperature for 2 hours, and immersing in water at room temperature for 22 hours. After twelve cycles of freezing and thawing, the weight gain, dimensional changes and flexural properties of the composites were evaluated. The results reveal that the dimensional stability of the developed biocomposites was enhanced after compatibilizers were incorporated in the composites. This may be due to the improved interfacial bonding between rice husk and polymers by the compatibilizers. After freezing and thawing cycle tests, the flexural strength and modulus of the composites decreased mainly due to the interfacial degradation caused by the moisture absorption. But, the addition of compatibilizers to the composites resulted in reduction in strength and stiffness to a certain extent. Similar findings have been presented where the flexural properties of wood flour-filled PVC composites decreased after freeze-thaw cyclic test. The authors mentioned that interfacial degradation due to moisture absorption is the main reason of defining such flexural behavior of developed composites. It is also worth mentioning that the loss in modulus is greater as compared to the strength of the composites when composites are subjected to freezing and thawing [43].

18.8 Biodegradation Behavior and Mechanical Characterization of Soil Buried Biocomposites

Biocomposites used for the development of any product need to be disposed of after the end of the product service life. Disposal of any material is one of the gravest environmental concerns. As the volume of disposed biocomposite parts is increasing day-by-day, additives are added to the biocomposite parts during their processing in order to accelerate the rate of biodegradability. In many countries there is no specific rule for disposing the used materials, and the used materials are simply discarded into the environment. In this case, the degradation of disposed materials only depends on nature. The biodegradation behavior of disposed materials is a serious matter of concern for sustainable ecological development. Several attempts had been made in the field of biodegradation characterization of biocomposites. A biodegradation study on flax fiber-reinforced PLA composites has been carried out where amphiphilic additives (benzilic acid, mandelic acid, zein, and DCP) were used as accelerators for biodegradation [44]. The prepared samples were buried in farmland soil for different periods of time and then weight loss was calculated. The results show that the neat PLA films degraded rapidly as compared to the PLA reinforced with flax fiber. After 20 days of burial in soil, it was found that the weight loss of neat PLA was about 4.2%. It was also realized that the samples of neat PLA turned out to be brittle in nature after biodegradation testing. Regarding the composite specimens, it was observed that the biodegradation rate was accelerated by 20-25% in the presence of mandelic acid when exposed for 50 to 60 days. But, in the presence of dicumyl peroxide (DCP), the biodegradation of biocomposites was found to be slow (only 5 to 10% loss in weight of the specimen) even after being exposed for 80 to 90 days. From the experimental results, it was also observed that the maximum weight loss of the composite specimens was experienced in the presence of mandelic acid among all the amphiphilic additives. The main objective of the study was to highlight the effect of different additives on the biodegradation of natural fiber-reinforced PLA composites. The study finally concluded that the use of additives may expedite or slow down the biodegradability of composite materials. It has also been stated that after the biodegradation test, the weight loss of the composite consisting of nonwoven flax fibers is higher as compared to the composite consisting of woven flax fibers. The results presented in the work address the biodegradation behavior of flax fiber-reinforced PLA composites when disposed of in the environment and also emphasize the effect of different additives on the biodegradation of developed biocomposites. The degradability study of starch/PLA/poly(hydroxyester-ether) composites was carried out under soil incubation conditions [45]. The fabrication of composites was followed by injection molding of 0 to 70% cornstarch, 13 to 100% polylactic acid, and 0 to 27% of poly(hydroxyester-ether) (PHEE). The time taken for the soil burial test was one year. The effect of starch and PHEE on the rates of biodegradation of developed composites was evaluated. It has been observed from the experimentation that the weight loss due to biodegradation for pure PLA is approximately 0% per year, for starch/PLA 0 to 15% per year, and for starch/PLA/PHEE 4 to 50% per year. The degradation rate was found to increase with an increase in the starch and PHEE contents. The water absorption behavior and tensile properties (tensile strength and modulus) of soil buried kenaf fiber-reinforced unsaturated polyester composites (KFRUPC) have been investigated [46]. It was observed that at the initial stage, the weight gain (due to the absorption of moisture) of the KFRUPC samples increases linearly and then reaches a saturation state with maximum moisture uptake after 960 hours (40 days). The results also showed that the weight of the composite specimen increases with an increase in the percentage of kenaf fibers. On the other hand, it has been found that there is no significant variation in the tensile properties of the KFRUPC after the composite specimens were buried in soil. This is due to the fact that degradation of the developed composites was found to be predominant when they are exposed for a long period of time. For the composite specimen consisting of 10 to 20% fiber content, degradation started almost after 120 days, but, for the 30% wt samples, the degradation started after 100 days. The weight loss due to the degradation of composite constituents was found to be very small after 120 days. This may be due to the structure of unsaturated polyester, which is not susceptible to microbial attack. Biodegradation characteristics of nettle fiber-reinforced PP composites were conducted to see the effect of exposure time on the mechanical properties of the developed composites [28]. The authors pointed out that there is very little change in tensile strength when composite specimens are buried in soil for 512 hours. The results show that there is only 1 to 3% decrease in tensile strength when composite specimens are buried in soil as compared to the unexposed composite specimen. This may be due to the fact that soil particles do not penetrate the composite specimen and only the moisture present in the soil is absorbed by the specimen. The absorption of moisture by the specimen fills the voids and gaps and results in a small drop of tensile strength. This means the effect of soil on the biodegradation of nettle fiber-reinforced PP composites is very insignificant for the period (0 to 512 hours) evaluated in the investigation. From the discussion it can be said that the biodegradation characterization of biocomposite is an important consideration as far as environmental issues are concerned. At the same time, from the literature it has been realized that the effect of soil on the tensile properties of biocomposites is insignificant when composites are buried in soil for a small duration of time.

18.9 Conclusions

Biocomposites are gradually replacing synthetic fiber composites in many indoor and outdoor applications. But, the higher rate of degradation of biocomposites as compared to the synthetic fiber composites becomes a barrier for complete commercialization of these attractive materials. The response of mechanical properties of biocomposites was found to change under different operating environments. A deep understanding of their mechanical response when subjected to different environments will certainly improve their durability in engineering applications. In the present chapter, the viability of potential utilization of biocomposites under different environments has been discussed. The chapter highlights the effect of water, soil, oil, diesel, sunlight, temperature, etc., on the mechanical properties of several types of developed biocomposites. The degradation mechanism corresponding to the environmental conditions has also been discussed for different types of biocomposites. Furthermore, the chapter also addressed the effect of fiber contents, exposure time, chemical treatment, etc., on the performance characteristics of the biocomposites when subjected to various environments.

Reference

1. I. Singh, K. Debnath, and A. Dvivedi, *Biomass based Bio-composites,* Smithers Rapra, U.K. (2013).
2. I. Singh, P.K. Bajpai, and V. Dhawan, *Biomass based Bio-composites,* Smithers Rapra, U.K. (2013).
3. P.K. Bajpai, I. Singh, and J. Madaan, *J. Rein. Plas. Compos.* 31, 1712 (2012).
4. P.K. Bajpai, I. Singh, and J. Madaan, *J. Thermoplastic Compos. Mat.* (2012) DOI: 10.1177/0892705712439571.
5. A.K. Mohanty, M. Misra, and G. Hinrichsen, *Macromol. Mat. Eng.* 276, 1 (2000).
6. Y. Xie, C.A.S. Hill, Z. Xiao, H. Militz, and C. Mai, *Compos. Part A* 41, 806 (2010).
7. Y-F. Shih, *Mat. Sci. Eng.* A 445–446, 289 (2007).
8. H. Chen, M. Miao, and X. Ding, *Compos. Part A* 40(12), 2013 (2008).
9. D.B. Dittenber, and H.V.S. GangaRao, *Compos. Part A* 43, 1419 (2012).
10. J.R. Araújo, W.R. Waldman, and M.A. De Paoli, *Polymer Degrad. Stability.* 93, 1770 (2008).
11. S. Kalia, B.S. Kaith, and I. Kaur, *Cellulose Fibers*: Bio- and Nano-polymer Composites Green Chemistry and Technology, Springer-Verlag, Berlin Heidelberg, UK (2011).
12. J. Holbery, and D. Houston, *JOM J Minerals Metals Mat. Soc.* 58, 80 (2006).
13. D. Puglia, J. Biagiotti, and J.M. Kenny. *J. Nat. Fibers* 1, 23 (2005).
14. A. Agarwal, S. Garg, P.K. Rakesh, I. Singh, and B.K. Mishra, *Indian J. Eng. Mat. Sci.* 17, 471 (2010).
15. K. Debnath, I. Singh, A. Dvivedi, and P. Kumar, Recent Advances in Composite Materials for Wind Turbine Blades, World Academic Publishing-Advances in Materials Science and Applications (2013).
16. M.J. John, and S. Thomas, *Carbohydr. Polymers* 71(3), 343 (2008).
17. K.J. Wong, B.F. Yousif, and K.O. Low, *Proc. Inst. Mechan. Eng. Part L* 224(3), 139 (2010).
18. K. Liu, H. Takagi, R. Osugi, and Z. Yang, *Compos. Sci. Technol.* 72(5), 633 (2012).
19. J. Summerscales, N.P.J. Dissanayake, A.S. Virk, and W. Hall, *Compos. Part A* 41, 1329 (2010).
20. P. Methacanon, U. Weerawatsophon, N. Sumransin, C. Prahsarn, and D.T. Bergado, *Carbohydr. Polym.* 82, 1090 (2010).
21. L.B. Manfredi, E.S. Rodrı´guez, M. Wladyka-Przybylak, and A. Va´zquez, *Polym. Degrad. Stability* 91, 255 (2006).
22. H.N. Dhakal, Z.Y. Zhang, and M.O.W Richardson, *Compos. Sci. Technol.* 67(7), 1674 (2007).
23. G.C. Yang, H.M. Zeng, J.J. Li, N.B. Jian, and W.B. Zhang, *Acta Sci. Nat. Univ. Sunyatseni.* 35, 53 (1999).
24. Q. Lin, X. Zhou, and G. Dai, *J. Appl. Polym. Sci.* 85(14), 2824 (2002).
25. A. Stamboulis, C.A. Baillie, S.K. Garkhail, H.G.H.V. Melick, and T. Peijs, *Appl. Compos. Mat.* 7(5–6), 273 (2000).
26. H.B. Daly, H.B. Brahim, N. Hfaied, M. Harchay, and R. Boukhili, *Polym. Compos.* 28(3), 355 (2007).
27. J.L. Thomason, *Composites* 26, 477 (1995).
28. P.K. Bajpai, D. Meena, S. Vatsa, and I. Singh, *J. Nat. Fibers* 10(3), 244 (2013).
29. H. Alamri, and I.M. Low, *Polym. Testing* 31, 620, (2012).
30. C. Karmaker, *J. Mat. Sci. Lett.* 16, 462 (1997).
31. I.M. Low, J. Somers, H.S. Kho, I.J. Davies, and B.A. Latella, *Comp. Interfaces* 16, 659 (2009).

32. H.M. Akil, L.W. Cheng, Z.A.M. Ishak, A.A. Bakar, and M.A.A. Rahman, *Compos. Sci. Technol.* 69, 1942 (2009).
33. S. Rassmann, R. Paskaramoorthy, and R.G. Reid, *Mat. Des.* 32, 1399 (2011).
34. C.V. Srinivasa, and K.N. Bharath, *Int. J. Mat. Biomat. Appl.* 2(2), 12 (2012).
35. K.A. Ajith Kumar, M.S. Sreekala, and S. Arun, *J. Biomat. Nanobiotechnol.* 3, 396 (2012).
36. L. Mathew, K.U. Joseph, and R. Joseph, *Bull. Mat. Sci.* 29(1), 91 (2006).
37. A. Campos, J.M. Marconcini, S.M. Martins-Franchetti, and L.H.C. Mattoso, *Polymer Degrad. Stability* 97, 1948 (2012).
38. J.L. Lopez, M. Sain, and P. Coope, *J. Appl. Polym. Sci.* 99, 2570 (2006).
39. M. Al-Shabanat, *Int. J. Chem.* 3(1), 129 (2011).
40. M.I. Popa, S. Pernevan, C. Sirghie, I. Spiridon, D. Chambre, D.M. Copolovici, and N. Popa, *J. Chem.* Article ID 343068 (2013).
41. T. Sen, and H.N. Jagannatha Reddy, *Adv. Mat. Sci. Eng.* Article ID 128158 (2013).
42. S. Panthapulakkal, S. Law, and M. Sain, *J. Appl. Polym. Sci.* 100(5), 3619 (2006).
43. J.M. Pilarski, L.M. Matuana, *J. Vinyl Additive Technol.* 11, 1 (2005).
44. R. Kumar, M.K. Yakubu, and R.D. Anandjiwala, *Express Polym. Lett.* 4(7), 423 (2010).
45. R.L. Shogrena, W.M. Doaneb, D. Garlottaa, J.W. Lawtona, and J.L. Willett, *Polym. Degrad. Stability* 79, 405 (2003).
46. A.A.A. Rashdi, S.M. Sapuan, M.M.H.M. Ahmad, and A. Khalina, *J. Food Agricult. Environ.* 7(3&4), 908 (2009).

Part IV

APPLICATIONS OF CELLULOSE/ POLYMER COMPOSITES

19

Cellulose Composites for Construction Applications

Catalina Gómez Hoyos* **and Analía Vazquez***

Institute of Polymer Technology and Nanotechnology (ITPN), Engineering Faculty, University of Buenos Aires, National Research Council (CONICET), Buenos Aires, Argentina.

Abstract

This chapter reflects the key tendencies of cellulose fiber composites for construction applications. These materials represent an alternative to traditional materials in the construction industry, its use will generate energy savings and emissions control, in addition to increasing recycling and reuse of building materials. Cellulosic fibers are commonly used to manufacture packaging bags for agro-industry products, ropes, threads and handicrafts. Further research development to extend the use of these fibers in the construction sector is an issue of great worldwide importance because the construction industry uses large quantities of raw materials.

Keywords: Natural fibers, natural fiber composites, micro and nanocellulose fibers, cement matrix, Portland cement

19.1 Polymers Reinforced with Natural Fibers for Construction Applications

Polymers reinforced with cellulose fibers have received much attention in recent years because of their low density, nonabrasive, combustible, nontoxic, low cost and biodegradable properties. Several authors have reviewed recent advances in the use of natural fibers in composites like flax [1], jute [2, 3], straw [4], kenaf [5, 6], coir [7–9], fique [10], among others. Natural fibers have been used to reinforce thermoplastics and thermosets polymers in automotive and aerospace applications [11]. The influence of surface treatments of natural fibers on interfacial characteristics was also studied [12–17], and Joshi *et al.* [18] compared the life-cycle environmental performance of natural fiber composites with glass fiber composites. In this study, natural fiber composites were found to be environmentally superior in most applications.

Fiber-reinforced plastics for building applications have been reported by several authors [10, 19–25], highlighting their advantages such as superior corrosion resistance, excellent thermomechanical properties and high strength-to-weight ratio. However

**Corresponding authors*: cgomez@fi.uba.ar; anvazque@yahoo.com.ar

Vijay Kumar Thakur (Ed.), Lignocellulosic Polymer Composites, (435–452) 2015 © Scrivener Publishing LLC

in these works, traditional petroleum-based resins and glass or carbon fibers were exclusively used. At the Sixth International Conference on Science and Technology of Composite Materials held in Durban, South Africa, in January 2007 [26, 27], it was established that currently, composites with cellulosic fibers and polymeric matrices represent an interesting option to develop new construction materials; it was reported that these materials are currently used in the production of roof coverings [20–22], thermal insulation materials [20, 22, 23], wall coverings and floors [24]. Additionally, continuous cross-sectional profiles are used in the production of extruded door frames [25], panels, interior woodwork [20-23], doors [10, 25], among others. Table 19.1 summarizes different investigations about polymers reinforced with natural fibers for construction applications; natural fiber, matrix, processing technique and applications used to elaborate composites are mentioned.

Current studies have shown that composite materials made from natural fibers and biopolymers have mechanical properties that allow their use as building materials. Christian *et al.* [22], conducted a mechanical testing, on two biobased composites made of hemp/acetate of cellulose and hemp/polyhydroxybutyrate, to evaluate their potential for replacing wood and engineered wood products in the construction industry. The authors realized tensile, shear and flexural tests according to ASTM standards and compared biobased composites responses with the properties of wood commonly used in construction, namely, Douglas fir (coastal), western hemlock, and ponderosa pine, and with engineered wood products like plywood, oriented strand board (OSB) and glue laminated timber (glulam). In flexure and shear tests, the biobased composite materials were comparable to wood used in construction. The flexural modulus of the biobased composites was roughly half that of wood parallel to the grain. In tension, biobased composites had strengths 4–5 times higher than wood loaded perpendicular to the grain and strengths comparable to wood loaded parallel to the grain. With respect to engineered wood products, biobased composites properties, resulted comparable to or better than properties measured in all cases taken into account, with exception of the flexural modulus of glulam. The authors mentioned that an advantage of biobased composites over wood is that biobased composites can easily be molded into structural shapes, including hollow sections.

Dweib *et al.* [21] developed a biobased composite by means of the vacuum-driven infusion process to elaborate structural panels or beams, using plant oil-based resin reinforced with different fiber combinations as follows: i) flax mats; ii) woven E-glass fiber; iii) recycled paper and chicken feather mats; iv) recycled paper and corrugated paper, and v) recycled paper and woven E-glass fiber. These composites were compared to conventional wood currently used in building: Douglas fir, spruce and cedar. The results showed that composite made of recycled paper beam with chicken feathers and composite made of recycled paper beam with corrugated cardboard have flexural rigidities comparable to the cedar. Moreover, all of composite beams were stronger than the weakest wood members, and in most cases, the composite beams had strengths nearly equal to or exceeding that of the strongest wood members. Moreover that composite beams made from recycled paper had strength and stiffness that make them suitable for use in structural applications where wood members would normally be used.

Table 19.1 Different published results of polymers reinforced with natural fibers for construction applications.

Fiber	Matrix	Processing Technique	Application	References
Hemp	Cellulose acetate and Polyhydroxybutyrate	Lamination	Replacing wood and engineered wood products in the construction industry	[22]
Flax, recycled paper chicken feathers and E-glass fiber	Soybean oil-based resin	Vacuum assisted resin transfer molding	Structural panels and unit beams	[21]
Hemp and flax natural fiber	Ortho unsaturated polyester	Casting pressure	Housing panel	[28]
Fique	Epoxy DGEBA	Pultrusion	Replacing wood and engineered wood products in the construction industry	[10]
Jute	Phenolic	Pultrusion	Door frames	[25]

Burgueño *et al.* [28] developed a cellular biocomposite from unsaturated polyester reinforced with natural and glass fibers by means of casting pressure. The authors compared the performance of the cellular biocomposite panel (biopanel), against conventional systems used for building and residential construction, named precast pre-stressed hollow core concrete slab (PC/PS HC slab), precast pre-stressed solid concrete slab (PC/PS solid slab), and oriented wood strand board insulated structural panel (ISP). Results showed that PC/PS HC supported 31% more pressure than the biopanel. However the biopanel effectively competes with the solid PC/PS slab and supports 72 % more pressure than the ISP board. In addition, the performance of the biopanel was also compared by means of allowable pressure with two commercial panels made of E-glass/vinylester sandwich: Durashield® and Composolite®. The biopanel outperformed the Composolite® by 28%, while the Durashield® outperformed biopanel by 17%. This evaluation shows that cellular biocomposite panels can effectively compete with commercial E-glass panels.

According to Van Wyk [20], all those polymer composites reinforced with natural fibers, have low negative environmental impact during manufacture and installation. Aditionally they offer high social impact in terms of growing, harvesting, processing and installation; they can outperform competing materials in use; and they are generally less toxic to manufacture, and install, than polymer composites reinforced with conventional fibers. In addition, they offer performance benefits in terms of toughness, thermal comfort, and indoor environmental quality. Inorganic fire retardants can be added to inhibit flaming and flameless combustion. Therefore, the future for natural fiber composites as building materials is bright, and several new natural fiber-based building materials are already making their mark in the building industry. According to Drzal *et al.* [29], the growth outlook for natural fiber composites with biobased resins between 2000 and 2005 varied from $30 million to $300 million.

19.1.1 Durability of Polymer Reinforced with Natural Fibers

The investigations mentioned above, are focused principally on the processing and mechanical response of polymers reinforced with natural fibers, without considering that their mechanical properties decrease after exposition to alkaline environmental of cement materials or weather. Natural fibers reduce their mechanical properties after exposition to alkaline environment of the cement matrix, nevertheless the use of polymeric matrix as a binder around the natural fibers provides protection for them. However, if the interface of composites is not good, and/or matrix is not alkaline resistant, hydration products like calcium hydroxide will migrate to interface, and polymer composite will deteriorate. Several studies have demonstrated that the mechanical properties of natural fibers decrease after exposure to alkaline environment of the cement matrix due to three different mechanisms [30–39]:

1. Chemical decomposition of hemicellulose and lignin, causing embrittlement and decreasing reinforce capacity of natural fibers; because break the link between elementary fibers [31, 40, 41]. Chemical decomposition of these compounds, is caused because they are solubilized in the calcium hydroxide and in other alkalis produced during hydration reaction.
2. Mineralization of natural fibers, caused by migration of hydration products into the lumen, the middle lamella and cell wall [31, 40, 41].
3. Fiber swelling and embrittlement by chemical and biological attack [42].

Several modifications have been developed on natural fibers and polymeric matrices, in order to increase durability of cellulosic fiber composites [10]. Alkaline treatments of natural fibers appears to be a low-cost alternative compared with other commonly used treatments such as acetylation or silanization. Alkaline treatment results in bleaching of the fiber, thereby removing impurities, waxy materials, lignin and hemicellulose [42–44]. In regard to matrix modification, homogeneous dispersion and exfoliation of montmorillonite comes out as an attractive alternative [10]. It has been reported that addition of montmorillonite reduces the coefficient of moisture diffusion, increases their mechanical properties, and significantly improves the mechanical strength of polymeric matrices [45, 46]. Gómez Hoyos *et al.* [10] elaborated epoxy fique composites for construction applications and compared them with conventional wood used in construction. The composites studied were made with fique fibers treated using Na(OH) solution at 18 w/v%; untreated fique fibers were also studied. The matrices were epoxy and epoxy with 5 wt% of chemically modified Cloisite 30B. Composites were aged in a calcium hydroxide solution, to simulate alkali environment of cement matrices. The authors evaluated the flexural properties loss which occurred over aging. The mechanical properties of composites were compared to conventional wood used in construction. The flexural properties of composite made with epoxy and alkali treated fiber, besides the composite made whit epoxy, cloisite 30B and alkali treated fiber; resulted comparable to or higher than properties reported for wood with the exception of the flexural strength of oriented strandboard. The flexural modulus of both epoxy fique composites was better than modulus of oriented strandboard and comparable to them after composites aging. This comparison suggests that epoxy fique composites with modified matrix and/or modified fiber, increased its durability and should be able to substitute or replace conventional wood used in construction before and after aging.

Singh *et al.* [25], processed pultruded jute/phenolic composites to elaborate door frames. Composites were aged in various environments: humidity, hydrothermal and weathering conditions. It was observed that dimensional change of the profiles was only up to 4% even in an accelerated water aging condition. Accelerated water aging increased the effect of absorbed moisture/water with respect to samples exposed to high humidity and alternate wetting and drying cycles. The properties of the jute profile door frame were satisfactory according to the requirements mentioned in IS: 4021–83 (Indian standard specification for timber door, window and ventilator frames). The performance of the installed door frame has shown no sign of dimensional instability after 3 years in terms of warping and bulging. It was suggested that jute door frames could be used as an alternative to the wooden door frames in buildings. Determination of mechanical properties and durability of natural fiber composite is crucial to the acceptability of the product in terms of the legislation governing building applications. Natural fiber composites may be classified according to their mechanical properties and durability behavior as structural and nonstructural. In order to determine what constitutes "acceptable" for natural fiber composites, their durability and properties should be compared with durability and properties of other composites likely to be used in construction [20].

19.1.2 Classification of Polymer Composites Reinforced with Natural Fibers

As we have seen, polymer reinforced with natural fibers may be classified according to matrix used as biocomposites or biobased composites. Biocomposites use traditional polymers from oil resources, while biobased composites use biopolymers from renewable resources. Also, they could be classified according to fiber used as a biocomposite or hybrid composite. Hybrid composite uses a mix of natural and synthetic fibers. According to Burgueño *et al.* [28], hybrid material arrangements overcome the limitations of a given constituent by enhancing the resulting properties with other materials.

On the other hand, polymer reinforced with natural fibers for construction applications could be classified according to its application as structural or nonstructural composite. A structural natural fiber composite can be defined as one that is needed to carry a load in use. Structural panels and unit beams [21], roof systems [22] and in some cases wood and engineered wood products [10, 22], are examples of structural composites. Performance of structural composites ranges broadly from high-performance to low-performance materials. As mentioned above, polymer reinforced with natural fibers has been designed, manufactured, and tested for suitability in structural beams, panels and wood and engineered wood products, yielding good results when tested.

A nonstructural composite can be defined as one that need not carry a load during service. Nonstructural composites are used for products such as ceiling tiles, furniture, windows and door frames [25]. Despite the environmental appeal of polymer reinforced with natural fibers, their use in the construction industry has been limited to nonstructural applications because of their lower strength and stiffness compared with polymer reinforced with synthetic fibers and other conventional structural materials. However, as can be observed in this work, recent developments have shown that the properties of polymer reinforced with natural fibers are comparable with polymer reinforced with synthetic fibers. Alkali treatment of natural fibers appears to be particularly promising, increasing the natural fiber surface area and roughness, causing surface fibrillation, improving fiber–matrix adhesion [10, 12, 15, 47], and thus increasing the mechanical properites and durability of composites materials.

19.2 Portland Cement Matrix Reinforced with Natural Fibers for Construction Applications

Natural fibers as building materials hold enormous potential and are crucial for achieving sustainability. Due to their low density and their cellular structure, natural fibers have very good acoustic and thermal insulation properties and have many advantageous properties over glass or rockwool fibers, like handling and disposal. A lot of research has been performed all over the world on the use of cellulosic fibers as a reinforcing material for the preparation of various types of composites like cement matrix composites [30, 35, 39–41, 48–57] or polymer-based composites [20–22, 28, 29, 58, 59]. The use of cellulosic fibers to reinforce cement matrix is not a new issue. There are records to show that for about 3500 years, man has used various types of natural fibers to reinforce fragile materials commonly used as building material [40, 60]. For instance, ancient civilizations used straw fibers to reinforce the clay used to elaborate bricks and reduce the cracking caused by the stress generated during the drying process [40]. However, over the years cellulosic fibers were replaced by asbestos mineral fibers, until 1973, when the Environmental Protection Agency (EPA) implemented regulations prohibiting the use of this material because it was detected to cause irreversible damage to the respiratory system [53]. From that moment on, research groups and cement industries started to thoroughly investigate the use of a wide variety of natural and synthetic fibers. Nowadays there are commercially existing cement materials reinforced with cellulosic fibers, polypropylene fibers and glass fibers; even the use of short fibers of polyvinyl alcohol (PVA) have been proposed as an asbestos fiber replacement. [30].

Moreover in recent years, several researchers around the world have invested great efforts in the use natural fibers as a replacement for asbestos fibers, in order to take advantage of their availability, low cost and low energy consumption during production [30, 31, 34, 35, 38–41, 48–57, 61–64]. The use of natural fibers to reinforce fragile matrices as cement mortar and concrete, represents an interesting possibility because it offers significant advantages over the use of other fibers such as lower production costs and lower energy consumption; additionally, natural fibers satisfy sustainability requirements and are a renovable resource [30, 31, 34, 35, 38–41, 48–57, 61–64]. However, as mentioned above, natural fibers decrease its mechanical properties after being embedded in cement matrix due to three different mechanisms. For this reason several modifications have been developed on natural fibers and on cement matrix, in order to increase the durability of cellulosic fiber composites. There are many studies that prove that both types of modifications are effective for increase durability. For instance, Tolêdo Filho *et al.* [30] demonstrated in cement materials reinforced with sisal fibers that decomposition mechanisms could be avoided using a cement matrix free of calcium hydroxide by means of a pozzolanic addition, and immersing natural fibers in slurried silica fume [32].

19.2.1 Modifications of Cement Matrix to Increase Durability

By mixing Portland cement with water, the hydration process begins. The hydration process is the combination of physical and chemical changes during which, the cement previously mixed with water becomes a porous solid. During this process a number of exothermic reactions give rise to cement matrix. In the case of Portland cement, the main reactions are based on interaction between two mineral components present in higher proportion: clinker and water. These reactions can be written in simplified form thus [30, 65, 66]:

$$2(3\ \mathrm{CaO{\cdot}SiO_2}) + 7\ \mathrm{H_2O} \rightarrow \mathrm{3CaO.2SiO_2.4H_2O} + \mathrm{3Ca(OH)_2} + \text{ heat} \quad \textbf{(19.1)}$$

$$2(2\ \mathrm{CaO{\cdot}SiO_2}) + 5\ \mathrm{H_2O} \rightarrow \mathrm{3CaO.2SiO_2.4H_2O} + \mathrm{Ca(OH)_2} + \text{ heat} \quad \textbf{(19.2)}$$

Taking into account their stoichiometry, both reactions are similar and produce the same products in different proportions [30, 65, 66]. Hydrated calcium silicate ($3CaO.2SiO_2.4H_2O$) is the principal hydration product and is the component that provides resistance to cement matrix and is formed on the surface of cement particles. The calcium hydroxide ($Ca(OH)_2$) is responsible for the pH 12 of the cement matrix; and is critical in protecting the steel reinforcement against corrosion; however it degradates the hemicellulose and lignin of natural fibers decreasing the properties of the composite material.

19.2.1.1 Pozzolanic Additions

Pozzolanic additions are siliceous minerals added during mixing of cement matrix to react with hydration products, especially with $Ca(OH)_2$. This reaction is exothermic and occurs between pozzolan and $Ca(OH)_2$ to form hydrated calcium silicate [30, 65, 66].

$$\text{Pozzolan} + \mathrm{Ca(OH)_2} + \mathrm{H_2O} \rightarrow \mathrm{CaO.SiO_2} + \text{ heat} \quad \textbf{(19.3)}$$

Pozzolanic reaction courses slower than reactions between clinker components and water; however its speed can be increased by varying the composition and surface area of the pozzolanic additive. Pozzolans more commonly used to increase durability of cement composites with cellulosic fibers are silica fume, metakaolin fly ash and slag.

In the literature, many studies show that some pozzolans like silica fume are highly effective, increasing durability of the cement matrix reinforced with cellulosic fibers [50, 62, 67, 68], because during pozzolanic reaction calcium hydroxide from hydration reaction is consumed. Gram [67] reduced alkalinity of cement matrix by adding silica fume, slag from cast iron, fly ash and natural pozzolans like rice husk ash, slug and silica gel. Results showed that replace 45 wt% of Portland cement by silica fume, entirely removed the loss of hardness of composite over time, caused by embrittlement of natural fibers. Natural pozzolans also increased the durability of composite materials; however fly ash and slag had no effect on the durability of composite material. In his thesis, Tolêdo Filho [68] studied durability of cement mortars reinforced with sisal fibers by monitoring the loss of flexural properties and observations of photomicrographs from scanning electron microscopy (SEM). Durability analysis was developed by exposing composite materials to the following environments: (i) cycles of wetting and drying, (ii) storage in water and weathering for long periods of time. Results from this investigation showed that cement mortar reinforced with short sisal fibers significantly decreased the mechanical properties after exposure to both environments for six months. This decrease in properties was attributed to the dissolution of non-cellulosic compound and mineralization of fibers. The author also studied the durability of cement mortar with pozzolanic additions like silica fume and slag, reinforced with sisal fibers. Results showed that silica fume increased durability significantly, while slag had no effect.

Gutierrez *et al.* [50] investigated the effect of replacing part of Portland cement by pozzolans as silica fume, fly ash and metakaolin on the mechanical properties of cement mortar reinforced with synthetic and natural fibers. The mechanical behavior and durability of composite materials was improved by replacing 15 wt% of cement with metakaolin or silica fume; however, because of its low pozzolanic power, fly ash had no effect on the durability of these materials. Khorami and Ganjian [62] increased flexural strength in 20% of cementitious materials reinforced with straw fibers and eucalyptus, by replacing 5 wt% of Portland cement with silica fume.

19.2.1.2 Carbonation of Cement Matrix

The alkaline pH of cement matrix can also be reduced by reaction between carbon dioxide (CO_2) and $Ca(OH)_2$. This reaction is known as carbonation of cementitious matrix and occurs in three stages: (i) Initially CO_2 diffuses in matrix through pores and is dissolved; (ii) subsequently CO_2 reacts with sodium hydroxide (NaOH) of cement matrix, decreasing the pH which favors $Ca(OH)_2$ formation; and (iii) finally a reaction occurs between CO_2 and $Ca(OH)_2$ that forms $Ca(HCO_3)_2$ and $CaCO_3$. Several authors have used this reaction to decrease pH of cement matrix and increase durability of cement composites reinforced with natural fibers [31, 37, 69]. To allow this reaction occurs, composite is cured in an environment rich in CO_2, which favors the formation of $CaCO_3$ and therefore decreases the pH in the matrix [32, 70–73]. However, cement composites cured using this method cannot be reinforced with steel, since as mentioned above, $Ca(OH)_2$ protects steel from oxidation.

Tolêdo Filho *et al.* [32] conducted a comprehensive study to determine specific modifications of cement matrix and natural fibers that were effective increasing durability. Results indicated that carbonation of the specimens for 109 days at conditions of 26.5°C is a promising alternative for increasing the durability of cement composites reinforced with cellulosic fibers.

19.2.2 Modifications of Natural Fibers to Increase Durability of Cement Composites

In the last decade, treatments have been developed for natural fibers that have significantly increased durability of cement composites, like immersing them in slurried silica fume [32], hornification [56] and alkaline treatments [37, 39], among others. Some treatments such as immersing natrual fibers in slurried silica fume were successful, but many others had no effect on durability of cement composites [67, 74]. Generally durability of cement composites is determined from the change of one or more properties, after subjecting composite material to one or more cycles of aging. Table 19.2 summarizes several investigations realized to determine durability of cement composites reinforced with natural fibers. It can be observed that mechanical properties, density, permeability and water absorption are properties commonly used to determine durability. Additionally, Table 19.2 also shows the effect of treatments realized on natural fibers on properties and durability of cement composites reinforced with natural fibers. Below are presented further details on the results of these and some other studies, which have been vital to develop new applications for these composites in the construction industry.

Gram [67] and Partek [74] thoroughly studied the possibility of using sisal fibers to reinforce concrete matrices and to develop tiles and sandwich elements in cellular materials. Durability studies were realized on these materials manufactured by machining processes adapted to the technology of concrete spraying. Durability studies were carried out by accelerated tests in air-conditioned environments and according to the fiber content. Before materials were manufactured, several modifications were applied to sisal fibers like acid mercerization, treatment with NH_3, treatment with calcium stearate, coating with polyvinylalcohol (PVA) and other polymers. However, in all cases, concrete reinforced with sisal fibers was more brittle than concrete without sisal fibers after exposure to different aging environments.

Additionally, to increase durability of cement matrix, Gram [67] also studied treatments of sisal fibers with different block agents like sodium silicate, sodium sulfate, magnesium sulfate, sulfite salts, barium salts and compounds of iron and copper, for the purpose of inhibiting the biological degradation process of natural fibers from microbial attack; but none of these treatments had a positive result on durability of composite material. Moreover, Gram studied the effect of impregnating natural fibers with hydrophobic compounds like stearic acid, forminas (group of proteins), polyvinylalcohol, ceramide, silicon oil, bitumen, rubber latex and asphalt, on durability. Results showed that only stearic acid and forminas increased the durability of composite materials. Finally, Gram [67] concluded that flexural strength of concrete reinforced with

Table 19.2 Effect of natural fiber treatments on properties and durability of Portland cement composites. (↑means that property increased, ↓means that property decreased).

Fiber	Fiber Content	Fibers Treatments	Processing Technique	Durability Test	Results	References
Bagasse from sugar cane	0 - 20 wt%	Soaked in water with a chemical admixture for 2 h.	Casting pressure 2-3 MPa	25 wetting and drying cycles.	Swelling: 0.765% Water absorption: ↑ 16.45% Internal bond strength: ↓14.5%	[41]
Kraft pulp (KF) and Cotton linters (CL)	4 wt%	Hornification*	Casting pressure 0.4 MPa	72 h drying with open air circulation at 60°C and 96 h soaking in water at room temperature.	**KF composite** Flexural strength: ↓ 20% Elastic Modulus: ↓ 8% Fracture energy: ↓ 48% **CL composite** Flexural strength: ↓ 16% Elastic Modulus: ↓ 9% Fracture energy: ↓ 7%	[56]
Fique	4 wt%	Immersion into a boiling lime solution for 5 min.	Hand casting	· 14 years under tropical weathering.· 14 years under protected environment in laboratory.	**Differences between Laboratory and tropical weathering:** Real density: ↑ 2% Water absorption: ↑ 16% Apparent void volume: ↑ 12% Samples aged under tropical weathering presented lower air permeability than tiles aged in laboratory, due to increased tortuosity of the pore structure.	[40]

sisal fibers remained after 4 years of outdoor exposure in Tanzania, where sisal fibers treated with stearic acid and barium nitrate were used, and between 40 to 50 wt% of Portland cement was replaced by silica fume [75].

Guimarães [76] has worked since 1980 on developing concrete reinforced with natural fibers and has deduced that durability is the most important property to consider in the development of these materials. To increase durability, Guimarães impregnated the natural fibers with PVA, solubilized in water and dimethyl formaldehyde. Subsequently, natural fibers were exposed to an alkaline environment for 28, 56, 84, 112 and 140 days, and then tested in tension. Results indicated that these impregnants did not prevent deterioration of natural fibers, since strength of these materials considerably decreased after having been exposed to an alkaline environment.

Le Troedec *et al.* [39] treated hemp fibers with sodium hydroxide solution 1.6 M pH 14 for 48 h at 20°C, in order to increase adhesion properties between natural fibers and cement matrix, increasing durability of the material. After treatment with NaOH, fibers were washed with distilled water with 1 v/v% of acetic acid, until reaching neutral pH. Results revealed that fiber treatment increased the maximal resistance of composite material by 60% with respect to composite material elaborated with natural fibers without treatment. Additionally, it was confirmed through acoustic analysis that natural fiber treatment delayed the start of damage mechanisms of fiber-matrix interface caused by degradation of natural cellulosic fibers.

Tolêdo Filho *et al.* [32] studied the effect of a series of treatments on natural fiber and cement matrix in order to increase durability of cement mortars reinforced with sisal fibers. Based on results from their investigation, it was concluded that incorporation of sisal fibers significantly reduced mechanical properties of composite material. However, the mechanical properties of composite were maintained over time if the sisal fibers were immersed in a suspension of silica fume. This is because the silica fume particles contact the fiber and their size allows them to be located in smaller defects in the fiber, preventing that compounds such as calcium hydroxide, come into contact with the fiber and destroy it by growth of crystals.

19.2.3 Application of Cement Composites Reinforced with Natural Fibers

The application of cement composites reinforced with cellulosic fibers in the building industry depends mainly on their mechanical properties and durability. And as we have seen, there is a growing interest in increasing these properties, especially durability. This interest is due among other things to the growing concern for the conservation of the environment, which has created new building regulations that optimize resources such as Leadership in Energy and Environmental Design (LEED) standards. Construction of buildings consumes 50% of worldwide energy resources; 45% of energy is used to heat, light and ventilate buildings and 5% is used during construction [26, 77]. Therefore, urgent changes are required relating to energy savings, emissions control, use of renewable resources, recycling and reuse of building materials. The development of new building materials from renewable resources and sustainable practices is of prime importance owing to growing environmental concerns. Natural fiber composites offer the opportunity for environmental gains by reducing energy consumption, because of their lighter weight, insulation and sound absorption properties; reduction in volatile organic emissions, and reduction in dependence on petroleum-based and forest product-based materials. Gallegos [78] reported that in Peru, fibers from cane reed are commonly used to elaborate composite material with cement matrices and used in construction of domes, cupolas for churches and drywall. Also, at an industrial level, pressed wood panels and fibrocement panels are manufactured with good thermal and acoustic properties, besides being light and easy to cut and fix.

Hashem [79] reported that 90% of Bangladesh's population lives in rural areas and uses bamboo, jute, sugar and palm cane to construct their houses; with a shelf life of between 2 and 3 years because of deterioration caused by external factors such as insects, microorganisms and fire. Bamboo is used to construct foundation columns in the preparation of walls, doors, windows, ceilings and pipes [80]. Jute and cane are used as fillers in bamboo structures that form walls and ceilings. Hashem also reported that re-covering fibers with copper sulfate, sodium dichromate, acetic acid and water, increased durability of these houses 3 or 4 times compared to houses elaborated with these materials without covering. Susuki and Yamamoto [81] proposed a composite material with natural fibers for fire-resistant constructions used in Japan. According to reports in their investigation, coating natural fibers with fly ash makes them fireproof and increases the durability of cement composites. Tolêdo Filho *et al.* [82] used sisal and coconut fibers to elaborate adobe blocks used to built homes in northeast of Brazil. Natural fibers were re-covered with hydrophobic agents to decrease their water absorption. Results showed that incorporing 4 wt% of fibers significantly decreased the brittleness of adobe blocks.

Tonoli *et al.* [40] elaborated corrugated sheets using a Portland cement composite reinforced with fique fibers, to be used as roofing tiles on buildings south of Colombia. Composite properties were evaluated after 14 years under tropical weathering and were compared with properties of the same tiles stored for 14 years in laboratory at controlled environment, as shown in Table 19.2. Tiles aged under tropical weathering showed higher water absorption and apparent increased void content more than those which remained in laboratory. Additionally, tiles aged under tropical weathering, showed lower air permeability and higher degree of hydration, than tiles aged under controlled environment at laboratory. The results contributed to understand fiber-cement degradation process, showing that water uptake and release under 14 years of tropical weathering led to damages in the fiber–matrix interface that may result in loss of strength of the tiles; and to fiber mineralization that might cause embrittlement of the fiber-cement tiles after tropical weathering exposition.

19.2.4 Micro and Nanofibers Used to Reinforce Cement Matrices

Until now, developments achieved in cement materials have shown great progress for humanity because of their usefulness in infrastructure, housing and transportation [66]. Cement is a combination of minerals especially formulated to react with water, following a series of chemical and physical processes known as hydration. Water is a substance strongly involved in preparation of cement materials, therefore the hydrophilic character and water holding capacity of micro and nanofibers of cellulose [83], is a property of great interest for cement materials. Properties of micro and nanofibers of cellulose, have been widely exploited in cosmetic, medical and food industries, for use as stabilizers for aqueous suspensions, flow controllers and as reinforcement matrix in final product [84, 85]. However applications of micro and nanocellulose fibers in construction industry it is a subject in the stage of laboratory research. Natural fibers commonly used to reinforce cement matrices have lengths between 1.5 and 10 mm [30, 31, 37, 39, 49, 53, 57, 61, 86]; therefore, reinforcements such as micro- and nanocellulose fibers extracted from natural fibers and vegetable matter for use in cement materials is a current topic.

Change in fiber size has a significant influence on properties of the cement composite. Changing conventional size fibers with nanofibers or microfibers significantly increases performance of the concrete; this is because fibers of micro and nanometric size have greater surface area and reactivity than conventional fibers. Additionally, reducing the fibers size increases the surface/volume ratio; i.e., for the same amount of fibers, nanofibers or microfibers have better distribution than conventional sized fibers in all volumes. Nano-binders or nano-engineered cement-based materials with nano-sized cement components like nanocellulose fibers, are the next innovative development. Below are mentioned progress by some of the most important research in this topic. Gómez Hoyos *et al.* [87] studied the effect of cellulose microcrystalline particles on mechanical properties and hydration process of cement composites by means of flexural test and thermogravimetric analysis respectively. Results showed that interactions between cellulose microcrystalline particles, cement particles, hydration products and water, decreased the workability and delayed the hydration reaction; however, these interactions increased temperature during curing process and increased the hydration degree of cement composites. It was concluded that cement-based composites with cellulose microcrystalline particles could be used to elaborate precast pieces. Until now, few investigations on cellulose microfibrils and cellulose nanofibrils added to cement materials were known [88–91], these investigations were focused on studying the mechanical properties of the composite material. Mohamed *et al.* [91] studied the effect of microcellulose addition to a self-compacting concrete mix. They found that adding 21 wt/v% of microcellulose fibers to cement composites increased their compression and flexural resistance after 7 days of curing. Moreover, Peters *et al.* [90] determined that adding 3 wt% of a mixture of micro- and nanocellulose fibers to a reactive powder concrete with high compressive strength, increased their impact strength by 53%. Additionally, Claramunt *et al.* [92] elaborated cement mortars with nanofibrilar cellulose to study its mechanical properties; they concluded that addition of 3 wt% of nanofibrilar cellulose, significantly increased flexural properties after 28 days of curing. Nilsson and Sargenius [88] developed Portland cement mortars with 0.11, 0.22 and 0.33 wt% of recycled cellulose microfibers from the paper industry in order to characterize changes in workability of fresh mixture and changes in mechanical properties, i.e., autogeneous shrinkage by drying and capillary absorption properties of cement mortars in hardened state. They concluded that addition of cellulose microfibers modifies the rheology, decreased the mechanical properties and had no effect on drying shrinkage properties. However, they found that the addition of cellulose microfibers had a positive effect on water absorption, strongly increasing absorption resistance, because cellulose microfibers modify pore structure. As consequence, addition of nanofibers and microfibers of cellulose to cement matrix is a new subject which needs future research.

References

1. B. Wang, S. Panigrahi, L. Tabil, and W. Crerar, "Pre-treatment of flax fibers for use in rotationally molded biocomposites," *J. Reinf. Plast.Compos.* 26(5), 447–463 (2007).
2. J.-M. Park, P.-G. Kim, J.-H. Jang, Z. Wang, B.-S. Hwang, and K.L. DeVries, "Interfacial evaluation and durability of modified/polypropylene (PP) composites using micromechanical test and acousticemission," *Compos. Part B Eng.* 39, (6), 1042–1061 (2008).
3. E. S. Rodriguez, P. M. Stefani, and A. Vazquez, "Effects of fibers'alkali treatment on the resin transfer molding processing and mechanical properties of jute-vinylester composites," *J. Compos. Mater.* 41(14),1729–1741 (2007).
4. S. Panthapulakkal, a Zereshkian, and M. Sain, "Preparation and characterization of wheat straw fibers for reinforcing application in injection molded thermoplastic composites," *Bioresour. Technol.* 97(2), 265–72(2006).
5. S. Ochi, "Mechanical properties of kenaf fibers and kenaf/PLAcomposites," *Mech. Mater.* 40(4–5), 446–452 (2008).
6. A. A. M. Mazuki, H. M. Akil, S. Safiee, Z. A. M. Ishak, and A. A.Bakar, "Degradation of dynamic mechanical properties of pultruded kenaf fiber reinforced composites after immersion in various solutions," *Compos. Part B Eng.* 42(1), 71–76 (2011).
7. S. Harish, D. P. Michael, A. Bensely, D. M. La, and A. Rajadurai, "Mechanical property evaluation of natural fiber coir composite," *Mater.Charact.* 60(1), 44–49 (2009).
8. N. Defoirdt, S. Biswas, L. De Vriese, L. Q. N. Tran, J. VanAcker, Q. Ahsan, L. Gorbatikh, A. Van Vuure, and I. Verpoest, "Assessment of the tensile properties of coir, bamboo, and jute fibre," *Compos. Part A Appl. Sci. Manuf.* 41(5), 588–595 (2010).
9. S. V. Prasad, C. Pavithran, and P. K. Rohatgi, "Alkali treatment of coir fibres for coir-polyester composites," *J. Mater. Sci.* 18(5),1443–1454 (1983).
10. C. Gómez Hoyos, and A. Vázquez, "Flexural properties loss of unidirectional epoxy/fique composites immersed in water and alkaline medium for construction application," *Compos. Part B Eng.* 43(8), 3120–3130 (2012).
11. D.U. Riedel, and D. J. Nickel, "Applications of natural fiber composites for constructive parts in aerospace, automobiles and other areas", *Biopolymers on-Line 10.* (2005)
12. W.Liu, A. K. Mohanty, L.T. Drzal, P. Askel, and M. Misra, "Effects of alkali treatment on the structure, morphology and thermal properties of native grassfibers as reinforcements for polymer matrix composites," *J. Mater. Sci.* 39,1051–1054 (2004).
13. N. Suizu, T. Uno, K. Goda, and J. Ohgi, "Tensile and impact properties of fully green composites reinforced with mercerized ramie fibers," *J. Mater Sci.* 44, 2477–2482 (2009).
14. K. Goda, M. S. Sreekala, A. Gomes, T. Kaji, and J. Ohgi, "Improvement of plant based natural fibers for toughening green composites–Effect of load application during mercerization of ramie fibers," *Compos. Part A Appl. Sci. Manuf.* 37(12), 2213–2220 (2006).
15. M. Z. Rong, M. Q. Zhang, Y. Liu, G. C. Yang, and H. M. Zeng, "The effect of fiber treatment on the mechanical properties of unidirectionalsisal-reinforced epoxy composites", *Compos. Sci. Technol.* 61(10),1437–1447 (2001).
16. S. Mohanty, "Mechanical and rheological characterization of treated jute-HDPE composites with a different morphology", *J. Reinf. Plast.Compos.* 25(13), 1419–1439 (2006).
17. W. G. Trindade, W. Hoareau, J. D. Megiatto, I. A T. Razera, A. Castellan, and E. Frollini, "Thermoset phenolic matrices reinforced with unmodified and surface-grafted furfuryl alcohol sugar cane bagasse and curaua fibers: properties of fibers and composites.," *Biomacromolecules* 6(5),2485–2496 (2005).

18. S. Joshi, L. Drzal, A. Mohanty, and S. Arora, "Are natural fibercomposites environmentally superior to glass fiber reinforced composites?," *Compos.Part A Appl. Sci. Manuf.* 35(3), 371–376 (2004).
19. C. Akis, L. Bank, V. Brown, E. Cosenza, J. Davalos, J. Lesko, A. Machida, A. S. Rizkalla, and T. Triantafillou, "Fiber-reinforced polymer composites for construction state-of-the-art review," *J. Compos. Constr.* 6(2), 73–87, (2002).
20. L. van Wyk, "The aplication of natural fibers composites in construction: A research case study," in *Sixth International Conference on Composite Sci. Technol.* (2007).
21. M. A. Dweib, B. Hu, A. O'Donnell, H. W. Shenton, and R. *P. Wool,*"All natural composite sandwich beams for structural applications," *Compos.Struct.* 63(2), 147–157 (2004).
22. S. J. Christian, and S. L. Billington, "Mechanical response of PHB- and cellulose acetate natural fiber-reinforced composites for construction applications," *Compos. Part B Eng.* 42(7), 1920-1928 (2011).
23. K. G. Satyanarayana, G. G. C. Arizaga, and F. Wypych, "Biodegradable composites based on lignocellulosic fibers—An overview," *Prog. Polym. Sci.* 34(9), 982–1021 (2009).
24. J. R. Burt, and K. A. Szabo, "Methods of making floor tile and wall covering from extruded hot recycled vinil thermoplastic membrane having discrete fibers ramdomly dispersed therein," Patent US5560797 (1995).
25. B. Singh, and M. Gupta, "Performance of pultruded jute fibre reinforced phenolic composites as building materials for door frame," *J.Polym. Environ.* 13(2), 127–137 (2005).
26. P. Joseph, and S. Tretsiakova-McNally, "Sustainable non-metallic building materials," *Sustainability* 2(2), 400–427 (2010).
27. B. Edwards, and P. Hyett, *Rough guide to sustainability,*1st ed., pp. 113, RIBA, London. (2001).
28. R. Burgueño, M. J. Quagliata, G. M. Mehta, A. K. Mohanty, M.Misra, and L. T. Drzal, "Sustainable cellular biocomposites from natural fibers and unsaturated polyester resin for housing panel applications," *J. Polym.Environ.* 13(2), 139–149 (2005).
29. L. T. Drzal, A. K. Mohanty, R. Burgueño, and M. Misra, "Biobasedstructural composite materials for housing and infrastructure applications :Opportunities and Challenges," in *NSF-PATH Housing Research Agenda* 129–140(2004).
30. R. D. Tolêdo Filho, F. de Andrade Silva, E. M. R. Fairbairn, and J. de A. Melo Filho, "Durability of compression molded sisal fiber reinforced mortar laminates," *Constr. Build. Mater.* 23(6), 2409–2420 (2009).
31. R. D. Tolêdo Filho, K. Scrivener, G. L. England, and K. Ghavami, "Durability of alkali-sensitive sisal and coconut fibres in cement mortar composites," *Cem. Concr. Compos.* 22(2), 127–143 (2000).
32. R. D. Tolêdo Filho, G. Khosrow, G. L. England, and K. Scrivener, "Development of vegetable fibre – mortar composites of improved durability," *Cem.Concr. Compos.* 25(2), 185–196 (2003).
33. F. de A. Silva, R. D. Tolêdo Filho, J. de A. Melo Filho, and E. de M. Rego Fairbairn, "Physical and mechanical properties of durable sisal fiber –cement composites," *Constr. Build. Mater.* 24(5), 777–785 (2010).
34. G. Ramakrishna, and T. Sundararajan, "Studies on the durability of natural fibres and the effect of corroded fibres on the strength of mortar," *Cem. Concr. Compos.* 27(5), 575–582 (2005).
35. G. Ramakrishna, and T. Sundararajan, "Impact strength of a few natural fibre reinforced cement mortar slabs: a comparative study," *Cem.Concr. Compos.* 27(5), 547–553 (2005).
36. P. Purnell, and J. Beddows, "Durability and simulated ageing of new matrix glass fibre reinforced concrete," *Cem. Concr. Compos.* 27(9–10), 875–884 (2005).

37. C. Onésippe, N. Passe-coutrin, F. Toro, S. Delvasto, K. Bilba, and M. Arsène, "Composites : Part A Sugar cane bagasse fibres reinforced cement composites : Thermal considerations," *Compos. Part A* 41(4), 549–556(2010).
38. Z. Li, L. Wang, and X. Ai Wang, "Cement composites reinforced with surface modified coir fibers," *J. Compos. Mater.* 41(12), 1445–1457(2007).
39. M. Le Troedec, P. Dalmay, C. Patapy, C. Peyratout, a. Smith, and T. Chotard, "Mechanical properties of hemp-lime reinforced mortars: influence of the chemical treatment of fibers," *J. Compos. Mater.* 45(22), 2347–2357(2011).
40. G. H. D. Tonoli, S. F. Santos, H. Savastano Jr., S. Delvasto, R. Mejia de Gutierrez, and M. del M. Lopez de Murphy, "Effects of natural weathering on microstructure and mineral composition of cementitious roofingtiles reinforced with fique fibre", *Cem. Concr. Compos.* 33(2), 225–232(2011).
41. L. K. Aggarwal, "Bagasse-reinforced cement composites," *Cem.Concr. Compos.* 17(2), 107–112 (1995).
42. M. J. John, B. Francis, K. T. Varughese, and S. Thomas, "Effect of chemical modification on properties of hybrid fiber biocomposites," *Compos. Part A Appl. Sci. Manuf.* 39(2), 352–363 (2008).
43. D. Ray, and B. K. Sarkar, "Characterization of alkali-treated jute fibers for physical and mechanical properties," *J. Appl. Polym. Sci.* 80(7), 1013–1020 (2001).
44. V. A. Alvarez, R. Ruscekaite, and A. Vazquez, "Mechanicalproperties and water absorption behavior of composites made from a biodegradable matrix and alkaline-treated sisal fibers," *J. Compos. Mater.* 37(17), 1575–1588 (2003).
45. A. Yasmin, J. J. Luo, J. L. Abot, and I. M. Daniel, "Mechanical and thermal behavior of clay/epoxy nanocomposites," *Compos. Sci. Technol.* 66(14), 2415–2422 (2006).
46. B. Qi, Q. X. Zhang, M. Bannister, and Y.-W. Mai, "Investigation of the mechanical properties of DGEBA-based epoxy resin with nanoclay additives," *Compos. Struct.* 75(1–4), 514–519 (2006).
47. Y. Liu, and H. Hu, "X-ray diffraction study of bamboo fibers treated with NaOH," *Fibers Polym.* 9, 735–739 (2008).
48. R. S. P. Coutts, and Y. Ni, "Autoclaved bamboo pulp fibre reinforced cement," *Cem. Concr. Compos.* 17(2), 99–106 (1995).
49. H. Savastano, P. G. Warden, and R. S. P. Coutts, "Brazilian waste fibres as reinforcement for cement-based composites," *Cem. Concr.Compos.* 22(5), 379–384 (2000).
50. R. M. de Gutierrez, L. N. Diaz, and S. Delvasto, "Effect of pozzolans on the performance of fiber-reinforced mortars," *Cem. Concr.Compos.* 27(5), 593–598 (2005).
51. H. Savastano Jr., and P. G. Warden, "Special theme issue: Natural fibre reinforced cement composites," *Cem. Concr. Compos.* 27(5),517 (2005).
52. R. S. P. Coutts, "A review of Australian research into natural fibre cement composites," *Cem. Concr. Compos.* 27(5), 518–526. (2005).
53. S. Delvasto, E. F. Toro, F. Perdomo, and R. Mejía de Gutiérrez, "An appropriate vacuum technology for manufacture of corrugated fique fibre reinforced cementitious sheets," *Constr. Build. Mater.* 24(2), 187–192(2010).
54. J. H. Morton, T. Cooke, and S. A. S. Akers, "Performance of slash pine fibers in fiber cement products," *Constr. Build. Mater.* 24(2),165–170. (2010).
55. G. H. D. Tonoli, H. Savastano Jr., E. Fuente, C. Negro, A. Blanco, and F. A. Rocco Lahr, "Eucalyptus pulp fibres as alternative reinforcement to engineered cement-based composites," *Ind. Crops Prod.* 31(2),225–232. (2010).

56. J. Claramunt, M. Ardanuy, J. A. Garcia-Hortal, and R. D. Tolêdo Filho, "The hornification of vegetable fibers to improve the durability of cement mortar composites," *Cem. Concr. Compos.* 33(5), 586–595. (2011).
57. R. D. Tolêdo Filho, K. Ghavami, M. A. Sanjuan, and G. L.England, "Free, restrained and drying shrinkage of cement mortar composites reinforced with vegetable fibres," *Cem. Concr. Compos.* 27(5), 537–546.(2005).
58. M. Mokhtar, A. Hassan, A. R. Rahmat, and S. Abd Samat, "Characterization and treatments of pineapple leaf fibre thermoplastic composite for construction application," Universiti Teknologi Malaysia Institutional Repository. (2005).
59. S. S. Pendhari, T. Kant, and Y. M. Desai, "Application of polymer composites in civil construction: A general review," *Compos. Struct.* 84(2), 114–124, (2008).
60. A. K. Bledzki, and J. Gassan, "Composites reinforced with cellulose based fibres," *Prog. Pol Sci.* 24(2), 221–274 (1999).
61. E. M. R. Fairbairn, B. B. Americano, G. C. Cordeiro, T. P.Paula, R. D. Toledo Filho, and M. M. Silvoso, "Cement replacement by sugar cane bagasse ash: CO2 emissions reduction and potential for carbon credits," *J.Environ. Manage.* 91(9), 1864–1871 (2010).
62. M. Khorami, and E. Ganjian, "Comparing flexural behaviour of fibre-cement composites reinforced bagasse: Wheat and eucalyptus," *Constr.Build. Mater.* 25(9), 3661–3667 (2011).
63. A. Kriker, A. Bali, G. Debicki, M. Bouziane, and M. Chabannet, "Durability of date palm fibres and their use as reinforcement in hot dry climates," *Cem. Concr. Compos.* 30(7), 639–648 (2008).
64. G. H. D. Tonoli, U. P. R. Filho, H. S. Jr., J. Bras, M. N.Belgacem, and F. A. R. Lahr, "Cellulose modified fibres in cement based composites," *Compos. Part A Appl. Sci. Manuf.* 40(12), 2046–2053 (2009).
65. V. S. Ramachandran, "Concrete science," in *Handbook of analytical techniques in concrete science and technology. Principles,Techniques, and Applications*, 1st ed., pp. 1–55, J. J. Beaudoin, Ed. Otawa.(2000).
66. L. Zongjin, *Advanced Concrete Technology*, 1st ed. Hoboken.(2011).
67. H. E. Gram, *"Durability of natural fibers in concrete CBIResearch No. 1–83,"* Swedish, 1983.
68. R. D. Tolêdo Filho, *"Natural fiber reinforced mortar composites:experimental characterization,"* Pontifical Catholic University of Rio deJaneiro, 1997.
69. N. Ukrainczyk, M. Ukrainczyk, J. Sipusie, and Ma. T., "XRD andTGA investigation of hardened cement paste," in *11. Conference on Materials, Processes, Friction and Wear* 22–24 (2006).
70. V. Agopyan, *"Developments on vegetable fibre – cement based materials in São Paulo, Brazil : an overview,"* 27, 527–536 (2005).
71. G. H. D. Tonoli, S. F. Santos, A. P. Joaquim, and H. J. Savastano, "Accelerated carbonation on vegetable fibre reinforced cementitious roofing tiles," in *10th Int. Inorganic- Bonded Fiber Composites Conference*. IIBCC 1980, 14–24 (2006).
72. Roma L.C. Jr, L. S. Martello, and H. J. Savastano, "Evaluation of mechanical, physical and thermal performance of cement-based tiles reinforced with vegetable fibers," 22(4), 668–674 (2008).
73. R. MacVicara, L. M. Matuana, and J. J. Balatinecza, "Aging mechanisms in cellulose fiber reinforced cement composites," *Cem. Concr. Compos.* 21(3), 189–196 (1999).
74. A. Partek, *Research Laboratory in Pargas*. Finland. (1979).
75. H. E. Gram, "Natural fiber concrete roofing," in *ConcreteTechnology and Desing Vol 5: Natural Fiber reinforced cement and concrete*, pp.143–172, Blackie and Son Ltd, Ed., London (1988).

76. S. S. Guimarães, "Some experiments in vegetable fibre-cement composites," in *Symposium on building materials, for low-income housing*, pp. 98–107 (1990).
77. B. Edwards, *Rough guide to sustainability*. London. (2002).
78. H. Gallegos, "Use of vegetable fibers as building material in Perú," in *Joint Symposium RILEM/ CIB/NCCL*, A25-A34 (1986).
79. A. Hashem, "Use of vegetables plants in rural housing in Bangladesh and their improvement," in Joint *Symposium RILEM/ CIB/NCCL*, C149-C183(1983).
80. I. S. Ghavami, "Ultimate load behaviour of bamboo-reinforced light weight Concrete Beams," *Practice* 17, 281–288 (1995).
81. T. Susuki and T. Yamamoto, "Fire resistan materials made with vegetables plants and fibers and inorganic particles," in *Secondinternational RILEM Symposium Proceedigs*, pp. 60–68 (1990).
82. R. D. Toledo Filho, N. P. Barbosa and K. Ghavami, "Application of sisal and cocconut fibers in adobe blocks," in *Second international RILEM*, pp. 139–149 (1990).
83. D. Fengel, and G. Wegener, *Wood—chemistry, ultrastructure, reactions.*, 1st ed., pp. 613, Wiley, Berlin. (1984).
84. G. E. Reier, *"Avicel PH microcrystalline cellulose, NF, Ph Eur., JP, BP."* on line: http://www.fmcbiopolymer.com/Portals/bio/content/Docs/PS-Section%2011.pdf
85. M. A. S. Azizi Samir, F. Alloin, and A. Dufresne, "Review ofrecent research into cellulosic whiskers, their properties and their application in nanocomposite field.," *Biomacromolecules* 6(2), 612–626(2005).
86. P. V. Joseph, M. S. Rabello, L. H. Mattosoc, K. Joseph, and S. Thomas,"Environmental effects on the degradation behaviour of sisal fibre reinforced polypropylene composites," *Compos. Sci. Technol.* 62(10–11), 1357–1372(2002).
87. C. Gómez Hoyos, E. Cristia, and A. Vázquez, "Effect of cellulose microcrystalline particles on properties of cement based composites," *Mater. Des.* 51, 810–818 (2013).
88. J. Nilsson, and P. Sargenius, "Effect of microfibrillar cellulose on concrete equivalent mortar fresh and hardened properties," KTH, ABE, Department of Civil and Architectural Engineering. (2011).
89. S. J. Peters, T. S. Rushing, E. N. Landis, and T. K. Cummins, "Nanocellulose and microcellulose fibers for concrete," *Transp. Res.Rec. J. Transp. Res. Board* 2142, 25–28 (2010).
90. S. J. Peters, "Fracture toughness investigations of micro and nano cellulose fiber reinforced ultra high performance concrete" Universityof Maine. (2009).
91. M. A. S. Mohamed, E. Ghorbel, and G. Wardeh, "Valorization of micro-cellulose fibers in self-compacting concrete," Constr. *Build. Mater.*24(12), 2473–2480, 2010.
92. J. Claramunt, M. Ardanuy, R. Arevalo, F. Pares, and R. D. deToledo Filho, "Mechanical performance of ductile cement mortar compositesreinforced with nanofibrillated cellulose," in *2nd International RILEM Conference. Strain Hardening Cementitious Composites*, pp. 131–138 (2011).

20

Jute: An Interesting Lignocellulosic Fiber for New Generation Applications

Murshid Iman and Tarun K. Maji*

Department of Chemical Sciences, Tezpur University, Assam, India

Abstract

The current state of polymers from renewable resources has attracted an increasing amount of attention due to their environmental friendliness and, most importantly, their contribution to thereduction of petroleum resources. The increasing awareness of the pressing need for greener and more sustainable technologies has focused attention on the use of biobased polymers instead of conventional petroleum-based polymers to fabricate biodegradable materials with high performance. Natural fibers such as hemp, ramie, jute, etc., are cheap, biodegradable and, most importantly, easily available worldwide. The biocomposites prepared by using natural fibers and varieties of natural polymers have evoked considerable interest in recent years due to their eco-friendly nature. Biocomposites offer the modern world an alternative solution to waste-disposal problems associated with conventional petroleum-based plastics. Therefore, the development of commercially feasible "green products" based on biofibers and polymers for a wide variety of applications is on the rise. Moreover, using nanotechnology for the production of biocomposites affords them better physical properties. This chapter discusses the efficacy of jute fibers for preparation of biocomposites to be used for a variety of applications.

Keywords: Polymers, biodegradable polymers, biofibers, jute, surface modification, nanotechnology, biocomposites.

20.1 Introduction

Rejuvenation of the civilized world is mainly dependent on an innovation of material science, basically "polymers." Perhaps they are the greatest thing chemistry has ever achieved, since the world would be a totally different place without polymers. However, the environmental and health effects of chemicals and chemical processes have begun to be considered. Therefore, in recent years there has been an expanding search for new materials with high performance at affordable costs. With growing environmental and health awareness, there has been an extraordinary focus within the scientific, industrial, and environmental societies on the use of eco-friendly materials or the use of renewable/recyclable/sustainable/triggered biodegradable materials. The development

*Corresponding author: tkm@tezu.ernet.in

Vijay Kumar Thakur (Ed.), Lignocellulosic Polymer Composites, (453–476) 2015 © Scrivener Publishing LLC

or selection of a material to meet the desired structural and design requirements calls for a compromise between conflicting objectives. This can be overcome by resorting to multi–objective optimization in material design and selection. Composite materials, which are prepared using lignocellulosic fiber such as jute and a variety of renewable matrix, are included in this chapter.

The development of biodegradable materials has not only been a big motivating factor for material scientists, but also an important source of opportunities to improve the living standard of people around the world. There is also the potential for economic development based on these materials, even though a major push for their use has been driven by the needs of industrialized countries [1-30]. It has been reported that increasing use of renewable materials would create or secure employment in rural areas, the distribution of which would be in agriculture, forestry, industry, etc. In the 20th century, extraordinary progress has been made in the improvement of practical procedures and products from natural polymers such as starch, cellulose, lactic acid, etc. The use of such materials was superseded due to the development of a wide range of synthetic polymers derived from petroleum-based or raw materials [31, 32]. However, since the 1990s, there has been a simultaneous and growing interest in developing biobased products and innovative process technologies that can reduce the dependence on fossil fuel and move towards a sustainable materials basis. The main reasons for the development of such materials are stated below: [33]

i. Mounting interest in reducing the environmental impact of synthetic polymers or composites due to increased eco-friendly consciousness;
ii. Dwindling petroleum resources, reducing the insistence for dependence on petroleum products, along with increased attention on maximizing the use of renewable materials; and
iii. Accessibility to enriched information on the properties and morphologies of natural materials such as lignocellulosic fibers, through the use of modern instruments at different levels, and hence a better understanding of their structure-property correlations.

Conventional polymer-based composite materials are now well established all over the world because of their high specific strength, modulus and long durability compared to conventional materials such as metals and alloys; these materials have created extensive applications. Built for the long haul, these polymers seem unsuitable for applications where plastics are used for short time periods and then disposed. The basic disadvantages of plastics are their non-degradability and environmentally hazardous nature. Plastic waste disposal management is now a great challenge due to the unavailability of free land for solid waste disposal in regions of high population density. Moreover, plastics are frequently soiled by food and other biological substances, making physical recycling of these materials unfeasible and generally objectionable. In contrast, biodegradable polymers disposed of in bioactive environments degrade by the enzymatic action of microorganisms such as bacteria, fungi, and algae. Their polymer chains may also be broken down by nonenzymatic processes such as chemical hydrolysis. Therefore, greater efforts have been triggered by scientists to find materials based on natural resources in view of their eco-friendly aspects. Such natural resources are organic in nature and also a source for carbon and a host of other useful materials and chemicals, particularly for the production of "green" materials [34-40].

Along with these, researchers have also focused their work on the processing of nanocomposites (materials with nanosized reinforcement) of biobased materials to improve their physical properties. Analogous to orthodox composites, nanocomposites also use a matrix where the nanosized reinforcement elements are dispersed. The reinforcement is currently achieved by a nanoparticle, where at least one of its dimensions is lower than 100 nm. This particular feature provides nanocomposites with unique and outstanding properties never found in conventional composites. Biobased nanocomposites are the next generation of materials.

20.2 Reinforcing Biofibers

In 1908, cellulose fiber-reinforced phenolic composites were made and later were extended to urea and melamine. In 1940, fiber-reinforced composites received commodity status with glass fiber in unsaturated polyesters. Composites are finding use in different fields ranging from guitars, tennis racquets, cars, microlight aircrafts and electronic components to artificial joints. Because of growing environmental awareness and demands of governmental authorities, the use of traditional composites, usually made of glass, carbon or aramid fibers reinforced with epoxy, unsaturated polyester resins, polyurethanes, or phenolics, are thought of unfavorably The most significant disadvantage of traditional composite materials is the problem of suitable disposal after the end of life. In the present polymer technology, it is essential that every material should exclusively be adapted to the environment. New fiber-reinforced materials called biocomposites were created by implanting natural reinforcing fibers, e.g., jute, flax, hemp, ramie, etc., into biobased polymer matrix derived from virgin derivatives of cellulose, starch, lactic acid, etc. [41-45]. This is a challenging field of research with unlimited future prospects. The research in this area is being pursued with great interest to develop newer biocomposites.

Over the last few decenniums, biofiber composites have been undergoing a notable transformation. These materials have become more and more satisfactory as new compositions and improvements have been intensively researched, developed and subsequently applied. The depletion of fossil fuels made biocomposites pointedly important and they have become engineering materials with a very diverse range of properties. The growth of natural fiber-thermoplastic composites, products that are progressively used in building materials and automotive interior parts, have been well recognized. These products have a number of key advantages in comparison with synthetic fiber-reinforced composites, which include low weight, low cost, lack of abrasiveness during processing, and the wide availability of the reinforcing fibers (wood or agricultural fiber) from renewable resources. This century could be called the cellulosic century, because more and more renewable plant resources for products are being discovered. It has been generally stated that natural fibers are renewable and sustainable, but they are, in fact, neither. The living plants from which the natural fibers are taken are renewable and sustainable, but not the fibers themselves[46-51].

Table 20.1 Commercially major fiber sources.

Fiber source	World production (10^3 ton)	Fiber source	World production (10^3 ton)
Bamboo	30,000	Coir	100
Jute	2300	Ramie	100
Kenaf	970	Abaca	70
Flax	830	Grass	700
Sisal	378	Hemp	214
Sugarcane bagasse	75,000		

Table 20.2 Examples of different types of fibers.

Fiber Type	Example
bast fibers	jute, flax, hemp, ramie and kenaf
leaf fibers	abaca, sisal and pineapple
seed fibers	coir, cotton and kapok
core fibers	kenaf, hemp and jute
grass and reed fibers	wheat, corn and rice
all other types	wood and roots

The plants from which the natural fibers are produced are classified as primary and secondary depending on their utilization. Primary plants are those grown for their fiber content, while secondary plants are plants in which the fibers are produced as a byproduct. Jute, hemp, kenaf and sisal are examples of primary plants, whereas pineapple, oil palm and coir are examples of secondary plants. Table 20.1 shows the main fibers used commercially in composites, which are now produced throughout the world [52]. There are six elementary types of natural fibers. The fiber classifications are shown in Table 20.2. The natural fibers are lignocellulosic in nature, which is generally the most ample renewable biomaterial of photosynthesis on earth. In terms of mass units, the net primary production of natural fiber per year is estimated to be 2×10^{11} tons as compared to synthetic polymers, which is 1.5×10^{8} tons [53]. Lignocellulosic materials are widely distributed in the biosphere in the form of plants, trees (wood) and crops. Cellulose, in its several forms, constitutes approximately half of all polymers utilized in the industry worldwide.

In this article, the attention is focused mainly on jute fiber and its composites and nanocomposites.

Jute (*Corchorus capsularis L.*) is an important tropical crop and grows in some Asian countries like India, Bangladesh, China, Thailand, Myanmar, Nepal and Indonesia. It occupies second place in terms of world production levels of cellulosic fibers. Jute, a rainy season crop, grows best in warm and humid climates with temperature between 24°C to 37°C. India, China and Bangladesh are the main producers of jute. To grow jute, farmers scatter the seeds on cultivated soil. About four months after planting, harvesting begins. The plants are usually harvested after they bloom, but before the blossoms go to seed. Jute plants are thinned out when they are about 6 inches (15 centimeters) tall. Jute is graded or rated according to its color, strength, and fiber length. There are two types of Jute fiber: (a) White Jute (*Corchorus capsularis*) and (b) Tossa Jute (*Corchorus olitorius*). The fibers are off-white to brown and 3 to 15 feet (0.9 to 4.5 meters) long. Jute is pressed into bales for shipment to manufacturers. Jute grows in alluvial soils and can survive in heavy flooding. It will only grow in areas with high temperatures, sand or loam soils, and having annual rainfall over 1,000 millimeters. Large-scale jute cultivation is virtually confined to northern and eastern Bengal, mostly in the floodplains of the Ganges and Brahmaputra Rivers. More than 97% of the world's jute is produced in Asia, including 65% in India and 28% in Bangladesh. The world's largest jute mill is in Bangladesh. In India, 4,000,000 families are involved in the cultivation of raw jute. There are 76 jute mills in India and a large number of people are employed in these mills. India is also self-sufficient in the production of jute seed. More than 90 percent of seeds are produced by the state seed corporation of Andhra Pradesh and Maharashtra. Jute has played a vital role in the socioeconomic development in some of these countries [54]. Jute and its products have played an important role in this country's work force. It is one of the inexpensive lignocellulosic fibers and is currently the bast fiber with the highest production volume.

20.2.1 Chemical Constituents and Structural Aspects of Lignocellulosic Fiber

The major constituents of lignocellulosic fibers are cellulose, hemicellulose and lignin. The amount of cellulose in the biofibers can vary depending on the species and the age of the plant or species. Cellulose is better described as a high molecular weight homopolymer of β-1,4-linked anhydro-D-glucose [55] units in which every unit is corkscrewed 180° with respect to its neighbors, and the repeat segment is frequently taken to be a dimer of glucose, known as cellobiose (Figure 20.1) [56]. It contains alcoholic hydroxyl groups. These hydroxyl groups form intramolecular hydrogen bonds inside the macromolecule itself and among other cellulose macromolecules as well as with hydroxyl groups from the air. Therefore, all of the natural fibers are hydrophilic in nature [57].

Even though the chemical assembly of cellulose from different natural fibers is the same, the degree of polymerization shows a discrepancy. The mechanical properties of a fiber are significantly related to degree of polymerization. Among various natural fibers, "Bast fibers" generally show the highest. Approximately, the degree of polymerization for bast fibers is around 10,000 [58].

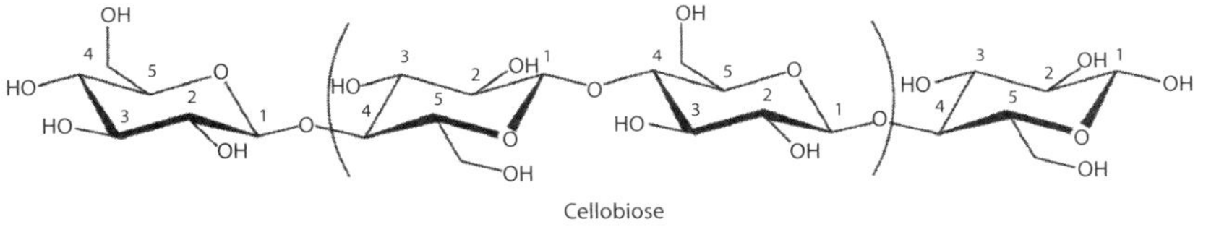

Figure 20.1 Structure of cellobiose (Chem. Rev., 2010, 110 (6), 3479–3500.).

Lignin is a phenolic compound commonly resistant to microbial degradation, but the pretreatment of fiber renders it susceptible to the cellulose enzyme [59, 60]. The actual chemical nature of the primary component of biofiber, the lignin, still remains vague [61, 62]. No method has so far been available by which it is possible to isolate the lignin in the native state from the fiber. Hence the exact structural formula of lignin in natural fiber has not yet been established, although most of the functional groups and units which make up the lignin molecule have been identified. It is highly unsaturated or aromatic in nature due to the high carbon and low hydrogen content; and is characterized by its related hydroxyl and methoxy groups. Ethylenic and sulfur-containing groups have also been found in lignin [63]. The chemical environment of lignin in lignocellulosic materials has been a challenging subject of research [64, 65]. Polysaccharides such as cellulose and hemicellulose are laid down first during synthesis of plant cell walls and lignin fills the spaces between the polysaccharide fibers, cementing them together. Thus it works as a structural support material in plants. This lignification process causes a hardening of cell walls, and the carbohydrate is protected from chemical and physical damage. The topology of lignin from different sources may be different but has the same basic composition. Lignin is believed to be linked with the carbohydrate moiety, though the exact nature of linkages in biofiber are not well known [66]. One of the linkages is alkali sensitive and formed by an ester-type combination between lignin hydroxyls and carboxyls of hemicellulose uronic acid. The other ether-type linkage occurs through the lignin hydroxyls combining with the hydroxyls of cellulose. The lignin, being polyfunctional, exists in combination with more than one neighboring chain molecule of cellulose and/or hemicellulose, making a crosslinked structure. The lignocellulosic material possesses many active functional groups [67] like primary and secondary hydroxyls, carbonyls, carboxyls (esters), carbon-carbon, ether, and acetal linkages.

Like all the lignocellulosic fibers, the physical and chemical properties of jute fiber is also dependent on the three chemical components, viz., cellulose, hemicellulose and lignin. The chemical composition and structural parameters of jute fibers are represented in Table 20.3.

20.2.2 Properties of Jute

The lignocellulosic fibers reveal substantial deviation in diameter along with the length of individual fibers. The qualities and properties of lignocellulosic fibers are generally dependent on features like size, maturity, as well as processing methods implemented for the extraction of fibers. The modulus of the fiber reduces with proliferation in diameter.

Table 20.3 Chemical composition and structural parameters of jute [57, 62, 66, 67].

Chemical composition	wt. (%)	Chemical composition	wt. (%)
Cellulose	61-71.5	Pectin	0.2
Lignin	12-13	Wax	0.5
Hemicellulose	13.6-20.4	Moisture content	12.6

The internal structure and chemical composition of fibers are responsible for properties such as density, electrical resistivity, ultimate tensile strength, initial modulus, etc. The strength and stiffness of fiber is generally correlated with the angle between axis and fibril of the fiber, i. e., the smaller this angle, the higher the mechanical properties; the chemical constituents and complex chemical structure of natural fibers also affect the properties significantly. Because of the very complex structure of lignocellulosic fibers, it is not possible to correlate the fiber strength exactly with cellulose content and micro-fibrillar angle. However, individual fiber properties can vary extensively depending on the source, separating technique, age, moisture content, speed of testing, etc. The lignin content of the fibers influences its structure [58], properties [51, 58, 66-77] and morphology [78]. The waxy substances of natural fibers generally influence their wettability and adhesion characteristics [79, 80].

Jute is 100% biodegradable and recyclable and thus environmentally friendly. It is a natural fiber with a golden and silky shine, and hence is called "The Golden Fiber." Jute is the world's cheapest vegetable fiber, procured from the bast or skin of the plant's stem. It is the second most important vegetable fiber after cotton in terms of usage, global consumption, production, and also availability. It has high tensile strength, low extensibility, and ensures better breathability of fabrics. Therefore, jute is a very suitable agricultural commodity for use in bulk packaging. It also helps to make the best quality industrial yarn, fabric, net and sacks. It is one of the most versatile natural fibers that has been used for packaging, textile, non-textile, construction, and agricultural sectors. Bulking of yarn results in a reduced breaking tenacity and hence an increased breaking extensibility when blended as a ternary blend. Unlike the other fiber, e.g., hemp, jute is not a form of *Cannabis*. Jute stem has a very high volume of cellulose, and hence it can also save the forests and meet the cellulose and wood requirements of the world. The best varieties of Jute are *Bangla Tosha - Corchorus olitorius (Golden shine)* and *Bangla White - Corchorus capsularis (Whitish Shine)*. Raw Jute and jute goods are interpreted as Burlap, Industrial Hemp, and Kenaf in some parts of the world. The best source of Jute in the world is the Bengal Delta Plain, which is occupied by Bangladesh and India [81]. An image of jute crops grown in India is shown in(Figure 20.2).

Figure 20.2 Image of jute crops.

The physical properties of jute are:

- The cells of jute fiber vary from 0.05-0.19 inch in length and 20-22μm in thickness.
- The aspect ratio (L/D) (mm) is about 152-365.
- Externally the fiber is smooth and glossy.
- It has specific gravity of 1.29. It is a very good insulator of heat and electricity.
- It is a highly hygroscopic fiber. Water absorption by jute takes place in two phases, i.e., molecular phase and capillary phase.
- The tenacity of jute varies from 3.5-4.5 (g/denier) at 4 cm test length. The tenacity may be as high as 6-7 g/denier.
- Its breaking elongation under normal atmospheric condition is 1-1.2%.

The chemical properties of jute are:

- In chemical composition, jute is different from linen and cotton. It is composed of a modified form of cellulose called lignocellulose (*bastose*).
- Jute develops its yellow color by iodine and sulphuric acid.
- Jute is more sensitive to the action of chemicals than cotton or linen. Therefore, it cannot be bleached, as treatment with alkalis and bleaching powder weakens, and breaks down the fiber significantly.
- The use of sodium silicate, soda ash, or sodium hydroxide is not recommended for treatment of jute. Lime water makes the fiber brittle. Ammonia provides a harsh feel and decreases its luster.

20.2.3 Cost Aspects, Availability and Sustainable Development

The world's supply of petroleum resources is dwindling; the demand for sustainable and renewable raw materials is high. So the use of available natural fibers has become an unavoidable task for scientists and industries. In order to ensure a reasonable profit for farmers, nontraditional outlets have to be explored for natural fibers. One such promising way is natural fiber-reinforced polymer composites. The price for lignocellulosic fibers, which are viable for different applications, differs a lot depending on the changing economies of the countries where such fibers are widely available. Jute is the so-called golden fiber from India and Bangladesh [62]. It is one of the cheapest natural fibers in the world. In recent years, prices of natural fibers were not stable, especially for flax fibers [72], which are about 30% more expensive than glass fibers. For these economic reasons, substitution of glass fibers with natural fibers does not seem easily realizable. However, lignocellulosic fibers offer several advantages; one of the most fascinating aspects about these natural fibers is their positive environmental impact. They also present safer handling and working conditions compared to synthetic reinforcements. The worldwide availability is an additional factor.

20.2.4 Surface Treatments

The fabrication of polymer composites containing lignocellulosic fibers will often result in fibers being physically spread in the polymeric environment. However, in the majority of cases, poor adhesion and subsequently insufficient mechanical properties result. Hence, the surface treatment of lignocellulosic fibers plays a dynamic role. Generally, surface treatment of lignocellulosic fibers is not mandatory to develop the bonding for the synthesis of biodegradable composites, in view of the analogous chemical scenery of both the biofiber and biopolymer matrix, which have a hydrophilic environment; unlike the situation with commodity polymers, which have an affinity to be hydrophobic. However, to improve many specific aspects of biodegradable composites, such as providing superior adhesion and diminished moisture sensitivity, surface treatment can be useful. Although better adhesion between the biopolymer matrix and lignocellulosic fibers is contributed to by the similar polarities of the two materials, this results in an increase in water absorption of the composite. Hence, these fibers need appropriate surface treatments.

Different surface treatment methods, namely chemical, physical, physical–chemical and physical–mechanical are employed. These modification methods are different for different types of fibers [57]. Dewaxing and alkali treatments are generally done for surface modification of jute fabrics. Jute is first treated with 2% liquid detergent at 70°C, followed by being washed with distilled water and finally dried in an oven. The washed fabrics are then dewaxed by treatment with a mixture of alcohol and benzene (1:2) for 72 h at 50°C, washed with distilled water and finally dried. The fabrics are then treated with 5% ($^{w}/_{v}$) NaOH solution for 30 min at room temperature and washed with distilled water several times to leach out the absorbed alkali. The fabrics are then kept immersed in distilled water overnight and washed repeatedly to avoid the presence of any trace amounts of alkali. The alkali-treated fabrics are dried in an oven stored at ambient temperature in a desiccator [82, 83]. This treatment has helped to improve its interaction with the matrix materials and increase adhesion of fibers with the matrix through surface roughness of fiber; leading to increased strength or other properties of composites through higher fiber incorporation and possibly providing greater durability of the composites.

20.2.5 Processing

A variety of molding methods can be used according to the end-item design requirements. The principal factors impacting the methodology are the nature of the chosen matrix and reinforcement materials. Another important factor is the gross quantity of material to be produced. Large quantities can be used to justify high capital expenditures for rapid and automated manufacturing technology. Small production quantities are accommodated with lower capital expenditures but higher labor and tooling costs at a correspondingly slower rate. The fabrication process may be one of two types based on the type of resins. The processes used for thermosetting resins are: compression molding, hand lay-up, resin transfer molding, vacuum-assisted resin transfer molding (VARTM), pultrusion, etc. The processes for thermoplastic resins include injection molding, extrusion molding, needle punched, direct long-fiber thermoplastic molding (D-LFT), etc. [84-89]

20.2.5.1 Compression Molding

Compression molding is a major technique for the construction of fiber-reinforced polymers. It is a high temperature and high-pressure process, which is widely used for all thermosetting polymers (polyester, epoxy, etc.). It is cost effective for high runs.

20.2.5.1.1 Sequence of Operation

- Lubrication of both top and bottom mold by petroleum jelly/silicon emulsion;
- Closing the mold and heating the charge;
- Filling the mold completely as judged by little flashing;
- Curing and ejection of the molded part.

Compression molding of composites based on a jute-phenolic system has been commonly practiced for a few decades. In this process, jute is impregnated with the phenolic resin by spraying process followed by drying under hot air drier. The pre-impregnated jute layers are placed together for desired thickness and compression molded at a high pressure of 700-800 kg/m^2 and at a temperature of around 120-140°C. Compression molding is relatively slow and labor intensive and is therefore a quite expensive method.

20.2.5.1.2 Hand Lay-Up Technique

In the hand lay-up technique the fibers are placed in a mold and the resin is later applied by rollers. One option is to cure using a vacuum bag. This removes the excess air and the atmospheric pressure exerts pressure to compact the part. The simplicity, low cost of tooling and flexibility of design are the main advantages of this procedure. On the other hand, the long production time, intensive labor, and low automation potential are some of the disadvantages.

In this process, the resin take up may go up to 300-400% on the basis of the jute fiber used, which is not economical. Some preprocessing or pretreatment of jute is required in order to enhance the interfacial interaction. Generally, when unsaturated polyester resin is used with glass fiber, the ratio maintained is 2.5:1. Whereas, for resin with jute, the ratio maintained is 3.5-4:1. Hybrid composite of jute and glass fiber can be made by two steps. Hand lay-up technique is used initially for making sheet moulding compound. This is followed by compression molding. In a typical process, a few plies of treated jute fiber are sandwiched between two outer plies of glass fiber and cured at approximately 80°C under a pressure of approx. 2 x 10^5 N/m^2 for 90 min. However, an increase in temperature increases the productivity. Even with unsaturated polyester resin, hot condition impregnation is usually done for higher productivity.

20.2.5.2 Resin Transfer Molding

Resin transfer moulding is used to replace the hand lay-up process for better productivity and quality. In this process, resin is injected into a closed cavity mold that is filled with reinforced fiber. It is a relatively low-pressure vacuum (100 psi) process that molds nearly complete shapes in 30–60 min. This is the inverse process of vacuum molding. This technique requires the fibers to be placed inside a mold consisting of two solid parts (close mold technique). A tube connects the mold with a supply of liquid resin, which is injected at low pressure through the mold, impregnating the fibers. The resultant part is cured at room temperature or higher till the completion of the curing reaction. The mold is opened and the product is removed. Parameters such as injection pressure, fiber content and mold temperature have a great influence on the development of the temperature profiles and the thermal boundary layers, especially for thin cavities. This technique has the advantage of rapid manufacturing of large, complex, and high performance parts. Several types of resins (epoxy, polyester, phenolic and acrylic) can be used for RTM as long as their viscosity is low enough to ensure a proper wetting of the fibers.

20.2.5.2.1 Sequence of Operation

- Laying of reinforcing jute in fiber/fabric/nonwoven form in the mold;
- Mating a matching mold half to the other half containing jute and clamping tightly;
- Forcing a pressurized resin system mixed with catalyst into the mold;
- Curing of resin and fiber system in the mold;
- Ejection of molded part.

20.2.5.3 Vacuum-Assisted Resin Transfer Molding (VARTM)

The use of the VARTM process in the manufacturing of high performance composite materials has been greatly recognized as one of the most important processes. The advantage of this technique is that it is possible to manufacture very large products with high mechanical properties. It also allows for the production of fiber contents in the range of 70% by weight and creates a void-free laminate. This process has brought lightweight, superior laminates to the average mold shop as well as to aerospace and high-tech fabricators. In a typical process, dry jute in unidirectional (UD) stitched form by a frame is put into the mold in several layers. After the tool is packed, a flexible bag is sealed around the perimeter and vacuum is applied which compacts the dry material, driving out the excess air. Resin is injected under vacuum through the dry jute. The pressure during injection is 1 bar. Veins and capillaries are used to distribute the resin to the entire part. After infusion at 50°C, the resin is allowed to cure at higher temperature (80°C). For post-cure, the composite samples are then dried in the oven at 80°C for 4 hours.

20.2.5.3.1 Pultrusion

Pultrusion is a modern technique used for producing continuous fiber-reinforced profile in which the orientation of the fiber is kept constant during cure. Although this process is utilized for both thermoplastic and thermoset resins, it is mainly suitable for thermosetting resins like polyester, epoxy and phenolic resin systems. Jute, available in continuous forms such as mat, roving, tapes, yarn, etc., is impregnated with resin, passed through a hot die, and shaped according to the desired cross-section of the product. Advantages of this process are the ability to build thin-walled structures, a large variety of cross-sectional shapes and the possibility for high degree of automation. The speed of pultrusion ranges from 0.4 meters to 1 meter per minute depending on the complexity of the products. The loading of jute is anywhere between 50-70%. Pultruded jute composites have good electrical insulation, corrosion and high resistance properties. They find applications in roofing sheets, cable trays, doors and window frames, paneling, sections for wardrobe, partitions, etc.

20.2.5.4 Injection Molding

The injection molding process is suitable for forming complex shapes and fine details with excellent surface finish and good dimensional accuracy at a high production rate and low labor cost. In injection molding, resin granules and short fibers are mixed into a heated barrel and transported to the mold cavity by a spindle. It is used for the manufacturing of plastics/composites and can produce from very small products such as bottle tops to very large car body parts.

20.2.5.4.1 Extrusion

Extrusion is the process where a solid plastic, usually in the form of beads or pallets, is continuously fed to a heated chamber and carried along by a feed screw within. As the solid resin is conveyed, it is compressed, melted and forced out of the chamber through a die. Through the cooling process the melt part results in a resolidification of the resin into a piece.

In the case of biocomposites, the basic extruder equipment works in a similar way or has few supported modifications. Generally, the bead or pallets of the thermoplastic resin are mixed together with 30 to 40% of short/long natural fibers, compressed, melted (wherein the natural fibers are impregnated with resin) and forced out through a die. The equipment which had supported modifications is intended to create raw materials for the injector.

20.2.5.5 Direct Long-Fiber Thermoplastic Molding (D-LFT)

One of the challenges posed by injection molding of natural fiber composites is to produce pellets of a consistent quality. This challenge has been addressed by North American and European injection molding equipment suppliers through a process called direct long-fiber thermoplastic molding (D-LFT). The D-LFT process consists of a twin-screw extruder where raw polypropylene and glass reinforcements are melted, resulting in a molten charge that is subsequently compression molded in a cold tool. Composite Products Inc. adopted this process to produce polypropylene reinforced with 40% natural fiber such as kenaf, flax and natural fiber/glass hybrids. Another company is Daimler Chrysler AG, which introduced the first large-scale application of natural fiber products (two-door vehicle of the 2005 Mercedes A-Class) resulting from the D-LFT process.

20.2.5.5.1 Filament Winding

In filament winding the continuous fibers are impregnated in a resin mix bath and then wound on a rotating mandrel. The successful cylinders with longitudinal or helical and hoop reinforcement were made with sisal epoxy and jute-polyester.

20.3 Biodegradable Polymers

Biodegradation is defined by Albertsson and Karlsson as an event which takes place through the action of enzymes and/or chemical decomposition associated with living organisms (bacteria, fungi, etc.) and their secretion products [90]. However, it is necessary to consider abiotic reactions like photodegradation, oxidation and hydrolysis, which may alter the polymer before, during or instead of biodegradation because of environmental factors. International organizations such as the American Society for Testing and Materials (ASTM) in connection with the Institute for Standards Research (ISR), the European Standardization Committee (CEN), the International Standardization Organization (ISO), the German Institute for Standardization (DIN), the Italian Standardization Agency (UNI), the Organic Reclamation and Composting Association (ORCA) are all actively involved in developing definitions and tests for biodegradability in different environments and compostability. A standard worldwide definition for biodegradable polymers is yet to be established. Therefore all of the already existing definitions correlate the degradability of a material in a specific disposal environment and to that in the simulating environment [91, 92].

Albertsson and Karlsson have also documented that nature can be used as a model in the scheme of degradable polymeric materials, because it can combine polymeric materials with different degradation times into a hierarchical system that optimizes both energy and material properties [93]. Naturally occurring biopolymers are susceptible to environmental degradation factors and their breakdown may be caused by a combination of these. Tailoring the properties of polymers to a wide range of uses and developing a predetermined service period for the materials have become increasingly significant and Albertsson and Karlsson suggest four different strategies:

1. The use of cheap, synthetic, bulk polymers with the addition of a biodegradable or photooxidizable component;
2. Chemical modification of the main polymer chain of synthetic polymers by the introduction of hydrolyzable or oxidizable groups;
3. The use of biodegradable polymers and their derivatives;
4. Tailor make new hydrolyzable structures, e.g., polyesters, polyanhydrides and polycarbonates.

Current research interest in biodegradable polymers is connected with well-defined areas of use. Biodegradable plastics offer one solution for managing packaging waste. However, biomedical applications of biodegradable and biocompatible polymers generate an enormous amount of research and interest. Uses in this field range from wound dressings, drug delivery applications and surgical implants to other different medical devices. The uses also include controlled-release fertilizers and pesticides for agricultural purposes, applications in the automotive industry and as surfactants [30].

20.4 Jute-Reinforced Biocomposites

Jute is one of the most common bast fibers having high tensile modulus and low elongation at break. If the low density (1.45 g/cm^3) of jute fiber is taken into consideration, then its specific stiffness and strength are comparable to those of glass fiber [94-97]. The specific modulus of jute fiber is better than glass fiber, and on a modulus per cost basis, jute is far superior. There are many reports that use jute as reinforcing agent for the synthesis of biodegradable composite. Mitra *et al.* [98] have reported on jute-reinforced composites, their limitations and some solutions through chemical modifications of fibers. Flexural strength, flexural modulus and the dynamic strength of chemically modified jute-poly(propylene) composites have been found to increase by 40%, 90% and 40% respectively compared to unmodified jute-poly(propylene) composites [99] due to the chemical modification of jute with maleic anhydride grafted polypropylene. The effect of different additives on the performance of biodegradable jute fabric-Biopol composites has been reported [100]. Biopol biodegradable polyester is a thermoplastic which has gained industrial consideration since it has a tensile strength comparable to that of isotactic poly(propylene) and is fully biodegradable. In the absence of any additive, both tensile strength and bending strength of composites were found to increase around 50%, whereas elongation at break was reduced by only 1% as compared to pure Biopol sheet. In order to study the effects of additives, the jute fabrics were soaked with several additive solutions of different concentrations. During such treatments dicumyl peroxide was used as the initiator. The effects of various surface modifications of jute on the performance of biodegradable jute-Biopol composites as prepared by hot-press technique have been reported by Mohanty *et al.* [83, 101]. The surface modification of jute, involving dewaxing, alkali treatment, cyanoethylation and grafting, is made with the aim of improving the hydrophobicity of the fiber in order to obtain decent fiber-matrix adhesion in the resulting composites. Differently chemically modified jute yarn-Biopol composites [78] showed maximum enhancement of mechanical properties like tensile strength, bending strength, impact strength and bending-modulus by 194%, 79%, 166% and 162% respectively in comparison to pure Biopol. With 10 and 25% acrylonitrile grafted yarns, the tensile strength of composites were enhanced by 102 and 84% in comparison to pure Biopol. Thus with the increase of grafting percent in yarn, the mechanical properties of the composite were found to decline. The composites made from alkali-treated yarns produced better mechanical properties than dewaxed and grafted yarns. Orientation of jute yarn played an important role in the properties. The enhancement of mechanical properties of composites is noticed only when the properties of the composites were measured along the yarn wrapping direction. Unlike jute yarn, the enhancement of mechanical properties of jute fabric-Biopol composites did not show any variation with the direction of measurement of properties [102]. An enhancement of more than 50% tensile strength, 30% bending strength and 90% impact strength of resulting composites as compared to pure Biopol sheets was observed under the experimental conditions used. Scanning electron microscopy also showed that the surface modifications improved the fiber matrix adhesion. The improvement in adhesion may be attributed to the formation of rough surface due to the removal of natural and artificial impurities from the surface of jute by the treatment with alkali [103]. In addition, alkali treatment leads to fiber fibrillation, i.e., breaking down of the fiber bundle of fabrics into smaller fibers, which improves the effective surface area available for contact with matrix polymer. Degradation studies showed that after 150 days of compost burial more than 50% weight loss of the jute/Biopol composites occurred.

Ray and other researchers have broadly investigated the properties of alkali-treated jute fiber reinforced with vinyl ester resin [104-106]. In their studies, they compared the mechanical, thermal, dynamic, and impact fatigue behavior of treated composites to those of untreated jute fiber–vinyl ester composites. Longer alkali treatment removed the hemicelluloses and improved the crystallinity, enabling better fiber dispersion. The dynamic, mechanical, thermal and impact properties were superior owing to the alkali treatment. The effects of hybridization [107] on the tensile properties of jute–cotton woven fabric-reinforced polyester composites were investigated as functions of the fiber content, orientation and roving texture. It was observed that tensile properties along the direction of jute roving alignment (transverse to cotton roving alignment) increased steadily with fiber content up to 50% and then showed a decreasing trend. The tensile strength of composites with 50% fiber content parallel to the jute roving is about 220% higher than pure polyester resin. Jute fiber-reinforced PP composites were evaluated in terms of the effect of matrix modification [83], the influence of gamma radiation [108], the effect of interfacial adhesion on creep and dynamic mechanical behavior [109], the influence of silane coupling agent [110, 111] and the effect of natural rubber [112]. The properties of jute/plastic composites were studied, including the thermal stability, crystallinity, modification, transesterification, weathering, durability, fiber orientation on frictional and wear behavior, eco-design of automotive components and alkylation [113-120]. Polyester resin was used as matrix for jute fiber-reinforced composites and the relationship between water absorption and dielectric behavior [121], the elastic properties, notched strength and fracture criteria [122], impact damage characterization [123], weathering and thermal behavior, and effect of silane treatment [124] were examined.

Iman and Maji have used the solution-induced intercalation method for the preparation of bionanocomposites based on surface-modified jute fabric and soy flour [125, 126] and starch [82]. The results of the various analyses suggest a strong interfacial interaction between the filler and the matrix. Furthermore, the mechanical strength, thermal stability, flame retardency, and dimensional stability of the nanoparticles- (such as clay, cellulose whiskers, etc.) filled composites were found to be much better than those of the unfilled composites. Behera *et al.* have developed jute-reinforced soy milk-based composites using nonwoven and woven jute fabrics. They have studied and reported the mechanical properties, viz., tensile strength, tensile and flexural modulus, flexural strength and elongation at break. Composites having 60 wt% jute felt or jute fabric possessed the best mechanical properties. Hydrophilicity of the composites was assessed by the measurement of contact angle and water absorption after immersing in water at ambient temperature as well as in boiling condition. Biodegradability of the composites was evaluated in compost soil burial condition. Fourier transform infrared and optical microscope analyses of the buried samples confirmed the degradation of the composites [127].

20.5 Applications

Jute-reinforced materials can find applications in the market segments like automotive industries (door panels, dashboard, and instrument panels); packaging—both for inland transport and export of agricultural products (crates, inlays for crates, pallets, boxes, and cases); and consumer products (housing for computer screens, refrigerators, etc.). Jute fibers as reinforcement and filler of injection-molded polypropylene products have been shown to possess excellent mechanical properties that can be molded on existing machinery into any shape or form such as crates and trays, pallets and boxes; but also films and straps for wrapping and taping. Natural fibers like jute can be pressed with resin to form laminates. Any foaming materials can be sandwiched between the laminates. These sandwiched-based laminates can be used as roofing, due to their improved thermal and acoustic insulating properties compared to corrugated iron roofing. Moreover, they do not cause zinc and rust pollution. Due to the use of local resources, the natural fiber sandwich is probably cheaper to manufacture. In comparison with traditional vegetable roofing, the sandwich is more durable. Similar panels can also be used as doors, table tops and shelves. Jute mat is an example of a nonwoven jute fiber composite. The material is composed of jute fiber, resin and a small amount of synthetic fiber. The manufacturing process creates a mat that can then be molded into creative shapes such as a car door panel. The result is a light but strong component. Building and construction technology trends worldwide establish the fact that the composites occupy a prominent position as building material, dislodging many conventional ones. Composites are an attractive proposition considering their embedded energy, especially against metals. Other important properties such as impact resistance, corrosion resistance, thermal and acoustic insulation all contribute favorably to composite, claiming its position as an ideal building material. Jute-coir composite boards, fibre reinforced plastic (FRP) sandwiched door shutters, FRP toilet blocks, etc., have also been developed. Regarding new applications of biocomposites, extensive work is currently being done by many companies and research institutes with lignocellulosic fiber and polymers in order to develop parts for different sectors like shipbuilding, the automotive industry, mass transit, street furniture and the energy industry.

20.6 Concluding Remarks

Motivated by economic crises, the depletion of petroleum resources together with rising environmental concerns have contributed to biocomposites gaining importance with time. Fibers like jute, hemp, coir, etc., are not only raw materials for biocomposite, but also meet the needs of the environment. Natural fiber like jute and biodegradable polymer used for making composites are considered as the most environmentally friendly materials due to their renewable and degradable nature. However, many problems like poor adhesion between matrix and fiber, difficulty of fiber orientation, the achievement of nanoscale size material and the evolution of truly green polymers, which will be environmentally friendly and renewable, must first be solved. Although the formation of composite using polymers and different nanofillers results in overall improvement in properties of the nanocomposites, fine tuning the ratio of different nanofillers is the scope for further reaserch. One of the major drawbacks of different natural polymers- (used as matrix) based composites are their limited long-term stability. Therefore, limited grafting of such polymers with other monomers may be tried in order to achieve superior properties of the composites. Ecobinders based on renewable furans and lignin, may fulfill the requirements for the selected product-market-combinations in the future.

Acknowledgement

M. I. is very grateful to the Council of Scientific and Industrial Research (CSIR), New Delhi, India, for the financial assistance under Senior Research Fellowship (SRF).

References:

1. K.G. Satyanarayana, L.P. Ramos, and Wypych, F. In *Biotechnology in Energy Management*; T.N. Ghosh, T. Chakrabarti, G. Tripathi, (Eds.), p. 583, APH Publishing Corporation, New Delhi (2005).
2. R.L. Lankey, and P.T. Anastas (Eds.), *Advancing Sustainability Through Green Chemistry and Engineering*, ACS Symposium Series 823, American Chemical Society, Washington, DC (2002).
3. A.K. Mohanty, M. Misra, and L.T. Drzal, Sustainable bio-composites from renewable resources: Opportunities and challenges in the green materials world. *J. Polym. Environ.* 10, 19–26 (2002).
4. A.K. Mohanty, L.T. Drzal, and M. Misra, Engineered natural fiber reinforced polypropylene composites: influence of surface modifications and novel powder impregnation processing. J. Adhes. Sci. Technol. 16, 999–1015 (2002).
5. M.K.Uddin, M.A. Khan, and K.M.I. Ali, Degradable jute plastic composites. *Polym. Degrad. Stabil.* 55, 1–7 (1997).
6. M. Wollerdorfer, and H. Bader, Influence of natural fibres on the mechanical properties of biodegradable polymers. *Ind. Crops Prod.* 8, 105–112 (1998).
7. S. Luo, and A.N. Netravali, Effect of 60Co γ-radiation on the properties of poly (hydroxybutyrate-co-hydroxyvalerate). *J. Appl. Polym. Sci.* 73, 1059–1067 (1999).
8. G.W. Coates, and M.A. Hillmyer, A virtual issue of macromolecules:"polymers from renewable resources. *Macromolecules* 42, 7987–7989 (2009).
9. M.A. DeWit, and E.R. Gillies, A cascade biodegradable polymer based on alternating cyclization and elimination reactions. *J. Am. Chem. Soc.* 131, 18327–18334 (2009).
10. Y.Z.Z. Ryberg, U. Edlund, and A.-C. Albertsson, Conceptual approach to renewable barrier film design based on wood hydrolysate. *Biomacromolecules* 12, 1355–1362 (2011).
11. J. Hartman, A.-C. Albertsson, M.S. Lindblad, and J. Sjöberg, Oxygen barrier materials from renewable sources: Material properties of softwood hemicellulose-based films. *J. Appl. Polym. Sci.* 100, 2985–2991 (2006).

12. H. Tian, Z. Tang, X. Zhuang, X. Chen, and X. Jing, Biodegradable synthetic polymers: Preparation, functionalization and biomedical application. *Prog. Polym. Sci.* 37, 237–280 (2012).
13. S.S. Ray, K. Yamada, M. Okamoto, and K. Ueda, Polylactide-layered silicate nanocomposite: a novel biodegradable material. *Nano Lett.* 2, 1093–1096 (2002).
14. D.S. Muggli, A.K. Burkoth, S.A. Keyser, H.R. Lee, and K.S. Anseth, Reaction behavior of biodegradable, photo-cross-linkable polyanhydrides. *Macromolecules* 31, 4120–4125 (1998).
15. A.T.Metters, K.S. Anseth, and C.N. Bowman, Fundamental studies of a novel, biodegradable PEG-b-PLA hydrogel. *Polymer* 41, 3993–4004 (2000).
16. K. Odelius, M. Ohlson, A. Höglund, and A.-C. Albertsson, Polyesters with small structural variations improve the mechanical properties of polylactide. *J. Appl. Polym. Sci.* 127, 27–33 (2013).
17. J.Y. Nam, S.S. Ray, and M. Okamoto, Crystallization behavior and morphology of biodegradable polylactide/layered silicate nanocomposite. *Macromolecules* 36, 7126–7131 (2003).
18. S.T. Lim, Y.H. Hyun, and H. Choi, Synthetic biodegradable aliphatic polyester/montmorillonite nanocomposites. *J. Chem. Mater.* 14, 1839–1844 (2002).
19. X. Tang, and S. Alavi, Structure and physical properties of starch/poly vinyl alcohol/laponite RD nanocomposite films. *J. Agric. Food Chem.* 60, 1954–1962 (2012).
20. S. Mishra, S.S. Tripathy, M. Misra, A.K. Mohanty, and S.K. Nayak, Novel eco-friendly biocomposites: Biofiber reinforced biodegradable polyester amide composites—Fabrication and properties evaluation. *J. Reinf. Plast. Compos.* 21, 55–70 (2002).
21. B.N. Dash, M. Sarkar, A.K. Rana, M. Mishra, A.K. Mohanty, S.S. Tripathy, A study on biodegradable composite prepared from jute felt and polyesteramide (BAK). *J. Reinf. Plast. Compos.* 21, 1493–503 (2002).
22. D. Plackett, T.L. Andersen, W.B. Pedersen, and L. Nielsen, Biodegradable composites based on polylactide and jute fibres. *Compos. Sci. Technol.* 63, 1287–1296 (2003).
23. P. Wambua, J. Ivens, I. Verpoest, Natural fibres: can they replace glass in fibre reinforced plastics? *Compos. Sci. Technol.* 63, 1259–1264 (2003).
24. A.N. Netravali, and S. Chabba, Composites get greener. *Mater. Today* 6, 22–29 (2003).
25. M. Geeta, A.K. Mohanty, K. Thayer, M. Misra, and L.T. Drzal, Novel biocomposites sheet molding compounds for low cost housing panel applications. *J. Polym. Environ.* 13, 169–175 (2005).
26. A.K. Mohanty, A. Wibowo, M. Misra, and L.T. Drzal, Effect of process engineering on the performance of natural fiber reinforced cellulose acetate biocomposites. *Composites A* 35, 363–370 (2004).
27. W. Liu, M. Misra, P. Askeland, L.T. Drzal, and A.K. Mohanty, 'Green'composites from soy based plastic and pineapple leaf fiber: Fabrication and properties evaluation. *Polymer* 46, 2710–2721 (2005).
28. R. Bhardwaj, A.K. Mohanty, L.T. Drzal, F. Pourboghrat, and M. Misra, Renewable resource-based green composites from recycled cellulose fiber and poly (3-hydroxybutyrate-co-3-hydroxyvalerate) bioplastic. *Biomacromolecules* 7, 2044–2051 (2006).
29. W. Liu, A.K. Mohanty, L.T. Drzal, and M. Misra, Novel biocomposites from native grass and soy based bioplastic: Processing and properties evaluation. *Ind. Eng. Chem. Res.* 44, 7105–7112 (2005).
30. M. Cheng, A.B. Attygalle, E.B. and Lobkovsky, G.W. Coates, Single-site catalysts for ring-opening polymerization: Synthesis of heterotactic poly (lactic acid) from rac-lactide. *J. Am. Chem. Soc.* 121, 11583–11584 (1999).
31. W. Amass, A. Amass, and B. Tighe, A review of biodegradable polymers: uses, current developments in the synthesis and characterization of biodegradable polyesters, blends of

biodegradable polymers and recent advances in biodegradation studies. *Polym. Int.* 47, 89–144 (1998).
32. C.K. Williams, and M.A. Hillmayer, Polymers from renewable resources: a perspective for a special issue of polymer reviews. *Polym. Rev.* 48, 1–10 (2008).
33. J.J. Maya, and S. Thomas, Biofibres and biocomposites. *Carbohydr. Polym.* 71, 343–364 (2008).
34. K.G. Satyanarayana, G.G.C. Arizaga, and F. Wypych, Biodegradable composites based on lignocellulosic fibers—An overview. *Prog. Polym. Sci.* 34, 982–1021 (2009).
35. O. Faruka, A.K. Bledzki, H-P. Fink, and M. Sain, Biocomposites reinforced with natural fibers: 2000–2010. *Prog. Polym. Sci.* 37, 1552–1596 (2012).
36. J. Summerscales, N.P.J. Dissanayake, A.S. Virk, and W. Hall, A review of bast fibres and their composites. Part 1–Fibres as reinforcements. *Compos. A* 41, 1329–1335 (2010).
37. []U. Witt, M. Yamamoto, U. Seeliger, R.-J. Müller, V. Warzelhan, Biodegradable polymeric materials—not the origin but the chemical structure determines biodegradability. *Angew. Chem. Int. Ed.* 38, 1438–1442 (1999).
38. D. Klemm, F. Kramer, S. Moritz, T. Lindström, M. Ankerfors, D. Gray, and A. Dorris, Nanocelluloses: A new family of nature-based materials. *Chem. Int. Ed.* 50, 5438–5466 (2011).
39. P. Maiti, C.A. Batt, and E.P. Giannelis, New biodegradable polyhydroxybutyrate/layered silicate nanocomposites. *Biomacromolecules* 8, 3393–3400 (2007).
40. T.L. Richard, Challenges in scaling up biofuels infrastructure. *Science* 329, 793–796 (2010).
41. H.W. Blanch, P.D. Adams, K.M. Andrews-Cramer, W.B. Frommer, B.A. Simmons, and J.D. Keasling, Addressing the need for alternative transportation fuels: The joint bioenergy institute. *ACS Chem. Biol.* 3, 17–20 (2008).
42. N. Venkateshwaran, and A. Elayaperumal, Banana fiber reinforced polymer composites – A review. *J. Reinf. Plast. Compos.* 29, 2387–2396 (2010).
43. S. Shinoja, R. Visvanathanb, S. Panigrahic, and M. Kochubabua, Oil palm fiber (OPF) and its composites: A review. *Ind. Crops Prod.* 33, 7–22 (2011).
44. A.K. Mohanty, M. Misra, and G. Hinrichsen, Biofibres, biodegradable polymers and biocomposites: An overview. *Macromol. Mater. Eng.* 276, 1–24 (2000).
45. A. Hassan, A.A. Salema, F.H. Ani, and A.A. Bakar, A review on oil palm empty fruit bunch fiber-reinforced polymer composite materials. *Polym. Compos.* 31, 2079–2101 (2010).
46. R.R. Franck (Ed.), *Bast and Other Plant Fibres,* p. 397, CRC Press, Boca Raton, FL (2005.
47. A.K. Bledzki, V.E. Sperber, and O. Faruk, Natural and wood fibre reinforcement in polymers, in *Rapra Review Reports,* Volume 13, pp. 1–144, iSmithers Rapra Publishing (2002).
48. C. Baillie (Ed.), *Green Composites: Polymer Composites and the Environment,* p. 308, Woodhead Publishing Limited, Cambridge, UK (2004).
49. A.K. Mohanty, M. Misra, and L.T. Drzal, (Eds.) *Natural Fibres, Biopolymers, and Biocomposites,* p. 896, Taylor & Francis Group, Boca Raton, Florida (2005).
50. K. Pickering (Ed.), *Properties and Performance of Natural-Fibre Composites, p. 557,* Woodhead Publishing Limited, Cambridge, UK (2008).
51. S. Thomas, and L.A. Pothan (Eds.), *Natural fibre Reinforced Polymer Composites: From Macro to Nanoscale,* p. 539, Old City Publishing, Philadelphia (2009).
52. A.K. Bledzki, and J. .Gassan, Composites reinforced with cellulose based fibres. *Prog. Polym. Sci.* , 24, 221–274 (1999).
53. M.P.Staiger, and N. Tucker, Natural-fibre composites in structural applications, in *Properties and Performance of Natural-Fibre Composites, K. Pickering, (Ed.), pp. 269–300, Woodhead Publishing Limited,* Cambridge, UK (2008).

54. Hon, D.N.S. Cellulose: A wonder material with promising future. *Polym. News* 13, 34–140 (1988).
55. W.M. Wang, Z.S. Cai, J.Y. Yu, and Z.P. Xia. Changes in composition, structure, and properties of jute fibers after chemical treatments. *Fibers Polym.* 10, 776–780 (2009).
56. H. Li, P. Zadorecki, and P. Flodin, Cellulose fiber-polyester composites with reduced water sensitivity (1)—chemical treatment and mechanical properties. *Polym. Compos.* 8, 199–202 (1987).
57. Y. Habibi, L.A. Lucia, and J.O. Rojas, Cellulose nanocrystals: chemistry, self-assembly, and applications. *Chem. Rev.* 110, 3479–3500 (2010).
58. A.K. Bledzki, S. Reihmane, J. Gassan, Properties and modification methods for vegetable fibers for natural fiber composites. *J. Appl. Polym. Sci.* 59, 1329–1336 (1996).
59. L.M. Lewin, and E.M. Pearce, *Handbook of Fiber Science and Technology, Volume IV Fiber Chemistry*, Marcel Dekker, New York (1985).
60. C. David, and R. Fornasier, Utilization of waste cellulose. 7. Kinetic study of the enzymatic hydrolysis of spruce wood pretreated by sodium hypochlorite. *Macromolecules* 19, 552–557 (1986).
61. M. Paillet, and A.Peguy, *J. Appl. Polym. Sci.* 40, 427–433 (1990).
62. A.K. Mohanty, and M. Misra, Studies on jute composites—A literature review. *Polym. Plast. Technol. Eng.* , 34, 729–792 (1995).
63. A.K. Mohanty, Graft copolymerization of vinyl monomers onto jute fibers. *J. Macromol. Sci. C Polym. Rev.* 27, 593–639 (1987.
64. D.N.-S. Hon, *Polym. News* 17, 102–107.
65. R.L. Crawford, *Lignin Biodegradation and Transformation*, John Wiley & Sons, New York (1981).
66. T.K. Kirk, T. Higuchi, and H.M. Chang, *Lignin Biodegradation: Microbiology, Chemistry and Potential Applications*, CRS Press, Boca Raton, Florida, 1980.
67. S.C.O. Ugbolue, Structure/property relationships in textile fibres. *Text. Prog.* 20, 1–43 (1990).
68. M. S. Kucharyov, M. S. *Tekst. Prom.* 8–9, 23–25 (1993).
69. Wuppertal, E.W. *Die Textilen Rohstoffe (Natur und Chemiefasern)*, Dr. Spohr-Verlag/ Deutscher Fachverlag, Frankfurt (1981).
70. R.M. Rowell, In *Emerging Technologies for Materials and Chemicals*, R.M. Rowell, T.P. Schultz, and R. Narayan, (Eds.), ACS Symposium Series 476, American Chemical Society, Washington, DC (1992).
71. D.S. Varma, M. Varma, and I.K. Varma, Coir fibers part I: Effect of physical and chemical treatments on properties. *Text. Res. Inst.* 54, 821–832 (1984).
72. P.S. Mukherjee, and K.G. Satyanarayana, Structure and properties of some vegetable fibres. *J. Mater. Sci.* 21, 51–56 (1986).
73. A.K. Bledzki, and J. Gassan, In *Handbook of Engineering Polymeric Materials*, N.P. Cheremisinoff, (Ed.), p. 787, Marcel Dekker Inc, New Jersey, (1997).
74. D.S. Varma, M. Varma, and I.K. Varma, Thermal behaviour of coir fibres. *Thermochim. Acta* 108, 199–210 (1986).
75. N. Chand, and P.K. Rohatgi, In *Natural Fibres and Composites*, p. 55, Periodical Experts Agency, Delhi, India (1994)
76. S.H. Zeronian, The mechanical properties of cotton fibers. *J. Appl. Polym. Sci.* 47, 445–461 (1991).
77. H. Saechtling, *International Plastics Handbook: For the Technologist, Engineer, and User; 2nd Ed.*, Hanser, Munich (1987).

78. B.C. Barkakaty, Some structural aspects of sisal fibers. *J. Appl. Polym. Sci.* 20, 2921–2940 (1976).
79. E.T.N. Bisanda, and M.P. Ansell, The effect of silane treatment on the mechanical and physical properties of sisal-epoxy composites. *Compos. Sci. Technol.* 41, 165–178 (1991).
80. S.H. Zeronian, H. Kawabata, and K.W. Alger, Factors affecting the tensile properties of nonmercerized and mercerized cotton fibers. *Text. Res. Inst.* 60, 179–183 (1990).
81. V.V. Safonov, *Treatment of Textile Materials,* Legprombitizdat, Moscow (1991).
82. http://textilelearner.blogspot.com/2012/03/features-oa-jute-fiber-properties-of.html#ixzz2qwXyya5u (January 7, 2014)
83. M. Iman, and T.K. Maji, Effect of crosslinker and nanoclay on starch and jute fabric based green nanocomposites. *Carbohydr. Polym.* 89, 290–297 (2012).
84. A.K. Mohanty, M.A. Khan, and G. Hinrichsen, Surface modification of jute and its influence on performance of biodegradable jute-fabric/Biopol composites. *Compos. Sci. Technol.* 60, 1115–1124 (2000).
85. A.C. Karmaker, and G. Hinrichsen, Processing and characterization of jute fiber reinforced thermoplastic polymers. *Polym. Plast. Technol. Eng.* 30, 609–629 (1991).
86. http://www.tifac.org.in/index.php?option = com_content&id = 541:development-of-natural-fibre-composites-in-india&catid = 85:publications&Itemid = 952 (January 7, 2014)
87. http://www.tifac.org.in/index.php?option = com_content&id = 545:jute-composite-technology-a-business-opportunities-&catid = 85:publications&Itemid = 952 (January 7, 2014)
88. A review on the tensile properties of natural fibre reinforced polymer composites, http://eprints.usq.edu.au/18884/1/Ku_Wang_Pattarachaiyakoop_Trada_AV.pdf (January 7, 2014)
89. K.V. de Velde, and P. Kiekens, Thermoplastic pultrusion of natural fibre reinforced composites. *Compos. Struct.* 54, 355–360 (2001).
90. D.N. Saheb, and J.P. Jog, Natural fiber polymer composites: a review. *Adv. Polym. Technol.* 18, 351–363 (1999).
91. A.-C. Albertsson, and S. Karlsson, *Chemistry and Technology of Biodegradable Polymers, p. 48,* Blackie, Glasgow, (1994).
92. ASTM standards on environmentally degradable plastics", ASTM publication code number (DCN): 03-420093-19 (1993).
93. CEN TC 261 SC4 W62 draft Requirements for packaging recoverable in the form of composting and biodegradation, test scheme for the final acceptance of packaging, CEN TC 261 SC4 W62 draft, *(March* 11, 1996).
94. A.-C. Albertsson, **and** S.J.Karlsson, Macromolecular architecture-nature as a model for degradable polymers. *Macromol. Sci. A* 33, 1565–1570 (1996).
95. S.R. Ranganathan, and P.K. Pal, Jute to Simplify Sheet Moulding Compound. *Popular Plast.* 31, 29–31 (1985).
96. P. Ghosh, and P.K. Ganguly, Plast. *Rubber Compos. Proc. Appl.* 20, 171 (1993).
97. P.J. Rae, and M.P. Ansell, Jute-reinforced polyester composites. *J. Mater. Sci.* 1985 20, 4015–4020 (1985).
98. A.C. Karmaker, and J. Youngquist, Injection molding of polypropylene reinforced with short jute fibers. *J. Appl. Polym. Sci.* 62, 1147–1151 (1996).
99. B.C. Mitra, R.K. Basak, and M. Sarkar, Studies on jute-reinforced composites, its limitations, and some solutions through chemical modifications of fibers. *J. Appl. Polym. Sci.* 67, 1093–1100 (1998).
100. J. Gassan, and A.K. Bledzki, The influence of fiber-surface treatment on the mechanical properties of jute-polypropylene composites. *Compos. A* 28, 1001–1005 (1997).

101. M.A. Khan, K.M.I. Ali, G. Hinrichsen, C. Kopp, and S. Kropke, Study on physical and mechanical properties of biopol-jute composite. *Polym. Plast. Technol. Eng.* 38, 99–112 (1999).
102. A.K. Mohanty, M.A. Khan, S. Sahoo, and G.J. Hinrichsen, Effect of chemical modification on the performance of biodegradable jute yarn-Biopol® composites. *Mater. Sci.* 35, 2589–2595 (2000).
103. D. Ray, B.K. Sarkar, S. Das, and A.K. Rana, Dynamic mechanical and thermal analysis of vinylester-resin-matrix composites reinforced with untreated and alkali-treated jute fibres. *Compos. Sci. Technol.* 62, 911–917 (2002).
104. D. Ray, B.K. Sarkar, and N.R. Bose, Impact fatigue behaviour of vinylester resin matrix composites reinforced with alkali treated jute fibres. *Compos. A* 33, 233–241 (2002).
105. B.K. Sarkar, and D. Ray, Effect of the defect concentration on the impact fatigue endurance of untreated and alkali treated jute–vinylester composites under normal and liquid nitrogen atmosphere. *Compos. Sci. Technol.* 64, 2213–2219 (2004).
106. K. Joseph, and L.H. Carvalho, Jute/cotton woven fabric reinforced polyester composites: effect of hybridization, In *Natural polymers and composites conference proceedings, L.H. Mattoso, A. Leao, and E. Frollini, (Eds.), pp. 454–459 (**2000**).*
107. T.T.L. Doan, S.L. Gao, and E. Mäder, Jute/polypropylene composites I. Effect of matrix modification. *Compos. Sci. Technol.* 66, 952–963 (2006).
108. Haydaruzzaman, R.A. Khan, M.A. Khan, A.H. Khan, and M.A. Hossain, Effect of gamma radiation on the performance of jute fabrics-reinforced polypropylene composites. *Radiat. Phys. Chem.* 78, 986–993 (2009).
109. B.A.Acha, M.M. Reboredo, and N.E. Marcovich, reep and dynamic mechanical behavior of PP–jute composites: Effect of the interfacial adhesion. *Compos. A* 38, 1507–1516 (2007).
110. X. Wang, Y. Cui, Q. Xu, B. Xie, and W.J. Li, Effects of alkali and silane treatment on the mechanical properties of jute-fiber-reinforced recycled polypropylene composites. *J. Vinyl Addit. Technol.* 16, 183–188 (2010).
111. C.K. Hong, I. Hwang, N. Kim, D. H. Park, B.S.D. HHwang, and C. Nah, Mechanical properties of silanized jute–polypropylene composites. *J. Ind. Eng. Chem.* 14, 71–76 (2008).
112. H.U. Zaman, R.A. Khan, M. Haque, M.A. Khan, A. Khan, T. Huq, N. Noor, M. Rahman, K.M. Rahman, D. Huq, and M.A. Ahmad, Preparation and mechanical characterization of jute reinforced polypropylene/natural rubber composite. *J. Reinf. Plas. Compos.* 29, 3064–3065 (2010).
113. S. Sarkar, and B. Adhikari, Jute felt composite from lignin modified phenolic resin. *Polym. Compos.* 22, 518–527 (2001).
114. A. Sampath, and G.C. Martin, Enhancement of natural fiber–epoxy interaction using bi-functional surface modifiers. In *SPE Conference. ANTEC 2000: Plastics the Magical Solution, pp. 2274–2278,* CRC Press, Boca Raton, FL (2000).
115. R.K. Samal, S. Acharya, M. Mohanty, and M.C. Ray, FTIR spectra and physico-chemical behavior of vinyl ester participated transesterification and curing of jute. *J. App. Polym. Sci.* 79, 575–581 (2001).
116. U.K. Dwivedi, and N. Chand, Influence of fibre orientation on friction and sliding wear behaviour of jute fibre reinforced polyester composite. *Appl. Compos. Mater.* 16, 93–100 (2009).
117. B. Singh, M. Gupta, and A. Verma, The durability of jute fibre-reinforced phenolic composites. *Compos. Sci. Technol.* 60, 581–589 (2000).
118. C. Alves, P.M.C. Ferrão, A.J. Silva, L.G. Reis, M. Freitas, L.B. Rodrigues, and D.E. Alves, Ecodesign of automotive components making use of natural jute fiber composites. *J. Clean. Prod.* 18, 313–327 (2010).

119. A. Mir, R. Zitoune, F. Collombet, and B. Bezzazi, Study of mechanical and thermomechanical properties of jute/epoxy composite laminate. *J. Reinf. Plas. Compos.* 29, 1669–1680 (2010).
120. M. Sarikanat, The influence of oligomeric siloxane concentration on the mechanical behaviors of alkalized jute/modified epoxy composites. *J. Reinf. Plas. Compos.* 29, 807–817 (2010).
121. A.N.Fraga, E. Frullloni, O. de la Osa, J.M. Kenny, and A. Vazquez, Relationship between water absorption and dielectric behaviour of natural fibre composite materials. Polym. Test. 25, 181–187 (2006).
122. A.K. Sabeel, S. Vijayarangan, and A.C.B. Naidu, Elastic properties, notched strength and fracture criterion in untreated woven jute–glass fabric reinforced polyester hybrid composites. *Mater. Des.* 28, 2287–2294 (2007).
123. C. Santulli, Post-impact damage characterisation on natural fibre reinforced composites using acoustic emission. *NDT E Int.* 34, 531–536 (2001).
124. B.N. Dash, A. Rana, H.K. Mishra, S.K. and Nayak, S.S. Tripathy, Novel low-cost jute–polyester composites. III. Weathering and thermal behavior. *J. App. Polym. Sci.* 78, 1671–1679 (2000).
125. K. Sever, M. Sarikanat, Y. Seki, G. Erkan, and Ü.H. Erdogan, The mechanical properties of -methacryloxypropyltrimethoxy silane-treated jute/polyester composites. *J. Compos. Mater.* 44, 1913–1924 (2010).
126. M. Iman, K.K. Bania, and T.K. Maji, Green jute-based cross-linked soy flour nanocomposites reinforced with cellulose whiskers and nanoclay. *Ind. Eng. Chem. Res.* 52, 6969–6983 (2013).
127. M.Iman, and T.K. Maji, Effect of crosslinker and nanoclay on jute fabric reinforced soy flour green composite. *J. Appl. Polym. Sci.* 127, 3987–3996 (2013).
128. A.K. Behera, S.. Avancha, R.K. Basak, R. Sen, and B. Adhikari, Fabrication and characterizations of biodegradable jute reinforced soy based green composites. B. *Carbohydr. Polym.* 88, 329–335 (2012).

21

Cellulose-Based Polymers for Packaging Applications

Behjat Tajeddin

Food Engineering and Post-Harvest Technology Research Department, Agricultural Engineering Research Institute, Karaj, Iran.

Abstract

Various synthetic materials have been and are still being used as packaging raw material, which may have many disadvantages. However, different polymers obtained from agricultural products are very suitable for the formulation of packaging materials. These polymers such as polysaccharides can be used in different forms like coatings, films, composites, nanocomposites, etc. Cellulose is one of the most abundant polysaccharides on earth, with an attractive combination of different properties. However, the efficient use of cellulose as a raw material for packaging seems rational and is a subject that should be taken into consideration.

This chapter provides a general overview of the various applications of cellulose in the packaging industry. In general, cellulose application in the packaging industry can be organized around three main topics: the preparation of composites, the production of coating materials, and the preparation of edible and non-edible films. Detailed discussions about each of these topics are presented in this chapter.

Keywords: Cellulose, coatings, films, composites, packaging

21.1 Introduction

Cellulose with the formula $(C_6H_{10}O_5)_n$ is one of the most abundant organic materials on earth, with an attractive combination of different properties. Although cellulose has long been known to mankind, its application in industries such as the packaging industry is still new and deserves more research. All natural fibers, whether wood or non-wood, are cellulosic in nature. This means that cellulose is a natural polymer that exists in the cell walls of all plants, both edible and inedible. Although humans are able to eat edible plants, the cellulose inside them is not digestible by the human digestive system and is rejected by the body. On the other hand, during the period of planting through harvesting of plants, parts of them may be normally or unintentionally discarded and turned into agricultural waste. Also, some plants naturally have short lives

**Corresponding author*: behjattajeddin@gmail.com; behjat.tajeddin@yahoo.com

Vijay Kumar Thakur (Ed.), Lignocellulosic Polymer Composites, (477–498) 2015 © Scrivener Publishing LLC

and after death may be left on the ground and therefore turn into waste. From another perspective, population growth and the transition of community life to the machine era have made the use of packaging inevitable, as packaged products are more available and easier to use. For years, different natural and synthetic materials have been and are still being used as packaging raw material, which may have many advantages and/or disadvantages.

In general, there are two process pathways for agro-materials: dry and wet methods [1, 2, 3].

The "dry process" is based on the biopolymers thermoplastic properties when plasticized and heated above their melting point to cause them to flow. Some good examples of dry processes are extrusion and heat pressing [2, 3]. The dry process is mainly used to form composites. Plant fibers, for example, as reinforcement for preparing the polymer matrix composites needed for extruder equipment.

The "wet process," "solvent process" or "casting" is based on the dispersion of the film-forming materials in the solvents followed by applying to flat surfaces, drying to eliminate the solvent and form a film structure. All the ingredients of materials dissolve or homogeneously disperse in the solvents to produce film-forming solutions. The casting process is used to form edible preformed films, or to apply coatings directly onto the products. To produce a homogeneous film structure avoiding phase separation, various emulsifiers can be added to the film-forming solution [2, 3]. For the wet process, the selection of solvents is one of the most important factors. If the film-forming solution is edible, only edible solvents like water, ethanol, and their mixtures will be appropriate.

Various polymers obtained from products or byproducts of agricultural origin are proposed for the formulation of biodegradable materials or edible films. These polymers, like cellulose polysaccharide, can be used in various forms (coatings, simple or multilayer films, three-dimensional items, simple materials, mixtures, blends and composites). The materials obtained from these agro-polymers are renewable and biodegradable (except when severe chemical modifications are applied). They are nontoxic to the soil and the environment, and when food-grade ingredients are used to formulate the materials, they can be edible. Furthermore, due to some original material properties, they can also be used as an "active bio-packaging" to modify and control food surface conditions (for example, gas selective materials, controlled release of specific functional agents such as flavor compounds, etc.).

However, in the food packaging industry, edible and biodegradable coatings or films produced from agro-polymers provide a supplementary and sometimes essential means of controlling physiological, microbiological and physicochemical changes in food products [3].

To summarize, cellulose applications in the packaging industry can be organized into three main topics. The first one is to extract cellulose from plants and use it directly to prepare composites. The second one is to produce cellulosic plastics like cellulose acetate, which are the best examples of biopolymers derived from renewable resources. The third one is to prepare cellulose coating materials, edible and non-edible films. Therefore, detailed discussions about each of these topics and processes are presented in this chapter along with many related subjects based on cellulose and its derivatives.

In addition, the topic of cellulose-based nanocomposites is a relatively new research field. Nowadays, there is a growing interest in biopolymer-based nanocomposites in

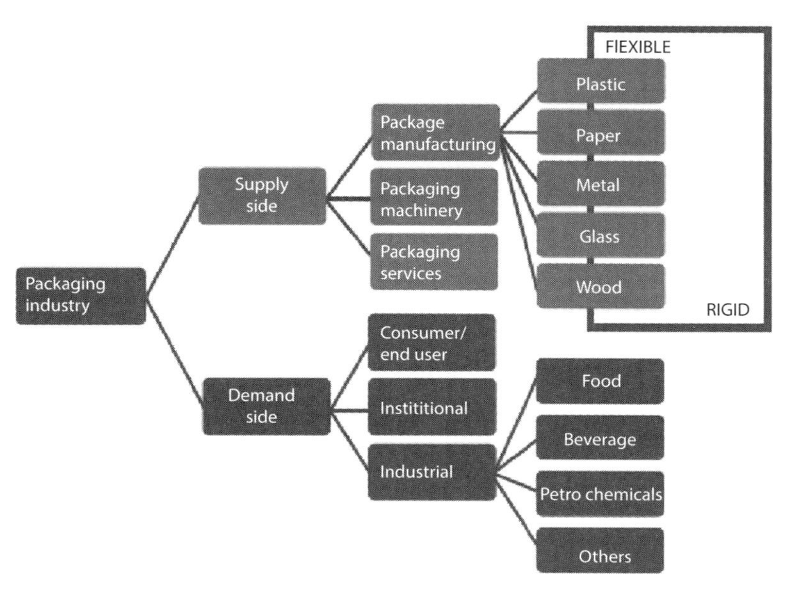

Figure 21.1 The Packaging industry [5].

both developed and developing countries, especially if the nanocomposites are based totally on renewable raw materials [4].

21.1.1 Packaging Materials

The supply side or providers and the demand side or end users are two major components of packaging products in the packaging industry (Figure 21.1). The supply side includes packaging machinery, packaging services and various package productions, from flexible to rigid materials [5].

As shown in Figure 21.1, plastic is the most flexible among all packaging materials, and this is the main reason why it has a very important role in food packaging.

21.1.2 Plastics

A polymer is a large and long-chain molecule constructed from many smaller structural units called monomers (repeating units of identical structure), covalently bonded together in any conceivable pattern. Certain polymers, such as proteins, cellulose, and silk, are found in nature, while many others, including polystyrene (PS), polyethylene (PE), polypropylene (PP), and polyamides, etc., are produced only by synthetic routes. Those commercial materials other than elastomers and fibers that are derived from synthetic polymers are called plastics, which are defined as processable materials [6, 7, 8]. Because of the diversity of function and structure found in the field of macromolecules, it is better presented in the form of a diagram. The diagram is shown in Figure 21.2.

Due to plastics' advantages in comparison with other materials, they have been extensively adopted in food packaging. These advantages are reflected in the physical,

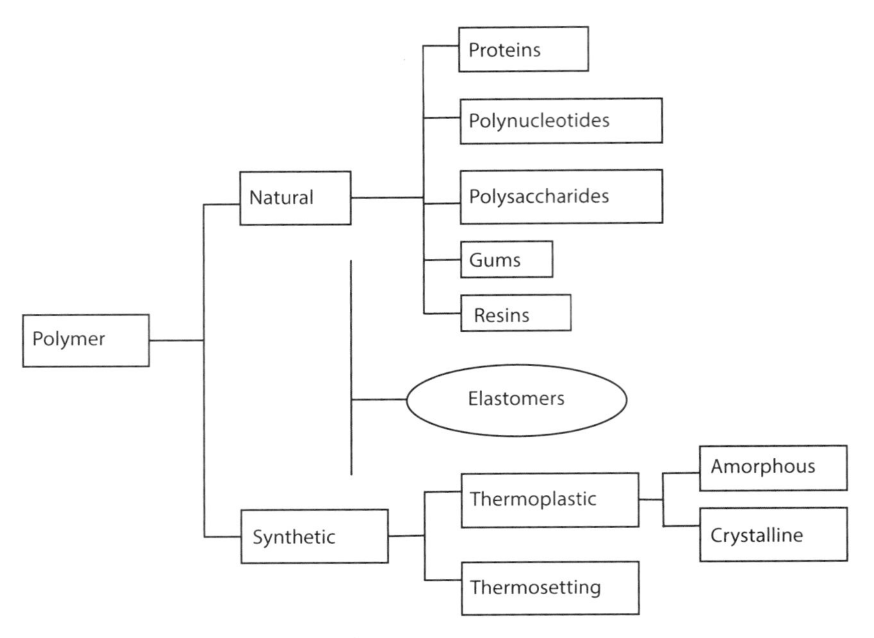

Figure 21.2 A classification of polymers [9].

mechanical, and chemical properties of plastics. They are stable in ambient and many hostile environments and not subject to degradation in normal use. It means that they will not change in properties or performance during the package life [6]. Plastics can be formed into end products such as bottles, containers, films, coatings, etc. The development of self-service stores with their large variety of products would be unimaginable without plastics.

21.1.3 Problems of Plastics

Nowadays, the limitation of petroleum resources, the shortage of landfill space, persistence of plastics in the environment, management of plastics wastes, and ingestion of packaging plastics by marine animals, birds, and others are some problems caused by plastics(Figure 21.3). Figure 21.4 is also a glimpse into the problems caused by plastic on a beach in Iran that may be found anywhere in the world.

Therefore, there is a considerable interest in replacing some or all of the synthetic plastics by natural or biodegradable materials in many applications. Since the food industry uses many plastics, even a small reduction in the amount of materials used for each package would result in a significant polymer reduction, and may improve solid waste problems [10]. It is clear that the use of biodegradable polymers for packaging offers an alternative and partial solution to the problem of gathering of solid waste composed of synthetic inert polymers [11].

The best-known renewable resources able to create biopolymer and biodegradable plastics are starch and cellulose [10,12]. Weber *et al.* believed that the only bio-based food packaging materials in use commercially on a major scale are based on cellulose [13].

Figure 21.3 Photo highlighting the harmful consequences to wildlife of plastics waste. (photo from Internet search).

Figure 21.4 Plastics accumulation on a beach in Chaloos, North of Iran.

Since starch is a source of energy and has an important role in human food, research should be focused on other subjects that are not food but can be used by people. Therefore, cellulose is a good subject to be used in the packaging industry.

According to regulations of the European Union (EU), cellulose is available in the list of allowed monomers and other starting substances, which shall be used for the manufacture of plastic materials, intended to be exposed to foodstuffs. In addition, according to Great Britain, cellulose is certified without time limit for use in the production of polymeric materials in contact with food or drinks or intended for such contact [14].

21.2 Cellulose as a Polymeric Biomaterial

Plant fibers have attracted more and more research interest owing to their advantages like renewability, environmental friendliness, low cost, light weight (low density), high specific mechanical performance (acceptable specific strength properties), easy recyclability, ease of separation, carbon dioxide sequestration, and biodegradability [15].

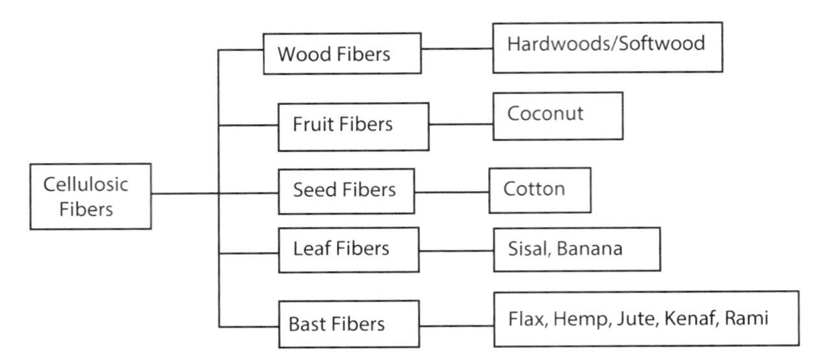

Figure 21.5 Plant fiber classification [16].

Cellulosic fibers originating from both hard and soft woods and from other cellulosic materials are shown in Figure 21.5 [16].

In general, plants include holocellulose (hemicellulose and cellulose), lignin and inorganic materials (ash). Cellulose is resistant to strong alkali (17.5 wt%) but is easily hydrolyzed by acids to water-soluble sugars. Cellulose is relatively resistant to oxidizing agents. Hemicelluloses form the supportive matrix for cellulose microfibrils. Hemicellulose is very hydrophilic and soluble in alkali and easily hydrolyzed in acids. Lignin is not hydrolyzed by acids, but soluble in hot alkali, readily oxidized, and easily condensable with phenol [17]. Cellulose with a degree of polymerization (DP) up to 650 [18,19] can be dissolved in dimethyl sulfoxide containing 10 to 20% (w/v) tetra butyl ammonium fluoride trihydrate without any pretreatment within 15 min at room temperature [19].

Cellulose is not plastic in its native form, but is converted into plastic through various approaches. However, to use cellulose as a polymeric material, it is often necessary to extract it from various plants. Cellulose from trees and plants is a substitute for petroleum feed stocks to make cellulosic plastics.

21.2.1 Cellulose Extraction

The amount of cellulose in lignocellulosic systems, can vary depending on the species and age of the plant. Table 21.1 shows the chemical composition of some common natural plant fibers.

However, there are some methods for extraction of cellulose from plants. Han and Rowell's (1996) method describes a procedure for extraction of holocellulose, hemicelluloses, cellulose and lignin [18]. The method consists of four principle steps: (1) preparation of sample (grinding of the wood), (2) removal of extractives, (3) preparation of holocellulose (removal of lignin), (4) preparation of α-cellulose (removal of hemicellulose). Several other procedures for α-cellulose extraction from wood samples have already been described during the last decades. Older methods used benzene-methanol instead of toluene-ethanol as organic solvent for the second step. Toluene-ethanol works as well as benzene-methanol mixture, and reduces health risks associated with the use of benzene and methanol.

Table 21.1 Chemical composition of some common natural vegetable fibers.

Type of Fiber	Cellulose (%)	Lignin (%)
Cotton	85-90	0.7-1.6
Seed flax	43-47	21-23
Hemp	57-77	9-13
Abaca	56-63	7-9
Sisal	47-62	7-9
Bamboo	26-43	21-23
Kenaf	44-57	15-19
Jute	45-63	21-26
Papyrus	38-44	16-19
Sugarcane bagasse	32-37	18-26
Cereal straw	31-45	16-19
Corn straw	32-35	16-27
Wheat straw	33-39	16-23
Rice straw	28-36	12-16

Source: [20, 21].

The method of Sheu and Chiu (1995) involves two main steps: (1) extraction with benzene-ethanol in a soxhlet extractor for 12 hours, followed by (2) bleaching and soaking in $NaClO_2$-CH_3COOH solution. At the end of the procedure, they also soaked their cellulose in distilled water at 70°C for 6 hours [22]. As cited in [22], Loader *et al.* (1997) also presented a method in which they cut samples into fine slivers (~40 mm) and use organic solvent (2:1 toluene-methylated spirit azeotrope) only for softwood. Finally, samples take place in an ultra-sonic bath, which enhances the removal extracts.

In addition, Borella *et al.* (2004) presented a method that is quite similar to Han and Rowell's (1996) method. They found that fine milling the samples is very important for (1) avoiding sample inhomogeneity leading to increased measurement uncertainty, and (2) permitting complete α-cellulose extraction.

21.2.2 Cellulosic Composites (Green Composites)

A composite material is a two-phase or multiphase compact material with its components (phases) separated by interfaces which can be formed naturally or be manmade. One of the composite material phases is the matrix (phase I). It exists in the solid (crystalline or amorphous) state of aggregation. Within the matrix, particles are distributed discretely. This is phase II or disperse phase [23]. Biocomposites are composite materials made from natural fiber and petroleum-derived nonbiodegradable polymers like PP, PE, and epoxies or biopolymers like poly lactic acid (PLA), cellulose esters. Composite materials derived from biopolymer and synthetic fibers such as glass and carbon come under biocomposites. Biocomposites derived from plant-derived fiber (natural/biofiber) and crop/bioderived plastics (biopolymer/bioplastic) are likely more ecofriendly, and such biocomposites are sometimes termed "green composites" [24].

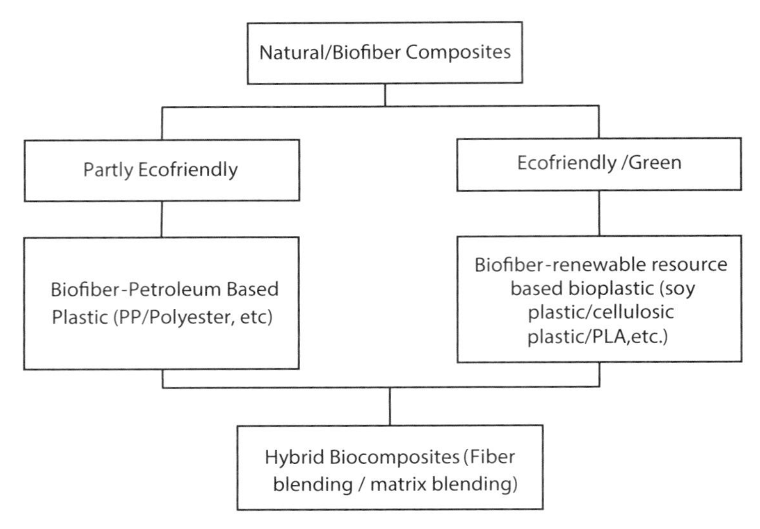

Figure 21.6 Classification of biobased composites [24].

Biocomposites have been the subject of international research since at least the mid-1990s and a number of practical applications are now emerging, including interior automotive components and housings for notebook computers [25]. Research conducted from the 1990s to the present has led to many new biobased products.

Biocomposites include biofibers and matrix polymers systems. They may be obtained from renewable resources and can be synthesized from petroleum-based chemicals. Blending of two or more biopolymers can produce a new biopolymer designed for specific requirements. As represented in Figure 21.6, biocomposites would come under the "partially biodegradable" type, in which the biofiber is biodegradable; and traditional thermoplastics such as PP or PE and thermosets such as unsaturated polyester are nonbiodegradable, or "completely biodegradable," in which matrix resin/polymer is biodegradable. "Hybrid biocomposites" are the result of two or more biofibers in combination with a polymer matrix [24].

Nowadays, the blending of biodegradable polymers with inert polymers has been accepted as a possible application in the waste disposal of plastics. In principal, the way of thinking behind this method is that if the biodegradable section is present in enough quantities and if microorganisms in the waste disposal environment degrade it, the plastic or film containing the residual inert component should lose its integrity, fall to pieces and fade away [10, 12].

Biocomposite products may be categorized as follows [24]:

1. Low biobased content product (20% or less biobased content);
2. Medium biobased content product (21-50% biobased content);
3. High biobased content product (51-90% biobased content).

Heat processing of agro-polymer-based materials like cellulose using techniques usually applied for synthetic thermoplastic polymers such as extrusion, injection, molding, etc., is more cost effective. This process is often used for making flexible films

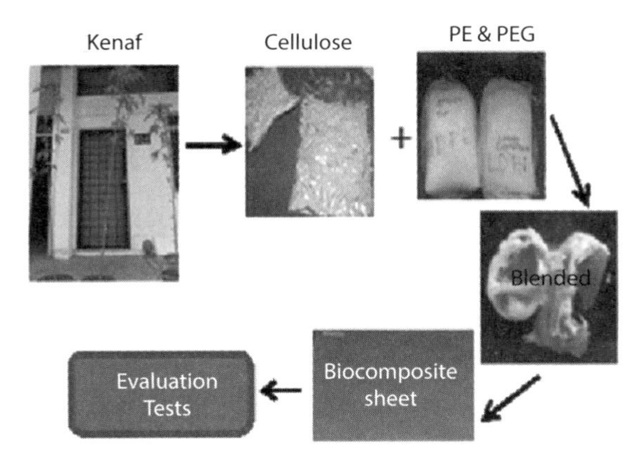

Figure 21.7 Plan for preparing kenaf cellulose/PE biocomposites [21].

(e.g., films for agricultural applications, packaging films, and cardboard coatings) or objects (e.g., biodegradable materials) that are sometimes reinforced with fibers (composite bioplastics for construction, automobile parts, etc.) [3].

The dry process of edible film production does not use liquid solvents such as water or alcohol [2]. In fact, the process pathway of cellulose composites is also a dry method. Molten casting, extrusion, and heat pressing are good examples of dry processes. Therefore, some equipment such as extruder, internal mixer, injection molder, hot press molder, etc., is necessary in this method. For the dry process, heat is applied to the materials to increase the temperature to above their melting point to cause the materials to flow. Thus, the thermoplastic properties of the materials should be identified in order to design composite or film manufacturing processes. It is necessary to determine the effects of compatabilizers, plasticizers and any other additives on the thermoplasticity of the composite or film-forming materials. Plasticizers decrease the glass transition temperature.

For example, Tajeddin (2009) extracted cellulose from Kenaf plant and prepared kenaf cellulose/polyethylene biocomposites (Figure 21.7). The Kenaf (*Hibiscus cannabinus* L.) plant completes its life cycle in one year. It consists of a large amount of cellulose and has many considerable economical and ecological advantages. With our increased population explosion and the quest to continue using cellulose crops from wood and cotton, more land is required to meet the global demand [21]. However, some plants like kenaf that have a short life cycle are good for this purpose as well as agricultural materials wastes.

As shown in Figure 21.7, at first, kenaf fiber was prepared from the bast part of kenaf stems. The cellulose extraction of kenaf fiber was then carried out using the Han and Rowell method [18]. Since cellulose is very hygroscopic material, the obtained cellulose was packed under vacuum packaging. Cellulose and the other materials were blended in the internal mixer and were then compressed using the hot press machine to be converted into sheet form. Finally, the sheet forms of treatments were evaluated by standard tests [21].

Although there is a possibility of using cellulose composites as a packing material to produce containers, its application as a raw material for packaging films is still not possible and requires further studies. However, cellulosic plastics like cellulose acetate are the best examples of biopolymers derived from renewable resources.

21.2.3 Cellulose Derivatives Composites

Due to infusibility and insolubility of cellulose, it is usually converted into derivatives to make it more processable. Some important derivatives of cellulose include ethers like methyl cellulose and hydroxyl-ethyl cellulose, and esters such as cellulose acetate, cellulose butyrate, cellulose acetate-butyrate, etc. However, little attention has been given to these materials in efforts to develop biodegradable plastics. Slow degradation rates, high cost, and processes that generate some noxious discharges are the likely reasons [10, 12].

The hydroxyl groups (-OH) of cellulose can be partially or fully reacted with various reagents to afford derivatives with useful properties like mainly cellulose esters and cellulose ethers.

21.2.3.1 Esterification

The commercially important thermoplastic materials are produced in esterification reactors. Esterification is "the process of combining an organic acid (RCOOH) with an alcohol (ROH) to form an ester (RCOOR) and water" [26]; or "a chemical reaction resulting in the formation of at least one ester product" [27]. Figure 21.8 follows the path of general esterification reactions.

Cellulose esterification does not basically differ from the path shown in Figure 21.8. In fact, the esterification reaction is occurred by the primary and secondary hydroxyl groups of cellulose instead of alcohols. A U.S. patent for the preparation of cellulose ester was issued to Kuo and Leonard in 1984 [28].

Edgar *et al.* (2001) have focused on the performance of cellulose esters in modern coatings, controlled release of actives, plastics (with particular focus on biodegradable plastics), composites and laminates, optical films, and membranes and related separation media [29].

The production of high-quality cellulose esters requires that special attention be paid to the selection of the starting materials. The cellulose bases generally consist of highly purified cotton linters with a cellulose content of over 99% and celluloses from wood pulp that contain between 90 and 97% cellulose.

There are good adhesion characteristics of the cellulose fiber-cellulose ester interface. Cellulose derivatives have been overlooked as potential components in composites with lignocellulosics; only a few studies have considered the use of cellulose esters as matrices in biocomposites. Cellulose esters are very well suited for use as matrix binders in natural fiber-based composites. Cellulose esters can be injection molded, extruded, blow and rotationally molded into structural components, thermoformed from sheet, and extruded in films. Typical processing temperatures lie between 180°

$$\text{Alcohol (ROH)} + \text{Organic Acid (RCOOH)} \xrightleftharpoons{\text{Catalyst}} \text{Ester (RCOOR)} + \text{Water } (H_2O)$$

Figure 21.8 General schematic of the equilibrium system of esterification reactions.

Table 21.2 Some cellulose ester derivatives.

Ester bases	Cellulose ester	Reagent
Organic esters	Cellulose acetate	Acetic acid and acetic anhydride
	Cellulose triacetate	Acetic acid and acetic anhydride
	Cellulose propionate	Propanoic acid
	Cellulose acetate propionate	Acetic acid and propanoic acid
	Cellulose acetate butyrate	Acetic acid and butyric acid
	Cellulose Xanthate	Xanthic acid
Inorganic esters	Nitrocellulose (cellulose nitrate) Cell-OH + $HNO_3 \rightarrow$ Cell-O-NO_2 + H_2O	Nitric acid or another powerful nitrating agent
	Cellulose sulfate Cell-OH + $H_2SO_4 \rightarrow$ Cell-O-SO_3H + H_2O	Sulfuric acid or another powerful sulfuring agent
	Cellulose phosphate Cell-OH + $H_2PO_4 \rightarrow$ Cell-O-PO_3H + H_2O	Phosphoric or another powerful phosphoring agent

and 240°C. Many cellulose ester formulations (e.g., acetates, propionates, butyrates) can be burned without generating toxic products or residues, thus allowing their use in the incineration of waste for recovery of heat value. Cellulose esters can be blended with other synthetic and natural components [30]. For example, cellulose acetate butyrate (CAB) is a cellulose ester that can be used as a film former, as a reactive polyol in curing coatings, and as an additive to other film formers. Due to its high glass transition temperature (Tg), CAB provides excellent hardness. Also, CAB increases productivity and prevents surface defects with reduction in dry time.

Table 21.2 shows some of the cellulose ester derivatives. There are two categories in this table: esters of organic acids are called organic esters and esters of inorganic acids are called inorganic esters. Cellulose nitrate, an ester of inorganic acid, was the first commercial cellulose derivative. It is usually prepared by nitrating using mixtures of nitric and sulfuric acid [28, 30]. There are some doubts about the safety of cellulose nitrate films in food packaging application [31].

21.2.3.2 Etherification

Etherification is defined as the process of converting a substance (as an alcohol or phenol) into an ether that is formed by treating alkali cellulose with a variety of reagents. Etherification requires ionisation of hydroxyl, i.e., C-OH → C-O^-.

A method for controlling the reaction between cellulose and an etherifying agent is provided in U.S. Patent 2492524, which is comprised of the steps: (a) combining cellulose with an alkylene oxide and alkyl halide in a reactor wherein an etherification reaction is initiated which includes a vapor phase, (b) periodically measuring the quantity of at least one of the reactants or products in the vapor phase of the reactor and therewith determining the rate of the etherification reaction taking place, and (c) regulating the reaction temperature continuously while the reaction is taking place by using the rate of reaction determined in step (b) to determine the amount of heating or cooling needed for the reactor [32].

Table 21.3 Some cellulose ether derivatives.

Ether bases	Cellulose ether	Reagent
Alkyl	Methylcellulose	Chloromethane
	Ethylcellulose	Chloroethane
	Ethyl methyl cellulose	Chloromethane and chloroethane
Hydroxyalkyl	Hydroxyethyl cellulose	Ethylene oxide
	Hydroxypropyl cellulose (HPC)	Propylene oxide
	Hydroxyethyl methyl cellulose	Chloromethane and ethylene oxide
	Hydroxypropyl methyl cellulose (HPMC)	Chloromethane and propylene oxide
	Ethyl hydroxyethyl cellulose	Chloroethane and ethylene oxide
Carboxyalkyl	Carboxymethyl cellulose (CMC)	Chloroacetic acid

Most ethers are water soluble and are used as thickeners in foods, cosmetics, pharmaceuticals, paints, etc. Critical properties of ethers include solubility, water-binding capacity, nontoxicity, and chemical stability. Degree of substitution (DS) determines properties of ethers that is the average number of –OHs substituted on anhydroglucose unit. Table 21.3 shows some of cellulose ether derivatives.

Carboxymethyl cellulose (CMC) is the most important cellulose ether commercially. Its DS is 0.4–1.4. DS; more than 0.6–0.8 gives a good water solubility. Carboxymethyl cellulose is one of the important modified celluloses that is widely used as an additive in industries. It possesses advantageous properties, especially solubility in water, immiscibility in oil and organic solvency, which makes it act as a multifunction agent. Therefore, it functions as stabilizer, thickener, binder and suspension agent in industries such as food and pharmaceutical. Hence, analysis of the CMC production process has been studied by researchers in order to enhance the efficiency of the process to achieve specific properties of CMC needed. There are many published works about CMC. For example, optimization of carboxymethyl cellulose production from etherification cellulose has been studied by Muei in 2010 [33].

The effect of methycellulose coating on the storage life of nectarine Var. Rafati was evaluated by Mizani *et al.* in 2009. The results showed that coating may increase overall acceptability, reduce weight loss in nectarine and increase the quality and shelf life of fruits [34].

21.2.3.3 Regenerated Cellulose Fibers

All forms of cellulose degrade before they melt but natural cellulose can be regenerated by the xanthate process to manufacture fibers, commonly called rayon or viscose, and film, commonly called by its earliest brand name Cellophane. Thus, cellulose can be turned into rayon, an important fiber that has been used for textiles since the beginning of the 20th century. Rayon and cellophane are known as "regenerated cellulose fibers." They are identical to cellulose in chemical structure and are usually made from dissolving pulp via viscose. Cellophane is used in the world for its performance in specialty markets including twist-wrapped confectionery, “breathable” packaging for baked goods, "live" yeast and cheese products, and ovenable and microwaveable packaging. However, there are many commercial products with different brand names such as Cellophane™, NatureFlex™, Rayophane™, etc.

The latter films are plasticized by glycols and water to overcome their brittleness and are transparent, colorless and of moderate crystallinity. They were very widely used for packaging but have been substantially replaced by synthetic thermoplastics, especially polypropylene. They have very high permeability to moisture and, especially when dry, very low permeability to permanent gases. Their moisture content varies greatly with their environment, reaching approximately 50% at 100% relative humidity. Not surprisingly, this causes many of their properties to vary considerably—so summarized values must be treated with considerable caution [35].

Uncoated regenerated cellulose films are highly permeable to water vapor but have a good barrier to flavors and aromas. They are suitable for twist wrap and lamination applications for products where protection from moisture is not required. The films are available in a range of thicknesses so they are suitable for single web or laminate applications [36].

21.2.3.4 Bacterial Cellulose (BC)

Bacterial cellulose is an organic compound produced from certain types of bacteria, principally of the genera *Acetobacter*, *Sarcina ventriculi* and *Agrobacterium*. Bacterial or microbial cellulose has different properties from plant cellulose and is characterized by high purity, strength, moldability and increased water holding ability. In natural habitats, the majority of bacteria synthesize extracellular polysaccharides, such as cellulose, which form protective envelopes around the cells. While bacterial cellulose is produced in nature, many methods are currently being investigated to enhance cellulose growth from cultures in laboratories as a large-scale process. By controlling synthesis methods, the resulting microbial cellulose can be tailored to have specific desirable properties. For example, attention has been given to the bacteria *Acetobacter xylinum* due to its cellulose's unique mechanical properties and applications to biotechnology, microbiology, and materials science. With advances in the ability to synthesize and characterize bacterial cellulose, the material is being used for a wide variety of commercial applications including textiles, cosmetics, and health food products, coatings, reinforcement for optically transparent films, as well as medical applications. Many patents have been issued in microbial cellulose applications and several active areas of research are attempting to better characterize microbial cellulose and utilize it in new areas [37].

Microbial cellulose as a new resource for some products has also been studied by Chawla *et al.*, in 2008 [38]. An advantage of using agricultural or industrial residual streams as feedstock for production of bacterial cellulose is the low cost of the raw material. When lignocellulosic feedstocks are pretreated at high temperature and high pressure, they give rise to inhibitory compounds due to the breakdown of polysaccharides and lignin [39]. In the studies of Hong and Qiu (2008) and Hong *et al.* (2011), the hydrolysates obtained through acid hydrolysis of konjak glucomannan and wheat straw had to be detoxified using overliming in order to enable bacterial growth and production of BC [40,41].

In addition, Cavka *et al.* (2013) investigated the appropriateness of waste fiber sludge for production of Bacterial cellulose (BC), and the possibility of combining the production of BC with the production of hydrolytic enzymes useful for degradation of lignocellulose. The fiber sludges were originated from a pulp mill using a sulfate-based process (kraft pulping) and from a lignocellulosic biorefinery using a sulfite-based process. The BC and enzymes were produced through sequential fermentations with the bacterium *Gluconacetobacter xylinus* and the filamentous fungus *Trichoderma reesei*. Fiber sludges from sulfate (SAFS) and sulfite (SIFS) processes were hydrolyzed enzymatically without prior thermochemical pretreatment and the resulting hydrolysates were used for BC production. Cellulase produced in this manner could tentatively be used to hydrolyze fresh fiber sludge to obtain medium suitable for production of BC in the same biorefinery. Fiber sludge consists mainly of cellulose and hemicellulose, and usually has a low content of lignin ($\leq 5\%$). Due to their composition and structure, fiber sludges are usually easy to be hydrolyzed enzymatically without prior thermochemical pretreatment, and could potentially yield hydrolysates with high glucose concentrations and low content of inhibitory compounds. A low content of inhibitory compounds should be advantageous for the bacterial strains used for production of BC. There are, however, drawbacks associated with enzymatic hydrolysis, especially the high cost of the hydrolytic enzymes used in the process [42].

The results of a 2013 study by Cavka *et al.* showed that there is great potential in utilizing waste fiber sludge for co-production of bacterial cellulose and enzymes. Fiber sludge serves as a low cost and abundant raw material that is easily hydrolyzed to sugars without pretreatment. The results also indicate that production of BC from fiber sludge hydrolysates gives a cellulose polymer displaying superior properties compared to the one produced from a glucose-based reference medium. Conditioning of hydrolysates and optimization of the cultivation conditions are likely to result in higher volumetric yields than what are reported here and deserve attention in future studies [42].

21.3 Cellulose as Coatings and Films Material

Coatings and edible films are used to protect food products and improve their quality and shelf life. There has been a great interest in production of edible films and coatings from biopolymers in recent years. The most widely used edible and biodegradable films are cellulose derivatives.

As cited in Guilbert and Gontard, 2005 [3], agro-polymers that have been proposed to formulate edible films or coatings are numerous (Cuq *et al.*, 1995; Guilbert and Cuq, 1998). Plant polysaccharides such as cellulose and derivatives are used in various forms (simple or composite materials, single-layer or multi-layer films). The formulation of bioplastic or edible films implies the use of at least one component able to form a matrix having sufficient cohesion and continuity. They are polymers, which under preparation conditions, have the property to form crystalline or amorphous continuous structures [3].

A few examples of applications of cellulose edible films or coatings are discussed below. They are used to increase product appearance or conservation (increasing shelf life).

21.3.1 Coatings

Edible coatings are thin layers of edible biopolymers applied on the surface of foods as protective coatings. One of the benefits of using cellulose esters in coating is the control of viscosity properties.

The evaluation of the effect of two methyl cellulose (MC)-based active coatings on the quality and storage life of tomatoes was studied by Sadeghipour *et al.* (2012). Fruits were dipped into active MC coating (MC, glycerol and potassium sorbate) and active MC-palmitic acid coating (MC, glycerol, potassium sorbate and palmitic acid) for 1 min at 20°C, then air-dried at room temperature and stored with uncoated samples at 15°C and 80-85% RH for 21 days. At one day intervals, coated and uncoated fruits were removed and evaluated for weight loss, total soluble solids, titrable acidity, pH, skin color, firmness, failure energy, ascorbic acid content, yeast and mold counts and overall acceptability. It was revealed that the coated fruits showed lower weight loss (up to 2.5 times), ripening rate, spoilage incidence (2-3 times), while the skin color, ascorbic acid content, sensory quality and overall acceptability were higher in coated tomatoes when compared with control during storage time. However, insignificant differences in the properties of tomatoes were observed by adding palmitic acid to the formulation of MC active coating [43].

Garcia *et al.* (2002) used methylcellulose (MC) and hydroxypropylmethylcellulose (HPMC) in coating formulations to reduce oil uptake in deep-fat fried potato strips and dough discs. The MC coatings were more effective in reducing oil uptake than HPMC ones. The effect of plasticizer addition (sorbitol) was also evaluated. The best formulations were 1% MC and 0.5% sorbitol for fried potatoes and 1% MC and 0.75% sorbitol for dough discs. For these formulations, oil uptake reduction was 40.6 and 35.2% for potato strips and dough discs compared to the uncoated samples; the increase in water content was 6.3 and 25.7%, respectively. Insignificant differences in the texture of coated and uncoated samples were observed. Although instrumental color differences were detected, all samples were accepted by the untrained panel [44].

The use of water-soluble cellulose acetate (WSCA) as a film coating material for tablets was investigated by Wheatley in 2007. Aspirin tablets were prepared by direct compression and coated with either WSCA or HPMC (hydroxypropyl methylcellulose) dispersions. Coatings of 1-3%, depending on the intended application, were applied to the model drug Aspirin tablets employing a side-vented coating pan. The WSCA has the capability of forming free films without plasticizers and the films dry at room temperature. Glass transition temperature (Tg) of WSCA is significantly higher relative to HPMC. Low viscosity WSCA was more soluble in water (25-30%) relative to medium viscosity WSCA (10-15%). Samples of coated (WSCA and HPMC) tablets and uncoated Aspirin cores were packaged for stability studies at room and elevated temperature storage. After three months at elevated temperature (35 and 45°C), there was no difference in moisture (weight) gain of Aspirin tablets coated with either WSCA or HPMC. The WSCA-coated tablets were not sticky or tacky, while the HPMC-coated tablets were tacky and stuck together [45].

21.3.2 Films

An edible film is essentially a dried and extensively interacting polymer network of a three-dimensional gel structure. Despite the film-forming process, film-forming materials should form a spatially rearranged gel structure with all incorporated film-forming agents such as biopolymers, plasticizers, other additives, and solvents in the case of wet casting. Biopolymer film-forming materials are generally gelatinized to produce film-forming solutions. Further drying of the hydrogels eliminates excess solvents from the gel structure. This does not mean that the film-forming mechanism during the drying process is only the extension of the wet-gelation mechanism. The film-forming mechanism during the drying process may differ from the wet-gelation mechanism, though wet gelation is the initial stage of the film-forming process [2].

In the film-forming mechanism (wet process), a wet gel should be converted into dry film in which the film-forming materials are dispersed in the solvents, followed by drying to remove the solvent and form a film structure. Therefore, there is a critical stage of a phase transition from a polymer-in-water (or other solvents) system to a water-in-polymer system.

As cited in [2], Peyron (1991) concluded that for the wet process, the selection of solvents is one of the most important factors. Since the film-forming solution should be edible and biodegradable, only water, ethanol, and their mixtures are appropriate as solvents. All the ingredients of film-forming materials should be dissolved or homogeneously dispersed in the solvents to produce film-forming solutions. The film-forming solution should be applied to flat surfaces using a sprayer, spreader or dipping roller, and dried to eliminate the solvent, forming a film structure. To produce a homogeneous film structure avoiding phase separation, various emulsifiers can be added to the film-forming solution.

Daraei *et al.* (2009) investigated the effects of adding four plastisizers, namely, glycerol (Gly), polyethylene glycol400 (PEG400), palmitic acid (PA) and GLY + PA, on some mechanical and physicochemical properties of edible methylcellulose (MC) films. The water vapor transmission rates (WVTR), tensile strength (TS), percent elongation (E), and percent soluble matter (SM) of the plastisized films were measured. The results indicate that pure MC films were very brittle. Plasticizers with low molecular weights (Gly) were most effective with regard to decreasing TS and increasing percent elongation of the MC films. On the other hand, plasticizers with higher molecular weights (PA) produced the greatest effect as regards decreasing WVTR of the MC films. The PA + GLY were more suitable for applications that require a lower permeability to water vapor, such as coatings on the surface of fruits [46].

21.4 Nanocellulose or Cellulose Nanocomposites

The idea of bionanocomposites—in which the reinforcing material has nanometer dimensions—is emerging to create the next generation of novel eco-friendly materials with superior performance and extensive applications in medicine, coatings, packaging, the automotive industry, etc.

Gindland and Keckes (2005) produced cellulose-based nanocomposite films with different ratio of cellulose I and II by means of partial dissolution of microcrystalline cellulose powder in lithium chloride/*N*, *N*-dimethylacetamide and subsequent film casting. The mechanical and structural properties of the films were characterized using tensile tests and X-ray diffraction. The films were isotropic, transparent to visible light, highly crystalline, and contain different amounts of undissolved cellulose I crystallites in a matrix of regenerated cellulose. The results show that by varying the cellulose I and II ratio, the mechanical performance of the nanocomposites can be tuned. Depending on the composition, a tensile strength up to 240 MPa, an elastic modulus of 13.1 GPa, and a failure strain of 8.6% were observed. Moreover, the nanocomposites clearly surpassed the mechanical properties of most comparable cellulosic materials, their greatest advantage being the fact that they are fully biobased and biodegradable, but also of relatively high strength [47].

Nanostructured composites, where both the reinforcement and the matrix are biobased, are discussed in the book by Oksman and Sain (2006). Cellulose combined with natural polymers led to the development of a class of biodegradable and environmentally-friendly bionanocomposites. This family of nanocomposites is expected to remarkably improve material properties when compared with the matrix polymers or conventional micro- and macro-composite materials. Such improvements in properties typically include a higher modulus and strength, improved barrier properties, and increased heat distortion temperature [4].

21.5 Quality Control Tests

In all materials used in different aspects of cellulose applications in food packaging, the material characteristics including cellulose, plasticizers, etc., and the fabrication procedures like composites, nanocomposites, casting of a film-forming solution, thermoforming, and so on, must be adapted to each specific food product and the conditions in which it will be used such as relative humidity, temperature, etc. Furthermore, edible and biodegradable films must meet a number of specific functional requirements like moisture barrier, gas barrier, water solubility, color and appearance, nontoxicity, etc.

Therefore, for quality control of composites or film-forming mechanisms, many tests like mechanical and rheological characteristics, thermal properties, biodegradability, etc., are necessary [48, 49].

In addition, several polymer chemistry laboratory techniques including X-ray diffraction, atomic force microscope (AFM), transmission electron microscope (TEM), scanning electron microscope (SEM), FTIR spectrometry, thermogravimetric analysis (TGA), differential scanning calorimetric (DSC), NMR spectrometry, electrophoresis, polarizing microscopy, and other polymer analysis methodologies are required. For example, to obtain the mechanical properties, tensile, elongation, flexural and Impact tests are some of the required tests. Because a major drawback of fiber-based material is its hydrophilicity, the degree of absorption of water is also necessary. It is required to determine the effects of plasticizers and any other additives on the thermoplasticity of the film-forming materials.

Han and Gennadios (2005) stated that for the determination of most physical chemistry parameters, cohesion of polymers is very important. If the composites or film-forming materials contain heterogeneous ingredients that are not compatible with the main biopolymers, the cohesion of the film-forming materials decreases and the film strength weakens. When the use of new biopolymers or additives is investigated, the compatibility of all film-forming ingredients should be maintained to obtain strong cohesion [2].

Owing to relatively low water-vapor barrier properties of agro-polymer-based materials and edible films, they can only be used as protective barriers to limit moisture exchange for short-term applications. However, they can be of considerable interest for numerous applications where very high water vapor permeability is required, such as in the case of modified atmosphere packaging (MAP) of fresh, minimally processed or fermented foods like fish, meat, fruits, vegetables, and cheeses. Agro-polymer-based materials have impressive gas-barrier properties in dry conditions, especially against oxygen. Increasing the water activity promotes both the gas diffusivity and the gas solubility, leading to a sharp increase in the gas permeability. With carbon dioxide, a sharp increase in the permeability is more important than with the oxygen permeability. The properties of amorphous or semicrystalline materials are seriously modified when the temperature of the compounds rises above the glass transition temperature (Tg). Generally, fully amorphous bioplastic applications are limited by the fact that a polymer's Tg is highly affected by the relative humidity (especially for hydrophilic polymers) [3].

The agro-polymer-based materials are generally fully biodegradable (apart from when some very severe chemical modifications are applied). Biodegradation kinetics are dependent on the type of polymer used (molecular weight, structure, crystallinity) and on the additives used such as plasticizers and fillers. As cited in [3], the methods used to evaluate biodegradability are generally based on Sturrn's procedure (Sturrn, 1973), i.e., International Standard IS0 14852, which measures the ultimate aerobic biodegradability of materials. Another approach consists of the evaluation of the biodegradability in soil.

In general, biodegradability of polymers can be tested using Figure 21.9. Screening tests by either enzymatic or aquatic means are inexpensive and fast, but real-life tests can be laborious and expensive. Neither of these two types of processes, however, actually simulates the condition truly present in landfills [50, 51].

Edible films and coatings must have organoleptic properties that are as neutral as possible (clear, transparent, odorless, tasteless, etc.) so as not to be detected when eaten. Enhancing the surface appearances (e.g., brilliance) and the tactile characteristics (e.g., reduced stickiness) can be required. It is possible to obtain materials with ideal organoleptic properties, but they must also be compatible with the food's filling. Films and coatings can also help to maintain desirable concentrations of coloring, flavor, spiciness, acidity, sweetness, saltiness, etc. The optical properties of films depend on the film formulation and fabrication procedures [3].

Cellulose nanocomposites also have proper thermal, barrier and mechanical properties compared to today's biomaterials and can be synthesized from biopolymer and nanosized reinforcements [4].

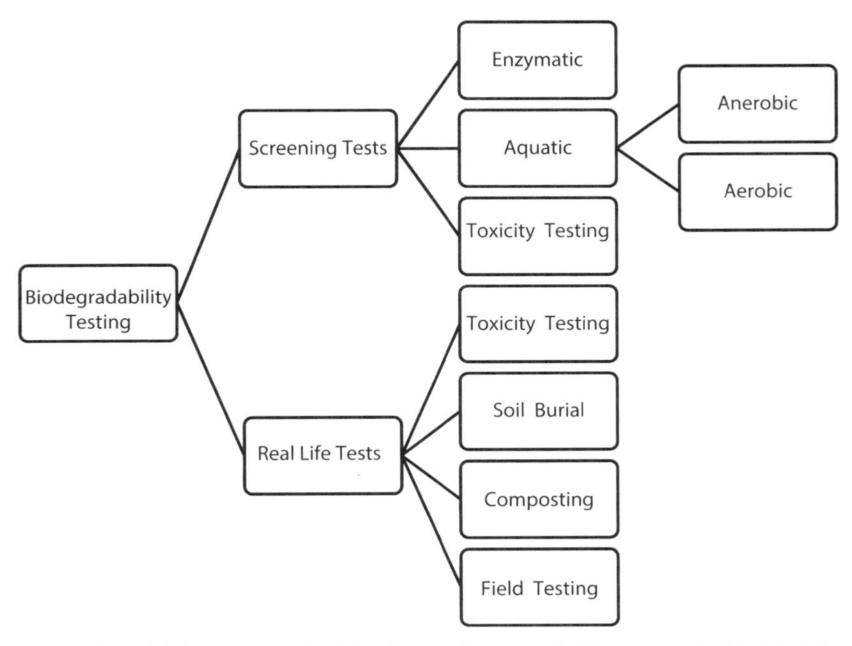

Figure 21.9 Biodegradability testing of solid polymers [Itavaara & Vikman, as cited in 50; 51].

21.6 Conclusions

Nowadays, our society faces many challenges due to environmental problems, which are the reason for the transition toward sustainability. The amount of nondegradable substances throughout the world is a fact, therefore, utilization of agricultural and forest resources are enormous.

The potential of cellulose as reinforcing agent has been demonstrated by its many properties. Different cellulose resources provide different mechanical characteristics. Cellulose is, for example, the major constituent of cotton (over 94%) and wood (over 50%). Cotton and wood are the major resources for all cellulose products such as paper, textiles, construction materials, cardboard, as well as such cellulose derivatives as cellophane, rayon, and cellulose acetate.

However, cellulose has an attractive combination of mechanical and thermal properties extracted from various plant sources. Extracted cellulose performs as a raw material in many packaging industry applications such as biocomposites, nanocomposites, coatings, films, etc. The process of converting cellulose to these products is quite complex. For example, to convert wood pulp to film, it undergoes a series of chemical processes to break the pulp down to a viscose liquid. This liquid is filtered in order to minimize waste and maximize the purity of the material to ensure the best film quality possible. The viscose is extruded and then cast along a series of rollers and baths, during which the film is cleaned and softened in order to ensure the right optical and mechanical properties.

Since uncoated cellulose films are highly permeable to water vapor, they are used for foods that do not require a complete moisture or gas barrier, including fresh bread and some types of confectionery. All cellulose films are heat resistant, heat sealable, machinable, and printable.

Cellulose derivatives are additive components for many thermoplastic and thermosetting coating systems. Cellulose esters as film formers provide fast drying and early hardness development, flooding and floating suppression, and crosslinking reactions.

References

1. S. Guilbert, B. Cuq, and N. Gontard. Recent innovations in edible and or biodegradable packaging materials. *Food Addit. Contam.* 14 (6–7), 741–751 (1997).
2. J.H. Han and A. Gennadios. Edible films and coatings: A review, in *Innovations in Food Packaging,* J. H. Han, (Ed.), pp. 239–262, Elsevier Academic Press, London (2005).
3. S. Guilbert and N. Gontard. Agro-polymers for edible and biodegradable films: review of agricultural polymeric materials, physical and mechanical, in *Innovations in Food Packaging,* J.H. Han, (Ed.), pp. 263–267, Elsevier Academic Press, London (2005).
4. K. Oksman and M. Sain (Ed.), *Cellulose Nanocomposites, Processing, Characterization and Properties,* ACS Symposium Series, Volume 938, American Chemical Society, Washington (2006).
5. N.M. Manalili, M.A. Dorado, and R. van Otterdijk. Appropriate food packaging solutions for developing countries, in *Food and Agriculture Organization of The United Nations (FAO),* pp. 1–37, FAO, Rome (2011).
6. W.E. Brown. *Plastics in food packaging: properties, design and fabrication.* Marcel Dekker, Inc., New York (1992).
7. J. R. Fri (Ed.), *Polymer Science & Technology,* Pearson Education, Inc (2003).
8. M. Sain and S. Panthapulakkal. Green fiber thermoplastic composites, in *Green Composites, Polymer Composites and the Environment,* C. Baillie, (Ed.), pp. 181–206, CRC Press, Boston (2004).
9. J.M.G. Cowie, *Polymers: Chemistry and Physics of Modern Materials,* Taylor & Francis, Boca Raton (2008).
10. R. Chandra and R. Rustgi. Biodegradable polymers program polymer science. *Prog. Polym. Sci.* 23, 1273– 1335 (1998).
11. R. Jayasekara, I. Harding, I. Bowater, G.B.Y. Christie, and G.T. Lonergan. Preparation, surface modification and characterization of Solution cast starch PVA blended films. *Polym. Test. 23,* 17–27 (2003).
12. S.E. Selke, Plastics recycling and biodegradable plastics", in *Modern Plastics Handbook,* C.A. Harper, (Ed.), Chapter 12, pp. 11–108, McGraw – Hill, New York (2000).
13. C.J. Weber, V. Haugaard, R. Festersen, and G. Bertelsen. Production and applications of biobased packaging materials for the food industry. *Food Addit. Contam.* 19 (4 Suppl.), 172 – 177 (2002).
14. O.V. Sheftel. *Indirect food additives and polymers: Migration and toxicology,* CRC Press, Boca Raton (2000).
15. M.Q. Zhang, M.Z. Rong, X. Lu. Fully biodegradable natural fiber composites from renewable resources: all-plant fiber composites. *Compos. Sci. Technol.* 65, 2514–2525 (2005).
16. S. Godavarti, Thermoplastic wood fiber composites, in *Natural Fibers, biopolymers, and Biocomposites,* A.M. Mohanty, M. Misra, and L.T. Drzal, (Eds.), pp. 347–389, CRC Press, Boca Raton (2005).

17. A. Bismarck, S. Mishra, and T. Lampke. Plant fibers as reinforcement for green composites, in *Natural fibers, biopolymers, and biocomposites,* A.M. Mohanty, M. Misra, and L.T. Drzal, (Eds.), pp. 37– 108, CRC Press, Boca Raton (2005).
18. J.S. Han, and J.S. Rowell, Chemical composition of fibers, in *Paper and Composites from Agro-Based Resources,* R.M. Rowell, and J. Rowell, (Eds.), pp. 83–130, CRC Press, Boca Raton (1996).
19. T. Heinze, R. Dicke, A. Koschella, A. H. Kull, E. A. Klohr, and W. Koch. Effective preparation of cellulose derivatives in a new simple cellulose solvent. Indirect food. In *Additives and Polymers,* V. Sheftel, (Ed.). Lewis, Boca Raton (2000).
20. F. Herrera, J. Pedro, and A. Valadez-Gonzalez. Fiber- matrix adhesion in natural fiber composites, in *Natural fibers, biopolymers, and biocomposites,* A.M. Mohanty, M. Misra, and L.T. Drzal, (Eds.), pp. 177– 230, CRC Press, Boca Raton (2005).
21. B. Tajeddin. Preparation and characterization of kenaf cellulose/polyethylene glycol/polyethylene biocomposites. Engineering Faculty, UPM, Malaysia, *PhD Thesis* (2009).
22. S. Borella, G. Menot, and M. Leuenberger. Sample homogeneity and cellulose extraction from plant tissue for stable isotope analyses, in *Handbook of stable isotope analytical techniques,* P.A. Groot, (Ed.), Volume 1, pp. 507–522, Elsevier (2004).
23. R.S. Saifullin. *Physical chemistry of inorganic polymeric and composite materials, (Series in polymer science & technology),* Ellis Horwood, New York (1992).
24. A.K. Mohanty, M. Misra, L.T. Drzal, S.E. Selke, B.R. Harte and G. Hinrichsen. Natural fibers, biopolymers, and biocomposites: an introduction, in *Natural fibers, biopolymers, and biocomposites,* A.M. Mohanty, M. Misra, and L.T. Drzal, (Eds.), pp. 1– 36, CRC Press, Boca Raton (2005).
25. D. Plackett and A. Vazquez, Natural polymers sources, in *Green Composites, Polymer Composites and the Environment,* C. Baillie, (Ed.), pp. 123–153, CRC Press, Boca Raton (2004).
26. Mosby, *Mosby's Medical Dictionary*, 8th Ed., Elsevier, London (2009).
27. Houghton Mifflin Company. *The American Heritage, Medical Dictionary,* Copyright © 2007, Houghton Mifflin Company (2004).
28. C.M. Kuo and A.P. Leonard. Process for esterification of cellulose using as the catalyst the combination of sulfuric acid, phosphoric acid, and a hindered aliphatic alcohol, US Patent 4480090 A, assigned to Eastman Kodak Company (October 30, 1984).
29. K.J. Edgar, C.M. Buchanan, J.S. Debenham, P.A. Rundquist, B.D. Seiler, M.C. Shelton, and D. Tindall. Advances in cellulose ester performance and application. *Prog. Polym. Sci.* 26(9), 1605–1688 (2001).
30. G. Toriz, P. Gatenholm, B.D. Seiler, and D. Tindall. Cellulose fiber-reinforsed cellulose esters: Biocomposites for the future, in *Natural Fibers, Biopolymers, and Biocomposites,* A.K. Mohanty, M. Misra, and L.T. Drzal, (Eds.), pp. 617–*638,* Boca Raton (2005).
31. Health and Safety Executive, The dangers of cellulose nitrate film. Series code: INDG469, http://www.hse.gov.uk/index.htm (2013).
32. Manufacture of cellulose ethers, US Patent 2492524, assigned to Hercules Powder Co Ltd, (September 18, 1945).
33. C.L. Muei. Optimization of carboxymethyl cellulose production from etherification cellulose, Universiti Malaysia Sabah, Malaysia, Master's thesis (2010).
34. M. Mizani, F. Darabi, F. Badiei, and A. Gerami. The influence of methylcellulose edible coating on the storage life of nectarine. *J. Food Technol. Nutr.* 6(3), (2009). (In Persian)
35. http://www.goodfellow.com/E/Cellulose-Film.html
36. http://www.innoviafilms.com/Our-Products/Packaging/Cellulose-Film-Families/Uncoated-Films.aspx

37. N. Petersen and P. Gatenholm. Bacterial cellulose-based materials and medical devices: Current state and perspectives. *Appl. Microbiol. Biotechnol.* 91(5), 1277–1286 (2011).
38. P.R. Chawla, I.B. Bajaj, S.A. Survase and R.S. Singhal. Microbial cellulose: Fermentative production and applications. *Food Technol. Biotechnol.* 47(2), 107–124 (2009).
39. L. Chen, F. Hong, X. Yang and S. Han. Biotransformation of wheat straw to bacterial cellulose and its mechanism. *Bioresour. Technol.* 135, 464–468 (2012).
40. F. Hong and K. Qiu. An alternative carbon source from konjac powder for enhancing production of bacterial cellulose in static cultures by a model strain Acetobacter aceti subsp. xylinus ATCC 23770. *Carbohydr. Polym.* 72, 545–549 (2008).
41. F. Hong, Y.X. Zhu, G. Yang, and X.X. Yang. Wheat straw acid hydrolysate as a potential cost-effective feedstock for production of bacterial cellulose. *J. Chem. Technol. Biotechnol.* 86, 675–680 (2011).
42. A. Cavka, X. Guo, S. J. Tang, S. Winestrand, L. J. Jönsson and F. Hong. Production of bacterial cellulose and enzyme from waste fiber sludge. *Biotechnol. Biofuels* 6 (25) (2013).
43. M. Sadeghipour, F. Badii, H. Behmadi and B. Bazyar. The effect of methyl cellulose based active edible coatings on the storage life of tomato. *J. Food Sci. Technol.* 9 (35), 89–98 (2012). (In Persian)
44. M.A García, C. Ferrero, N. Bértola, M. Martino, and N. Zaritzky. Edible coatings from cellulose derivatives to reduce oil uptake in fried products. *Innov. Food Sci. Emerg. Technol.* 3(4), 391–397 (2002).
45. T.A. Wheatley. Water soluble cellulose acetate: A versatile polymer for film coating. *FMC Biopolym.* 33(3), 281–90 (2007).
46. F. Daraei, F. Badii, M. Mizani, and A. Gerami. Effects of addition of plasticizers on mechanical and physicochemical properties of edible methylcellulose films. *Iran. J. Nutr. Sci. Food Technol.* 4(3) 47–54 (2009). (In Persian)
47. W. Gindland, and J. Keckes. All-cellulose nanocomposite. *Polymer* 46(23), 10221–10225 (2005).
48. B. Tajeddin, R.A. Rahman, and L.C. Abdulah. The effect of polyethylene glycol on the characteristics of kenaf cellulose/low-density polyethylene biocomposites. *Int. J. Biol. Macromol.* 47, 292–297 (2010).
49. B. Tajeddin and L. C. Abdulah. Thermal properties of high density polyethylene-kenaf cellulose composites. *Polym. Polym. Compos.* 18(5), 195–199 (2010).
50. W. Sridach, K.T. Hodgson, and M.M. Nazhad, Biodegradation and recycling potential of barrier coated paperboards. *Bioresources* 2(2), 179–192 (2006).
51. B. Tajeddin, A.R. Russly, C.A. Luqman, A.Y. Yus and I. Nor Azowa. Effect of PEG on the biodegradability studies of Kenaf cellulose -polyethylene composites. *Int. Food Res. J.* 16, 243–247 (2009).

22

Applications of Kenaf-Lignocellulosic Fiber in Polymer Blends

Norshahida Sarifuddin[1] and Hanafi Ismail*,[1,2]

[1]*School of Material and Mineral Resources Engineering, USM Engineering Campus, Penang, Malaysia*
[2]*Clusters for Polymer Composites (CPC), Science and Engineering Research Centre (SERC), USM Engineering Campus, Penang, Malaysia*

Abstract
Increasing environmental concerns have encouraged researchers worldwide to expand the invention of new materials that may replace conventional materials in various applications. Among the available lignocellulosic fibers, kenaf has recently emerged as an interesting reinforcing component for the development of natural fiber-based composites. The unique attributes of kenaf fibers, which are comparable to those of synthetic fibers, have driven the interest in exploiting their full potentiality in polymer composites and their diverse use. Yet, their compatibility with polymer matrices and inherent moisture absorption have become crucial concerns. Thus, this chapter addresses various aspects of kenaf fiber-reinforced polymer composites with respect to the characteristics of natural fibers. Also discussed are Malaysian kenaf, properties of kenaf-composites and possible treatments for fiber.

Keywords: Kenaf, Malaysian cultivation, sago starch

22.1 Introduction

Growing dependency on petroleum-derived plastic materials has raised some concerns from the environmental perspective. Since they are inherently inert to microorganisms or chemicals in the environment, petroleum-based products can pose significant issues regarding waste disposal to some extent. Great attention has been paid to the biodegradable or environmentally acceptable materials over the past decades due to environmental pressure derived from the consumption of petroleum-based materials [1]. This has prompted the urgent development of new materials allied with environmental preservation. Attempts have been undertaken to study the viability of using a natural fiber as a reinforcing agent in polymer matrices [2].

**Corresponding author*: hanafi@eng.usm.my

Vijay Kumar Thakur (Ed.), Lignocellulosic Polymer Composites, (499–522) 2015 © Scrivener Publishing LLC

The biological nature of cellulose-based materials offers the prospect of integrating environmentally-friendly aspects into many products [3]. Natural fiber-based products have shown their biodegradability under controlled conditions after a certain lifetime, which suggests they are a reliable approach for technically degradable composite. In fact, they are relatively abundant, low in cost, low in density, nonabrasive to equipment, non-irritable to skin, renewable and durable [2, 4, 5]. Also, they encompass specific tensile properties and impart fewer health risks. Certainly, natural fibers have become preferable in various applications, and accordingly, efforts are underway to produce a composite material with superior environmental and mechanical performance with a relative cost balance [3]. Despite their attractive characteristics, several major technical considerations must be taken into account. Challenges include a full understanding of the degree of polymerization and fiber distribution, adhesion between the fiber and matrix, as well as moisture-resistant properties.

This chapter is based on present state-of-the-art research undertaken to develop such a composite. A review on the available natural fibers and polymer matrix is given in brief for a comparative study. Given the broad scope of this chapter, it will inevitably be incomplete, but will hopefully provide an overview of the topic.

22.2 Natural Fibers

As previously mentioned, the incorporation of reinforcements such as natural fibers is one of the interesting approaches to enhance the properties of polymer composites, which can address the significant requirements of most engineering applications. For this reason, the demand for natural fiber-reinforced composites has increased dramatically over the past few years for various commercial applications [6]. The increasing number of publications in recent years reflects the growing importance of this type of composite. Figure 22.1 shows the number of journals published about natural fibers and polymer composites over the past five years. Up to now, studies have established a rough indication of the increased interest of researchers on this topic. The natural fibers

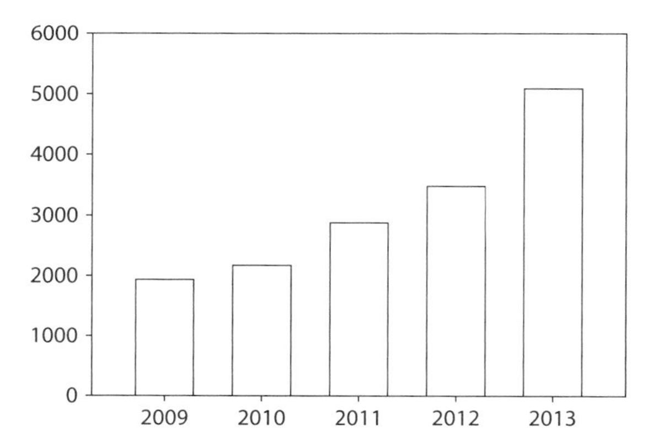

Figure 22.1 Number of journals published on natural fibers and polymer composites. Source: www.sciencedirect.com; keywords: natural fiber, polymer composites.

Table 22.1 Recently reported work on natural fibers and polymer composites.

Fiber	Matrix	References
Baggase	Polypropylene	[15]
Banana	Poly (lactic acid)	[16]
Doum	Polypropylene	[17]
Flax	Polypropylene	[18]
Hemp	Polypropylene	[19]
Henequen	High density polyethylene (HDPE)	[20]
Sisal	Polypropylene	[21]
Sugarcane	Low density polyethylene (LDPE)	[22]

which have been reported in literature to reinforce different polymer matrices are listed in Table 22.1.

In view of recent global environmental issues, researchers worldwide have begun to show an interest in exploiting the full potentiality of natural fibers as reinforcing agents in polymer composites and their diverse use. Efforts are underway to replace synthetic materials with natural materials. These natural fibers demonstrate interesting attributes in fulfilling the particular requirements of the composites, particularly in most engineering applications, owing to their advantages over conventional glass and carbon fibers. Utilization of these fibers as reinforcing components in polymer composites is an effective way to produce lightweight, low cost, eco-friendly and naturally degradable composites without affecting their rigidity [7]. Accordingly, fibers are known to provide strength and rigidity to the weak and brittle matrices due to their relatively high strength and stiffness. Apart from the aforementioned advantages, natural fibers have attracted increasing research interest owing to their renewability, low density, ease of separation, carbon dioxide sequestration and non-abrasiveness to equipment [2, 4, 5]. These studies emphasize the use of natural fibers derived from annually renewable resources as reinforcing fibers in both thermoplastic and thermoset matrix composites, which seem to confer positive environmental benefits pertaining to their disposability [8].

Basically, natural fibers are derived from plants, animals or minerals. In numerous applications, natural fibers from plants are commonly considered as reinforcements for polymeric composites [9]. As shown in Figure 22.2, the plant fibers (cellulose or lignocellulose) can generally be classified accordingly into bast, leaf, fruit, seed, grass and stalk fibers. The selection of fibers for various applications depends on their compositions and structures. A sound knowledge of these properties helps in understanding their physical and mechanical properties in detail. Chemical compositions of some common cellulosic or lignocellulosic fibers are illustrated in Table 22.2. Natural fibers basically consist of cellulose, lignin and hemicelluloses, while pectins, pigments and extractives can be found in smaller amounts. Natural fibers are also referred to as cellulosic or lignocellulose fibers because each fiber is essentially a composite where rigid cellulose microfibrils are embedded in a softer matrix mainly composed of lignin and hemicelluloses [10].

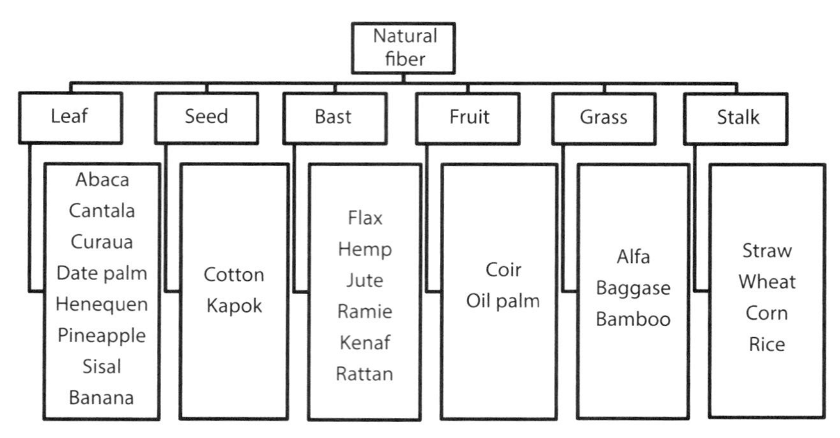

Figure 22.2 Classification of natural fibers [5].

Natural fiber has an intricate layered structure which contains a primary cell wall and three secondary cell walls. Figure 22.3 shows the structural constitution of natural fiber cell. Each cell wall consists of three key components which are mainly cellulose, hemicelluloses and lignin. Lignin-hemicellulose acts as a matrix while microfibrils (made up of cellulose molecules) serve as fibers [9]. In plant fiber, cellulose can be described as a natural polymer where each repeating unit contains three hydroxyl groups. Cellulose is resistant to hydrolysis, strong alkali and oxidizing agents, but to some extent, it is degradable when exposed to chemical and solution treatments. Hemicellulose is a polysaccharide with a lower molecular weight that functions as a cementing matrix between cellulose microfibrils, forming the main structural component of the fiber cell. It is hydrophilic and can be easily hydrolyzed by dilute acids and bases. Lignin is a complex hydrocarbon polymer that gives the plant rigidity and assists in the transportation of water. It is hydrophobic, resists acid hydrolysis and most microbial attacks, is soluble in hot alkali, readily oxidized and easily condensable with phenol [9].

Apart from chemical composition, the properties of cellulosic fibers are greatly influenced by their morphological factors such as internal fiber structure, microfibril angle, dimensions and defects of the cell [10]. It is believed that microfibrillar angles are responsible for the mechanical properties of fibers. IAdditionally, natural fibers with higher cellulose content, higher degree of polymerization of cellulose, longer cell length and lower microfibrillar angle possess higher mechanical strength [9]. Table 22.3 compares the mechanical and physical properties of various natural fibersreported on in the literature. The properties of natural fibers differ among cited works, because different fibers were used, different moisture conditions were present, and different testing methods were employed [11].

The application of natural fibers as a reinforcement in composite materials is a continuing development. This is because, there are some challenges associated with the use of natural fibers as reinforcements in polymer composites. These include the incompatibility between fiber and polymer matrices, poor moisture resistance and limited processing temperature. The main cause for these shortcomings is the presence of hydroxyl and other polar groups in natural fibers, which makes them hydrophilic in nature. This hydrophilicity results in incompatibility with the hydrophobic polymer matrix [12].

Table 22.2 Chemical composition of some common natural fibers [7, 11].

Fiber	Cellulose (wt.%)	Hemicellulose (wt.%)	Lignin (wt.%)	Waxes (wt.%)
Baggase	55.2	16.8	25.3	
Bamboo	26-43	30	21-31	
Flax	71	18.6-20.6	2.2	1.5
Kenaf	72	20.3	9	
Jute	61-71	14-20	12-13	0.5
Hemp	68	15	10	0.8
Ramie	68.6-76.2	13-16	0.6-0.7	
Abaca	56-63	20-25	1-9	3
Sisal	65	12	9.9	2
Coir	32-43	0.15-0.25	40-45	
Oil Palm	65		29	
Pineapple	81		12.7	
Curaua	73.6	9.9	7.5	
Wheat Straw	38-45	15-31	12-20	
Rice Husk	35-45	19-25	20	14-17
Rice Straw	41-57	33	8-19	8-38
Alfa	45.4	38.5	14.9	2
Banana	63-67.6	10-19	5	
Coconut	36-43	0.15-0.25	41-45	
Corncob	26.1	45.9	11.3	
Cotton	82.7-90	5.7	<2	0.6
Henequen	60-77.6	4-28	8-13.1	0.5
Isora	74		23	1.09
Nettle	86	10		4
EFB	65		19	
Olive husk	40.56	18.1	23.43	
Piassava	28.6	25.8	45	
PALF	70-83		5-12.7	
Soy hulls	56.4	12.5	18	
Eucalyputs wood	37.6	32.9	19.1	
Hardwood	31-64	25-40	14-34	
Softwood	30-60	20-30	21-37	

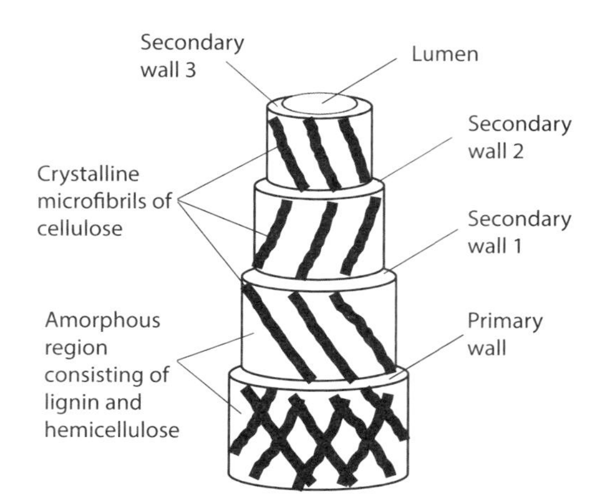

Figure 22.3 Schematic representation of the structural constitution of natural fiber cell (Figure inspired by studies in [23]).

Table 22.3 Mechanical and physical properties of various natural fibers [7, 24].

Fiber	Tensile Strength (MPa)	Young's Modulus (GPa)	Elongation at break (%)	Density (g/cm^{-1})
Abaca	400	12	3-10	1.5
Alfa	350	22	5.8	0.89
Baggase	290	17		1.25
Bamboo	140-230	11-17		0.6-1.1
Banana	500	12	5.9	1.35
Coir	175	4-6	30	1.2
Cotton	287-597	5.5-12.6	7-8	1.5-1.6
Curaua	500-1150	11.8	3.7-4.3	1.4
Date palm	97-196	2.5-5.4	2-4.5	1-1.2
Flax	345-1035	27.6	2.7-3.2	1.5
Hemp	690	70	1.6	1.48
Henequen	430-570	10.1-16.3	3.7-5.9	1.2
Isora	500-600		5-6	1.2-1.3
Jute	393-773	26.5	1.5-1.8	1.3
Kenaf	930	53	1.6	
Nettle	650	38	1.7	
Oil palm	248	3.2	25	0.7-1.55
Piassava	134-143	1.07-4.59	7.8-21.9	1.4
Pineapple	1.44	400-627	14.5	0.8-1.6
Ramie	560	24.5	2.5	1.5
Sisal	511-635	9.4-22	2-2.5	1.5

Therefore, focusing on those issues may improve the performance of the composite and widen the applications for such materials. Several researchers have proved that fiber treatment is necessarily done to increase the compatibility between polymer and natural fibers [13].

For these circumstances, kenaf has recently gained a lot of attention as a reinforcing material [14]. With regards to technical utilization of natural fibers, kenaf-reinforced composites are one of the most important areas. They have successfully been widely incorporated in various applications. In the following section of this chapter, the overall characteristics of kenaf fiber as a reinforcing agent in polymer composites will be reviewed and discussed in detail [11].

22.3 Kenaf: Malaysian Cultivation

Kenaf is an annual plant and a member of the Hibiscus family (Hibiscus cannabinus L. Malvaceae). It has a high growth rate and can reach a height of 3.7-5.5 m with a stem diameter of 25-51 mm within 4-5 months in suitable temperature and rainfall conditions [24, 25]. Kenaf, a word of Persian origin, refers to a warm season, short-day and annual herbaceous plant. It is a dicotyledon, which represents three main layers in the stalk: an outer cortical (or bast) tissue layer (phloem), an inner woody (called the core) tissue layer (xylem) and thin central pith layer [26].

It is the third world crop, after wood and bamboo, which has been introduced as a new annually renewable resource for industrial purposes [24]. Recently, kenaf is being explored as a useful raw material for a layer in papermaking [26], where previously it was used mainly for cordage, canvas and sacking [27].

At present, kenaf fiber is selected as a reinforcing material in composites for the following reasons. First, kenaf is a crop that grows commercially and it has already been used to reinforce common synthetic polymers such as polyolefin and polyester. Second, due to its high cellulose content (about 72.8%) in comparison with other fibers, kenaf fibers have a high reported modulus [1]. A single fiber of kenaf can have a tensile strength and a modulus as high as 11.9 and 60.0 GPa respectively. Third, natural fibers like kenaf have some excellent attributes over traditional reinforcing materials such as glass fibers in terms of cost, density, renewability, recyclability, abrasiveness and biodegradability [28]. Interestingly, it also has a high ability to consume CO_2 and its rate of photosynthesis is at least three times higher than that of common plants [14]. In fact, kenaf grows quickly; it can produce two crops per year in tropical areas and requires relatively little care [26].

Basically, kenaf has two distinctive stem regions: the outer part, or so-called bast, constitutes around 34% of the weight of the stem and the inner, woody core is about 66%. The long bast fibers are usually used to produce paper, protective packaging, and composite boards and are also used in textile industries. On the other hand, the short fibers are used to manufacture products like animal bedding and horticultural mixtures [25].

Based on research findings about its technical and commercial potential, kenaf has been recognized as the new prospective crop of Malaysia. This is associated with its fast growing characteristics compared to other plants [27]. Kenaf was first introduced in the early 1970s and in the 1990s it was identified as an alternative fibrous material for the production of panel products such as fiberboard and particleboard [29]. In the year 2000, kenaf was cultivated in the eastern part of Peninsular Malaysia as a trial plantation for research purposes [30]. Currently, the Malaysian government has promoted the planting of kenaf for social and economic reasons. Realizing the diverse possibilities of products derived from kenaf, the National Kenaf and Tobacco Board (formerly known as the National Tobacco Board) envisaged the development of kenaf cultivation in order to replace the current tobacco cultivation [29]. For this reason, the Malaysian government has allocated a huge amount of funding for this program. Recognizing its immense potential and interest, the aim is to develop the kenaf industry as the country's new source of growth in conjunction with the palm oil and rubber industries. With such abundance, relatively low capital investment and renewable sources of plant fibers, it is convenient to explore the potential of the vast economic performance of kenaf cultivation [31]. Considering these objectives, the main characteristics of the plant fiber are typically described in terms of their microscopic features, chemical composition and physical properties. To date, the chemical composition, anatomy, microfibril angle, mechanical properties and structure of the cell wall of kenaf plants in Malaysia are widely available in open literature [31]. In fact, some research works have already been conducted on locally planted kenaf. For instance, Abdul Khalil *et al.* [31] studied the anatomical and chemical nature of kenaf cultivated in Malaysia, and the chemical and morphological nature of kenaf were examined by Ashori *et al.* [26].

Generally, some varieties of kenaf can be found in Malaysia. These types may differ in terms of color of stem, shape of the leaf, color of the flower and seed, as well as its suitability in different environmental conditions. Kenaf sp. V36 is a kenaf variety that is widely planted in Malaysia. It is found to be the most applicable type of kenaf for commercial use [24]. H'ng *et al.* have examined nine varieties of kenaf including the V36 type, especially in terms of their anatomical appearance as well as cell morphology [32]. They have found that all kenaf varieties have slight differences in their length, fiber and lumen diameter. The mean values of fiber length, fiber diameter and fiber cell wall thickness of V36-type kenaf are shown in Table 22.4. Recent data reported by Nayeri *et al.* [33] are comparable to those published by H'ng *et al.* [32].

Besides, they also greatly differ in terms of chemical compositions. It has been reported that their chemical components vary according to their different cultivate, years, climates and soil conditions [30]. It is a fibrous plant consisting of an inner core fiber (60-75%) and an outer bast fiber (25-40%) in the stem [31]. As previously mentioned, lignocellulose material from wood or non-wood plants consists of cellulose,

Table 22.4 Mean value of fiber length, fiber diameter and fiber cell wall thickness of V36-type kenaf [32].

Properties	Bast	Core
Fiber length (mm)	2.89 ± 0.61	0.89 ± 0.17
Fiber diameter (μm)	24.09 ± 4.41	26.38 ± 5.58
Fiber lumen diameter (μm)	11.74 ± 1.83	17.43 ± 3.32
Fiber cell wall thickness (μm)	6.18 ± 1.02	4.47 ± 0.73

Table 22.5 Chemical composition of different fractions of kenaf fibers [31].

Chemical components	Kenaf whole	Kenaf core	Kenaf bast
Extractive (%)	6.4	4.7	5.5
Holocellulose (%)	87.7	87.2	86.8
α-cellulose (%)	53.8	49.0	55.0
Lignin (%)	21.2	19.2	14.7
Ash (%)	4.0	1.9	5.4

hemicelluloses, lignin, extractive and a minor part of inorganic matter. Thus, many extensive studies have been carried out to understand the chemical composition of natural fibers like kenaf.

Lignin is an amorphous, three-dimensionally branched network polymer that serves as a binder that holds the fiber together. Cellulose molecules consist of long, linear chains of homopolysaccharide composed of β-D-glucopyranose units which are lined by (1-4)-glycosidic bonds. It may exist in crystalline or amorphous phase. Hydrogen bonding between cellulose molecules provides strength to the fiber. Unlike cellulose which is a homopolysaccharide, hemicelluloses are heteropolysaccharides. Extractives are a heterogeneous group of lipophilic and hydrophilic compounds including terpenes, fatty acids, esters, tannins, volatile oils, polyhydric alcohols and aromatic compounds. The inorganic constituent of lignocellulosic material is usually called ash. The ash consists of various metal salts such as silicates, carbonates, oxalates and phosphates of potassium, magnesium, calcium, iron and manganese as well as silicon [25].

Analyses conducted by Ashori *et al.* [26] showed that there is a significant variation in the chemical compositions of the kenaf in comparison to the ones reported by Abdul Khalil *et al.* [31]. Kenaf bast and core are quite different in respect to their chemical components. Bast fiber is composed with a higher percentage of alpha-cellulose content which provides a stronger layer [26]. The chemical compositions of the whole kenaf fiber (core and bast) are illustrated in Table 22.5. According to the table, the core fiber of kenaf indicates higher percentages of holocellulose and lignin, while bast fibers of kenaf reveal higher percentages of α-cellulose, extractive and ash content [31]. Recent work by Ibrahim *et al.* [27] on the chemical composition of Malaysian cultivated kenaf was comparable to those published by Abdul Khalil *et al.* [31] and Ashori *et al.* [26], considering the natural variations in chemical characteristics of kenaf fibers.

In addition, the observations made by Ashori *et al.* [26] indicate that there are significant differences in morphological characteristics of bast and core fibers. The average length of bast fiber is 2.48 mm, while the length of core fiber is a 0.72 mm layer [26]. The average lumen diameters of the kenaf core and bast fibers are 6.7 and 2.8 μm, respectively. This showed that the core fibers have a higher lumen diameter compared to the bast fiber layer [26]. Ibrahim *et al.* [27] have reported that the average fiber lengths of bast and core fibers are 1.94 and 0.7 mm respectively, which are comparable to those reported previously.

22.4 Kenaf Fibers and Composites

At present, kenaf fibers are being used as reinforcing agents in thermoplastics. This has led to more assessments in order to acquire a better understanding of their mechanical and physical performance. Numerous reports are available on kenaf fiber-reinforced polymer composites. Table 22.6 summarizes some of the reported work on these types of composites.

Yusuff *et al.* [34] have studied the influence of kenaf fiber on the mechanical, thermal and biodegradable properties of poly(lactic acid) (PLA)/kenaf composites. They compounded 20 wt% of kenaf (approximately 70-250 μm in diameter) with PLA granules using a twin-screw extruder. The pelletized mixtures were then injection-molded in order to prepare samples for characterization. From the study they have found that with the addition of kenaf, the flexural modulus of the composites was increased up to 4 GPa. The flexural strength was obtained at 90 MPa. Despite the increment in modulus and strength, this finding was expected since kenaf fiber has higher cellulose content in comparison to other fibers, which mainly contributes to the better mechanical properties of the composites. From thermogravimetric analysis (TGA) results, they observed a decline of thermal stability of polymer matrix with addition of kenaf fibers. In this work, they also reported on the properties of PLA-kenaf composites after being subjected to a soil burial test. It was found that PLA-kenaf composites gradually degraded with time. This correlates to the cellulose content of kenaf, which leads to higher water absorption and certainly affects the rate of degradation.

Another work by Ahmad Thirmizir *et al.* reported on the effect of kenaf loadings (10-40 wt%) on the mechanical properties of water absorption of kenaf bast fiber

Table 22.6 The development of kenaf fiber-reinforced polymer composites in chronological order.

Year of publication	Reinforcement	Matrix	References
2004	Kenaf Hemp	Polyester	[8]
2006	Kenaf bast	Poly (lactic acid)	[14]
2007	Kenaf	Soy flour	[1]
2008	Kenaf	Poly (lactic acid)	[38]
2009	Kenaf cellulose	Low density polyethylene (LDPE)	[39]
2010	Kenaf bast	Poly (butylene succinate)	[30]
2010	Kenaf Rice Husk	Poly (lactic acid)	[34]
2010	Kenaf	Chitosan	[40]
2010	Kenaf core	High density polyethylene (HDPE)/Soya powder	[36]
2012	Kenaf Dust	Polycaprolactone/thermoplastic sago starch	[37]

(KBF)-filled poly(butylene succinate) [30]. The composite samples were prepared by compounding the kenaf bast fiber and poly(butylene succinate) using an internal compounder with co-rotating double winged rotors. They found that increasing the KBF loading from 10 to 30 wt% resulted in an increase in the flexural strength, whereas, further increase of the KBF loading reduced the strength. This can be explained by the reinforcing mechanism provided by the fibers, which allows bending forces to be transferred to the fibers efficiently. A similar finding was also reported by Prachayawarakon *et al.* [35]. Apart from that, the authors have also reported on the augmentation of water uptake of composite with increasing KBF contents. This is contributed by the hydroxyl groups of cellulose, hemicelluloses, and lignin constituents in the fiber, which tend to form hydrogen bonds with water molecules and subsequently result in a higher water absorption rate. This is in accordance with the findings reported by Ismail *et al.* [36] on kenaf core-reinforced HDPE/soya powder composites.

Recent work by Chang *et al.* discussed the properties of polycaprolactone (PCL)/thermoplastic sago starch (TPSS) blend reinforced with kenaf dust [37]. PCL pellets, TPSS and kenaf dust were melt-mixed using an internal mixer. The amounts of PCL and TPSS were fixed at 1:1 of weight ratio, whereas, kenaf dust loadings were varied from 0 to 30 phr. Then, the PCL/TPSS/kenaf dust composites were subjected to compression molding. From the study, they obtained an optimum tensile strength at 30 phr kenaf dust loading. Clearly, kenaf dust imparted a reinforcing effect to the PCL/TPSS blend. This was further justified by the SEM micrographs of the composite's fractured surfaces. The fibers were found to be well dispersed in the polymer blends, which contributed to the efficient stress transfer from matrix to the fiber. Similar to previous findings, the authors have noticed that the incorporation of kenaf dust in the polymer blend would significantly boost the capacity of water absorption. This is due to the hydrophilicity of kenaf and the presence of voids between matrix and fiber.

From the previous work, it can be concluded that the major limitation of using these fibers as reinforcements in such matrices is poor interfacial adhesion between polar, hydrophilic fibers and a non-polar, hydrophobic matrix. Thus, in order to overcome this problem, modification of fibers is required. Chemical modification can be an interesting approach to alter the cell wall structure and surface chemistry of the fibers and consequently improve the adhesion between fiber and matrix [24]. This will be discussed in a later section.

22.5 Kenaf Fiber Reinforced Low Density Polyethylene/Thermoplastic Sago Starch Blends

As mentioned previously, great attention has been paid to the development of kenaf fiber-reinforced polymer composites. This is in conjunction with the tremendous global awareness associated with the environmental impact of the disposal of conventional materials, which are certainly nondegradable. Therefore, recent work by the authors is an attempt to produce a material which is comparable to the conventional one together with the benefit of biodegradability, which will then at least lessen the environmental problems.

In view of the fact that polyethylene (PE) is one of the most commonly used polymers, especially in packaging products, in this section, PE blended with biodegradable polymers will be our subject of discussion. Thus, an endeavour has been undertaken to use natural polymer (i.e., starch) [41]. Up-to-date studies on the blend of PE with various starches (i.e., maize, tapioca, rice, corn, etc.) have been established. Since Malaysia is one of the largest exporters of sago, its excessive production can be used as a blending component with synthetic polymers. Starch turns into thermoplastic starch (TPS) in the presence of plasticizers at high temperature and under shear, allowing its use similar to most conventional synthetic thermoplastic polymers. In order to preserve renewability and biodegradability, as well as to improve mechanical properties of the final products, different researchers have discovered the prospective use of natural fiber-reinforced synthetic polymer/biopolymer [42, 43]. This trend suggests that natural fibers like kenaf can be preferable in light of the fact that they are one of the crops that have recently been actively and commercially cultivated in Malaysia, and they offer excellent attributes in terms of strength and modulus due to their high cellulose content [1]. The other attractive features of kenaf fibers such as their low cost, lightweight, renewability, biodegradability and high specific mechanical properties have made them suitable to be used as reinforcement in polymer composites.

Up to now there has been little literature available regarding the combination of starch and kenaf in polyolefin. Thus, this work focuses on the investigation of the mechanical and morphological properties of this type of composite. Likewise, the properties were determined by means of fiber loading. In this work, thermoplastic sago starch (TPSS) was prepared by mixing sago starch with glycerol. After that, they were melt-blended with low-density polyethylene (LDPE) and kenaf core fiber (KCF) using an internal mixer. Different amounts of kenaf fiber were loaded into the mixer (i.e., 10-40 phr). The mixtures were then compression molded using a hot press before being subjected to mechanical testing and characterizations.

Figure 22.4(a)-(b) illustrates the plots of tensile strength and Young's modulus of LDPE/TPSS blend reinforced with different amounts of KCF. It was found that the addition of 10% (by weight) of KCF into the LDPE/TPSS blend raises the tensile strength up to 6.704 MPa. This is an indication of good degree of adhesion between fibers and matrix. Thus, it is believed that the amount of fiber loading is sufficient to provide the reinforcing effect to the LDPE/TPSS blend. This also can be attributed to the chemical similarity of both TPSS and KCF, which formed a hydrogen bond and consequently enhanced the interaction. The results were in accordance with the findings

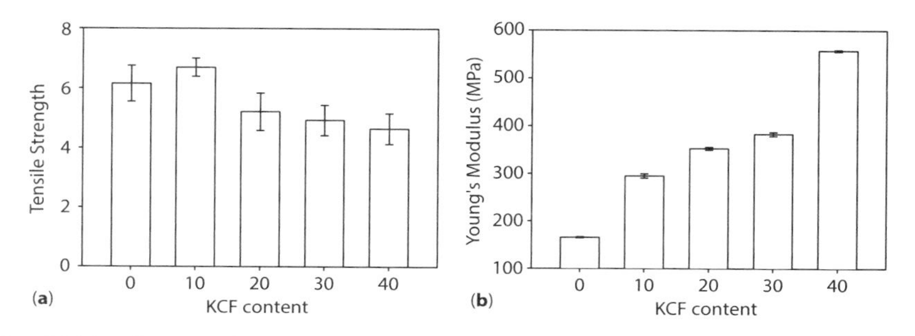

Figure 22.4 (a) Tensile strength and (b) Young's modulus of LDPE/TPSS blends reinforced by KCF.

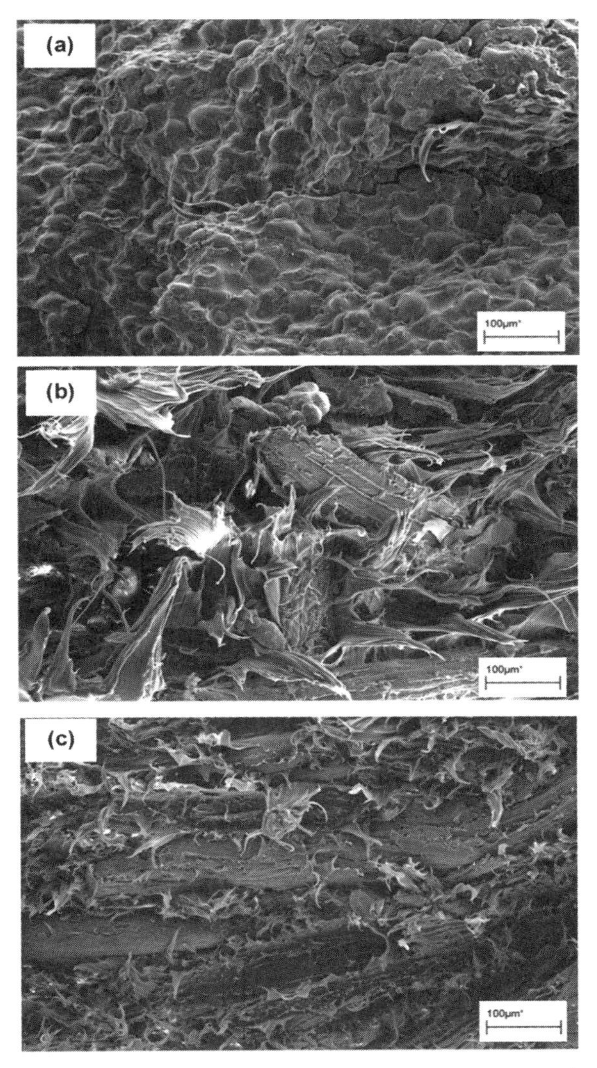

Figure 22.5 Tensile fractured surfaces of LDPE/TPSS/KCF composites at (a) 0 phr, (b) 10 phr and (c) 40 phr KCF loadings (magnification of 150X).

reported by Prachayawarakon *et al.* [35]. However, further increase of KCF incorporated into the composite system caused a reduction in tensile strength. This is because fiber-fiber interaction becomes more prominent than fiber-matrix interaction with increasing KCF content. Reduction in the effective surface area of fiber phase results in efficient stress transfer from the matrix. This is justified by the SEM micrograph in Figure 22.5. The moduli are increased from 172.5 MPa until 557.75 MPa when 10-40% (by weight) of KCF is loaded in the composite system. The addition of KCF is expected to increase the modulus resulting from the inclusion of rigid filler particles in the soft matrix [44].

The SEM micrographs of the tensile fractured surface of LDPE/TPSS blend and morphological effects of KCF incorporation are depicted in Figure 22.5. As presented in

Figure 22.5 (a), without KCF loading, the tensile fractured surface of LDPE/TPSS blend shows the spots of starch granules within the LDPE continuous phase. The immiscibility between the two components results in weak interfacial adhesion; starch granules tend to agglomerate resulting in low strength. The tensile properties obtained in the previous section are corroborated with this morphological structure. Whereas Figure 22.5 (b) illustrates the tensile fractured surface of LDPE/TPSS blend reinforced by 10% (by weight) KCF. It can be observed that fiber is embedded in the matrix and the surface of fiber is wetted by the matrix. There are no obvious starch granular figures. The good dispersion signifies the presence of chemical bonding between KCF and TPSS, probably due to the formation of hydrogen bonds and good stress transfer from matrix to fiber, which is evidence of yielded tensile strength. On the other hand, Figure 22.5 (c) illustrates the tensile fractured surface of LDPE/TPSS/KCF composite when KCF s loaded at 40% (by weight). The micrographs of fractured surface show poor wettability between fiber and matrix in which fiber agglomeration becomes more prominent in the composites. The non-uniform distribution of fiber gives rise to the formation of stress concentrated areas. This situation is reflected in the mechanical properties.

These findings clearly indicate that the presence of kenaf fiber in the polymer blend has by some means improved the overall properties of the composite system. The uniform distribution of kenaf fibers in the microstructure of polymer composites, particularly at an optimum amount of fiber loading, is the major factor responsible for the improvement of the mechanical properties.

However, it has some disadvantages, mainly; the incompatibility between the hydrophilic fibers and hydrophobic thermoplastic matrices, which can negatively affect the mechanical properties. This requires appropriate treatments to enhance the adhesion between the fiber and matrix [13]. Fiber modification is necessary to increase the adhesion between the hydrophilic natural fibers and the hydrophobic polymer matrix at the interface. Many investigations have been carried out by a number of researchers. However, different methods and chemicals were used. Accordingly, this crucial factor will be discussed in detail later.

22.6 The Effects of Kenaf Fiber Treatment on the Properties of LDPE/TPSS Blends

As mentioned previously, the main bottleneck in the broad use of these fibers in thermoplastics is the poor compatibility between the fibers and the matrix. The inherent high moisture sorption of lignocellulosic fibers certainly has an effect on their dimensional stability [28]. This may lead to the microcracking of the composites and degradation of mechanical properties [28]. Like other natural fibers, kenaf absorbs moisture due to its hydrophilicity. The key issue related to the development and production of natural fiber-reinforced composites is the interfacial adhesion between the fiber and polymer matrix. Because of their inherent dissimilarities, natural fibers/polymer matrix composites are not compatible and interfacial adhesion in these composites tends to be poor. The weak bonding at the interfaces between natural fibers and polymer matrix is surely a critical cause of the reduction of useful properties and performance of the composites. Therefore, a number of studies on physical and chemical surface modification on a variety of natural fibers have been devoted to enhancing the interfacial strength and to increasing the wettability between the natural fiber and polymer matrix [45, 46].

As reported in previous studies, modification of fiber surface and the use of compatibilizers or coupling agents are found to be suitable to improve the compatibility between the fiber and matrix [47]. The commonly used coupling agents include maleic anhydride, silane, etc. [48]. Modification of fiber surface can be done with chemical treatments such as alkylation, acylation, mercerization, and peroxide treatment. A few researchers also have reported on the modification of fiber by ultrasound and enzymatic treatments as an alternative technique to remove lignin and to modify cellulose in fiber [47]. Normally there are two ways to treat kenaf fibers to make them compatible with matrix; one is alkaline treatment, by using an alkaline solution where the fiber results in better fiber matrix adhesion, and the second is using a coupling agent. The coupling agent has two functions: reacting with OH groups of the cellulose and reacting with the functional groups of the matrix with the goal of facilitating stress transfer between the fibers and matrix [48].

Natural fibers are hydrophilic in nature as they are derived from lignocelluloses which contain highly polarized hydroxyl groups [5]. In addition, the presence of impurities on the surface of fibers hinders them from forming mechanical interlocking with matrix [49]. Attempts at improving the performance of the composites are to lessen the number of hydroxyl groups and to create new chemical bonds between the fiber and matrix. It is essential to modify the surface of fiber to render it more hydrophobic and also more compatible with matrices [50]. Generally, chemical treatments remove the lignin from the surface of natural fibers so that it becomes rough. It also reduces the number of free hydroxyl groups of the cellulose, which results in the reduction of the polarity of the cellulose molecules and enhances the compatibility with the hydrophobic polymer matrices [51]. One of the most reliable techniques to improve compatibility of fibers and widen their application is grafting. It is a facile method for incorporating desired functional groups onto natural cellulosic fibers [52]. Graft copolymerization may be achieved by treating the fibers with an ester agent and other agents or by grafting various monomers onto the fibers. Some monomers grafted onto fibers by copolymerization are glycidyl methacrylate and butyl acrylate [52, 53].

Extensive research work has been carried out on graft copolymerization of various monomers onto jute fibers [54-56]. But very limited research is available in open literature on grafting of kenaf. Hence, modification of kenaf fibers by graft copolymerization of hydrophobic vinyl monomer is an option to increase the compatibility of fiber and matrix. Investigations demonstrate that acrylic monomers, methyl methacrylate (MMA), grafted onto fibers yield an improvement in the mechanical properties [57]. It has been chosen for modification of fiber surface because of its compatibility with other common polymers [58]. The grafting mechanism would be the free-radical initiator (e.g., benzoyl peroxide) which dissociates and subsequently abstracts hydrogen from the active hydroxyl groups in KCF backbone to form microradicals which then react with MMA to form the grafted KCF [59].

Several studies have modified the properties of natural fibers by grafting of monomer onto the fiber by the radiation method [56, 60]. However, the purpose of this

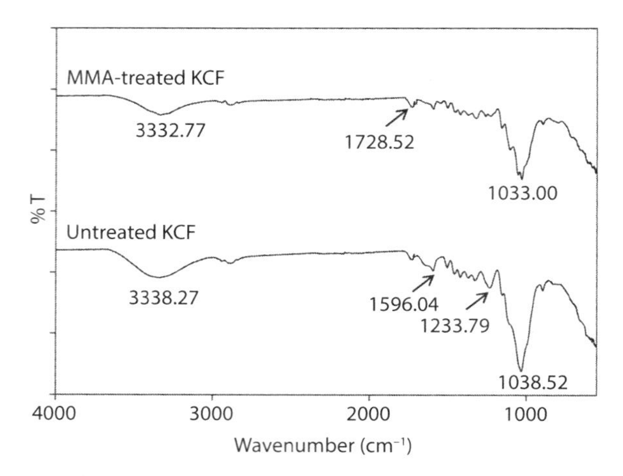

Figure 22.6 FTIR spectra of untreated and MMA-treated KCF.

present work is to investigate the effect of the graft polymerization of MMA onto KCF using benzoyl peroxide (BPO) as the initiator. Given that the hydroxyl groups in fiber are not readily accessible, the fiber is pretreated with sodium hydroxide to allow the chemical reaction to take place. Basically, the extent of grafting may be influenced by the various reaction parameters such as concentration of monomer, initiator ratio and reaction condition. However, the following discussion will reflect only on the strength of the overall system as well as on the morphology of the fractured surfaces. The system incorporated a fixed amount of monomer in order to determine the optimum fiber loading in the matrices.

Fourier transform infrared (FTIR) analysis was conducted on raw KCF as well as on MMA-treated KCF to identify the useful information on the related functional groups of the fibers. The FTIR spectra of both untreated KCF and MMA-treated KCF are shown in Figure 22.6. As seen, both spectra present basically the same pattern of absorption band. There is not much difference in the spectra of raw and treated KCF. Both KCF and MMA-treated KCF show a broad peak at 3332.7 cm^{-1} due to O-H stretching absorption in lignin, cellulose and hemicelluloses. However, it can be observed that the stretching vibration band for MMA-treated KCF is slightly shifted towards lower band intensity compared to the untreated one. These observations imply that the number of hydroxyl groups in treated KCF is reduced. In addition, for both untreated and treated IR spectra, it is observed that the peaks around 2896, 1596 and 1044.04 cm^{-1} arise from $-CH_2$, C = C and C-O stretching, respectively [61]. It can be recognized that the MMA-treated KCFs exhibit a strong band at 1734.04 cm^{-1} corresponding to the ester stretching vibration of carbonyl group in comparison to the untreated one. The intensification of this band confirms the occurrence of grafting of MMA monomer onto KCF through covalent bonds [52]. These findings also have been postulated in other reports elsewhere [58]. On top of that, this is another possible explanation for the rise of strength observed when the MMA-treated KCFs are incorporated in the composites.

Figure 22.7 depicts the effect of KCF loading on the tensile properties of LDPE/TPSS blends. The effects are illustrated as a function of MMA-treated KCF in comparison to the untreated one. Apparently, the tensile strength increases gradually with the increase

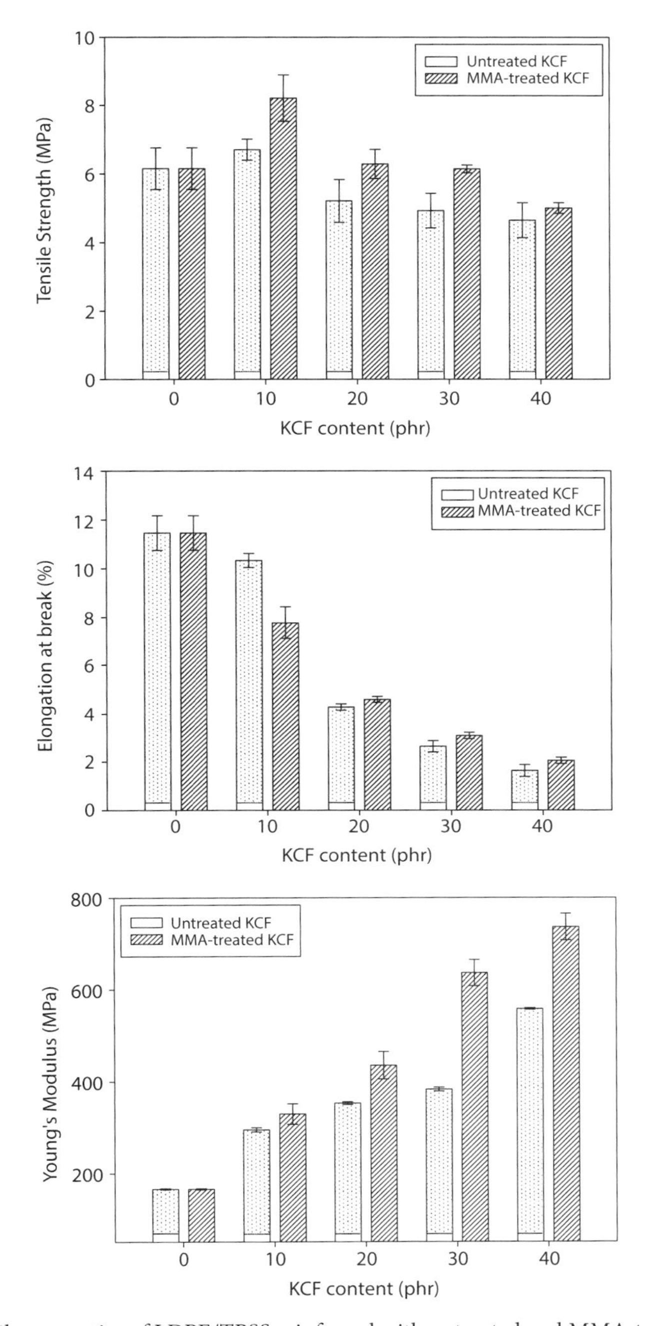

Figure 22.7 Tensile properties of LDPE/TPSS reinforced with untreated and MMA-treated KCF.

of both untreated KCF and MMA-treated KCF loading up to 10 phr. However, higher values of tensile strength are obtained in the case of MMA-treated KCF-loaded composites. It is found that the tensile strength of MMA-treated KCF-reinforced LDPE/TPSS blends is increased by 22.5% compared to the untreated one. The enhanced tensile strength is likely due to the efficient stress transfer from the matrix to the fiber, as the load applied results in improved adhesion between fiber and matrix [62]. This mainly is attributed to the presence of MMA-treated KCF in the composites which are eventually compatible with the hydrophobic matrix. The MMA molecules grafted onto the surface of KCF tend to lower the polarity of KCF and form ester carbonyl groups, which are assumed to link both fiber and matrix [60]. This occurrence is in accordance to the IR spectra presented in Figure 22.6. It is assumed that the MMA polymerized onto the surface of fiber causes it to be rougher. This eventually leads to better mechanical interlocking between fiber and matrix and improves the interfacial contact between both components, which is shown later in SEM results. However, both MMA-treated and untreated KCF result in a gradual reduction of tensile strength with the addition of above 10 phr of fiber in the composites. A further increase in fiber loadings has a dentrimental effect on the strength. This is because an optimum level is attained, whereby, further increase in the fiber loading results in the agglomeration of the fibers, which raises the number of stress-concentrated areas and hence results in discontinuity in the matrix. The addition of both untreated and MMA-treated KCF is found to reduce the elongation break drastically. This phenomenon is common for thermoplastic matrix reinforced with fibers. Generally, the inclusion of fibers imparts the rigidity of composites and restrains the deformability of interface between fiber and matrix, consequently causing a reduction in degree of ductility. A similar trend was also reported by Ismail *et al.* [63]. However, the blend shows a tendency for lower elongation at break with the addition of an optimum amount of treated KCF (10 phr) as compared to the untreated one. It is believed that improving adhesion between matrix and fiber augments the stiffness of composites and thus imparts reduction of percentage of elongation at break. It can be seen that the Young's modulus shows a gradual increase with KCF loading for both systems (untreated and MMA-treated). This is in accordance with other reports elsewhere [63]. Such trends indicate that incorporation of more fibers into the blends increases stiffness of the composites. Higher values of Young's modulus were observed for MMA-treated KCF-reinforced LDPE/TPSS blends in comparison to the untreated one. This evidences that MMA-treated KCF exhibits a better bonding with matrix which is attributed to an increase in the degree of the stiffness of the composites.

The tensile fractured surfaces of LDPE/TPSS blend loaded with untreated and MMA-treated KCF were analyzed by SEM to evaluate the degree of interfacial adhesion between fibers and matrix as shown in Figure 22.8. In relation to the tensile fracture surfaces, a typical micrograph of surface characteristic of MMA-treated KCF is also presented.

According to Figure 22.8 (a), the MMA-treated KCF shows the presence of MMA on the smooth surface of KCF. The deposition of MMA chains allows the surface of KCF to seem rougher [64]. The pits and surface impurities observed for raw KCF are no longer present in grafted fiber. The surface becomes more uniform owing to deposition of MMA and results in an appreciable increase in the strength when it is incorporated in the LDPE/TPSS blend, as explained in the previous section [54]. Apparently, the tensile fractured surfaces of the composites incorporated with MMA-treated fibers shown in Figure 22.8 (b) reveal less distinct interfacial areas, which imply improved interfacial adhesion between fibers and matrices. As mentioned previously, the grafting of polymer on the surface of fiber causes an increase in the roughness of the surface.

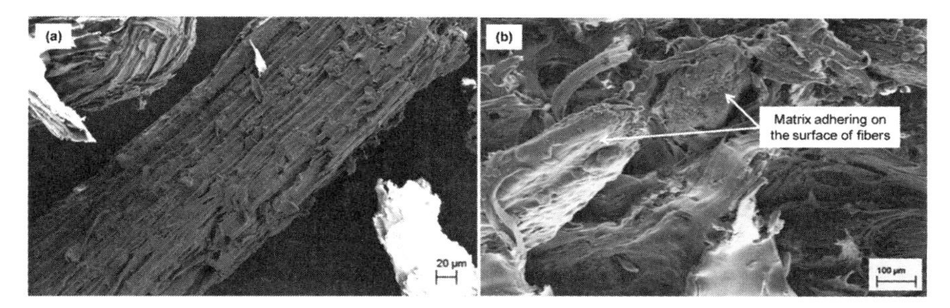

Figure 22.8 SEM micrographs of (a) MMA-treated KCF and (b) LDPE/TPSS/10phr MMA-treated KCF composites.

The rougher surface of the fiber would increase the mechanical interlocking with polymer matrix leading to a better interfacial contact whereby it seems that those matrices are adhered on the fiber surface. Clearly, the morphological studies support the results on tensile strength of the composites. Therefore, it is evident that improved adhesion between fiber and the polymer matrix is achieved upon treatment.

22.7 Outlook and Future Trends

Natural fiber-reinforced polymer composites have been significantly developed over the past years because of their advantages in processing, mechanical properties and biodegradability. These types of composites are predicted to have more and more applications in the near future. Adhesion and compatibility between natural fiber and polymer matrix will remain a critical issue concerning the overall properties of the composites. Hence, further research is necessary to overcome this shortcoming. Current scientific research has been focused on the selection of the most suitable materials and the optimization of all parameters.

Acknowledgement

The authors wish to acknowledge the financial support provided by Research University Grant (Grant No. 1001/PKT/8640014) from Universiti Sains Malaysia (USM) and the fellowship scheme of International Islamic University Malaysia (IIUM) which helped to make this research work possible. The authors also would like to thank the National Kenaf and Tobacco Board, Malaysia, for supplying kenaf used in this study.

References

1. W. Liu, , L.T. Drzal, A.K. Mohanty, and M. Misra, Influence of processing methods and fiber length on physical properties of kenaf fiber reinforced soy based biocomposites. *Compos. B Eng.* 38(3), 352–359 (2007).

2. T. Mukherjee, and N. Kao, PLA based biopolymer reinforced with natural fibre: A review. *J. Polym. Environ.* 19(3), 714–725 (2011).
3. J.K. Pandey, S.H. Ahn, C.S. Lee, A.K. Mohanty, and M. Misra, Recent advances in the application of natural fiber based composites. *Macromol. Mater. Eng.* 295(11), 975–989 (2010).
4. H. Ku, H. Wang, N. Pattarachaiyakoop, and M. Trada, A review on the tensile properties of natural fiber reinforced polymer composites. *Compos. B Eng.* 42(4), 856–873 (2011).
5. S. Kalia, , B.S. Kaith, and I. Kaur, Pretreatments of natural fibers and their application as reinforcing material in polymer composites—A review. *Polym. Eng. Sci.* 49, 1253–1272 (2009).
6. H.M. Akil, M.F. Omar, A.A.M. Mazuki, S. Safiee, Z.A. M. Ishak, and A. Abu Bakar, Kenaf fiber reinforced composites: A review. *Mater. Des.* 32, 4107–4121 (2011).
7. K. Majeed, M. Jawaid, A. Hassan, A. Abu Bakar, H.PS. Abdul Khalil, A.A. Salema, and I. Inuwa, Potential materials for food packaging from nanoclay/natural fibres filled hybrid composites. *Mater. Des.* 46, 391–410 (2013).
8. S.H. Aziz, and M.P. Ansell, The effect of alkalization and fibre alignment on the mechanical and thermal properties of kenaf and hemp bast fibre composites: Part 1 - polyester resin matrix. *Compos. Sci. Technol.* 64, 1219–1230 (2004).
9. Z.N. Azwa, B.F. Yousif, A.C. Manalo, and W. Karunasena, A review on the degradability of polymeric composites based on natural fibres. *Mater. Des.* 47, 424–442 (2013).
10. C. Johansson, J. Bras, I. Mondragon, P. Nechita, D. Plackett, P. Šimon, D.G. Svetec, S. Virtanen, M.G. Baschetti, C. Breen, F. Clegg, and S. Aucejo, Renewable fibers and bio-based materials for packaging applications-a review of recent developments. *Bioresource* 7(2), 2506–2552, 2012.
11. O. Faruk, A.K. Bledzki, H.P. Fink, and M. Sain, Biocomposites reinforced with natural fibers: 2000–2010. *Prog. Polym. Sci.* 37, 1552–1596 (2012).
12. H.D. Rozman, G.S. Tay, R.N. Kumar, A. Abusamah, H. Ismail, and Z.A. Mohd, The effect of oil extraction of the oil palm empty fruit bunch on the mechanical properties of polypropylene–oil palm empty fruit bunch–glass fibre hybrid composites. *Polym. Plast. Technol. E ng.* 40(2), 103–115 (2001).
13. H. Ismail, A.M. Norjulia, and Z. Ahmad, The effects of untreated and treated kenaf loading on the properties of kenaf fibre-filled natural rubber compounds. *Polym. Plast. Technol. Eng.* 49(5), 519–524 (2010).
14. S. Serizawa, , K. Inoue, , and M. Iji, Kenaf-fiber-reinforced poly (lactic acid) used for electronic products. *J. Appl. Polym. Sci.* 100(1), 618–624 (2006).
15. O. Agunsoye, and V.S. Aigbodion, Bagasse filled recycled polyethylene bio-composites: Morphological and mechanical properties study. *Results Phys.* 3, 187–194 (2013).
16. P.J. Jandas, S. Mohanty, and S.K. Nayak, Surface treated banana fiber reinforced poly (lactic acid) nanocomposites for disposable applications. *J. Clean. Prod.* 52, 392–401 (2013).
17. H. Essabir, A. Elkhaoulani, K. Benmoussa, R. Bouhfid,F.Z. Arrakhiz, and A. Qaiss, Dynamic mechanical thermal behavior analysis of doum fibers reinforced polypropylene composites. *Mater. Des.* 51, 780–788 (2013).
18. A. Bourmaud, G. Ausias, G. Lebrun, M.L. Tachon, and C. Baley, Observation of the structure of a composite polypropylene/flax and damage mechanisms under stress. *Ind. Crops Prod.* 43, 225–236 (2013).
19. A. Elkhaoulani, F.Z. Arrakhiz, K. Benmoussa, R. Bouhfid, and A. Qaiss, Mechanical and thermal properties of polymer composite based on natural fibers: Moroccan hemp fibers/polypropylene. *Mater. Des.* 49, 203–208 (2013).
20. A. May-Pat, A. Valadez-González, and P.J. Herrera-Franco, Effect of fiber surface treatments on the essential work of fracture of HDPE-continuous henequen fiber-reinforced composites. *Polym. Test.* 32 (6), 1114–1122 (2013).

21. S. Kaewkuk, W. Sutapun, and K. Jarukumjorn, Effects of interfacial modification and fiber content on physical properties of sisal fiber/polypropylene composites. *Compos. B Eng.* 45 (1), 544–549 (2013).
22. A. Moubarik, N. Grimi, and N. Boussetta, Structural and thermal characterization of Moroccan sugar cane bagasse cellulose fibers and their applications as a reinforcing agent in low density polyethylene. *Compos. B Eng.* 52, 233–238 (2013).
23. M.Z. Rong, M.Q. Zhang, Y. Liu, G.C. Yang, and H.M. Zeng, The effect of fiber treatment on the mechanical properties of unidirectional sisal-reinforced epoxy composites. *Compos. Sci. Technol.* 61, 1437–1447 (2001).
24. H.P.S. Abdul Khalil, and N.L. Suraya, Anhydride modification of cultivated kenaf bast fibers: morphological, spectroscopic and thermal studies. *Bioresources* 6(2), 1122–1135 (2011).
25. A.A. Mossello, J. Harun, S.R.F. Shamsi, H. Resalati, P.M. Tahir, , R. Ibrahim, and A.Z. Mohmamed, A review of literatures related to kenaf as a alternative for pulpwoods. *Agric. J.* 5, 131–138 (2010).
26. A. Ashori, J. Harun, W.D. Raverty, and M.N.M. Yusoff, Chemical and morphological characteristics of Malaysian cultivated kenaf (Hibiscus cannabinus) fiber. *Polym. Plast. Technol . Eng.* 45(1), 131–134 (2006).
27. M. Ibrahim, W.R. Wan Daud, W. Rosli, and K.N. Law, Comparative properties of soda pulps from stalk, bast, and core of Malaysian grown kenaf. *Bioresources* 6(4), 5074–5085 (2011).
28. A.M.M. Edeerozey, H.M. Akil, A.B. Azhar, and M.I. Zainal Ariffin, Chemical modification of kenaf fibers. *Mater. Lett.* 61(10), 2023–2025 (2007).
29. L.S. Ang, C.P. Leh, and C.C. Lee, Effects of alkaline pre-impregnation and pulping on malaysia cultivated kenaf (Hibiscus cannabinus). *Bioresources* 5(3), 1446–1462 (2010).
30. M.Z. Ahmad Thirmizir, Z.A. MohdIshak, R. Mat Taib, R. Sudin, and Y.W. Leong, Mechanical, water absorption and dimensional stability studies of kenaf bast fibre-filled poly (butylene succinate) composites. *Polym. Plast. Technol. Eng.* 50, 339 – 348 (2011).
31. H.P.S. Abdul Khalil, A.F. IreanaYusra, A.H. Bhat, and M. Jawaid, Cell wall ultrastructure, anatomy, lignin distribution, and chemical composition of Malaysian cultivated kenaf fiber. *Ind. Crops Prod.* 31, 113–121 (2010).
32. P.S. H'ng, B.N. Khor, N. Tadashi, A.S.N. Aini, and M.T. Paridah, Anatomical structures and fiber morphology of new kenaf varieties. *Asian J. Sci. Res.* 2, 161–166 (2009).
33. M.D. Nayeri, P.M. Tahir, J. Harun, L.C. Abdullah, E.S. Bakar, M. Jawaid, and F. Namvar, Effects of temperature and time on the morphology, pH, and buffering capacity of bast and core kenaf fibres. *Bioresources*, 8(2), 1801–1812 (2013).
34. A.A. Yussuf, I. Massoumi, and A. Hassan, Comparison of polylactic acid/kenaf and polylactic acid/rise husk composites: The influence of the natural fibers on the mechanical, thermal and biodegradability properties. *J. Polym. Environ.* 18(3), 422–429 (2010).
35. J. Prachayawarakorn, L. Hommanee, D. Phosee, and P. Chairapaksatien, Property improvement of thermoplastic mung bean starch using cotton fiber and low-density polyethylene. *Starch–Stärke*, 62(8), 435–443 (2010).
36. H. Ismail, A.H. Abdullah, and A.A. Bakar, The effects of a silane-based coupling agent on the properties of kenaf core-reinforced high-density polyethylene (HDPE)/soya powder composites. *Polym. Plast. Technol. Eng.* 49(11), 1095–1100 (2010).
37. Y. Chang, H. Ismail, and Q. Ahsan, Effect of maleic anhydride on kenaf dust filled polycaprolactone/thermoplastic sago starch composites. *Bioresources* 7(2), 1594–1616 (2012).
38. S. Ochi, Mechanical properties of kenaf fibers and kenaf/PLA composites. *Mech. Mater.* 40, 446–452 (2008).

39. B. Tajeddin, R.A. Rahman, and L.C. *Abdulah, Mechanical and morphological properties of kenaf cellulose/LDPE biocomposites. J. Agric. Environ. sci.* 5(6), 777–785 (2009).
40. N.M. Julkapli, and M.H. Akil, Influence of a plasticizer on the mechanical properties of kenaf-filled chitosan bio-composites. *Polym. Plast. Technol. Eng.* 49(9), 944–951 (2010).
41. D. Bikiaris, and C. Panayiotou, LDPE/starch blends compatibilized with PE-g-MA copolymers. *J. Appl. Polym. Sci.* 70, 1503–1521 (1998).
42. L. Averous, C. Fringant, and L. Moro, Plasticized starch–cellulose interactions in polysaccharide composites. *Polymer* 42, 6565–6572 (2001).
43. M.D.R. Rahman, M.D.M. Huque, M.D.N. Islam, and M. Hasan, Mechanical properties of polypropylene composites reinforced with chemically treated abaca. *Compos. A Appl. Sci. Manuf.* 40, 511– 517 (2009).
44. X.V. Cao, H. Ismail, A.A. Rashid, T. Takeichi, and T.V. Huu, Mechanical properties and water absorption of kenaf powder filled recycled high density polyethylene/natural rubber biocomposites using MAPE as a compatibilizer. *Bioresources* 6(3), 3260–3271 (2011).
45. []D. Cho, and H.S. Lee, Effect of fiber surface modification on the interfacial and mechanical properties of kenaf fiber-reinforced thermoplastic and thermosetting polymer composites. *Compos. Interface* 16(7–9), 711–729 (2009).
46. N. Le Moigne, M. Longerey, J.M. Taulemesse, J.C. Bénézet, and A. Bergeret, Study of the interface in natural fibres reinforced poly (lactic acid) biocomposites modified by optimized organosilane treatments. *Ind. Crops Prod.* 52, 481–494 (2014).
47. M.R. Islam, M.D.H. Beg, and A. Gupta, Characterization of laccase-treated kenaf fibre reinforced recycled polypropylene composites. *Bioresources* 8(3), 3753–3770 (2013).
48. Y. Xu, S. Kawata, K. Hosoi, T. Kawai, and S. Kuroda, Thermomechanical properties of the silanized-kenaf/polystyrene composites. *Expr. Polym. Lett.* 3(10), 657–664 (2009).
49. X. Wang, Y. Cui, H. Zhang, and B. Xie, Effects of methyl methacrylate grafting and polyamide coating on the interfacial behavior and mechanical properties of jute-fiber-reinforced polypropylene composites. *J. Vinyl Addit. Technol.* 18, 113–119 (2012).
50. A. Khan, T. Huq, M. Saha R.A. Khan, and M.A. Khan, Surface modification of calcium alginate fibers with silane and methyl methacrylate monomers. *J. Reinf. Plast. Compos.* 29 (20), 3125–3132 (2010).
51. V.K. Kaushik, A. Kumar, and S. Kalia, Effect of mercerization and benzoyl peroxide treatment on morphology, thermal stability and crystallinity of sisal fibers. *Int. J. Text. Sci.* 1(6), 101–105 (2012).
52. V.K. Thakur, M.K. Thakur, and A.S. Singha, Free radical–induced graft copolymerization onto natural fibers. *Int. J. Polym. Anal. Char.* 18(6), 430–438 (2013).
53. Y. Li, X. Dong, Y. Liu, J. Li, and F. Wang, Improvement of decay resistance of wood via combination treatment on wood cell wall: Swell-bonding with maleic anhydride and graft copolymerization with glycidyl methacrylate and methyl methacrylate. *Int. Biodeter. Biodegrad.* 65 (7), 1087–1094 (2011).
54. J. Rout, M. Misra, and A.K. Mohanty, Surface modification of coir fibers I: studies on graft copolymerization of methyl methacrylate on to chemically modified coir fibers. *Polym. Adv. Technol.* 10, 336–344 (1999).
55. I.H. Mondal, *J. Eng. Fibers Fabr.* 8(3), 42–50 (2013).
56. F. Khan, and S.R. Ahmad, Graft copolymerization reaction of water-emulsified methyl methacrylate with preirradiated jute fiber. *J. Appl. Polym. Sci.* 65, 459–468 (1997).
57. T.J. Madera-Santana, and F. Vazquez Moreno, Graft polymerization of methyl methacrylate onto short leather fibers. *Polym. Bull.* 42, 329–336 (1999).

58. Z. Wan, Z. Xiong, H. Ren, Y. Huang, H. Liu, H. Xiong, Y. Wu, and J. Han, Graft copolymerization of methyl methacrylate onto bamboo cellulose under microwave irradiation. *Carbohydr. Polym.* 83 (1), 264–269 (2011).
59. Q. Peng, Q. Xu, D. Sun, and Z. Shao, Grafting of methyl methacrylate onto Antheraea pernyi silk fiber with the assistance of supercritical CO2. *J. Appl. Polym. Sci.* 100, 1299–1305 (2006).
60. H.U. Zaman, R.A. Khan, M.A. Khan, and M.D.H. Beg, Physico-mechanical and degradation properties of biodegradable photografted coir fiber with acrylic monomers. *Polym. Bull.* 70, 2277–2290 (2013).
61. J. Sharif, S.F. Mohamad, F. Othman, N.A. Bakaruddin, H.N. Osman, and G. Olgun, Graft copolymerization of glycidyl methacrylate onto delignified kenaf fibers through pre-irradiation technique. *Radiat. Phys. Chem.* 91, 125–131 (2013).
62. G. Raju, C.T. Ratnam, N.A. Ibrahim, M.Z.A. Rahman, and W.M.Z.W. Yunus, Enhancement of PVC/ENR blend properties by poly (methyl acrylate) grafted oil palm empty fruit bunch fiber. *J. Appl. Polym. Sci.* 110, 368–375 (2008).
63. H. Ismail, A.H. Abdullah, and A.A. Bakar, Influence of acetylation on the tensile properties, water absorption, and thermal stability of (High-density polyethylene)/(soya powder)/(kenaf core) composites. *J. Vinyl Addit. Technol.* 17, 132–137 (2011).
64. A.S. Singha, and R.K. Rana, Natural fiber reinforced polystyrene composites: Effect of fiber loading, fiber dimensions and surface modification on mechanical properties. *Mater. Des.* 41, 289–297 (2012).

23

Application of Natural Fiber as Reinforcement in Recycled Polypropylene Biocomposites

Sanjay K Nayak[*,1] **and Gajendra Dixit**[2]

[1]*Laboratory for Advanced Research in Polymeric Materials (LARPM), Central Institute of Plastics Engineering & Technology (CIPET), Bhubaneswar, India*
[2]*Maulana Azad National Institute of Technology (MANIT), Bhopal, Madhya Pradesh, India*

Abstract
The new environmental, economic, and petroleum crises have induced the scientific community to deal with a sustainable recycling practice for polymers. Numerous academic and industrial studies have been focusing on incorporation of natural fibers into recycled polypropylene (RPP) to obtain eco-friendly and economical materials with unique combinations of physical, chemical, mechanical, thermal and electrical properties. Therefore, this investigation is centered on mechanical, thermal, and morphological properties of Sisal Fibers (SF)-reinforced RPP composites in the presence of clay and compatibilizer (MA-g-PP). Short, randomly oriented RPP/SF composites with different fiber loading were prepared. The RPP composite with 40 wt% of chemically treated fiber content and 5% MA-g-PP (compatibilizer) has shown an improved mechanical property. This enhancement in the properties may be attributed to the improved interfacial adhesion between fiber and matrix, compact macromolecular packing arrangement of fibers within RPP matrix, and nucleation of clay. The glass transition temperature (T_g) and thermal degradation temperature of these modified composites have also been observed to be improved. The SEM micrographs have confirmed the compatibility between SF and RPP along with better interfacial bonding.

Keywords: Natural fibers, recycled PP, SEM, thermal

23.1 Introduction

The use of natural fiber reinforcements in recycled thermoplastic polymers has generated considerable research interest due to the increasing environmental threat created by the high consumption rate of petroleum-based plastics. Therefore, there is a growing interest in the development of green, environmentally-friendly composites for agro-based structural materials. Apart from this, the renewed interests in recyclability and environmental sustainability have also arisen in the last few years. The best possible way to develop such composites are to incorporate natural fibers into recycled

**Corresponding author*: drsknayak@gmail.com

Vijay Kumar Thakur (Ed.), Lignocellulosic Polymer Composites, (523–550) 2015 © Scrivener Publishing LLC

polypropylene (RPP) instead of traditional glass, carbon or other synthetic fiberbased PP [1,2]. RPP is emerging as a material of choice for diverse applications in a range of single-use packaging products such as bottles, yogurt containers, medicine bottles, caps, straws, plastic cups, and food packaging. A uniform code for the recycling of plastics was first introduced in 1988 by the Society of the Plastics Industry. The code was designed to make the identification of certain plastic resins easier to enable a more efficient and precise sorting stage.

The combination of natural fibers with RPP can achieve environmentally-friendly composite materials with comparable properties to those of traditional synthetic fiber-reinforced PP. Natural fibers such as bamboo [3], flax [4], hemp [5], jute [6], henequen [7], ramie [7], and pineapple [8], etc., have been incorporated into RPP to form composites with improved properties. Natural fibers are the fibrous plant materials, which are sometimes referred to as vegetable, biomass, photo mass, agro mass, solar mass, and/or photosynthetic fibers. Another term used is lignocelluloses fibers, which means containing lignin and cellulose. Natural organic fibers have been around for a very long time, from the beginning of life on earth. The archeological artifacts suggest that human beings used these materials in fabrics many thousands of years ago. Many kinds of textiles, ropes, canvas and paper produced from natural fibers are in use today. It may seem surprising, but the first natural fiber composites were prepared more than 100 years ago. In 1896, for example, airplane seats and fuel tanks were made of natural fibers with a small content of polymeric binders. As early as 1908, the first composite materials were applied for the fabrication of large quantities of sheets, tubes and pipes for electronic purposes. However, these attempts were without recognition of the composite principles and the importance of fibers as the reinforcing part of composites. The use of natural fibers was suspended due to low cost and growing performance of technical plastics. A renaissance in the use of natural fibers as reinforcements in technical applications began in the 20th century in the 1990s. The first composite materials were applied for the fabrication of large quantities of sheets, tubes and pipes for electronic purposes made of phenol- or melamine-formaldehyde resins [9]. In the 1960s, fiber-reinforced polymer composites became one of the focal research areas. Since then, many conventional polymer composites have been prepared. These composites possess better chemical and physical properties such as high strength and modulus, good dimensional stability, improved chemical and corrosion resistance, and various functionalities compared to pure polymeric materials. The composites from natural fibers and conventional polyolefin, that is, polypropylene and polyethylene, have been extensively used [10-13]. The huge production of these composites created problems such as shortage of landfill space for disposal, carbon emissions during incineration, depletion of petroleum resources, and environmental pollution. Therefore, to solve these problems, various efforts have made to develop eco-friendly natural fiber-reinforced green thermoplastic composites [5, 14-17]. One of the most promising thermoplastic polymers is RPP. Moreover, the high rate of urbanization demands the usage of lightweight disposable materials, which has forced scientific communities to develop eco-friendly materials with improved performance characteristics. Also, growing environmental awareness and new rules and regulations are forcing industries and researchers to seek more ecologically-friendly polymer materials. These can be based on renewal biobased plant and agriculture products that can compete in the markets currently dominated by petroleum-based products. As a result, new types of composites based on plant fibers have been developed in recent years. Demand for natural fibers in plastic composite is forecast to grow 15–20% annually, with a growth rate of 15–20% in automotive applications, and 50% or more in selected building applications. Other emerging markets are industrial and consumer applications such as permanent formworks, facades, tanks, pipes, long-span roofing elements, strengthening of existing structures, and structural components of building. At the moment, three quarters of all agricultural fibers used in composite is wood fiber, with the remaining one quarter kenaf, jute, hemp and sisal, etc. Complete matrix fusion and formation of strong fiber/matrix interface are vital requirements for the manufacture of reliable, eco-friendly natural composites with better physico-mechanical properties. Like other countries, India is continuously using natural fibers such as jute fibers for the preparation of composites. The government of India promoted large projects where jute-reinforced polyester resins were used in buildings. Nowadays, natural fiber-reinforced composites and nanocomposites are finding applications in the automobile and packaging industries. The national research agencies in India are continuously developing natural fiber-reinforced composites, however, their commercialization has been limited so far. In order to improve upon the laboratory-industry linkage towards application development and commercialization, the Advanced Composites Mission was launched by the Department of Science and Technology, Government of India. The commercial application of jute-based composites in automobile interiors has been achieved by Birla Jute Industries Ltd., Kolkata, India.

Natural fiber-reinforced composites have been found to have optimum mechanical and physical properties due to improved adhesion between fiber and matrix. The major drawback of natural fibers is their moisture sensitivity, which can be reduced through chemical modification such as acetylation, silane, alkaline treatment and other methods. Among natural fibers, sisal is one of the most versatile agro-based fibers, mostly grown in Tanzania and northeast Brazil. Due to its superior mechanical properties and recyclable nature, sisal fiber can be used as potential input material for making composites for applications in buildings, automobiles, railways, geotextiles, packaging industries, etc. Sisal fiber-reinforced composite building materials like wood substitute products, panels, doors, corrugated roofing sheets and instant houses suitable for disaster-prone areas (floods/tsunamis/earthquakes), would attract prospective entrepreneurs and stakeholders due to its durability and cost effectiveness. Since asbestos fibers are carcinogenic, sisal fiber cement corrugated roofing sheets, which are eco-friendly, can be an effective alternative. Moreover, sisal fiber has enormous potential for the manufacture of green composites due to its abundant availabilty, low cost, and renewable, non-abrasive, viscoelastic, biodegradable, combustible, compostable nature; and also its better insulation against noise and heat due to its hollow structure.

23.1.1 Natural Fibers – An Introduction

Natural fibers such as sisal, flax, jute and wood-fibers possesses good reinforcing capability when precisely compounded with polymers. These fibers have attracted attention because of their low cost [19-21]. The advantages of natural fibers over synthetic fibers are low density, renewability, favorable values of specific strength and specific modulus, excellent chemical resistance and significant processing benefits which require minor changes to equipment [9]. However, the hydrophilic nature of natural fibers is a major drawback for their application as reinforcement in composites. The poor moisture resistance of natural fibers leads to incompatibility and poor wettability with hydrophobic polymers [22,23]. Since 1990s, natural fiber-based composites are emerging as a good alternative to synthetic fiber such as glass fiber-reinforced composites. Natural fiber-reinforced composites, such as hemp fiber-epoxy, flax fiber-polypropylene (PP), and china reed fiber-PP are particularly attractive in automotive applications due to lower cost and lower density. Glass fibers used for composites have density of 2.6 g/cm^3 and cost between \$1.30 and \$2.00/kg, whereas, the density of flax fibers was found to be 1.5 g/cm^3 and its cost between \$0.22 and \$1.10/kg [24]. Natural fiber composites are also claimed to offer environmental advantages such as reduced dependence on non-renewable sources, lower pollutant emissions, lower greenhouse gas emissions, and enhanced energy recovery. Environmental and economical concerns are stimulating research to design eco-friendly renewable natural fibers for construction, furniture, packaging and automotive industries. These fibers are often considered only for markets that require low costs, high production rates and can accept low performance. However, these fibers have many advantages such as light weight, non-abrasive, and low energy requirements for processing. Therefore, researchers are looking forward to use these natural fibers to reinforce thermoplastic polymers [25, 26].

Due to the exponential growth of petroleum-based polymeric materials and environmental concern, there is an urgent need to search the best possible alternative to replace these conventional materials. In this contest, natural fiber-reinforced thermoplastic composites are proving their suitability to replace conventional fillers like glass and carbon fiber-based composites [27, 28]. There are many advantages of using natural fibers composites such as renewable, biodegradable, low pollution level during processing of composites, less carbon emission, and comparatively less energy is required for production. Automotive industries are now looking for natural fiber-based composites [29]. For an automotive concern, a car made from grass may not sound strong and healthy; however plant-based cars are the wave of the future. Researchers at Michigan State University are working on developing materials from plants like hemp, kenaf, sisal, corn straw, and grass to replace plastics and metal-based car components.

23.1.2 Chemical Composition of Natural Fiber

Climatic conditions, age and the digestion process influence the structure and chemical composition of the fiber. The major components of natural fiber are cellulose, hemicellulose, and lignin. Cellulose is the essential component of natural fiber. It is a linear condensation polymer consisting of D-anhydroglucopyranose units (glucose units) joined together by 1, 4-glycosidic bonds [30, 31]. The pyranose rings are in the $4C_1$ conformation meaning that $-CH_2OH$, $-OH$ groups and glycosidic bonds, are equatorial position of the rings. The supramolecular structure of cellulose is responsible for chemical and physical properties. In the fully extended molecule, their mean planes at an angle of 180° to each other oriented to adjacent chain units. Thus, the repeating unit in cellulose is the anhydro cellulobiose unit and the number of repeating units per molecule is half of the degree of polymerization. This may be as high as 14,000 in native cellulose, but purification procedures usually reduce it to in the order of 2,500. The mechanical properties of natural fibers depend on cellulose type, since each type of cellulose has its own cell geometry.

The another component of natural fiber is hemi-cellulose which comprise of a group of polysaccharides. The hemi-cellulose differs from cellulose in three important aspects. First, they contain several different sugar units whereas cellulose contains only 1, 4-D-glucopyranose units. Secondly, they exhibit a considerable degree of chain branching, whereas cellulose is a strictly linear polymer. Finally, the degree of polymerization of hemi-cellulose is ten to one hundred times lower than that of native cellulose. Unlike cellulose, the constituents of hemi-cellulose differ from plant to plant [32]. The another main component of natural fiber is lignin which is a complex hydrocarbon polymer with both aliphatic and aromatic constituent [33]. Lignin fills the spaces in the cell wall between cellulose, hemicellulose, and pectin components, especially in xylem tracheids, vessel elements and sclereid cells. It is covalently linked to hemicellulose and, therefore, crosslinks different plant polysaccharides, conferring mechanical strength to the cell wall and by extension the plant as a whole [34]. It is particularly abundant in compression wood but scarce in tension wood,which are types of reaction wood. Lignin plays a crucial part in conducting water in plant stems. The polysaccharide components of plant cell walls are highly hydrophilic and thus permeable to water, whereas lignin is more hydrophobic. The crosslinking of polysaccharides by lignin is an obstacle for water absorption to the cell wall. Thus, lignin makes it possible for the plant's vascular tissue to conduct water efficiently [35]. The mechanical properties of lignin are lower than that of cellulose. The internal cross-section of natural fiber is shown in Figure 23.1 and a comparative study of different natural fibers in terms of chemical composition is given in Table- 23.1.

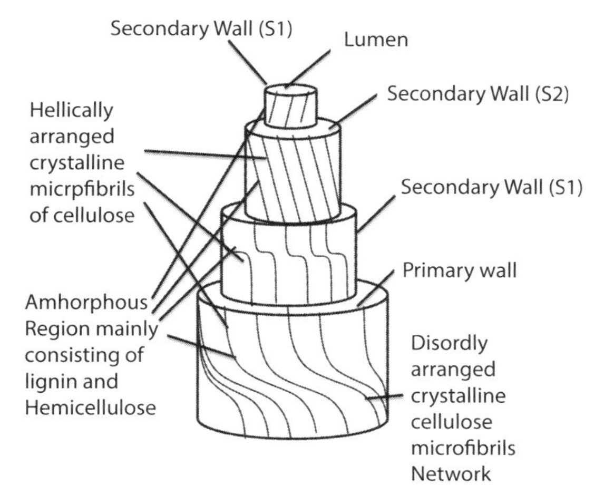

Figure 23.1 Structure of natural fiber.

Table 23.1 Comparison of different natural fibers in terms of chemical composition [Mohanty *et al.* 2000].

Fiber	Cellulose (%)	Hemicellulose (%)	Lignin (%)	Ash (%)	Pectin (%)	Wax (%)	Mocrofibril angle (°)	Moisture (%)
Jute	61-71.5	13-20.4	12-13		0.2	0.5	8.0	12.6
Flax	71-78.5	18-20.6	2.2	1.5	2.2	1.7	10.0	10.0
Hemp	70-74.4	17-22.4	3-5.7	2.6	0.9	0.8	6.2	10.8
Ramie	68-76.2	13-16.7	.6-.7		1.9	0.3	7.5	8.0
Kenaf	31-39	15-19	21.5	4.7				
Sisal	67-78	10-14.2	8-11		10	2.0	20.0	11.0
PALF	70-82		5-12				14.0	11.8
Henequen	77.6	4-8	13.1					
Cotton	82.7	5.7			0.6	0.6		
Coir	36-43	0.15-0.25	41-45				41-45	8.0
Wood (Soft)	40-44	25-29	25-31	0.2				
Wood (Hard)	43-47	25-35	16-24	0.4				

23.1.3 Classification of Natural Fibers

The classification of natural fibers is given in Figures 23.2 and 23.3. All plant fibers are composed of cellulose while animal fibers consist of proteins (hair, silk, and wool). Plant fibers include bast (or stem) fibers, leaf or hard fibers, seed, fruit, wood, cereal straw, and other grass fibers. Over the last few years, a number of researchers have been involved in investigating the exploitation of natural fibers as load bearing constituents in composite materials [2, 36]. It is important to know that, some plants yields more than one type of fiber for example jute, flax, hemp, and kenaf have both bast and core fibers. Cerials grains have both stem and hull fiber. Plant fiber can themselves be classified as either wood or non-wood. Non-wood fibers may be further subdivided into bast, leaf or seed hair fibers, depending on their origin in the plant, while wood fiber

Figure 23.2 Different types of natural fibers.

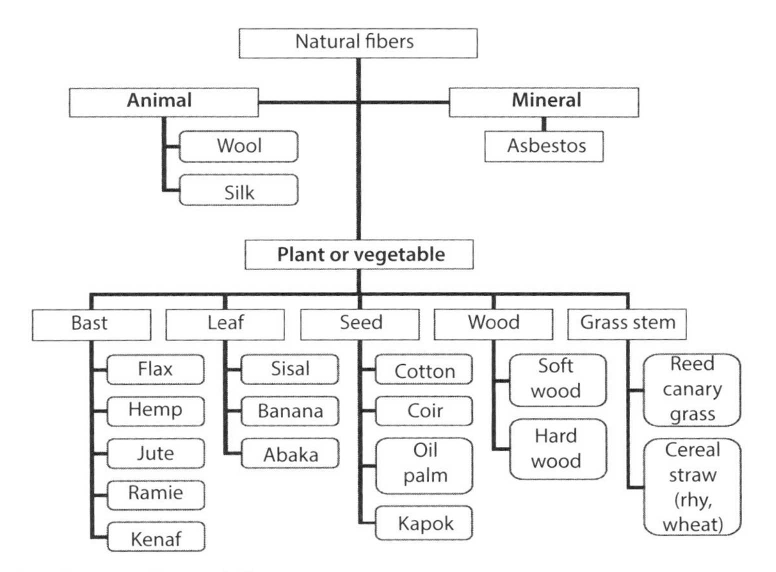

Figure 23.3 classification of natural fibers.

can, for convenience, are subdivided into softwood and hardwood fiber. All vegetable fibers, whether from wood or non-wood origin, are, however, composed of three main cell wall polymers: cellulose, lignin and matrix.

23.1.4 Surface Modification of Natural Fibers

The strength of the fiber is very important for natural fiber-reinforced composites. Polar natural fibers are inherently incompatible with hydrophobic polymers. Therefore, the surface modification of natural fibers is required to improve its physical and chemical properties. Various chemical treatment methods have been used to modify the surface properties of the natural fibers. Few methods of surface modification are described below:

23.1.4.1 Alkali Treatment

Alkaline treatment or mercerization is one of the best used chemical treatment for natural fibers.

Due to alkali treatment there is an increase in the amount of amorphous cellulose at the expense of crystalline cellulose. By this treatment there is a removal of hydrogen bonding in the network structure. Reaction which takes place during this treatment is shown below:

$$\text{Fiber-OH} + \text{NaOH} \rightarrow \text{Fiber-O-Na}^{+} + H_2O \quad (23.1)$$

The type of alkali treatment such as KOH, LiOH, NaOH and its concentration will influence the degree of swelling and degree of lattice transformation into cellulose [37]. Alkali solution not only affects the cellulosic components inside the plant fiber but also affect the non cellulosic components such as hemicellulose, lignin and pectin [38]. Jacob *et al.* examined the effect of NaOH conc. (0.5, 1, 2, 4 and 10%) for treating sisal fiber-reinforced composites and concluded that maximum tensile strength resulted from the 4% NaOH treatment at room temperature [39]. Mishra *et al.* reported that 5% treated NaOH fiber-reinforced polyester composites having better tensile strength than 10% NaOH treated composites [40]. Because at high concentration there is delignification of natural fiber taking place and as a result damage of fiber surface. The tensile strength of composite decreased drastically after certain optimum NaOH concentration.

23.1.4.2 Silane Treatment (SiH4)

The scheme for silane treatment for natural fiber is given in Figure 23.4. Silanes are used as coupling agents to let glass fibers adhere to a polymer matrix, stabilizing the composite materials. Silane coupling agents may reduce the number of hydroxyl groups in the fiber matrix interface. In the presence of moisture, hydrolysable alkoxy group leads to the formation of silanols. The silanols reacts with hydroxyl group of the fiber, forming stable covlantbonds to the cell wall that are chemisorbed on to the fiber surface. As a result the hydrocarbon chains provided by the application of silane restrain the swelling of the fiber by creating a crosslinked network due to covalent bonding between the matrix and the fiber. The reaction schemes are given as follows [41].

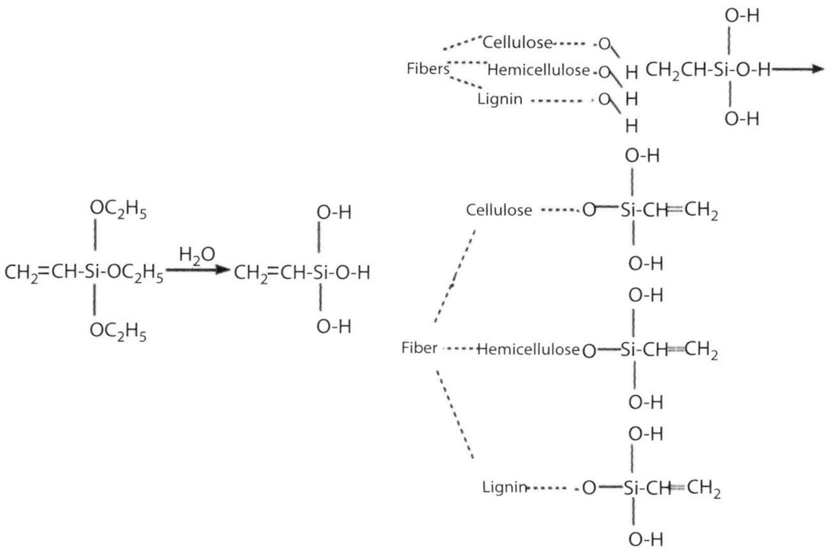

Figure 23.4 Scheme for silane treatment of natural fiber.

$$CH_2CHSi\ (OC_2H_5)_3 \rightarrow CH_2CHSi\ (OH)_3 + 3C_2H_5OH \qquad (23.2)$$

$$CH_2CHSi\ (OH)_3 + Fiber - OH = CH_2CHSi\ (OH)_2O\text{-}Fiber + H_2O \qquad (23.3)$$

Silane treatment in surface modification of glass fiber composites have been applied by various researchers [42-45]. They have found that Silane coupling agents are effective in modifying natural fiber–polymer matrix interface and increasing the interfacial strength.

23.1.4.3 Acetylation of Natural Fibers

During Acetylation there is an introduction of an acetyl functional group (CH_3COO–) into an organic compound. Esterification is a well known method for acetylation of natural fibers and it causes plasticization of cellulosic fibers. The reaction which involves the generation of acetic acid (CH_3COOH) as by-product which must be removed from the lignocellulosic material before the fiber is used. Chemical modification with acetic anhydride (CH_3-C(= O)-O-C(= O)-CH_3) substitutes the polymer hydroxyl groups of the cell wall with acetyl groups, modifying the properties of these polymers so that they become hydrophobic. The reaction of acetic anhydride with fiber is shown as [2].

$$Fiber\ \text{-}OH + CH_3 -C(= O)\text{-}\ O - C(= O)\text{-}\ CH_3 \rightarrow Fiber\ \text{-}OCOCH_3 + CH_3COOH \qquad (23.4)$$

Acetylation can reduce the hygroscopic nature of natural fibers and increases the dimensional stability of composites. Acetylation was used in surface treatments of fiber for use in fiber-reinforced composites [46, 47].

Out of the above mentioned methods, alkali treatment process (mercerization) has been adopted for the surface modification of Sisal Fiber (SF). The treatment of the fiber was carried out by immersing the fibers in 1N sodium hydroxide (NaOH) solution for 1 hour at room temperature. Then, these fibers were washed with distilled water containing few drops of acetic acid, followed by thorough washing under continuous

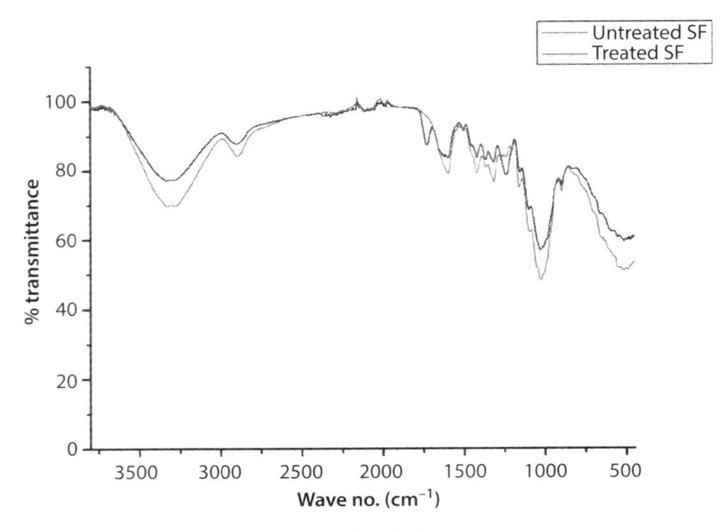

Figure 23.5 FTIR spectra for treated and untreated sisal fiber.

stream of water to remove residual NaOH. Subsequently, the mercerized fibers were dried at room temperature for 24 hours, and then in a vacuum oven at 80°C for 24 hours and finally chopped to obtain fibers with length of 2-3 mm. The FTIR analysis has been performed on untreated and treated SF and shown in Figure 23.5. It is inferred from the spectra, that the absorbance between 1030-1150 cm^{-1} corresponds to the symmetric stretching of primary and secondary C-O-C and C-O. Stretch at 1654 cm^{-1} and 3400 cm^{-1} confirmed the presence of –C = O and –OH groups of aliphatic carboxylic acid and ketonic groups, respectively. The absence of characteristic peak at 1643 and 1156 cm^{-1} in the spectra confirmed the modification of SF. The disappearance of these peaks may be due to the decomposition of hemicellulose and partial leaching out of lignin. The shifting of the peaks in the spectra indicates the participation of free hydroxyl groups in the chemical reaction, where –OH group of fiber is replaced with –ONa.

23.1.5 Properties of Natural Fibers

The properties of some natural fibers are summarized in Table 23.2. Generally, tensile strength and Young modulus of fibers increases with increasing cellulose content. The microfibrillar angle determines the stiffness of the fibers. Plant fibers are more ductile if the microfibrils have a spiral orientation to the fiber axis. If the microfibrils are oriented parallel to the fiber axis, the fibers will be rigid, inflexible, and have high tensile strength [48, 49]. The criteria of chosing suitable reinforcing fibers are elongation at failure, thermal stability, adhesion of fibers and matrix, dynamic and long-term behavior, price and processing costs. The inherently polar and hydrophilic nature of lignocellulosic fibers and the non-polar characteristics of most thermoplastics result in compounding difficulties leading to non-uniform dispersion of fibers within the matrix. These properties possess serious problems during shipping, storage, and composite processing. The non uniformity and variation of dimensions and of their mechanical properties (even between individual plants in the same cultivation) possess another serious problem.

Table 23.2 Properties of natural fibers [Mohanty *et al.* 2000].

Fiber	Tensile strength (MPa)	Young's modulus (GPa)	Elongation at break (%)	Density (g/cm^3)
Abaca	400	12	3–10	1.5
Alfa	350	22	5.8	0.89
Bagasse	290	17	–	1.25
Bamboo	140–230	11–17		0.6–1.1
Banana	500	12	5.9	1.35
Coir	175	4–6	30	1.2
Cotton	287–597	5.5–12.6	7–8	1.5–1.6
Curaua	500–1,150	11.8	3.7–4.3	1.4
Date palm	97–196	2.5–5.4	2–4.5	1–1.2
Flax	345–1,035	27.6	2.7–3.2	1.5
Hemp	690	70	1.6	1.48
Henequen	500 6 70	3.2 6 3.1	8 6 1.1	1.2
Isora	500–600	–	5–6	1.2–1.3
Jute	393–773	26.5	1.5–1.8	1.3
Kenaf	930	53	1.6	–
Nettle	650	38	1.7	–
Oil palm	248	3.2	25	0.7–1.55
Piassava	134–143	1.07–4.59	1.9–7.8	1.4
Pineapple	1.44	400–627	14.5	0.8–1.6
Ramie	560	24.5	2.5	1.5
Sisal	511–635	9.4–22	2.0–2.5	1.5

It is quite clear that the advantages, disadvantages and most of the shortcomings have remedial measures in the form of chemical treatments.

23.2 Recycled Polypropylene (RPP) - A matrix for Natural Fiber Composites

Polypropylene is one of the most widely used thermoplastics, not only because of its balance of physical and mechanical properties, but also due to its environmental friendliness such as recyclability and low cost. Polypropylene (PP) possesses excellent resistance to organic solvents, degreasing agents and electrolytic attack. It is the lightest of the widely used thermoplastics with the exception of plastic foams. With a specific gravity of less than one, polypropylene will float on water. In the latest years industries are attempting the recycling practices on the thermoplastic wastes to reduce the dependence on petroleum-based fuels and products due to the dwindling availability of landfill sites, rising oil prices and increased environmental consciousness. This is leading to the need to investigate environmentally friendly, sustainable materials to replace existing ones. The tremendous increase of production and use of plastics in every sector of our life lead to huge plastic wastes. Disposal problems, as well as strong regulations and criteria for cleaner and safer environment, have directed great part of the scientific research toward the use of recycled materials. Among the different types of eco-composites recycled polypropylene (RPP) containing natural fibers (NF) and natural polymers have a key role. Currently the most viable way toward eco-friendly composites is the use of natural fibers as reinforcement in RPP. Natural fibers represent a traditional class of renewable materials which, nowadays, are experiencing a great revival. In the latest years there have been many researches developed in the field of natural fiber-reinforced recycled plastics [50, 51]. Most of them are based on the study of the mechanical and thermal properties of recycled composites reinforced with NF.

23.3 Natural Fiber-Based Composites – An Overview

Research on natural fiber composites has existed since the early 1900s but has not received much attention until late in the 1980's. Composites, primarily glass but including natural fiber-reinforced composites, are found in countless consumer products including: boats, skis, agricultural machinery and cars. A major goal of natural fiber composites is to alleviate the need to use expensive glass fiber ($3.25/kg) which has a relatively high density (2.5 g/cm^3) and is dependent on nonrenewable sources [52-54]. Recently, automobile companies have been interested in incorporating natural fiber composites into both interior or and exterior parts. This serves a two-fold goal of the companies; to lower the overall weight of the vehicle thus increasing fuel efficiency and to increase the sustainability of their manufacturing process. Many companies such as Mercedes Benz, Toyota and Daimler Chrysler have already accomplished this and are looking to expand the uses of natural fiber composites. Natural fibers such as sisal, cotton, flax, hemp, etc., primarily consist of: cellulose, hemicelluloses, pectin and lignin. The individual percentage of these components varies with the different types of fibers. This variation can also be effected by growing and harvesting conditions. Cellulose is a semicrystalline polysaccharide and is responsible for the hydrophilic nature of natural fibers. Hemicellulose is a fully amorphous polysaccharide with a lower molecular weight compared to cellulose. The amorphous nature of hemicelluloses results in it being partially soluble in water and alkaline solutions [54]. Pectin, whose function is to hold the fiber together, is a polysaccharide like cellulose and hemicellulose. Lignin is an amorphous polymer but unlike hemicellulose, lignin is comprised mainly of aromatics and has little effect on water absorption [55, 56]. The largest advantages to using natural fibers in composites are the cost of materials, their sustainability and density. Natural fibers can cost as little as $0.50/kg, and can be grown in just a few months [12]. They are also easy to grow and have the potential to be a cash crop for local farmers. Natural fibers are also significantly lighter than glass, with a density of 1.15-1.50 g/cm^3 versus 2.4g/cm^3 for E-glass [12]. Two major factors currently limit the large scale production of natural fibers composites. First, the strength of natural fiber composites is very low compared to glass. This is often a result of the incompatibility between the fiber and the resin matrix. The wettability of the fibers is greatly reduced compared to glass and this constitutes a challenge for scale up productions. Though when comparing specific strengths, natural fibers are not much less than glass fiber composites. The second factor limiting large scale production of natural fiber composites is water absorption. Natural fibers absorb water from the air and direct contact from the environment. This absorption deforms the surface of the composites by swelling and creating voids. The result of these deformations is lower strength and an increase in mass. The treatment of fibers is currently an area of research receiving significant attention. The absorption of water is commonly thought to occur at the free hydroxyl groups on the cellulose chains. With a ratio of 3 hydroxyl groups per glucose repeat unit the amount of water that can be absorbed is substantial. By capping the hydroxyl groups this ratio can be reduced. There are several promising techniques that have been studied by various groups [57-60]. Among all treatments, treatments mercerization (alkaline) treatment has had the most reviews [57, 59]. Utilizing silanes as coupling agents is a treatment commonly used in glass composite production and is starting to find uses in natural fiber composites [54]. Acetylation is another treatment that is common with cellulose to form a hydrophobic thermoplastic and has the potential to have the same results on natural fibers [57-60].

23.3.1 Sisal Fiber–Based Recycled Polypropylene (RPP) Composites

A number of investigations have been conducted on several types of natural fibers such as sisal, kenaf, hemp, flax, bamboo, and jute to study the effect of these fibers on the mechanical properties of composite materials [61, 62]. Among the various natural fibers, sisal fibers possess moderately high specific strength and stiffness and can be used as a reinforcing material in polymeric resin matrices to make useful structural composite materials [63]. Sisal fiber is a lignocellulosic material extracted from the plant Agave Sisalana and is available in quantity in the southern parts of India. The incorporation of sisal fiber into plastics and elastomers to obtain cost reduction and reinforcement have been reported by various researchers. Parameswaran and Abdulkalam [64] investigated the feasibility of developing polymer-based composites using sisal fiber. Pavithran *et al.* [65] reported the impact properties of a unidirectionally oriented sisal-fiber composite. Thomas and coworkers reported on the use of sisal and pineapple fibers as a potential reinforcing agent in polyethylene, thermosets (epoxy resin, phenol-formaldehyde, polyester), polystyrene and natural rubber [66-70]. However, very limited studies have been reported in the literature on the use of sisal fiber as a reinforcement in recycled polypropylene thermoplastic (RPP). A detailed investigation in terms of mechanical, thermal, morphological, and effect of weathering has therefore been carried out on sisal-fiber-reinforced RPP composites.

Prior to compounding, untreated and treated sisal fibers as well as recycled polypropylene (RPP) were predried at 80°C in a vacuum for 12 hours. Different compositions viz- 10%, 20%, 30%, 40% wt% of sisal fibers were incorporated in RPP and the mixture were melt blended in a batch mixer (Haake Rheomix OS, Germany). The mixing was carried out at 190°C for 12 minutes with rotor speed of 40 rpm. The molten materials were cooled to room temperature, granulated and conditioned at 80°C for 2 hours prior to specimen preparation. Finally, the samples were prepared using twin screw micro-compounder (Haake Rheomix 9000) with a capacity of just a few grams of material. In the first stage, RPP/SF containing different weight percentage of sisal fiber (10%, 20%, 30% & 40%) were prepared, and then 40 wt% of fiber loading RPP/alkali-treated SF/MA-g-PP was prepared. The optimized sample of RPP and TSF in the ratio 60:40 with 3 wt% of clay loading has also been prepared under identical condition. All compositions were prepared at a temperature of 190⁰C, mixing speed of 40 rpm for duration of 12 minutes.

23.3.1.1 Mechanical and Dynamic Mechanical Properties of Sisal RPP Composites

The results of mechanical properties of SF, RPP, and their composites are summarized in Table 23.3 (a) and (b), and Figures 23.6 (a), (b), (c), and (d), respectively. It is evident from the test results that percentage elongation of Untreated Sisal Fiber (UTSF)-reinforced RPP at all composition was found to be lower than that of RPP. An Approximate decrement of 24.27%, 30.1%, 39.8%, and 51.5% has been observed for 10, 20, 30, and 40 wt% UTSF loaded RPP composites, respectively. Similarly, UTSF could not result an appreciable enhancement in tensile strength of RPP. However, an increasing trend of tensile strength with values of 0.23%, 1.07% and 3.72% for 20%, 30%, and 40 wt% fiber-filled RPP composites has been observed. This marginal improvement may be an indicative of suitability of sisal fiber as reinforcement in RPP class of thermoplastic. Moreover, UTSF with different composition (10, 20, 30, and 40 wt%) results in an increase in the tensile modulus by 38%, 55%, 67%, and 74%, as compared to the RPP. The optimum modulus value of 2233 MPa was observed for 40 wt% of UTSF loading, which was 73.89% higher than that of RPP alone. This enhancement may probably due to the sufficiently good adhesion between the fibers and the matrix leading to the compact packing arrangement. Furthermore, the similar properties have been evaluated for Treated Sisal Fiber (TSF) incorporated RPP. The tensile modulus (2337 MPa) of 40 wt% TSF loaded RPP was observed to be far better than similar composition of RPP/UTSF (2233 MPa) as well as RPP alone (583 MPa). This phenomenon of an enhancement may be attributed to the

Table 23.3 (a) Mechanical properties of RPP/UTSF composites.

S. no.	Sample (wt %)	Tensile Strength (MPa)	Tensile Modulus (MPa)	% Elongation	Impact strength (J/m)
1.	RPP	21.1 ± 1	583 ± 62	60.06 ± 33	103 ± 2
2.	RPP/UTSF (90/10)	20.7 ± 0.41	941 ± 51	11.78 ± 2	78 ± 5
3.	RPP/UTSF (80/20)	21.25 ± 0.27	1289 ± 68	7.2 ± 1	72 ± 3
4.	RPP/UTSF (70/30)	21.43 ± 0.36	1765 ± 78	3.67 ± 1	62 ± 5
5.	RPP/UTSF (60/40)	22.02 ± 1	2233 ± 158	2.37 ± 0.06	50 ± 5

Table 23.3 (b) Mechanical properties of V-RPP, RPP/UTSF, RPP/TSF RPP/UTSF/MA-g-PP, RPP/TSF/MA-g-PP, RPP/TSF/MA-g-PP/C15A.

S. no.	Samples	Tensile Strength (MPa)	Tensile Modulus (MPa)	Elongation at break (%)	Notched Impact (J/m)
1.	V-RPP	21.1 ± 1	583 ± 62	60.06 ± 33	103 ± 2
2.	RPP/UTSF (60/40)	22.02 ± 1	2233 ± 158	2.37 ± 0.06	50 ± 5
3.	RPP/TSF (80/20)	22.19 ± 0.39	2337 ± 62	2.02 ± 0.08	53 ± 3
4.	RPP/UTSF/ MA-g-PP (55/40/5)	24.83 ± 1	2502 ± 118	1.74 ± 0.14	50 ± 0.7
5.	RPP/TSF/ MA-g-PP (55/40/5)	26.37 ± 0.25	2563 ± 57	1.72 ± 0.11	48 ± 3
6.	RPP/TSF/ MA-g-PP/ C15A	23.85 ±0.57	2743 ± 41	1.74 ± 0.03	46 ± 5

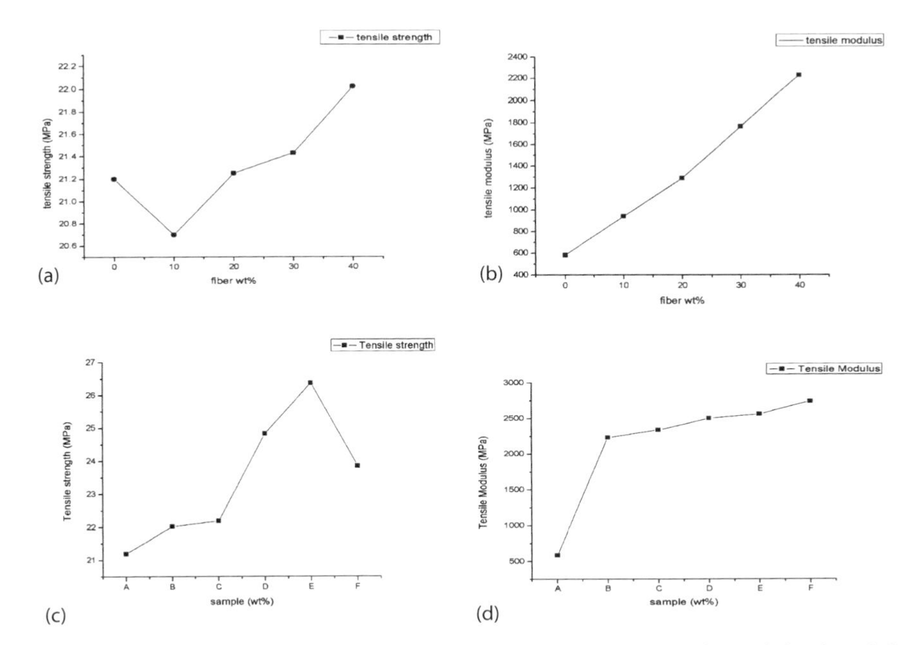

Figure 23.6 (a) Tensile strength (MPa) graph for different fiber wt% (b) Tensile modulus (MPa) data for different fiber wt% (c) Tensile strength (MPa) of A = V-RPP, B = RPP/UTSF, C= RPP/TSF, D = RPP/UTSF/MA-g-PP, E = RPP/TSF/MA-g-PP and F = RPP/TSF/MA-g-PP/C15A (d) Tensile modulus (MPa) of A = V-RPP, B = RPP/UTSF, C= RPP/TSF, D = RPP/UTSF/MA-g-PP, E = RPP/TSF/MA-g-PP and F = RPP/TSF/MA-g-PP/C15A.

surface uniformity of TSF, which provides mechanical interlocking between the fiber and the matrix at the interface. The SEM image of TSF has also confirmed the above fact, where a more rigid and uniform surface obtained [Figure 23.7 (a)]. Moreover, the impact strength of 10, 20, and 30 wt% incorporated UTSF was observed to be higher than that of 40 wt% loaded UTSF and TSF composites. This deterioration at higher fiber loading may be due to the incapability of the filler to absorb sufficient energy necessary to stop crack formation. On the other hand, the addition of small amount of MAPP (5%) into RPP/UTSF (55:40) and RPP/TSF (55:40) drastically improves their tensile strength and modulus values as compared to the RPP as well as other composites. This is assumed that the polar groups present in the SF, which are the active sites for incompatibility and crack propagation were replaced by the anhydride group of MAPP. Hence, a compatibilized structure with modified interface between the SF and RPP may be the determinant of this appreciable enhancement. The SEM images clearly indicating sufficiently good adhesion between fiber and matrix provided by MAPP compatibilizer [Figures 23.7 (b) and (c)]. In addition to this, mechanical properties of well intercalated nanoclay-filled RPP composite has also been evaluated and the highest value of tensile modulus (2743 MPa) among all composites was found in RPP/TSF/MA-g-PP/C15A. It is expected that the exfoliated clay layer might have filled the available gaps in the interfacial region thereby providing more compact structure with improved interfacial strength, which leads to an increase in the stiffness.

Furthermore, the variation of storage modulus loss modulus, and mechanical loss factor of RPP and RPP/TSF/MA-g-PP composite as a function of temperature has also determined and shown in the Figures 23.8 (a), (b), and (c), respectively. It is evident from the results that SF also contributes towards an increase in the storage modulus of RPP. The RPP/TSF/MA-g-PP composite containing 40 wt% fiber showed a storage modulus value of around 6000MPa, which was almost doubled as compared to the RPP (3000MPa). This considerable increase in the storage modulus may be due to the more efficient stress transfer from the matrix to the fiber as well as stiffness induced by the fibers and the compatibilizer, which restricts the segmental motion of the RPP matrix. However, the storage modulus of both samples was observed to be decreases with increasing temperature. This may happened due to the softening of polymer matrix as temperature increases, which decreases the intermolecular forces between fiber and matrix. A sudden decrease in the storage modulus after β transition took place in both samples indicating a glassy/rubbery transition phase. Moreover, the β and α transitions occurred at around 10°C and 50-80°C, respectively. Whereas, the onset of the β transition indicates the storage modulus glass transition and the peak point

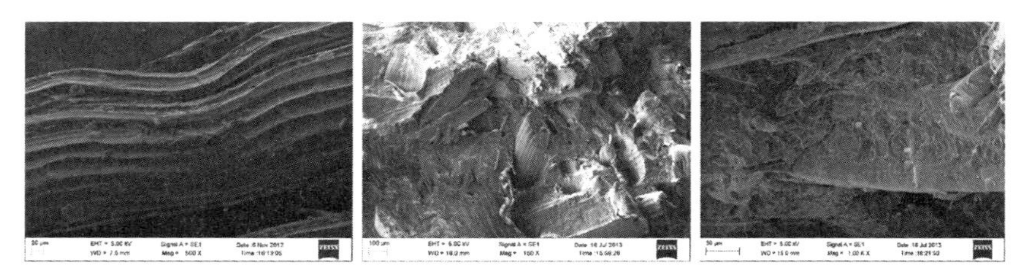

Figure 23.7 (a) SEM of TSF (b) SEM image of RPP/UTSF/MA-g-PP (c) SEM image of RPP/TSF/MA-g-PP.

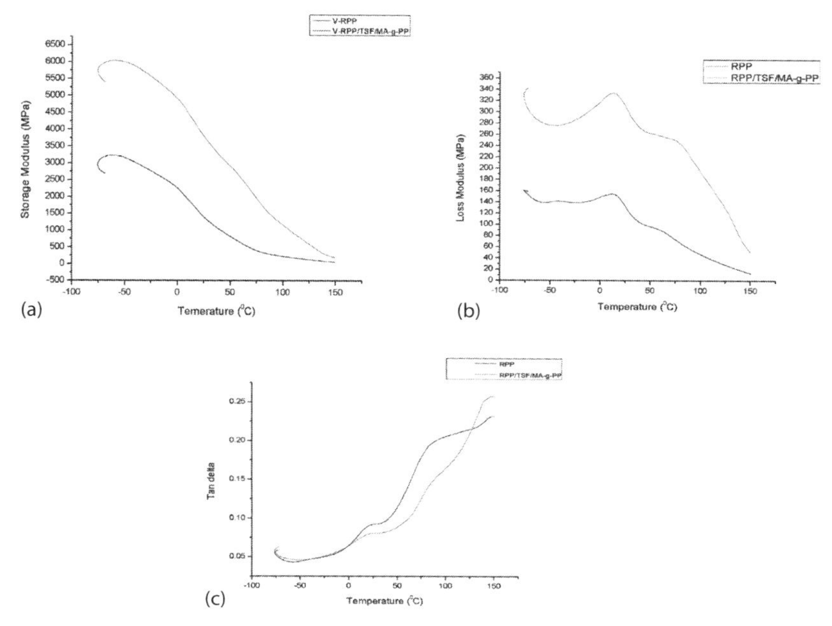

Figure 23.8 (a) Variation of Storage modulus (E′) as a function of temperature (b) Variation of Loss modulus (E″) as a function of temperature (c) Variation of tan δ (mechanical loss factor) as a function of temperature.

of the β transition is the loss modulus glass transition of RPP and RPP/TSF/MA-g-PP composite. The loss modulus value of the composite was found to be higher than that of RPP, which may be probably an indication of good interfacial adhesion between the fiber and matrix [71-73]. On the other hand, the highest value of mechanical loss factor (tan δ) was observed for RPP, which is due to the free movement of polymeric chain during relaxation process. But, the addition of fiber in to the host matrix lowers the tan δ value, which may be due to the elastic deformation and lack of the free movement of polymeric chains.

23.3.1.2 Thermal Properties Sisal RPP Composites

The DSC values of virgin RPP and its composits are summarized in Table 23.4. The glass transition temperature (T_g) of virgin RPP was observed at around 0°C, which was increased by 18°C at 40 wt% UTSF loading. This increase in the T_g value may be attributed to the compact packing of fibers within the matrix [74]. T_g value of 40 wt% TSF loaded RPP exhibits similar trend of enhancement as compared to the RPP alone. Moreover, the addition of MAPP contributes to the further enhancement in the T_g of UTSF and TSF-filled composites. This additional increment may be due to the effect of MAPP which compatibilized the blends of fiber and matrix. On the other hand, a decreasing pattern of crystallinity (%) values in all the composites has been observed as compared to the virgin RPP. The lowest observed crystallinity value (14.1) was found in RPP/TSF (60/40) which may be due to the hydrophilic character of fibers. Moreover, there was no effect of fibers on peak maximum and the crystallization peaks of both RPP/UTSF (60/40) and RPP/TSF (60/40) composites was observed to be more or less

Table 23.4 DSC analysis of RPP and its composites.

S. no.	Sample	T_g (°C)	T_c		% crystal linity	T_m	
			T_{c1}	T_{c2}		T_{m1}	T_{m2}
1.	V-RPP	0	121.68	-	33.91	166.77	128.44
2.	RPP/UTSF (60/40)	18	120.8	-	22.53	165.87	127.94
3.	RPP/TSF (80/20)	18	121.12	-	14.1	165.34	127.57
4.	RPP/UTSF/ MA-g-PP (55/40/5)	19	126.99	114.5	14.47	166.57	128.87
5.	RPP/TSF/ MA-g-PP (55/40/5)	19	127.12	114.01	14.08	166.69	128.75
6.	RPP/TSF/ MA-g-PP/ C15A	17	126.42	113.32	14.4	165.62	128.07

similar to the virgin RPP. However, the addition of MAPP leads to an appreciable increment in the crystallization peaks as compared to virgin RPP as well as other composites. The peak value of 126.99 °C and 127.12 °C has been determined for RPP/UTSF/ MA-g-PP RPP/TSF/MA-g-PP composites, respectively. This improvement may likely be due to the sufficiently good interfacial bonding between the fiber and host matrix. Moreover, nanoclay has also been incorporated to see its effect on the similar DSC parameters. Almost, same trend of increase in the crystallization peak of RPP/TSF/ MA-g-PP/C15A composite (126.42 °C) has been observed and attributed to the nucleating effects of nanoclay layers, which requires comparatively more amount of energy for phase transition.

The Thermogravimetric Analysis (TGA) has also been performed on the virgin RPP and its composites and summarized in Table- 23.5, and also represented in Figure 23.9. It is inferred from the TGA thermogram, that virgin RPP shows single step degradation with an initial degradation temperature (IDT) of 314°C, maximum degradation temperature of 446°C and final degradation temperature of 473°C. The 40 wt% UTSF incorporated RPP composite shows lower value of IDT and started to degrade at 220 °C with 50% weight loss. This deterioration of thermal stability may be due to the loss of hydroxyl groups present in the UTSF and depolymerisation of cellulose to anhydroglucose [75]. A small shoulder observed at a region of 225-275 °C corresponds to the degradation of hemicelluloses part of the UTSF may also be the cause of this deterioration. However, a marginal increase of IDT has been achieved by the addition of MAPP into RPP/UTSF composite. On the other hand, an appreciable increment in the thermal stability of 40 wt% TSF loaded RPP has been observed, which may be attributed to the improvement in the surface property of sisal leading to a toughned composites, which

Table 23.5 TGA analysis of RPP and its composites.

S. no.	Sample (wt %)	Initial degradation temp (°C)	Maximum degradation temp (°C)	Final degradation temp (°C)
1.	V-RPP	314	446	473
2.	RPP/UTSF (60/40)	220	446	448
3.	RPP/TSF (80/20)	329	373	406
4.	RPP/UTSF/ MA-g-PP (55/40/5)	230	456	614
5.	RPP/TSF/ MA-g-PP (55/40/5)	206	457	633
6.	RPP/TSF/ MA-g-PP/ C15A	327	365	411

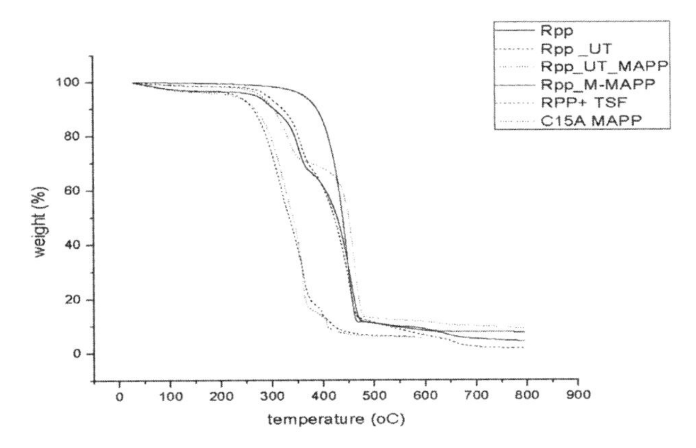

Figure 23.9 TGA thermogram of RPP and composites.

can bear comparatively more heat as compared to the virgin RPP. A similar enhancement in the IDT of nanoclay-filled RPP/TSF/MA-g-PP composite has been reported. This is probably due to the well dispersion of sufficiently high thermal stable silicate particles within the host matrix.

23.3.1.3 Weathering and Its Effect on Mechanical Properties of Sisal RPP Composites

The results obtained for the mechanical properties of RPP and its composites before and after weathering are shown in **Figures 23.10 (a), (b), and (c)**, respectively. The

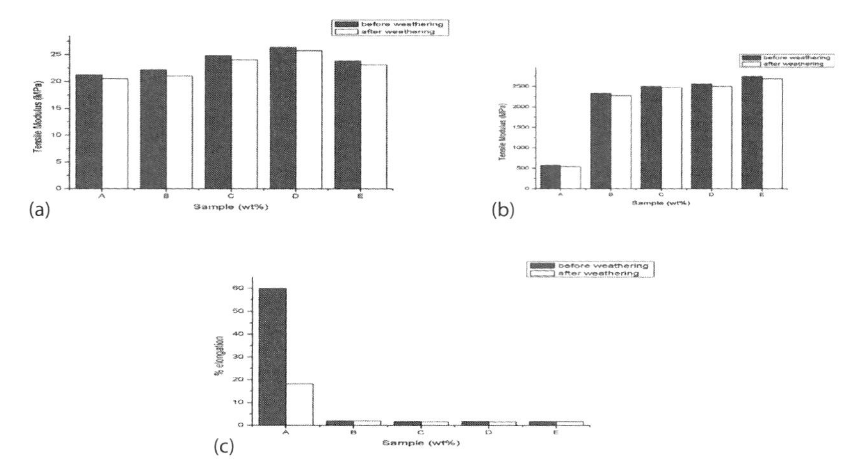

Figure 23.10 (a) Tensile strength of samples before and after weathering (A = RPP, B = RPP/TSF, C = RPP/UTSF/MA-g-PP, D = RPP/TSF/MA-g-PP, E = RPP/TSF/MA-g-PP/C15A) (b) Tensile modulus of samples before and after weathering (A = RPP, B = RPP/TSF, C = RPP/UTSF/MA-g-PP, D = RPP/TSF/MA-g-PP, E = RPP/TSF/MA-g-PP/C15A) (c) % elongation of samples before and after weathering (A = RPP, B = RPP/TSF, C = RPP/UTSF/MA-g-PP, D = RPP/TSF/MA-g-PP, E = RPP/TSF/MA-g-PP/C15A).

overall mechanical properties of all the composites were found to be greater than virgin RPP. This behavior is in accordance with the data reported for all composites in section 23.3.1.1 (Mechanical and Dynamic Mechanical Properties of Sisal RPP Composites) of this article. The degradation processes due to the weathering affect the overall mechanical properties of the virgin RPP and composites. Although, the sisal fiber provide stiffness to the RPP composites but it reduces due to fiber- matrix debonding weathering process. This loss of tensile properties was found to be due to the degradation of main components such as cellulose, hemicelluloses, lignin of the sisal fiber. Also, the PP chain scission leading to the formation of sufficient amount of carboxyl and vinyl groups with the concomitant decrease in the molecular weight is the determinant of the degradation [76, 77]. The elongation at break value of the RPP is significantly higher than that of the SF loaded RPP composite due to the increase in the rigidity and brittleness. The elongation at break value for all the RPP composites has been drastically declined after exposure to UV for 170 h. This loss in elongation of SF loaded RPP composites are again be attributed to the degradation of the major components and extensive chain scission of RPP. Moreover, weathering also induced the microcracks formation on the surface of RPP and composites and causes fiber-matrix debonding, which breaks the samples even at a lower load. The another possible cause of deterioration may be the presence of humidity inside the accelerated weathering chamber. This humidity increases the hydrophilicity of fibers leading to the decrease of interfacial strength of the SF-filled RPP composites [78]. The SEM images confirmed the presence of microcracks on the surface of the composites. Although, surface cracks are evident in all samples after being weathered for 170 h, the severity of cracks varies from one another. It can be seen that the surface of RPP/TSF has more deteriorated with large, continuous surface cracks [Figure 23.11 (a)] The least damage done by weathering is on compatibilized samples, i.e., RPP/TSF/MA-g-PP and RPP/UTSF/MA-g-PP [Figures 23.11 (b) and (c)]

Figure 23.11 (a) SEM image of weathered RPP/TSF (b) SEM image of weathered RPP/TSF/MA-g-PP (c) SEM image of weathered RPP/UTSF/MA-g-PP.

that contain comparatively small cracks, suggested that the ultimate failure is determined by the strength of interface between fibers and matrix.

23.3.1.4 Fracture Analysis of RPP and its Composites

As said earlier, that the recycled polypropylene (RPP) composites are increasing attention for many different application fields due to depletion of petroleum resources and environmental concern. However, owing to its poor fracture resistivity, especially under extreme condition such as high strain rates, the usefulness of RPP as an engineering thermoplastic is still limited. In order to overcome this problems, attempts have been carried out through the addition of a fibrous phase within the polymer matrix. The sisal fiber (SF) has been chosen as reinforcement in RPP due to having appreciable mechanical properties. A very little research and development is done on SF-filled recycled RPP. Before commercially employing SF-based RPP composites for the end use application, the determination of their fracture behavior is critical. Several features of the SF have a decisive influence on the values of the fracture toughness of RPP composites. The fracture behavior of RPP is strongly affected by the addition of SF. RPP composites with enhanced tensile, stiffness, impact, and flexural strength with lower fracture can be achieved by the incorporation of SF. The fracture properties of SF filled composites can be evaluated by component properties, composition, structure and interaction between phases.

The low impact strength is a major disadvantage of natural fiber (NF)-reinforced thermoplastic because NF provides points of stress concentrations, thus providing sites for crack initiation and potential composite failure. The impact properties of composites depend strongly on the interfacial adhesion between components. This can be maximized by improving the interaction and adhesion between the two phases in final composites. There are mainly two approaches to improve the interfacial adhesion:

polymeric matrix and NF modification. Different coupling agents have been used to modify the polymeric matrix and improve the interfacial strength and subsequently the impact properties of the composites. Maleic anhydride grafted styrene-ethylene-butylene-styrene (SEBS-g-MA) [79], maleic anhydride grafted polyolefin such as HDPE-g- MA [80], PP-g-MA [81], and LDPE-g-MA [82] are the most common examples of reported works in the literature. Another approach for enhancement of interfacial adhesion in natural fiber-reinforced thermoplastic matrix is fiber treatment before mixing with polymer. Some of these treatments have physical nature and some of them are of chemical nature. Alkali treatment [37, 38], silane treatment , acetylation [2], etc., are some chemical methods for modification of fibers. Their results show enhanced polymer-matrix adhesion.

In this study, the Izod impact test has been performed on the RPP and SF loaded composites and represented in [Figure 23.12 (a) and (b)]. The Izod impact test was conducted with all samples pre-notched with a notch-cutter. The Resulting fracture surfaces were analyzed by SEM analysis. The SEM images are taken from the previous study done by Mohanty *et al.* [83]and represented in [Figures 23.13 (a), (b), (c), (d), and (e)], respectively. It can be seen from the images that treated SF exhibit smooth surface as compared to the untreated SF [Figure 23.13 (a) and (b)]. Figure 23.3 (c), shows evidence of interfacial deboning of untreated SF with RPP. This behavior corroborates a mechanism of fracture through cracks that spread preferentially between the fiber and matrix due to low interfacial resistance. In this case the fibers suffered rupture at one end, while other fibers have been pulled out from the matrix leaving behind cracks. This may be due to the poor adhesion between fiber and matrix. However, the RPP/UTSF/MA-g-PP (55/40/5) composites showed improved impact properties as compared to RPP/UTSF (60/40) which is attributed to a complex mechanism of individual interaction of the MAPP and UTSF with the RPP matrix during crack initiation and further propagation [Figure 3.1 (i)]. Moreover, RPP/TSF (60/40) and RPP/TSF/MA-g-PP (55/40/5) shows far better properties than that of UTSF loaded RPP composite. The modified features obtained by the treatment of SF could be responsible for this significant improvement, which increases the stress concentration at the fiber/matrix interface. Hence, it is possible to imagine that the improved adhesion between fiber and matrix is necessary, since rupture of the composite takes place from interface

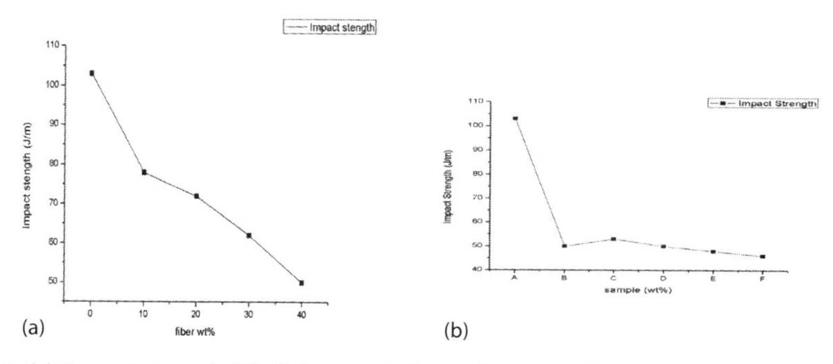

Figure 23.12 (a) Impact strength (J/m) data for different fiber wt% (b) Impact strength (J/m) of (A = V-RPP, B = RPP/UTSF, C = RPP/TSF, D = RPP/UTSF/MA-g-PP, E = RPP/TSF/MA-g-PP, F = RPP/TSF/MA-g-PP/C15A).

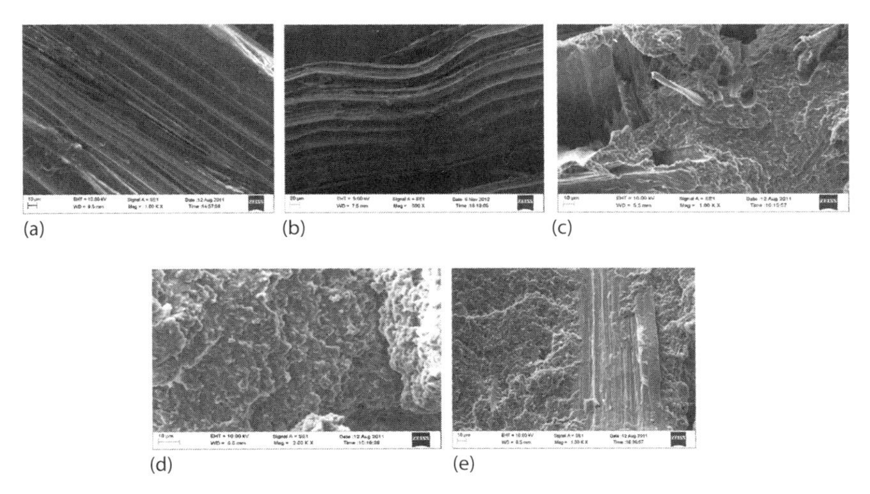

Figure 23.13 (a) SEM of UTSF (b) SEM of TSF (c) SEM of RPP/UTSF (d) SEM of RPP/UTSF/MA-g-PP (e) SEM of RPP/TSF.

through debonding of the fibers. It is verified that during the application of load, some of the fibers can break while others remain intact. It is seen that, in case of TSF loaded RPP the fibers are well embedded into the RPP matrix, which are difficult to separate [Figure 23.13 (e)]. This efficient adhesion and ability to transfer mechanical load from the matrix to the entire fiber is the determinant of improvement of the fracture behavior of RPP composite.

23.4 Conclusion

A number of reviews have been studied on the potential of natural fibers such as sisal, kenaf, hemp, flax, bamboo, and jute for the preparation of thermoplastic composites. In this work, however sisal fiber (SF) has been used as reinforcement due to easily availability and comparatively low cost. The untreated and treated SF-reinforced RPP composites have been prepared and investigated their thermal, mechanical, morphological, weathering and impact properties. An improved mechanical, thermal, and morphological property has been observed for chemical treated SF as well as clay loaded RPP. The analysis revealed that SF-reinforced RPP composites with enhanced properties can be successfully achieved which warrants to replace the synthetic fillers-based conventional thermoplastic composites. These SF-based RPP composites can be the material of choice in the field of aeronautic, automobiles, civil engineering, etc., due to its low cost, low density, non-toxicity, recyclability, acceptable strength, high specific properties, and minimum waste disposal problems.

References

1. K. Jayaraman, Manufacturing sisal–polypropylene composites with minimum fibre degradation. *Compos. Sci. Technol.* 63, 367–374 (2003).

2. M.Z. Rong, M.Q. Zhang, Y. Liu, G.C. Yang, and H.M. Zeng, The effect of fiber treatment on the mechanical properties of unidirectional sisal-reinforced epoxy composites. *Compos. Sci. Technol.* 61, 1437–1447 (2001).
3. K. Okubo, and T. Fujii, Eco-composites using bamboo and other natural fibers and their mechanical properties, *Proceedings of the international workshop on "Green" composites*, pp. 17–21 (2002).
4. M. Wollerdorfer, and H. Bader, Influence of natural fibers on the mechanical properties of biodegradable polymers. *Ind. Crops Prod.* 8, 105–112 (1998).
5. A.K.Mohanty, A. Wibowo, M. Misra, and L.T.Drzal, Effect of process engineering on the performance of natural fiber reinforced cellulose acetate biocomposites. *Compos. A Appl. Sci. Manuf.* 35, 363–370 (2004).
6. D. Plackett, T.L. Andersen, W.B. Pedersen, and L. Nielsen, Biodegradable composites based on polylactide and jute fibers. *Compos. Sci. Technol.* 63, 1287–1296 (2003).
7. P.J. Herrera-Franco, and A. Valadez-Gonzalez. Mechanical properties of continuous natural fibre-reinforced polymer composites. *Compos. A Appl. Sci. Manuf.* 35, 339–345 (2004).
8. S. Luo, and A.N. Netravali. Interfacial and mechanical properties of environment-friendly green composites made from pineapple fibers and poly (hydroxybutyrate-co-valerate) resin. *J. Mater. Sci.* 34, 3709–3719 (1999).
9. T.P. Nevell and S.H. Zeronian, Wiley, New York (1985).
10. P.O. Olesen and D.V. Plackett, Perspectives on the Performance of Natural Plant Fibres in IENICA EVENTS, *Natural Fibres Performance Forum*, Copenhagen, May 27–28, Volume 28, http://www.ienica.net/fibresse minar/fibresindex.htm.
11. S. Hattallia, A. Benaboura, F. Ham-Pichavant, A. Nourmamode, and A. Castellan, *Polym. Degrad. Stabil.* 75, 259 (2002)
12. X. Li, L.G. Tabil, and S. Panigrahi, Chemical treatments of natural fiber for use in natural fiber-reinforced composites: A review. *J. Polym. Environ.* 15, 25–33 (2007).
13. M. Sain and S. Panthapulakkal, Chitosan–starch composite film: Preparation and characterization. *Ind. Crops. Prod.* 21, 185–192 (2005).
14. S.V. Joshi, L.T. Drzal, A.K. Mohanty, and S. Arora, Are natural fiber composites environmentally superior to glass fiber reinforced composites? *Compos. A* 35, 371–376 (2004).
15. M. Brahmakumara, C. Pavithran, and R.M. Pillai, Coconut fibre reinforced polyethylene composites: effect of natural waxy surface layer of the fibre on fibre/matrix interfacial bonding and strength of composites. *Compos. Sci. Technol.* 65, 563–569 (2005).
16. K.G. Satyanarayana, K. Sukumaran, P.S. Mukherjee, C. Pavithran and S.G.K. Pillai. Natural fiber–polymer composites. *J. Cem. Concr. Compos.* 12(2), 117–136 (1990).
17. K.G. Satyanarayana, K. Sukumaran, A.G. Kulkarni, S.G.K. Pillai, and P.K. Rohatgi, Fabrication and properties of natural fiber-reinforced polyester composites. *J. Compos.* 17(4), 329–333 (1986).
18. M.A. Mansur and M.A. Aziz, Study of bamboo-mesh reinforced cement composites. *Int. Cem. Compos. Lightweight Concr.* 5(3), 165–171 (1983).
19. M.A. Maleque, F.Y. Belal, and S.M. Sapuan, Mechanical properties study of pseudo-stem banana fiber reinforced epoxy composite. *Arab. J. Sci. Eng.* 32 (2B) 359–364 (2007).
20. T.M. Gowda, A.C.B. Naidu, and R. Chhaya, Some mechanical properties of untreated jute fabric-reinforced polyester composites. *J. Compos. A Appl. Sci. Manuf.* 30(3), 277–284 (1999).
21. L.A. Pothana, Z. Oommenb, and S. Thomas, Dynamic mechanical analysis of banana fiber reinforced polyester composites. *Compos. Sci. Technol.* 63(2), 283–293 (2003).
22. T. Corbière-Nicollier, B.G. Laban, L. Lundquist, Y. Leterrier, J.-A.E. Månson, and O. Jolliet, Life cycle assessment of biofibers replacing glass fibers as reinforcement in plastics. *Resour. Conserv. Recy.* 33(4), 267–287 (2001).

23. L.A. Pothan, T. Sabu, and Neelakantan, Short banana fiber reinforced polyester composites: mechanical, failure and aging characteristics. *J. Reinf. Plast. Compos.* 16(8), 744–765 **(1997)**.
24. S. Joseph, M.S. Sreekala, Z. Oommen, P. Koshy, and T. Sabu, A comparison of the mechanical properties of phenol formaldehyde composites reinforced with banana fibers and glass fibers. *Compos. Sci. Technol.* 62(14), 1857–1868 (2002).
25. B. Braun, J.R. Dorgan, and D.M. Knauss, Reactively compatibilized cellulosic polylactide microcomposites. *J. Polym. Environ.* 14, 49–58, (2006).
26. K. Oksman, M. Skrifvars, and J.F. Selin, Natural fibres as reinforcement in polylactic acid (PLA) composites. *Comp. Sci. Technol.* 63, 1317–1324 (2003).
27. P. Pan, B. Zhu, W. Kai, S. Serizawa, M. Iji, and Y. Inoue, Crystallization behavior and mechanical properties of bio-based green composites based on poly (L-lactide) and kenaf fiber. *J. Appl. Polym. Sci.* 105, 1511–1520 (2007).
28. S.H. Lee, and S. Wang, Biodegradable polymers/bamboo fiber biocomposite with bio-based coupling agent. *Compos. A* 37, 80–91 (2006).
29. N. Reddy, and Y. Yang, Properties and potential applications of natural cellulose fibers from cornhusks. *Green Chem.* 4, 190–195 (2005).
30. Y. Nishiyama, P. Langan, and H. Chanzy, Crystal structure and hydrogen-bonding system in cellulose Iβ from synchrotron X-ray and neutron fiber diffraction. *J. Am. Chem. Soc* 124 (31), 9074–9082 (2002).
31. R.L. Crawford, *Lignin biodegradation and transformation,* John Wiley and Sons, New York (1981).
32. H. Onggo and S. Pujiastuti, Effect of weathering on functional group and mechanical properties of polypropylene-keneaf composites. *Indonesian J. Mater. Sci.* 11(2), 1–6 (2010).
33. L.M. Matuana, J.S. Jin, and N.M. Stark, Ultraviolet weathering of HDPE/wood-flour composites coextruded with a clear HDPE cap layer. *Polym. Degrad. Stabil.* 96(1), 97–106 (2011).
34. M. Chabannes, K. Ruel, A. Yoshinaga, B. Chabbert,A. Jauneau, J.P. Joseleau, and A.M. Boudet, In situ analysis of lignins in transgenic tobacco reveals a differential impact of individual transformations on the spatial patterns of lignin deposition at the cellular and subcellular levels. *Plant J.* 28 (3), 271–282 (2001).
35. K.V. Sarkanen and C.H. Ludwig, *Lignins: Occurrence, Formation, Structure, and Reactions.* Wiley Intersci, New York (1971).
36. T.P. Nevell and S.H. Zeronian, *Cellulose Chemistry and its Applications,* Wiley, New York (1985).
37. D. Fengel and G. Wegener, *Wood: Chemistry, Ultrastructure, Reactions*, pp. 482, de Gruyter, Berlin (1983).
38. I.V. Weyenberg, T.C. Truong, B. Vangrimde, and I. Verpoest, Improving the properties of UD flax fibre reinforced composites by applying an alkaline fibre treatment.*Compos. A* 37, 1368–1376 (2006).
39. M. Jacob, S. Thomas, and K.T. Varughese, Mechanical properties of sisal/oil palm hybrid fiber reinforced natural rubber composites. *Compos. Sci. Technol.* 64, 955–965 (2004).
40. S. Mishra, A.K. Mohanty, L.T. Drzal, M. Misra, S. Parija, S.K. Nayak, and S.S. Tipathy, Studies on mechanical performance of biofibre/glass reinforced polyester hybrid composites. *Compos. Sci. Technol.* 63, 1377–1385 (2003)
41. P.J. Herrera-Francoand, and A. Valdez Gonzalez, *Effect of fiber treatment on the mechanical properties of LDPE-henequen cellulosic fiber composites. J. Appl. Polym. Sci.* 65, 197–207 (1997).
42. J.K. Kim, M.L. Sham, and J. Wu, Nanoscale characterisation of interphase in silane treated glass fibre composites. *Compos. A Appl. Sci. Manuf.* 32, 607–618 (2001).

43. Z.A. Mohd Ishak, A. Ariffin, and R. Senawi Effects of hygrothermal aging and a silane coupling agent on the tensile properties of injection molded short glass fiber reinforced poly (butylene terephthalate) composites. *Eur. Polym. J. 37,* 1635–1647 (2001).
44. G.W. Lee, N.J. Lee, J. Jang, K.J. Lee, and J.D. Nam, Effects of surface modification on the resin-transfer moulding (RTM) of glass-fibre/unsaturated-polyester composites. *Compos. Sci. Technol.* 62, 9–16 (2002).
45. S. Debnath, S.L. Wunder, J.I. McCool, and G.R. Baran, Silane treatment effects on glass/resin interfacial shear strengths. *Dent. Mater.* 19(5), 441–448 (2003).
46. A.S.C. Hill, H.P.S. Abdul Khalil, and M.D. Hale, A study of the potential of acetylation to improve the properties of plant fibres. *Ind. Crops Prod.* 8(1), 53–63 (1998).
47. M.S. Sreekala, and S. Thomas, Effect of fibre surface modification on water-sorption characteristics of oil palm fibres. *Compos. Sci. Technol.* 63(6), 861–869 (2003).
48. P.K. Shin, M.H. Kim, and J.M. Kim, Biodegradability of degradable plastics exposed to anaerobic digested sludge and simulated landfill conditions. *J. Environ. Polym. Degrad.* 5(1), 33–39 (1997).
49. J.-D. Gu, Microbiological deterioration and degradation of synthetic polymeric materials: recent research advances. *Int. Biodeter. Biodegr.* 52, 69–91 (2003).
50. A.K. Bledzki, and J. Gassan, Composites reinforced with cellulose based fibres. *J. Prog. Polym. Sci.* 24, 221–274 (1999).
51. A. Magurno, Vegetable fibres in automotive interior components. *Angew. Makromol. Chem.* 272, 99–107 (1999).
52. J. Holbery, and D. Houston, Natural-fiber-reinforced polymer composites in automotive applications. *JOM* 58(11), 80–86 (2006).
53. A.K. Bledzki, O. Faruk, and V.E. Sperber, Cars from Bio-Fibres, *Macromol. Mater. Eng.* 291, 449–457 (2006).
54. A.K. Mohantya, M. Misraa, and G. Hinrichsen, Biofibres, Biodegradable PolymerBiocomposites: An Overview. *Macromol. Mater. Eng.* 276/277, 1–24 (2000).
55. A.K. Bledzki, and J. Gassan, Composites reinforced with cellulose based fibres. *Prog. Polym. Sci.* 24, 221–274 (1999).
56. D.N. Saheb, and J.P. Jog, Natural fiber polymer composites: A review. *Adv. Polym. Technol.* 18 (4), 351–363 (1999).
57. G. Bogoeva-Gaceva, Natural fiber eco-composites. *Polym. Compos.* 28(1), 98–107 (2007).
58. K.L. Pickering, *Properties and Performance of Natural-Fibre Composites,* CRC Press, Florida, (2008).
59. R.M. Rowell, Property enhanced natural fiber composite materials based on chemical modification, in *Science and Technology of Polymers and Advanced Materials*, pp. 717–732, Springer, US (1998).
60. J. George, M.S. Sreekala, and S. Thomas, A review on interface modification and characterization of natural fiber reinforced plastic composites. *Polym. Eng. Sci.* 41 (9), 1471–1485 (2001).
61. S.L. Bai, C.M.L. Wu, Y.W. Mai, H.M. Zeng, and R.K.Y. LI, Failure mechanisms of sisal fibres in composites. *Adv. Compos. Lett.* 8(1), 13– 17 (1999).
62. B.C. Barkakaty, Some structural aspects of sisal fibres. *J. Appl. Polym. Sci.* 20, 2921–2940 (1976).
63. S.S. Bhagavan, D.K. Tripathy, and S.K. De, Stress relaxation *in short jute fibre-reinforced nitrile rubber composites. J. Appl. Polym. Sci.* 33, 1623–1634 (1987).
64. T. Paramasivan, and A.P.J. Abdulkalam, On the study of indigenous natural-fibre composites. *Fiber. Sci. Technol.* 7, 85–88 (1974).

65. C. Pavithran, P.S. Mukherjee, M. Brahmakumar, and A.D. Damodaran, Impact properties of natural fibre composites. *J. Mater. Sci. Lett.* 6, 882–884 (1987).
66. K. Joseph, S. Thomas, C. Pavithran, and M. Brahmakumar, Tensile properties of short sisal fiber-reinforced polyethylene composites. *J. Appl. Polym. Sci.* 47,1731–1739 (1993).
67. K. Joseph, C. Pavithran, and S. Thomas, *Eur. Polym. J. Nat. Rubber Res.* 4, 55 (1991).
68. S. Varghese, B. Kuriakose, S. Thomas, and A.T. Koshy, Studies on natural rubbershort sisal fiber composites. *Indian J. Nat. Rubber Res.* 4, 55 (1991).
69. K.C. Manikandan Nair, S.M. Diwan, and S. Thomas, Tensile properties of short sisal fiber reinforced polystyrene composites. *J. Appl. Polym. Sci.* 60, 1483–1497 (1996).
70. J. George, S.S. Bhagawan, N. Prabhakaran, and S.Thomas, Short pineapple-leaf-fiber-reinforced low-density polyethylene composites. *J. Appl. Polym. Sci.* 57, 843–854 (1995).
71. N. Sgriccia, and M.C. Hawley, Thermal, morphological, and electrical characterization of microwave processed natural fiber composites. *Compos. Sci. Technol.* 67, 1986–1991 (2007).
72. T.A. Bullions, D. Hoffman, R.A. Gillespie, J. Price- O'Brien, and A.C. Loos, Contributions of feather fibers and various cellulose fibers to the mechanical properties of polypropylene matrix composites. *Compos. Sci. Technol.* 66, 102–114 (2006).
73. M. Wollerdorfer, and H. Bader, Influence of natural fibres on the mechanical properties of biodegradable polymers. *Ind. Crops Prod.* 8, 105–112 (1998).
74. N. Islam, R. Rahman, M.. Haque, and M. Huque, Physico-mechanical properties of chemically treated coir reinforced polypropylene composites. *Compos. A* , 41, 192–198 (2010).
75. M. Pracella, D. Chionna, I. Anguillesi, Z. Kulinski, and E. Piorkowska, Functionalization, compatibilization and properties of polypropylene composites with Hemp fibres. *Compos. Sci. Technol.* 66(13), 2218–2230 (2006).
76. P.V. Joseph, K. Joseph, and S. Thomas, Effect of processing variables on the mechanical properties of sisal-fiber-reinforced polypropylene composites. *Compos. Sci. Technol.* 59(11), 1625–1640 (1999).
77. C. Hill and M. Hughes, Natural fibre reinforced composites opportunities and challenges. *J. Biobased Mater. Bioenerg.* 4(2), 148–158, (2010).
78. A. Shahzad, Hemp fiber and its composites—A review. *J. Compos. Mater.* 46(8), 973–986 (2012).
79. M. Bengtsson, P. Gatenholm, and K. Oksman, The effect of crosslinking on the properties of polyethylene/wood flour composites. *Compos. Sci. Technol.* 65(10), 1468–1479 (2005). ISSN 0266-3538
80. I. Polec, P.J. Hine, M.J. Bonner, I.M. Ward, and D.C.Barton, Die drawn wood polymer composites. I. mechanical properties. *Compos. Sci. Technol.* 70(1), 45–52 (2010). ISSN 0266-3538
81. Farsi, M. Wood–plastic composites: influence of wood flour chemical modification on the mechanical performance. *J. Reinf. Plast. Compos.* 29(24), 3587–3592 (2010). ISSN 0731-6844
82. M. Tasdemir, H.Biltekin, G. Caneba, and T. Gerald, Preparation and characterization of LDPE and ppwood fiber composites, *J. Appl. Polym. Sci.* 112(5) 3095–3102 (2009). ISSN 0021-8995.
83. A.K. Gupta, M. Biswal, S. Mohanty, and S.K. Nayak, Mechanical, thermal degradation, and flammability studies on surface modified sisal fiber reinforced recycled polypropylene composites. *Adv. Mech. Eng.* 2012, Article ID 418031, http://dx.doi.org/10.1155/2012/418031 (2012).

Index

Accelerated weathering, 335, 342, 345–347, 349–351, 353, 355–357, 359–362
Acetic anhydride, 24, 341
Acetoguaiacon, 111
Acetylation, 263, 341, 358, 362, 529–530
Acid hydrolysis, 211
Acoustic emission spectroscopy, 33–34
  fibre breakage, 34
Acoustic properties, 369, 372, 374–75, 380–81, 383
Acrylic, 336, 359, 363
Acrylonitrile treatment, 317
  hydrophobicity, 317
  kenaf fiber, 317
Activation energies, 339–340
Adhesion, 18–19, 42, 44–48, 56–58, 103, 105, 112, 115, 122, 272–278, 371, 376, 378, 380–81, 386, 392–395, 399, 500, 509, 510, 512, 513, 517, 518
Advanced materials, 42
Aerospace, 43
AFM, 32–33
Agglomerate, 512
Aging, 45
Agricultural field, 43
Agricultural product, 41, 46
Agro-residual fibers, 233
Aldehyde, 104, 108, 111
Aliphatic, 108, 111
Alkali, 341, 346, 360–361
Alkali treatment, 44, 52, 55–57, 216, 221, 303, 528–529,
  FTIR, 303
  SEM, 304
  static Young's modulus, 306
  wood polymer composites, 303
  XRD, 305
Alkaline treatment, 21
Alkalization, 260, 263
Alkyl halide treatment, 320
Alkyl methacrylate (AMA), 345
Aluminum-reinforced polyester resin, 387
Amino propyl triethoxy silane, 277
Amphiphilic additives, 428
Anti-shielding, 403
Applications of natural fiber compostites,
  polymers reinforced with natural fibers, 435–438
  Portland cement matrix reinforced with natural fibers, 445–447
Aramid, 335
Arenga pinnata, 44, 46
Artificial weathering, 345, 357
Ash, 110, 111, 118
Atomic force microscopy, 32–33
  surface roughness, 33
Automotive, 43
Automotive applications, 531–532

Bagasse fibers, 238
Banana bunch fibers, 236
Banana leaf fibers, 236
Banana stem fibers, 236
Bast fibers, 328, 456, 457, 466
Benzene diazonium salt treatment, 306
  2,6-diazo cellulose, 308
  compressive modulus, 307
  dynamic Young's modulus, 307
  modulus of elasticity, 307
Benzophenone, 158–159, 335
  absorption of benzophenone, 164
Benzotriazoles, 335, 355
Bi-linear cohesive law, 405
Biobased composites, 436–437
Biocomposite, 483–486, 495–498

Biocomposites, 44
fully biodegradable, 418, 419
partially biodegradable, 418, 419
Biodegradability, 385
Biodegradable, 272, 275, 277, 278, 453, 454, 459, 461, 465 466, 468, , 478, 480, 484–486, 491–494, 496, 508, 510, 522
biodegradability, 481, 493–495, 498
biodegradability test, 495
completely biodegradable, 484
nonbiodegradable, 483–484
partially biodegradable, 484
Biodegradable polymers, 330, 345, 347, 360
Biodiversity, 43
Biological fiber extraction, 240–241
Biomass, 46
Biopolymer, 478, 480, 483–484, 486, 490–492, 494, 496–497
Biopolymers, 44, 104, 125, 126, 129
Biorefinery, 490
Bleaching, 263
Bond dissociation energies, 332–333
Boundary Element Method (BEM), 407
Brittleness, 389

Calcium carbonate, 350, 363
Carbon black, 334, 355–356
Carbon fiber, 43
Carbon fiber composite, 392
Carbonyl group, 514
Carbonyl groups, 332–334, 346
Carbonyl index, 346, 348–349, 355–356, 362
Carbonyls, 111
Casting, 478, 485, 492–493
Castor,
oil, 104–105, 108–109, 115, 117, 122
seeds, 105
Catalysts, 395
Cellophane, 488–489, 495
Cellulose, 41, 43–45, 54, 125–145, 247–248, 271–273, 275, 276, 328–330, 332, 334–338, 341–342, 358, 420, 421, 500–502, 504–509, 513, 514, 520–522
fiber, 328, 334, 336–337, 340–341, 345–346, 351, 357, 360–363
fibers, 105, 108, 117,
microfribils, 121
Cellulose extraction, 482–483, 485, 497
Cellulose fibres, 18–19
Cellulose microcrystalline particles, 447–448
Cellulose-based fiber, 41, 43
Cellulose-based polymer, 41, 43, 44, 57, 58
Cellulosic fibers, 271, 272
Cement, 45, 49, 50
Cement hidration process, 441
Cerium dioxide (CeO_2), 359
Chain scission, 332–335, 339, 346, 349, 351, 353–355, 358, 362
Chemical, 45, 47, 48, 54
bonding, 393, 395
modification, 394
pre-treatment, 394
Chemical composition, 502, 506, 507
Chemical composition of agro-residual fibers, 250
Chemical fiber separation, 241
Chemical functionalization of cellulosic fibers,
alkali treatment, 283
benzoylation, 283
composites fabrication, 283
evaluation of physico-chemical properties, 290–295
FTIR analysis, 288–289
Limiting Oxygen Index (LOI) test, 295–297
mechanical properties, 284–288
SEM analysis, 289–290
thermogravimetric analysis, 290
Chemical modification, 21–26, 127–128, 130–131, 133–134, 141, 144
MDI, 19–20
Chemical properties of plant fibers, 247–250
Chemical treatment, 274, 275, 277
acetylation, 274, 277, 278
admicellar treatment, 278
benzoylation, 274, 277, 278
mercerization, 276–278, 280
polystyrene maleic anhydrie, 277, 278
silane treatment, 274, 276, 277
Chemical treatments, 260–263
Chromophores, 333, 357
Clustering, 397403
Clusters, 403
Coalescence of voids, 402
Coating, 477–478, 480, 485–492, 494–498
Coconut,
fibers, 104–105, 110, 112–113, 115–116, 119–122
Coconut husk fibers, 239
Co-crystallization, 394
Coextrusion, 335–336

Cohesive,
  crack, 408
  element, 404, 405, 407
  failure, 395–406
  finite element (CFE) method, 405
  stress, 406, 407
  Zone Model (CZM), 404–406, 408, 409
Coir fiber, 239, 369–375, 376, 378–80, 383
Compact tension (CT), 398
Compatibility, 499, 505, 512, 513, 518
Compatibilizer, 64, 68–69, 105, 330, 357–358
Compatibilizers,
Composite, 477–480, 482–486, 490, 493–494, 496–498
Composite films, 125, 129, 130–131, 136–140, 144
Composite materials, 369–71, 373, 378, 380
Composites, 103, 104, 105, 106, 108–109, 110, 112–113, 114, 115, 117, 118, 120, 121, 122
Composition, 45, 53
Compression molding, 272, 273, 462
Computational chemistry,
  molecular mechanics methods, 149–150
  semi-empirical methods, 150–153
Condensation, 341–342, 345–346, 350–351, 355, 357–360, 362
Conductivity, 49–51
Continuum mechanics, 403
Core, 505–510, 520, 522
Core fibers, 456
Corn cob fibers, 235
Corn husk fibers, 235, 242, 254, 255
Corn stalk fibers, 235
Coupling agent, 216, 219, 341–342, 362, 393, 394, 513, 520
Crack,
  behavior, 407
  branching, 397
  deflection, 390, 397
  growth, 390, 391, 402, 403
  initiation, 400, 401, 404, 408
  opening displacement (COD), 406, 407
  path, 396, 397, 404, 408
  pinning, 390, 396, 397
  propagation, 400
  resistance, 387–398
Critical energy release rate, 385, 405
Crystallinity, 329, 334, 338, 346, 349, 351, 354, 361
Cure, 47, 48
Cyclic anhydride treatments, 321

Damage,
  analysis, 386
  evolution, 403
  initiation and propagation, 385
  mechanism, 392, 397, 401, 402
Debonding, 385–398, 403, 404, 407, 409
Decay, 47
Decohesion, 402–404
Defects, 85
  too high strain in tows, 85, 89, 94–98
  tow buckles, 85, 89, 98
Deformability, 176
Degradability, 392
Degradation, 49, 51, 53, 54
Degree of substitution, 130–131, 143
Delamination, 397, 398, 403, 404
Dewaxing, 461, 466
Differential scanning calorimetry (DSC), 51, 54–56, 109, 120, 121, 122
Digital image correlation (DIC), 385, 395, 401, 406, 408
Digital speckle photography, 401
Dimensional stability, 254–255, 369, 372, 374, 377–79, 383
Direct methods, 408
Discoloration, 345, 350, 355–357, 359, 359–360, 363
Dissipation factor, 188
Dissociation energies, 332–333
D-LFT, 462, 464
Double cantilever beam (DCB), 398
Double-edge notched tension (DENT), 398
Drawbacks, 43
Ductile polymer composites, 399400
Ductility, 378–79
Durability,
  polymers reinforced with natural fibers, 438–439
  Portland cement matrix reinforced with natural fibers, 443–445
Durian (Durio ziberthinus), 44, 49–51
Durian seed flour, 51
Dynamic mechanical analysis, 186

Eco-friendly, 453, 454
E-glass/epoxy braided composites, 392
Elasticity, 47, 50, 51
Elastomeric matrices, 66

Electrical conductivity, 188
Electronic speckle pattern interferometry (ESPI), 394
Electrostatic potential, 160, 163, 167, 170–171
Element Free Galerkin (EFG) method, 407
Element-failure method (EFM), 404
Elongation, 379, 47, 51, 52
Elongation at break, 517, 187
Embedded localization method, 408
Empty fruit brunch (EFB), 47, 48
Enau, 44
Encapsulation, 194
End notched flexure (ENF), 398
Energy absorption, 386
Environment, 46
Environmental, 499–501, 506, 509
Environmental condition,
 biodegradation, 428, 429
 humid environment, 423, 424
 oil absorption, 424, 425
 soil buried, 428, 429
 thermal loading, 426, 427
 UV-irradiation, 425, 426
Environmental impact, 63, 64, 67
Environmental threat, 523
Environmentally friendly material, 41, 43
Enzyme treatments, 264–265
Epoxides treatment, 320
Epoxidised natural rubber, 194
Epoxidised natural rubber (ENR), 47
Epoxy, 44, 45, 47, 48
Epoxy novolac, 369, 371–76, 378, 381, 383
Epoxy–silica nanocomposites, 387
Equilibrium moisture content (EMC), 337, 340,
Essential work of fracture (EWF), 400
Esterification, 486, 497
 ester, 483, 486–487, 491, 496–497
Etherification, 487–488, 497
 ether, 486–488, 497
Ethylacrylate, 346
Ethylene dimethylacrylate, 278
Ethylene–propylene copolymer, 358
Exfoliation, 397
Experimental, 105
Explicit dynamic simulations, 404
Extended finite element method (XFEM), 408
Extensibility, 369, 382
Extensometer, 400
Extractives, 110, 112, 115, 117, 122
Extrinsic, 405406
Extruded WPC, 392
Extrusion, 272, 273, 464

Fatigue, 47
Ferric oxide, 355
Fiber, 479, 482–486, 488, 490, 493, 496–498, 499–522
 biofiber, 483–484
 breakage, 398399403
 cellulosic fiber, 482, 497
 fiber sludges, 490
 natural fibers, 477, 483, 496–497
 plant fiber, 478, 481–483, 496, 497
 pullout, 403
 synthetic fiber, 483
Fiber content, 45, 47, 51, 54, 55
Fiber extraction, 239–246
Fiber saturation point (FSP), 337
Fiber treatment, 505, 512, 519
Fiber-matrix interface, 372, 375, 379, 382
Filament winding, 465
Fillers, 42, 46–49, 51
Films, 477–478, 480, 484–487, 489–498
 edible films, 477–478, 490, 494, 496
Flax, 328, 337
Flax fabric, 84, 86–87
Flax tows, 84
Flextural,
 modulus, 190, 193, 347, 351–352, 360, 388–394
 strength, 190, 193, 199, 347, 350–351, 353, 356–357, 360–361, 388–394
 stress, 192
 toughness, 190, 193
 properties, 391–394
Flexural modulus, 347, 351–352, 360
Food packaging, 478–480, 487, 493, 496
 packaging industry, 477–479, 481, 495
 packaging materials, 477, 479–480, 496
Fourier Transform Infra-Red (FTIR), 47, 48, 52
Fourier Transformed Infrared Spectroscopy (FTIR), 105, 108, 113, 122
Fracture,
 analysis, 385, 386, 402

behavior, 386–439, 495, 498, 400, 401, 404, 409
criterion, 404
energy, 387, 392, 404
initiation, 400, 409
mechanism, 385, 386, 388–394, 396, 398, 399, 401, 403, 404, 409
toughness, 385–387, 389–392, 395–398, 400, 406
Freezing, 426, 427
FTIR, 160, 161–163, 164–167, 168–170
FTIR spectroscopy, 344–346, 132, 133, 140
Functionalized polymers, 394

Galerkin/cohesive zone model (DG/CZM), 406
Gel, 47, 48
General properties,
ageing, 68, 72
flexural, 68, 70–75
impact, 64, 66–67, 69–74
shear, 70–71, 75
tensile, 64, 66, 68–75
thermal, 64, 67–70
toughness, 64, 72–74
Generalized Method of Cells (GMC), 403
Genetic algorithms, 408
Geometry optimization, 160–161, 163–164, 167–168
Glass fiber, 48, 360
Glass transition temperature, 336, 340
Glass-filled epoxy, 391
Glibenclamide,
absorption of glibenclamide, 167
mechanics of action, 159
medical uses, 160
Glucans, 107, 111
Glycerol, 125, 128–132, 140, 142–144
Gomuti, 43–46
Graft copolymeraization, 265
Grafted maleic anhydride groups, 393
Grafting, 513, 514, 517, 521
Grass and reed fibers, 456
Green composite, 132, 144
Green composites, 523–524

Halpin-Tsai (H-T) model, 402
Hand layup, 272
Hand Lay-Up, 462
Hardness, 46, 47
Hashin-Shtrikman (H-S) model, 402
Hemicellulose, 45, 54, 55, 248, 328–330, 336–338, 341, 362, 420, 421
Hemicelluloses, 501, 507, 509, 514
Hemp, 328, 337–339, 347, 360–361
Hemp fiber polypropylene composites, 392
Heterogeneous, 386, 394, 395, 402, 405
High density polyethylene, 216
High density polyethylene (HDPE), 47, 51, 53–58, 338, 344, 346–350, 352–356, 358–359, 363
High impact polystyrene (HIPS), 358
High-density polyethylene (HDPE), 390
Hindered amine light stabilizers (HALS), 335–336, 353, 357, 362
Hot strand stretching, 390
Hybrid composites, 369–76, 378–79, 381, 383
Hybridization, 369, 371–72, 383
Hydrogen bonding, 337–338, 341
Hydrogen bonds, 509, 512
Hydrolysis reaction, 423
Hydrophilic, 44, 45, 53, 56, 330, 337–338, 340–342, 359, 418, 502, 507, 509, 512, 513
Hydrophobic, 330, 337, 340, 353, 359, 363, 418, 502, 509, 512, 513, 517
Hydrophopic, 272
Hydroxyl, 502, 509, 513, 514
Hydroxyl groups, 330, 332, 337–338, 341–342, 359, 361

Impact, , 45–48
energy, 400
resistance, 392
strength, 386388
Impact modification, 20
PEAA, 20
Impact modifiers, 64
Impact strength, 333, 347, 350–351, 357, 360, 373, 375, 379–80, 383
Impregnation, 46
Incompatibility, 502, 512
Indonesia, 44, 46, 49, 51
Industries, 43
Initiator, 513, 514
Injection molding, 272, 273, 458

Inorganic, 507
fiber, 392
fillers, 395
particles, 386389
Interaction, 510, 511
Intercalation, 397
Interfacial,
adhesion, 394,395, 399
area, 392, 398
interactions, 386, 392
parameters, 405, 408
properties, 385, 408
resistance, 388, 389
shear strength, 388
strength, 386, 395, 398
stress criterion, 387
Interfacial adhesion, 216, 227, 509, 512, 516
Interfacial bonding, 44
Interlaminar shear failure mechanism, 392
Intermolecular cleavage, 398
Interphase, 388, 398, 402
Intrinsic, 404, 405, 408
Introduction, 104
Inverse method, 408
Irradiation, 47, 48, 345, 351–352, 355–357, 360
Izod or Charpy impact energy, 400

J-integral, 400, 406, 407
Jute, 328, 337–338, 340, 347, 351, 358, 360, 363, 453–469

Kabung, 44
Kaolinite, 193
Kenaf, 328, 345, 347, 499, 505–510, 512, 513, 518–522
Kenaf core, 507–510, 519, 521
Kupiah, 44

Laminates,
delamination, 74–75
stacking sequence, 72–74
symmetry, 72–74
Layering pattern, 369, 374, 376, 378
Leaf fibers, 456
Lignin, 45, 54, 103, 104, 105, 106–114, 117–122, 248–249, 271, 275, 328–330, 332–333, 337, 350, 357–359, 362, 420, 421, 457–459, 469, 501, 502, 504, 507, 509, 513, 514, 520
Lignocellulose fibers, 524
Lignocellulosic, 1, 4–13, 103–105, 111, 117, 120, 122, 271, 272,
Lignocellulosic fiber, 328–330, 332, 336, 342, 345, 362, 454, 457–460, 468, 499, 501, 512
Lignocellulosic materials,
rice husk, 154–155
wheat gluten husk, 155–158
Lignocellulosic polymer composites (LPCs), 330, 334, 336, 338, 340–346, 347, 352, 359, 362–362
Linear Fickian, 421
Local saturation, 407
Local shielding, 403
Localized shear bands, 387, 397, 398
Low density polyethylene, 520
Luffa fiber, 369–70, 372–75, 378–80, 383
Lumen, 420, 421

Macroscale modeling, 402
Makassar, 51
Malaysia, 44, 46, 49
Maleic anhydride (MAH), 47
Maleic anhydride polyethylene (MAPE), 393, 394
Maleic anhydride polypropylene (MAPP), 393–395, 400
Maleic anhydride PP (MAPP), 351, 357
Maleic anhydride treatment, 318
Manufacturing process, 43
Marine, 43
Materials, 41–43, 49, 50
Matrices, 45, 49, 103–106, 113
Matrix microcracking, 392
Matrix-fibre adhesion, 20
PLA-wood flour, 18–19
Mechanical damping parameter, 187
Mechanical fiber separation, 241–242
Mechanical interlocking, 393–395
Mechanical properties, 104, 105, 106, 121, 125, 127–129, 139, 143–144, 218–220, 225–226, 328, 330, 332, 338, 344, 346, 347, 351–353, 355, 357–362, 369, 371–72, 374, 376, 378–79, 381–83, 480–481, 489, 492–496, 498, 501, 502, 506, 508, 510, 512, 517–521,
Mechanical properties of agro-residual fibers, 251–252, 253
Mechanical testing, 86–97
biaxial tensile test, 87, 94
flax fabric testing, 86, 94

flax tow testing, 86, 90–93, 95–97
tow tensile strain, 86, 90–93, 95–97
Melt blending, 51
Mercerization, 341, 363, 182
Metal oxide nanoparticle reinforced epoxy composites, 396
Methyl acrylate, 47, 48
Methyl methacrylate, 513, 521, 522
Microbial resistance, 257
Microcrystalline cellulose, 131–132, 142
Microfibril angle, 329
Microfibrillar angle, 420, 421
Microfibrils, 420, 421, 501, 502
Microfibrils of cellulose, 447–448
Micromechanical deformation, 392, 394
Micro–meso–macro-approaches, 404
Microorganisms, 454
Microwave curing, 273, 274, 278
Mode-I,
dynamic stress intensity factor, 391
fracture behavior, 391
Mode-II,
fracture energy, 392
fracture toughness, 392400
Modification, 509, 512, 513, 520–522
Modification of agro-residual fibers, 258–266
Modification of cement matrix, 441
Modification of natural fibers, 443
Modulus, 42, 44–48, 50–52, 505, 508, 510, 511, 517
storage, 186
tensile, 191
Modulus of elasticity (MOE), 391
Modulus of rupture, 392
Moiré interferometry, 401
Moisture, 106, 107, 108, 111, 115, 117, 328–330, 335–338, 339–344, 347, 354, 358–359, 361–363
sensitivity, 330
Moisture absorption, 44, 45, 252–254, 421–423
Molecular dynamics, 403
Molecular entanglements, 395
Monomer, 513, 514
Mori–Tanaka Method (MTM), 403
Morphology, 217–218, 223–226, 369, 371, 375, 378, 381, 383, 391, 397, 400, 402, 403
Multi-scale, 403, 405, 407

Nanocellulose, 492
Nanoclay treatment, 318
dynamic mechanical thermal analysis, 319
flame retardant, 319
glass transition, 320
impregnation, 319
melamine-urea-formaldehyde, 319
thermal stability, 319
water absorption, 319
wood polymer nanocomposites, 320
Nanocomposite, 492–493, 495–496, 498
nanobiocomposite, 492–494
Nanofibrils of cellulose, 447–448
Nanoparticle, 387, 390, 391, 396, 397
Natural fiber, 44, 53
Natural fiber composites,
polymers reinforced with natural fibers, 435
portland cement matrix reinforced with natural fibers, 440
Natural fiber-reinforced composites, 387, 500, 512
Natural fibers, 84, 103, 104–105, 108, 115, 132, 142, 369–70, 372, 376, 380–81, 499–507, 510, 512, 513, 519–521, 525–526,
animal fibers, 272, 419, 420
bamboo, 66, 68, 74
cellulose, 526–527
chemical composition, 526–527
classification, 528
flax, 66, 68, 70, 73
fracture analysis, 538–539
grass, 74
hemicellulose, 527
hemp, 65–66, 68, 70
jute, 67–69, 72, 75
kenaf, 66, 70–71, 73
lignin, 527
loofa, 73
mechanical properties, 533–535
mineral fibers, 272, 419, 420
palm, 66, 68, 74
plant fibers, 272, 419, 420
sisal, 66, 68–70, 73
thermal properties, 535, 537
Natural fibre composites, 17
fibre treatment, 21–26
interfacial adhesion, 30
PLA, 18
Natural rubber, 47, 52
Natural weathering, 342, 345–346, 349–350, 356–358, 360–362

Necking, 396, 397
Nelder-Mead algorithm, 408
Nettle fibers, 237, 243
Non-destructive test (NDT), 391
Non-polar (hydrophobic) molecules, 393
Norton–Bailey creep law, 392
Nucleation or coalescence of the microvoids, 401

o-hydroxybenzene Diazonium Salt Treatment, 310
  2,6-diazocellulose, 311
  agglomerations, 312
  coir fiber, 311
  interfacial adhesion, 312
  micro-voids, 312
Oil palm, 176
  biocomposites, 176
  empty fruit bunch, 176
  empty fruit bunch fibre, 177
  fibre, 177
Oil palm (Elaeis guineensis), 44, 46–49
Oil palm bunch fibers, 238
Oil palm fibre, 179
  arabinose, 180
  dielectric constant, 181, 187
  galactose, 180
  lignin, 180
  mannose, 180
  pentosan, 180
  rigidity index, 181
  true density, 179
  xylose, 180
Okra stem fibers, 236, 251, 256
Opacity, 129, 132, 136–137
Optimization technique, 408
Organic acid, 125, 128–131, 133–134, 136–137, 140, 144–145
Organofunctionality, 395
Oxidation, 333, 336, 339, 345–346, 350, 352, 355, 357–359, 361–362
Oxidation of natural iber, 321

Packaging products, 524
Palm fibers, 214
Palm sugar, 44
Paperboard, 49
Particle,
  bridging, 390
  clustering, 403
  debonding, 387, 390, 396
Particulate,
  composites, 386–389, 391, 395, 402, 403, 407, 408
  polymer composites (PPCs), 385, 386, 391, 396, 417
Pectin, 249
Phenolic, 107, 108, 111
Photo degradation, 257
Photodegradation, 333–335, 353–355, 359
Photometer, 343
Photostabilizer, 334, 352–354
Physical penetration, 393
Physical properties, 369, 372–73, 376, 383
Physical properties of agro-residual fibers, 249–251
Physical structures, 222
Physical treatments, 258
Pigment, 335–336, 352, 354–357, 362
Pineapple fibers, 220
Pineapple leaf fibers, 238
Plant fibers, 329
Plasma modification, 27–30
  natural fibres, 27
  plasma induced chemical grafting, 28
  wood flour, 28
Plasma-treated glass fiber-reinforced nylon-6,6 composites, 392
Plastic,
  deformation, 396, 397, 403, 407
  void growth, 387, 396, 398
Plastic void growth, 387396398
Plasticization, 336, 340
Plastics, 478, 479–484, 486, 496
Polar, 502, 509
Polar (hydrophilic) molecules, 393
Polar groups, 418
Pollution, 43
Poly(vinyl chloride) (PVC), 47–48
Polyester, 43, 46–49, 52, 53, 335–336, 338–340, 360
Polyethylene, 333, 338, 340, 352
Polyethylene terephthalate (PET), 390, 391
Poly-hydroxybutyrate (PHB), 361
Polymer binder tearing, 401
Polymer Bonded Explosives (PBX), 391
Polymer composites, 1, 3–5, 7, 9–13, 370, 376, 499–502, 505, 508–510, 512, 518, 519
Polymer matrices, 499, 501, 502, 513
Polymerization, 500, 502, 514, 522
Polymethylenepolyphenyl isocyanate (PMPPIC), 48

Polyoses/ hemicellulose, 110, 111–113, 117–118, 120, 122
Polypropylene, 46–48, 51, 103, 105, 109, 112–114, 119–122, 222, 338, 340, 342, 346, 357
Polystyrene, 277, 278, 280
Polyurethane, 103, 104–105, 108–109, 115, 117, 118, 122
Polyvinyl chloride (PVC), 347, 351, 356, 359
Potassium hydroxide (KOH), 341
Pozzolans, 441–443
Prevailing dissipation mechanisms, 398
Primary cell wall, 419–421
Processing, 41, 42, 44
Pseudo-Fickian, 422
Pultrusion, 272, 463

Quasi-static, 387, 391, 392, 408

Radical initiators, 395
Ramie, 328, 337, 338
Rayon, 488, 495
Reactivity, 258
Recycled polypropylene, 531
Recycling, 67–68, 76
Reed stalk fibers, 237
Regulations, 43, 47
Reinforcement, 43, 103, 104–106, 109, 115, 117, 122, 370–72, 378, 455, 460, 461, 464, 465, 468
Reinforcements, 500–502, 509
Relative humidity (RH), 335–337, 357, 361
Renewable, 453–456, 460, 468, 469, 500, 501, 505, 506
Representative volume element (RVE), 405407
Residual bending strength, 392
Resin transfer molding, 272, 273, 462
Resins, 43, 48
Resources, 43, 51, 57, 58
RGB color segmentation techniques, 388
Rice husk, 50
Ricinoleic acid, 105
Rigidity, 501, 502, 517
Roofing, 44
Rupture, 47

Sago starch, 508–510, 519
Scanning electron, 44, 46, 40
Scanning electron microscope (SEM), 401
Scanning electron microscopy, 31
  interfacial adhesion, 32
Scanning Electron Microscopy (SEM), 109, 114, 115, 116, 122
Scorch time, 47
Secondary cell wall, 419–421
Seed fibers, 456
SEM, 369, 375, 380–82,
Semi-analytical micromechanical methods, 403
Semi-empirical approach, 402
Sheet molding, 272
SIFT–EFM approach, 404
Silane, 341, 346, 362, 394, 395
Silane treatment, 263, 529
  amino, 187
  fluro, 187
  vinyl, 187
Silica,
  filled epoxies, 387
  graphene oxide nanoplatelets (GONP) composites, 396
  reinforced epoxy adhesive, 392
Single Leg Bending (SLB) joints, 408
Single-edge notched tension (SENT), 398
Single-edged notched three point bending (SENB), 398
Sisal, 328, 337–338, 345, 347, 357
Sisal fiber composites, 532–533
Sliding mode, 388, 392
Smooth Particle Hydrodynamics (SPH), 407
Sodium hydroxide (NaOH), 341, 358
Solution-induced intercalation method, 467
Sound absorption, 369–70, 375, 380–82, 46
Soundproof material, 46
Sources of agro-residual fibers, 235–239, 243–244
Southeast Asia, 41, 43, 44, 46, 49, 57, 58
Species, 43, 44, 49
Specimen, 44, 48, 53
Speckled patterns, 401
Spectroscopic techniques, 401
Starch, 126–129, 358, 361, 363, 480–481, 496, 508–510, 512, 520, 521
Starch microparticles, 125, 129–132, 136–137, 140, 144
Stiffness, 328, 329, 336, 338, 340, 350, 356, 361, 501, 517, 44
Strain at failure, 45
Strain gauge, 400
Strain Invariant Failure Theory (SIFT), 404

Strain rate, 391, 392, 400
Strength, 328, 333, 336, 339, 345–347, 350–351, 353, 356–358, 360–361, 501, 502, 505, 507–518,
  izod impact, 201
  tear, 186
  tensile, 186, 189, 199, 205
Stress,
  concentration, 394, 395, 397
  evolution, 407
  intensity, 386, 391, 400, 401
Stress-wave loading, 391
Structural composite materials, 533
Structure of plant fiber, 419–421
Sub-micro-fibril, 390
Succinic anhydride treatment, 313
  adhesion, 313
  biocomposites, 313
  compatibility, 313
  hydrophilic, 316
  impurities/waxy substances, 314
  jute fiber, 314
  polypropylene, 313
  tensile strength, 315
Sugar palm, 43–45, 49
Sugarcane bagasse, 103, 104–106, 108, 110, 111, 114, 122
Sugarcane fibers, 238
Sulawesi, 51
Sulfonation, 265
Surface, 44–46, 51, 52
Surface modification of natural fibers, 528
Surface treatment, 330, 334, 346, 352, 357, 105, 106
Surface treatments, 64, 69, 72, 74, 76
Swelling, 197
Synthetic fiber, 1, 4, 5, 417
Synthetic polymers, 505, 510

Talc, 350–351, 363
Tater absorption, 508, 509, 520–522
Tear strength, 47
Technology, 42
Temperature, 44, 47, 49, 51–58
Tensile, 44–48, 51, 52
Tensile properties, 371–73, 375, 378–79
Tensile strength, 510, 516
Textile forming, 85, 94–98
  sheet forming device, 85, 87–90
Thailand, 46
Thawing, 426, 427
Thermal,
  decomposition, 386
  insulation, 385
Thermal analysis, 139
Thermal properties, 45, 51, 53
Thermal stability, 45, 52–57, 255–257, 258
Thermogravimetric analysis (TGA), 51, 53, 54, 109, 117, 118, 119, 122
Thermoplastic, 330, 336, 345–347, 359, 362, 385, 389, 393, 478, 480, 484–486, 489, 493, 496, 501, 508–510, 512, 517, 520, 521
Thermoplastic matrices,
  polar matrices, 69
  polycarbonate, 68
  polyethylene, 68, 69
  polypropylene, 68, 69
  polystyrene, 68, 69
  polyvinylchloride, 68, 69
Thermoplastic starch, 510, 127–129, 142–144
Thermoplastics, 103–105
Thermoset, 330, 336, 345, 347, 359, 389, 390, 395
Thermoset matrices,
  epoxy, 70–71
  polyester, 64, 70–72, 74
Thermosetting, 103–106, 108, 113
Thickness swelling, 369, 372, 374, 376–78, 383
Time–temperature superposition, 392
Titanium dioxide, 355–356
Toluene, 278
Topography, 275
Toughening,
  agents, 389
  mechanisms, 387, 390, 396–398
Toughness, 371–72, 379, 42, 45–47
Traction separation law (TSL), 404, 405
Traditional application, 44
Treatment, 505, 513, 518, 521
  alkali, 183
  benzoylation, 195, 202
  titanate, 182
  toluene diisocyanate, 182
Tropical countries, 41, 43, 46, 51
Twin screw microcompounder, 533
Two-stage diffusion, 422

Ultraviolet (UV), 329–339, 342–347, 349–363
Ultraviolet absorbers (UVAs), 334–336, 352, 354–355, 362
Unidirectional oriented sisal fiber composites, 533
Unit cell, 402407
Unplasticized poly (vinyl chloride) (UPVC), 351
UV radiation, 278
UV-A, 331
UV-B, 331
UV-C, 331

Variability, 257–258
VARTM, 463
Vinyl index, 346, 348–349, 362
Viscoelastic, 391392
Voronoi cell approach, 403

Waste, 46, 49
Water absorption, 46, 49, 51, 52, 197, 369, 371–72, 374, 376–78, 381, 383
Water hyacinth (Eichhornia crassipes), 51–54
Water uptake, 129, 132, 137–139
Waxes, 249
Weathering effect, 537–538
Weathering resistance, 330, 350, 352, 357–358, 362–363
Wettability, 512, 513
Wheat straw fibers, 235
Whisker, 402
Wood fiber, 328–329, 340, 347, 349–352, 357, 359
Wood fibers, 126
Wood flour, 46–47
Wood/recycled plastic composite (WRPC), 388
Woodflour, 329, 336, 338, 344, 346–347, 352–356
Wood-plastic composites (WPCs), 385, 386, 388, 391–395, 398–403, 409
Wood-polymer composites (WPCs), 330, 335, 359

XPS,
X-ray diffraction, 132–134, 136, 140
X-ray photoelectron spectroscopy, 401
Xylans, 107, 111

Yamada-Sun failure criterion, 403
Young's modulus, 351, 517

β- propiolactone treatments, 320

Also of Interest

Check out these published related titles from Scrivener Publishing

Journal of Renewable Materials
Editors: Alessandro Gandini and Ramaswamy Nagarajan
www.scrivenerpublishing.com

Lignocellulosic Polymer Composites
Processing, Characterization, and Properties
Edited by Vijay Kumar Thakur
Published 2015. ISBN 978-1-118-77357-4

Nanocellulose Polymer Nanocomposites
Fundamentals and Applications
Edited by Vijay Kumar Thakur
Published 2015. ISBN 978-1-118-87190-4

Handbook of Cellulosic Ethanol
By Ananda S. Amarasekara
Published 2014. ISBN 978-1-118-23300-9

The Chemistry of Bio-based Polymers
By Johannes Karl Fink
Published 2014. ISBN 978-1-118-83725-2

Biofuels Production
Edited by Vikash Babu, Ashish Thapliyal and Girijesh Kumar Patel
Published 2014. ISBN 978-1-118-63450-9

Handbook of Bioplastics and Biocomposites Engineering Applications
Edited by Srikanth Pilla
Published 2011. ISBN 978-0-470-62607-8

Biopolymers
Biomedical and Environmental Applications
Edited by Susheel Kalia and Luc Avérous
Published 2011. ISBN 978-0-470-63923-8

Renewable Polymers: Synthesis, Processing, and Technology
Edited by Vikas Mittal
Published 2011. ISBN 978-0-470-93877-5

Plastics Sustainability
Towards a Peaceful Coexistence between Bio-based and Fossil Fuel-Based Plastics
Michael Tolinski
Published 2011. ISBN 978-0-470-93878-2

Green Chemistry for Environmental Remediation
Edited by Rashmi Sanghi and Vandana Singh
Published 2011 ISBN 978-0-470-94308-3